Flora of Maine

a Manual for Identification
of Native and Naturalized Vascular Plants
of Maine

Arthur Haines

Thomas F. Vining

V. F. Thomas Co.
P. O. Box 281
Bar Harbor, Maine 04069-0281

cover illustration (*Salix arctophila*) by Arthur Haines
Maine has the single United States station of this plant. Represented by a lone, staminate individual, this dwarf willow is found in a remote ravine on Katahdin, Piscataquis Co.

Library of Congress Catalog Card Number: 98-60742

ISBN 0-9664874-0-0

DEDICATION

A. H. — Thanks to Les Eastman for providing the inspiration for beginning my botanical study and to Tom Vining for taking care of the obstacles that stood in the way.

T. F. V. — To my parents who, by their example and guidance, helped me understand the importance of learning.

TABLE OF CONTENTS

PREFACE

If you are reading these words (which clearly you are), you may, perhaps probably, have read the prefaces to other, earlier manuals covering our area. Several features of those works are, without apology, repeated here.

Arthur Cronquist found it appropriate to abandon in the 2nd edition of his Manual of Vascular Plants a feature of previous works that, although it had been useful in the past, had, in his opinion, lost its value. Similarly, we have chosen to discontinue using the plethora of names that have been assigned by workers in various groups to the same plant part; for example, we are calling a fern leaf a *leaf* and its divisions *leaflets* and *leafules* instead of *pinnae* and *pinnules*.

Merritt Lyndon Fernald, in his preface to the 8th edition of Gray's Manual of Botany, emphasized the importance of field work. Agreeing with his view, we have tried, insofar as possible, to examine plants in their natural habitats. This effort has enabled us to constantly refine and sometimes completely redo portions of the keys and/or species descriptions. Even as this book goes to press, there are changes that, given another few weeks, we might be able to make. However, like Fernald, we recognize that there "has to be a deadline!" Consequently, we present this Flora of Maine to be, as the Flora of New Brunswick was described, "a baseline from which we can move forward."

Finally, like other authors who have preceded us, we recognize those who have contributed in a significant way to help us complete this project. Of course, we accept full responsibility for our final decisions in taxonomy and other matters.

TAXONOMIC ASSISTANCE

George Argus (*Salix*)
Peter Ball (*Carex, Salicornia*)
Harvey Ballard Jr. (*Viola*)
Mary Barkworth (Poaceae)
Kåre Bremer (family ordering)
Luc Brouillet (Asteraceae)
Mark Chase (family ordering)
Arthur Gilman (pteridophytes)
C. Barre Hellquist (*Stuckenia*)
Harold Hinds (Polygonaceae)
Walter Judd (family circumscriptions, Ericaceae)

Eric Lamont (Asteraceae)
Michel Lelong (*Panicum*)
Lawrence Magrath (Orchidaceae)
James Phipps (*Crataegus*)
Gregory Plunkett (Apiaceae)
Robert Price (Brassicaceae)
Anton Reznicek (Cyperaceae)
John Semple (Asteraceae)
S. Galen Smith (*Eleocharis, Schoenoplectus*)
Todd Yetter (Lamiaceae)
Frederick Utech (Asparagales, Liliales)

OTHER HELP

David Boufford
Christopher Campbell
William Crins
Garrett Crow
Craig Freeman
Walter Kittredge
Les Mehrhoff

Sally Rooney
University herbaria at:
  Harvard
  Maine
  New Hampshire
Jill Weber
Woodlot Alternatives, Inc.

BRIEF HISTORY OF THE FLORA OF MAINE

John Josselyn, namesake of the Josselyn Botanical Society of Maine, compiled in 1672 a list of plants observed in the Black Point section of Scarborough, Maine. His interest was primarily in plants with medicinal or cultural uses, and his list reflected that interest.

Not until 1848 was an attempt made at describing the entire flora of the state. Aaron Young, state botanist, produced a single volume of an anticipated 20 or so. In this work a plant was affixed to a page, covered by a protective sheet, and accompanied by information on a second sheet. For reasons unknown to us, no further volumes were produced.

In 1862, George Goodale, curator of botany for the Portland Society of Natural History and botanist of the Maine Scientific Survey, compiled the first extensive list of Maine's plants. This list was reworked and printed in 1868 as *The Portland Catalog of Maine Plants*. This catalog prevailed until 1892 when Merritt Lyndon Fernald, who largely rewrote *Gray's Manual of Botany* for its 8th edition (1950), revised Goodale's list as the *Second Edition* of the Portland catalog. Just three years later, because of rapid advances in botanical knowledge, Fernald revised his revision excluding 77 species and varieties and adding 155.

During the first half of the 20th century, many new plant discoveries were made. In 1948, the first *Check-list of the Vascular Plants of Maine* was produced. This checklist was "intended to be a forerunner to a Flora of the State." Indeed, Professor Eugene Ogden of the University of Maine and lead author of the checklist assigned to his taxonomy students the task of writing keys for various families of plants. These keys were to be worked into the Flora of Maine he spoke of in his introduction. However, Dr. Ogden soon moved to New York. Since then, three revisions of the checklist were made, but still no flora.

For several years prior to 1996, Arthur Haines, a graduate student at the University of Maine, had been studying Maine's flora and writing keys for seminars and labs he taught. In 1996, he asked Thomas Vining, also a graduate student at the University, if he would like to join him in writing a Flora of Maine. Two years and two months later (coincidentally the amount of time Thoreau, who himself botanized in the Maine woods, spent at Walden Pond) and 150 years after Aaron Young's single volume, they delivered this book to the printer.

INTRODUCTION

Purpose. This book is principally, as its subtitle indicates, a manual for identification. We have, therefore, placed primary emphasis on the keys; other information is designed to illuminate them. Genus notes, for example, provide suggestions for field observation and/or clarify important features, and species information includes description of habit, statewide frequency, and habitat.

Scope. We have chosen to restrict our work to vascular plants that are native or naturalized to Maine. Therefore, the keys may not be applicable elsewhere when other species are present. This geographic limitation has enabled us to use characters not appropriate for books covering a larger range. Although hybrids are common in many genera, as noted in the species descriptions, they are often included in species descriptions, but not in keys.

Taxonomy and nomenclature. Families are arranged in phylogenetic order, such that all monophyletic groups of families are kept together, as far as is known (Angiosperm Phylogeny Group in prep., Bremer *et al.* 1998, Källersjö *et al.* 1998, and references therein). Genera within families and species within genera are alphabetical. Family circumscription follows Judd *et al.* (in prep.), who used monophyly as their primary criterion. Nomenclature is that of authors of Flora of North America, when available. Otherwise, primarily Kartesz (1994) was followed. If you need more information, please contact Arthur Haines in care of V. F. Thomas Co, P. O. Box 281, Bar Harbor, Maine 04069-0281.

Keys. Keys in this book are dichotomous and generally like others of that kind. There are, however, some features to be noted. Semicolons divide each stanza into descriptions of individual characters. Commas separate character states in addition to normal, grammatical use. We would rather err on the side of being too wordy than assuming the reader understands what we are thinking. Information in ( ) is an exception to stated ranges or character states; *e.g.*, sepals 3(4) or (2.0–)3.0–4.0 mm or corolla red (white). Material in [ ] is additional information meant to clarify a statement; *e.g.*, flowers precocious [appearing earlier than the leaves]. Decimals are used instead of fractions. A color given as "green-yellow" is intermediate between green and yellow, whereas "green to yellow" means the object of interest may be green or yellow or any shade between.

Voucher specimens. Insofar as possible, all species of Maine's flora have been verified from voucher specimens. Exceptions are *Actinidia arguta* and *Aristolochia macrophylla* which are not known to be vouchered, but have been observed by Linda Gregory, Acadia National Park botanist (pers. comm.).

Species descriptions.
synonym(s) - Names following a subspecies that is enclosed in < > are synonyms of that subspecies. A semi-colon separates synonymy of different subspecies.
common name(s) - Names following a subspecies that is enclosed in < > are common names of that subspecies. A hyphen indicates that an implied relationship is not true; *e.g.*, cotton-grass is not a grass.
habit - Trees and shrubs are deciduous unless otherwise noted.
range - For non-native species, origin/native range is given first, followed by areas of introduction and/or naturalization.

state frequency - The frequency given for species that are naturalized in Maine applies only to the naturalized individuals, not to the planted, but non-escaped, ones. The following guidelines were used when determining state frequency:

common - grows throughout the state in a variety of habitats, often in both pristine and disturbed areas (*e.g.*, *Cornus canadensis*, *Pinus strobus*, *Ranunculus acris*). Some extremely abundant coastal species are included in this rank.

occasional - found throughout much of the state and may be locally abundant, but only in specific ecosystems (*e.g.*, *Chamaedaphne calyculata*, *Cypripedium acaule*, *Sorbus decora*)

uncommon - infrequently found throughout state, restricted to a portion of the state, or growing in specific community types, but with generally more than 30 occurrences statewide (*e.g.*, *Adiantum pedatum*, *Betula lenta*, *Pogonia ophioglossoides*)

rare - generally fewer than 30 occurrences (*e.g.*, *Diapensia lapponica*, *Eriocaulon parkeri*, *Panax quinquefolius*)

very rare - limited to a single geographic feature (*e.g.*, mountain or river) or generally with fewer than 5 occurrences statewide (*e.g.*, *Cryptogramma stelleri*, *Pedicularis furbishiae*, *Rhododendron lapponicum*)

habitat - Habitat is for plants growing in Maine. Plants are terrestrial unless otherwise noted. The word *and* was generally used to connect habitats modified by the same adjective; *e.g.*, "sandy fields and waste places" means that *both* the fields and waste places are sandy, whereas "sandy fields, waste places" means that only the fields are necessarily sandy.

notes - Rankings of rare plants follows the 3 March 1997 list of the Maine Natural Areas Program. The following are explanations of those rankings.

S1 = critically imperiled in Maine because of extreme rarity (5 or fewer occurrences or very few remaining individuals or hectares) or because some aspect of its biology makes it especially vulnerable to extirpation from the State of Maine.

S2 = imperiled in Maine because of rarity (6–20 occurrences or few remaining individuals or hectares) or because of other factors making it vulnerable to further decline.

S3 = rare in Maine (on the order of 20–100 occurrences).

SH = occurred historically in Maine, and could be rediscovered; not known to have been extirpated.

SU = possibly in peril in Maine, but status uncertain; need more information.

SX = apparently extirpated in Maine (historically occurring species for which habitat no longer exists in Maine).

S? = Probably rare or historic in Maine, based on status elsewhere in New England, but not reviewed or documented by the Maine Natural Areas Program.

CHANGES IN MAINE'S FLORA

Many additions to and exclusions from Maine's flora have occurred since the publication of the third revision of the *Checklist of the Vascular Plants of Maine* (1995).

Additions

The following taxa have been added to Maine's flora based on the existence of a voucher specimen.

Actinidia arguta
Agrostis elliottiana
Aristolochia macrophylla
Axyris amaranthoides
Cardamine impatiens
Carex blanda
Carex pensylvanica
Carex tonsa var. *rugosperma*
Carex vacillans
Centaurium pulchellum
Chenopodium berlandieri var. *macrocalycium*
Chenopodium foggii
Elaeagnus umbellata
Helenium autumnale
Humulus lupulus var. *lupuloides*
Knautia arvensis
Laburnum ×watereri
Ligustrum amurense
Ligustrum vulgare
Malus baccata

Malus prunifolia
Morus alba
Nymphaea odorata var. *tuberosa*
Parthenocissus tricuspidata
Petunia axillaris
Pinguicula vulgaris
Pinus nigra
Ranunculus hispidus var. *caricetorum*
Rosa cinnamomea
Rosa gallica
Rosa rubrifolia
Rosa spinosissima
Salix cinerea ssp. *oleifolia*
Sanicula odorata
Setaria faberi
Sorbaria sorbifolia
Sporobolus asper
Symphoricarpos albus
Ulmus glabra

Exclusions

The following taxa have been excluded from Maine's flora for the reasons given in parentheses after the taxon name and coded as follows:

    1 = not naturalized in Maine
    2 = outside known range of species/subspecies
    3 = voucher specimen(s) misidentified
    4 = voucher specimen unknown

Ageratina altissima var. *roanensis* (2)
Allium schoenoprasum var. *schoenoprasum* (3)
Amberboa moschata (1)
Ambrosia psilostachya (2)
Antennaria rosea ssp. *pulvinata* (3)
Aster dumosus (3)
Aster praealtus var. *praealtus* (2)

Aster solidagineus (3)
Arabis ×divaricarpa (3)
Brachyactis frondosa (4)
Brassica oleracea (1)
Cakile maritima (3)
Calendula officinalis (1,4)
Carex albolutescens (3)

Carex complanata (2)
Carex debilis var. *debilis* (2,3)
Carex inops (2)
Carex lapponica (2)
Carex macloviana (2,4)
Carex magellanica ssp. *magellanica* (2)
Carex viridula ssp. *oedocarpa* (3)
Centaurea montana (1,4)
Cerastium biebersteinii (4)
Chenopodium watsonii (1,2)
Cichorium endivia (1)
Crataegus ×*anomala* (4)
Crataegus mollis (2)
Eleocharis flavescens (2)
Eleocharis uniglumis (2,4)
Eupatorium purpureum (3)
Festuca rubra ssp. *arenaria* (2,3)
Hieracium praealtum var. *praealtum* (2)
Iris pumila (1)
Iris sibirica (1)
Lepidium hirtum (4)
Lupinus angustiflorus (1)
Luzula congesta (2)

Lyonia ligustrina var. *foliosiflora* (2)
Machaeranthera gracilis (1)
Madia gracilis (1)
Neobeckia aquatica (2,4)
Poa pratensis ssp. *alpigena* (2)
Polemonium cuspidatum (1)
Potamogeton diversifolius (2,4)
Pycnanthemum flexuosum (2,3)
Ranunculus hispidus var. *nitidus* (2)
Rubus arcticus ssp. *acaulis* (2,4)
Rubus argutus (2,4)
Sagittaria stagnorum (2,4)
Salix babylonica (1,2)
Salix elaeagnos (1,4)
Scabiosa atropurpurea (1,4)
Schizanthus pinnatus (1)
Solidago canadensis ssp. *elongata* (2)
Solidago canadensis ssp. *salebrosa* (2)
Stellaria corei (1,4)
Valeriana dioica var. *sylvatica* (2,3)
Viburnum dentatum var. *scabrellum* (2)

GLOSSARY

a - Latin prefix meaning not, without.

abaxial - on the side away from the axis, usually refers to the underside of a leaf (*cf.* adaxial).

acaulescent - without an upright, leafy stem.

accrescent - increasing in size with age.

accumbent cotyledons - cotyledons oriented such that one of their edges is set against the radicle (*cf.* incumbent cotyledons).

achene - a dry, indehiscent, usually one-seeded, fruit.

acroscopic - toward the distal end (*cf.* basiscopic).

actinomorphic - radially symmetric, usually refers to the perianth of a flower (*cf.* zygomorphic).

acuminate - gradually tapering to a narrow point, more tapering than acute, less than attenuate.

acute - condition of an apex with more or less straight sides that meet to form an angle of less than 90° (*cf.* obtuse).

adaxial - on the side toward the axis, usually refers to the top side of a leaf (*cf.* abaxial).

adhere - to stick together, but not be actually fused, applies to unlike organs (*cf.* cohere, connivent).

adnate - fused with a structure different from itself, as when stamens are adnate to petals (*cf.* connate, distinct).

adpressed - laying close to and pressed against a surface pointing in the proximal direction (*cf.* appressed).

adventive - not intentionally introduced, but occurring sporadically here and there.

aerenchyma - a type of plant tissue that is filled with large air spaces.

aerial - above ground.

aggregate fruit - a class of fruits in which 2 or more ovaries from an apocarpous gynoecium are united to form a single fruit (*cf.* multiple fruit, simple fruit).

alveolate - with alveoli.

alveolus (plural: **alveoli**) - one of many pronounced, angular cavities that collectively provide a honeycomb appearance.

ament - a type of inflorescence, usually with an elongate, non-fleshy axis, composed of flowers with a reduced or no perianth, subtended by bracts, usually anemophilous.

amphibious - capable of living in water and land; often referring to plants capable of surviving when the water source has disappeared.

amphistomic - with stomata on both surfaces of the leaves.

androecium - the stamens taken collectively.

androgynous - a spike with both staminate and carpellate flowers, the carpellate located at the base, below the staminate (*cf.* gynaecandrous).

anemophilous - wind pollinated.

annular - with the form of a ring.

annulus - a ring of specialized cells encircling a sporangium that aids in sporangia dehiscence.

ante - in front of, as with antepetalous stamens where the stamens are positioned directly in front of or opposite the petals, as opposed to alternating with them.

anther - pollen-bearing structure usually at the end of a stalk called a filament.

anthocarp - a compound unit found in the Nyctaginaceae consisting of the fruit, an achene or utricle, and the enclosing, indurate calyx.

anthocyanin - a pigment ranging in color from blue to purple to red.

antrorse - upward or forward oriented (*cf.* retrorse).

aphyllopodic - having basal sheaths without blades; with new shoots arising laterally from parent shoot (*cf.* phyllopodic).

apical placentation - a type of placentation in which the ovule is borne at the apex of the locule.

apiculate - having an apiculus.

apiculus - an abrupt, short, very small, projected tip.

apo - prefix meaning the parts are distinct.

apophysis (plural: **apophyses**) - portion of a scale of a megasporangiate strobilus, or seed cone, that is visible when the strobilus is closed.

appressed - laying close to and pressed against a surface pointing in the distal direction (*cf.* adpressed).

aquatic - living in fresh water.

areole - a bounded space, such as the area surrounded by veinlets of leaves with reticulate venation.

aril - a specialized appendage on a seed, derived from the funiculus or seed coat.

aristate - tipped with a slender bristle.

armed - bearing a sharp projection such as a prickle, spine, or thorn (*cf.* unarmed).

article - an individual segment of a loment fruit.

articulated - joined to something, often at a place of eventual separation.

ascending - diverging from an axis at an angle of 15–45°, less upward than erect, less outward than spreading.

attenuate - tapering very gradually to a prolonged tip.

auricle - lobe-like appendage at the base of an organ.

awn - a terminal or sometimes abaxial bristle as on the lemmas of some grasses or the scales of some *Carex* species.

axil - the position between an axis and a lateral organ from that axis (*e.g.*, a leaf axil).

axile placentation - a kind of placentation in which the ovules are borne on the central axis formed by the meeting of the partitions, found only in ovaries with 2 or more locules.

axillary - borne in an axil.

axis - the main stem of a structure, such as a plant or inflorescence.

basal placentation - a type of placentation in which the ovule is borne at the base of the locule.

basifixed - attached at the base.

basiscopic - toward the proximal end (*cf.* acroscopic).

beak - a slender, terminal appendage on a 3-dimensional organ.

berry - an indehiscent, several-seeded, fleshy fruit.

bi - prefix meaning 2.

bifacial - a term used to describe leaves that have both the ab- and adaxial surface visible (*cf.* unifacial).

bifid - cleft into two more or less equal parts.

bilabiate - condition of a calyx or corolla with 2 lips, usually an upper and a lower one.

bisexual - containing both pollen- and seed-producing organs.

blade - the expanded, distal portion of an organ, such as a leaf or petal, as opposed to the narrow, basal portion.

bloom - a white or white-blue powdery or waxy coating that can be rubbed away.

bract - a modified leaf that subtends an inflorescence or flower that is not part of the flower proper, commonly of reduced size composed with the foliage leaves.

bracteal - with the form and position of bracts.

bracteate - having bracts.

bracteolate - having bracteoles.

bracteole - diminutive of bract; a bract of the second order.

bractlet - same as bracteole.

branchlet - an ultimate segment of a branch.

bristle - a stiff hair.

bud - an undeveloped shoot, inflorescence, or flower, in woody plants often covered by scales and serving as the overwintering stage.

bulb - a group of modified leaves serving as a food-storage organ, borne on a short, vertical, underground stem (*cf.* corm).

bulbiferous - bearing or having bulbs.

bulblet - diminutive of bulb; a new bulb arising from a parent bulb; a new bulb borne in the leaf axils or in place of flowers.

bulbil - same as bulblet (*q.v.*).

bulbous - swollen.

bundle scar - one of the marks found within a leaf scar that indicate the number and position of the vascular bundles that passed from stem to leaf.

ca. - about, approximately (Latin = *circa*).

caducous - falling off early, as stipules that leave behind a scar.

callosity - a hardened thickening.

callous - with a thick, firm texture.

callus - a firm, thickened portion of an organ; the firm base of the lemma in the Poaceae.

calyculate - possessing a set of small bracts at the base of an involucre.

calyptra - a lid or cap that encloses or sits atop a structure and is eventually pushed off by the structure as it develops.

calyptrate - possessing a calyptra.

calyx - sepals taken collectively.

campanulate - a concavely flaring tube; bell-shaped.

caniculate - channeled longitudinally.

capillary - very fine, hair-like, not-flattened.

capitate - abruptly expanded at the apex, thereby forming a knob-like tip.

capitulescence - the arrangement of capitula on a plant.

capitulum - a type of inflorescence, found mostly in the Asteraceae, composed of many flowers borne on an expanded receptacle and subtended by a common involucre.

capsule - a dry, dehiscent, fruit derived from more than 1 carpel (therefore dehiscing by 2 or more valves) and lacking a replum.

carpel - one type of modified leaves of a flower, bearing the ovules and usually composed of an ovary, style, and stigma.

carpellate - bearing carpels.

carpophore - a stalk-like prolongation of the receptacle forming a central axis of the flower to which the base of the petals and filaments may be adnate.

cartilaginous - nearly as firm as bone, but flexible.

caruncle - an appendage at or near the hilum of some seeds.

caryopsis - a specialized type of achene in which the seed coat is adnate to the pericarp, found in the Poaceae.

cataphyll - a leaf sheath that does not bear a blade

caudate - having a tail-like appendage; a leaf apex drawn out into a very long tapering point, more so than acuminate.

caudex - firm, hardened, summit of a root mass that functions as a perennating organ.

caudicle - the slender stalk of a pollinium in the Orchidaceae.

caulescent - possessing a stem.

cauline - of or pertaining to the above ground portion of the stem.

cespitose - growing in a compact cluster with closely-spaced stems; tufted.

cf. - compare [Latin = *confer*].

chaff - thin, dry scales borne on the receptacle in the Asteraceae.

channeled - with a deep, rounded groove.

chartaceous - with the texture of waxed paper.

chasmogamous - flowers that are open and cross-pollinate (*cf.* cleistogamous).

chlorophyll - the essential green pigment found in some plant organs.

cilia - hairs found at the margin of an organ.

ciliate - provided with cilia.

ciliolate - diminutive of ciliate.

circinate - a type of vernation with the tip curled up, such that the apex is the center of the coil.

circumboreal - an encircling region of boreal habitats north of the temperate region and south of the arctic region.

circumscissile - dehiscing by an encircling, transverse line so that the top comes off like a lid or cap.

clathrate - with a pattern of parallel areoles; lattice-like.

clavate - widened in the distal portion, like a baseball bat.

claw - the narrow, basal portion of perianth parts (*cf.* blade).

cleistogamous - flowers that remain closed and self-pollinate (*cf.* chasmogamous).

cohere - to stick together, but not actually be fused, applied to like organs (*cf.* adhere, connivent).

coiled raceme - an indeterminate inflorescence where all the pedicels arise on the same side of the axis and the entire inflorescence is arched or coiled near the apex.

coma - tuft of hairs that aids in wind dispersal of seeds.

commissure - the surface where 2 carpels cohere, normally applied to schizocarpic fruits.

comose - with a coma.

compound - composed of multiple parts.

compound leaf - a leaf that is divided to the midrib, with distinct, expanded portions called leaflets.

concolored - with the parts or sides of similar or nearly similar color.

concrescent - grown together.

conduplicate - folded lengthwise into nearly equal parts.

conic - cone-shaped, used to describe a 3-dimensional object.

connate - describing 2 like parts that are fused (*cf.* adnate, distinct).

connivent - converging and touching but not actually fused, applied to like organs (*cf.* adhere, cohere).

convolute - arranged such that one edge is covered and the other is exposed, usually referring to petals in bud.

cordate - with a rounded lobe on each side of a central sinus; heart-shaped.

coriaceous - with a firm, leathery texture.

coralloid - coral-like.

corm - a short, vertical, enlarged, underground stem that serves as a food storage organ (*cf.* bulb).

corpusculum - a sticky gland to which a pollinium is attached by way of a translator arm that aids in the pollinia's dispersal, found in the Apocynaceae.

corymb - an indeterminate inflorescence, somewhat similar to a raceme, that has elongate lower branches that create a more or less flat-topped inflorescence.

costa (plural: **costae**) - a prominent midvein or midrib of a leaflet.

costule - diminutive of costa; a second order costa, found on a leafule.

crenate - with rounded teeth.

crenulate - finely crenate.

crest - an elevated, often complex, appendage or rib on an organ.

crisped - a leaf margin that is undulate or crimped perpendicular to the plane of the leaf surface.

crown - a form that pappus can take, in which the ovary is surrounded by a lobed or toothed rim.

cucullate - an organ that is arched and enfolding; hood-shaped.

culm - a stem, often used for graminoid species.

cuneate - tapering to the base with relatively straight, non-parallel margins; wedge-shaped.

cupulate - cup-shaped.

cupule - a diminutive, cupulate structure.

cyathiescence - the arrangement of cyathia on a plant, used in the Euphorbiaceae.

cyathium (plural: **cyathia**) - a type of pseudanthial inflorescence in which carpellate and staminate flowers lacking perianth are subtended by a set of bracts that are connate into a cup-like structure and sometimes bearing petaloid appendages, the entire unit resembling a single flower, found in the Euphorbiaceae.

cylindric - with the 3-dimensional shape of a cylinder.

cyme - a class of determinate inflorescences, in which the terminal flower of the inflorescence develops first and then each flower enters anthesis in succession to the base.

cymule - same as dichasium.

cypsela - a special type of achene that has an adnate pappus, found in the Asteraceae and Caprifoliaceae.

cystolith - concretions of calcium carbonate, appearing as dots on the leaf surfaces in some Urticaceae.

deciduous - not persistent, falling off after the normal function.

declined - curved downward.

decompound - repeatedly compound into numerous segments, often in an irregular fashion.

decumbent - a stem that is prostrate at the base and curves upward to have an erect or ascending, apical portion.

decurrent - possessing an adnate line or wing that extends down the axis below the node, usually referring to leaves on a stem (*cf.* surcurrent).

decurved - curving downward.

decussate - arrangement of organs with 2 organs produced at each node, and each successive node with organs set at right angles to the orientation of those of the previous node, usually referring to branches or leaves.

deflexed - abruptly bent downward.

dehiscent - condition of a structure that splits open, thereby releasing contents such as seeds or pollen.

deltate - triangle-shaped.

dentate - provided with outward oriented teeth (*cf.* crenate, serrate, undulate).

denticulate - diminutive of dentate; with small, outward oriented teeth.

depauperate - poorly developed due to unfavorable conditions.

determinate - growing to a fixed size, with the apex or apical portion developing first, then successively developing to the base (*cf.* indeterminate).

diadelphous - the condition of stamens that have filaments connate in 2 groups, often one distinct and the other fused.

dichasial cyme - a cyme that is composed of 2 or more dichasia or branches in a dichasial fashion.

dichasium (plural: **dichasia**) - a special type of cyme that bears 3 flowers—a central, earlier blooming flower and 2 lateral, later developing flowers positioned opposite one another (*cf.* monochasium).

dichlamydeous - condition of a perianth composed of 2 series of parts.

dichotomously branched - splitting into 2 essentially equivalent branches at each node.

dimorphic - having 2 forms.

dioecious - having only unisexual flowers of the same type (*e.g.*, only carpellate) on a given plant (*cf.* monoecious, polygamous, synoecious).

distal - toward, or at the end away from, the point of attachment (*cf.* proximal).

distichous - arranged along a stem in 2 ranks.

distinct - not connate, separate from other similar organs (*cf.* adnate, connate).

dithecal - an anther with 2 locules.

divaricate - horizontally spreading.

divergent - spreading, not parallel.

divided - condition of an organ with sinuses extending all the way to the axis, the organ thus considered to be compound (*cf.* lobed).

drupe - a fleshy fruit with a firm, often bony, endocarp that encloses the usually solitary seed. The endocarp may be composed of 1 unit or 2 or more closely associated units called pyrenes.

e, ex - Latin prefix meaning without.

elaiosome - an oily or fatty body or appendage, usually on seeds, offering food to ants.

ellipsoid - 3-dimensional equivalent of an ellipse.

elliptic - more or less shaped like an ellipse, widest in the middle and tapering to more or less rounded ends.

emarginate - provided with a small notch at the apex.

emergent - growing with roots and possibly lower stem in the water but with upper portions of the plant in the air.

endocarp - the inner portion of the pericarp.

ensiform - more or less linear with a sharply pointed apex; sword-shaped .

entire - condition of a leaf-margin that lacks teeth of any kind.

entomophilous - insect pollinated.

epicalyx - a set of bracts that are closely spaced and resemble the true calyx, creating the illusion of more sepals than are actually present.

epigynous - a flower with an inferior ovary, with or without a hypanthium (*cf.* hypogynous, perigynous).

epipetalous - attached to the petals.

epistomic - with stomata on only the adaxial surface.

equilateral - describing the basal portion of leaf blades when each side is similar size and shape (*cf.* inequilateral).

equitant - with the edge, rather than the face, of conduplicate, unifacial leaves set toward the stem.

erect - diverging from an axis at an angle of up to 15°.

erose - with a ragged edge.

evanescent - fading away and not reaching the margin or apex.

excurrent - with the central rib or axis continuing or projecting beyond the organ.

exfoliating - peeling (*e.g.*, birch bark).

exserted - projecting beyond an enclosing structure (*cf.* included).

extrorse - curved or directed outward (*cf.* introrse).

false indusium - a modified tooth or reflexed margin of a fern leaf that covers the sorus (*cf.* indusium).

farina - inflated hairs, commonly white or ochroleucous, that collectively create a mealy coating on a surface.

farinose - provided with farina.

fascicle - a bundle or compact cluster.

fenestrate - with openings or window-like slits.

few - 2 to 4 (*cf.* several, many).

fibrillose - the condition of leaf sheaths that are breaking down into fibers; appearing fibrous.

fibrovascular bundle - a group of specialized conducting cells in the petioles of ferns.

fid - suffix meaning deeply divided.

filament - stalk supporting an anther.

filiform - thread-like.

flabellate - prominently dilated apically with a truncate to rounded apex; fan-shaped.

flaccid - limp, lax, not retaining its form when removed from the water or supporting terrain.

flexuous - condition of an elongate axis that arches or bends in alternating directions in a zig-zag fashion.

floral leaves - leaves associated with flowers or found on the same branch.

floricane - the second year flowering stem of *Rubus* (*cf.* primocane).

floriferous - bearing flowers.

foliaceous - more or less resembling the texture and color and often shape of a leaf.

follicle - a dehiscent, unicarpellate fruit that splits along a single suture on one side of the mature carpel.

fornix - a scale or appendage in the corolla tube of some Boraginaceae.

free - the condition of a plant organ not attached or fused to another type of plant organ (*cf.* adnate, distinct).

free-central placentation - arrangement of ovules attached to a column or axis extending from the base of a compound, unilocular ovary.

fruit - the mature or ripened ovary and any associated structures that ripen and unite with it.

funnelform - constantly flaring tube; funnel-shaped.

fusiform - narrowly tapering at both ends and thickest in the middle; spindle-shaped.

galea - the concave, hood-like, upper lip of some corollas, as in the Veronicaceae.

galeate - possessing or having the form of a galea.

gamo - Greek prefix meaning the parts are connate (*cf.* apo).

gamophyllous - having connate leaf bases, such that the stem appears to pierce through the leaves.

gemma (plural: **gemmae**) - a small vegetative bud that separates from the parent plant and forms a new plant.

gemmiferous - bearing or producing gemmae.

gibbous - conspicuously swollen on one side.

girdle - an encircling mark or ridge; a zone on the immediate, distal side of the *Isoetes* megaspore.

glabrate - nearly glabrous or becoming so.

glabrous - lacking hairs.

gland - a protuberance or depression on an organ or at the summit of a hair that produces a sticky or greasy substance.

glandular - bearing glands (*cf.* eglandular).

glaucous - covered with a bloom.

globose - spherical, shaped like a globe.

glomerule - a compact cluster of 2 or more flowers.

glume - scale that is empty and does not subtend flowers, found at the base of Poaceae spikelets, where it is usually paired, or at the base of *Eleocharis* spikes.

glutinous - covered with a sticky or tacky substance.

graminoid - with the form of grasses, specifically with narrow, linear leaves, often used to refer to the Cyperaceae, Juncaceae, and Poaceae.

gynaecandrous - a spike with both staminate and carpellate flowers, the staminate located at the base, below the carpellate (*cf.* androgynous).

gynobase - a prolonged portion of the receptacle in the Boraginaceae and Lamiaceae.

gynobasic style - a style attached to the gynobase and individual carpels.

gynoecium - the carpels of a flower taken collectively.

gynophore - a stalk that elevates the gynoecium above the base of the flower.

gynostegium - a short-cylindric structure formed from adnate filament, anther, and style material, found in some Apocynaceae.

halophyte - a plant adapted to growing in a salty substrate.

hastate - more or less triangular in outline with outward oriented basal lobes; halberd-shaped.

haustorium - a specialized, root-like connection to a host plant that a parasite uses to extract nourishment.

hemi - Greek prefix meaning half.

hemiparasite - a partially parasitic plant that possesses chlorophyll (*cf.* holoparasite).

herb - a plant that dies back to the ground at the end of each growing season.

herbaceous - herb-like, not woody; having the texture of leaves.

herbage - the above ground vegetative parts of a plant (*e.g.*, stems, leaves).

heterophyllous - having foliage leaves that are larger than bracteal leaves.

hilum - the scar at the point of attachment of seeds.

hirsute - pubescent with coarse, somewhat stiff, usually curving hairs, coarser than villous but softer than hispid.

hispid - pubescent with coarse, stiff hairs that may be uncomfortable to the touch, coarser than hirsute but softer than bristly.

hispidulous - diminutive of hispid.

holoparasite - a completely parasitic plant that lacks chlorophyll (*cf.* hemiparasite).

homophyllous - have foliage leaves and bracteal leaves of similar size (*cf.* heterophyllous).

hood - a tubular-shaped portion of the corona in some Apocynaceae.

horn - a slender process borne at the base of the hoods of the corona in some Apocynaceae.

hydrophilous - water pollinated.

hydrophyte - a plant adopted to growing in the water.

hypanthium - a ring or cup-like section of tissue in a flower formed from the fusion of the basal portion of the sepals, petals, and stamens.

hypogynous - a flower with a superior ovary that lacks a hypanthium (*cf.*, epigynous, perigynous).

hypostomic - with stomata only on the abaxial surface.

imbricate - overlapping, as shingles on a roof; arrangement of bud scales with the outer/lower ones overlying and partially concealing the inner/upper ones.

included - shorter than and contained within an enclosing structure (*cf.* exserted).

incumbent cotyledons - cotyledons oriented such that the flat surface of one is set against the radicle (*cf.* accumbent cotyledons).

indehiscent - not separating or splitting open at maturity.

indeterminate - a structure that develops from the base and continues to elongate or expand throughout the growing season (*cf.* determinate).

indument - the epidermal outgrowths of a plant taken collectively, this term does not include the ramentum of ferns.

indurate - hard, firm.

indusium - an epidermal outgrowth of tissue that covers a sorus, found in the ferns (*cf.* false indusium).

inequilateral - describing the basal portion of leaf blades when one side is a size and/or shape that is different from the tissue on the other size of the midvein (*cf.* equilateral).

inferior - occurring below a particular structure(s), as an inferior ovary that is below the points of attachment of sepals, petals, and stamens.

inflorescence - arrangement of flowers on a stem.

infructescence - arrangement of fruits on a stem.

internode - portion of a stem between two successive nodes.

introrse - curved or directed inward (*cf.*, extrorse).

invest - to enclose or envelop.

involucel - diminutive of involucre; an involucre of the second order.

involucral bract - one bract of an involucre.

involucre - a structure that basally surrounds another structure, used to describe a series of bracts at the base of an inflorescence.

involute - rolled inward/upward toward the adaxial surface (*cf.* revolute).

isodiametric - approximately the same size in all dimensions.

keel - a sharp, longitudinal ridge where two surfaces join; the 2 lower, united petals in the corolla of the Fabaceae.

labellum - the modified, usually lowest, petal of the Orchidaceae.

lacerate - appearing torn, with an irregular and jagged margin.

lacuna (plural: **lacunae**) - a space or gap located within tissues.

lanceolate -widest below the middle and tapering at both ends, narrower than ovate but wider than linear; lance-shaped.

lateral - on the side(s), neither adaxial nor abaxial.

latex - a clear or colored liquid produced in some plants, containing colloidal particles of terpenes in water.

latitudinal - perpendicular to the main axis (*cf.* longitudinal).

leaflet - one of the expanded portions of a compound leaf; an ultimate segment of a once-divided leaf.

leafule - one of the expanded portions of a compound leaflet; an ultimate segment of a twice-divided leaf.

legume - a dehiscent fruit that splits along one suture and is derived from a unicarpellate ovary, found in the Fabaceae.

lemma - one of the pair of bracts that subtend the florets in the Poaceae oriented so its adaxial surface faces the axis of the spikelet (*cf.* glume, palea).

lentibule - a hypothetical structure at or near the apex of an amical, often without glands.

lenticel - a slightly raised portion, often of different color, on the bark of woody plants without a corky texture.

lenticular - lens-shaped.

lepidote - covered with small scales that are attached near their center.

liana - a woody plant that trails over the ground or climbs on/over other vegetation.

ligule - an appendage produced at the junction of the sheath and blade of the adaxial surface of a leaf in many Poaceae and Cyperaceae; the flattened portion of the ray flower in the Asteraceae.

limb - the expanded portion of a gamopetalous corolla, distal to the throat.

linear - very narrow, with more or less parallel sides, narrower than lanceolate.

lip - a projecting portion of the calyx or corolla.

lobe - a projecting portion of an organ too large to be referred to as a tooth.

lobed - possessing lobes, with the sinuses between the lobes not extending all the way to the axis (*cf.* divided).

locule - a seed-containing chamber in an ovary; a chamber in any structure that contains one or more items.

loculicidal - a type of dehiscence in which a capsule splits along the midrib or outer median line of each carpel such that the capsule dehisces through each carpel (*cf.* septicidal).

lodicule - a tiny scale found in the flowers of the Poaceae that is the vestigial perianth.

loment - a legume composed of 1-seeded segments that may disarticulate from one another.

long shoot - a typical branch or shoot with remote leaves (*cf.* short shoot).

longitudinal - parallel to the main axis (*cf.* latitudinal).

lunate - crescent-shaped.

lustrous - shiny, possessing a luster.

maculate - spotted or splotched.

many - 11 or more (*cf.* few, several).

marcescent - withering but persistent and not falling.

marginal - pertaining to or borne on or near the margin.

marginal placentation - a type of placentation in which the ovules are borne at the connate margins of a simple carpel.

megaspore - a spore that gives rise to a female gametophyte, usually much larger than a microspore.

megasporophyll - a leaf, often modified (reduced and/or without chlorophyll), that bears megaspores in its axils.

membranaceous - thin, flexible, almost translucent; like a membrane.

mericarp - each half- or single-carpellate segment of a schizocarpic fruit.

merous - Greek suffix meaning members of a set or cycle, used to indicate the numbers comprising each set of the perianth.

mesic - moist, neither dry nor wet (*cf.* hydric, xeric).

microspore - a spore that gives rise to a male gametophyte, usually much smaller than a megaspore.

mineral soil - soils composed mostly of degraded parent rock material such as clay, silt, and sand.

mixed inflorescence - an inflorescence with an indeterminate, central axis and determinate branches.

monadelphous - the condition of stamens that have all of the filaments connate into a single group.

moniliform - arrangement of a set of spherical or ovoid bodies that are separate and non-overlapping on an elongate axis; bead-like.

monocarpic - flowering and fruiting only once and then dying.

monochasium (plural: **monochasia**) - a 2-flowered cymule, a terminal flower and a single, lateral flower (*cf.* dichasium).

monochlamydeous - perianth comprised of 1 series of parts.

monoecious - having unisexual flowers, with both the carpellate and staminate flowers borne on the same plant (*cf.* dioecious, polygamous, synoecious).

monomorphic - having one form.

monopodial - with a simple axis from which branches or appendages originate (*cf.* sympodial).

monothecal - an anther with a single locule.

mucro - a short, abrupt, projecting tip of an organ.

mucronate - having a mucro.

mucronulate - diminutive of mucronate.

multiple fruit - a fruit composed of 2 or more ovaries from separate flowers (*cf.*, aggregate fruit, simple fruit).

muricate - with small, pointed projections.

nectary - a structure that produces nectar.

nectariferous - bearing nectar.

nerve - a conspicuous vein of an organ.

node - point of attachment of a leaf or leaves to a stem.

nutlet - diminutive of nut.

ob - Latin prefix indicating a reverse in direction or orientation (*e.g.*, obovate - egg-shaped in cross-section with the wider end near the apex).

oblique - emerging or joining at an angle other than parallel or perpendicular.

obloid - 3-dimensional equivalent of oblong.

oblong - shaped similar to a rectangle that is longer than wide.

obsolete - reduced to the point of being undetectable or virtually absent, implies presence in ancestral forms (*cf.* rudimentary, vestigial).

obtuse - bluntly pointed, with the margins forming an angle of greater than 90° (*cf.* acute).

ocrea (plural: **ocreae**) - the tubular, sheathing stipule located apically to the petiole, found in the Polygonaceae.

ocreola (plural: **ocreolae**) - the tubular bracteole subtending the flowers in the Polygonaceae.

ochroleucous - a pale yellow-white color.

oligotrophic - lakes that lack considerable amounts of aquatic vegetation, usually deep, clear, and with sandy or rocky bottoms.

opposite - positioned directly across from one another; a node that bears 2 similar organs.

orbicular - circular in outline.

organic soil - soils composed mostly of partially degraded plant material, such as peat and muck.

orifice - an opening into a cavity.

ornamented - with elaborate processes or patterns.

oval - broad-elliptic.

ovary - the expanded basal portion of a carpel that bears the ovules.

ovate - 2 dimensional equivalent of egg-shaped, widest below the middle and broadly tapering to each end, wider than lanceolate.

ovoid - the 3-dimensional equivalent of ovate; egg-shaped.

ovule - an immature seed.

palate - a raised area on the lower lip of a bilabiate corolla that can obscure or close off the opening.

palea - one of a pair of bracts subtending a single grass floret, oriented so its adaxial surface faces away from the axis of the spikelet (*cf.* glume, lemma).

palmate - with structures radiating from a common point, like fingers or parts of a fan.

pandurate - with a constriction near the middle of the organ.

panicle - a branched, indeterminate inflorescence.

papillae - short, rounded projections.

papillate - covered with papillae.

papillose - same as papillate.

pappus - the modified calyx in the Asteraceae, found crowning the ovary, and appearing as bristles, scales, or awns or a mixture of these.

parasite - a plant that acquires its food and water through specialized attachments to a host plant.

parietal placentation - a type of placentation in which the partitions extend only part way toward the center of the ovary and the ovules are borne on the partitions, found in ovaries composed of 2 or more carpels.

pectinate - the condition of an unbranched axis with a single row of narrow appendages emitting from one or both sides.

pedicel - stalk of a single flower or fruit in an inflorescence or infructescence (*cf.* peduncle).

pedicellate - borne on a pedicel.

peduncle - a stalk that supports an inflorescence composed of one or more flowers (*cf.* pedicel).

pedunculate - borne on a peduncle.

pellucid - transparent or translucent.

peltate - attachment of an organ at the center rather than at the base; umbrella-like.

pendulous - limply hanging from near the base.

penicillate - provided with a tuft of hairs at the apex.

pentadelphous - the condition of stamens that have filaments connate into 5 groups.

penultimate - next to last.

pepo - a specialized type of berry with an inseparable, often leathery, rind, found in the Cucurbitaceae.

perennate - to survive the winter season, often in a form different from that of the summer phase of the plant (*e.g.* buds, seeds, corms).

perennial - a plant that lives for more than 2 years.

perfoliate leaf - a leaf that has the basal margins connate, creating the illusion that the stem pierces the middle of the leaf.

perianth - the calyx, corolla, and corona taken collectively, one or more of which may be absent.

pericarp - the mature ovary wall of a fruit.

perigynium (plural: **perigynia**) - a specialized bract that subtends the carpellate flowers in *Carex*, whose margins are united thereby forming a sac-like structure with an orifice at the apex through which the stigmas protrude.

perigynous - condition of a flower with a superior ovary and a hypanthium (*cf.* epigynous, hypogynous).

persistent - remaining attached through the year or after the normal function.

petal - one of the modified leaves of a flower, usually pigmented and thereby attracting pollinators (*cf.* epigynous, hypogynous).

petaliferous - producing or possessing petals.

petaloid - resembling a petal (usually in shape and color).

petiole - non-expanded, basal portion of a leaf.

petiolate - having a petiole.

petiolulate - having a petiolule.

petiolule - non-expanded, basal portion of a leaflet.

photosynthetic - capable of photosynthesizing due to the presence of chlorophyll.

phyllary - one bract of the involucre that subtends a capitulum.

phyllopodic - having the basal sheaths blade-bearing; with new shoots arising from the center of parent shoot (*cf.* aphyllopodic).

phyllotaxis - arrangement of nodes on the twig in alternate-leaved species. Phyllotaxis is given as 2 numbers separated by a slash (*e.g.*, 2/5). The first number (2 in the example) refers to the number of revolutions around the twig that must be made to reach a position on exactly the same side of the twig that one started from. The second number refers to the number of nodes (5 in the example) that must be counted in a given direction to reach the same position that one started from. Example: Given a plant with a phyllotaxis of 2/5, an observer, starting at a node or winter bud, would have to make 2 revolutions around the twig and count 5 nodes or winter buds in the process to reach a node or winter bud oriented in the same direction as the one started from.

phytomelin - a black crust found on seeds.

pilose - with sparse, straight, and spreading hairs.

pin - one type of heterostylic flower in which the style is longer than the stamens (*cf.* thrum).

pinnate - arranged along both sides of an elongate axis, such as a feather.

pinnatifid - pinnately lobed.

pistil - the outward appearance of the carpels of a flower.

pith - the internal cylinder of parenchymous tissue in a stem or root.

placentation - arrangement of ovules in an ovary, specifically, how they are attached.

planoconvex - cross-sectional shape, for objects with 2 surfaces, that is flat on one surface and convexly rounded on the other.

plumose - like a tuft of hairs or feathers; specifically applied to hairs or bristles that have lateral branches arising in a pinnate fashion.

pollen - the microgametophyte of seed plants, contained in anthers.

pollinium - a mass of pollen grains cohering as a single unit, as in the Orchidaceae and some Apocynaceae.

polygamous - with both unisexual and bisexual flowers on the same plant (*cf.* dioecious, monoecious, synoecious).

pome - a special type of berry found in some Rosaceae; specifically, a fleshy fruit derived from a compound, inferior ovary, with a papery, cartilaginous, or bony endocarp and usually 2 or more seeds.

praemorse - abruptly ending in a truncated fashion.

precocious - the condition of the flowers' developing before the leaves (*cf.* coaetaneous, serotinous).

prickle - small, more or less sharp, epidermal outgrowth (*cf.* spine, thorn).

primocane - the first-year, vegetative stem in *Rubus* (*cf.* floricane).

process - a slender, protruding feature.

propagule - a body with the function of reproduction, such as a seed, turion, or bulb.

prophyll - one of a pair of bracteoles that subtends the perianth in *Juncus*.

proximal - toward or at the point of attachment (*cf.*, distal).

pseudanthium (plural: **pseudanthia**) - a cluster of small flowers that appear to be a single flower.

pseudoterminal bud - the distalmost bud on an indeterminate branch, identified by the presence of a twig scar adjacent to the bud (*cf.* terminal bud).

pseudowhorl - falsely appearing to be whorled.

puberulent - diminutive of pubescent.

pubescence - an epidermal outgrowth, not as stiff to be a prickle, not flattened as a scale; a hair.

pubescent - with pubescence.

pulvinate - densely matted together in clumped colonies; cushion-like.

punctate - dotted with small glands or pits.

pustule - a blister.

pustulose - having or with the form of pustules.

pyrene - one of the units comprising the endocarp of a drupe, used when the endocarp is composed of 2 or more units.

pyriform - pear-shaped.

pyxis - a dehiscent fruit that is circumscissile around the middle, the top falling off and exposing the seed(s).

q.v. - abbreviation of Latin *quod vide*, "which see", a cross-reference.

raceme - an indeterminate inflorescence with a central axis, often elongate, with pedicellate flowers.

rachilla - diminutive of rachis; the axis of the flowers in a grass spikelet or a sedge flower.

rachis - the central or main axis of a structure.

ramentum - the flattened scales of the leaves and stems of ferns.

rank - a longitudinal position on a stem.

ray - the flat, usually elongate, portion of a ray flower corolla.

ray flower - the zygomorphic, ray-bearing flower in the Asteraceae.

receptacle - an expanded portion at the apex of the pedicel or peduncle to which the parts of a flower are attached.

reflexed - abruptly bent backward or downward.

remote - separated to some extent from others of the same kind (*e.g.*, remote leaves, remote spikes).

reniform -wider than long, with a central, basal sinus; kidney-shaped.

replum - a frame-like placenta positioned between the valves of a silique or a silicle.

resupinate - possessing a half-twist in the pedicel or ovary so that the flower is upside down.

reticulate - forming a network due to a pattern of splitting and rejoining.

retrorse - downward or backward oriented (*cf.* antrorse).

retuse - with a terminal notch in an otherwise blunt or rounded apex.

revolute - rolled under, toward the abaxial surface (*cf.* involute).

rhizomatous - bearing rhizomes.

rhizome - a horizontal, underground stem.

rhizophore - modified leafless shoots that produce shoots.

rhombic - diamond-shaped.

rostellum - a cap-like or beak-like, sterile projection of the stigma.

rotate - the shape of a corolla that is nearly circular in outline with widely spreading lobes and lacking a tubular, basal portion; saucer-shaped.

rotund - circular in outline.

rudimentary - poorly developed and reflecting an earlier evolutionary stage, not a reduction of the feature (*cf.* obsolete, vestigial).

rugose - wrinkled.

rugulose - diminutive of rugose.

runner - an ultimate branch segment of a stolon or rhizome.

saccate - bearing, or in the shape of, a pouch or bag.

sagittate - arrow-shaped, specifically applied to structures with backward oriented basal lobes.

salverform - describing a corolla with a slender, basal tube and a more or less horizontally spreading limb.

samara - a specialized type of achene that bears a wing or wings that aid in wind dispersal.

scabrous - rough textured due to short, stiff hairs or minute prickles.

scabrule - a single hair or prickle that collectively provides a scabrous texture.

scale - a small, thin, flat structure, not resembling a leaf.

scape - a peduncle that arises from near the ground and lacks leaves (although it may possess small bracts), found in plants that lack aerial stems.

scapose - with the form of a scape.

scarious - thin and dry, often chartaceous and translucent.

schizocarp - a simple fruit composed of 2 or more carpels that separates at maturity into half- or unicarpellate segments called mericarps.

scurfy - beset with small scales, providing a rough, irregular texture.

secund - with flowers or branches oriented or directed to one side of the axis, one-sided.

seed - a mature ovule and perennating organ of many plants.

senesce - to die back, as in the autumn.

sepal - one of the modified leaves of a flower, commonly the outermost series and usually green to brown, lacking bright pigment.

sepaloid - appearing like a sepal usually in color, texture, and shape.

septate - possessing septa.

septum (plural: **septa**)- a partition; a partition formed by the walls of adjacent carpels in an ovary.

septicidal - splitting through the septa, so that the capsule dehisces by separating the individual carpels (*cf.* loculicidal).

sericeous - silky, due to the presence of long, slender, neatly appressed hairs.

serotinous - late, the flowers developing after the formation of the leaves (*cf.* coaetaneous, precocious).

serrate - a margin of a leaf provided with forward pointing, sharp teeth (*cf.* crenate, dentate).

serrulate - diminutive of serrate, the teeth smaller and finer.

sessile - attached at the base, without a stalk of some form.

setaceous - bristle-like.

setose - covered with bristles.

several - 5 to 10 (*cf.* few, many).

sheath - the basal portion of leaves of bracts that surrounds and/or encloses the stem or branches.

short shoot - a peg- or knob-like shoot with closely crowded leaves (*cf.* long shoot).

shrub - a woody plant lacking a tree-like form, usually shorter than 6.0 m and with many stems at the base.

silicle - similar to a silique, but up to 3 times as long as wide.

silique - a fruit that dehisces by 2 valves, leaving the persistent replum to which the seeds are attached, found only in the Brassicaceae, this name is reserved for fruits greater than 3 times as long as wide (*cf.* silicle).

simple - undivided.

simple fruit - a fruit derived from a single ovary of a flower (*cf.* aggregate fruit, multiple fruit).

sinus - the indented area between 2 lobes of an organ.

sobol - a stem arising from axillary buds, near the ground, that are horizontal for a very short distance and then turn to become upright, aerial stems.

soboliferous - bearing sobols.

sordid - with a dingy, gray coloration.

sorus (plural: **sori**) - aggregation of spore-bearing structures of ferns.

spadix - a spike-like inflorescence with flowers crowded on a fleshy axis.

spathe - a relatively large, solitary bract that subtends and commonly encloses the inflorescence.

spatulate - abruptly widened from a narrow base into a somewhat rounded apex; spoon-shaped.

spicule - a minute, stiff, spine-like process on the margins of an organ.

spike - an indeterminate inflorescence consisting of an elongate axis with sessile flowers.

spikelet - diminutive of spike; the second order designation of a compound spike; one or more florets of a grass flower subtended by a pair of glumes.

spine - a slender, firm, sharp structure derived from a leaf or a portion of a leaf (*cf.* prickle, thorn).

spinescent - with the form of or bearing a spine.

spinose - bearing spines.

spinule - diminutive of spine.

spinulose - diminutive of spinose.

sporangiaster - modified sporangia, forming an indument that partly covers the sorus.

sporangiophore - a stalk that bears sporangia.

sporangium - a structure that houses or contains spores.

spore - a one-celled reproductive structure in ferns and fern allies, giving rise to the gamete-bearing plant.

sporocarp - a hardened, capsule-like structure derived from a modified leaflet that encloses the sorus in the Marsileaceae.

sporophore - the fertile, spore-bearing portion of the leaf blade in the Ophioglossaceae (*cf.* trophophore).

sporophyll - a specialized leaf that bears a sporangium.

spreading - diverging from an axis by an angle of 45–90°.

spur - a hollow, often elongate, process borne on the calyx or corolla; a slender process, often resembling a bristle, borne on the stamens of some species.

squarrose - abruptly curving outward near the apex.

stamen - one series of modified leaves in a flower that bear pollen; anther and filament collectively.

staminate - describing a structure or plant bearing stamens.

staminode - a sterile, modified stamen.

stellate - compound hairs that branch 3 or more times at the base.

sterigma (plural: **sterigmata**) - small, woody projection of a twig of some conifers to which the leaf is attached.

stigma - pollen-receptor surface and associated tissue at the distal end of a style (when the latter is present).

stipe - a stalk of a structure, used when more precise terms (such as petiole, peduncle, style, *etc.*) are inappropriate.

stipel - an appendage, often paired, found at the base of petiolules in some species.

stipitate - borne on a stipe.

stipule - an appendage, often paired, found at the base of the petiole in some species.

stolon - a horizontal stem at or just above the surface of the ground.

stoloniferous - producing and/or possessing stolons.

stomate - an opening in the epidermal tissue of a leaf through which gas exchange occurs.

stramineous - light yellow-brown; straw-colored.

striate - ornamented with parallel lines.

strigose - pubescent with straight, neatly appressed hairs.

strobilus (plural: **strobili**) - the reproductive organ of the ferns, fern allies, and conifers, consisting of a cluster of sporophylls on an axis; a cone.

strophiole - an appendage at the hilum of some seeds.

style - the stalk that connects the stigma(s) to the ovary.

stylopodium (plural: **stylopodia**) - a disk-like or cylindrical enlargement of the basal portion of the style.

sub - Latin prefix meaning under or nearly so.

submersed - below the surface of the water.

subtend - to be positioned at the base of, as with bracts of a flower.

subulate - narrowly tapering from the base to the apex; awl-shaped.

succulent - fleshy and/or juicy.

suffrutescent - firmer than herbaceous, somewhat woody or woody only near the base.

suffruticose - somewhat woody.

superior ovary - an ovary attached to the receptacle above the insertion of all other flower parts (*cf.* inferior ovary).

surcurrent - possessing an adnate line or wing that extends up the axis above the node (*cf.* decurrent).

suture - a line of fusion that will eventually split in a dehiscent fruit.

sympodial - with a compound axis composed of short, axillary branches that supersede or overgrow the original, terminal bud of each axis segment (*cf.* monopodial).

syncarpous - describing a gynoecium composed of connate carpels.

synoecious - having only bisexual flowers (*cf.* dioecious, monoecious, polygamous).

taproot - the large, central root.

tendril - a specialized leaf, branch, or inflorescence that aids a vine or liana in climbing.

tepals - petals and sepals that are indistinguishable from each other due to overall similarity.

terete - circular in cross-section.

terminal - the distal-most position.

terminal bud - position of the distalmost bud on a determinate twig, identified by the absence of a twig scar adjacent to the bud.

ternate - divided into 3 parts.

terrestrial - pertaining to non-aquatic, land environments.

testa - mature seed coat derived from mature integuments of the ovule.

tetradynamous - description of a 6-stamened androecium in which 4 stamens are longer than the other 2, found commonly in Brassicaceae.

tetragonous - 4-angled, more or less square in cross-section.

thalloid - appearing as a thallus.

thallus - a body that is not differentiated into a stem and leaf.

theca - a locule of an anther.

thorn - a short, stiff, pointed process that is a modified branch.

throat - the narrow, basal portion of a sympetalous corolla, proximal to the limb.

thrum - one type of heterostylic flower in which the stamens nearly equal to exceed the length of the style(s) (*cf.* pin).

thyrse - a mixed inflorescence with racemosely arranged cymes.

tomentose - pubescence consisting of curled, tangled, often matted, hairs; woolly.

tomentulose - diminutive of tomentose.

torus - the receptacle of a flower or capitulum.

trifid - cleft into 3 more or less equal parts.

trigonous - triangular in cross-section.

triquetrous - 3-angled with each angle projecting in a thin, wing-like rib.

trophophore - the sterile, photosynthetic portion of the leaf in the Ophioglossaceae (*cf.* sporophore).

truncate - with the base or apex of an organ latitudinally straight, appearing to have been cut off.

tubercle - a swelling or projection borne on an organ, usually of different color and/or texture.

tuberculate - bearing tubercles.

tuber - a thickened food-storage portion of a rhizome.

tuberiferous - bearing tubers.

tuberous - tuber-like.

turion - a small, bulb-like, perennating organ.

twining - growing in a curling, encircling fashion, thereby gaining support from other vegetation or structures.

umbel - an indeterminate inflorescence, somewhat similar to a raceme, which has a greatly reduced central axis, such that the flowers appear to originate from the same point, usually resulting in a more or less flat-topped inflorescence.

umbellet - diminutive of umbel; the ultimate unit of a compound umbel.

umbo - a raised portion of the apophysis of a seed cone.

unarmed - without any prickles, spines, or thorns.

undulate - wavy-margined, without prominent teeth.

unifacial - leaves that have margins united so that only the abaxial surface is visible (*cf.* bifacial).

unisexual - a flower that bears only stamens or only carpels; a flower that bears both stamens and carpels but one of which is non-functional or abortive and the flower is functionally unisexual (*cf.* bisexual).

urceolate - constricted at a point just before an opening; urn-shaped.

utricle - a specialized type of achene in which the pericarp is loosely attached and often readily removed.

vallecular cavity - a set of longitudinal cavities in the stems of *Equisetum* external to the central cavity and internal to the carinal cavities.

valvate - the condition of a structure that opens by valves; a position of petals in bud or bud scales that have the margins of each touching but not overlapping (*cf.* imbricate).

valve - one portion of an ovary wall that separates when a capsule dehisces; the inner set of accrescent tepals in *Rumex*.

vascular bundle - a strand of xylem, phloem, and associated tissue, usually used in reference to fern petiole cross-sections or leaf scars in woody plants.

vein - a visible vascular bundle passing through an organ, same as nerve.

veinlet - the ultimate branch of a vein.

velum - the thin, membranous tissue that covers the sporangium in *Isoetes*.

velutinous - pubescent with fine, short, spreading hairs; velvety feeling.

venation - the pattern of veins on an organ.

vernation - the orientation of leaves in bud.

versatile - an arrangement of anthers in which the filament is connected to the center of an anther (*cf.* basifixed).

verticil - one whorled cycle of organs.

verticillate - arranged in whorls.

vestigial - poorly developed due to evolutionary reduction (*cf.* obsolete, rudimentary).

villous - pubescent with long, soft, bent hairs, the hairs not crimped or tangled.

vine - an herbaceous plant that does not fully support its own weight but rather trails along the ground or climbs over other vegetation often by means of special climbing appendages or growth form.

virgate - a tall, slender, erect form.

viscid - sticky, glutinous.

viscidulous - diminutive of viscid.

viscidulum (plural: **viscidula**) - a sticky gland to which the pollinia are attached that aids in the pollinia's dispersal, found in the Orchidaceae.

wanting - absent or nearly so.

weeds - rapidly spreading plants of disturbed habitats.

whorled - leaf arrangement in which a node has 3 or more leaves.

wing - one of the 2 lateral petals in flowers of the Fabaceae; one of the 2 large, petaloid sepals in flowers of the Polygalaceae; a thin, flat projection from the side, top, or back o a structure.

xeric - dry conditions (*cf.* hydric, mesic).

zygomorphic - bilaterally symmetric, a shape that can be divided into equal halves by only plane (*cf.* actinomorphic).

VASCULAR METAPHYTA OF MAINE
(139 families, 699 genera, 2096 species)

KEY TO THE FAMILIES

1a. Plants typically reproducing by spores; seeds and fruits not produced; gametophyte independent of sporophyte; ferns and fern-like plants .... **Group 1**

1b. Plants typically reproducing by seeds, the seeds borne within a fruit or not; gametophyte dependent on sporophyte; seed plants

  2a. Plants not producing true flowers; seeds commonly borne in strobili on the surface of a scale (embedded in a fleshy aril in *Taxus*) never enclosed in an ovary; styles and stigmas absent; trees and shrubs with narrow, scale- or needle-like, usually persistent, leaves .... **Group 2**

  2b. Plants usually producing true flowers; seeds [ovules] enclosed in an ovary; stigma(s) and usually style(s) present, elevated above the ovary; trees, shrubs, and herbs, commonly with broader, flat leaves

    3a. Leaves usually parallel-veined (or the plants thalloid in some Araceae); seeds with 1 cotyledon; perianth typically 3- or 6-merous; vascular bundles scattered throughout the stem; secondary growth absent .... **Group 3**

    3b. Leaves usually pinnately veined; seeds with 2 cotyledons; perianth typically 4-, 5-, or more, -merous; vascular bundles arranged in a ring around a central pith; secondary growth absent or present

      4a. Plants definitely woody; secondary growth present

        5a. Each node with 2 or more leaves or leaf scars [leaves opposite or whorled] .... **Group 4**

        5b. Each node with 1 leaf or leaf scar [leaves alternate] (often subopposite in *Rhamnus cathartica* and *Salix purpurea*) .... **Group 5**

      4b. Plants herbaceous or suffrutescent; secondary growth absent

        6a. Flowers epigynous or partly so [ovary inferior] .... **Group 6**

        6b. Flowers hypogynous or perigynous [ovary superior] or the flowers lacking a perianth

          7a. Flowers zygomorphic .... **Group 7**

          7b. Flowers actinomorphic or without a perianth

            8a. Gynoecium composed of either a single carpel or 2 or more connate carpels, the ovary thereby superficially appearing as 1

              9a. Corolla gamopetalous .... **Group 8**

              9b. Corolla apopetalous, absent, or very inconspicuous .. **Group 9**

            8b. Gynoecium apocarpous, appearing as 2 or more distinct ovaries ... . **Group 10**

Group 1

1a. Stems conspicuously jointed, bearing at each joint a small whorl of dark leaves that are united at the base; sporangia aggregated in a terminal strobilus, borne on the underside of peltate sporangiophores .... **Equisetaceae**

1b. Stems not conspicuously jointed, the leaves not arranged in distinct and remote whorls; sporangia aggregated in a terminal strobilus or not, but never borne on peltate sporangiophores

2a. Leaves 4-parted, clover-like, usually floating on the surface of water; sporangia borne in hardened sporocarps .... **Marsileaceae**

2b. Leaves neither conspicuously 4-parted nor floating on the surface of water; sporangia borne on the surface of leaves, sometimes concealed

   3a. Leaves narrow, with unexpanded, grass-like blades; sporangia embedded in the leaf bases near the surface of the substrate; plants aquatic .... **Isoetaceae**

   3b. Leaves various, but not grass-like; sporangia variously arranged, but usually elevated above the ground; plants terrestrial, though sometimes found in wet environments

      4a. Leaves small, up to 1.5 cm long, scale-like, with a single midvein; sporangia commonly aggregated in terminal strobili

         5a. Strobili terete in cross-section; leaves without a ligule; spores of 1 size, less than 0.05 mm wide .... **Lycopodiaceae**

         5b. Strobili quadrangular in cross-section (terete in *Selaginella selaginoides*); leaves with a small, adaxial ligule; spores of 2 sizes, the larger more than 0.30 mm wide .... **Selaginellaceae**

      4b. Leaves much larger, foliaceous, prominently veined; sporangia various, but not borne in terminal strobili

         6a. Sporangia thick-walled, lacking an annulus, commonly bearing more than 1000 spores; leaves not circinate in bud .... **Ophioglossaceae**

         6b. Sporangia thin-walled, with an annulus (poorly developed in the Osmundaceae), bearing up to 512 spores; leaves circinate in bud

            7a. Sporangium with a poorly developed annulus composed of a group of apical, thick-walled cells, bearing 256–512 spores; spore-bearing leaves brown at maturity, conspicuously different from the vegetative leaves .... **Osmundaceae**

            7b. Sporangium with a well developed annulus that forms a complete or nearly complete ring around the sporangium, usually bearing only 16–64 spores; spore-bearing leaves normally green and similar to, or slightly different from, the vegetative leaves (except in *Matteuccia* and *Onoclea*) .... **Polypodiaceae**

Group 2

1a. Seeds borne singly, partly concealed by a red, fleshy aril; abaxial surface of the flat leaves bearing longitudinal, pale yellow, stomatal bands .... **Taxaceae**

1b. Seeds borne 1–400 in woody or fleshy strobili, lacking a fleshy, red aril; abaxial surface of the flat to tetragonous leaves without yellow bands

   2a. Leaves opposite or whorled, imbricate and concealing the twig (the twig visible in *Juniperus communis*), persistent on dried specimens; strobili with peltate scales, or with opposite, basifixed scales, or the scales fleshy and coalesced into a berry-like form; each scale of a strobilus bearing 1–20 ovules, without subtending bracts ....
   .... **Cupressaceae**

   2b. Leaves alternate or fascicled, not concealing the twig, deciduous from dried specimens; strobili with flat, alternately arranged, basifixed scales; each scale of a strobilus bearing 2 ovules, subtended by a bract .... **Pinaceae**

Group 3

1a. Plants thalloid, not differentiated into stems and leaves, 0.8–15.0 mm long, floating on the surface of water .... **Araceae**
1b. Plants not thalloid, differentiated into stems and leaves; habit various
  2a. Flowers replaced by, or modified into, vegetative propagules
    3a. Plants from a conspicuous bulb, with a strong odor of onion; stems scapose ....
    .... **Alliaceae**
    3b. Plants without a conspicuous, bulbous base, lacking the odor of onion; stems with leaves
      4a. Leaves terete and septate (*Juncus*) .... **Juncaceae**
      4b. Leaves flat, with or without septa
        5a. Plants submersed aquatics, with flaccid leaves (*Potamogeton*) ....
        .... **Potamogetonaceae**
        5b. Plants terrestrial, with firmer leaves
          6a. Inflorescence a panicle; leaves arranged in 2 ranks; plants of alpine habitats (*Festuca*) .... **Poaceae**
          6b. Inflorescence a spike, with flowers aggregated in an umbel-like cluster; leaves arranged in 3 ranks; plants of sandy or gravelly shores (*Cyperus*) .... **Cyperaceae**
  2b. Flowers usually produced normally, though not always present
    7a. Gynoecium composed of 2 or more distinct carpels [each carpel with a single stigma]; plants commonly aquatic
      8a. Perianth present, showy, differentiated into sepals and petals; petals white ......
      .... **Alismataceae**
      8b. Perianth, if present, reduced, not differentiated into sepals and petals; petals [inner tepals], if present, usually green, yellow-green, or brown
        9a. Plants emersed; leaves terete in cross-section
          10a. Each carpel with 2 ovules; inflorescence bracteate; each leaf with a terminal pore .... **Scheuchzeriaceae**
          10b. Each carpel with 1 ovule; inflorescence without bracts; leaves without terminal pores .... **Juncaginaceae**
        9b. Plants submersed (except sometimes the inflorescence); leaves flat or capillary
          11a. Leaves alternate; flowers bisexual, with 2 or 4 stamens ....
          .... **Potamogetonaceae**
          11b. Leaves opposite; flowers unisexual; staminate flowers with 1 stamen
          .... **Zannichelliaceae**
    7b. Gynoecium composed of 1 carpel or 2 or more connate carpels [then the ovary surmounted by 2 or more stigmas]; plants aquatic or terrestrial
      12a. Plants aquatic or marine, submersed, with flaccid, often translucent, leaves (or with thick, emergent leaves in some *Sparganium*); flowers unisexual
        13a. Flowers [and fruits] in dense, spherical clusters, each flower with 1 stigma; leaves flat or commonly keeled on 1 surface, especially near the base (*Sparganium*) .... **Typhaceae**
        13b. Flowers not in spherical clusters, each flower with 2–6 or more stigmas; leaves flat

14a. Flowers bearing a perianth composed of 3 sepals and 3 petals; androecium usually with 2 or 9 stamens; gynoecium composed of 3–6 weakly united carpels; leaves whorled (at least in part) or all basal .... **Hydrocharitaceae**

14b. Flowers without a perianth; androecium with 1 stamen; gynoecium composed of a single, unilocular, uni- or bicarpellate ovary; leaves alternate or opposite

    15a. Pollen filiform; gynoecium with 2 stigmas; leaves mostly alternate; plants perennial, with rhizomes, marine ..... **Zosteraceae**

    15b. Pollen globose; gynoecium with 2–4 stigmas; leaves opposite; plants annual, without rhizomes, of fresh water habitats .... **Najadaceae**

12b. Plants usually terrestrial, though sometimes of wet habitats, commonly with firmer stems and leaves, mostly emersed; flowers unisexual or, more commonly, bisexual

16a. Inflorescence a spadix

    17a. Spathe narrow, resembling the foliage leaves; leaves ensiform, unifacial, parallel-veined; ovary 2- or 3-locular; anthers introrse .... **Acoraceae**

    17b. Spathe broad, wrapped around and concealing the spadix (not concealing the spadix in *Calla*); leaves broad and flat, bifacial, pinnately veined; ovary unilocular; anthers extrorse .... **Araceae**

16b. Inflorescence various, but not a spadix

    18a. Flowers clearly entomophilous, with an evident, showy perianth

        19a. Ovary superior

            20a. Flowers borne in the axils of firm, imbricate bracts, aggregated in a dense, ellipsoid to ovoid, terminal spike; leaves grass-like; petals yellow .... **Xyridaceae**

            20b. Flowers not from the axils of imbricate bracts, commonly in more open inflorescences; leaves various; petal color various

                21a. Sepals sepaloid, green

                    22a. Leaves grass-like, with parallel venation; inflorescence composed of more than 1 flower; fruit a capsule .... **Commelinaceae**

                    22b. Leaves broad, with pinnately branched, secondary veins; inflorescence composed of a solitary, terminal flower; fruit a berry .... **Trilliaceae**

                21b. Sepals petaloid, of a color similar to that of the petals

                    23a. Perianth connate at the base, thereby forming a tube, zygomorphic or actinomorphic; stamens 3 or 6, adnate to the corolla tube; plants aquatic .... **Pontederiaceae**

                    23b. Perianth distinct or connate, actinomorphic; stamens (4)6, free or adnate to the corolla; plants terrestrial

                        24a. Leaves parallel-veined, generally lacking a well defined petiole and blade, without tendrils

25a. Actual leaves scale-like; upper branches filiform, green, simulating leaves, clustered in the leaf axils .... **Asparagaceae**

25b. Leaves foliaceous; upper branches not clustered in the leaf axils

  26a. Tepals evidently connate; filaments adnate to the tepals

    27a. Tepals 5.0–13.0 cm long, orange or yellow; anthers versatile; leaves mostly 50.0–100.0 cm long .... **Hemerocallidaceae**

    27b. Tepals 0.6–2.2 cm long, white, green-white, or yellow-green; anthers basifixed; leaves 4.0–20.0 cm long

      28a. Perianth minutely roughened on the abaxial [outside] surface; inflorescence a raceme; pedicels ascending; fruit a capsule .... **Nartheciaceae**

      28b. Perianth not roughened on the abaxial surface; inflorescence axillary or a secund raceme; peduncles or pedicels drooping; fruit a berry .... **Convallariaceae**

  26b. Tepals distinct or nearly so; filaments free or adnate

    29a. Flowers in a many-flowered umbel at the summit of a scape; seeds with phytomelin; plants with the odor of onion or garlic .... **Alliaceae**

    29b. Flowers arranged otherwise (forming an umbel with 3–6 flowers in *Clintonia*) at the summit of a scape or leafy stem; seeds without phytomelin (except *Ornithogalum*) plants without the odor of onion or garlic

      30a. Scape viscid-pubescent; leaves equitant and distichous .... **Tofieldiaceae**

      30b. Stem or scape glabrous or pubescent, but not viscid; leaves not set edge to stem, commonly not distichous

31a. Leaves in 1 or 2 whorls; styles elongate, capillary, exceeding the tepals; fruit a purple or black berry (*Medeola*) .... **Liliaceae**
31b. Leaves arranged otherwise; styles shorter; fruit a capsule or berry, the berry blue, red, or becoming red
   32a. Leaves conspicuously plicate-veined; tepals yellow-green; styles 3, separate to the base; fruit a septicidal capsule .... **Melanthiaceae**
   32b. Leaves not plicate-veined; tepals white, yellow, red, orange, or yellow-green in *Clintonia*; styles 1, divided or lobed only in the apical portion; fruit a berry or loculicidal capsule
      33a. Stems from a bulb or corm; fruit a capsule, circular or bluntly angled in cross-section
         34a. Tepals white, with a green stripe on the abaxial surface; leaves with a white stripe on the adaxial surface ....
.... **Hyacinthiaceae**
         34b. Tepals yellow, orange, or red; leaves without a white stripe .... **Liliaceae**
      33b. Stems from rhizomes; fruit a berry, or a capsule with triquetrous cross-section in *Uvularia*
         35a. Tepals white; inflorescence a terminal raceme or panicle (*Maianthemum*) .... **Convallariaceae**

35b. Tepals yellow, green-white, yellow-green, or pink; inflorescence axillary, an umbel, or a solitary, terminal flower .... **Uvulariaceae**

24b. Leaves pinnately veined, with a well defined petiole and blade, with stipular tendrils .... **Smilacaceae**

19b. Ovary inferior

36a. Flowers zygomorphic, with 1 or 2 stamens; pollen cohering in pollinia; seeds dust-like .... **Orchidaceae**

36b. Flowers actinomorphic, with 3 or 6 stamens; pollen separate; seeds larger

37a. Flowers with 3 stamens; leaves equitant; seeds without phytomelin .... **Iridaceae**

37b. Flowers with 6 stamens; leaves not equitant; seeds usually with phytomelin

38a. Perianth pubescent on the abaxial [outside] surface; leaves pubescent; plants from corms .... **Hypoxidaceae**

38b. Perianth glabrous; leaves glabrous; plants from bulbs .... **Amaryllidaceae**

18b. Flowers anemophilous or self-pollinated, without a perianth or with an inconspicuous perianth of scales, bristles, or sepaloid tepals

39a. Flowers arranged in a solitary, terminal, involucrate, white to gray, pseudanthial cluster; roots conspicuously septate .... **Eriocaulaceae**

39b. Flowers arranged otherwise; roots lacking conspicuous septa

40a. Inflorescence consisting of a dense, elongate, cylindrical spike, the carpellate portion below, the staminate portion above; fruit a wind-dispersed, 1-seeded follicle (*Typha*) .... **Typhaceae**

40b. Inflorescence otherwise; fruit a capsule, achene, utricle, or caryopsis

41a. Fruit a 3- to many-seeded capsule; perianth of 6 sepaloid tepals .... **Juncaceae**

41b. Fruit indehiscent; perianth of 2–6 inconspicuous tepals, scales, bristles, or absent

42a. Ovary with 1 stigma (2 stigmas in *Sparganium eurycarpum*); perianth of 2–6 tepals; inflorescence a dense, spherical cluster; plants aquatic (*Sparganium*) .... **Typhaceae**

42b. Ovary with 2 or 3 stigmas; perianth of scales, bristles, or absent; inflorescence variously open to compact; plants terrestrial or aquatic

    43a. Leaves arranged in 3 ranks or absent and represented by a bladeless sheath; stems usually solid, very often triangular in cross-section; flowers spirally arranged (distichous in *Cyperus* and *Dulichium*); fruit an achene .... **Cyperaceae**

    43b. Leaves arranged in 2 ranks; stems usually hollow, almost never triangular in cross-section; flowers distichously arranged [with reference to the floral scales]; fruit a caryopsis or rarely a utricle .... **Poaceae**

Group 4

1a. Plants 0.5–3.0 cm tall, parasitic on *Picea*, *Larix*, and *Pinus*; anthers sessile .... **Viscaceae**

1b. Plants much taller or longer at maturity, not parasitic; anthers with filaments

    2a. Leaves compound

        3a. Leaves palmately compound; flowers zygomorphic; fruit a large capsule 3.0–5.0 cm thick (*Aesculus*) .... **Sapindaceae**

        3b. Leaves pinnately compound; flowers actinomorphic; fruit an achene, schizocarp, drupe, or samara

            4a. Plants vines; flowers with numerous stamens; fruit an achene terminated by an elongate, plumose style (*Clematis*) .... **Ranunculaceae**

            4b. Plants upright shrubs or trees; flowers with 2–12 stamens; fruit otherwise

                5a. Fruit a samaroid schizocarp; ovary bilobed, compressed; androecium composed of 4–12 stamens, commonly 8 (*Acer*) .... **Sapindaceae**

                5b. Fruit a samara or drupe; ovary not bilobed, not compressed; androecium composed of 2 or 5 stamens

                    6a. Inflorescence an umbel; leaves 2-times pinnately compound; gynoecium with 5 carpels (*Aralia*) .... **Apiaceae**

                    6b. Inflorescence a cyme, fascicle, raceme, or panicle; leaves 1-time pinnately compound; gynoecium with 2–5 carpels

                        7a. Perianth absent; androecium composed of 2 stamens; fruit a samara; flowers usually unisexual; trees at maturity (*Fraxinus*) .... **Oleaceae**

                        7b. Perianth present and gamopetalous; androecium composed of 5 stamens; fruit a drupe; flowers bisexual; shrubs at maturity (*Sambucus*) .... **Adoxaceae**

    2b. Leaves simple

        8a. Apical portions of the stem succulent; leaves minute and scale-like, 1.0–3.0 mm long; stems jointed; plants of Atlantic coast shores (*Sarcocornia*) .... **Amaranthaceae**

        8b. Apical portions of the stem herbaceous to woody; leaves foliaceous, mostly longer; stems not jointed; plants of inland habitats

            9a. Leaves and leaf scars whorled; plants of wetlands

10a. Inflorescence a dense cyme; flowers each with 8 or 10 stamens; petals pink-purple, 10.0–15.0 mm long; stems woody in the basal portion, surrounded by loose, spongy tissue, the buds barely emerging through the tissue (*Decodon*) .... **Lythraceae**

10b. Inflorescence a dense, spherical cluster; flowers each with 4 stamens; petals white, 5.0–8.0 mm long; stems woody throughout, not surrounded by spongy tissue, the buds clearly visible (*Cephalanthus*) .... **Rubiaceae**

9b. Leaves and leaf scars opposite; plants of various habitats

11a. Leaves and twigs conspicuously white and brown lepidote-scaly; plants dioecious (*Shepherdia*) .... **Elaeagnaceae**

11b. Leaves and twigs without lepidote scales; plants synoecious, polygamous, or dioecious

12a. Flowers zygomorphic; ovary 4-lobed; fruit a schizocarp that separates into 4 mericarps; plants mat-forming (*Thymus*) .... **Lamiaceae**

12b. Flowers actinomorphic (except in some *Lonicera*); ovary unlobed or with fewer that 4 lobes; fruit not a schizocarp (a winged schizocarp in *Acer*); plants not mat-forming (except *Diapensia*)

13a. Leaves 0.6–1.5 cm long, crowded and imbricate; plants of exposed, alpines areas above 1000 m elevation ..... **Diapensiaceae**

13b. Leaves longer than 3.0 cm, not crowded; plants of various habitats usually less than 1000 m (except *Viburnum edule*)

14a. Stems woody only in the basal portion, surrounded by loose, spongy tissue, the buds barely emerging through the tissue; stamens dimorphic, of 2 different lengths (*Decodon*) .... **Lythraceae**

14b. Stems woody throughout, not surrounded by spongy tissue, the buds clearly visible; stamens monomorphic

15a. Fruit a samaroid schizocarp; ovary bilobed and compressed (*Acer*) .... **Sapindaceae**

15b. Fruit a capsule or drupe; ovary not, or inconspicuously, lobed, not compressed

16a. Perianth 4-merous

17a. Leaves punctate with translucent dots; petals yellow (*Hypericum*) .... **Clusiaceae**

17b. Leaves not punctate; petals white, yellow, green, blue, or purple

18a. Flowers with 2 stamens; corolla gamopetalous .... **Oleaceae**

18b. Flowers with 5–many stamens; corolla apopetalous

19a. Fruit a drupe; leaves entire, with arcuate venation (*Cornus*) .... **Cornaceae**

19b. Fruit a capsule; leaves toothed, with pinnate venation

> 20a. Flowers with 20 or more stamens; perianth 2.5–3.5 cm wide in living material; styles present, elongate; twigs terete .... **Hydrangeaceae**
>
> 20b. Flowers with 4 stamens; perianth 0.6–1.0 cm wide in living material; styles absent; twigs more or less quadrangular (*Euonymus*) .... **Celastraceae**

16b. Perianth 5-merous

> 21a. Style absent or very short; corolla actinomorphic, white, small, the fertile flowers up to 8.0 mm long (*Viburnum*) .... **Adoxaceae**
>
> 21b. Style elongate; corolla actinomorphic or zygomorphic, white, pink, yellow, red, or purple, 3.0–50.0 mm .... **Caprifoliaceae**

Group 5

1a. Inflorescence a capitulum; fruit a cypsela; leaves often white-tomentose on one or both surfaces (*Artemisia*) .... **Asteraceae**

1b. Inflorescence not a capitulum; fruit not a cypsela; leaves variously glabrous or pubescent

2a. Leaves compound

> 3a. Perianth zygomorphic; filaments conspicuously connate (only at the base in *Amorpha*); fruit a legume .... **Fabaceae**
>
> 3b. Perianth actinomorphic; filaments distinct (connate at the base in Rutaceae); fruit not a legume
>
>> 4a. Leaves palmately compound; plants vines, with tendrils produced opposite the leaves (*Parthenocissus*) .... **Vitaceae**
>>
>> 4b. Leaves pinnately compound; plants not vines, without tendrils (plants climbing by means of aerial roots in *Toxicodendron radicans*)
>>
>>> 5a. Inflorescence a unisexual ament, the staminate drooping; flowers small, with an inconspicuous perianth; fruit a nut, sometimes enclosed in a husk .... **Juglandaceae**
>>>
>>> 5b. Inflorescence not an ament, uni- or bisexual, none drooping; flowers with an evident perianth; fruit a drupe, samara, follicle, pome, or aggregate
>>>
>>>> 6a. Leaves punctate with pellucid dots; fruit a fleshy follicle or an orbicular samara .... **Rutaceae**
>>>>
>>>> 6b. Leaves not punctate; fruit otherwise
>>>>
>>>>> 7a. Leaflets entire except for 1 or more coarse teeth near the base; fruit a samara .... **Simaroubaceae**
>>>>>
>>>>> 7b. Leaflets toothed to lobed; fruit not a samara
>>>>>
>>>>>> 8a. Ultimate unit of the inflorescence an umbel; styles swollen at the base forming a stylopodium .... **Apiaceae**
>>>>>>
>>>>>> 8b. Ultimate unit of the inflorescence not an umbel; styles without a stylopodium

9a. Perianth composed of a single series [the sepals]; wood
    yellow (*Xanthorhiza*) .... **Ranunculaceae**
9b. Perianth composed of 2 series [sepals and petals both
    present]; wood not yellow
   10a. Flowers with a hypanthium, bisexual; plants with a
    watery sap; stems armed or unarmed .... **Rosaceae**
   10b. Flowers without a hypanthium, commonly unisexual;
    plants with a milky or poisonous sap; stems unarmed ...
    .... **Anacardiaceae**

2b. Leaves simple
  11a. Plants lianas
    12a. Calyx gamosepalous, 3-lobed, with a strongly curved basal tube, petaloid,
    yellow-brown marked with red-brown; flowers with 6 stamens (*Aristolochia*)
    .... **Aristolochiaceae**
    12b. Calyx gamosepalous, aposepalous, or obsolete, without a strongly curved
    tube, sepaloid; flowers with 5 stamens, when stamens are present
     13a. Corolla gamopetalous; anthers dehiscing by terminal pores or clefts;
    some of the leaves with a distinct pair of small lobes or leaflets (*Solanum*)
    .... **Solanaceae**
     13b. Corolla apopetalous (cohering at the summit in *Vitis*); anthers dehiscing
    by longitudinal slits; leaves simple or lobed, the lobes, when present, not
    confined to the base of the leaf
      14a. Leaves palmately veined or palmately lobed; tendrils and
    inflorescences produced opposite the leaves; fruit a berry, red to
    purple to black (*Vitis*) .... **Vitaceae**
      14b. Leaves pinnately veined; tendrils absent; inflorescences terminal or
    axillary; fruit a capsule or green-yellow berry
       15a. Flowers with 5 stamens; fruit an orange capsule up to 1.0 cm
    long; pith continuous (*Celastrus*) .... **Celastraceae**
       15b. Flowers with many stamens; fruit a green-yellow berry *ca.*
    2.5 cm long; pith diaphragmed .... **Actinidiaceae**
  11b. Plants neither climbing nor twining
    16a. Plants with a milky latex; fruit a fleshy multiple (*Morus*) .... **Urticaceae**
    16b. Plants with a watery sap; fruit simple or aggregate
     17a. Flowers unisexual
      18a. Leaves palmately 3- to 5-lobed; flowers in dense, spherical clusters;
    fruit a linear-shaped achene .... **Platanaceae**
      18b. Leaves not palmately lobed (except in some 3-lobed leaves of
    Sassafras); flowers not in dense, spherical clusters; fruit otherwise
       19a. Plants up to 0.5 m tall, with numerous, crowded, linear leaves
    3.0–8.0 mm long; fruit a drupe .... **Ericaceae**
       19b. Plants usually taller, with larger, broader leaves; fruit various
        20a. Inflorescence an ament; perianth absent or tiny and
    inconspicuous

21a. Winter buds covered by a single, cap-like scale or with the lowest scale centered over the leaf scar; fruit a capsule, containing comose seeds .... **Salicaceae**

21b. Winter buds with 2 or more scales, the lowest scale not centered over the leaf scar; fruit an achene, samara, or nut

22a. Foliage very fragrant; plants dioecious ... **Myricaceae**

22b. Foliage not fragrant; plants monoecious

23a. Ovary 3- or 6-locular; carpellate flowers solitary or in small clusters; fruit a nut .... **Fagaceae**

23b. Ovary 2-locular; carpellate flowers commonly in aments; fruit an achene, samara, or nut ....
.... **Betulaceae**

20b. Inflorescence not an ament; perianth present, evident

24a. Pith diaphragmed; staminate flowers with 8–15 stamens (*Nyssa*) .... **Cornaceae**

24b. Pith continuous; staminate flowers with 4–9 stamens

25a. Staminate flowers with 9 stamens; anthers dehiscing by uplifting valves; plants with aromatic foliage and wood .... **Lauraceae**

25b. Staminate flowers with 4–8 stamens; anthers dehiscing by longitudinal slits; plants without aromatic foliage or wood

26a. Ovary with 1 stigma, the style unbranched; stamens antesepalous

27a. Leaves and twigs covered with silver lepidote scales; plants usually armed with thorns; petals absent; styles elongate (*Elaeagnus*) ....
.... **Elaeagnaceae**

27b. Leaves and twigs without lepidote scales; plants unarmed; petals present; styles very short, nearly absent .... **Aquifoliaceae**

26b. Ovary with 2–4 stigmas, the style 2- to 4-lobed; stamens antepetalous .... **Rhamnaceae**

17b. Flowers bisexual

28a. Leaves entire, covered with silver lepidote scales; plants usually armed with thorns (*Elaeagnus*) .... **Elaeagnaceae**

28b. Leaves entire or toothed, not lepidote-scaly; plants unarmed or armed

29a. Flowers zygomorphic; stamens monadelphous; fruit a legume (*Genista*) .... **Fabaceae**

29b. Flowers actinomorphic; stamens distinct or shortly connate at the base; fruit not a legume

30a. Corolla of 4 yellow, strap-shaped petals; plants flowering in the fall during and after the falling of the leaves ....
.... **Hamamelidaceae**

30b. Corolla of 4–10 variously colored petals, the petals not strap-shaped with parallel margins; plants not flowering after leaf fall

    31a. Fruit an elliptic to suborbicular samara; flowers tiny, anemophilous, precocious; leaves doubly serrate .... **Ulmaceae**

    31b. Fruit not a samara; flowers various, generally entomophilous, precocious to serotinous; leaves usually not doubly serrate

        32a. Peduncle of the inflorescence adnate to a conspicuous, elongate bract; the numerous stamens commonly connate into 5 antepetalous groups (*Tilia*) .... **Malvaceae**

        32b. Peduncle of the inflorescence without bracts or, when bracteate, free from the bract; stamens 4–many, distinct

            33a. Pith diaphragmed; gynoecium composed of a single carpel (very rarely 2 carpels) (*Nyssa*) .... **Cornaceae**

            33b. Pith continuous; gynoecium composed of 1–10 carpels

                34a. Perianth monochlamydeous

                    35a. Flowers with a conspicuous hypanthial tube; stamens 8 .... **Thymelaeaceae**

                    35b. Flowers with a short, inconspicuous hypanthial tube, or lacking one altogether; stamens 4 or 5

                        36a. Stamens antepetalous; leaves commonly subopposite; branches often ending in a short, spine-like process (*Rhamnus*) .... **Rhamnaceae**

                        36b. Stamens antesepalous; leaves alternate; branches without spines (*Nemopanthus*) .... **Aquifoliaceae**

                34b. Perianth composed of 2 or more whorls, both the sepals and petals present

                  37a. Flowers perigynous

                    38a. Stamens as many as the petals

                      39a. Leaves palmately veined or palmately lobed, without stipules .... **Grossulariaceae**

                      39b. Leaves pinnately veined or pinnately lobed, with stipules or stipule scars adjacent to the leaf scar

                        40a. Plants usually armed with stout thorns; stamens antesepalous (*Crataegus*) .... **Rosaceae**

40b. Plants unarmed or with a slender, spine-like process at the apex of some branches; stamens antepetalous .... **Rhamnaceae**

38b. Stamens numbering more than the petals .... **Rosaceae**

37b. Flower hypogynous

41a. Plants armed with spines; perianth composed of 6 sepals [in 2 cycles] and 5 petals [in 2 cycles]; anthers dehiscing by uplifting valves (*Berberis*) .... **Berberidaceae**

41b. Plants unarmed or with thorns in *Lycium*; perianth composed of fewer parts; anthers dehiscing by longitudinal slits or apical pores

42a. Sepals dimorphic, the 2 outer smaller than the 3 inner; petals yellow .... **Cistaceae**

42b. Sepals monomorphic; petals white, pink, red, purple, green, or green-white

43a. Stamens numbering the same as the petals and antepetalous, often enclosed by the petals [especially in bud], or the petals absent and the stamens alternating with the sepals .... **Rhamnaceae**

43b. Stamens fewer than, as many as, or more than the petals, alternating with the petals, not enclosed by the petals (except in *Kalmia*)

44a. Stamens fewer than or as many as the petals or inner tepals

45a. Ovary 2-locular; fruit a red berry 1.0–2.0 cm in diameter; plants usually with thorns (*Lycium*) .... **Solanaceae**

45b. Ovary 4- to 8-locular; fruit otherwise; plants unarmed

46a. Styles short, the stigma nearly sessile on the summit of the ovary; sepals present, though small and inconspicuous .... **Aquifoliaceae**

46b. Styles elongate, the stigma elevated above the summit of the ovary; sepals usually evident ..... **Ericaceae**

44b. Stamens numbering more than the petals

47a. Plants evidently pubescent with compound hairs; ovary 3-locular; leaves flat, deciduous .. **Clethraceae**

47b. Plants glabrous or with simple hairs; ovary 2- to 9-locular; leaves commonly revolute-margined and/or evergreen .... **Ericaceae**

Group 6

1a. Inflorescence a capitulum (the capitula with only 1 flower in *Echinops*); anthers in the disk flowers connate in a ring around the style; fruit a cypsela .... **Asteraceae**

1b. Inflorescence not a capitulum (except in *Knautia*); anthers not connate in a ring; fruit otherwise

2a. Flowers unisexual

3a. Plants vines, with tendrils; corolla gamopetalous .... **Cucurbitaceae**

3b. Plants not vines, without tendrils; corolla apopetalous (very inconspicuous and reduced to a rim of sepals in *Hippuris*)

4a. Plants aquatic, with thin, flaccid, submersed leaves; flowers anemophilous, with an inconspicuous perianth

5a. Leaves pectinately divided into narrow segments or the leaves reduced to minute scales or bumps on the stem, mostly alternate or whorled; flowers with 4 or 8 stamens (*Myriophyllum*) .... **Haloragaceae**

5b. Leaves simple, whorled; flowers with a single stamen (*Hippuris*) .... **Veronicaceae**

4b. Plants terrestrial, with thick, firm leaves; flowers entomophilous, with well developed perianths

6a. Leaves simple; fruit a drupe; inflorescence an axillary cymule of 3 flowers—the 2 lateral flowers usually staminate, the central bisexual (*Geocaulon*) .... **Santalaceae**

6b. Leaves compound or evidently lobed; fruit not a drupe; inflorescence otherwise

    7a. Leaves 1- to 3-times pinnately compound; fruit an achene enclosed in an indurate hypanthium or a follicle .... **Rosaceae**

    7b. Leaves palmately divided into 3–7 broad segments; fruit a schizocarp (*Sanicula*) .... **Apiaceae**

2b. Flowers bisexual

  8a. Leaves 4–6 per node, entire, with arcuate venation; flowers pseudanthial, small, collectively subtended by 4 large, white, petaloid bracts (*Cornus*) ..... **Cornaceae**

  8b. Leaves 1–12 per node, without arcuate venation; flowers not pseudanthial, lacking petaloid bracts

    9a. Calyx large, 2.0–4.0 cm long, 3-merous, gamosepalous, petaloid, red-brown; flower solitary, on a stout peduncle, arising between a pair of cordate leaves (*Asarum*) .... **Aristolochiaceae**

    9b. Calyx mostly smaller, not 3-merous, connate or not, sepaloid (petaloid tepals in Santalaceae), not red-brown; flowers arranged otherwise

      10a. Plants aquatic

        11a. Leaves with a prominent, sheathing base; inflorescence an umbel (*Sium*) .... **Apiaceae**

        11b. Leaves lacking a prominent, sheathing base; inflorescence axillary flowers or a raceme

          12a. Leaves compound, all basal, with 3 thick, fleshy leaflets; fruit a 2-valved capsule (*Menyanthes*) .... **Menyanthaceae**

          12b. Leaves simple or compound, not confined to the base of a plant, not with 3 thick leaflets; fruit an achene-like drupe or a 4- or 5-valved capsule

            13a. Leaves whorled, simple; flowers with a single stamen (*Hippuris*) .... **Veronicaceae**

            13b. Leaves alternate, simple, lobed or compound; flowers with 3, 5, or 8 stamens

              14a. Leaves opposite; perianth composed of 4 sepals; fruit a 4-valved capsule (*Ludwigia*) .... **Onagraceae**

              14b. Leaves alternate (with a basal tuft in *Samolus*); perianth and fruit otherwise

                15a. Leaves pinnately lobed or pinnately compound (merely toothed in the emersed leaves of *Proserpinaca palustris*); perianth composed of only 3 sepals; flowers with 3 antesepalous stamens (*Proserpinaca*) .... **Haloragaceae**

                15b. Leaves simple; perianth composed of 5 sepals and 5 petals; flowers with 5 antepetalous stamens (*Samolus*) .... **Primulaceae**

      10b. Plants terrestrial, though sometimes of wet habitats

        16a. Inflorescence a capitulum; calyx composed of 8–12 setaceous appendages (*Knautia*) .... **Caprifoliaceae**

        16b. Inflorescence and calyx otherwise

17a. Perianth 3- or 4-merous
 18a. Corolla gamopetalous; stamens 3 or 4; leaves opposite and with interpetiolar stipules or whorled .... **Rubiaceae**
 18b. Corolla apopetalous above the hypanthium or absent in *Ludwigia*; stamens 2, 4, or usually 8; leaves without stipules, alternate or opposite .... **Onagraceae**
17b. Perianth 5-merous or absent (sepals numbering 2 in the Portulacaceae, petals rarely 4 or 6 in *Portulaca*)
 19a. Leaves opposite; flowers with 3 or 4 stamens (5 in *Triosteum*) .... **Caprifoliaceae**
 19b. Leaves alternate or all basal; flowers with 5–many stamens
  20a. Inflorescence usually an umbel; fruit a schizocarp or, less commonly, a berry; leaves compound (simple), usually with a sheathing base; styles swollen at the base forming a stylopodium; calyx usually a series of small teeth around the rim of the ovary or absent .... **Apiaceae**
  20b. Inflorescence otherwise (except in *Primula*); fruit a capsule, pyxis, or drupe; leaves simple (trifoliate in *Menyanthes*), mostly without a sheathing base; gynoecium without a stylopodium; calyx of evident sepals (or apparently absent in the Santalaceae)
   21a. Leaves compound, with 3 thick, fleshy leaflets; petals pubescent on the adaxial [inner] surface (*Menyanthes*) .... **Menyanthaceae**
   21b. Leaves simple; petals glabrous on the adaxial surface (with a tuft of hairs near the base in *Comandra*)
    22a. Flowers with mostly 6–10 stamens; petals yellow; prostrate weeds (*Portulaca*) .... **Portulacaceae**
    22b. Flowers with 5 stamens; petals (or tepals) white, red, green-purple, blue, or purple; plants not prostrate weeds
     23a. Perianth monochlamydeous; fruit a drupe (dryish in *Comandra*) .... **Santalaceae**
     23b. Perianth dichlamydeous; fruit a capsule
      24a. Corolla 4.0–50.0 mm long; style present, evident; placentation axile .... **Campanulaceae**
      24b. Corolla 1.3–2.5 mm long; style very short or absent; placentation free-central (*Samolus*) .... **Primulaceae**

Group 7

1a. Perianth with a nectary spur
 2a. Plants insectivorous by means of bladders or viscid leaves; flowers with 2 stamens; placentation free-central .... **Lentibulariaceae**
 2b. Plants not insectivorous; flowers with 4–6 stamens; placentation axile or parietal

3a. Stamens numerous; both the calyx and corolla spurred, the 2 upper petals prolonged backward into a spur that extends into the upper, spurred sepal; leaves palmately divided (*Delphinium*) .... **Ranunculaceae**

3b. Stamens 4–6; either the calyx or corolla spurred, but not both; leaves not palmately divided (palmately lobed in some *Viola*)

    4a. Calyx spurred; capsule dehiscing elastically; gynoecium with 5 carpels ....
.... **Balsaminaceae**

    4b. Corolla spurred; capsule not dehiscing elastically; gynoecium with 2 or 3 carpels

        5a. Calyx of 2, often caducous, sepals; flowers with 6 stamens; leaves decompound .... **Papaveraceae**

        5b. Calyx of 5 sepals; flowers with 4 or 5 stamens; leaves variously simple to lobed, but not many-times dissected

            6a. Corolla apopetalous; flowers with 5 stamens; gynoecium with 3 carpels, therefore, the capsule with 3 valves .... **Violaceae**

            6b. Corolla gamopetalous; flowers with only 4 functional stamens; gynoecium with 2 carpels, therefore, the capsule with 2 valves ....
.... **Veronicaceae**

1b. Perianth lacking a nectary spur

  7a. Upper sepal much larger than the other 4, petaloid, arched, forming a helmet-shaped structure that conceals the 2 petals; leaves palmately divided, with pinnately lobed segments (*Aconitum*) .... **Ranunculaceae**

  7b. Calyx without a single, larger, helmet-shaped sepal; petals 3–8, not concealed by the calyx; leaves otherwise

    8a. Calyx with 2 sepals; androecium with 3 stamens; fruit a capsule with 1–3 seeds (*Montia*) .... **Portulacaceae**

    8b. Calyx with 4–6 sepals; androecium with 2–25 stamens, but not 3; fruit a capsule, legume, loment, or schizocarp

        9a. Fruit a legume or loment; gynoecium with 1 carpel; stamens monadelphous or diadelphous (distinct in *Thermopsis* and *Baptisia*); lower 2 petals fused into a keel (only the banner petal present in *Amorpha*) .... **Fabaceae**

        9b. Fruit a capsule; gynoecium with 2 or more carpels; stamens distinct (monadelphous in *Polygala*); lower 2 petals variously distinct or connate

            10a. Corolla apopetalous; apical portion of the petals bearing a fringe of linear to oblanceolate segments that are more or less equal to the length of the petal; androecium with 10–25 stamens .... **Resedaceae**

            10b. Corolla gamopetalous; apical portion of the petals lacking a fringe, though sometimes pubescent; androecium with 2–8 stamens

                11a. Calyx composed of 2 small sepals and 2 large, petaloid sepals; flowers with 6 or 8 monadelphous stamens .... **Polygalaceae**

                11b. Calyx composed of sepaloid sepals; flowers with 2–5 distinct stamens

                    12a. Carpels 2, each carpel lobed, often evidently so, therefore, the gynoecium appearing to be composed of 4 carpels, separating at maturity into 4 half-carpellate segments; fruit a schizocarp

13a. Inflorescence a coiled raceme; leaves alternate; stems terete (*Myosotis*) .... **Boraginaceae**

13b. Inflorescence not a coiled raceme; leaves opposite; stems commonly square

14a. Style terminal, with a conspicuously lobed stigma; carpels weakly lobed; inflorescence indeterminate [a spike or raceme]; placentation marginal; plants not aromatic .... **Verbenaceae**

14b. Style gynobasic (terminal in *Ajuga*, *Teucrium*, and *Trichostema*), usually with an inconspicuous stigma; carpels usually evidently lobed; inflorescence mixed, with cymose lateral branches, sometimes the inflorescence condensed in dense verticils of flowers, only rarely completely indeterminate [a raceme]; ovules attached to lateral partitions and not carpel margins; plants commonly with aromatic foliage .... **Lamiaceae**

12b. Carpels 2, each carpel unlobed, therefore, the gynoecium appearing to be composed of 2 carpels (appearing unicarpellate in *Phryma* but the stigmas 2); fruit a capsule or achene

15a. Fruit an achene; ovary unilocular and uniovulate; calyx tightly reflexed in fruit .... **Phrymaceae**

15b. Fruit a capsule; ovary bilocular or unilocular in holoparasitic genera, with 2 or more ovules; calyx not reflexed in fruit

16a. Anthers with 1 locule, dehiscing by a single, distal slit, not sagittate at the base .... **Scrophulariaceae**

16b. Anthers with 2 locules, dehiscing by 2 openings or the locules connate near the apex and opening by a distal, V-shaped slit, usually sagittate at the base

17a. Plants holoparasitic [without chlorophyll] or hemiparasitic [with chlorophyll], producing haustoria, commonly darkening in drying .... **Orobanchaceae**

17b. Plants autotrophic [with chlorophyll], not producing haustoria, usually not darkening in drying .... **Veronicaceae**

Group 8

1a. Plants parasitic, with haustoria, lacking chlorophyll; stems twining, pink-yellow to orange, with scale-like leaves (*Cuscuta*) .... **Convolvulaceae**

1b. Plants autotrophic, without haustoria, with chlorophyll; stems upright or twining, usually green, with foliaceous leaves (scale-like in *Bartonia*)

2a. Depressed, pulvinate, evergreen, alpine herbs of elevation exceeding 1000 meters; inflorescence a solitary, peduncled flower .... **Diapensiaceae**

2b. Plants of various habit, but commonly upright or vining, with deciduous leaves of non-alpine habitats; inflorescence various, but commonly composed of more than 1 flower

3a. Leaves all basal (opposite and borne on a stem in *Plantago psyllium*); calyx and corolla inconspicuous, the corolla scarious, 3- or, more commonly, 4-merous (*Plantago*) .... **Veronicaceae**
3b. Leaves borne on a stem (all basal in *Limosella*, *Menyanthes*, and *Primula*); calyx and/or corolla evident, at least one of which is petaloid and not scarious
    4a. Plants with a milky latex; carpels connate at the summit only, with a common stigma; fruit a pair of follicles .... **Apocynaceae**
    4b. Plants with a watery sap; carpels connate throughout or only in the basal portion; fruit an achene, berry, schizocarp, capsule, or pyxis
        5a. Leaves with evident, sheathing stipules [ocreae]; pedicels of the flowers subtended by sheathing bracteoles [ocreolae]; perianth monochlamydeous, composed of more or less petaloid tepals; fruit an achene .... **Polygonaceae**
        5b. Leaves without sheathing stipules; pedicels without sheathing bracteoles; perianth dichlamydeous; fruit otherwise
            6a. Stems prostrate, emitting at nodes a tuft of 5–10 narrow leaves and 1 or more, 1-flowered peduncles; plants of tidal shores (*Limosella*) .... **Veronicaceae**
            6b. Stems upright or vining, the leaves normally 1 or 2 from a node, sometimes more; inflorescence various; usually with more than 1 flower; plants not of tidal shores (of coastal shores in *Mertensia maritima* and *Heliotropium*)
                7a. Plants aquatic, with densely crowded, dissected leaves; inflorescence an umbel or raceme, the peduncle and axis of each individual raceme inflated and hollow, constricted at the nodes and bearing a whorl of flowers (*Hottonia*) .... **Primulaceae**
                7b. Plants aquatic or terrestrial, the leaves not both densely crowded and dissected; inflorescence without inflated peduncles and axes
                    8a. Foliage leaves opposite or whorled
                        9a. Gynoecium appearing to be composed of 4 carpels [actually 2, each weakly to evidently lobed], the carpels maturing as a schizocarp; flowers with 4 stamens; stems quadrangular
                            10a. Style terminal; plants not aromatic .... **Verbenaceae**
                            10b. Style gynobasic; plants aromatic (*Mentha*) .... **Lamiaceae**
                        9b. Gynoecium composed of 1–3 carpels, the carpels maturing as a capsule or pyxis; flowers with 4–7 stamens; stems mostly terete or angled
                            11a. Stamens antepetalous; carpels with 1 stigma; placentation free-central .... **Primulaceae**
                            11b. Stamens antesepalous; carpels usually with a 2- or 3-lobed stigma; placentation parietal or axile

12a. Gynoecium with 3 carpels and 3 stigmas; placentation axile; calyx with scarious intervals between the green, longitudinal ribs (*Phlox*) .... **Polemoniaceae**

12b. Gynoecium with 2 carpels and 2 stigmas; placentation parietal; calyx without scarious intervals between longitudinal ribs .. **Gentianaceae**

8b. Foliage leaves alternate or all basal (appearing opposite in *Datura*, *Nicandra*, and *Physalis* due to forking of the stem)

13a. Inflorescence a coiled raceme; style gynobasic (terminal in *Heliotropium*); fruit a schizocarp .... **Boraginaceae**

13b. Inflorescence otherwise; style terminal; fruit a capsule, pyxis, or berry (the berry sometimes dry in the Solanaceae)

14a. Leaves reduced to scales on the stem; flowers 2.5–4.0 mm long (*Bartonia*) .... **Gentianaceae**

14b. Leaves foliaceous; flowers larger

15a. Plants aquatic; leaves either floating or with 3 fleshy leaflets .... **Menyanthaceae**

15b. Plants terrestrial; the leaves neither floating nor with 3 fleshy leaflets

16a. Leaves confined to the base of the flowering stems; stamens antepetalous; placentation free-central (*Primula*) .... **Primulaceae**

16b. Leaves, at least in part, borne on the flowering stem; stamens antesepalous; placentation basal, parietal, or axile

17a. Ovary with 4 or 6 ovules; fruit a 4- or 6-seeded capsule; plants climbing or trailing (mostly upright in *Calystegia spithamaea*) .... **Convolvulaceae**

17b. Ovary usually with more than 6 ovules; fruit usually a many-seeded capsule, berry, or pyxis; plants upright (climbing or trailing in *Solanum dulcamara*)

18a. Gynoecium composed of 2 carpels (3–5 in *Nicandra*); fruit a berry, 2-valved capsule, or pyxis

19a. Fruit a capsule; anthers with 1 locule, opening by a single, distal slit (*Verbascum*) ...... **Scrophulariaceae**

19b. Fruit a berry, spiny capsule, or pyxis; anthers with 2 locules, opening by a pair of pores or longitudinal slits .... **Solanaceae**

18b. Gynoecium composed of 3 carpels;
fruit a 3-valved capsule ....
.... **Polemoniaceae**

Group 9

1a. Plants submerged aquatics, attaching to the substrate by fleshy disks; flowers without perianth, borne singly in the axils of leaves, subtended by a tubular spathe ....
.... **Podostemaceae**
1b. Plants aquatic or terrestrial, but not attaching to the substrate by fleshy disks; flowers with at least 1 whorl of perianth (lacking a perianth in *Callitriche* and *Euphorbia*), not subtended by a tubular spathe
  2a. Placentation laminar; plants aquatic, with conspicuous elliptic or orbicular floating leaves with 2 basal lobes, arising directly from a rhizome .... **Nymphaeaceae**
  2b. Placentation otherwise; plants terrestrial or aquatic; leaves various, but not at once floating, basally lobed, and arising from an underwater rhizome
    3a. Inflorescence a cyathium, with a single carpellate flower and 2 or more staminate flowers borne in a cupulate involucre, the margin of the involucre sometimes subtended by petaloid glands [commonly 4], the entire arrangement resembling a single flower; plants with a milky sap (*Euphorbia*) .... **Euphorbiaceae**
    3b. Inflorescence not pseudanthial; plants with a watery sap (milky or colored latex in the Papaveraceae)
      4a. Calyx of 2, often caducous, sepals; plants with a milky or colored latex ....
.... **Papaveraceae**
      4b. Calyx of 3 or more sepals (2 sepals in the Portulacaceae and some Elatinaceae), or the perianth monochlamydeous and with 3 or more tepals, or the perianth absent; plants without latex, the sap watery
        5a. Plants insectivorous by means of viscid hairs or pitcher-shaped leaves, of organic soils; leaves all basal
          6a. Leaves with viscid, glandular-hairs, flat, the margins not connate; inflorescence a raceme; styles 3, distinct .... **Droseraceae**
          6b. Leaves without glandular hairs, the margins connate, thereby forming a tube that is retrorsely pubescent on the adaxial [inside] surface; inflorescence a single flower; styles 5, connate, modified into a 5-rayed, umbrella-shaped body .... **Sarraceniaceae**
        5b. Plants not insectivorous, of mineral or organic soils; leaves various
          7a. Leaves opposite or whorled (sometimes the upper alternate in *Chrysosplenium*)
            8a. Plants free-floating, rootless, aquatic; leaves 2- to 4-times dichotomously forked; perianth composed of 7–15 basally connate sepals; fruit an achene with 2–15 marginal spines ....
.... **Ceratophyllaceae**
            8b. Plants usually rooted, terrestrial or aquatic; leaves not dichotomously forked; perianth otherwise; fruit various, but not an achene with spines

9a. Basal leaves peltate, 3.0–4.0 dm wide; fruit a yellow berry 4.0–5.0 cm wide (*Podophyllum*) .... **Berberidaceae**

9b. None of the leaves peltate, narrower; fruit an achene, capsule, silique, or schizocarp

    10a. Leaves punctate with pellucid dots; petals yellow or flesh-colored to pink .... **Clusiaceae**

    10b. Leaves not punctate; petals variously colored, but not yellow or flesh-colored (yellow in *Portulaca* and *Linum medium*)

        11a. Leaves scale-like; stems jointed; flowers sunken into the fleshy stem; plants coastal halophytes (*Salicornia*) .... **Amaranthaceae**

        11b. Leaves foliaceous; stems without joints; flowers not sunken into a fleshy stem; plants of various habitats

            12a. Plants aquatic; androecium with 1–3 stamens

                13a. Leaves crowded at the apex forming a floating rosette; flowers unisexual, without perianth; fruit a schizocarp (*Callitriche*) ... **Veronicaceae**

                13b. Leaves not crowded at the apex; flowers bisexual, with sepals and petals; fruit an achene .... **Elatinaceae**

            12b. Plants terrestrial, though sometimes of wet habitats; androecium with 4–10 stamens

                14a. Perianth monochlamydeous

                    15a. Perianth petaloid

                        16a. Leaves whorled, 3–8 per node; placentation axile .... **Molluginaceae**

                        16b. Leaves opposite; placentation basal or free-central

                            17a. Flowers in a cyme, with a sepaloid involucre subtending 1–5 flowers and a petaloid calyx; fruit an achene, enclosed in an indurate calyx [collectively called an anthocarp]; plants introduced .... **Nyctaginaceae**

                            17b. Flowers axillary, with only a petaloid calyx; fruit a 5-valved capsule; coastal halophyte (*Glaux*) .... **Primulaceae**

                    15b. Perianth sepaloid or inconspicuous

                        18a. Stamens 4–8, set in the notches of an 8-lobed disk that fills much of the center of the flower; fruit a 2-valved capsule (*Chrysosplenium*) .... **Saxifragaceae**

                        18b. Stamens 1–5, not set on a disk; fruit an achene or 3- to 10-valved capsule

19a. Flowers unisexual; fruit an achene

  20a. Ovary with 2 styles; leaves palmately lobed or divided; staminate flowers each with 5 stamens .... **Cannabaceae**

  20b. Ovary with 1 style; leaves simple, without lobes or divisions; staminate flowers each with 4 stamens or 5 in *Laportea* .... **Urticaceae**

19b. Flowers bisexual; fruit a capsule (a utricle in *Scleranthus*)

  21a. Ovary with 3–5 locules; placentation axile; leaves whorled .... **Molluginaceae**

  21b. Ovary with a single locule; placentation free-central; leaves opposite .... **Caryophyllaceae**

14b. Perianth dichlamydeous

  22a. Calyx of 2 sepals; stamens antepetalous .... **Portulacaceae**

  22b. Calyx of 3 or more sepals; stamens antesepalous

    23a. Flowers 4-merous

      24a. Leaves palmately lobed or palmately divided; androecium composed of 6 tetradynamous stamens; fruit a silique (*Cardamine*) .... **Brassicaceae**

      24b. Leaves simple; androecium composed of 4 or 8 stamens; fruit a capsule

      25a. Anthers dehiscing by a single, terminal pore; filaments twisted, thereby bringing all the stamens to one side of the flower at anthesis; leaves 2.0–7.0 cm long .... **Melastomaceae**

      25b. Anthers dehiscing by longitudinal slits; filaments not twisted to one side; leaves 0.2–2.0 cm long

        26a. Sepals 2.0–2.5 mm, entire at the apex; leaves linear-subulate; capsules

2.0–3.0 mm long (*Sagina*) ...
.... **Caryophyllaceae**

26b. Sepals *ca.* 1.0 mm long,
3-cleft at the apex; leaves
ovate to oblong; capsules
ca. 1.0 mm long
(*Radiola*) .... **Linaceae**

23b. Flowers 5- or 6-merous

27a. Anthers dehiscing by 2 terminal
pores; stigmas large, peltate,
orbicular, with 4 or 5 lobes ....
.... **Ericaceae**

27b. Anthers dehiscing by longitudinal
slits; stigmas smaller, linear to
capitate, without lobes

28a. Leaves palmately lobed or
pinnately divided; fruit a
schizocarp, with 5 1-seeded
mericarps that elastically dehisce
from a persistent, central column
.... **Geraniaceae**

28b. Leaves simple; fruit a capsule

29a. Flowers perigynous, with
intersepalar appendages;
ovary 2-locular ....
.... **Lythraceae**

29b. Flowers hypogynous
(perigynous in *Scleranthus*),
without intersepalar
appendages; ovary 1- or 10-
locular

30a. Capsule with 10 locules;
placentation apical-axile;
stamens commonly
alternating with small
staminodes; calyx
aposepalous (*Linum*) ......
.... **Linaceae**

30b. Capsule with a single
locule; placentation free-
central; androecium
without staminodes; calyx
apo- or gamosepalous .....
.... **Caryophyllaceae**

7b. Leaves alternate or all basal

31a. Leaves compound, 3-foliate, with obcordate leaflets; petals yellow or white with pink stripes .... **Oxalidaceae**

31b. Leaves simple to compound, but not as in the Oxalidaceae; petals variously colored

    32a. Flowers with 4 petals (very rarely 0 or 2) and 6 tetradynamous stamens; fruit a silique or silicle .... **Brassicaceae**

    32b. Flower with 0–6 petals and a various number of stamens, but not 6 and tetradynamous; fruit otherwise

        33a. Stamens monadelphous, the filaments forming a conspicuous tube surrounding the style; gynoecium composed of 5–20 carpels .... **Malvaceae**

        33b. Stamens distinct or connate only at the very base; gynoecium composed of 1–6(–7) carpels

            34a. Leaves with tubular, sheathing stipules [ocreae]; flowers subtended by sheathing bracteoles [ocreolae] ... . **Polygonaceae**

            34b. Leaves without tubular, sheathing stipules; flowers not subtended by sheathing bracteoles

                35a. Leaves compound

                    36a. Perianth composed of 6 petaloid sepals and 6 petals that are modified into shorter, flabellate nectaries opposite the sepals; anthers dehiscing by flaps; fruit an exposed, blue seed (*Caulophyllum*) .... **Berberidaceae**

                    36b. Perianth composed of 4 or 5 petaloid sepals; anthers dehiscing by longitudinal slits; fruit an achene or follicle

                        37a. Flowers with numerous stamens; perianth caducous; fruit a follicle; leaves 2- to 3-times divided (*Cimicifuga*) .... **Ranunculaceae**

                        37b. Flowers with 4 stamens; perianth not caducous; fruit an achene enclosed in a indurate hypanthium; leaves 1-time divided (*Sanguisorba*) .... **Rosaceae**

                35b. Leaves simple, toothed, or lobed, but not divided

                    38a. Perianth sepaloid

                        39a. Stamens 10 per flower; fruit a 5- to 7-lobed and -locular capsule ... **Penthoraceae**

                        39b. Stamens 1–5 per flower (or 8–16 in *Acalypha*, which has palmately lobed bracts); fruit an achene, utricle, pyxis, or 3-locular capsule

40a. Flowers perigynous, bisexual; leaves palmately lobed (*Alchemilla*) .... **Rosaceae**

40b. Flowers hypogynous, unisexual or sometimes bisexual in the Amaranthaceae; leaves simple to pinnately lobed

    41a. Ovary 3-locular; styles 3, each one branched or irregularly lacerate; fruit a 3-valved capsule (*Acalypha*) .... **Euphorbiaceae**

    41b. Ovary 1-locular; styles mostly 2 or 3, without additional branching; fruit an achene, utricle, or pyxis

        42a. Sepals usually dry and scarious; filaments distinct or, more commonly, connate near the base; ovary with 3 stigmas (*Amaranthus*) .... **Amaranthaceae**

        42b. Sepals membranaceous to herbaceous (hyaline in *Axyris*); filaments distinct; ovary usually with 1 or 2 stigmas

            43a. Ovary with 1 stigma; flowers unisexual; plants sometimes with stinging hairs .... **Urticaceae**

            43b. Ovary with 2(–5) stigmas; flowers unisexual or bisexual; plants without stinging hairs .... **Amaranthaceae**

38b. Perianth, at least in part, petaloid

    44a. Plants lacking chlorophyll, the stems white to pink (to red) or yellow to brown; leaves reduced and scale-like (*Monotropa*) .... **Ericaceae**

    44b. Plants with chlorophyll, the stems and/or leaves green; leaves foliaceous

        45a. Pollen dehiscing by 2 apical pores; leaves all basal; fruit a 5-valved capsule (*Pyrola*) .... **Ericaceae**

        45b. Pollen dehiscing by longitudinal slits; leaves and fruits various

46a. Gynoecium with 10 styles; fruit a dark purple berry ... **Phytolaccaceae**

46b. Gynoecium with 2–7 styles; fruit a capsule, pyxis, or utricle

47a. Flowers perigynous

48a. Perianth 6-merous; sepals alternating with appendages at the rim of the hypanthium .... **Lythraceae**

48b. Perianth 5-merous; sepals not alternating with appendages

49a. Androecium with 5 stamens and 5 trifid, white staminodes; petals 10.0–18.0 mm long; flowers with 4 carpels; fruit a 4-locular capsule .... **Parnassiaceae**

49b. Androecium with 10 stamens; petals 2.0–6.0 mm long; flowers with 2 carpels; fruit a follicle or a 1- or 2-locular capsule .... **Saxifragaceae**

47b. Flowers hypogynous

50a. Calyx definitely gamosepalous; halophytic plants of Atlantic coast shores ...... **Plumbaginaceae**

50b. Calyx aposepalous or nearly so, connate only at the very base, if at all; plants not halophytic

51a. Calyx composed of 2 sepals; stems prostrate; fruit a pyxis (*Portulaca*) .... **Portulacaceae**

51b. Calyx composed of 5 sepals; stems upright; fruit a capsule

52a. Gynoecium of 5 carpels; capsule 10-valved; leaves borne

on a stem (*Linum*) .....
.... **Linaceae**

52b. Gynoecium of 2 or 4 carpels; capsule 2- or 4-valved; leaves all basal or essentially so

53a. Androecium with 5 stamens and 5 trifid, white staminodes; petals 10.0–18.0 mm long; flowers with 4 carpels; fruit a 4-locular capsule .... **Parnassiaceae**

53b. Androecium with 10 stamens; petals 2.0–6.0 mm long; flowers with 2 carpels; fruit a follicle or a 1- or 2-locular capsule .... **Saxifragaceae**

Group 10

1a. Leaves peltate; plants aquatic

    2a. Leaves orbicular, unlobed; carpels individually sunken in the spongy receptacle; fruit a nut, loosely borne in cavities of the firm, accrescent receptacle .... **Nelumbonaceae**

    2b. Leaves elliptic; carpels free from the receptacle; fruit a leathery follicle, not enclosed by the receptacle (*Brasenia*) .... **Nymphaeaceae**

1b. Petiole, if present, attached at the base of the leaf blade; plants terrestrial or sometimes aquatic in *Caltha, Comarum, Mentha,* and *Ranunculus*

    3a. Corolla gamopetalous; carpels 2, each carpel evidently lobed, therefore the gynoecium appearing to be composed of 4 carpels, separating at maturity into 4 half-carpellate segments

        4a. Inflorescence a coiled raceme; leaves alternate; stems terete .... **Boraginaceae**

        4b. Inflorescence not a coiled raceme; leaves opposite; stems commonly square .... **Lamiaceae**

    3b. Corolla apopetalous or absent; carpels 1–many, not schizocarpic (except Malvaceae)

        5a. Stamens monadelphous; anthers unilocular; fruit a schizocarp .... **Malvaceae**

        5b. Stamens distinct; anthers bilocular; fruit otherwise

            6a. Flowers lacking a hypanthium

                7a. Plants succulent with fleshy, simple leaves; carpels with a small, scale-like, nectariferous appendage at the base, stamens usually 10 or fewer per flower (up to 80 in *Sempervivum*); fruit a follicle .... **Crassulaceae**

7b. Plants not succulent, with thinner, simple, lobed, or compound leaves; carpels without a scale-like, nectariferous appendage; stamens usually many per flower; fruit an achene, berry, or follicle .... **Ranunculaceae**

6b. Flowers with a hypanthium

8a. Leaves simple, without stipules; flowers with 2 or 4 carpels; fruit a capsule .... **Saxifragaceae**

8b. Leaves simple to compound, with stipules (without stipules in *Aruncus*); flowers with 2–many carpels; fruit an achene, drupe, or follicle, either solitary or in an aggregate .... **Rosaceae**

LYCOPODIACEAE
(4 genera, 15 species)

KEY TO THE GENERA

1a. Sporophylls borne on upright stems in zones alternating with sterile leaves; gemmiferous branches present; horizontal stem absent; upright stems clustered .... *Huperzia*
1b. Sporophylls aggregated in a terminal strobilus; gemmiferous branches absent; horizontal stem present (but may be underground); upright stems separate, produced along the horizontal stem
   2a. Sporophylls yellow to brown, shorter and broader than vegetative leaves; sterile stems upright; peduncles with reduced and remote leaves; plants mostly of upland habitats
      3a. Leaves in 4 or 5 ranks; branches flat (terete in *Diphasiastrum sitchense*); peduncles, if present, dichotomously branched; ultimate branches, including the leaves, 2.0–6.0 mm in diameter .... *Diphasiastrum*
      3b. Leaves in 6–8 ranks; branches terete (flat in *Lycopodium obscurum*); peduncles, if present, pseudomonopodial; ultimate branches, including the leaves, 5.0–12.0 mm in diameter .... *Lycopodium*
   2b. Sporophylls green, similar in shape and size to vegetative leaves; sterile stems prostrate or arching along the ground; peduncles leafy; plants mostly of wetlands ....
.... *Lycopodiella*

Diphasiastrum
(4 species)

Fertile hybrids are produced within this genus, but detection is difficult because abortive spores, usually a good indicator of hybridization in clubmosses, are absent. Further, these hybrids are often disjunct from their parents' ranges, which makes determining parental taxa difficult. Presence or absence of annual constrictions on the branches is used to separate *Diphasiastrum complanatum* and *D. digitatum*. However, absence of these constrictions does not necessarily mean the plant in question is *D. digitatum*. It may be that the collection is from first-year growth. Reproductive plants allow for the most reliable use of the annual demarcation character.

KEY TO THE SPECIES

1a. Leaves mostly 5-ranked, adnate to the stem less than half their length; peduncles absent (up to 1.0 cm tall); strobili solitary; branches terete; plants commonly shorter than 12.0 cm .... *D. sitchense*
1b. Leaves 4-ranked, those of the lateral ranks adnate to the stem at least half their length; peduncles 0.5–15.0 cm tall; strobili 1–7; branches flat; plants commonly taller than 12.0 cm
   2a. Leaves of the lower rank 1.0–1.5 mm long, similar in length to the leaves of the lateral and upper ranks; branches narrow, 1.2–2.0 mm wide, blue-green ....
.... *D. tristachyum*
   2b. Leaves of the lower rank up to 1.0 mm long, shorter than the leaves of the lateral and upper ranks; branches wider, 2.0–3.0 mm wide, green

3a. Branches without annual constrictions; lateral branches of the upright shoots forming regular and planar, fan-shaped splays; strobili mostly 20.0–40.0 mm long, often each with a slender, sterile tip .... ***D. digitatum***

3b. Branches with annual constrictions; lateral branches of the upright shoots forming irregular, spreading splays; strobili 10.0–25.0 mm long, each without a sterile tip .... ***D. complanatum***

Diphasiastrum complanatum (L.) Holub

synonym(s): *Lycopodium complanatum* L., *Lycopodium complanatum* L. var. *canadense* Victorin
common name(s): northern running-pine
habit: stoloniferous, perennial herb
range: Labrador and Newfoundland, s. to NH and NY, w. to WA, n. to British Columbia and AK
state frequency: uncommon; absent from the southern part of Maine
habitat: dry, upland forests with acidic soils
notes: Hybrids with *Diphasiastrum alpinum* (L.) Holub, called *D.* ×*issleri* (Rouy) Holub, *D. digitatum*, and *D. tristachyum*, called *D.* ×*zeilleri* (Rouy) Holub, occur in Maine.

Diphasiastrum digitatum (Dill. *ex* A. Braun) Holub

synonym(s): *Lycopodium digitatum* Dill. *ex* A. Braun, *Lycopodium complanatum* L. var. *flabelliforme* Fern.
common name(s): southern running-pine
habit: stoloniferous, perennial herb
range: Newfoundland, s. to GA, w. to TN, n. to MI and MN
state frequency: common
habitat: dry uplands, coniferous forests, open fields
notes: Hybrids with *Diphasiastrum complanatum* and *D. tristachyum*, called *D.* ×*habereri* (House) Holub, occur Maine. The sterile tips of the strobili of *D. digitatum* are found in about 50% of the specimens. Annual constrictions may rarely be produced in deeply shaded plants.

Diphasiastrum sitchense (Rupr.) Holub

synonym(s): *Lycopodium sitchense* Rupr., *Lycopodium sabinaefolium* Willd. var. *sitchense* (Rupr.) Fern.
common name(s): Sitka clubmoss, Alaskan clubmoss
habit: stoloniferous, perennial herb
range: Labrador, s. to ME, w. to OR, n. to AK
state frequency: very rare
habitat: exposed alpine meadows and barrens
notes: This species is ranked S1 by the Maine Natural Areas Program. Hybrids with *Diphasiastrum tristachyum*, called *D.* ×*sabinifolium* (Willd.) Holub, occur in Maine. This fertile hybrid does not necessarily grow with its parents, is more common than *D. sitchense*, and can be distinguished from *D. sitchense* by its mostly 4-ranked leaves, adnation of lateral leaves for more than half their length, and flat branches.

Diphasiastrum tristachyum (Pursh) Holub

synonym(s): *Lycopodium tristachyum* Pursh
common name(s): blue ground-cedar, wiry ground-cedar
habit: rhizomatous, perennial herb
range: Newfoundland, s. to SC, w. to IN, n. to Manitoba
state frequency: occasional
habitat: dry, upland forests of acidic soils
notes: Hybrids with *Diphasiastrum complanatum* (q.v.), *D. digitatum* (q.v.), and *D. sitchense* (q.v.) occur in Maine.

Huperzia
(3 species)

Hybrids within this genus grow in Maine in the limited areas where ranges of parental taxa overlap. They produce *ca.* 25–50% abortive spores and exhibit morphological intermediacy between the parental species. *Huperzia appalachiana* and *H. selago* differ subtly from one another, so it is important to collect specimens that show all the characters in the key. Be sure the gemmae have not fallen from the gemmiferous branches.

KEY TO THE SPECIES

1a. Leaves 7.0–12.0 mm long, obovate, serrate with 1–8 teeth, hypostomic .... ***H. lucidula***
1b. Leaves 3.0–8.0 mm long, lanceolate to triangular, entire or with 1–3 irregular teeth, amphistomic
    2a. Shoots determinate, the entire plant dying after several years, without annual constrictions; gemmiferous branches in 1–3 pseudowhorls at the end of annual growth; leaves in the mature [distal] portion 2.0–3.5 mm long; leaves in the juvenile [proximal] portion ascending to spreading, narrow-triangular .... ***H. appalachiana***
    2b. Shoots indeterminate, the bases of plant dying and rotting away, with weak annual constrictions; gemmiferous branchlets in 1 pseudowhorl at the end of annual growth; leaves in the mature portion 3.5–5.0 mm long; leaves in the juvenile portion often reflexed, triangular .... ***H. selago***

Huperzia appalachiana Beitel & Mickel
synonym(s): —
common name(s): mountain firmoss
habit: perennial herb
range: Greenland, s. along Appalachian mountains to SC; disjunct on the shore of Lake Superior
state frequency: rare
habitat: exposed high-elevation, rocky, often igneous, areas or cliffs near Atlantic coast
notes: This species is ranked S2 by the Maine Natural Areas Program. Hybrids with *Huperzia selago* occur in Maine.

Huperzia lucidula (Michx.) Trevisan
synonym(s): *Lycopodium lucidulum* Michx.
common name(s): shining clubmoss
habit: perennial herb
range: Newfoundland, s. to NC, w. to IL, n. to MN and Saskatchewan
state frequency: occasional
habitat: shady, mossy forests
notes: Hybrids with *Huperzia selago*, called *H.* ×*buffersii* (Abbe) Kartesz & Gandhi, occur in Maine.

Huperzia selago (L.) Bernh. *ex* Mart. & Schrank
synonym(s): *Lycopodium selago* L., *Lycopodium selago* L. var. *appressum* Desv., *Lycopodium selago* L. var. *patens* Desv.
common name(s): northern firmoss
habit: perennial herb
range: Newfoundland, s. to RI, w. to MN, n. to Alberta and Yukon
state frequency: rare; possibly more common than thought
habitat: ditches, swales, swamps, rarely into alpine zone
notes: This species is ranked S1 by the Maine Natural Areas Program. Hybrids with *Huperzia appalachiana* and *H. lucidula* (*q.v.*) occur in Maine.

Lycopodiella
(2 species)

Maine's 2 species hybridize with each other and are thought to produce fertile hybrids.

KEY TO THE SPECIES

1a. Upright, reproductive shoots 3.5–10.0 cm tall; sporophylls spreading; leaves of horizontal stem 0.5–0.7 mm wide; stems slender, [excluding the leaves] 0.5–0.9 mm in diameter .... *L. inundata*

1b. Upright, reproductive shoots 4.0–45.0 cm tall; sporophylls appressed; leaves 0.8–1.1 mm wide; stems thick, [excluding the leaves] 1.5–2.0 mm in diameter ....
.... *L. appressa*

Lycopodiella appressa (Chapman) Cranfill
synonym(s): *Lycopodium appressum* (Chapman) Lloyd & Underwood, *Lycopodium inundatum* L. var. *appressum* Chapman, *Lycopodium inundatum* L. var. *bigelovii* Tuckerman
common name(s): appressed bog clubmoss
habit: prostrate, perennial herb
range: Newfoundland, s. to FL, w. to TX, n. to KS
state frequency: very rare
habitat: acidic shores and bogs, damp, sandy soil
notes: This species is ranked SH by the Maine Natural Areas Program. Known to hybridize with *Lycopodiella inundata*.

Lycopodiella inundata (L.) Holub
synonym(s): *Lycopodium inundatum* L.
common name(s): bog clubmoss
habit: prostrate, perennial herb
range: Newfoundland, s. to WV, w. and n. to WI and MN; disjunct in the Pacific northwest
state frequency: occasional
habitat: bogs, shores, wet meadows, damp, sandy soil
notes: Known to hybridize with *Lycopodiella appressa*. Large individuals approach the morphology of *L. appressa* but retain the spreading orientation of leaves and sporophylls.

Lycopodium
(6 species)

Hybridization is thought to be extremely rare. Leaf position and orientation are important characters in the *Lycopodium obscurum* complex but are often lost or obscured during pressing, so make field notes.

KEY TO THE SPECIES

1a. Leaves with a caducous bristle-tip (persistent on shoot apices); strobili pedunculate
  2a. Peduncles with 1 strobilus or sometimes with a second, non-pedicellate strobilus; leaves 3.0–5.0 mm long, ascending to appressed; upright stems with 2 or 3 ascending branches .... *L. lagopus*
  2b. Peduncles with 2–5 pedicellate strobili; leaves 4.0–6.0 mm long, spreading; upright stems with 3–6 spreading branches .... *L. clavatum*

1b. Leaves lacking a bristle-tip, acute or acuminate at the apex; strobili sessile [*i.e.*, lacking peduncles]

3a. Horizontal stems above ground surface; upright stems mostly simple, occasionally with 1 or 2 branches; leaves 5.0–11.0 mm long; strobilus solitary at the terminus of an upright stem .... ***L. annotinum***

3b. Horizontal stems below ground surface; upright stems many times branched; leaves 2.5–5.5 mm long; strobili 1–7 per upright stem

    4a. Leaves of the upright stem's main axis widely spreading [45–90°], light green; leaves of the branches in 2 upper, 2 lower, and 2 lateral ranks ... ***L. dendroideum***

    4b. Leaves of the upright stem's main axis appressed, not spreading more than 30°, dark green; leaves of the branches in 1 upper, 1 lower, and 4 lateral ranks

        5a. Branches flat; leaves of the lower rank smaller than those of other ranks; leaves of the lateral ranks twisted [the adaxial leaf surface faces up] .... ***L. obscurum***

        5b. Branches terete; leaves of the lower rank similar in length to those of the other ranks; leaves of the lateral ranks not twisted [the adaxial surface faces the stem] .... ***L. hickeyi***

Lycopodium annotinum L.

synonym(s): *Lycopodium annotinum* L. var. *acrifolium* Fern., *Lycopodium annotinum* L. var. *pungens* (La Pylaie) Desv.
common name(s): stiff clubmoss, bristly clubmoss
habit: stoloniferous, perennial herb
range: circumboreal; s. to NJ and TN
state frequency: common
habitat: coniferous forests, extending into subalpine zone
notes: —

Lycopodium clavatum L.

synonym(s): —
common name(s): wolf-claw clubmoss, running-pine
habit: trailing, perennial herb
range: circumboreal; s. to NJ and TN
state frequency: occasional
habitat: dry fields, open woods
notes: —

Lycopodium dendroideum Michx.

synonym(s): *Lycopodium obscurum* L. var. *dendroideum* (Michx.) D. C. Eat.
common name(s): prickly tree clubmoss
habit: erect, perennial herb
range: Newfoundland, s. to WV, w. to MN and British Columbia, n. to AK
state frequency: common
habitat: forests, commonly with dry soil
notes: —

Lycopodium hickeyi W. H. Wagner, Beitel, & Moran

synonym(s): *Lycopodium obscurum* L. var. *isophyllum* Hickey
common name(s): Hickey's tree clubmoss
habit: erect, perennial herb
range: Newfoundland, s. to TN, w. to IN, n. to MN
state frequency: uncommon
habitat: hardwood forests
notes: —

Lycopodium lagopus (C. Hartman) G. Zinserling *ex* Kuzeneva-Prochorova
synonym(s): *Lycopodium clavatum* L. var. *lagopus* Laestad. *ex* C. Hartman, *Lycopodium clavatum* L. var. *megastachyon* Fern. & Bissell, *Lycopodium clavatum* L. var. *monostachyon* Hook. & Grev.
common name(s): one-cone clubmoss
habit: trailing, perennial herb
range: Greenland, s. to ME, w. to British Columbia, n. to AK
state frequency: occasional; less common in the southern part of Maine
habitat: fields, open woods
notes: —

Lycopodium obscurum L.
synonym(s): —
common name(s): princess-pine, ground-pine
habit: erect, perennial herb
range: New Brunswick, s. to NJ and mountains of GA, w. to TN, n. to MN and Ontario
state frequency: occasional
habitat: forests
notes: —

ISOETACEAE
(1 genus, 6 species)

Isoetes
(6 species)

Megaspore morphology is an important character for distinguishing species in this genus. Megaspores are found at the base of the outer ring of leaves [called megasporophylls]. Specimens collected too early in the season contain underdeveloped, "mealy" megaspores and are not easily identified. If collecting must be done early in the season, try to find the rotting leaf bases from the previous year as these may still contain usable megaspores. It is best to collect late in the season, dry the megaspores thoroughly, and use a 30× lens for at least some of the species. Hybrids, not uncommon in this genus, can be identified by the presence of malformed megaspores. An important feature of the megaspores is the girdle. This is a zone immediately distal to the equatorial ridge, in the opposite hemisphere as the radial ridges. When the girdle is similar in texture to the remaining portion of the distal hemisphere, it is termed obscure. When the girdle is of a texture different from that of the remaining portion of the distal hemisphere and is called evident.

KEY TO THE SPECIES
1a. Megaspores echinate with thin, sharp spines; leaves deciduous .... ***I. echinospora***
1b. Megaspores smooth or variously ornamented, but without a spiny texture; leaves persisting more than 1 year
    2a. Leaves with abundant stomates, bright green; plants aquatic to amphibious, occasionally emergent; girdle of megaspores obscure .... ***I. riparia***
    2b. Leaves with few or no stomates, dark green to red-green or red-brown; plants aquatic, usually submerged; girdle evident (obscure in *I. prototypus*)
        3a. Megaspores 0.55–0.75 mm in diameter, averaging more than 0.6 mm; leaves abruptly tapering to the tip; plants submerged 1.0–3.0 m .... ***I. lacustris***
        3b. Megaspores 0.4–0.65 mm in diameter, averaging less than 0.6 mm; leaves gradually tapering to the tip; plants submerged to 1.0 m deep

4a. Velum covering the entire sporangium; leaves very rigid and straight to the tip; sporangium wall unpigmented; girdle of the megaspore obscure .... *I. prototypus*

4b. Velum covering less than half the sporangium; leaves pliant and curling at the tips; sporangium wall brown-streaked; girdle evident

5a. Megaspores with a papillate girdle and irregular, roughened crests .... *I. tuckermanii*

5b. Megaspores with a smooth girdle and smooth, rounded crests .... *I. acadiensis*

Isoetes acadiensis Kott

synonym(s): —
common name(s): Acadian quillwort
habit: evergreen, aquatic, perennial herb
range: Newfoundland, s. to MA and NY
state frequency: very rare
habitat: oligotrophic, often acidic, lakes and streams
notes: This species is ranked S1 by the Maine Natural Areas Program. Known to hybridize with *Isoetes lacustris*.

Isoetes echinospora Durieu

synonym(s): *Isoetes echinospora* Durieu var. *braunii* (Durieu) Engelm., *Isoetes echinospora* Durieu ssp. *muricata* (Durieu) A. & D. Löve, *Isoetes muricata* Durieu
common name(s): spiny-spored quillwort
habit: deciduous-leaved, aquatic, perennial herb
range: Greenland, s. to NJ, w. to CA, n. to AK
state frequency: occasional
habitat: oligotrophic, often acidic, lakes and streams
notes: Known to hybridize with *Isoetes lacustris*, *I. riparia*, and *I. tuckermanii*. Only *I. echinospora* and *I. riparia* have stomates on the leaves. In *I. echinospora*, the stomates are found only near the tip.

Isoetes lacustris L.

synonym(s): *Isoetes hieroglyphica* A. A. Eat., *Isoetes macrospora* Durieu
common name(s): lake quillwort
habit: evergreen, aquatic, perennial herb
range: Greenland, s. to NY, w. to MN, n. to Saskatchewan; disjunct in midwest and southeast United States
state frequency: occasional
habitat: oligotrophic, often acidic, lakes and streams
notes: This is Maine's deepest water quillwort. It is known to hybridize with all quillworts in Maine except *Isoetes prototypus*. Hybrids with *I. tuckermanii*, called *I.* ×*harveyi* A. A. Eat., occur in Maine.

Isoetes prototypus D. M. Britt.

synonym(s): —
common name(s): prototype quillwort
habit: evergreen, aquatic, perennial herb
range: New Brunswick, Nova Scotia, and ME
state frequency: very rare
habitat: oligotrophic, often acidic, lakes and streams
notes: This species is ranked S1 by the Maine Natural Areas Program.

Isoetes riparia Engelm. *ex* A. Braun
synonym(s): —
common name(s): shore quillwort
habit: evergreen, aquatic or emergent, perennial herb
range: ME, s. to NC, w. and n. to Quebec
state frequency: very rare
habitat: margins of lakes and streams
notes: This species is ranked SH by the Maine Natural Areas Program. Known to hybridize with
  Isoetes echinospora, I. lacustris, and *I. tuckermanii.*

Isoetes tuckermanii A. Braun *ex* Engelm.
synonym(s): —
common name(s): Tuckerman's quillwort
habit: evergreen, aquatic, perennial herb
range: Newfoundland, s. to NJ, w. to PA, n. to Quebec; disjunct to Ontario
state frequency: occasional
habitat: lakes and streams, often acidic
notes: Known to hybridize with *Isoetes echinospora, I. lacustris (q.v.),* and *I. riparia.*

SELAGINELLACEAE
(1 genus, 3 species)

Selaginella
(3 species)

Maine's representatives of this genus are easily separated by morphology or habitat or, for the 2 rare species, state distribution.

KEY TO THE SPECIES
1a. Leaves monomorphic, spirally arranged, not in distinct ranks; axillary leaves absent at branching points
  2a. Leaves thin, spinulose-toothed, acuminate; strobili cylindric; sporophylls spreading, mostly 10-ranked; rhizophores absent .... *S. selaginoides*
  2b. Leaves thicker, ciliolate, bristle-tipped; strobili tetragonous; sporophylls appressed, 4-ranked; rhizophores present .... *S. rupestris*
1b. Leaves dimorphic, arranged in 4 ranks, the 2 lateral ranks with larger leaves than the 2 median ranks; axillary leaves present at branching points .... *S. apoda*

Selaginella apoda (L.) Spring *ex* Mart., Eichler, & Urban
synonym(s): —
common name(s): meadow spikemoss, creeping spikemoss
habit: trailing, evergreen, perennial herb
range: ME, s. to FL, w. to TX and OK, n. to IL
state frequency: very rare; known only from southern Maine
habitat: meadows, lawns, stream banks
notes: This species is ranked S1 by the Maine Natural Areas Program.

Selaginella rupestris (L.) Spring
synonym(s): —
common name(s): rock spikemoss

habit: trailing, evergreen, perennial herb
range: Greenland, s. to GA, w. to OK, n. to SD and Alberta
state frequency: occasional
habitat: exposed rocks, rocky barrens
notes: —

Selaginella selaginoides (L.) Beauv. *ex* Mart. & Schrank
synonym(s): —
common name(s): northern spikemoss, low spikemoss
habit: trailing, evergreen, perennial herb
range: circumboreal; s. to ME, w. to MN; also s. to Alberta, w. to British Columbia; disjunct to CO
    and NV
state frequency: very rare; known only from northern Maine
habitat: calcareous wetlands
notes: This species is ranked S1 by the Maine Natural Areas Program.

OPHIOGLOSSACEAE
(2 genera, 10 species)

KEY TO THE GENERA
1a. Leaf blades simple, entire; venation reticulate; sporangia borne in a compact, linear
    sporophore .... ***Ophioglossum***
1b. Leaf blades lobed or compound, entire or toothed; venation forking; sporangia borne in
    pinnately branched sporophores .... ***Botrychium***

Botrychium
(9 species)

Botrychium is a highly variable genus with noticeably different morphologies from sun to shade and even within the same population. Single-plant collections sometimes cannot be identified. Several samples are required to insure that collections capture this variability. Multiple collections will not harm the plants if the below-ground portions are left intact.

KEY TO THE SPECIES
1a. Leaf blades small, up to 10.0 cm long, seldom longer than 4.0 cm, glabrous; leaflets
    entire to lobed
   2a. Leaflets obovate to flabellate, the apex rounded to truncate, the margins entire or
       distally few-lobed; trophophore [leaf blade] and sporophore erect in bud
      3a. Trophophore shorter than the stalk of the sporophore and often not reaching the
          base of the sporophore; basal pair of leaflets evidently largest in well developed
          plants .... ***B. simplex***
      3b. Trophophore equaling or exceeding the length of the stalk of the sporophore;
          basal pair of leaflets not evidently larger than the adjacent pair
         4a. Plants yellow-green to glaucous green; summer leaves not planar, at least the
             basal leaflets somewhat clasping the stalk of the sporophore; leaflets remote,
             usually narrowly flabellate; lower margin of the basal leaflets extending from
             the axis at an angle of 90° or less [i.e., ascending]; apical leaf segments

slightly reduced and more or less of similar shape as medial segments; medial leaflets 1.0–9.0 mm wide .... *B. minganense*

4b. Plants green to blue-green; summer leaves planar; leaflets approximate to remote, usually broadly flabellate; lower margin of the basal leaflets extending from the axis at an angle greater than 90° [*i.e.*, descending]; apical leaf segments greatly reduced in size and of different shape from medial segments; medial leaflets 6.0–18.0 mm wide.................................. *B. lunaria*

2b. Leaflets linear to narrowly ovate, the apex obtuse to acuminate, the margins shallowly to deeply lobed; at least the tip of the trophophore reflexed in bud

5a. Trophophore oblong to ovate, usually stalked; sporophore with 1 main axis or with 1 larger and 2 smaller branches, mainly erect in bud ..... *B. matricariifolium*

5b. Trophophore deltate, usually sessile; sporophore usually divided into equally long branches, reflexed in bud .... *B. lanceolatum*

1b. Leaf blades larger, 3.0–25.0 cm long, seldom shorter than 5.0 cm, pubescent, at least in bud; leaflets 1- to 3-times compound

6a. Trophophore [leaf blade] herbaceous, deciduous, appearing sessile on the common stem; leaf sheath open .... *B. virginianum*

6b. Trophophore coriaceous, persisting more than 1 year, stalked; leaf sheath closed

7a. Leafules regularly dissected, divided to the tip; terminal leaflet segment equal to or slightly longer than adjacent, lateral leaf segments; ultimate segments of leaves obtuse to rounded; leaves green in fall and winter .... *B. multifidum*

7b. Leafules irregularly dissected, not divided in the apical portion; terminal leaflet segment noticeably larger than adjacent, lateral leaf segments; ultimate segments of leaves rounded to acuminate; leaves green or bronze to purple in fall and winter

8a. Leafules ovate, with crenulate to denticulate margins and usually rounded apices; leaves green in fall and winter .... *B. oneidense*

8b. Leafules lanceolate to ovate, with denticulate to lacerate margins and usually acute apices; leaves bronze to purple in fall and winter .... *B. dissectum*

Botrychium dissectum Spreng.
synonym(s): *Botrychium obliquum* Muhl. *ex* Willd.
common name(s): dissected grapefern
habit: perennial herb
range: Nova Scotia and New Brunswick, s. to FL, w. to TX, n. to Lake Superior
state frequency: occasional
habitat: various habitats from open areas to woodlands
notes: *Botrychium dissectum* is difficult to distinguish from *B. oneidense*. Leafules of the latter have blunter teeth and an apex that is more rounded.

Botrychium lanceolatum (Gmel.) Angstr. ssp. *angustisegmentum* (Pease & Moore) Clausen
synonym(s): —
common name(s): lance-leaved moonwort, triangle moonwort
habit: perennial herb
range: Newfoundland, s. to NJ and TN, w. and n. to MN; disjunct to MT, Alberta, and British Columbia
state frequency: uncommon
habitat: shaded woods
notes: *Botrychium lanceolatum* ssp. *lanceolatum* was reported from Maine in *Rhodora* 32:133–136, but this specimen has not been verified.

Botrychium lunaria (L.) Sw.

synonym(s): —
common name(s): moonwort
habit: perennial herb
range: Greenland, s. to PA, w. to CA, n. to AK
state frequency: very rare
habitat: fields and woods
notes: This species is ranked S1 by the Maine Natural Areas Program. In deep shade *Botrychium lunaria* approaches the morphology of *B. minganense*, especially in the remoteness of the leaflets. Noted to hybridize with other species of *Botrychium*.

Botrychium matricariifolium (Döll) A. Braun *ex* Koch

synonym(s): —
common name(s): daisy-leaved moonwort
habit: perennial herb
range: Newfoundland and Labrador, s. to NJ and TN, w. and n. to MN
state frequency: occasional
habitat: fields and deciduous forests
notes: —

Botrychium minganense Victorin

synonym(s): *Botrychium lunaria* (L.) Sw. var. *minganense* (Victorin) Dole
common name(s): Mingan moonwort
habit: perennial herb
range: Newfoundland and Labrador, s. to ME and NY, w. to CA, n. to AK
state frequency: very rare
habitat: fields and woods
notes: Deep-shade forms of this species show a glaucous-green color and more strongly folded leaves.

Botrychium multifidum (Gmel.) Rupr.

synonym(s): *Botrychium multifidum* (Gmel.) Rupr. var. *intermedium* (D. C. Eat.) Farw.
common name(s): leathery grapefern
habit: perennial herb
range: Greenland, s. to WV, w. to CA, n. to AK
state frequency: occasional
habitat: various habitats, mostly open
notes: —

Botrychium oneidense (Gilbert) House

synonym(s): *Botrychium dissectum* Spreng. var. *oneidense* (Gilbert) Farw.
common name(s): blunt-lobed grapefern
habit: perennial herb
range: New Brunswick, s. to NC, w. and n. to MN
state frequency: uncommon
habitat: moist woods, swamps
notes: This species is difficult to distinguish from *Botrychium dissectum* and *B. multifidum*.

Botrychium simplex E. Hitchc.

synonym(s): *Botrychium simplex* E. Hitchc. var. *tenebrosum* (A. A. Eat.) Clausen
common name(s): little grapefern, least moonwort
habit: perennial herb

range: Greenland, s. to NC, w. and n. to SD and ND; also MT, s. to UT, w. to CA, n. to British Columbia
state frequency: occasional
habitat: moist or dry open areas, ditches
notes: Morphologically, this species is extremely variable. In deep shade the specimens possess reduced, undivided trophophores and tall, slender stalks.

Botrychium virginianum (L.) Sw.

synonym(s): *Botrychium virginianum* (L.) Sw. var. *europaeum* Angstr.
common name(s): rattlesnake fern, common grapefern
habit: perennial herb
range: Labrador and Newfoundland, s. to FL, w. to AZ, n. to AK
state frequency: occasional
habitat: shaded woods
notes: —

Ophioglossum
(1 species)

Ophioglossum pusillum Raf.

synonym(s): *Ophioglossum vulgatum* L. var. *pseudopodum* (Blake) Farw.
common name(s): northern adder's-tongue
habit: perennial herb
range: New Brunswick and Nova Scotia, s. to NC, w. to NE, n. to ND; also British Columbia, s. to CA
state frequency: very rare
habitat: open fields, fens, marshes
notes: This species is ranked S2 by the Maine Natural Areas Program. It is, however, likely overlooked due to its inconspicuousness.

EQUISETACEAE
(1 genus, 8 species)

Equisetum
(8 species)

Hybrids are found in this genus and can be identified by their white, malformed spores (*vs.* green and spherical in non-hybrids). Important are notes of the dimensions of the internal cavities of the stem, as these dimensions will be distorted in pressing. *Equisetum ×ferrissii* Clute is reported to occur in Maine by the Flora of North America (1993, vol. 2, p. 83), but voucher specimens are unknown.

KEY TO THE SPECIES FOR USE WITH VEGETATIVE MATERIAL
1a. Stems persisting for more than 1 year, mostly simple [without regularly whorled branches]; stomates sunken, arranged linearly on the stem
    2a. Stems 3- to 12-ridged; leaves not articulated to the sheaths and persistent on them; central cavity up to 0.35 times the diameter of the stem

3a. Stems flexuous, 6-ridged, 0.5–1.0 mm thick; sheaths with 3 leaves; central cavity
   absent; vallecular cavities 3 .... *E. scirpoides*
3b. Stems erect and straight, 3- to 12-ridged, 1.0–4.5 mm thick; leaves numbering
   the same as the ridges; central cavity evident; vallecular cavities 5–12 ....
   .... *E. variegatum*
2b. Stems 14- to 50-ridged; leaves articulated to and deciduous from the sheaths; central
   cavity greater than 0.5 times the stem diameter .... *E. hyemale*
1b. Stems not persisting for more than 1 year, simple or with regular whorls of branches;
   stomates not sunken, scattered on the stem or in broad bands or absent
   4a. Length of the first internode of each branch shorter than its subtending stem sheath
      (or sometimes the branches absent in *E. fluviatile*); grooves of branches rounded
      [*i.e.*, U-shaped]; plants of wetlands
      5a. Leaves more than 11 per sheath, 1.5–3.0 mm long, dark throughout or with a
         narrow, white band; stems shallowly 9- to 25-ridged; central cavity 0.75 or more
         times the stem diameter .... *E. fluviatile*
      5b. Leaves fewer than 11 per sheath, 2.0–5.0 mm long, with prominent white
         margins; stems deeply 5- to 10-ridged; central cavity less than 0.35 times the
         stem diameter .... *E. palustre*
   4b. Length of the first internode of each branch equal to or exceeding its subtending stem
      sheath; grooves of branches channeled [*i.e.*, C-shaped]; plants of moist woods, fields,
      roadsides, and forested wetlands
      6a. Leaves red-brown, cohering into 3 or 4 large groups, papery; stem branches
         again branched .... *E. sylvaticum*
      6b. Leaves brown to black, separate or in 4 or more small groups, firm; stem
         branches unbranched
         7a. Leaves of the branches attenuate; central cavity 0.25 times the stem diameter;
            stem internodes smooth or sparsely scabrous .... *E. arvense*
         7b. Leaves of the branches deltate; central cavity 0.35–0.5 times the diameter
            of the stem; stem internodes scabrous .... *E. pratense*

KEY TO THE SPECIES FOR USE WITH REPRODUCTIVE MATERIAL
1a. Stems persisting for more than 1 year, mostly simple [*i.e.*, without regularly whorled
   branches]; strobilus apex pointed; stomates sunken, arranged linearly on the stem
   2a. Strobili 0.2–1.0 cm long; stems 3- to 12-ridged; leaves not articulated to and
      persistent on the sheaths; central cavity up to 0.35 times the diameter of the stem
      3a. Strobili 2.0–5.0 mm long; stems flexuous, 6-ridged, 0.5–1.0 mm thick; sheaths
         with 3 leaves; central cavity absent; vallecular cavities 3 .... *E. scirpoides*
      3b. Strobili 5.0–10.0 mm long; stems erect and straight, 3- to 12-ridged, 1.0–4.5 mm
         thick; leaves numbering the same as the ridges; central cavity evident; vallecular
         cavities 5–12 .... *E. variegatum*
   2b. Strobili 1.0–2.0 cm long; stems 14- to 50-ridged; leaves articulated to and deciduous
      from the sheaths; central cavity greater than 0.5 times the stem diameter ....
      .... *E. hyemale*
1b. Stems not persisting for more than 1 year, simple or with regular whorls of branches;
   strobilus apex rounded; stomates not sunken, scattered on the stem or in broad bands or
   absent

4a. Strobili appearing in spring, borne on soft, pink to light brown stems, the green photosynthetic stems forming or appearing later [*i.e.*, stems dimorphic]
    5a. Leaves red-brown, cohering into 3 or 4 large groups, papery .... ***E. sylvaticum***
    5b. Leaves brown to black, separate or in 4 or more small groups, firm
        6a. Leaves brown throughout; stems without stomates, promptly senescing ....
        .... ***E. arvense***
        6b. Leaves black with a pale margin; stems with stomates, becoming green and branched, persisting through the season .... ***E. pratense***
4b. Strobili appearing in summer, borne on firmer, green stems that persist throughout the season and are photosynthetic [*i.e.*, stems monomorphic]
    7a. Leaves more than 11 per sheath, 1.5–3.0 mm long, dark throughout or with a narrow, white band; stems shallowly 9- to 25-ridged; central cavity 0.75 or more times the stem diameter .... ***E. fluviatile***
    7b. Leaves fewer than 11 per sheath, 2.0–5.0 mm long, with prominent, white margins; stems deeply 5- to 10-ridged; central cavity less than 0.35 times the stem diameter .... ***E. palustre***

Equisetum arvense L.
synonym(s): *Equisetum arvense* L. var. *boreale* (Bong.) Rupr.
common name(s): common horsetail, field horsetail
habit: perennial herb
range: Greenland, s. to SC and MS, w. to NV and CA, n. to AK; also Eurasia
state frequency: common
habitat: damp to dry open areas, fields, roadsides
notes: Hybrids with *Equisetum fluviatile*, called *E. ×litorale* Kühlewein *ex* Rupr., occur in Maine. *Equisetum ×litorale* can be identified by a central cavity 0.5–0.65 times the diameter of the stem, solid branches, presence of vallecular cavities, and abortive spores. Also known to hybridize with *E. palustre*, and *E. pratense*.

Equisetum fluviatile L.
synonym(s): —
common name(s): river horsetail, water horsetail
habit: perennial herb
range: Labrador, s. to VA, w. to WA, n. to AK
state frequency: occasional
habitat: standing, still, or slow-moving water
notes: Hybrids with *Equisetum arvense* (*q.v.*) occur in Maine. Also known to hybridize with *E. palustre*.

Equisetum hyemale L. ssp. *affine* (Engelm.) Calder & Taylor
synonym(s): —
common name(s): common scouring-rush
habit: evergreen, perennial herb
range: Newfoundland, s. to FL, w. to CA, n. to AK
state frequency: occasional
habitat: moist banks, shores, and roadsides
notes: Hybrids with *Equisetum variegatum*, called *E. ×mackaii* (Newman) Brichan, occur in Maine. This hybrid resembles small forms of *E. hyemale*. It differs in its 7–16 persistent leaves and stems 20.0–86.0 cm tall *vs. E. hyemale* with 14–50 deciduous leaves and stems up to 220.0 cm tall.

Equisetum palustre L.

synonym(s): —
common name(s): marsh horsetail
habit: perennial herb
range: Labrador, s. to PA, w. to WA, n. to AK; also Eurasia
state frequency: uncommon
habitat: marshes, swamps, stream banks
notes: Known to hybridize with *Equisetum arvense* and *E. fluviatile.*

Equisetum pratense Ehrh.

synonym(s): —
common name(s): meadow horsetail
habit: perennial herb
range: Newfoundland, s. to NY, w. to MN and British Columbia, n. to AK; also Eurasia
state frequency: uncommon
habitat: meadows, moist woods
notes: Known to hybridize with *Equisetum arvense.*

Equisetum scirpoides Michx.

synonym(s): —
common name(s): dwarf scouring-rush
habit: perennial herb
range: Greenland, s. to NY, w. to MN and British Columbia, n. to AK
state frequency: uncommon
habitat: moist and/or mossy, often coniferous, woods
notes: Known to hybridize with *Equisetum variegatum.*

Equisetum sylvaticum L.

synonym(s): *Equisetum sylvaticum* L. var. *pauciramosum* Milde
common name(s): woodland horsetail
habit: perennial herb
range: Greenland, s. to VA, w. to MN and OR, n. to AK
state frequency: occasional
habitat: moist woods
notes: The reproductive stems turn green and become branched, unlike *Equisetum arvense*, whose
    reproductive stems senesce soon after spore dispersal.

Equisetum variegatum Schleich. *ex* F. Weber & D. M. H. Mohr

synonym(s): —
common name(s): variegated scouring-rush
habit: evergreen, perennial herb
range: circumboreal; s. to NY
state frequency: rare
habitat: shores, banks, ditches, moist woods
notes: This species is ranked S3 by the Maine Natural Areas Program. Hybrids with *Equisetum
    hyemale (q.v.)* occur in Maine. Also known to hybridize with *E. scirpoides.*

OSMUNDACEAE
(1 genus, 3 species)

Osmunda
(3 species)

Hybridization within this genus is extremely rare.

KEY TO THE SPECIES

1a. Fertile leaflets borne on a separate blade; axil of leaflets bearing a tuft of white, turning light brown with age, hairs on the abaxial side .... *O. cinnamomea*
1b. Fertile leaflets borne on the same blade as the sterile leaflets; axil of leaflets without a tuft of hairs
    2a. Fertile leaflets produced at the apex of the blade; leaflets divided .... *O. regalis*
    2b. Fertile leaflets produced in the middle of the blade with sterile leaflets basal and apical; leaflets lobed, but not divided to the costae .... *O. claytoniana*

Osmunda cinnamomea L.
synonym(s): —
common name(s): cinnamon fern
habit: tall, perennial herb
range: Labrador, s. to FL, w. to TX, n. to Ontario
state frequency: common
habitat: moist areas, often acidic
notes: This species is vegetatively similar to *Osmunda claytoniana*. In addition to the hair tufts in the axils of the leaflets, *O. cinnamomea* has leafules each with an obtuse to acute apex, whereas *O. claytoniana* lacks the hair tufts and has leafules each with a rounder or blunt apex.

Osmunda claytoniana L.
synonym(s): —
common name(s): interrupted fern
habit: tall, perennial herb
range: Labrador, s. to NC, w. to MO, n. to Ontario
state frequency: common
habitat: moist woods and margins
notes: Known to produce a sterile hybrid with *Osmunda regalis*. This species is vegetatively similar to *O. cinnamomea* (*q.v.*).

Osmunda regalis L. var. *spectabilis* (Willd.) Gray
synonym(s): *Osmunda spectabilis* Willd.
common name(s): royal fern
habit: tall, perennial herb
range: Newfoundland, s. to FL, w. to TX, n. to Ontario
state frequency: occasional
habitat: wet places, often acidic
notes: Known to hybridize with *Osmunda claytoniana* (*q.v.*).

MARSILEACEAE
(1 genus, 1 species)

Marsilea
(1 species)

Marsilea quadrifolia L.
synonym(s): —
common name(s): water-clover
habit: floating to emergent, aquatic herb
range: Eurasia; introduced to ME, s. to MD, w. to MO, n. to IA and MI
state frequency: very rare
habitat: shallow water
notes: Thought to have been introduced to Maine.

POLYPODIACEAE
(18 genera, 41 species)

KEY TO THE GENERA

1a. Sori located at the leaf margin; indusium formed, at least in part, by a revolute leaf margin or portion of the margin; spores tetrahedral

    2a. Indument of stems and leaves composed of hairs only 1 cell wide; stems usually moderately thick and green

        3a. Leaves 2-times divided, lanceolate; base of the petiole with fewer than 10 vascular bundles; ultimate leaf segments toothed; sori spherical, discrete, each covered by a modified tooth of the leaf margin .... **Dennstaedtia**

        3b. Leaves 4-times divided, broadly deltate; base of the petiole with 10 or more vascular bundles; ultimate leaf segments entire; sori linear, confluent, covered by reflexed margin of leaf segment .... **Pteridium**

    2b. Indument of stems and leaves composed of scales 2 or more cells wide; stems usually thin and dark

        4a. Leaves 3.0–15.0 cm long, with 1 main rachis; veins of leaf segments obscure, pinnately branched, divergent; sori confluent as a marginal band, concealed by the reflexed margin of the leaf segment [called a false indusium] ....
.... **Cryptogramma**

        4b. Leaves 15.0–110.0 cm long, appearing to have 2 recurved-spreading rachises; veins of leaf segments prominent, dichotomously branched, essentially parallel; sori discrete, borne on the reflexed margin of the ultimate leaf segments ....
.... **Adiantum**

1b. Sori not located at the margin (except *Dryopteris marginalis*); indusium not formed by a revolute leaf margin (formed by a highly modified, involute leaf margin in *Matteuccia* and *Onoclea*); spores oblong, reniform, bilateral, or winged

    5a. Some of the sporangia modified to form sporangiasters; indusium absent; leaves evergreen, coriaceous, once-pinnate .... **Polypodium**

    5b. Sporangiasters absent; indusium present, sometimes inconspicuous (absent); leaves evergreen or deciduous, herbaceous to coriaceous, variously divided

6a. Sori in 2 rows, 1 row on each side of, and immediately adjacent to, the midvein, covered by an elongate, flap-like indusium that opens toward the midvein .... ***Woodwardia***

6b. Sori various, when elongate and flap-like, opening toward the margin

    7a. Scales of the stem clathrate-reticulate; indusia elongate and flap-like; petiole with 2 vascular bundles, these united into 1 X-shaped bundle distally .... ***Asplenium***

    7b. Scales of the stem not clathrate; indusia elongate and flap-like, circular, reniform, lacerate, or absent; petiole with 2–7 vascular bundles, when 2, united into 1 U-shaped bundle distally

        8a. Plants pubescent with needle-like, transparent hairs; grooves of the adaxial rachis surface not continuous with the grooves of the costae

            9a. Leaf blades triangular; some of the basal leaflets connected by wings along the rachis; costae not grooved adaxially; indusium absent .... ***Phegopteris***

            9b. Leaf blades narrowly lanceolate; none of the basal leaflets connected by wings; costae grooved adaxially; indusium present .... ***Thelypteris***

        8b. Plants usually not pubescent, but if so, the hairs usually pigmented, glandular, or flattened [the latter are actually microscales]; grooves of the adaxial rachis surface, if present, continuous with the grooves of the costae except in *Deparia*, *Gymnocarpium*, and *Matteuccia*

            10a. Leaves strongly dimorphic—fertile leaves bearing sporangia hidden by highly modified, indurate, revolute leaf segments and sterile leaves either deeply lobed and with reticulate venation or with leaflets reduced in size very gradually to the base and abruptly to the apex; spores green

                11. Sterile leaves deeply lobed, divided only in the basal portion if at all; blade deltate, widest at the base; venation reticulate; sori enclosed in spherical leafules .... ***Onoclea***

                11b. Sterile leaves once-divided; blade elliptic to narrowly oblanceolate, widest above the middle; venation forking and not rejoining; sori enclosed in linear leaflets .... ***Matteuccia***

            10b. Leaves monomorphic to weakly dimorphic; sterile leaves with simple or forking veins that do not rejoin, at least once-divided, and with basal leaflets that do not very gradually reduce in size; spores light brown to brown

                12a. Indusium inferior, composed of lacerate or filamentous segments; veins not reaching margin of leaf segments; petioles articulated near base (not articulated in *Woodsia obtusa*) .... ***Woodsia***

                12b. Indusium absent or present, sometimes inconspicuous, superior when present, attached centrally or laterally, not divided; veins reaching margin of leaf segments; petioles not articulated near base

                    13a. Indusium round to reniform; petiole base with 3–7 vascular bundles; teeth of leaf segments often bristle-tipped

14a. Indusium peltate [attached at the center by a stalk]; leaflets or leafules with a basal auricle .... *Polystichum*

14b. Indusium attached laterally at a distinct sinus; leaflets and leafules lacking a basal auricle .... *Dryopteris*

13b. Indusium ovate-linear, hooked, or horseshoe-shaped, or absent; petiole base with 2 vascular bundles; teeth of leaf segments not bristle-tipped

15a. Indusium absent; leaf blades deltate to pentagonal; rhizome long and slender; leaflets weakly articulated to rachis; basal leaflets noticeably longer than next-apical pair .... *Gymnocarpium*

15b. Indusium present (inconspicuous in *Cystopteris*); leaf blades lanceolate to delate; rhizome short; leaflets not articulated to rachis; basal leaflets not evidently longer than next-apical pair

16a. Indusium linear, hooked at one end, or horseshoe-shaped, elongate, evident; vascular bundles at the base of the petiole lunate in cross-section; petiole swollen and with 2 rows of marginal teeth at the very base

17a. Leaf blade once-divided; rachis with septate hairs, its grooves not continuous with those of costae; adaxial grooves of costae shallow; sori not crossing veinlet .... *Deparia*

17b. Leaf blade twice-divided; rachis without septate hairs, its grooves continuous with those of costae and costules; adaxial grooves of costae deep; sori often crossing veinlet .... *Athyrium*

16b. Indusium ovate to lanceolate, short, inconspicuous; vascular bundles at the base of the petiole round or oblong in cross-section; petioles slender, without basal teeth .... *Cystopteris*

Adiantum
(2 species)

KEY TO THE SPECIES

1a. Leaves lax and arching; leaf segment stalks 0.5–1.5(–1.7) mm long; ultimate leaf segments about 3.0 times as long as broad, the acroscopic margin with rounded lobes separated by shallow sinuses mostly 0.1–2.0 mm deep; leaflets larger; plants of non-serpentine habitats .... *A. pedatum*

1b. Leaves stiffly erect; leaf segment stalks 0.2–0.9(–1.3) mm long; ultimate leaf segments 2.5–4.0 times as long as wide, the acroscopic margin with angular lobes separated by deep sinuses mostly 0.6–4.0 mm deep; leaflets smaller; plants of serpentine habitats .... *A. aleuticum*

Adiantum aleuticum (Rupr.) Paris
synonym(s): *Adiantum pedatum* L. var. *aleuticum* Rupr., *Adiantum pedatum* L. ssp. *calderi* Cody
common name(s): western maidenhair, Aleutian maidenhair
habit: perennial herb
range: Newfoundland and several disjunct localities in the east; also AK, s. to CA
state frequency: very rare
habitat: serpentine boulder fields
notes: This species is ranked S1 by the Maine Natural Areas Program. *Adiantum aleuticum* varies subtly in many ways from *A. pedatum*.

Adiantum pedatum L.
synonym(s): —
common name(s): northern maidenhair
habit: perennial herb
range: Quebec, s. to GA, w. to LA and OK, n. to MN
state frequency: uncommon
habitat: rich, deciduous forests
notes: —

Asplenium
(4 species)

These ferns are identifiable by their small, simple to once-divided leaves. They hybridize to produce sterile hybrids of intermediate morphology.

KEY TO THE SPECIES

1a. Leaves simple, narrowly deltate to linear-lanceolate, rooting at the tip; venation reticulate; sori irregularly scattered over the blade .... *A. rhizophyllum*
1b. Leaves once-divided [pinnate], linear to narrowly oblanceolate, not rooting at the tip; venation free [not rejoining]; sori in paired rows on each leaflet
   2a. Leaves dimorphic—the fertile erect and tall, the sterile prostrate or arching and short; leaflets with a basal auricle .... *A. platyneuron*
   2b. Leaves monomorphic; leaflets without a basal auricle
      3a. Rachis dark brown to red-brown; leaflets oblong to oval, the middle ones 2.5–4.0 mm wide; petioles 0.15–0.25 times the length of the blade .... *A. trichomanes*
      3b. Rachis green; leaflets deltate to rhombic, the middle ones 4.0–5.0 mm wide; petioles 0.25–0.5(–1.0) times the length of the blade .... *A. trichomanes-ramosum*

Asplenium platyneuron (L.) B. S. P.
synonym(s): *Asplenium platyneuron* (L.) B. S. P. var. *incisum* (Howe *ex* Peck) B. L. Robins.
common name(s): ebony spleenwort
habit: rhizomatous, perennial herb
range: ME, s. to FL, w. to TX, n. to KS, IA, and WI; also CO, AZ, and NM
state frequency: rare; known from the southern portion of Maine
habitat: dry, circumneutral soil of woods, rocky slopes, and ledges
notes: This species is ranked S1 by the Maine Natural Areas Program. Known to hybridize with *Asplenium rhizophyllum* and *A. trichomanes*.

Asplenium rhizophyllum L.

synonym(s): *Camptosorus rhizophyllus* (L.) Link
common name(s): walking fern
habit: evergreen, perennial herb
range: ME, s. to GA, w. to OK, n. to MN
state frequency: very rare
habitat: calcareous ledges and boulders
notes: This species is ranked SX by the Maine Natural Areas Program. Known to hybridize with
   Asplenium platyneuron and *A. trichomanes*.

Asplenium trichomanes L.

synonym(s): —
common name(s): maidenhair spleenwort
habit: rhizomatous, perennial herb
range: New Brunswick and Nova Scotia, s. to GA, w. to OK, n. to MN and Ontario; also WY and SD,
   s. to NM and AZ; also British Columbia, s. to CA
state frequency: occasional
habitat: shaded, usually calcareous, rock
key to the subspecies:
   1a. Spores 0.027–0.032 mm in diameter; plants of acidic habitats .... *A. t.* ssp. *trichomanes*
   1b. Spores 0.037–0.043 mm in diameter; plants of calcareous habitats ....
   .... *A. t.* ssp. *quadrivalens* D. E. Mey.
notes: Known to hybridize with all other members of this genus that grow in Maine. Also, the
   subspecies of *Adiantum trichomanes* are known to hybridize with each other.

Asplenium trichomanes-ramosum L.

synonym(s): *Asplenium viride* Huds.
common name(s): green spleenwort
habit: rhizomatous, perennial herb
range: Greenland, s. to ME, w. to Ontario; also AK, sporadically south to UT and CO
state frequency: very rare
habitat: shaded, calcareous rock
notes: This species is ranked S1 by the Maine Natural Areas Program. Known to hybridize with
   Asplenium trichomanes.

Athyrium
(1 species)

Leaves often weakly dimorphic, the reproductive ones more erect and contracted.

Athyrium filix-femina (L.) Roth *ex* Mertens var. *angustum* (Willd.) Lawson

synonym(s): *Athyrium filix-femina* (L.) Roth *ex* Mertens var. *michauxii* (Spreng.) Farw.
common name(s): lady fern
habit: rhizomatous, perennial herb
range: Greenland, s. to NC, w. to NE, n. to Saskatchewan
state frequency: common
habitat: moist woods, stream banks
notes: The light brown to red petioles are sparsely beset with brown to brown-black scales. The leaves
   are variable in division, ranging from 1.5- to 2.5-times divided. Plants sometimes glandular.

Cryptogramma
(1 species)

Vegetative specimens superficially resemble some species of *Cystopteris*. They can be separated by examining the leafules and petiole cross-section—*Cryptogramma* has more confluent leafules and 2 vascular bundles near the base of the petiole; *Cystopteris* has relatively distinct leafules and 3 or more vascular bundles.

Cryptogramma stelleri (Gmel.) Prantl *ex* Engler
synonym(s): —
common name(s): slender cliff-brake
habit: perennial herb
range: Labrador, s. to NJ, w. to IA, n. to MN; also AK, s. to CO
state frequency: very rare
habitat: moist, shaded, calcareous cliffs and ledges
notes: This species is ranked S1 by the Maine Natural Areas Program.

Cystopteris
(3 species)

Plants show tremendous variation in leaf morphology, and often difficult to assign to species, especially when stunted forms are found in stressful habitats. Hybrids are known and can be detected by misshapen spores.

KEY TO THE SPECIES
1a. Plants often with bulblets; rachises, costae, and indusia with glandular hairs; blade ovate to broadly or narrowly deltate [widest near base] .... *C. bulbifera*
1b. Plants lacking bulblets; rachises, costae, and indusia without glandular hairs; blade lanceolate to elliptic [widest near middle]
    2a. Indusia up to 1.0 mm long; basal leafules sessile or nearly so, broadly tapering or rounded at the base, approximate, emerging nearly perpendicular to the costa [*i.e.*, spreading]; apical portion of blade with deltate to ovate leaflets; margins of leafules serrate; acroscopic margin of leaflet convex due to the fact that the costa of the leaflet is nearly straight .... *C. fragilis*
    2b. Indusia up to 0.5 mm long; basal leafules stalked, mostly cuneate at the base, remote, emerging at an acute angle to the costa [*i.e.*, ascending]; apical portion of blade with ovate to narrowly elliptic leaflets; margins of leafules crenate; acroscopic margin of leaflet concave due to the fact that the costa of the leaflet curves slightly upward ....
.... *C. tenuis*

Cystopteris bulbifera (L.) Bernh.
synonym(s): —
common name(s): bulblet fern
habit: glandular, rhizomatous, perennial herb
range: Newfoundland, s. to GA, w. to AZ, n. to Manitoba
state frequency: uncommon
habitat: rich, rocky slopes, forests, and cliffs
notes: Known to hybridize with *Cystopteris fragilis*.

Cystopteris fragilis (L.) Bernh.

synonym(s): —
common name(s): fragile fern
habit: rhizomatous, perennial herb
range: Greenland, s. to CT, w. to CA, n. to AK
state frequency: occasional
habitat: rocky forests, cliffs
notes: Known to hybridize with *Cystopteris bulbifera* and *C. tenuis*. Habitat is helpful for distinguishing *C. fragilis* and *C. tenuis*. The former is more likely to be found at higher elevations and on cliffs, whereas the latter is more often at lower elevations and growing in soil. Hybrids with *C. tenuis* are more or less intermediate (although often larger than both parents), but are usually confused with *C. fragilis*.

Cystopteris tenuis (Michx.) Desv.

synonym(s): *Cystopteris fragilis* (L.) Bernh. var. *mackayi* Lawson
common name(s): Mackay's brittle fern, slender fragile fern
habit: rhizomatous, perennial herb
range: Nova Scotia, s. to NC, w. to OK, n. to Ontario; also in southwestern United States
state frequency: occasional
habitat: rocky forests, shaded cliffs
notes: Difficult to distinguish from *Cystopteris fragilis* (*q.v.*) and known to hybridize with it.

Dennstaedtia
(1 species)

These ferns have septate hairs and glandular hairs, useful characters for separating vegetative or dwarf forms from other species.

Dennstaedtia punctilobula (Michx.) T. Moore

synonym(s): —
common name(s): hay-scented fern, boulder fern
habit: perennial herb
range: New Brunswick and Nova Scotia, s. to GA and TN, w. to MO, n. to Quebec
state frequency: occasional
habitat: clearings, open, rocky slopes, fields
notes: Aromatic leaves.

Deparia
(1 species)

The petiole and rachis are pubescent with septate hairs.

Deparia acrostichoides (Sw.) M. Kato

synonym(s): *Athyrium acrostichoides* (Sw.) Diels, *Athyrium thelypterioides* (Michx.) Desv., *Diplazium acrostichoides* (Sw.) Butters
common name(s): silvery spleenwort, silvery glade fern
habit: rhizomatous, perennial herb
range: Quebec, New Brunswick, and Nova Scotia, s. to GA, w. to AR, n. to Ontario
state frequency: uncommon
habitat: rich, often rocky, forests
notes: —

Dryopteris
(9 species)

Infrequently hybridizes but many parental combinations possible. Hybrids are sterile and can be identified by abortive spores and intermediate morphology.

KEY TO THE SPECIES
1a. Leaves 15.0–120.0 cm long, mostly more than 25.0 cm, lacking aromatic glands, with few or no abaxial scales; indusia separate
  2a. Leaves 2.5- to 3.5-times divided; leaf segments with bristle-tipped teeth
    3a. First basiscopic [lower] leafule of basal leaflet no longer than adjacent basiscopic leafule and not noticeably larger than first acroscopic [upper] leafule of basal leaflet; plant axes and indusia with glandular hairs .... *D. intermedia*
    3b. First basiscopic leafule of basal leaflet longer than adjacent basiscopic leafule and 2.0–5.0 times as long as first acroscopic leafule of basal leaflet; plants without glandular hairs
      4a. First pair of leafules of basal leaflet nearly opposite, not more than 4.0 mm apart, the basiscopic leafule 2.0(–3.0) times the length of the acroscopic one; blade ovate-lanceolate .... *D. carthusiana*
      4b. First pair of leafules of basal leaflet offset 4.0–15.0 mm, the basiscopic leafule (2.0–)3.0–5.0 times the length of the acroscopic one; blade ovate-deltate .... *D. campyloptera*
  2b. Leaves 1- to 2-times divided; leaf segments with or without bristle-tipped teeth
    5a. Sori borne at or near margins of leaf segments; plants blue-green; petioles with a dense tuft of white-brown scales at the base .... *D. marginalis*
    5b. Sori borne between margins and midrib of leaf segments; plants green; petioles with scattered light to dark brown scales
      6a. Plants with scales of 2 distinct types—one broad, one hair-like; scales abundant on petiole, rachis, and costae; leaves with 20–30 pairs of leaflets; petiole up to 0.25 times the length of the blade .... *D. filix-mas*
      6b. Plants with scales of various sizes, mostly broad, but not distinctly of 2 sizes; scales abundant on only the petiole; leaves with 10–25 pairs of leaflets; petiole 0.25–0.35 times the length of the blade
        7a. Scales of the petiole dark brown to black; sori borne close to midrib; basal leaflets ovate; teeth of leaf segments not bristle-tipped ....
.... *D. goldiana*
        7b. Scales of the petiole light brown; sori borne midway between margin and midrib; basal leaflets deltate to elongate-deltate; teeth of leaf segments bristle-tipped
          8a. Fertile leaves deciduous, their leaflets twisted, parallel to the horizon, noticeably larger and more erect than the persistent sterile leaves; basal leaflets deltate to broadly deltate .... *D. cristata*
          8b. Fertile leaves persistent, their leaflets not twisted, parallel to the leaf axis, only slightly different from the sterile leaves; basal leaflets narrowly deltate to oblong-deltate .... *D. clintoniana*
1b. Leaves 6.0–25.0 cm long, with aromatic glands, densely scaly abaxially; indusia often overlapping .... *D. fragrans*

Dryopteris campyloptera (Kunze) Clarkson

synonym(s): *Dryopteris spinulosa* (Swarz) Watt var. *americana* (Fisch. *ex* Kunze) Fern.
common name(s): mountain wood fern
habit: robust, deciduous, rhizomatous, perennial herb
range: Newfoundland, s. to NC and TN
state frequency: occasional
habitat: cool, moist forests, often of higher elevation than closely related species
notes: Known to hybridize with *Dryopteris intermedia* and *D. marginalis.*

Dryopteris carthusiana (Vill.) H. P. Fuchs

synonym(s): *Dryopteris spinulosa* (O. F. Muell.) Watt
common name(s): spinulose wood fern
habit: deciduous, rhizomatous, perennial herb
range: Newfoundland, s. to NC, w. to WA, n. to British Columbia; also Yukon and Northwest
    Territory
state frequency: uncommon
habitat: moist woods and slopes, forested wetlands, stream shores
notes: Hybrids with *Dryopteris cristata*, called *D.* ×*uliginosa* (A. Br.) Druce, and *D. intermedia* (*q.v.*)
    occur in Maine. Also known to hybridize with *Dryopteris clintoniana, D. goldiana*, and *D.*
    marginalis.

Dryopteris clintoniana (D. C. Eat.) Dowell

synonym(s): *Dryopteris cristata* (L.) Gray var. *clintoniana* (D. C. Eat.) Underwood
common name(s): Clinton's wood fern
habit: dimorphic-leaved, rhizomatous, perennial herb
range: New Brunswick, s. to PA, w. to IN, n. to MI
state frequency: rare; known only from southern Maine
habitat: forested wetlands
notes: Hybrids with *Dryopteris goldiana* and *D. intermedia* occur in Maine. Also known to hybridize
    with *D. carthusiana, D. cristata*, and *D. marginalis.* The fertile leaves of this species die back to
    the ground, while the sterile leaves remain green through the winter. *Dryopteris clintoniana* is an
    allohexaploid derivative of *D. cristata* and *D. goldiana.*

Dryopteris cristata (L.) Gray

synonym(s): —
common name(s): crested wood fern
habit: dimorphic-leaved, rhizomatous, perennial herb
range: Newfoundland, s. to NC, w. to IL, n. to British Columbia
state frequency: occasional
habitat: forested wetlands
notes: Hybrids with *Dryopteris carthusiana* (*q.v.*), *D. intermedia*, called *D.* ×*boottii* (Tuckerman)
    Underwood, and *D. marginalis*, called *D.* ×*slossoniae* Wherry *ex* Lellinger, occur in Maine. Also
    known to hybridize with *D. clintoniana.* The fertile leaves die back to the ground, while the sterile
    leaves form a rosette and remain green through the winter.

Dryopteris filix-mas (L.) Schott

synonym(s): —
common name(s): male fern
habit: rhizomatous, perennial herb
range: Greenland, s. to ME; also in the Great Lakes region and western United States
state frequency: very rare
habitat: rich, rocky forests
notes: This species is ranked S1 by the Maine Natural Areas Program. Known to hybridize with
    Dryopteris marginalis.

Dryopteris fragrans (L.) Schott

synonym(s): —
common name(s): fragrant wood fern
habit: aromatic, rhizomatous, perennial herb
range: Greenland, s. to ME, w. to WI, n. to AK
state frequency: rare
habitat: rich cliffs and talus, often shaded and moist
notes: This species is ranked S3 by the Maine Natural Areas Program. Known to hybridize with
     Dryopteris marginalis.

Dryopteris goldiana (Hook. *ex* Goldie) Gray

synonym(s): —
common name(s): Goldie's fern
habit: robust, rhizomatous, perennial herb
range: Quebec, s. to GA, w. to AL, n. to MN
state frequency: rare
habitat: rich, rocky, forests and slopes
notes: This species is ranked S2 by the Maine Natural Areas Program. Hybrids with *Dryopteris*
     clintoniana occur in Maine. Also known to hybridize with *D. carthusiana*, *D. intermedia*, and *D.*
     marginalis.

Dryopteris intermedia (Muhl. *ex* Willd.) Gray

synonym(s): *Dryopteris spinulosa* (O. F. Muell.) Watt var. *intermedia* (Muhl. *ex* Willd.) Underwood
common name(s): evergreen wood fern
habit: evergreen, rhizomatous, perennial herb
range: Newfoundland, s. to GA, w. to AR, n. to Ontario
state frequency: common
habitat: forests, forested wetlands
notes: Hybrids with *Dryopteris carthusiana*, called *D.* ×*triploidea* Wherry, *D. clintoniana*, and *D.*
     cristata (*q.v.*) occur in Maine. Also known to hybridize with *D. campyloptera*, *D. goldiana*, and
     D. marginalis.

Dryopteris marginalis (L.) Gray

synonym(s): —
common name(s): marginal wood fern
habit: coriaceous-leaved, evergreen, rhizomatous, perennial herb
range: Greenland, s. to GA, w. to OK, n. to Ontario
state frequency: occasional
habitat: rocky slopes and forests
notes: Hybrids with *Dryopteris cristata* (*q.v.*) occur in Maine. Also known to hybridize with all other
     species of *Dryopteris* found in Maine.

Gymnocarpium

(1 species)

Gymnocarpium jessoense (Koidzumi) Koidzumi ssp. *parvulum* Sarvela is reported to occur in Maine by
the Flora of North America (1993, vol. 2, pp. 261–262), but voucher specimens are unknown.

Gymnocarpium dryopteris (L.) Newman

synonym(s): —
common name(s): oak fern
habit: rhizomatous, perennial herb

range: Greenland, s. to WV, w. to WA, n. to AK; also several scattered locations in western United States
state frequency: occasional
habitat: forests, rocky slopes
notes: —

Matteuccia
(1 species)

Matteuccia struthiopteris (L.) Todaro var. ***pensylvanica*** (Willd.) Morton
synonym(s): —
common name(s): ostrich fern, fiddle-head fern
habit: dimorphic, rhizomatous, perennial herb
range: Newfoundland, s. to VA, w. to SD, n. to AK
state frequency: occasional
habitat: rich and/or alluvial forests
notes: The sporangia do not dehisce until the following spring before the new leaves are produced.

Onoclea
(1 species)

Onoclea sensibilis L.
synonym(s): —
common name(s): sensitive fern
habit: dimorphic, rhizomatous, perennial herb
range: Newfoundland, s. to FL, w. to TX, n. to Manitoba
state frequency: common
habitat: wetlands, shores, ditches
notes: The sporangia do not dehisce until the following spring before the new leaves are produced.

Phegopteris
(2 species)

KEY TO THE SPECIES
1a. Basal leaflets mostly 10.0–20.0 cm long, connected to the adjacent pair by a wing; scales on costae narrowly lanceolate, white to light brown, usually 3–5 cells wide at the base; larger leaflets with evidently lobed segments; middle leaflet veins forked or pinnate .... *P. hexagonoptera*
1b. Basal leaflets mostly 3.0–10.0 cm long, not connected to next pair; scales on costae ovate-lanceolate, brown, usually 6–12 cells wide at the base; larger leaflets with entire or crenate segments; middle leaflet veins often simple .... *P. connectilis*

Phegopteris connectilis (Michx.) Watt
synonym(s): *Dryopteris phegopteris* (L.) C. Christens., *Thelypteris phegopteris* (L.) Slosson
common name(s): long beech fern, northern beech fern
habit: rhizomatous, perennial herb
range: Greenland, s. to NC, w. to WA, n. to AK
state frequency: occasional
habitat: moist woods, often rocky
notes: Known to hybridize with *Phegopteris hexagonoptera.*

Phegopteris hexagonoptera (Michx.) Fée
synonym(s): *Dryopteris hexagonoptera* (Michx.) C. Christens., *Thelypteris hexagonoptera* (Michx.) Nieuwl.
common name(s): broad beech fern, southern beech fern
habit: rhizomatous, perennial herb
range: ME, s. to FL, w. to TX, n. to WI
state frequency: rare
habitat: rich, shady woods
notes: This species is ranked S2 by the Maine Natural Areas Program. Known to hybridize with *Phegopteris connectilis*.

Polypodium
(2 species)

The sporangiasters of the genus appear as small, stipitate-glandular, bulbous heads mixed together with the sporangia of a sorus. Hybridization is reported to be frequent in this genus. The morphologically intermediate triploid hybrids can be recognized, in part, by their misshapen, translucent spores *vs.* yellow and kidney-bean-shaped ones in diploid and tetraploid plants [*Polypodium appalachianum* and *P. virginianum*, respectively].

KEY TO THE SPECIES
1a. Sporangiasters more than 40 per sorus; scales of rhizome and petiole golden brown; spores averaging smaller than 0.052 mm in diameter; leaflet apex usually pointed; leaves tending to be more deltate, widest near the base .... *P. appalachianum*
1b. Sporangiasters fewer than 40 per sorus; scales of rhizome and petiole golden brown with a dark brown central strip; spores averaging larger than 0.052 mm in diameter; leaflet apex usually rounded; leaves tending to be more oblong, widest near the middle ....
.... *P. virginianum*

Polypodium appalachianum Haufler & Windham
synonym(s): —
common name(s): Appalachian polypody
habit: evergreen, rhizomatous, perennial herb
range: Newfoundland, s. to GA, w. to AL, n. to Quebec
state frequency: common
habitat: cliffs, talus slopes, rocky woods
notes: —

Polypodium virginianum L.
synonym(s): *Polypodium vulgare* L. var. *americanum* Hook.
common name(s): rock polypody, common polypody
habit: evergreen, rhizomatous, perennial herb
range: Newfoundland, s. to GA, w. to AR, n. to Northwest Territory
state frequency: common
habitat: cliffs, talus slopes, rocky woods
notes: —

Polystichum
(2 species)

KEY TO THE SPECIES

1a. Leaves once-divided, the fertile leaflets contracted; sori often confluent and nearly covering abaxial leaflet surface .... *P. acrostichoides*
1b. Leaves more or less twice-divided, the fertile leaflets like the sterile and not contracted; sori distinct, not covering the entire abaxial leafule surface .... *P. braunii*

Polystichum acrostichoides (Michx.) Schott
synonym(s): —
common name(s): Christmas fern
habit: evergreen, rhizomatous, perennial herb
range: Quebec and Nova Scotia, s. to FL, w. to TX, n. to MN
state frequency: occasional
habitat: shaded and/or rocky forests
notes: Hybrids with *Polystichum braunii*, called *P.* ×*potteri* Barrington, occur in Maine.

Polystichum braunii (Spenner) Fée
synonym(s): —
common name(s): Braun's holly fern
habit: scaly, evergreen, perennial herb
range: Newfoundland, s. to CT, w. to MN, n. to Ontario; also in northwestern North America
state frequency: rare
habitat: rich, rocky forests
notes: Known to hybridize with *Polystichum acrostichoides* (*q.v.*).

Pteridium
(1 species)

The large, ternately branched leaves makes this one of the more easily recognized ferns in Maine. It also possesses septate hairs as in *Dennstaedtia*.

Pteridium aquilinum (L.) Kuhn *ex* Decken var. *latiusculum* (Desv.) Underwood *ex* Heller
synonym(s): —
common name(s): bracken fern, hogbrake
habit: erect, rhizomatous, perennial herb
range: Newfoundland, s. to FL, w. to TX, n. to Manitoba
state frequency: common
habitat: fields, open woods, barrens
notes: —

Thelypteris
(3 species)

Thelypteris possess thin hairs and 2 vascular bundles in the petiole, characters useful for separating small, vegetative specimens from members of the Polypodiaceae.

KEY TO THE SPECIES

1a. Basal leaflets greatly reduced, much shorter than middle leaflets; leaves with 23–46 pairs of leaflets; leaf margin ciliate .... *T. noveboracensis*

1b. Basal leaflets only slightly reduced compared with middle leaflets; leaves with 8–31 pairs of leaflets; leaf margin lacking cilia
    2a. Veins of leaf segments mostly simple; costae lacking scales; blades with sessile, red to orange glands on the abaxial surface .... *T. simulata*
    2b. Veins of leaf segments mostly forked; costae with light brown scales; blades lacking glands .... *T. palustris*

Thelypteris noveboracensis (L.) Nieuwl.
synonym(s): *Dryopteris noveboracensis* (L.) Gray
common name(s): New York fern
habit: rhizomatous, perennial herb
range: Newfoundland, s. to GA, w. to LA, n. to Ontario; also AR and OK
state frequency: occasional
habitat: woods, margins of streams and swamps
notes: —

Thelypteris palustris Schott var. pubescens (Lawson) Fern.
synonym(s): *Dryopteris thelypteris* (L.) Gray, *Dryopteris thelypteris* (L.) Gray var. *pubescens* (Lawson) Weatherby
common name(s): marsh fern
habit: rhizomatous, perennial herb
range: Newfoundland, s. to FL, w. to TX, n. to Manitoba
state frequency: occasional
habitat: marshes, swamps, bogs, wet ditches
notes: —

Thelypteris simulata (Davenport) Nieuwl.
synonym(s): *Dryopteris simulata* Davenport
common name(s): Massachusetts fern
habit: rhizomatous, perennial herb
range: New Brunswick and Nova Scotia, s. to VA and WV; also WI
state frequency: uncommon; more common in southern Maine
habitat: acid soils of swamps and bogs
notes: —

Woodsia
(4 species)

KEY TO THE SPECIES
1a. Petioles articulated [therefore the persistent petiole bases mostly of equal length], the articulation point noticeable as a slightly swollen node; leafules entire to crenate; indusium composed of narrow filaments 1 cell wide; leaves 2.5–25.0 cm long
    2a. Plants glabrous, with sessile glands; basal leaflets flabellate; leaves 1.0–1.5 cm wide; petioles green to light brown .... *W. glabella*
    2b. Plants with hairs, scales, and stalked glands; basal leaflets narrowly ovate to deltate; leaves 1.0–3.0 cm wide; petioles brown
        3a. Leaflets glabrous or nearly so on the abaxial surface, the larger leaflets with 2 or 3 pairs of leafules; indusium composed of few filaments .... *W. alpina*
        3b. Leaflets with scales on the abaxial surface, the larger leaflets with 4–7 pairs of leafules; indusium composed of many filaments .... *W. ilvensis*

1b. Petioles not articulated [therefore the persistent petiole bases of various lengths]; leafules dentate with pointed teeth; indusium composed of broad filaments 2 or more cells wide; leaves 8.0–60.0 cm long .... ***W. obtusa***

Woodsia alpina (Bolton) S. F. Gray

synonym(s): —
common name(s): alpine cliff fern
habit: cespitose, rhizomatous, perennial herb
range: Greenland, s. to ME and NH, w. to MN, n. to AK
state frequency: very rare
habitat: calcareous ledges and cliffs
notes: This species is ranked S1 by the Maine Natural Areas Program. Known to hybridize with *Woodsia ilvensis*. *Woodsia alpina* is an allotetraploid derivative of *W. glabella* and *W. ilvensis*.

Woodsia glabella R. Br. *ex* Richards.

synonym(s): *Woodsia alpina* (Bolton) S. F. Gray var. *glabella* (R. Br. *ex* Richards.) D. C. Eat.
common name(s): smooth cliff fern
habit: cespitose, rhizomatous, perennial herb
range: Greenland, s. to ME and NH, w. to British Columbia, n. to AK
state frequency: very rare
habitat: calcareous ledges and cliffs
notes: This species is ranked S1 by the Maine Natural Areas Program.

Woodsia ilvensis (L.) R. Br.

synonym(s): —
common name(s): rusty cliff fern
habit: cespitose, rhizomatous, perennial herb
range: Greenland, s. to NC, w. to IL, n. to Northwest Territory and AK
state frequency: occasional
habitat: cliffs, talus slopes, rocky woods
notes: Known to hybridize with *Woodsia alpina*.

Woodsia obtusa (Spreng.) Torr.

synonym(s): —
common name(s): blunt-lobed cliff fern
habit: cespitose, rhizomatous, perennial herb
range: Quebec and ME, s. to FL, w. to TX, n. to MN
state frequency: very rare
habitat: cliffs, ledges, talus slopes
notes: This species is ranked S1 by the Maine Natural Areas Program.

Woodwardia

(2 species)

Large ferns of acidic wetlands.

KEY TO THE SPECIES

1a. Leaves monomorphic, with 15–20 pairs of distinct, lobed leaflets; veins forming 1 row of areoles adjacent to the costae or costules .... ***W. virginica***
1b. Leaves dimorphic, with 7–10 pairs of leaflets, the sterile leaflets more or less confluent and serrulate; veins forming 2 or more rows of areoles adjacent to the costae or costules .... ***W. areolata***

Woodwardia areolata (L.) T. Moore
synonym(s): —
common name(s): netted chain fern
habit: rhizomatous, perennial herb
range: ME, s. along coastal plain to FL, w. to TX
state frequency: very rare
habitat: bogs and acidic, wet woods
notes: This species is ranked SH by the Maine Natural Areas Program.

Woodwardia virginica (L.) Sm.
synonym(s): —
common name(s): Virginia chain fern
habit: rhizomatous, perennial herb
range: Nova Scotia, s. to FL, w. to TX; also PA, w. to IL
state frequency: uncommon
habitat: acid soils of marshes, swamps, bogs, and wet ditches
notes: —

PINACEAE
(5 genera, 13 species)

KEY TO THE GENERA

1a. Branches dimorphic—leaves borne in fascicles or spirally arranged tufts on short shoots, and singly and remote on long shoots
    2a. Leaves all deciduous, those of the short shoots in spirally arranged tufts of 10–60; each scale of a strobilus without an apophysis; bracts exserted in basal portion of strobilus .... ***Larix***
    2b. Leaves persisting more than 1 year, those of the short shoots in fascicles of 2–5; each scale of a strobilus with an apophysis; bracts hidden by scales .... ***Pinus***
1b. Branches monomorphic; leaves borne singly, alternate
    3a. Leaves attached to peg-like projections [sterigmata] from twig, the leaf bases narrowed to the attachment; buds not, or slightly, resinous; strobili pendant at maturity, with persistent scales
        4a. Leaves flat, petiolate, rounded at apex, dimorphic—longer, spreading ones and shorter, often appressed, ones; terminal shoot nodding .... ***Tsuga***
        4b. Leaves tetragonous, sessile, pointed at apex, monomorphic; terminal shoot erect
        .... ***Picea***
    3b. Leaves attached directly to twig, expanded to a circular base; buds very resinous; strobili erect, with deciduous scales .... ***Abies***

Abies
(1 species)

Abies is often identified by its somewhat 2-ranked leaves, although this character, which is dependent solely on the branches' exposure to sun, is unreliable.

Abies balsamea (L.) P. Mill.
synonym(s): *Abies balsamea* (L.) P. Mill. var. *phanerolepis* Fern.
common name(s): balsam fir
habit: evergreen tree to depressed shrub
range: Labrador, s. to NJ and mountains of VA and WV, w. to MN, n. to Alberta
state frequency: common
habitat: boreal forests, moist woods
notes: —

Larix
(1 species)

Larix laricina (Du Roi) K. Koch
synonym(s): —
common name(s): larch, hackmatack, tamarack
habit: deciduous tree
range: Labrador, s. to NJ, w. to IN, n. to British Columbia and AK
state frequency: occasional
habitat: bogs, swamps, low, wet woods
notes: —

Picea
(4 species)

KEY TO THE SPECIES FOR USE WITH VEGETATIVE MATERIAL
1a. Twigs pubescent
    2a. Pubescence of twigs glandular; leaves gray-green, dull, glaucous; plants of organic soils .... *P. mariana*
    2b. Pubescence of twigs not glandular; leaves yellow to dark green, lustrous, not glaucous; plants commonly of mineral soils
        3a. Branches drooping; apex of bud scales without awns .... *P. abies*
        3b. Branches spreading; apex of some bud scales with conspicuous awns ....
.... *P. rubens*
1b. Twigs glabrous
    4a. Twigs glaucous, light brown; branches only slightly drooping .... *P. glauca*
    4b. Twigs not glaucous, red-brown; branches drooping .... *P. abies*

KEY TO THE SPECIES FOR USE WITH REPRODUCTIVE MATERIAL
1a. Strobili 0.6–6.0 cm long; scales flabellate, widest at the apex
    2a. Strobili 3.0–6.0 cm long; scale apex entire .... *P. glauca*
    2b. Strobili 0.6–4.5 cm long; scale apex slightly to evidently irregularly toothed
        3a. Strobili mostly 1.5–2.5 cm long, persisting on plant for many years; scale apex evidently erose .... *P. mariana*
        3b. Strobili mostly 2.5–4.5 cm long, usually falling from plant by first winter; scale apex nearly entire to erose .... *P. rubens*
1b. Strobili 12.0–16.0 cm long; scales elliptic to rhombic, widest near the middle ....
.... *P. abies*

Picea abies (L.) Karst.
synonym(s): —
common name(s): Norway spruce
habit: evergreen tree
range: Europe; escaped from cultivation
state frequency: uncommon
habitat: woods
notes: —

Picea glauca (Moench) Voss
synonym(s): —
common name(s): white spruce, cat spruce
habit: evergreen tree
range: Labrador, s. to ME, w. to British Columbia, n. to AK
state frequency: occasional
habitat: upland forests, exposed headlands
notes: —

Picea mariana (P. Mill.) B. S. P.
synonym(s): —
common name(s): black spruce
habit: evergreen tree to depressed shrub
range: Labrador, s. to NJ, w. to MN and British Columbia, n. to AK
state frequency: uncommon
habitat: bogs, subalpine regions
notes: Hybrids with *Picea rubens* occur in Maine.

Picea rubens Sarg.
synonym(s): —
common name(s): red spruce
habit: evergreen tree
range: New Brunswick, s. to NJ and along mountains to NC and TN, w. to NY, n. to Quebec
state frequency: common
habitat: upland forests
notes: Hybrids with *Picea mariana* occur in Maine.

Pinus
(6 species)

KEY TO THE SPECIES FOR USE WITH VEGETATIVE MATERIAL
1a. Leaves in fascicles of 5, with 1 fibrovascular bundle in cross-section; bud scales with
    entire margins .... *P. strobus*
1b. Leaves in fascicles of 2 or 3, with 2 fibrovascular bundles in cross-section; bud scales
    fringed on margin
    2a. Leaves in fascicles of 3 .... *P. rigida*
    2b. Leaves in fascicles of 2
        3a. Leaves 9.0–16.0 cm long, straight or only slightly twisted
            4a. Leaves brittle, tending to break when bent; buds red-brown; smaller branches
                orange-brown .... *P. resinosa*
            4b. Leaves pliable, tending not to break when bent; buds gray; smaller branches
                brown to gray-brown .... *P. nigra*
        3b. Leaves 2.0–7.0 cm long, noticeably twisted

5a. Leaves 3.0–7.0 cm long; larger branches orange-brown .... *P. sylvestris*
5b. Leaves 2.0–3.5 cm long; larger branches gray-brown .... *P. banksiana*

KEY TO THE SPECIES FOR USE WITH REPRODUCTIVE MATERIAL

1a. Strobili 10.0–15.0 cm long; apophyses terminal .... *P. strobus*
1b. Strobili 3.0–9.0 cm long; apophyses dorsal
  2a. Strobili symmetric, with a straight axis and without unequal development of apophyses
    3a. Strobili, when falling, leaving some of the basal scales on the branch [hollow-based]; umbo unarmed .... *P. resinosa*
    3b. Strobili falling with all the scales intact; umbo armed with a slender prickle, the prickle often deciduous in *P. nigra*
      4a. Basal, abaxial portion of scale aging red-brown .... *P. rigida*
      4b. Basal, abaxial portion of scale aging black .... *P. nigra*
  2b. Strobili asymmetric, with a curved axis and/or unequal development of apophyses on the abaxial side
    5a. Strobili serotinous, persisting on plant for many years; apophyses all more or less of equal size .... *P. banksiana*
    5b. Strobili usually falling after second winter; apophyses elongating toward base of strobilus on abaxial side .... *P. sylvestris*

Pinus banksiana Lamb.
synonym(s): —
common name(s): jack pine
habit: evergreen tree
range: Quebec, s. to ME, w. to WI and MN, n. to Alberta and Northwest Territories; also PA
state frequency: uncommon
habitat: sandy, acid soils
notes: —

Pinus nigra Arnold
synonym(s): —
common name(s): black pine, Austrian pine
habit: evergreen tree
range: Europe; introduced to our area
state frequency: uncommon
habitat: roadsides, edges, near establishments
notes: —

Pinus resinosa Ait.
synonym(s): —
common name(s): red pine, Norway pine
habit: evergreen tree
range: Newfoundland, s. to PA, w. to SD, n. to Ontario
state frequency: occasional
habitat: dry, sandy soil
notes: —

Pinus rigida P. Mill.
synonym(s): —
common name(s): pitch pine

habit: evergreen tree
range: ME, s. to GA
state frequency: uncommon
habitat: dry, sandy soil and ledges
notes: —

Pinus strobus L.

synonym(s): —
common name(s): eastern white pine
habit: evergreen tree
range: Newfoundland, s. to GA, w. to IA, n. to MN and Manitoba
state frequency: common
habitat: sandy, well drained soils
notes: —

Pinus sylvestris L.

synonym(s): —
common name(s): Scotch pine
habit: evergreen tree
range: Eurasia; introduced to southern Canada and northern United States
state frequency: rare
habitat: abandoned fields, agricultural regions, roadsides
notes: —

Tsuga
(1 species)

Tsuga canadensis (L.) Carr.

synonym(s): —
common name(s): eastern hemlock
habit: evergreen tree
range: Quebec, s. to GA, w. to AL, n. to MN
state frequency: common
habitat: moist ravines, hillsides
notes: —

TAXACEAE
(1 genus, 1 species)

Taxus
(1 species)

Taxus canadensis Marsh.

synonym(s): —
common name(s): American yew
habit: low, straggling, evergreen shrub
range: Newfoundland, s. to VA, w. to IA, n. to Ontario
state frequency: occasional
habitat: forests
notes: —

CUPRESSACEAE
(3 genera, 5 species)

KEY TO THE GENERA

1a. Strobili woody, the scales opening and exposing the axis; seeds winged; leaves of the lateral rank evidently folded

2a. Branchlets terete or tetragonous; strobilus scales peltate; strobili nearly spherical .... *Chamaecyparis*

2b. Branchlets flat; strobilus scales basifixed; strobili ellipsoid .... *Thuja*

1b. Strobili fleshy and resembling a berry, the scales not opening; seeds not winged; leaves flat or keeled, but not folded .... *Juniperus*

Chamaecyparis
(1 species)

Chamaecyparis thyoides (L.) B. S. P.
synonym(s): —
common name(s): Atlantic white-cedar
habit: evergreen tree
range: ME, s. to FL and LA
state frequency: rare
habitat: forested wetlands along the coastal plain
notes: This species is ranked S2 by the Maine Natural Areas Program.

Juniperus
(3 species)

Habit of the plant is vital for assessing species identification in this group, especially when distinguishing between *Juniperus horizontalis* and *J. virginiana*. Include this information in field notes and on herbarium labels. *Juniperus communis* L. var. *montana* Ait. is reported to grow in Maine by Gleason and Cronquist (1991), but valid voucher specimens are unknown.

KEY TO THE SPECIES

1a. Leaves whorled, not concealing the branchlet, monomorphic; strobili axillary .... *J. communis*

1b. Leaves opposite, closely imbricate and concealing the branchlet, dimorphic—longer, subulate ones and shorter, scale-like ones; strobili borne terminally

2a. Plants prostrate shrubs with erect branches, up to 0.3 m high; strobilus with a curved stalk, containing 3–5 seeds; apex of scale-like leaves apiculate .... *J. horizontalis*

2b. Plants upright shrubs or trees, to 20.0 m tall; strobilus with a straight stalk, containing 1 or 2 seeds; apex of scale-like leaves merely acute .... *J. virginiana*

Juniperus communis L. var. **depressa** Pursh
synonym(s): —
common name(s): common juniper
habit: low, evergreen shrub
range: Newfoundland, s. to GA, irregularly w. to CA, n. to AK
state frequency: occasional
habitat: rocky slopes, open woodlands
notes: —

Juniperus horizontalis Moench
synonym(s): —
common name(s): creeping Juniper
habit: low, trailing, evergreen shrub
range: Newfoundland, s. to MA, irregularly w. to MT, n. to AK
state frequency: uncommon
habitat: cliffs, rock outcrops, open, dry fields
notes: Known to hybridize with *Juniperus virginiana*.

Juniperus virginiana L.
synonym(s): —
common name(s): eastern redcedar
habit: erect, evergreen shrub to small tree
range: ME, s. to FL, w. to TX, n. to MN
state frequency: uncommon; known from the southern portion of the state
habitat: woods, abandoned fields
notes: Known to hybridize with *Juniperus horizontalis*.

Thuja
(1 species)

Thuja occidentalis L.
synonym(s): —
common name(s): northern white-cedar, arbor vitae
habit: small to medium-sized, evergreen tree
range: Quebec, s. to NC, w. to IL, n. to Manitoba
state frequency: common
habitat: moist or wet soil, swamps, uplands, cliffs
notes: —

CERATOPHYLLACEAE
(1 genus, 2 species)

Ceratophyllum
(2 species)

Submersed, commonly free-floating, aquatic plants with whorled, dichotomously forked branches.

KEY TO THE SPECIES
1a. Leaves mostly once- or twice-forked, serrate, ultimate segments flat; styles 4.5–6.0 mm long; achenes with 2 basal spines .... *C. demersum*
1b. Leaves mostly 3- or 4-times forked, mostly entire, ultimate segments capillary; styles 5.0–10.0 mm long; achenes with 4–15 spines [2 basal and 2–13 lateral] ... *C. echinatum*

Ceratophyllum demersum L.
synonym(s): —
common name(s): hornwort
habit: submerged, aquatic herb
range: nearly cosmopolitan; throughout most of North America

state frequency: uncommon
habitat: quiet water
notes: —

Ceratophyllum echinatum Gray
synonym(s): —
common name(s): prickly hornwort
habit: submerged, aquatic herb
range: New Brunswick, s. to FL, w. to Mexico, n. to British Columbia
state frequency: rare
habitat: quiet water
notes: This species is ranked SH by the Maine Natural Areas Program.

NYMPHAEACEAE
(3 genera, 6 species)

KEY TO THE GENERA

1a. Leaves without a sinus; perianth composed of 3 sepals and 3 petals; carpels 4–8, not submersed in the receptacle .... *Brasenia*

1b. Leaves with a pronounced, basal sinus; perianth composed of numerous tepals that progressively become petaloid in centripetal fashion; carpels numerous, sunken into a large, spongy receptacle
   2a. Leaves with angular basal lobes; petioles with 4 large air cavities; flowers white (pink); sepals sepaloid .... *Nymphaea*
   2b. Leaves with rounded basal lobes; petioles with many small air cavities; flowers yellow; sepals petaloid .... *Nuphar*

Brasenia
(1 species)

Submerged portions of *Brasenia* are covered with a mucilaginous jelly.

Brasenia schreberi J. F. Gmel.
synonym(s): —
common name(s): water-shield
habit: aquatic herb
range: Prince Edward Island and Nova Scotia, s. to FL, w. to TX, n. to British Columbia
state frequency: uncommon
habitat: ponds, slow streams, quiet water
notes: —

Nuphar
(3 species)

KEY TO THE SPECIES

1a. Leaf blades 3.5–10.0 cm long, with a sinus greater than 0.4 times the length of the blade; petioles very slender; sepals 1.0–2.0 cm long, usually 5 per flower; stigmatic disk red, 3.0–6.0 mm in diameter, with 6–10 lines; fruit *ca.* 15.0 cm tall, the petals and stamens deciduous from it .... *N. microphylla*

1b. Leaf blades 7.0–40.0 cm long, with a sinus less than 0.4 times the length of the blade; petioles thicker; sepals 2.5–5.0 cm long, 6 per flower; stigmatic disk green to yellow (red), 5.0–15.0 mm in diameter, with 8–25 lines; fruit 2.0–4.0 cm tall, the petals and stamens persistent at its base

    2a. Petioles strongly flattened on the adaxial surface, often wing-margined; leaves with a narrow, closed, basal sinus, floating; sepals red at the base of the abaxial surface; fruit often purple .... *N. variegata*

    2b. Petioles terete to oval in cross-section, not wing-margined; leaves with a broad, open, basal sinus, often emergent; sepals green (red) at the base of the abaxial surface; fruit green .... *N. advena*

Nuphar advena (Ait.) Ait. f.

synonym(s): *Nuphar lutea* (L.) Sm. ssp. *advena* (Ait.) Kartesz & Gandhi, *Nuphar lutea* (L.) Sm. ssp. *macrophylla* (Small) E. O. Beal, *Nymphaea advena* Soland., *Nymphaea macrophylla* Small
common name(s): cow-lily, large yellow pond-lily
habit: erect, often emergent, rhizomatous, aquatic, perennial herb
range: ME, s. to FL, w. to TX, n. to WI
state frequency: rare; locally common in Merrymeeting Bay
habitat: shallow, still or slow-moving waters, fresh tidal shores
notes: This species is ranked S2 by the Maine Natural Areas Program.

Nuphar microphylla (Pers.) Fern.

synonym(s): *Nymphaea microphylla* Pers.
common name(s): small yellow pond-lily
habit: rhizomatous, floating-leaved, aquatic, perennial herb
range: Quebec, s. to NJ, w. and n. to MN and Manitoba
state frequency: uncommon
habitat: still or slow-moving, shallow water
notes: Hybridizes with *Nuphar variegata* (*q.v.*).

Nuphar variegata Dur.

synonym(s): *Nuphar lutea* (L.) Sm. ssp. *variegata* (Dur.) E. O. Beal
common name(s): yellow water-lily
habit: rhizomatous, floating-leaved, aquatic, perennial herb
range: Newfoundland, s. to MD, w. to ID, n. to Yukon
state frequency: occasional; abundant on some lakes and rivers
habitat: still or slow-moving, shallow water
notes: Hybrids with *Nuphar microphylla*, called *N.* ×*rubrodisca* (Morong) Greene, occur in Maine and can be identified by leaf blades 5.0–25.0 cm long, anthers shorter than the filaments, red stigmatic disk 5.0–10.0 mm in diameter, and 5 or 6 sepals 2.0–2.5 cm long.

Nymphaea
(2 species)

KEY TO THE SPECIES

1a. Flowers 4.0–8.0 cm wide; stamens 20–40; stigmatic disk with 6–9 lines; leaf blades elliptic, 4.0–9.0 cm wide, with a wide sinus, mottled when young .... *N. leibergii*

1b. Flowers 7.0–20.0 cm wide; stamens 40–100; stigmatic disk with 10–25 lines; leaf blades orbicular, 10.0–30.0 cm wide, with a narrow sinus, not mottled .... *N. odorata*

Nymphaea leibergii Morong
synonym(s): *Nymphaea tetragona* Georgi ssp. *leibergii* (Morong) Porsild
common name(s): pygmy water-lily
habit: rhizomatous, aquatic, perennial herb
range: Quebec, s. to ME, w. to WA, n. to Manitoba
state frequency: very rare
habitat: cold ponds, lakes, and swamps
notes: This species is ranked S1 by the Maine Natural Areas Program. *Nymphaea leibergii* has been
reported to hybridize with *N. odorata* ssp. *odorata*. The hybrid is very rare, known only from
northern Maine and Saskatchewan.

Nymphaea odorata Ait.
synonym(s): <*N. o.* ssp. *odorata*> = *Castalia odorata* (Ait.) Wood; <*N. o.* ssp. *tuberosa*> = *Castalia
tuberosa* (Paine) Greene
common name(s): <*N. o.* ssp. *odorata*> = pond-lily, fragrant water-lily; <*N. o.* ssp. *tuberosa*> =
tuberous white water-lily
habit: rhizomatous, aquatic, perennial herb
range: Newfoundland, s. to FL, w. to TX, n. to Manitoba
state frequency: occasional; abundant on some lakes and rivers
habitat: ponds, quiet waters
key to the subspecies:
    1a. Petals acute to narrow-rounded at the apex; abaxial surface of the sepals and leaves green or,
more commonly, purple; inner filaments narrower than the anthers; seeds 1.5–2.5 mm long;
branches of the rhizome not constricted at the base; petioles faintly, if at all, striped; flowers
strongly fragrant .... *N. o.* ssp. *odorata*
    1b. Petals broad-rounded at the apex; abaxial surface of the sepals and leaves green; all the
filaments broader than the anthers; seeds 2.8–4.5 mm long; branches of the rhizome
constricted at the base, breaking into tuber-like segments; petioles striped with brown-purple;
flowers weakly or not at all fragrant .... *N. o.* ssp. *tuberosa* (Paine) Weisema & Hellquist
notes: *Nymphaea odorata* ssp. *tuberosa* is a relatively new addition to the flora. It is still uncertain
whether it is native or was introduced. As with other northern populations, Maine plants may be
difficult to assign to subspecies due to blurred distinctions among subspecies. *Nymphaea odorata*
ssp. *odorata* is reported to hybridize with *N. leibergii* (*q.v.*).

ARISTOLOCHIACEAE
(2 genera, 2 species)

KEY TO THE GENERA
1a. Plants herbaceous, without aerial stems; calyx actinomorphic, the basal, tubular portion
straight; flowers with 12 stamens; leaves mostly 8.0–14.0 cm wide .... ***Asarum***
1b. Plants woody, with twining, aerial stems; calyx zygomorphic, the basal, tubular portion
strongly bent; flowers with 6 stamens; leaves mostly 10.0–35.0 cm wide .... ***Aristolochia***

Aristolochia
(1 species)

Aristolochia macrophylla Lam.
synonym(s): *Aristolochia durior* Hill
common name(s): Dutchman's pipe
habit: deciduous, woody, twining liana

range: Ontario, s. to GA; escaped from cultivation to New England and NY
state frequency: very rare
habitat: forests, rocky slopes
notes: —

Asarum
(1 species)

Asarum canadense L.
synonym(s): —
common name(s): wild ginger
habit: rhizomatous herb
range: New Brunswick, s. to NC, w. to LA, n. to MN and Ontario
state frequency: rare
habitat: rich woods, calcareous ledges
notes: This species is ranked S1/S2 by the Maine Natural Areas Program.

LAURACEAE
(2 genera, 2 species)

KEY TO THE GENERA

1a. Leaves often lobed, palmately veined; inflorescence produced at the tips of the previous year's branchlets, peduncled; tepals accrescent; drupe blue .... *Sassafras*
1b. Leaves simple, pinnately veined; inflorescence produced in the axils of the previous year's branchlets, sessile; tepals deciduous; drupe red .... *Lindera*

Lindera
(1 species)

Lindera benzoin (L.) Blume
synonym(s): *Benzoin aestivale* (L.) Nees
common name(s): spicebush
habit: medium-sized shrub
range: ME, s. to FL, w. to TX, n. to Ontario
state frequency: rare
habitat: damp, rich woods and brook margins
notes: This species is ranked S3 by the Maine Natural Areas Program. Plants with aromatic leaves.

Sassafras
(1 species)

Sassafras albidum (Nutt.) Nees
synonym(s): *Sassafras albidum* (Nutt.) Nees var. *molle* (Raf.) Fern.
common name(s): sassafras
habit: small tree
range: ME, s. to FL, w. to TX, n. to MI
state frequency: rare
habitat: woods, thickets, fields, roadsides
notes: This species is ranked S2 by the Maine Natural Areas Program. Plants with aromatic leaves.

ACORACEAE
(1 genus, 1 species)

Acorus
(1 species)

Acorus americanus (Raf.) Raf.
synonym(s): *Acorus calamus* L.
common name(s): sweet flag, flagroot
habit: emergent, rhizomatous, perennial herb
range: New Brunswick, s. to FL, w. to TX and CO, n. to OR, MT, and British Columbia
state frequency: uncommon
habitat: swamps, shallow water
notes: —

NARTHECIACEAE
(1 genus, 1 species)

Aletris
(1 species)

Aletris farinosa L.
synonym(s): —
common name(s): unicorn root
habit: erect, rhizomatous, perennial herb
range: ME, s. to FL, w. to TX, n. to MN
state frequency: very rare
habitat: open woods, sandy, gravelly, or peaty soil
notes: This species is ranked SX by the Maine Natural Areas Program.

TOFIELDIACEAE
(1 genus, 1 species)

Triantha
(1 species)

Triantha glutinosa (Michx.) Baker
synonym(s): *Tofieldia glutinosa* (Michx.) Pers.
common name(s): sticky false asphodel
habit: erect, simple, perennial herb
range: Newfoundland, s. to mountains of GA, w. to CA, n. to AK
state frequency: rare
habitat: fens, circumneutral ledges and shores
notes: —

SCHEUCHZERIACEAE
(1 genus, 1 species)

Scheuchzeria
(1 species)

Scheuchzeria palustris L. spp. **americana** (Fern.) Hultén
synonym(s): —
common name(s): podgrass
habit: rhizomatous, sometimes aquatic, perennial herb
range: circumboreal; s. to NJ, w. to CA
state frequency: uncommon
habitat: organic soil wetlands
notes: —

NAJADACEAE
(1 genus, 3 species)

Najas
(3 species)

KEY TO THE SPECIES
1a. Each margin of the leaf blade with 13–17 minute, unicellular spinules; base of leaves
    abruptly expanded into conspicuous auricles; staminate flowers 1.0–1.5 mm long ....
    .... *N. gracillima*
1b. Each margin of the leaf blade with 20–100 minute, unicellular spinules; base of leaves
    somewhat expanded into sloping shoulders; staminate flowers 2.5–3.2 mm long
    2a. Seeds pitted, 1.2–2.5 mm long, fusiform, widest near the middle; anthers 4-locular;
        leaves 0.2–2.0 mm wide .... *N. guadalupensis*
    2b. Seeds smooth, (1.2–)2.5–3.7 mm long, more or less obovate, widest above the
        middle; anthers 1-locular; leaves 0.2–0.6 mm wide .... *N. flexilis*

Najas flexilis (Willd.) Rostk. & Schmidt
synonym(s): —
common name(s): northern water nymph
habit: monoecious, branched, aquatic, annual herb
range: Newfoundland, s. to VA, w. to CA, n. to British Columbia
state frequency: occasional
habitat: fresh to brackish, shallow water
notes: —

Najas gracillima (A. Braun *ex* Engelm.) Magnus
synonym(s): —
common name(s): thread-like naiad, slender water nymph
habit: monoecious, slender, sparsely branched, aquatic, annual herb
range: Nova Scotia, s. to AL, w. to MO, n. to MN
state frequency: rare
habitat: ponds, shores
notes: This species is ranked S3 by the Maine Natural Areas Program.

Najas guadalupensis (Spreng.) Magnus
synonym(s): —
common name(s): southern naiad
habit: monoecious, branched, aquatic, annual herb
range: ME, s. to PA, w. to IA, n. to Manitoba; also ID and OR
state frequency: very rare
habitat: shallow water
notes: This species is ranked SH by the Maine Natural Areas Program.

ALISMATACEAE
(2 genera, 7 species)

KEY TO THE GENERA

1a. Androecium composed of 6 stamens; gynoecium with a single whorl of carpels on a
small, flat receptacle; all the flowers bisexual .... ***Alisma***
1b. Androecium composed of 7–40 stamens; gynoecium with several whorls of carpels on a
large receptacle, forming a subglobose cluster; flowers all bisexual or, in some species,
all the flowers unisexual .... ***Sagittaria***

Alisma
(2 species)

KEY TO THE SPECIES

1a. Achenes 1.8–3.0 mm long; sepals with broad, conspicuous, scarious margins; petals
3.8–4.5 x 3.0–3.9 mm, noticeably longer than the sepals; styles 0.4–0.6 mm long ....
.... ***A. triviale***
1b. Achenes 1.5–2.2 mm long; sepals with narrow, less conspicuous, scarious margins;
petals 1.8–2.5 x 1.4–2.0 mm, scarcely, if at all, longer than the sepals; styles 0.2–0.4 mm
long .... ***A. subcordatum***

Alisma subcordatum Raf.
synonym(s): *Alisma parviflorum* Pursh, *Alisma plantago-aquatica* L. var. *parviflorum* (Pursh) Torr.
common name(s): southern water-plantain
habit: rhizomatous, aquatic herb
range: ME, s. to FL, w. to NE, n. to MN and Ontario
state frequency: occasional
habitat: shallow water, muddy shores
notes: —

Alisma triviale Pursh
synonym(s): *Alisma brevipes* Greene, *Alisma plantago-aquatica* L. var. *americanum* J. A. Schultes
common name(s): northern water-plantain
habit: rhizomatous, aquatic herb
range: Quebec and Nova Scotia, s. to MD, w. to NE, n. to British Columbia; also Mexico, NM, AZ,
and CA
state frequency: occasional
habitat: shallow water, muddy shores, streams
notes: —

Sagittaria
(5 species)

Leaf shape is extremely variable in this genus, with most species capable of producing leaves with or without basal lobes. The identification key gives the common expression of the leaves. Like *Eriocaulon*, members of this genus produce conspicuously septate roots. *Sagittaria*, however, usually have a flattened, apical portion of the leaf.

KEY TO THE SPECIES

1a. Sepals appressed in fruit; pedicels spongy, recurved in fruit; bisexual flowers with 9–12 stamens; plants of tidal shores .... *S. calycina*
1b. Sepals reflexed in fruit; pedicels not spongy, normally ascending in fruit; bisexual flowers with 12–40 stamens; plants commonly of non-tidal shores
   2a. Leaves usually with basal lobes; filaments smooth
      3a. Beak of the achene 0.1–0.5 mm long, erect; petals 7.0–10.0 mm long; achenes 1.8–2.6 mm long .... *S. cuneata*
      3b. Beak of the achene 0.6–1.8 mm long, horizontally spreading; petals 10.0–20.0 mm long; achenes 2.5–4.0 mm long .... *S. latifolia*
   2b. Leaves simple, without basal lobes; filaments rough, beset with minute scales
      4a. Carpellate flowers borne on pedicels 1.0–3.0 cm long; flowering stem straight, without a distinct bend; achenes [rarely produced] 1.5–2.0 mm long, with a short beak 0.1–0.3 mm long; leaves linear-lanceolate to broad-lanceolate .... *S. graminea*
      4b. Carpellate flowers sessile or on short pedicels to 0.5 cm long; flowering stem often with a conspicuous bend at the lowest whorl of flowers; achenes [usually produced] 2.5–4.0 mm long, with a beak 1.0–1.5 mm long; leaves lanceolate to ovate .... *S. rigida*

Sagittaria calycina Engelm. var. *spongiosa* Engelm.
synonym(s): *Lophotocarpus spongiosus* (Engelm.) J. G. Sm., *Sagittaria montevidensis* Cham. & Schlecht. ssp. *spongiosa* (Engelm.) Bogin, *Sagittaria spatulata* (J. G. Sm.) Buch.
common name(s): tidal arrowhead
habit: aquatic, perennial herb
range: New Brunswick, s. to VA, w. to Mexico, n. to SD and MN
state frequency: rare
habitat: brackish to nearly fresh tidal shores
notes: This species is ranked S3 by the Maine Natural Areas Program.

Sagittaria cuneata Sheldon
synonym(s): *Sagittaria arifolia* Nutt. *ex* J. G. Sm.
common name(s): northern arrowhead, wapato
habit: emergent to submerged, aquatic, tuberiferous, perennial herb
range: Quebec and Nova Scotia, s. to NY, w. to TX and CA, n. to British Columbia
state frequency: uncommon
habitat: shallow water, muddy shores
notes: This species commonly possesses relatively broad, cordate, floating leaves.

Sagittaria graminea Michx.
synonym(s): —
common name(s): grass-leaved sagittaria, grass-like arrowleaf
habit: submerged to emergent, short-rhizomatous, perennial herb

range: Labrador, s. to FL, w. to TX, n. to SD, MN, and Ontario
state frequency: occasional
habitat: shallow water, mud
notes: —

Sagittaria latifolia Willd.
synonym(s): *Sagittaria latifolia* Willd. var. *obtusa* (Engelm.) Wieg.
common name(s): common arrowhead, broad-leaved arrowleaf, wapato
habit: more or less erect, emergent, tuberiferous, perennial herb
range: coastal Quebec and New Brunswick, s. to SC and LA, w. to CA, n. to British Columbia
state frequency: occasional
habitat: swamps, ponds, streams
notes: —

Sagittaria rigida Pursh
synonym(s): —
common name(s): sessile-fruited arrowhead, swamp arrowhead
habit: erect to lax, emergent to submersed, perennial herb
range: ME, s. to VA, w. to MO, n. to MN
state frequency: very rare
habitat: shallow water, mud
notes: This species is ranked S1 by the Maine Natural Areas Program.

HYDROCHARITACEAE
(2 genera, 3 species)

KEY TO THE GENERA
1a. Leaves borne on a stem, at least the upper leaves whorled, mostly 0.6–1.7 cm long, without a prominent lacunar band; staminate flowers with 3–9 stamens, usually 9 .... *Elodea*
1b. Leaves all basal, 15.0–200.0 cm long, with a prominent lacunar band nearly the entire width of the leaf; staminate flowers with 2 stamens .... *Vallisneria*

Elodea
(2 species)

KEY TO THE SPECIES
1a. Leaves 1.0–5.0 mm wide, averaging 2.0 mm, firm, blunt at the apex, the upper densely crowded and overlapping; staminate flowers elevated on a long, slender, pedicel-like hypanthium 2.0–30.0 cm long, with a spathe 7.0–10.0 mm long, with sepals 3.0–4.0 mm long; carpellate flowers with sepals 2.0–3.5 mm long .... *E. canadensis*
1b. Leaves 0.3–1.75 mm wide, averaging 1.3 mm, weak, pointed at the apex, the upper not densely crowded; staminate flowers sessile, with a spathe 2.0–4.0 mm long, with sepals *ca.* 2.0 mm long; carpellate flowers with minute sepals up to 1.5 mm long .... *E. nuttallii*

Elodea canadensis Michx.
synonym(s): *Philotria canadensis* (Michx.) Britt.
common name(s): common waterweed, Canada waterweed
habit: submerged, aquatic, perennial herb
range: Quebec, s. to NC and AL, w. to CA, n. to Saskatchewan

state frequency: uncommon

habitat: quiet water, often basic

notes: The staminate flowers of this species reach the surface of the water by elongation of the pedicel-like hypanthium.

Elodea nuttallii (Planch.) St. John

synonym(s): *Philotria nuttallii* (Planch.) Rydb.

common name(s): Nuttall's waterweed, free-flowered waterweed

habit: aquatic, perennial herb

range: Quebec, s. to NC and MS, w. to NM, n. to ID

state frequency: uncommon

habitat: shallow, fresh or brackish, often acidic, water, tidal estuaries

notes: The staminate flowers of this species must break free and float to the surface due to the absence of an elongate hypanthium.

Vallisneria
(1 species)

Vallisneria americana Michx.

synonym(s): *Vallisneria spiralis auct. non* L.

common name(s): tapegrass, eelgrass

habit: submerged, aquatic, perennial herb

range: New Brunswick, s. to FL, w. to AZ, n. to ND; naturalized to WA

state frequency: uncommon

habitat: quiet water

notes: —

JUNCAGINACEAE
(1 genus, 3 species)

Triglochin
(3 species)

KEY TO THE SPECIES

1a. Leaves usually equaling the height of the flowering stem, arching outward from the sheath; spike 2.0–7.0 cm tall .... *T. gaspense*

1b. Leaves usually shorter than the flowering stem, erect; spike 6.0–45.0 cm tall

    2a. Flowers with 6 carpels and 6 stigmas; mature follicles 2.0–3.0 mm wide, *ca.* 2.0 times as long as wide, the cluster of 6 follicles on a slender, non-wing-angled axis .... *T. maritimum*

    2b. Flowers with 3 carpels and 3 stigmas; mature follicles 1.0 mm wide, *ca.* 5.0–7.0 times as long as wide, the cluster of 3 follicles on a triquetrous axis .... *T. palustre*

Triglochin gaspense Lieth & D. Löve

synonym(s): —

common name(s): Gaspe arrow-grass

habit: erect, rhizomatous, graminoid, perennial herb

range: Newfoundland, s. to Prince Edward Island, w. to ME, n. to Quebec

state frequency: very rare

habitat: salt marshes
notes: This species is ranked SH by the Maine Natural Areas Program.

Triglochin maritimum L.
synonym(s): —
common name(s): common arrow-grass
habit: erect, rhizomatous, graminoid, perennial herb
range: Labrador, s. to DE, w. to CA, n. to AK
state frequency: uncommon
habitat: brackish shores, sometimes inland wetlands
notes: —

Triglochin palustre L.
synonym(s): —
common name(s): slender arrow-grass, marsh arrow-grass
habit: slender, erect, rhizomatous, perennial herb
range: circumboreal; s. to RI, w. to NM and CA
state frequency: uncommon
habitat: coastal marshes
notes: —

ZANNICHELLIACEAE
(1 genus, 1 species)

Zannichellia
(1 species)

Zannichellia have hydrophilous flowers, the carpellate ones possessing and odd, peltate-funnelform stigma. The linear, 1-nerved leaves, occasionally vary in the number produced at a node [ranging 1–4, usually 2].

Zannichellia palustris L.
synonym(s): *Zannichellia palustris* L. var. *major* (Hartman) W. D. J. Koch
common name(s): horned-pondweed
habit: slender, branched, submerged, rhizomatous, aquatic, perennial herb
range: Newfoundland, s. to FL, w. to TX and Mexico, n. to AK
state frequency: rare
habitat: fresh to brackish water
notes: This species is ranked S2 by the Maine Natural Areas Program.

ZOSTERACEAE
(1 genus, 1 species)

Zostera
(1 species)

Marine herbs with elongate, linear, parallel-veined leaves.

Zostera marina L. var. ***stenophylla*** Aschers. & Graebn.
synonym(s): —
common name(s): eelgrass, grass-wrack
habit: monoecious, branched, submerged, rhizomatous, perennial herb
range: circumboreal; s. along the coast to FL
state frequency: occasional
habitat: shallow seawater
notes: —

POTAMOGETONACEAE
(3 genera, 25 species)

KEY TO THE GENERA
1a. Each flower with 0 tepals and 2 stamens; fruits long-stipitate; stipules completely
    connate to the leaves; plants of brackish water .... ***Ruppia***
1b. Each flower with 4 tepals and 5 stamens; fruits sessile; stipules free from the leaves or
    basally connate and with a free tip; plants mostly of fresh water
    2a. Leaves narrow-linear, 0.2–1.5 mm wide, all submersed, cross-septate throughout;
        stipules connate to the blade for a distance of 0.4–3.0 cm, commonly more than
        1.0 cm; spikes with well separated lower whorls of flowers, on thin, flexuous
        peduncles, submersed [*i.e.*, the flowers hydrophilous]; drupes not compressed, with
        an indistinctly coiled embryo .... ***Stuckenia***
    2b. Leaves filiform to orbicular, 0.1–75.0 mm wide, all submersed or some floating,
        cross-septate only in the lacunar bands (if present); stipules distinct from the blade or
        connate for a distance of less than 1.0 cm (except *Potamogeton robbinsii*); spikes
        usually with contiguous whorls of flowers, on thick, stiff peduncles, emersed [*i.e.*, the
        flowers anemophilous]; drupes compressed, with a distinctly coiled embryo (except
        P. robbinsii) .... ***Potamogeton***

Potamogeton
(22 species)

The floating leaves in *Potamogeton* can be identified on herbarium specimens by their position [species
with floating leaves produce them at the shoot apex], and are noticeably thicker, firmer, and less translucent
than the submersed leaves. Species of *Potamogeton* are known to hybridize, with hybridization events
more common in the wide-leaved species. Leaf measurements are taken from leaves of the primary shoot
and not from short, axillary shoots.

KEY TO THE SPECIES
1a. Submersed leaves less than 4.0 mm wide
    2a. Submersed leaves with the stipule basally fused to the leaf blade, only the tip of the
        stipule distinct from the blade
        3a. Apex of submersed leaves acute to long-tapering; fused portion of the stipule
            usually shorter than the distinct portion; drupes often with an entire or toothed
            lateral keel on each side of the dorsal keel .... ***P. bicupulatus***
        3b. Apex of submersed leaves obtuse; fused portion of the stipule usually longer than
            the distinct portion; drupes without a pair of lateral keels .... ***P. spirillus***

2b. Submersed leaves with stipules that are distinct from the blade for their entire length
    4a. Floating leaves [greatly expanded in width] produced and generally present on some of the plants in each colony
        5a. Floating leaves with blades 0.6–1.5 cm long, 5- to 9-veined; submersed leaves 0.2–1.0 mm wide, thin, transparent, lacking lacunar cells; spikes 3.0–8.0 mm long; drupes 1.5–2.5 mm long .... *P. vaseyi*
        5b. Floating leaves with blades 1.5–12.0 cm long, 7- to 37-veined; submersed leaves 0.24–10.0 mm wide, either with abundant lacunar cells or with thick and nearly opaque texture; spikes 10.0–50.0 mm long; drupes (2.0–)2.5–4.0 mm long
            6a. Submersed leaves 1.0–10.0 mm wide, thin, transparent, with prominent bands of lacunar cells 1.0–2.0 mm wide on each side of the midrib; drupes with dorsal and lateral keels .... *P. epihydrus*
            6b. Submersed leaves 0.25–2.0 mm wide, thick and nearly opaque, petiole-like, lacking lacunar cells; drupes with only a dorsal keel
                7a. Submersed leaves 0.8–2.0 mm wide; floating leaves with blades 5.0–10.0 x 2.5–6.0 cm, cordate to rounded at the base; petioles pale at the apex; stipules 4.0–10.0 mm long; drupes 3.5–5.0 mm long, deeply wrinkled on each side, with an inconspicuous dorsal keel .... *P. natans*
                7b. Submersed leaves 0.25–1.0 mm wide; floating leaves with blades 2.5–6.0 x 1.0–3.0 cm, cuneate to rounded at the base; petioles usually lacking a whitish band at the apex; stipules 2.5–4.0 mm long; drupes 2.0–3.5 mm long, not wrinkled on the sides, with a prominent dorsal keel .... *P. oakesianus*
    4b. Floating leaves not produced, all the leaves submerged
        8a. Rhizome elongate; peduncles (5.0–)10.0–25.0 cm long; leaves up to 0.5 mm wide, 1-veined .... *P. confervoides*
        8b. Rhizome scarcely developed (elongate in *P. foliosus*); peduncles 0.5–6.0 cm long; leaves 0.2–5.0 mm wide, 1- to 35-veined
            9a. Stems conspicuously compressed; leaves 15- to 25(35)-veined; spikes with 7–11 whorls of flowers; drupes 4.0–4.5 mm long .... *P. zosteriformis*
            9b. Stems terete or slightly compressed; leaves 1- to 13-veined; spikes with 1–5 whorls of flowers; drupes 1.5–3.5 mm long
                10a. Nodal glands absent; rhizome elongate; spikes 4.0–7.0 mm long, with 1 or 2 whorls of flowers; peduncles 0.5–1.5 cm long; drupes with a conspicuous dorsal keel .... *P. foliosus*
                10b. Nodal glands present; rhizome scarcely developed; spikes 2.0–18.0 mm long, with 1–5 whorls of flowers; peduncles 0.5–6.0 cm long; drupes without a dorsal keel (sometimes with a low keel in *P. obtusifolius*)
                    11a. Stipules white and [including those of the winter buds] coarsely fibrous
                        12a. Leaves rounded at the apex, sometimes with an apiculus, 1.5–3.0 mm wide, 5- to 7(9)-veined; winter buds with inner leaves oriented at a right angle to the outer leaves, which are not corrugated at the base .... *P. friesii*

12b. Leaves obtuse to attenuate at the apex, 0.5–2.0 mm wide, 3- to 5(7)-veined; winter buds with the inner and outer leaves oriented in the same plane, the outer leaves corrugated at the base .... *P. strictifolius*

11b. Stipules [including those of the winter buds] green, brown, or white, delicate, not fibrous

13a. Leaves 0.2–2.5 mm wide, acute to obtuse at the apex, green; drupe with a beak 0.1–0.6 mm long, the body 1.5–2.2 mm long .... *P. pusillus*

13b. Leaves 1.0–3.5 mm wide, obtuse to rounded at the apex, sometimes with an apiculus, commonly suffused with red; drupe with a beak 0.6–0.7 mm long, the body 2.5–3.5 mm long .... *P. obtusifolius*

1b. Submersed leaves 4.0 mm wide or wider (those of the axillary shoots may be narrower in some species)

14a. Stipules basally fused to the leaf, only the tip of the stipule distinct from the leaf; leaves with a pair of auricles at the base, mostly minutely serrulate, at least in the apical portion .... *P. robbinsii*

14b. Stipules of submersed leaves distinct from the leaf their entire length; leaves without auricles (though cordate-clasping in some species), entire

15a. Floating leaves not produced; leaves all submersed, sessile and conspicuously cordate-clasping at the base

16a. Leaf apex cucullate; rhizome spotted with red-brown; stipules persistent; drupes 4.0–5.7 mm long, with a prominent, dorsal keel .... *P. praelongus*

16b. Leaf apex flat; rhizome unspotted; stipules disintegrating and remaining as fibers or completely absent; drupes 1.6–4.2 mm long, with an inconspicuous keel or the keel absent

17a. Leaf blades ovate to suborbicular, 1.0–5.0(–8.0) cm long, delicately 7- to 15-veined; stipules disintegrating and absent on the lower part of the stem; drupes 1.6–3.0 mm long .... *P. perfoliatus*

17b. Leaf blades lance-ovate to narrow-lanceolate, (3.0–)5.0–12.0 cm long, coarsely 13- to 21-veined; stipules disintegrating into persistent fibers; drupes 2.2–4.2 mm long .... *P. richardsonii*

15b. Floating leaves produced (except in *P. zosteriformis*), though not always present; submersed leaves sessile or petiolate, not cordate-clasping

18a. Stems and petioles black-spotted; submersed leaves with undulate margins .... *P. pulcher*

18b. Stems and petioles unspotted; submersed leaves with straight margins

19a. Submersed leaves arched and conduplicate, 2.0–7.0 cm wide; achenes 4.0–5.5 mm long .... *P. amplifolius*

19b. Submersed leaves neither arched nor conduplicate, 0.1–3.0 cm wide; achenes 3.0–4.5 mm long

20a. Stems conspicuously compressed; floating leaves not produced; leaves 15- to 25(35)-veined .... *P. zosteriformis*

20b. Stems terete to somewhat compressed; floating leaves produced, though not always present; submersed leaves 3- to 15-veined

21a. Sheath-like stipules 4.0–10.0 cm long; submersed leaves with petioles 2.0–13.0 cm long .... *P. nodosus*
21b. Sheath-like stipules 1.0–3.0 cm long; submersed leaves sessile
  22a. Submersed leaves distichous, linear, with prominent lacunar bands 1.0–2.0 mm wide on each side of the midrib; each lacunar band composed of 9–18 rows of cells ..... *P. epihydrus*
  22b. Submersed leaves not distichous, linear to narrow-obovate, with inconspicuous lacunar bands up to 0.5 mm wide on each side of the midrib; each lacunar band composed of 0–4 rows of cells
    23a. Submersed leaves 5.0–20.0 cm long, strictly entire, red-tinged, especially in drying; drupes depressed in the center of the 2 sides .... *P. alpinus*
    23b. Submersed leaves 3.0–8.0 cm long, provided with minute, caducous, 1-celled spicules along the margin, green; drupes not depressed in the center of each side .... *P. gramineus*

Potamogeton alpinus Balbis
synonym(s): *Potamogeton alpinus* Balbis var. *subellipticus* (Fern.) Ogden, *Potamogeton alpinus* Balbis var. *tenuifolius* (Raf.) Ogden
common name(s): red pondweed
habit: submerged, rhizomatous, aquatic herb
range: circumboreal; s. to MA, w. to CA
state frequency: uncommon
habitat: ponds and slow streams
notes: Hybrids with *Potamogeton gramineus* occur in Maine.

Potamogeton amplifolius Tuckerman
synonym(s): —
common name(s): bigleaf pondweed
habit: thick, simple or uncommonly branched, rhizomatous, aquatic, perennial herb
range: Newfoundland, s. to VA and mountains of GA, w. to CA, n. to British Columbia
state frequency: uncommon
habitat: deep water of lakes and streams
notes: —

Potamogeton bicupulatus Fern.
synonym(s): *Potamogeton diversifolius* Raf. var. *trichophyllus* Morong
common name(s): snail-seed pondweed
habit: branched, aquatic, perennial herb
range: ME, s. to TN
state frequency: very rare
habitat: quiet water
notes: This species is ranked S1 by the Maine Natural Areas Program.

Potamogeton confervoides Reichenb.
synonym(s): —
common name(s): alga pondweed
habit: slender, branched, submersed, aquatic, perennial herb
range: Newfoundland, s. to NJ, w. to PA, n. to WI

state frequency: rare
habitat: shallow water, often acidic
notes: This species is ranked S2 by the Maine Natural Areas Program.

Potamogeton epihydrus Raf.
synonym(s): *Potamogeton epihydrus* Raf. var. *nuttallii* (Cham. & Schlecht.) Fern.
common name(s): ribbonleaf pondweed
habit: slender, branched, long-rhizomatous, aquatic, perennial herb
range: Newfoundland, s. to GA, w. to CA, n. to AK
state frequency: occasional
habitat: lakes, ponds, slow streams
notes: —

Potamogeton foliosus Raf.
synonym(s): *Potamogeton foliosus* Raf. var. *macellus* Fern.
common name(s): leafy pondweed
habit: usually branched, slender- and long-rhizomatous, submersed, aquatic, perennial herb
range: New Brunswick and Nova Scotia, s. to FL, w. to CA, n. to AK
state frequency: uncommon
habitat: streams and ponds, often alkaline
notes: —

Potamogeton friesii Rupr.
synonym(s): —
common name(s): Fries's pondweed
habit: simple or forked, aquatic, perennial herb
range: Newfoundland, s. to NJ, w. to IA and UT, n. to AK
state frequency: very rare
habitat: calcareous water
notes: This species is ranked S1 by the Maine Natural Areas Program.

Potamogeton gramineus L.
synonym(s): *Potamogeton gramineus* L. var. *maximus* Morong, *Potamogeton gramineus* L. var.
     myriophyllus J. W. Robbins
common name(s): variable pondweed
habit: slender, branched, aquatic, perennial herb
range: circumboreal; s. to NJ, w. to IA
state frequency: uncommon
habitat: lakes, ponds, streams
notes: Hybrids with *Potamogeton alpinus*, *P. nodosus*, *P. perfoliatus*, and *P. pusillus* var. *tenuissimus*
     occur in Maine.

Potamogeton natans L.
synonym(s): —
common name(s): floating pondweed
habit: simple or forked, aquatic, perennial herb
range: circumboreal; s. to PA, w. to CA
state frequency: occasional
habitat: lakes, ponds, quiet streams
notes: The shape of the base of floating leaves is dependent on stream velocity. Plants growing in fast
     streams tend to lose the cordate base and have a rounded base instead.

Potamogeton nodosus Poir.

synonym(s): —
common name(s): longleaf pondweed
habit: branched, aquatic, perennial herb
range: Labrador, s. to VA, w. to CA, n. to British Columbia
state frequency: uncommon
habitat: slow to fast, deep to shallow water of streams and ponds
notes: Hybrids with *Potamogeton gramineus* occur in Maine.

Potamogeton oakesianus J. W. Robbins

synonym(s): —
common name(s): Oakes's pondweed
habit: simple or branched, aquatic, perennial herb
range: Newfoundland, s. to NJ, w. to PA, n. to WI and Ontario
state frequency: uncommon
habitat: quiet water
notes: —

Potamogeton obtusifolius Mert. & Koch

synonym(s): —
common name(s): bluntleaf pondweed
habit: slender, branched, submersed, aquatic, perennial herb
range: Newfoundland, s. to NJ, w. to WY, n. to Yukon
state frequency: uncommon
habitat: cold ponds and streams
notes: —

Potamogeton perfoliatus L.

synonym(s): *Potamogeton perfoliatus* L. var. *bupleuroides* (Fern.) Farw.
common name(s): perfoliate pondweed, redhead-grass
habit: slender, branched, submersed, aquatic, perennial herb
range: interruptedly circumboreal; s. in eastern North America to NC and OH; also disjunct to the gulf
    states
state frequency: occasional
habitat: calcareous water of ponds and slow streams
notes: Hybrids with *Potamogeton gramineus*, *P. pusillus* var. *pusillus*, called *P.* ×*mysticus* Morong,
    P. pusillus var. *tenuissimus*, and *P. richardsonii* occur in Maine.

Potamogeton praelongus Wulfen

synonym(s): —
common name(s): whitestem pondweed
habit: branched or simple, submersed, stout-rhizomatous, aquatic, perennial herb
range: circumboreal; s. to MD, w. to CO and CA
state frequency: uncommon
habitat: cold, usually deep, water
notes: Many herbarium specimens will show some of the leaves with a split apex. This is caused by
    pressing the plant, which causes the cucullate leaf tip to tear as it is forced to lay flat.

Potamogeton pulcher Tuckerman

synonym(s): —
common name(s): spotted pondweed
habit: simple, pale-rhizomatous, aquatic, perennial herb
range: Nova Scotia and ME, s. to FL, w. to OK, n. to MN

state frequency: very rare
habitat: peaty or muddy shores
notes: This species is ranked S1 by the Maine Natural Areas Program.

Potamogeton pusillus L.
synonym(s): <*P. p.* var. *gemmiparus*> = *Potamogeton gemmiparus* (J. W. Robbins) J. W. Robbins *ex* Morong; <*P. p.* var. *tenuissimus*> = *Potamogeton berchtoldii* Fieber, *Potamogeton berchtoldii* Fieber var. *acuminatus* Fieber, *Potamogeton berchtoldii* Fieber var. *lacunatus* (Hagstr.) Fern., *Potamogeton berchtoldii* Fieber var. *polyphyllus* (Morong) Fern., *Potamogeton berchtoldii* Fieber var. *tenuissimus* (Mert. & Koch) Fern.
common name(s): slender pondweed
habit: slender, branched, submersed, aquatic, perennial herb
range: Greenland, s. to FL, w. to Mexico, n. to AK
state frequency: occasional
habitat: acidic to basic water
key to the subspecies:
    1a. Leaves 0.2–0.7 mm wide, acute at the apex, 1-veined, with a single row of lacunae on each side of the midvein; peduncles 1.0–3.5 cm long .... *P. p.* var. *gemmiparus* J. W. Robbins
    1b. Leaves 0.2–2.5 mm wide, acute to obtuse (rarely apiculate) at the apex, 3- or 5-nerved, with 0–5 rows of lacunae on each side of the midvein; peduncles 0.5–6.2 cm long
        2a. Leaves with up to 2 rows of lacunae on each side of the midvein, acute at the apex (rarely apiculate); stipules with united margins, forming a tube surrounding the stem; spikes mostly terminal, 1–3 per plant, with peduncles 1.0–6.2 cm long, with 2–4 distinct whorls of flowers .... *P. p.* var. *pusillus*
        2b. Leaves with up to 5 rows of lacunae on each side of the midvein, acute to obtuse at the apex; stipules convolute, but without united margins; spikes terminal and/or axillary, often more than 3 per plant, with peduncles 0.5–3.0(–4.5) cm long, with 1–3 crowded whorls of flowers .... *P. p.* var. *tenuissimus* Mert. & Koch
notes: *Potamogeton pusillus* var. *pusillus* hybridizes with *P. perfoliatus* (*q.v.*), and *Potamogeton pusillus* var. *tenuissimus* hybridizes with *P. gramineus*. Both hybrids occur in Maine.

Potamogeton richardsonii (Benn.) Rydb.
synonym(s): —
common name(s): redhead pondweed
habit: usually branched, submersed, aquatic, perennial herb
range: Labrador, s. to CT, w. to NV and CA, n. to AK
state frequency: uncommon
habitat: ponds, streams
notes: Hybrids with *Potamogeton perfoliatus* occur in Maine.

Potamogeton robbinsii Oakes
synonym(s): —
common name(s): fern pondweed, Robbins' pondweed
habit: rhizomatous, submersed, aquatic herb
range: Quebec, s. to DE and AL, w. to OR, n. to British Columbia
state frequency: uncommon
habitat: slow, muddy water
notes: —

Potamogeton spirillus Tuckerman
synonym(s): —
common name(s): northern snail-seed pondweed
habit: branched, often submersed, aquatic, perennial herb

range: Newfoundland, s. to VA, w. to IA, n. to SD and Manitoba
state frequency: occasional
habitat: shallow, quiet water
notes: —

Potamogeton strictifolius Benn.

synonym(s): —
common name(s): straight-leaved pondweed
habit: slender, simple or branched, submersed, aquatic, perennial herb
range: New Brunswick and ME, s. to CT, w. to IN, n. to MN and Saskatchewan
state frequency: very rare
habitat: calcareous ponds and streams
notes: This species is ranked SH by the Maine Natural Areas Program. Known to hybridize with
   Potamogeton zosteriformis.

Potamogeton vaseyi J. W. Robbins

synonym(s): —
common name(s): Vasey's pondweed
habit: filiform, branched, aquatic, annual herb
range: New Brunswick and ME, s. to MA, w. to IL, n. to MN and Ontario
state frequency: very rare
habitat: shallow, still or slow-moving water
notes: This species is ranked S1 by the Maine Natural Areas Program.

Potamogeton zosteriformis Fern.

synonym(s): —
common name(s): flatstem pondweed
habit: branched, submersed, aquatic, perennial herb
range: New Brunswick and Nova Scotia, s. to VA, w. to CA, n. to AK
state frequency: uncommon
habitat: quiet water of ponds and slow streams
notes: Known to hybridize with *Potamogeton strictifolius*.

Ruppia
(1 species)

Ruppia maritima L.

synonym(s): *Ruppia maritima* L. var. *longipes* Hagstr., *Ruppia maritima* L. var. *obliqua* (Schur)
   Aschers. & Graebn., *Ruppia maritima* L. var. *rostrata* Agardh, *Ruppia maritima* L. var.
   subcapitata Fern. & Wieg.
common name(s): ditch-grass
habit: submersed, aquatic herb
range: interruptedly circumboreal; s. along the east coast to FL
state frequency: uncommon
habitat: saline to brackish pools and ditches
notes: —

Stuckenia
(2 species)

KEY TO THE SPECIES

1a. Leaves acute at the apex, with an apiculus on young plants; stigma borne on a short style, the style persistent on the achene and forming a tiny beak on mature drupes; drupes mostly 3.0–4.5 mm long .... ***S. pectinata***

1b. Leaves blunt to obtuse at the apex, sometimes retuse, and only rarely apiculate; stigma broad and sessile, not forming a beak on mature drupes; drupes mostly 2.0–3.0 mm long .... ***S. filiformis***

Stuckenia filiformis (Pers.) Boerner
synonym(s): <*S. f.* ssp. *alpinus*> = *Coleogeton filiformis* (Pers.) D. Les & R. Haynes ssp. *alpinus* (Blytt) D. Les & R. Haynes, *Potamogeton filiformis* Pers. var. *alpinus* (Blytt) Aschers. & Graebn., *Potamogeton filiformis* Pers. var. *borealis* (Raf.) St. John, *Potamogeton filiformis* Pers. var. *macounii* Morong; <*S. f.* ssp. *occidentalis*> = *Coleogeton filiformis* (Pers.) D. Les & R. Haynes ssp. *occidentalis* (J. W. Robbins) D. Les & R. Haynes, *Potamogeton filiformis* Pers. var. *occidentalis* (J. W. Robbins) Morong
common name(s): threadleaf pondweed
habit: rhizomatous, tuberiferous, submersed, aquatic herb
range: circumboreal; s. in the east to PA
state frequency: very rare; known only from northern Maine
habitat: shallow, circumneutral waters
key to the subspecies:
    1a. Plants larger, 3.0–10.0 dm tall; leaves 0.5–2.0 mm wide; stipules of lower portion of stem loose and disintegrating early .... *S. f.* ssp. *occidentalis* (Robbins) Haynes, Les, & Kral
    1b. Plants smaller, 1.0–3.0 dm tall; leaves 0.2–0.8 mm wide; stipules of lower portion of stem tightly clasping and persistent .... *S. f.* ssp. *alpinus* (Blytt) Haynes, Les, & Kral
notes: *Stuckenia filiformis* ssp. *occidentalis* is ranked S1 by the Maine Natural Areas Program. *Stuckenia filiformis* ssp. *alpinus* is ranked S2.

Stuckenia pectinata (L.) Borner
synonym(s): *Coleogeton pectinatus* (L.) D. Les & R. Haynes, *Potamogeton pectinatus* L.
common name(s): sago pondweed
habit: rhizomatous, tuberiferous, submersed, aquatic herb
range: cosmopolitan; throughout the east
state frequency: rare
habitat: shallow, circumneutral water
notes: —

ARACEAE
(7 genera, 8 species)

KEY TO THE GENERA

1a. Plants thalloid, not differentiated into stems and leaves, 0.8–15.0 mm long, floating on the surface of water
    2a. Thallus broad-ellipsoid to globose, without roots; flowers without a spathe ... ***Wolffia***
    2b. Thallus flat, with roots; flowers with a spathe

3a. Each thallus with 2 or more roots [usually 5–12], the abaxial [lower] surface red-purple, the adaxial surface with 3–11 nerves .... *Spirodela*

3b. Each thallus with a single root, the abaxial surface green to somewhat purple, the adaxial surface with 1–3 nerves .... *Lemna*

1b. Plants not thalloid, differentiated into stems and leaves, much larger, terrestrial or emergent aquatics

    4a. Leaves compound, with usually 3 leaflets; flowers concealing only the basal portion of the axis of the spadix .... *Arisaema*

    4b. Leaves simple; flowers concealing most or all of the axis of the spadix

        5a. Spathe white, open, nearly flat, not concealing the spadix; fruit a cluster of distinct, red berries; plants from a long, creeping rhizome .... *Calla*

        5b. Spathe green or green to brown and marked with purple, convolute, concealing the basal portion of the spadix; fruit a cluster of green to brown berries or mature carpels submerged in the fleshy spadix; plants from erect rhizomes or fibrous roots

            6a. Mature leaf blades rounded or cordate at the base, lacking a prominent vein than extends into the short, basal lobes (if present); spadix subglobose; flowers with 4 tepals; fruit of mature carpels submerged in the fleshy spadix; plants malodorous .... *Symplocarpus*

            6b. Leaf blades sagittate to hastate, with pronounced basal lobes and a conspicuous vein extending into each lobe; spadix elongate; flowers without a perianth; fruit a green to brown berry; plants not malodorous .... *Peltandra*

Arisaema
(1 species)

Arisaema triphyllum (L.) Schott
synonym(s): *Arisaema atrorubens* (Ait.) Blume
common name(s): Jack in the pulpit
habit: erect, perennial herb
range: Nova Scotia, s. to FL, w. to TX, n. to ND and Manitoba
state frequency: common
habitat: moist to wet, rich woods
notes: —

Calla
(1 species)

Calla palustris L.
synonym(s): —
common name(s): wild calla
habit: emergent, rhizomatous, perennial herb
range: circumboreal; s. to MD, w. to IN, MN, and AK
state frequency: occasional
habitat: shallow water of swamps, bogs, and pond margins
notes: —

Lemna
(2 species)

KEY TO THE SPECIES

1a. Thallus 6.0–15.0 mm long, tapering to a long stalk; lateral thalli remaining attached to the parent thallus, forming tangled colonies suspended below the water .... *L. trisulca*

1b. Thallus 2.5–6.0 mm long, sessile or nearly so; lateral thalli separating from the parent thallus or remaining attached and forming floating rosettes .... *L. minor*

Lemna minor L.
synonym(s): —
common name(s): lesser duckweed
habit: floating, thalloid, aquatic herb
range: nearly cosmopolitan; in North America from Labrador, s. to FL, w. to CA, n. to AK
state frequency: uncommon
habitat: quiet water
notes: —

Lemna trisulca L.
synonym(s): —
common name(s): star duckweed
habit: submerged to floating, thalloid, aquatic, herb
range: nearly cosmopolitan; in North America from New Brunswick, s. to FL, w. to CA, n. to AK
state frequency: rare
habitat: shallow, still or slow-moving, circumneutral water
notes: —

Peltandra
(1 species)

Peltandra virginica (L.) Schott
synonym(s): —
common name(s): tuckahoe
habit: scapose, emergent when in water, perennial herb
range: ME, s. to FL, w. to TX, n. to MI and Ontario
state frequency: uncommon
habitat: swamps, pond margins, slow streams, shallow water
notes: —

Spirodela
(1 species)

Spirodela polyrrhiza (L.) Schleid.
synonym(s): —
common name(s): water flaxseed, greater-duckweed
habit: floating, thalloid, aquatic herb
range: nearly cosmopolitan; in North America from Quebec, s. to FL, w. to TX and Mexico, n. to also British Columbia
state frequency: uncommon
habitat: quiet margins of ponds and streams
notes: —

Symplocarpus
(1 species)

Symplocarpus foetidus (L.) Salisb. *ex* Nutt.
synonym(s): *Spathyema foetida* (L.) Raf.
common name(s): skunk-cabbage
habit: malodorous, perennial herb
range: New Brunswick, s. to GA, w. to IA, n. to MN and Manitoba
state frequency: occasional
habitat: swampy woods and thickets, wet meadows
notes: —

Wolffia
(1 species)

Wolffia columbiana Karst.
synonym(s): —
common name(s): Columbia water-meal
habit: floating to submerged, thalloid, aquatic herb
range: ME, s. to FL, w. to LA, n. to MN and Ontario
state frequency: very rare
habitat: quiet water
notes: This species is ranked S1 by the Maine Natural Areas Program.

TRILLIACEAE
(1 genus, 4 species)

Trillium
(4 species)

KEY TO THE SPECIES
1a. Petals white, with a red-purple crescent-shaped marking near the base; leaves with a
    short, but distinct, petiole; ovary 3-lobed .... *T. undulatum*
1b. Petals white, pink, or red-purple; leaves sessile; ovary 6-angled or -winged
    2a. Petals 4.0–8.0 cm long, ascending in the basal portion and spreading in the apical
        portion; stigma straight, of uniform diameter throughout its length, usually more than
        0.5 times as long as the ovary .... *T. grandiflorum*
    2b. Petals 1.5–6.0 cm long, spreading from near the base; stigma conspicuously
        recurved, tapering in the apical portion, usually *ca.* 0.5 times as long as the ovary
        3a. Petals red-purple, 2.5–6.0 cm long; ovary red-purple to purple; peduncles
            3.0–8.0 cm long, erect to spreading, the flowers only rarely hidden below the
            leaves .... *T. erectum*
        3b. Petals white, 1.5–2.5 cm long; ovary white to pink; peduncles 1.0–5.0 cm long,
            recurved, hiding the flowers under the leaves .... *T. cernuum*

Trillium cernuum L.
synonym(s): —
common name(s): nodding trillium

habit: erect, simple, rhizomatous, perennial herb
range: Newfoundland, s. to MD and mountains of GA, w. to IA, n. to WI
state frequency: uncommon
habitat: damp woods and thickets
notes: —

Trillium erectum L.

synonym(s): —
common name(s): purple trillium, stinking Benjamin, wet dog trillium
habit: erect, simple, rhizomatous, perennial herb
range: Quebec and New Brunswick, s. to MD and mountains of GA, w. to IL, n. to Manitoba
state frequency: occasional
habitat: rich, moist woods
notes: This species can rarely have white, yellow, or green petals instead of the normal red-purple
ones. Plants with these lighter petal colors will also have a lighter colored ovary.

Trillium grandiflorum (Michx.) Salisb.

synonym(s): —
common name(s): big white trillium
habit: erect, simple, thick-rhizomatous, perennial herb
range: ME and NH, s. to mountains of GA, w. to AR, n. to MN and Ontario
state frequency: very rare
habitat: rich woods and thickets
notes: This species is ranked SX by the Maine Natural Areas Program.

Trillium undulatum Willd.

synonym(s): —
common name(s): painted trillium
habit: erect, simple, thick-rhizomatous, perennial herb
range: Quebec and New Brunswick, s. to NJ and mountains of GA and TN, w. and n. to WI and
Ontario
state frequency: occasional
habitat: moist woods and stream banks
notes: —

MELANTHIACEAE
(1 genus, 1 species)

Veratrum
(1 species)

Veratrum viride Ait.

synonym(s): —
common name(s): false hellebore, Indian poke
habit: stout, erect, simple, thick-rhizomatous, perennial herb
range: Ontario and Quebec, s. to NC; also AK, s. to OR
state frequency: occasional
habitat: swamps, moist to wet forests, meadows, and low areas
notes: —

SMILACACEAE
(1 genus, 2 species)

Smilax
(2 species)

KEY TO THE SPECIES

1a. Plants herbaceous and annual; stems unarmed; carpels with 2 ovules; flowers malodorous; berry with 2–6 seeds; peduncles 5.0–8.0 times as long as the petioles of the subtending leaves .... *S. herbacea*

1b. Plants woody and perennial; stems with stout, flattened prickles; carpels with 1 ovule; flowers not malodorous; berry with 2 or 3 seeds; peduncles about as long as the petiole of the subtending leaves .... *S. rotundifolia*

Smilax herbacea L.
synonym(s): *Nemexia herbacea* (L.) Small
common name(s): carrion flower, Jacob's ladder
habit: dioecious, climbing, annual herb
range: Quebec and ME, s. along the Appalachians to GA
state frequency: uncommon
habitat: roadsides, thickets
notes: —

Smilax rotundifolia L.
synonym(s): —
common name(s): common greenbriar
habit: dioecious, armed, woody vine
range: Nova Scotia, s. to FL, w. to TX and OK, n. to MI and Ontario
state frequency: rare
habitat: thickets, roadsides
notes: —

UVULARIACEAE
(3 genera, 5 species)

KEY TO THE GENERA

1a. Leaves all basal; inflorescence a 3- to 6-flowered umbel; fruit a blue berry ...... *Clintonia*

1b. Leaves borne on a stem; inflorescence axillary or a solitary, terminal flower; fruit a red berry or a capsule

    2a. Tepals yellow, 12.0–25.0 mm long; fruit a capsule; peduncle jointed or geniculate above the base .... *Uvularia*

    2b. Tepals green-white or pink, 6.0–13.0 mm long; fruit a red berry; peduncle neither jointed nor geniculate .... *Streptopus*

Clintonia
(1 species)

Clintonia borealis (Ait.) Raf.
synonym(s): —
common name(s): bluebead-lily, corn-lily, clintonia
habit: slender-rhizomatous, scapose, perennial herb
range: Labrador, s. to NJ and mountains of NC and TN, w. to IN, n. to MN and Manitoba
state frequency: common
habitat: woods, thickets
notes: —

Streptopus
(2 species)

KEY TO THE SPECIES
1a. Leaves sessile, but not clasping; nodes of the stem pubescent; tepals pink, with darker red stripes .... *S. lanceolatus*
1b. Leaves cordate-clasping; nodes of the stem glabrous; tepals green-white ....
.... *S. amplexifolius*

Streptopus amplexifolius (L.) DC.
synonym(s): *Streptopus amplexifolius* (L.) DC. var. *americanus* J. A. & J. H. Schultes
common name(s): twisted stalk, white mandarin
habit: branched or simple, rhizomatous, perennial herb
range: circumboreal; s. to mountains of NC, w. to mountains of AZ
state frequency: uncommon
habitat: rich, moist woods, thickets
notes: Hybrids with *Streptopus lanceolatus* occur in Maine. This sterile hybrid differs from both parents in possessing deep red flowers.

Streptopus lanceolatus (Ait.) Reveal
synonym(s): *Streptopus roseus* Michx. var. *perspectus* Fassett
common name(s): rose twisted stalk, rose mandarin
habit: simple or forked, rhizomatous, perennial herb
range: Labrador, s. to NJ and mountains of NC and GA, w. and n. to the Pacific northwest
state frequency: uncommon
habitat: rich woods, moist thickets
notes: —

Uvularia
(2 species)

KEY TO THE SPECIES
1a. Leaves sessile; ovary and capsule stipitate; capsule ellipsoid, narrowed at both ends; tepals not glandular-papillose .... *U. sessilifolia*
1b. Leaves perfoliate; ovary and capsule sessile at the end of the peduncle; capsule truncate, narrowed only in the basal portion; tepals glandular-papillose on the adaxial [inner] surface .... *U. perfoliata*

Uvularia perfoliata L.
synonym(s): —
common name(s): bellwort, wild-oats
habit: slender, erect, forked, short-rhizomatous, perennial herb
range: ME and NH, s. to FL, w. to TX, n. to Ontario
state frequency: very rare
habitat: dry forests and woodlands
notes: —

Uvularia sessilifolia L.
synonym(s): —
common name(s): wild-oats, little merrybells
habit: slender, erect, forked, rhizomatous, perennial herb
range: New Brunswick and Nova Scotia, s. to FL, w. to LA and AR, n. to ND and Manitoba
state frequency: common
habitat: upland and alluvial forests, woodlands, fields, edges
notes: —

LILIACEAE
(3 genera, 5 species)

KEY TO THE GENERA

1a. Leaves in 1 or 2 whorls; tepals green-yellow, 0.6–1.1 cm long; fruit a berry; stems arising from tuberous rhizomes .... *Medeola*

1b. Leaves all basal, alternate, or whorled, when whorled, with more than 2 nodes; tepals yellow, orange, or red, 1.5–10.0 cm long; fruit a capsule; stems arising from bulbs or corms

    2a. Leaves all basal, light-mottled; tepals 1.5–5.0 cm long; anthers basifixed .... *Erythronium*

    2b. Leaves borne on a stem, not mottled; tepals 5.0–10.0 cm long; anthers versatile .... *Lilium*

Erythronium
(1 species)

Erythronium americanum Ker-Gawl.
synonym(s): —
common name(s): trout-lily, dog's-tooth-violet, fawn-lily
habit: colonial, perennial herb
range: New Brunswick and Nova Scotia, s. to FL, w. to OK, n. to MN and Ontario
state frequency: occasional
habitat: moist, rich woods
notes: —

Medeola
(1 species)

Medeola virginiana L.
synonym(s): —
common name(s): Indian cucumber-root
habit: slender, erect, simple, perennial herb
range: New Brunswick and Nova Scotia, s. to FL, w. to LA and MO, n. to MN and Ontario
state frequency: occasional
habitat: rich woods
notes: —

Lilium
(3 species)

KEY TO THE SPECIES
1a. Leaves alternate throughout, with bulblets in the upper axils; tepals conspicuously
    recurved .... *L. lancifolium*
1b. Leaves mostly or all whorled, with axillary bulblets; tepals spreading to slightly recurved
    2a. Flowers erect; tepals with a distinct, abruptly narrowed, basal portion, red-orange
        .... *L. philadelphicum*
    2b. Flowers nodding; tepals gradually narrowed in the basal portion, yellow or orange-
        yellow .... *L. canadense*

Lilium canadense L.
synonym(s): —
common name(s): Canada lily, wild yellow lily
habit: tall, slender, erect, perennial herb
range: Quebec and New Brunswick, s. to MD and mountains of VA, w. to IN, n. to Ontario
state frequency: uncommon
habitat: moist to wet woods and meadows, low thickets
notes: —

Lilium lancifolium Thunb.
synonym(s): *Lilium tigrinum* Ker-Gawl.
common name(s): tiger lily
habit: tall, stout, perennial herb
range: Asia; escaped from cultivation to New Brunswick and ME, s. to VA, w. and n. to ND
state frequency: uncommon
habitat: old homesites, dry thickets, roadsides
notes: —

Lilium philadelphicum L.
synonym(s): —
common name(s): wood lily, wild orange-red lily
habit: erect, perennial herb
range: ME, s. to MD and mountains of VA, w. to KY, n. to MN and Ontario
state frequency: uncommon
habitat: dry thickets, clearings, and open woods
notes: —

HEMEROCALLIDACEAE
(1 genus, 2 species)

Hemerocallis
(2 species)

KEY TO THE SPECIES
1a. Flowers red-orange, not fragrant; tepals reticulate-veined; capsules rarely maturing ....
.... *H. fulva*
1b. Flowers yellow, fragrant; tepals parallel-veined; capsules maturing ... *H. lilioasphodelus*

Hemerocallis fulva (L.) L.
synonym(s): —
common name(s): orange day-lily
habit: tall, scapose, perennial herb
range: Eurasia; escaped from cultivation
state frequency: uncommon
habitat: old homesites, fields, roadsides, waste places
notes: —

Hemerocallis lilioasphodelus L.
synonym(s): *Hemerocallis flava* (L.) L.
common name(s): yellow day-lily
habit: tall, scapose, perennial herb
range: Asia; escaped from cultivation to New Brunswick and Nova Scotia, s. to PA, w. and n. to MI
state frequency: rare
habitat: old homesites
notes: —

CONVALLARIACEAE
(3 genera, 7 species)

KEY TO THE GENERA
1a. Leaves all basal; inflorescence a secund raceme with drooping pedicels .... *Convallaria*
1b. Leaves borne on a stem; inflorescence otherwise
    2a. Tepals evidently connate forming a tube, 10.0–22.0 mm long; inflorescence of
    axillary flowers; berry dark blue to black; rhizome thick and knotty ..... *Polygonatum*
    2b. Tepals distinct, 1.5–6.0 mm long; inflorescence a terminal raceme or panicle; berry
    initially, or ultimately becoming, red; rhizome slender .... *Maianthemum*

Convallaria
(1 species)

Convallaria majalis L.
synonym(s): —
common name(s): lily-of-the-valley
habit: colonial, rhizomatous, perennial herb
range: Europe; escaped from cultivation

state frequency: uncommon
habitat: thickets, roadsides, waste places, areas or cultivation
notes: —

Maianthemum
(4 species)

KEY TO THE SPECIES

1a. Flowers with 4 tepals; leaves cordate at the base .... *M. canadense*
1b. Flowers with 6 tepals; leaves narrowed near the base
    2a. Inflorescence a raceme; tepals 4.0–6.0 mm long, longer than the stamens
        3a. Leaves 7–12 per stem, sessile and somewhat clasping at the base, pubescent on
        the abaxial surface; peduncle of the raceme short, often less than 2.0 cm long ......
        .... *M. stellatum*
        3b. Leaves 2–4 per stem, narrowed to a subpetiolar base, glabrous; peduncle or
        raceme 2.0–6.0 cm long .... *M. trifolium*
    2b. Inflorescence a panicle; tepals 1.5–3.0 mm long, shorter than the stamens ....
    .... *M. racemosum*

Maianthemum canadense Desf.
synonym(s): *Unifolium canadense* (Desf.) Greene
common name(s): Canada mayflower, wild lily-of-the-valley
habit: low, erect, slender-rhizomatous, perennial herb
range: Labrador, s. to DE and mountains of GA, w. to IA and SD, n. to Manitoba
state frequency: common
habitat: forests
notes: —

Maianthemum racemosum (L.) Link
synonym(s): *Smilacina racemosa* (L.) Desf., *Smilacina racemosa* (L.) Desf. var. *cylindrata* Fern.,
    Vagnera racemosa (L.) Morong
common name(s): false spikenard, false Solomon's seal
habit: ascending, fleshy-rhizomatous, perennial herb
range: New Brunswick and Nova Scotia, s. to VA and mountains of NC, w. to MO and AZ, n. to
    British Columbia
state frequency: occasional
habitat: rich woods, slopes, clearings
notes: —

Maianthemum stellatum (L.) Link
synonym(s): *Smilacina stellata* (L.) Desf., *Vagnera stellata* (L.) Morong
common name(s): star-flowered false Solomon's seal
habit: ascending to erect, rhizomatous, perennial herb
range: Newfoundland, s. to NJ and mountains of VA, w. to AZ and CA, n. to British Columbia
state frequency: occasional
habitat: woods, thickets, meadows, shores
notes: —

Maianthemum trifolium (L.) Sloboda
synonym(s): *Smilacina trifolia* (L.) Desf., *Vagnera trifolia* (L.) Morong
common name(s): three-leaved false Solomon's seal

habit: slender, erect, slender-rhizomatous, perennial herb
range: Labrador, s. to NJ, w. to IL, n. to Yukon
state frequency: occasional
habitat: wet woods and meadows, bogs
notes: —

Polygonatum
(2 species)

KEY TO THE SPECIES
1a. Leaves glabrous, with 7–19 prominent veins; flowers 14.0–22.0 mm long ... *P. biflorum*
1b. Leaves pubescent on the veins of the abaxial surface, with 3–9 prominent veins; flowers
10.0–13.0 mm long .... *P. pubescens*

Polygonatum biflorum (Walt.) Ell. var. **commutatum** (J. A. & J. H. Schultes) Morong
synonym(s): —
common name(s): Solomon's seal
habit: slender to stout, erect to arching, rhizomatous, perennial herb
range: ME, s. to FL, w. to TX, n. to ND and Ontario
state frequency: very rare
habitat: woods, thickets, roadsides
notes: It is unclear whether this species was introduced or if the South Thomaston (Knox County)
station is a native occurrence.

Polygonatum pubescens (Willd.) Pursh
synonym(s): —
common name(s): Solomon's seal
habit: slender, erect, rhizomatous, perennial herb
range: New Brunswick and Nova Scotia, s. to MD and mountains of SC and GA, w. to IA, n. to MN
and Manitoba
state frequency: uncommon
habitat: moist, rocky woods, thickets
notes: —

HYPOXIDACEAE
(1 genus, 1 species)

Hypoxis
(1 species)

Hypoxis hirsuta (L.) Coville
synonym(s): —
common name(s): common stargrass, yellow stargrass
habit: ascending to reclining, scapose, perennial herb
range: ME, s. to FL, w. to TX, n. to ND and Manitoba
state frequency: very rare
habitat: open woods, meadows
notes: This species is ranked SX by the Maine Natural Areas Program.

ASPARAGACEAE
(1 genus, 1 species)

Asparagus
(1 species)

Asparagus officinalis L.
synonym(s): —
common name(s): asparagus
habit: branched, short-rhizomatous, perennial herb
range: Europe; escaped from cultivation
state frequency: uncommon
habitat: fields, waste places
notes: —

AMARYLLIDACEAE
(2 genera, 2 species)

KEY TO THE GENERA
1a. Tepals wide-spreading; perianth with a conspicuous corona; flowers usually solitary ....
.... *Narcissus*
1b. Tepals ascending; perianth without a corona; flowers in a 2- to 8-flowered umbel ....
.... *Leucojum*

Leucojum
(1 species)

Leucojum aestivum L.
synonym(s): —
common name(s): summer snowflake
habit: erect to arching, scapose, perennial herb
range: Europe; escaped from cultivation to Nova Scotia and ME, s. to VA
state frequency: very rare
habitat: low woods, meadows
notes: —

Narcissus
(1 species)

Narcissus poeticus L.
synonym(s): —
common name(s): poet's narcissus
habit: erect, scapose, perennial herb
range: Europe; escaped from cultivation
state frequency: rare
habitat: fields, meadows
notes: —

ALLIACEAE
(1 genus, 5 species)

Allium
(5 species)

KEY TO THE SPECIES

1a. Leaves flat, 2.0–8.0 cm wide, produced in early season and soon withering, not present during anthesis; capsule prominently 3-lobed .... *A. tricoccum*

1b. Leaves flat or terete, 0.1–1.5 cm wide, not early withering, present during anthesis; capsule only slightly lobed

   2a. Leaves definitely flat; bulbs covered by a fibrous outer layer .... *A. canadense*

   2b. Leaves terete and hollow (except at the very base); bulbs covered by a membranaceous-scaly outer layer

      3a. Pedicels 3.0–7.0 mm long, shorter than to equaling the flowers; filaments monomorphic, not widened near the base; umbels without bulblets ....
.... *A. schoenoprasum*

      3b. Pedicels 10.0 mm or longer, longer than the flowers; filaments dimorphic—the inner widened and flat, at least near the base, the outer slender; some or all of the flowers commonly replaced by bulblets

         4a. Stems hollow; leaves 5.0–15.0 mm wide; filaments lacking hair-like appendages .... *A. cepa*

         4b. Stems solid; leaves commonly less than 5.0 mm wide; filaments terminating in 2 hair-like appendages .... *A. vineale*

Allium canadense L.
synonym(s): —
common name(s): wild garlic
habit: erect, bulbiferous, perennial herb
range: New Brunswick and Nova Scotia, s. to FL, w. to TX, n. to ND and Ontario
state frequency: rare
habitat: woods, thickets, banks, meadows, rocky shores
notes: This species is ranked S2 by the Maine Natural Areas Program.

Allium cepa L.
synonym(s): —
common name(s): common onion
habit: coarse, inflated-stemmed, bulbiferous, perennial herb
range: Iran and Pakistan; escaped from cultivation
state frequency: very rare
habitat: areas of cultivation
notes: —

Allium schoenoprasum L. var. *sibiricum* (L.) Hartman
synonym(s): *Allium sibiricum* L.
common name(s): chives
habit: stout, erect, bulbiferous, perennial herb
range: circumboreal; s. to ME, w. to CO and WA

state frequency: uncommon
habitat: rocky or gravelly shores
notes: —

Allium tricoccum Ait.

synonym(s): —
common name(s): wild leek, ramps
habit: erect, bulbiferous, perennial herb
range: New Brunswick and Nova Scotia, s. to MD and mountains of GA, w. to AL, n. to ND
state frequency: rare
habitat: rich woods, alluvial bottomlands
notes: This species is ranked S2 by the Maine Natural Areas Program. Similar to *Allium canadense* in that the bulb is covered by a fibrous outer layer.

Allium vineale L.

synonym(s): —
common name(s): scallions, field garlic
habit: erect, bulbiferous, perennial herb
range: Europe; naturalized to ME, s. to GA, w. to AR, n. to IL and MI
state frequency: very rare
habitat: fields, meadows, lawns
notes: —

HYACINTHIACEAE
(1 genus, 1 species)

Ornithogalum
(1 species)

Ornithogalum umbellatum L.

synonym(s): —
common name(s): star of Bethlehem, nap at noon
habit: scapose, perennial herb
range: Europe; naturalized to Newfoundland, s. to SC, w. to UT, n. to British Columbia
state frequency: very rare
habitat: woods, thickets, fields, roadsides
notes: —

IRIDACEAE
(2 genera, 9 species)

KEY TO THE GENERA

1a. Tepals 10.0–70.0 mm long, the outer recurved to spreading, the inner spreading to erect; styles flattened and petaloid, covering and concealing the stamens; leaves 3.0–30.0 mm wide .... ***Iris***

1b. Tepals 7.0–12.0 mm long, all of similar orientation; styles filiform, not petaloid, not concealing the stamens; leaves 1.0–6.0 mm wide .... ***Sisyrinchium***

Iris
(5 species)

KEY TO THE SPECIES

1a. Outer tepals [sepals] pubescent .... *I. germanica*
1b. Outer tepals glabrous
    2a. Tepals yellow to ochroleucous; capsules 5.0–8.0 cm tall .... *I. pseudoacorus*
    2b. Tepals largely blue or purple (white); capsules 2.5–5.5 cm tall
        3a. Leaves 0.2–0.7 cm wide; capsule sharply trigonous; rhizome at or near the
            surface, 0.2–0.5 cm wide .... *I. prismatica*
        3b. Leaves 5.0–30.0 cm wide; capsule bluntly 3-angled; rhizome at or well below the
            surface, 0.5–3.0 cm wide
            4a. Inner tepals [petals] 2.0–5.0 cm long, 0.5–0.65 times as long as the outer
                tepals, flat, without a bristle tip; capsule tipped by a beak; seeds D-shaped .....
                .... *I. versicolor*
            4b. Inner tepals reduced, 1.0–2.0 cm long, less than 0.5 times as long as the outer
                tepals, involute, bristle-tipped; capsule blunt at the apex, without a beak;
                seeds pyriform-D-shaped .... *I. setosa*

Iris germanica L.
synonym(s): —
common name(s): German iris
habit: stout, erect, perennial herb
range: Europe; escaped from cultivation
state frequency: very rare
habitat: old gardens, roadsides, waste places
notes: —

Iris prismatica Pursh *ex* Ker-Gawl.
synonym(s): —
common name(s): slender blue flag
habit: erect, rhizomatous, perennial herb
range: Nova Scotia, s. to GA
state frequency: rare
habitat: brackish to fresh marshes, swamps, and shores, meadows
notes: This species is ranked S2 by the Maine Natural Areas Program.

Iris pseudacorus L.
synonym(s): —
common name(s): yellow iris, yellow flag, water flag
habit: erect, rhizomatous, perennial herb
range: Europe; escaped from cultivation
state frequency: uncommon
habitat: marshes, swamps, stream- and brooksides
notes: —

Iris setosa Pallas *ex* Link var. **canadensis** M. Foster *ex* B. L. Robins. & Fern.
synonym(s): *Iris hookeri* Penny *ex* D. Don
common name(s): beachhead iris, arctic blue flag
habit: coarse, cespitose, stout-rhizomatous, perennial herb
range: Labrador, s. to ME

state frequency: rare
habitat: rocky headlands, beaches, shores
notes: —

Iris versicolor L.
synonym(s): —
common name(s): northern blue flag
habit: erect to arching, stout-rhizomatous, perennial herb
range: Labrador, s. to VA, w. to OH, n. to MN and Manitoba
state frequency: common
habitat: marshes, swamps, meadows, shores, ditches
notes: —

Sisyrinchium
(4 species)

Measurement of stem width should be taken at the mid-portion of the stem.

KEY TO THE SPECIES
1a. Spathes sessile, usually solitary at the apex of the stem; stems appearing unbranched
  2a. Stems slender, 0.5–1.5 mm wide; leaves 1.0–2.0 mm wide; outer bract of the spathe
    with margins united only at the very base; pedicels spreading to recurving in fruit ......
    .... ***S. mucronatum***
  2b. Stems winged, 1.5–3.0 mm wide; leaves 2.0–3.0 mm wide; outer bract of the spathe
    with basal margins united for 2.0–6.0 mm; pedicels erect to ascending in fruit ....
    .... ***S. montanum***
1b. Spathes with a conspicuous peduncle, usually 2–5 per stem; stems appearing branched
  due to long peduncles
  3a. Stems broadly winged, 2.0–4.0 mm wide; leaves 2.0–6.0 mm wide, commonly
    darkening in drying; outer bract of the spathe mostly 2.0–4.0 cm long ....
    .... ***S. angustifolium***
  3b. Stems narrow, scarcely winged, 0.5–2.0 mm wide; leaves 1.0–3.0 mm wide,
    commonly remaining light green in drying; outer bract of the spathe mostly
    1.5–2.0 cm long .... ***S. atlanticum***

Sisyrinchium angustifolium P. Mill.
synonym(s): —
common name(s): —
habit: loosely tufted, ascending to spreading, perennial herb
range: Newfoundland, s. to FL, w. to TX, n. to MN and Ontario
state frequency: uncommon
habitat: damp woods, thickets, meadows, and shores
notes: —

Sisyrinchium atlanticum Bickn.
synonym(s): —
common name(s): —
habit: slender, spreading to erect, perennial herb
range: Nova Scotia, s. to FL, w. to MS and MO, n. to MN
state frequency: rare
habitat: damp woods and meadows, marshes, swales
notes: —

Sisyrinchium montanum Greene var. *crebrum* Fern.
synonym(s): —
common name(s): —
habit: tufted, erect, perennial herb
range: Newfoundland, s. to NJ and mountains of NC, w. and n. to Ontario
state frequency: occasional
habitat: meadows, shores, sandy open areas
notes: —

Sisyrinchium mucronatum Michx.
synonym(s): —
common name(s): —
habit: tufted, erect, perennial herb
range: ME, s. to NC, w. to ND, n. to Manitoba
state frequency: uncommon
habitat: meadows, fields, open woods
notes: This species is ranked S1 by the Maine Natural Areas Program.

ORCHIDACEAE
(18 genera, 46 species)

KEY TO THE GENERA
1a. Labellum basally or entirely inflated and saccate, forming a pouch
  2a. Labellum saccate only in the basal portion, 3.0–14.0 mm long; inflorescence a 4- to many-flowered raceme or spike
    3a. Principal leaves all basal, usually white-striped or -reticulated; flowers white or green .... *Goodyera*
    3b. Principal leaves borne on a stem, alternate, without white markings; flowers purple-green .... *Epipactis*
  2b. Labellum saccate throughout, 15.0–60.0 mm long; inflorescence a solitary flower or a 2- or 3-flowered raceme
    4a. Labellum pubescent; plants with 1 leaf; flowers with 1 anther; stem bulbous at the base .... *Calypso*
    4b. Labellum glabrous; plants with 2 or more leaves; flowers with 2 anthers; stem without a bulbous swelling at the base .... *Cypripedium*
1b. Labellum neither inflated nor saccate, not forming a pouch
  5a. Labellum basally prolonged into a conspicuous spur
    6a. Perianth with 2 distinct colors (very rarely uniform in color); plants with 1 or 2 leaves; caudicles of pollinia convergent, the viscidula enclosed in a single pouch
      7a. Sepals and lateral petals pink to pale purple; labellum white, unspotted, and unlobed; plants usually with 2 leaves .... *Galearis*
      7b. Sepals and lateral petals purple to white; labellum white, spotted with pale purple, 3-lobed, the central lobe notched at the apex; plants usually with 1 leaf .... *Amerorchis*
    6b. Perianth essentially uniform in color; plants with 1–many leaves; caudicles of pollinia divergent, the viscidula not enclosed in a pouch

8a. Floral bracts 2.0–4.0 times as long as the flowers; labellum with lateral lobes longer than the midlobe; viscidula covered by a thin membrane .... ***Coeloglossum***

8b. Floral bracts shorter than to equaling the length of the flowers (sometimes the lowest bract longer than the flowers); labellum with lateral lobes equal to or shorter than the midlobe; viscidula not covered by a thin membrane .... ***Platanthera***

5b. Labellum lacking a spur (flowers with a small, inconspicuous, adnate spur in *Corallorhiza* formed by the 2 lateral sepals)

9a. Flowers resupinate, the labellum the lowest petal; labellum glabrous or pubescent on the adaxial [inner] surface

10a. Stem leaves whorled; sepals and lateral petals conspicuously unalike .... ***Isotria***

10b. Stem leaves alternate or opposite; sepals and lateral petals similar or not

11a. Labellum pubescent on the adaxial surface, 17.0–33.0 mm long

12a. Leaves flat, 1.0–2.5 cm wide, numbering 1–3 on the stem, present at anthesis; sepals and lateral petals distinct, the outer spreading .... ***Pogonia***

12b. Leaves plicate, 0.2–0.4 cm wide, solitary on the stem, absent or not fully formed at anthesis; sepals and lateral petals connate at the base, the outer erect .... ***Arethusa***

11b. Labellum glabrous, 2.0–12.0 mm long

13a. Principal leaves in a pair and appearing opposite

14a. Labellum without a prominent notch at the apex and without a pair of auricles at the base; leaves near the base of the stem .... ***Liparis***

14b. Labellum 2-lobed due to a prominent notch at the apex, bearing a pair of auricles or teeth near the base; leaves near the middle of the stem .... ***Listera***

13b. Principal leaves not in pairs, alternately arranged or absent

15a. Plants with scale-like leaves and coralloid rhizomes, lacking chlorophyll [the stems yellow-green, brown, or purple] .... ***Corallorhiza***

15b. Plants with well developed leaves and fleshy, tuberous roots or bulbs, possessing chlorophyll

16a. Labellum with 3 prominent, longitudinal, green ridges; inflorescence a few-flowered raceme .... ***Triphora***

16b. Labellum lacking ridges; inflorescence a several- to many-flowered raceme or spike

17a. Sepals and lateral petals similar; labellum 3.5–12.0 mm long; stem with bracts between the principal leaves and the flowers; inflorescence a spike .... ***Spiranthes***

17b. Sepals and lateral petals dissimilar; labellum 2.0–3.0 mm long; stem without bracts between the principal leaves and the flowers; inflorescence a raceme .... ***Malaxis***

9b. Flowers not resupinate, the labellum the uppermost petal; labellum pubescent on
the adaxial surface .... *Calopogon*

Amerorchis
(1 species)

Amerorchis rotundifolia (Banks *ex* Pursh) Hultén
synonym(s): *Orchis rotundifolia* Banks *ex* Pursh
common name(s): one-leaf orchis, small round-leaved orchis
habit: short-rhizomatous, perennial herb
range: Greenland, s. to ME and NY, w. to MT, n. to AK
state frequency: very rare
habitat: swamps, wet woods
notes: This species is ranked S1 by the Maine Natural Areas Program.

Arethusa
(1 species)

Arethusa bulbosa L.
synonym(s): —
common name(s): dragon's mouth, swamp-pink
habit: erect, perennial herb
range: Newfoundland, s. to DE, MD, and mountains of NC and SC, w. to IN, n. to MN and Ontario
state frequency: rare
habitat: *Sphagnum* bogs
notes: —

Calopogon
(1 species)

Calopogon tuberosus (L.) B. S. P.
synonym(s): *Calopogon pulchellus* (Salisb.) R. Br. *ex* Ait. f.
common name(s): grass-pink
habit: erect, perennial herb
range: Newfoundland, s. to FL, w. to TX, n. to MN and Manitoba
state frequency: uncommon
habitat: *Sphagnum* bogs
notes: —

Calypso
(1 species)

Calypso bulbosa (L.) Oakes var. **americana** (R. Br. *ex* Ait. f.) Luer
synonym(s): *Cytherea bulbosa* (L.) House
common name(s): calypso, Venus' slipper
habit: erect, perennial herb
range: circumboreal; s. to NY, w. to NM and CA
state frequency: rare
habitat: cool, mossy woods, *Thuja* swamps
notes: —

Coeloglossum
(1 species)

Coeloglossum viride (L.) Hartman var. ***virescens*** (Muhl. *ex* Willd.) Luer
synonym(s): *Habenaria viridis* (L.) R. Br. *ex* Ait. f. var. *bracteata* (Muhl. *ex* Willd.) Reichenb. *ex* Gray
common name(s): frog orchid, bracted orchid, long-bracted green orchid
habit: erect, perennial herb
range: Newfoundland, s. to NJ and mountains of NC, w. to NE and CO, n. to AK
state frequency: uncommon
habitat: rich woods, thickets, meadows, shores, often calcareous
notes: —

Corallorhiza
(3 species)

KEY TO THE SPECIES
1a. Labellum with a lateral lobe or tooth on each side, 3.5–8.0 mm long; lateral sepals spreading to down-curved; plants flowering mid-May to late-August
    2a. Labellum 6.0–8.0 mm long; sepals and lateral petals with 3(–5) nerves; spur present, though inconspicuous; stem commonly purple to brown; plants flowering in early July to late August; raceme usually with 10–30 flowers .... *C. maculata*
    2b. Labellum 3.5–5.0 mm long; sepals and lateral petals with 1 nerve; spur absent; stem commonly yellow-green; plants flowering in mid-May to mid-June; raceme usually with 2–12 flowers .... *C. trifida*
1b. Labellum without lateral lobes, entire to erose, 3.0–4.0 mm long; lateral sepals up-curved; plants flowering mid-August to late September .... *C. odontorhiza*

Corallorhiza maculata (Raf.) Raf.
synonym(s): —
common name(s): spotted coralroot
habit: erect, rhizomatous, perennial herb
range: Labrador, s. to MD and mountains of NC, w. to Mexico and CA, n. to British Columbia
state frequency: uncommon
habitat: dry woods
notes: —

Corallorhiza odontorhiza (Willd.) Nutt.
synonym(s): —
common name(s): autumn coralroot
habit: erect, rhizomatous, perennial herb
range: ME, s. to FL, w. to Mexico, n. to MN
state frequency: very rare
habitat: dry, open woods
notes: This species is ranked S1 by the Maine Natural Areas Program.

Corallorhiza trifida Chatelain
synonym(s): *Corallorhiza trifida* Chatelain var. *verna* (Nutt.) Fern.
common name(s): northern coralroot, pale coralroot, early coralroot
habit: erect, rhizomatous, perennial herb
range: circumboreal; s. to NJ, w. to NM

state frequency: uncommon
habitat: bogs, wet woods
notes: —

Cypripedium
(5 species)

KEY TO THE SPECIES

1a. Plants with 2 basal leaves; labellum with a cleft along its entire length on the adaxial [upper] surface .... *C. acaule*
1b. Plants with few to several leaves alternately arranged on the stem; labellum with a round opening near the base on the adaxial surface
    2a. All 3 sepals distinct; labellum 1.5–2.5 cm long, strongly red-veined, prolonged downward near the apex into a conical pouch .... *C. arietinum*
    2b. Lateral sepals [2] connate, thereby forming a single blade; labellum 2.0–5.0 cm long, neither red-veined nor prolonged downward to form a protruding sac
        3a. Labellum white (pink) and streaked with pink or purple; upper sepal and the flat, lateral petals obtuse to rounded at the apex; 2 lateral sepals completely connate ... . *C. reginae*
        3b. Labellum yellow, usually with purple veins; upper sepal and the usually spirally twisted lateral petals acute to acuminate at the apex; 2 lateral sepals connate except at the very tip and forming a 2-lobed apex
            4a. Lateral petals 5.0–8.0 cm long; labellum 3.0–5.0 cm long; stems with 4–6 leaves; sepals usually green-yellow; plants commonly of mesic forests .... *C. pubescens*
            4b. Lateral petals 3.0–5.0 cm long; labellum 2.0–3.0 cm long; stems with 3 or 4(–5) leaves; sepals usually red-purple; plants commonly of organic or mineral soil wetlands, shores, and wet rocks .... *C. parviflorum*

Cypripedium acaule Ait.
synonym(s): *Fissipes acaulis* (Ait.) Small
common name(s): pink lady's slipper, moccasin flower
habit: erect, scapose, perennial herb
range: Newfoundland, s. to GA, w. to AL, n. to Alberta
state frequency: occasional
habitat: dry, acid woods
notes: The labellum in this species is pink or white.

Cypripedium arietinum Ait. f.
synonym(s): —
common name(s): ram's head lady's slipper
habit: erect, perennial herb
range: Quebec, s. to MA, w. to MN, n. to Manitoba
state frequency: very rare
habitat: acid woods, *Thuja* swamps
notes: This species is ranked S1 by the Maine Natural Areas Program.

Cypripedium parviflorum Salisb.
synonym(s): *Cypripedium calceolus* L. var. *parviflorum* (Salisb.) Fern.
common name(s): small yellow lady's slipper, small yellow moccasin flower

habit: erect, perennial herb
range: Newfoundland, s. to NJ and mountains of GA, w. to TX and NM, n. to WA and British Columbia
state frequency: uncommon
habitat: moist woods, fens, wet shores
notes: —

Cypripedium pubescens Willd.
synonym(s): *Cypripedium calceolus* L. var. *pubescens* (Willd.) Correll
common name(s): large yellow lady's slipper, large yellow moccasin flower
habit: erect, perennial herb
range: New Brunswick, s. to GA, w. to AL and MO, n. to MN
state frequency: uncommon
habitat: mesic woods
notes: —

Cypripedium reginae Walt.
synonym(s): —
common name(s): showy lady's slipper
habit: tall, erect, perennial herb
range: Newfoundland, s. to NJ and mountains of GA, w. to AR, n. to ND and Manitoba
state frequency: rare
habitat: fens
notes: This species is ranked S2/S3 by the Maine Natural Areas Program.

Epipactis
(1 species)

Epipactis helleborine (L.) Crantz
synonym(s): *Serapias helleborine* L.
common name(s): helleborine
habit: erect, short-rhizomatous, perennial herb
range: Europe; naturalized to New Brunswick, s. to NJ, w. to MO, n. to Ontario
state frequency: uncommon
habitat: woods, thickets, roadsides
notes: —

Galearis
(1 species)

Galearis spectabilis (L.) Raf.
synonym(s): *Orchis spectabilis* L.
common name(s): showy orchis
habit: erect, short-rhizomatous, perennial herb
range: New Brunswick, s. to GA, w. to KS, n. to MN and Ontario
state frequency: very rare
habitat: rich, often calcareous, woods
notes: This species is ranked S1 by the Maine Natural Areas Program.

Goodyera
(4 species)

KEY TO THE SPECIES

1a. Spike dense and cylindric, the spiral hardly discernible; labellum lacking fleshy callosities on the adaxial [inner] surface, with a reflexed tip, the apical portion less than 0.5 times the length of the basal, saccate portion; scape with 4–14 bracts .... ***G. pubescens***

1b. Spike less dense and open, secund or spirally secund; labellum with fleshy callosities on the adaxial surface, with a spreading or recurved tip, the apical portion 0.5–1.0 times the length of the basal, saccate portion; scape with 2–7 bracts

   2a. Labellum 5.0–8.0 mm long; leaves typically white only on the midvein; beak of the rostellum 2.3–3.6 mm long .... ***G. oblongifolia***

   2b. Labellum 3.0–5.5 mm long; leaves typically white on the primary veins; beak of the rostellum 0.2–1.7 mm long

      3a. Spike secund; leaves usually with 5 nerves; labellum deeply saccate, the pouch as deep as long; beak of the rostellum 0.2–0.6 mm long; anthers blunt ..... ***G. repens***

      3b. Spike spirally secund; leaves usually with 5–9 nerves; labellum shallowly saccate, the pouch longer than deep; beak of the rostellum 0.6–1.7 mm long; anthers acuminate .... ***G. tesselata***

Goodyera oblongifolia Raf.
synonym(s): *Peramium decipiens* (Hook.) Piper
common name(s): giant rattlesnake-plantain, western rattlesnake-plantain
habit: coarse, stout, erect, scapose, perennial herb
range: Quebec, s. to ME, w. to NM and CA, n. to AK
state frequency: very rare
habitat: dry to moist, often coniferous, woods
notes: This species is ranked S1 by the Maine Natural Areas Program.

Goodyera pubescens (Willd.) R. Br. *ex* Ait. f.
synonym(s): *Peramium pubescens* (Willd.) MacM.
common name(s): downy rattlesnake-plantain
habit: erect to repent, scapose, stout, perennial herb
range: New Brunswick, s. to FL, w. to MO, n. to MN and Ontario
state frequency: uncommon
habitat: dry to moist woods
notes: —

Goodyera repens (L.) R. Br. *ex* Ait. f.
synonym(s): *Goodyera repens* (L.) R. Br. *ex* Ait. f. var. *ophioides* Fern., *Peramium ophioides* (Fern.) Rydb.
common name(s): dwarf rattlesnake-plantain, lesser rattlesnake-plantain
habit: slender, stoloniferous, scapose, perennial herb
range: circumboreal; s. to mountains of NC, w. to British Columbia
state frequency: uncommon
habitat: moist to dry, cold woods
notes: —

Goodyera tesselata Lodd.
synonym(s): *Peramium tessalatum* (Lodd.) Heller
common name(s): checkered rattlesnake-plantain, alloploid rattlesnake-plantain
habit: erect, scapose, perennial herb
range: Newfoundland, s. to MD, w. to MN, n. to Manitoba
state frequency: uncommon
habitat: dry to moist woods
notes: —

Isotria
(2 species)

Species in this genus vegetatively resemble *Medeola virginica*, which has a solid, pubescent stem arising from a white tuber. *Isotria* have glabrous, hollow stems and lack white tubers.

KEY TO THE SPECIES
1a. Sepals 3.5–6.5 cm long, usually green-yellow in the basal portion and red-purple in the apical portion; lateral petals 1.5–2.5 cm long; labellum 2.0–2.5 cm long; peduncle 2.0–5.5 cm long, at least as long as the ovary; stems commonly 2.0–4.0 dm tall .... ***I. verticillata***
1b. Sepals 1.2–2.5 cm long, usually green-yellow throughout; petals 1.3–1.7 cm long; labellum 1.0–1.5 cm long; peduncle 1.0–1.5 cm long, shorter than the ovary; stems commonly 1.5–2.5 dm tall .... ***I. medeoloides***

Isotria medeoloides (Pursh) Raf.
synonym(s): —
common name(s): small whorled-pogonia
habit: erect, perennial herb
range: ME and NH, s. to NC, w. to MO, n. to MI and Ontario
state frequency: rare
habitat: early successional forests
notes: This species is ranked S2 by the Maine Natural Areas Program.

Isotria verticillata Raf.
synonym(s): —
common name(s): large whorled-pogonia
habit: erect, perennial herb
range: ME, s. to FL, w. to TX, n. to MI and Ontario
state frequency: very rare
habitat: acid woods
notes: This species is ranked SX by the Maine Natural Areas Program.

Liparis
(1 species)

Liparis loeselii (L.) L. C. Rich.
synonym(s): —
common name(s): Loesel's bog twayblade, yellow twayblade
habit: erect, perennial herb
range: New Brunswick, s. to MD and mountains of AL, w. to AR and KS, n. to ND and
    Saskatchewan

state frequency: rare
habitat: damp to wet woods, thickets, and meadows, bogs
notes: —

Listera
(3 species)

KEY TO THE SPECIES
1a. Labellum 3.0–5.0 mm long, divided half of its length or more into 2 linear segments; column *ca.* 0.5 mm tall; axis of the raceme glabrous; floral bracts about 0.7–2.0 mm long .... *L. cordata*
1b. Labellum 6.0–11.0 mm long, divided a third its length into 2 oblong or ovate segments; column 2.5–4.0 mm tall; axis of the raceme glandular-pubescent; floral bracts 2.0–5.0 mm long
   2a. Labellum narrowed at the base, usually bearing a pair of small teeth at the base; pedicels and ovary glandular-pubescent .... *L. convallarioides*
   2b. Labellum maintaining its width and not narrowed at the base, bearing a pair of retrorse auricles at the base; pedicels and ovary glabrous .... *L. auriculata*

Listera auriculata Wieg.
synonym(s): *Ophrys auriculata* (Wieg.) House
common name(s): auricled twayblade
habit: erect, perennial herb
range: Newfoundland, s. to NH, w. to MN, n. to Ontario
state frequency: very rare
habitat: wet woods, thickets, and banks, *Thuja* swamps
notes: This species is ranked S1 by the Maine Natural Areas Program. Hybrids with *Listera convallarioides*, called *L.* ×*veltmanii* Case, occur in Maine.

Listera convallarioides (Sw.) Nutt. *ex* Ell.
synonym(s): *Ophrys convallarioides* (Sw.) W. Wight
common name(s): broad-lipped twayblade, broad-leaved twayblade
habit: stout, erect, perennial herb
range: Newfoundland, s. to MA and mountains of NC, w. to AZ, n. to AK
state frequency: uncommon
habitat: damp to wet woods and thickets, swamps, fens
notes: Known to hybridize with *Listera auriculata* (*q.v.*).

Listera cordata (L.) R. Br. *ex* Ait. f.
synonym(s): *Ophrys cordata* L.
common name(s): heart-leaved twayblade
habit: slender, erect, perennial herb
range: circumboreal; s. to mountains of NC, w. to NM and CA
state frequency: uncommon
habitat: wet woods and banks, *Sphagnum* bogs
notes: —

Malaxis
(2 species)

KEY TO THE SPECIES

1a. Labellum entire, acuminate at the apex; pedicels 1.0–2.0 mm long; inflorescence less crowded and tapering to the apex .... *M. monophyllos*
1b. Labellum with 2 lobes at the apex and with a central tooth in the sinus between the lobes; pedicels 4.0–8.0 mm long; inflorescence more crowded and rounded or flattened at the apex .... *M. unifolia*

Malaxis monophyllos (L.) Sw. ssp. *brachypoda* (Gray) A. & D. Löve
synonym(s): *Malaxis brachypoda* (Gray) Fern., *Microstylis monophyllos* Lindl.
common name(s): white adder's mouth
habit: slender, erect, perennial herb
range: Newfoundland, s. to FL, w. to CA, n. to Manitoba
state frequency: very rare
habitat: damp woods, bogs
notes: This species is ranked S1 by the Maine Natural Areas Program.

Malaxis unifolia Michx.
synonym(s): *Microstylis unifolia* (Michx.) B. S. P.
common name(s): green adder's mouth
habit: slender, erect, perennial herb
range: Newfoundland, s. to FL, w. to TX, n. to Saskatchewan
state frequency: uncommon
habitat: damp woods, swamps, bogs
notes: —

Platanthera
(12 species)

Hybrids occur between closely related species.

KEY TO THE SPECIES

1a. Labellum simple, with entire, toothed, or conspicuously fringed margins
   2a. Stems scapose, without leaves or with 1–5 minute, bract-like leaves; leaves all basal; pollen sacs divergent
      3a. Spur 13.0–45.0 mm long; scape with 0–5 bract-like leaves; basal leaves 2.5–25.0 cm wide, broad-elliptic or broad-obovate to orbicular; ovary straight
         4a. Scape with 0(1) bract-like leaves, 2.0–4.0 dm tall; ovary sessile; spike 2.0–4.0 cm wide; flowers green to green-yellow; labellum upcurved .... *P. hookeri*
         4b. Scape with 1–5 bract-like leaves, 3.0–6.0 dm tall; ovary on a stipe 5.0–10.0 mm long; spike 4.0–5.0(–8.0) cm wide; flowers white to green-white; labellum downward oriented .... *P. orbiculata*
      3b. Spur 3.0–9.0 mm long; scape with 0(1) bract-like leaves; basal leaves 1.0–5.5 cm wide, oblanceolate to obovate; ovary arcuate .... *P. obtusata*
   2b. Stems with 1–several leaves; leaves not confined to the base of the plant; pollen sacs mostly parallel
      5a. Margin of the labellum entire

6a. Flowers green to green-white or green-yellow; labellum gradually widened at the base .... *P. hyperborea*

6b. Flowers white; labellum abruptly widened at the base .... *P. dilatata*

5b. Margin of the labellum toothed, erose, or conspicuously fringed

7a. Labellum conspicuously fringed, 8.0–11.0 mm long; spur 10.0–25.0 mm long .... *P. blephariglottis*

7b. Labellum with basal or apical teeth or lobes, sometimes with crenations or irregular teeth, but definitely not fringed, 3.0–7.0 mm long; spur 3.0–12.0 mm long

8a. Labellum without a protuberance, with 3 teeth or very shallow lobes at the apex; stem with 1 primary leaf; spur 7.0–12.0 mm long .... *P. clavellata*

8b. Labellum with a prominent, erect protuberance, usually with 2 basal teeth, often crenate or irregularly toothed over the margin; stem with 1–3 primary leaves; spur 3.0–6.0 mm long .... *P. flava*

1b. Labellum deeply 3-lobed, with erose or conspicuously fringed margins

9a. Flowers green-white, yellow-white, yellow-green, or white (sometimes tinged with bronze or rose); margin of the labellum conspicuously long-fringed, with some fringe segments exceeding 0.5 times the length of the lobes of the labellum; labellum with cuneate to broad-cuneate lobes; largest lower leaves 1.0–3.5 cm wide; viscidulum linear to oval

10a. Sepals 4.5–7.0 mm long, the lateral ones deflexed behind the labellum; labellum 7.0–15.0 mm long, its segments cuneate; spur 1.4–2.1 cm long; lateral petals entire .... *P. lacera*

10b. Sepals 7.0–13.0 mm long, the lateral ones merely divergent; labellum 15.0–20.0 mm long, its lobes broad-cuneate; spur 2.0–4.0 cm long; lateral petals toothed .... *P. leucophaea*

9b. Flowers pink to rose-purple, very rarely white; margin of the labellum erose to short-fringed, with fringe segments up to 0.5 times the length of the lobes of the labellum; labellum with broad-flabellate lobes; largest lower leaves 2.0–9.0 cm wide; viscidulum suborbicular

11a. Labellum 7.0–12.0 x 8.0–15.0 mm; lateral petals 5.0–7.0 x 3.0–6.0 mm; lateral sepals 5.0–7.0 x 3.0–4.0 mm; segments of the fringe less than 0.35 times the total length of the lobes of the labellum; raceme mostly 1.0–3.0(–4.5) cm wide .... *P. psycodes*

11b. Labellum 10.0–18.0 x 10.0–25.0 mm; lateral petals 6.0–10.0 x 5.0–6.0 mm; lateral sepals 6.0–10.0 x 4.0–6.0 mm; segments of the fringe 0.35–0.5 times the total length of the lobes of the labellum; raceme mostly 4.0–5.0(–9.0) cm wide .... *P. grandiflora*

Platanthera blephariglottis (Willd.) Lindl.
synonym(s): *Habenaria blephariglottis* (Willd.) Hook.
common name(s): white-fringed orchid
habit: erect, perennial herb
range: Newfoundland, s. to FL, w. to TX, n. to MI and Ontario
state frequency: uncommon
habitat: *Sphagnum* bogs, swamps, wet meadows
notes: —

Platanthera clavellata (Michx.) Luer
synonym(s): *Habenaria clavellata* (Michx.) Spreng., *Habenaria clavellata* (Michx.) Spreng. var.
   ophioglossoides Fern.
common name(s): green woodland orchid, club-spur orchid
habit: slender, erect, perennial herb
range: Newfoundland, s. to FL, w. to TX, n. to MN and Ontario
state frequency: uncommon
habitat: wet woods, thickets, shores
notes: —

Platanthera dilatata (Pursh) Lindl. *ex* Beck
synonym(s): *Habenaria dilatata* (Pursh) Hook.
common name(s): leafy white orchid, tall white bog orchid, bog candle
habit: tall, erect, perennial herb
range: Greenland, s. to NJ, w. to NM and CA, n. to AK
state frequency: uncommon
habitat: bogs, wet meadows and woods
notes: Known to hybridize with *Platanthera hyperborea*.

Platanthera flava (L.) Lindl. var. *herbiola* (R. Br. *ex* Ait. f.) Luer
synonym(s): *Habenaria flava* (L.) R. Br. var. *herbiola* (R. Br. *ex* Ait. f.) Ames & Correll
common name(s): pale green orchid, tubercled orchid
habit: slender, erect, perennial herb
range: New Brunswick, s. to FL, w. to TX, n. to MN and Ontario
state frequency: rare
habitat: swampy woods, wet shores
notes: This species is ranked S2 by the Maine Natural Areas Program.

Platanthera grandiflora (Bigelow) Lindl.
synonym(s): *Habenaria fimbriata* (Ait.) R. Br. *ex* Ait. f.
common name(s): large purple-fringed orchid
habit: erect, perennial herb
range: Newfoundland, s. to DE and mountains of NC, w. to TN, n. to Ontario
state frequency: uncommon
habitat: wet thickets, meadows, and rich woods
notes: Known to hybridize with *Platanthera psycodes* (*q.v.*), which it closely resembles.

Platanthera hookeri (Torr. *ex* Gray) Lindl.
synonym(s): *Habenaria hookeri* Torr. *ex* Gray
common name(s): Hooker's orchid
habit: erect, perennial herb
range: Newfoundland, s. to NJ, w. to IA, n. to MN and Manitoba
state frequency: uncommon
habitat: dry to moist woods
notes: —

Platanthera hyperborea (L.) Lindl. var. *huronensis* (Nutt.) Luer
synonym(s): *Habenaria hyperborea* (L.) R. Br. *ex* Ait. f. var. *huronensis* (Nutt.) Farw.
common name(s): leafy northern green orchid, tall northern bog orchid
habit: erect, perennial herb
range: Greenland, s. to NJ, w. to NM and CA, n. to AK
state frequency: uncommon
habitat: bogs, wet woods, thickets, meadows, and ditches
notes: Known to hybridize with *Platanthera dilatata*.

Platanthera lacera (Michx.) G. Don
synonym(s): *Habenaria lacera* (Michx.) R. Br.
common name(s): ragged orchid, ragged-fringed orchid
habit: erect, perennial herb
range: Newfoundland, s. to FL, w. to TX, n. to Manitoba
state frequency: uncommon
habitat: moist woods, meadows, and roadsides
notes: Hybrids with *Platanthera psycodes*, called *P.* ×*andrewsii* (M. White) Luer, occur in Maine.

Platanthera leucophaea (Nutt.) Lindl.
synonym(s): *Habenaria leucophaea* (Nutt.) Gray
common name(s): prairie white-fringed orchid
habit: erect, perennial herb
range: Nova Scotia and ME, s. to LA, w. to OK, n. to Ontario
state frequency: very rare
habitat: circumneutral fens
notes: This species is ranked S1 by the Maine Natural Areas Program.

Platanthera obtusata (Banks *ex* Pursh) Lindl.
synonym(s): *Habenaria obtusata* (Banks *ex* Pursh) Richards.
common name(s): blunt-leaf orchid, one-leaf rein orchid
habit: slender, erect, perennial herb
range: circumboreal; s. to MA, w. to CO and OR
state frequency: uncommon
habitat: moist bogs and woods, often coniferous
notes: —

Platanthera orbiculata (Pursh) Lindl.
synonym(s): <*P. o.* var. *macrophylla*> = *Habenaria macrophylla* Goldie; <*P. o.* var. *orbiculata*> = *Habenaria orbiculata* (Pursh) Torr.
common name(s): large round-leaved orchid
habit: erect, perennial herb
range: Labrador, s. to NJ and mountains of SC, w. to OR, n. to AK
state frequency: uncommon
habitat: dry to moist woods
key to the subspecies:
    1a. Spur 0.8–2.7 cm long; labellum 1.0–1.5 cm long; capsules 1.5–2.0 cm long; upper sepal suborbicular; lateral sepals drooping to reflexed .... *P. o.* var. *orbiculata*
    1b. Spur 3.0–4.5 cm long; labellum 1.5–2.2 cm long; capsules 1.8–2.5 cm long; upper sepal ovate; lateral sepals ascending .... *P. o.* var. *macrophylla* (Goldie) Luer
notes: —

Platanthera psycodes (L.) Lindl.
synonym(s): *Habenaria psycodes* (L.) Spreng.
common name(s): small purple-fringed orchid, soldier's plume
habit: stout, erect, perennial herb
range: Newfoundland, s. to MD and mountains of GA, w. to NE, n. to Manitoba
state frequency: uncommon
habitat: wet woods, borders, meadows, roadsides, and rocky shores
notes: Known to hybridize with *Platanthera grandiflora* and *P. lacera* (*q.v.*). Where *P. psycodes* grows adjacent to the similar *P. grandiflora*, the former normally flowers 10–15 days later.

Pogonia
(1 species)

Pogonia ophioglossoides (L.) Ker-Gawl.
synonym(s): —
common name(s): rose pogonia, snakemouth
habit: erect, short-rhizomatous, perennial herb
range: Newfoundland, s. to FL, w. to TX, n. to MN and Ontario
state frequency: uncommon
habitat: *Sphagnum* bogs, wet meadows
notes: —

Spiranthes
(5 species)

The inflorescence in this genus is a raceme composed of flowers that gradually decrease in size from the base to the apex. Measurements of sepals and petals need to be taken from the lower third of the raceme. Hybrids do occur in this genus and should be sought in mixed populations.

KEY TO THE SPECIES

1a. Inflorescence spirally secund, the flowers not so crowded as to obscure the single, long spiral; labellum white with a large, central, green spot .... *S. lacera*
1b. Inflorescence crowded, the spiral obscured, the flowers appearing in 3 or 4 vertical ranks; labellum white with yellow, yellow-white, green, or green-yellow markings
  2a. Sepals and lateral petals connivent, thereby forming a tubular hood; labellum pandurate, strongly constricted near the middle and dilated to the apex ....
.... *S. romanzoffiana*
  2b. Only the upper sepal and lateral petals connivent, the lateral sepals free and not forming part of a tubular hood; labellum not conspicuously pandurate
    3a. Labellum bright yellow near the apex, oblong; perianth 5.0–6.0 mm long; sepals connate 0.6–0.8 mm at their base; basal auricles of the labellum shorter than 1.0 mm and glabrous; viscidulum 0.3–0.6 mm long; leaves oblanceolate to oblong; plants flowering late June–mid-July .... *S. lucida*
    3b. Labellum with green-yellow or yellow-white markings, narrow-ovate to ovate-oblong; perianth 8.0–12.0 mm long; sepals distinct or nearly so; basal auricles of the labellum longer than 1.0 mm and pubescent; viscidulum 1.0–2.0 mm long; leaves linear-lanceolate to narrow-oblanceolate; plants flowering late August–late September
      4a. Labellum usually with yellow-white markings, strongly decurved, narrow-ovate; basally narrowed portion of the labellum 0.8–1.5 mm long ....
.... *S. ochroleuca*
      4b. Labellum usually with green-yellow markings, moderately arcuate-recurved, ovate to ovate-oblong; basally narrowed portion of the labellum 0.3–0.8 mm long .... *S. cernua*

Spiranthes cernua (L.) L. C. Rich.
synonym(s): *Ibidium cernuum* (L.) House
common name(s): common ladies' tresses, nodding ladies' tresses
habit: erect, simple, perennial herb

range: Nova Scotia, s. to FL, w. to TX, n. to ND and Ontario
state frequency: uncommon
habitat: moist, sandy areas, bogs
notes: —

Spiranthes lacera (Raf.) Raf.
synonym(s): <*S. l.* var. *gracilis*> = *Ibidium gracile* (Bigelow) House, *Spiranthes gracilis* (Bigelow) Beck
common name(s): slender ladies' tresses
habit: slender, erect, simple, perennial herb
range: New Brunswick, s. to FL, w. to TX, n. to MN and Saskatchewan
state frequency: uncommon
habitat: sandy thickets, clearings, and open areas
key to the subspecies:
    1a. Leaves persistent during anthesis, thin and nearly translucent; plants flowering in July–early August; inflorescence less crowded or with only a few spirals; sepals and lateral petals forming a slender tube 1.0–1.75(–2.2) mm in diameter; apex of the labellum with a thick, white border .... *S. l.* var. *lacera*
    1b. Leaves senesced prior to anthesis, thick and opaque; plants flowering in August; inflorescence more crowded, with several spirals; sepals and lateral petals forming a tube 1.5–2.5 mm in diameter; apex of the labellum with a thin, white border .... *S. l.* var. *gracilis* (Bigelow) Luer
notes: *Spiranthes lacera* var. *gracilis* is ranked SH by the Maine Natural Areas Program. Hybrids with *S. romanzoffiana* occur in Maine.

Spiranthes lucida (H. H. Eat.) Ames
synonym(s): *Ibidium plantagineum* (Raf.) House
common name(s): shining ladies' tresses, broad-leaved ladies' tresses
habit: slender, erect, simple, perennial herb
range: New Brunswick, s. to NJ and mountains of NC, w. to NE, n. to MN and Ontario
state frequency: very rare
habitat: damp thickets, slopes, meadows, and shores, marshes
notes: This species is ranked S1 by the Maine Natural Areas Program.

Spiranthes ochroleuca (Rydb.) Rydb.
synonym(s): *Spiranthes cernua* (L.) L. C. Rich. var. *ochroleuca* (Rydb.) Ames
common name(s): —
habit: erect, simple, perennial herb
range: Nova Scotia, s. to mountains of NC, w. to KY, n. to MI
state frequency: rare
habitat: rocky slopes, open woods, fields
notes: —

Spiranthes romanzoffiana Cham.
synonym(s): *Ibidium strictum* (Rydb.) House
common name(s): hooded ladies' tresses, Romanzoff's ladies' tresses
habit: erect, simple, perennial herb
range: Labrador, s. to MA, w. to CA, n. to AK
state frequency: uncommon
habitat: swamps, bogs, wet thickets, meadows, fields, and shores
notes: Hybrids with *Spiranthes lacera* occur in Maine.

Triphora
(1 species)

Triphora trianthophora (Sw.) Rydb.
synonym(s): —
common name(s): nodding-pogonia, three birds orchid
habit: small, nodding, perennial herb
range: ME, s. to FL, w. to TX, n. to WI
state frequency: very rare; found predominantly in the western portion of the state
habitat: hardwood forests and slopes, especially *Fagus*
notes: This species is ranked S1/S2 by the Maine Natural Areas Program.

PONTEDERIACEAE
(2 genera, 2 species)

KEY TO THE GENERA

1a. Leaves linear, without a midrib or petiole, narrow at the base; flowers pale yellow, with 3
    stamens, all exserted from the perianth; anthers basifixed; fruit with 7–30 seeds ....
    .... ***Zosterella***
1b. Leaf blades lanceolate to deltate-ovate, with a midrib and petiole, cordate at the base;
    flowers blue-purple (white), with 6 stamens—3 exserted and 3 included within the
    perianth; anthers versatile; fruit with 1 seed .... ***Pontederia***

Zosterella
(1 species)

Zosterella dubia (Jacq.) Small
synonym(s): *Heteranthera dubia* (Jacq.) MacM.
common name(s): water stargrass
habit: submersed to emersed or stranded, aquatic, perennial herb
range: New Brunswick, s. to FL, w. to CA, n. to OR and ID
state frequency: very rare
habitat: streams, quiet water, mud flats
notes: This species is ranked S1 by the Maine Natural Areas Program.

Pontederia
(1 species)

Pontederia cordata L.
synonym(s): —
common name(s): pickerelweed
habit: erect, emergent, rhizomatous, perennial herb
range: New Brunswick, s. to FL, w. to OK, n. to MN and Ontario
state frequency: occasional
habitat: marshes, shallow water, muddy shores
notes: —

COMMELINACEAE
(2 genera, 3 species)

KEY TO THE GENERA

1a. Corolla weakly zygomorphic—one of the petals usually smaller than the other 2; inflorescence subtended by a folded spathe; flowers with 3 fertile stamens and usually 3 sterile stamens with cruciform anthers .... *Commelina*

1b. Corolla actinomorphic; inflorescence subtended by leaves; flowers with 6 fertile stamens .... *Tradescantia*

Commelina
(1 species)

Commelina communis L.
synonym(s): —
common name(s): common dayflower
habit: erect to decumbent, simple or branched, annual herb
range: Asia; naturalized to ME, s. to GA, w. to TX, n. to ND
state frequency: rare
habitat: roadsides, ditches, gardens
notes: —

Tradescantia
(2 species)

KEY TO THE SPECIES

1a. Pedicels pubescent; sepals 10.0–15.0 mm long, pubescent .... *T. virginiana*

1b. Pedicels glabrous; sepals 8.0–12.0 mm long, glabrous or pubescent near the apex .... *T. ohiensis*

Tradescantia ohiensis Raf.
synonym(s): —
common name(s): smooth spiderwort
habit: slender, often branched, perennial herb
range: MA, s. to FL, w. to TX, n. to MN; escaped from cultivation
state frequency: very rare
habitat: thickets, meadows
notes: —

Tradescantia virginiana L.
synonym(s): —
common name(s): Virginia spiderwort
habit: erect to ascending, perennial herb
range: CT, s. to GA, w. to MO, n. to MN; escaped from cultivation to ME, s. to MA
state frequency: rare
habitat: woods, thickets, meadows, roadsides
notes: —

TYPHACEAE
(2 genera, 9 species)

KEY TO THE GENERA

1a. Inflorescence spherical; fruit an achene with a persistent stigma beak; leaves 2.0–12.0 mm wide, flaccid to firm, parallel-veined and checkered with rectangular areoles; stems rarely exceeding 1.0 m in height .... *Sparganium*
1b. Inflorescence cylindric; fruit a wind dispersed follicle; leaves 5.0–23.0 mm wide, firm, parallel-veined; stems usually exceeding 1.0 m in height .... *Typha*

Sparganium
(7 species)

The leaves are variable in cross-section depending on the water level; submersed and floating leaves tend to be thinner, with a faint keel on the abaxial surface. Emersed leaves are thicker and have a conspicuous keel. Submersed leaves may resemble the leaves of *Vallisneria americana* but are easily separable. *Vallisneria* have a central band of lacunae whereas *Sparganium* are checkered throughout with rectangular areoles. The staminate spikes often disarticulate before maturation of the carpels. Their numbers can still be determined late in the season by examining the axis of the inflorescence above the most distal carpellate spike. There is a visible node where they were attached.

KEY TO THE SPECIES

1a. Flowers and fruits with 2 stigmas; ovary with 2 locules; achenes broad-obpyramidal, truncate at the apex, 4.0–8.0 mm wide .... *S. eurycarpum*
1b. Flowers and fruits with 1 stigma; ovary with 1 locule; achenes fusiform to ellipsoid, tapering to the apex, up to 4.0 mm wide
    2a. Flowering stems with 1 staminate spike; fruiting spikes 0.8–1.2 cm in diameter; beak of achene 0.5–1.5 mm long .... *S. nutans*
    2b. Flowering stems with 2–20 staminate spikes; fruiting spikes 1.2–3.5 cm in diameter; beak of achene 0.5–6.0 mm long
        3a. Anthers and stigmas oblong to ovoid, 0.4–0.8 mm long; tepals inserted on the basal half of the stipe; beak of mature fruit flattened and strongly curved .... *S. fluctuans*
        3b. Anthers and stigmas linear, 0.6–4.0 mm long; tepals inserted mostly on the apical half of the stipe; beak of mature fruit terete, slightly curved
            4a. Both the sessile carpellate spikes of the main axis and the peduncle bases of the lateral branches borne directly in the axils of leaves or bracts
                5a. Fruiting spikes 2.5–3.5 cm in diameter; stigmas 2.0–3.2 mm long; achenes usually lustrous, with a beak 4.5–6.0 mm long; anthers 1.0–1.6 mm long .... *S. androcladum*
                5b. Fruiting spikes 1.5–2.5 cm in diameter; stigmas 0.8–1.5(–2.0) mm long; achenes dull, with a beak 1.5–5.0 mm long; anthers 0.8–1.2 mm long .... *S. americanum*
            4b. Some or all of the sessile carpellate spikes of the main axis and/or the peduncle bases of the lateral branches borne above the axil of leaves or bracts

6a. Achene green near the base, with a beak 2.0–4.3 mm long, the beak nearly as long as the body of the achene; stigmas 2.0–4.3 mm long; staminate portion of the inflorescence usually 4.0–10.0 cm long; stems and leaves usually erect and emersed; leaves up to 85.0 cm long .... *S. erectum*

6b. Achene red near the base, with a beak 0.5–2.0 mm long, the beak much shorter than the body of the achene; stigmas 0.6–1.5 mm long; staminate portion of the inflorescence usually 1.0–4.0 cm long; stems and leaves usually flaccid, floating; leaves longer, up to 120.0 cm long .... *S. angustifolium*

Sparganium americanum Nutt.
synonym(s): —
common name(s): —
habit: stout to slender, erect, rhizomatous, perennial herb
range: Newfoundland, s. to FL, w. to TX, n. to ND and Ontario
state frequency: occasional
habitat: muddy or peaty shores, shallow water
notes: —

Sparganium androcladum (Engelm.) Morong
synonym(s): *Sparganium lucidum* Fern. & Eames
common name(s): —
habit: stout, erect, rhizomatous, perennial herb
range: Quebec, s. to VA; also northern midwest states
state frequency: very rare
habitat: muddy shores, shallow water
notes: —

Sparganium angustifolium Michx.
synonym(s): *Sparganium chlorocarpum* Rydb. var. *acaule* (Beeby) Fern., *Sparganium emersum* Rehmann, *Sparganium multipedunculatum* (Morong) Rydb.
common name(s): —
habit: slender, rhizomatous, submersed, aquatic, perennial herb
range: Newfoundland, s. to NJ, w. to CA, n. to AK
state frequency: uncommon
habitat: deep or shallow water, wet shores
notes: —

Sparganium erectum L.
synonym(s): *Sparganium chlorocarpum* Rydb.
common name(s): —
habit: slender, erect, rhizomatous, perennial herb
range: Newfoundland, s. to WV, w. to IA, n. to Ontario; also ID
state frequency: uncommon
habitat: swamps, muddy or peaty shores, shallow water
notes: —

Sparganium eurycarpum Engelm. *ex* Gray
synonym(s): —
common name(s): giant bur-reed
habit: stout, erect, rhizomatous, perennial herb

range: Newfoundland, s. to VA, w. to CA, n. to British Columbia
state frequency: uncommon
habitat: muddy shores, shallow water
notes: —

Sparganium fluctuans (Morong) B. L. Robins.

synonym(s): —
common name(s): —
habit: slender, rhizomatous, submersed, aquatic, perennial herb
range: Newfoundland, s. to CT, w. to PA and MN, n. to British Columbia
state frequency: uncommon
habitat: cold lakes and ponds
notes: —

Sparganium nutans L.

synonym(s): *Sparganium minimum* (Hartman) Wallr.
common name(s): —
habit: very slender, rhizomatous, submersed to emergent, perennial herb
range: circumboreal; s. to NJ, w. to NM
state frequency: rare
habitat: shallow water of ponds or brooks
notes: —

Typha
(2 species)

KEY TO THE SPECIES

1a. Leaves 5.0–11.0 mm wide; carpellate portion and staminate portion of the inflorescence
usually separated by a distance of more than 5.0 mm, the carpellate portion light red-
brown, mostly 1.0–2.0 cm wide at maturity; stigmas linear; stems 0.75–1.5 m tall; plants
mostly of tidal shores .... *T. angustifolia*
1b. Leaves (8.0–)10.0–23.0 mm wide; carpellate portion and staminate portion of the
inflorescence normally contiguous, not separated by more than 5.0 mm, the carpellate
portion dark brown, mostly 2.0–3.0 cm wide at maturity; stigmas broader, spatulate;
stems 1.0–3.0 m tall; plants of inland, mostly non-tidal, shores .... *T. latifolia*

Typha angustifolia L.

synonym(s): —
common name(s): narrow-leaved cat-tail
habit: tall, erect, simple, perennial herb
range: nearly cosmopolitan; in eastern North America from Nova Scotia, s. to SC
state frequency: uncommon
habitat: coastal marshes
notes: Hybrids with *Typha latifolia*, called *T.* ×*glauca* Godr., occur in Maine and can be identified by
their separate carpellate and staminate portions of the inflorescence [the length of the separation
intermediate between the lengths of the parental species], mature carpellate spikes 1.6–2.0 cm
wide, and leaves 6.5–11.0 mm wide. However, they are larger than either parent, with stems
2.0–3.5 m tall and carpellate portions of the spikes measuring 18.0–50.0 cm long.

Typha latifolia L.
synonym(s): —
common name(s): common cat-tail, broad-leaved cat-tail
habit: tall, erect, simple, perennial herb
range: Newfoundland, s. to FL, w. to CA, n. to AK
state frequency: common
habitat: marshes, shallow water
notes: Known to hybridize with *Typha angustifolia* (*q.v.*).

XYRIDACEAE
(1 genus, 3 species)

Xyris
(3 species)

Flowers of *Xyris* have 3 sepals. The upper sepal is markedly different from the other 2. It is larger, thin, somewhat encloses the petals, and is deciduous with the maturation of the capsule. The other 2 sepals, called lateral sepals, are smaller, scarious, keeled, and persistent in fruit.

KEY TO THE SPECIES
1a. Floral scales lacking a central green portion; leaves 1.0–2.5 mm wide, commonly less than 2.0 mm; keel of lateral sepals entire or nearly so; inflorescence narrow-ellipsoid; seeds 0.8–1.0 mm long .... *X. montana*
1b. Floral scales with a central green portion; leaves 0.5–10.0 mm wide, usually the larger exceeding 2.0 mm; keel of lateral sepals lacerate; inflorescence broad-ellipsoid; seeds 0.5–0.7 mm long
    2a. Apex of lateral sepals somewhat concealed by the subtending floral bracts; seeds *ca.* 0.5 mm long; leaves commonly less than 3.0 mm wide; inflorescence 0.5–1.5 cm tall .... *X. difformis*
    2b. Apex of lateral sepals visible, exceeding the subtending floral bracts; seeds 0.6–0.7 mm long; leaves 3.0–10.0 mm wide; inflorescence 1.0–2.5 cm tall .... *X. smalliana*

Xyris difformis Chapman
synonym(s): *Xyris caroliniana* Walt. *sensu* Fernald (1950)
common name(s): yellow-eyed-grass
habit: erect, tufted, perennial herb
range: ME, s. to FL, w. to TX, n. to WI
state frequency: uncommon
habitat: damp to wet peat and sand, river swamps
notes: —

Xyris montana Ries
synonym(s): —
common name(s): yellow-eyed-grass
habit: erect, densely tufted, perennial herb
range: Newfoundland, s. to NJ, w. to PA, n. to MN and Ontario
state frequency: uncommon
habitat: wet peat and sand, *Sphagnum* bogs
notes: —

Xyris smalliana Nash
synonym(s): *Xyris congdonii* Small
common name(s): yellow-eyed-grass
habit: erect, coarse, tufted, perennial herb
range: ME, s. to FL, w. to MS
state frequency: very rare
habitat: peaty shores, and wet, low areas, usually coastal
notes: This species is ranked S1 by the Maine Natural Areas Program.

ERIOCAULACEAE
(1 genus, 2 species)

Eriocaulon
(2 species)

The conspicuously septate roots and rosette of tapering, pointed leaves allow this genus to be identified easily in the vegetative state.

KEY TO THE SPECIES

1a. Inflorescence subglobose, gray to white, 4.0–5.0 mm wide; bractlets evidently pubescent with clavate hairs; flowering stem (4)5- to 7-ridged; plants of fresh water habitats .... *E. aquaticum*
1b. Inflorescence hemispherical, gray to light yellow-brown, 3.0–4.0 mm wide; bractlets sparsely pubescent with clavate hairs; flowering stem 4- or 5-ridged; plants of tidal shores .... *E. parkeri*

Eriocaulon aquaticum (Hill) Druce
synonym(s): *Eriocaulon septangulare* Withering
common name(s): pipewort, white buttons
habit: scapose, submerged to emergent, aquatic, perennial herb
range: Newfoundland, s. to VA and mountains of NC, w. to IN, n. to MN and Ontario
state frequency: occasional
habitat: shallow water of ponds and streams
notes: —

Eriocaulon parkeri B. L. Robins.
synonym(s): —
common name(s): Parker's pipewort
habit: scapose, submerged to emergent, aquatic, perennial herb
range: Quebec, s. along the coast to NC
state frequency: rare
habitat: estuaries, tidal flats
notes: This species is ranked S3 by the Maine Natural Areas Program.

JUNCACEAE
(2 genera, 35 species)

KEY TO THE GENERA

1a. Plants glabrous; capsules uni- to trilocular; seeds many per capsule, lacking a caruncle; stamens 3 or 6 .... *Juncus*

1b. Plants pubescent, at least when young; capsules unilocular; seeds 3 per capsule, with a caruncle; stamens 6 .... *Luzula*

Juncus
(28 species)

Plants are most easily identified when capsules are mature, but this is not essential for many of the species. Stamen remnants [usually the filaments] still exist in fruiting individuals and are found behind the tepals. Species with 3 stamens will have stamen remnants behind only the outer 3 tepals. Species with 6 stamens will have remnants behind both the outer and inner tepals. Therefore, it is necessary to search only behind the inner tepals because absence of stamens indicates a 3-stamen species, while presence indicates a 6-stamen species.

KEY TO THE SPECIES

1a. Flowers bracteolate, each flower with a pair of bracts at the base of the pedicel and an additional pair of bracteoles [called prophylls] subtending the flower; flowers all pedicellate; leaves various but never septate

  2a. Involucral bract with similar cross-sectional shape as the culm and seeming to be a continuation of it, giving the appearance of a lateral inflorescence; leaves bladeless

    3a. Stamens 6; anthers 3.0–5.0 times the length of the filaments; tepals with a conspicuous dark stripe on each side of the mid-vein; stems smooth or irregularly wrinkled; plants mostly of saline habitats .... *J. arcticus*

    3b. Stamens 3 or 6; anthers 0.3–0.5 times the length of the filaments; tepals lacking conspicuous dark stripes; stems coarsely to finely grooved; plants mostly of non-saline habitats

      4a. Stamens 3; stems 1.0–5.0 mm thick above basal sheaths; inflorescence commonly with more than 20 flowers; placentation axile; involucral bract commonly less than 0.5 times the length of the stem .... *J. effusus*

      4b. Stamens 6; stems 0.5–1.0 mm thick above basal sheaths; inflorescence with fewer than 20 flowers, commonly 8 or fewer; placentation parietal; involucral bract commonly more than 0.5 times the length of the stem .... *J. filiformis*

  2b. Involucral bract flat in cross-section, not appearing as a continuation of the stem and the inflorescence evidently terminal; leaves with well formed blades

    5a. Plants annual; inflorescence occupying 0.35–0.90 of the height of the plant; leaf sheaths lacking auricles

      6a. Inner tepals [petals] acute to acuminate; capsules exceeded by the inner tepals, mostly acute to subacute (truncate); inflorescence relatively open .... *J. bufonius*

      6b. Inner tepals rounded to acute; capsules equaling or exceeding the inner tepals, mostly truncate at the apex; inflorescence contracted .... *J. ambiguus*

5b. Plants perennial; inflorescence occupying less than 0.25 of the height of plant; leaf sheaths with auricles
    7a. Auricles erose; inflorescence with 1–3 flowers; capsules with a beak up to 0.7 mm long; margin of the leaf blade minutely serrulate .... *J. trifidus*
    7b. Auricles entire; inflorescence with many flowers; capsules beakless; margin of the leaf blade entire
        8a. Leaves terete or caniculate; tepals appressed, shorter than the capsules
            9a. Capsules 4.1–5.4 mm long, golden brown; seeds with distinct, slender, slightly curved, white appendages 0.2–0.4 mm long; involucral bract up to 5.0(7.0) cm long, usually not exceeding the inflorescence; inner tepals 3.2–4.0 mm long .... *J. vaseyi*
            9b. Capsules 3.0–4.0 mm long, dark brown; seeds with short, blunt, white appendages less than 0.2 mm long; involucral bract 2.0–15.0(–20.0) cm long, usually taller than the inflorescence; inner tepals 1.9–3.4 mm long .... *J. greenei*
        8b. Leaves flat or involute (terete in some *J. dichotomus*); tepals ascending to spreading, exceeding the length of the capsules (except *J. compressus*)
            10a. Leaves not confined to the base of the stem, some borne on the upper half of the stem; stems arising singly from elongate rhizomes; tepals with red-purple margins, the outer [sepals] with an obtuse (acute) apex; plants mostly of saline habitats
                11a. Stamens 1.5–2.2 mm long; anthers 2.0–6.0 [mean = 3.0] times as long as the filaments; capsules shorter than to equaling the length of the tepals; lowest bract 1.0–5.0 cm long, usually not surpassing the inflorescence .... *J. gerardii*
                11b. Stamens 0.8–1.0 mm long; anthers less than 2.0 [mean = 1.0] times as long as the filaments; capsules usually longer than the sepals; lowest bract 2.0–7.5(–13.0) cm long, usually surpassing the inflorescence .... *J. compressus*
            10b. Leaves mostly basal, borne on only the lower third of the stem; stems cespitose from short rhizomes; tepals green to brown at the margin, the outer with an acuminate apex; plants mostly of non-saline habitats
              12a. Placentation axile [*i.e.*, partitions of capsule extending to center and meeting]; leaves not extending past midpoint of stem .... *J. secundus*
              12b. Placentation parietal [*i.e.*, partitions of capsule not extending to center]; leaves extending past midpoint of stem
                13a. Auricles scarious, 1.0–5.0 mm long; sheath margin pliable
                    14a. Inflorescence with widely scattered flowers; branches of the inflorescence erect to ascending, often inwardly curved; ultimate branches of the cyme mostly 3.0–5.0 cm long; tepals 2.5–3.5(–4.0) mm long; capsules less than 0.75 times as long as the tepals; plants mostly 7.0–9.0 dm tall ... *J. anthelatus*
                    14b. Inflorescence with crowded to remote flowers; branches of the inflorescence erect to divergent; ultimate branches of

the cyme 1.0–2.0 cm long; tepals (2.8–)3.5–4.5 mm long; capsules greater than 0.75 times as long as the tepals; plants usually shorter than 7.0 dm .... *J. tenuis*

13b. Auricles scarious to coriaceous, 0.2–0.6 mm long; sheath margin brittle

15a. Capsules dark brown; inner leaf sheath pink; leaves flat or terete; auricles membranaceous or cartilaginous, usually green to brown .... *J. dichotomus*

15b. Capsules light brown; inner leaf sheath brown; leaves flat; auricles cartilaginous, yellow .... *J. dudleyi*

1b. Flowers lacking bracteoles, with only the pair of bracts at the base of the pedicel; flowers in glomerules (except *J. pelocarpus* and *J. subtilis*); leaves terete and septate (except *J. marginatus* and *J. stygius*)

16a. Capsules 6.0–7.0 mm long; seeds 3.0–3.5 mm long; leaves terete, lacking septa .... *J. stygius*

16b. Capsules 1.5–5.7 mm long; seeds 0.5–1.9 mm long; leaves both terete and septate or flat

17a. Leaves flat, without septa; capsules globose, with red speckles ... *J. marginatus*

17b. Leaves terete and septate; capsules oblong to ellipsoid, without red speckles

18a. Leaves dimorphic—those of the rhizome submersed and capillary, those of the upright stem emergent and thicker, 4.0–6.0 mm in diameter; inflorescence exceeded by the lowest cauline leaf .... *J. militaris*

18b. Leaves monomorphic, no capillary leaves produced on the rhizome, 0.5–1.5 mm in diameter, the lowest not exceeding the inflorescence

19a. Flowers occurring singly on a pedicel or in glomerules of 2, sometimes replaced by vegetative propagules [bulbils]

20a. Stems creeping or floating; flowers 1 or 2 per plant; outer tepals [sepals] 2.2–2.8 mm long .... *J. subtilis*

20b. Stems upright; flowers many per plant; outer tepals 1.6–2.3 mm long .... *J. pelocarpus*

19b. Flowers occurring in glomerules of 2 or more, not replaced by vegetative propagules

21a. Seeds with tail-like appendages, the appendages commonly white; leaf sheaths with ribs that curve near the summit; flowers usually with 3 stamens

22a. White appendages of seeds 0.1 times the length of the seed body; outer tepals obtuse to subacute; capsules 2.4–3.8 mm long .... *J. brachycephalus*

22b. White appendages of seeds 0.3–0.7 times the length of the seed body; outer tepals acuminate; capsules 3.3–4.8 mm long

23a. Inflorescence strictly ascending; white appendages of the seed 0.3–0.5 times the length of the seed body; capsules noticeably exceeding the tepals; glomerules with 2–5 flowers .... *J. brevicaudatus*

23b. Inflorescence ascending to spreading; white appendages of the seed 0.5–0.7 times the length of the seed body; capsules

equaling or slightly exceeding the tepals; glomerules with
5–50 flowers .... *J. canadensis*
21b. Seeds without tail-like appendages; leaf sheaths with nerves that do
not curve at the summit; flowers with 6 stamens (3 in *J. acuminatus*)
24a. Stamens 3; auricles 1.0–1.5 mm long .... *J. acuminatus*
24b. Stamens 6; auricles up to 1.0 mm long (2.0–5.0 mm long in *J. torreyi*)
25a. Glomerules obpyramidal to hemispherical; valves of capsules
separate after dehiscence; principal involucral bract usually
shorter than the inflorescence
26a. Capsules acute to rounded, with a mucronate tip; outer
tepals obtuse to acute at the apex; inflorescence strictly
ascending .... *J. alpinoarticulatus*
26b. Capsules acute to obtuse; outer tepals acute to acuminate;
inflorescence spreading and open .... *J. articulatus*
25b. Glomerules spherical; valves of capsules cohering at the
summit after dehiscence; principal involucral bract usually
longer than the inflorescence
27a. Auricles 0.5–1.0 mm long, cartilaginous, yellow;
glomerules composed of 6–20 flowers; inner and outer
tepals of same length .... *J. nodosus*
27b. Auricles 2.0–5.0 mm long, scarious, light brown;
glomerules composed of 25–100 flowers; inner tepals
shorter than outer tepals .... *J. torreyi*

Juncus acuminatus Michx.

synonym(s): —
common name(s): sharp-fruited rush
habit: cespitose, perennial herb
range: Nova Scotia and ME, s. to FL, w. to TX and Mexico, n. to MN and Ontario; also British
Columbia, s. to OR
state frequency: uncommon
habitat: damp meadows and shores, wet woods
notes: This species has glomerules with 10–50 flowers.

Juncus alpinoarticulatus Chaix ex Vill.

synonym(s): *Juncus alpinoarticulatus* Chaix *ex* Vill. ssp. *nodulosus* (Wahlenb.) Hämet-Ahti, *Juncus alpinus* Vill., *Juncus alpinus* Vill. var. *rariflorus* Hartman
common name(s): alpine rush
habit: ascending to erect, perennial herb
range: circumboreal; s. in the east to PA
state frequency: rare
habitat: wet meadows and shores, both fresh and tidal
notes: This species is ranked S2 by the Maine Natural Areas Program. Known to hybridize with
Juncus articulatus, J. brevicaudatus, J. nodosus, and *J. torreyi.*

Juncus ambiguus Guss.

synonym(s): *Juncus bufonius* L. var. *halophilus* Buch. & Fern.
common name(s): —

habit: fleshy, simple or branched, annual herb
range: Labrador, s. to CT, w. to CO, n. to Saskatchewan
state frequency: very rare
habitat: Atlantic coast shores
notes: —

Juncus anthelatus (Wieg.) R. E. Brooks

synonym(s): *Juncus tenuis* Willd. var. *anthelatus* Wieg.
common name(s): —
habit: slender, cespitose, perennial herb
range: Quebec and New Brunswick, s. to GA, w. to TX, n. to WI
state frequency: uncommon
habitat: swamps, wet ditches
notes: Similar to *Juncus tenuis*, but growing in wetter habitats and flowering 10–15 days earlier.

Juncus arcticus Willd. var. *balticus* (Willd.) Trautv.

synonym(s): *Juncus arcticus* Willd. var. *littoralis* (Engelm.) Boivin, *Juncus balticus* Willd. var. *littoralis* Engelm.
common name(s): wire rush
habit: simple, long-rhizomatous, perennial herb
range: circumboreal; s. to PA, w. to MO and British Columbia
state frequency: uncommon; more common in coastal areas
habitat: brackish to fresh shores
notes: —

Juncus articulatus L.

synonym(s): *Juncus articulatus* L. var. *obtusatus* Engelm.
common name(s): jointed rush
habit: erect, loosely cespitose, rhizomatous, perennial herb
range: Newfoundland, s. to NC, w. to OR, n. to British Columbia
state frequency: uncommon
habitat: bogs, wet meadows and ditches, sandy or muddy shores
notes: Known to hybridize with *Juncus alpinoarticulatus, J. brevicaudatus, J. canadensis*, and *J. nodosus*. The glomerules are composed of 3–10 flowers in this species.

Juncus brachycephalus (Engelm.) Buch.

synonym(s): —
common name(s): —
habit: slender, spreading to ascending, densely cespitose, perennial herb
range: New Brunswick, s. to NJ, w. to IL, n. to WI and Ontario; also GA
state frequency: very rare
habitat: wet meadows, calcareous shores, marshes
notes: The glomerules are composed of 2–5 flowers in this species.

Juncus brevicaudatus (Engelm.) Fern.

synonym(s): —
common name(s): short-tailed rush
habit: slender, ascending to erect, densely cespitose, perennial herb
range: Labrador, s. to NC, w. to IL, n. to Alberta
state frequency: common
habitat: marshes, wet meadows and shores
notes: Known to hybridize with *Juncus alpinoarticulatus* and *J. articulatus*.

Juncus bufonius L.

synonym(s): —
common name(s): toad rush
habit: simple or branched, annual herb
range: Greenland, s. to FL, w. to CA, n. to AK
state frequency: occasional
habitat: moist, open areas, sometimes of coastal shores
notes: —

Juncus canadensis J. Gay *ex* Laharpe

synonym(s): *Juncus canadensis* J. Gay *ex* Laharpe var. *sparsiflorus* Fern.
common name(s): Canada rush
habit: erect, cespitose, perennial herb
range: Newfoundland, s. to GA, w. to LA, n. to NE, MN, and Ontario
state frequency: occasional
habitat: swamps, marshes, wet shores and ditches, fresh to saline areas
notes: Known to hybridize with *Juncus articulatus*.

Juncus compressus Jacq.

synonym(s): —
common name(s): compressed rush
habit: erect, cespitose, rhizomatous, perennial herb
range: Europe; naturalized to Newfoundland, s. to NY, w. to CO and WY, n. to Manitoba
state frequency: very rare
habitat: disturbed, saline soil
notes: —

Juncus dichotomus Ell.

synonym(s): *Juncus platyphyllus* (Wieg.) Fern., *Juncus tenuis* Willd. var. *dichotomus* (Ell.) Wood
common name(s): forked rush
habit: densely cespitose, short-rhizomatous, perennial herb
range: ME, s. to FL
state frequency: uncommon
habitat: dry to moist soils
notes: —

Juncus dudleyi Wieg.

synonym(s): *Juncus tenuis* Willd. var. *dudleyi* (Wieg.) F. J. Herm., *Juncus tenuis* Willd. var. *uniflorus* Farw.
common name(s): Dudley's rush
habit: cespitose, short-rhizomatous, perennial herb
range: Newfoundland, s. to NJ and mountains of VA, w. to AZ, n. to British Columbia
state frequency: rare
habitat: damp to dry soils
notes: This species is ranked S2 by the Maine Natural Areas Program.

Juncus effusus L.

synonym(s): *Juncus effusus* L. var. *compactus auct. non* Lej. & Court., *Juncus effusus* L. var. *costulatus* Fern., *Juncus effusus* L. var. *decipiens* Buch., *Juncus effusus* L. var. *pylaei* (Laharpe) Fern. & Wieg., *Juncus effusus* L. var. *solutus* Fern. & Wieg.
common name(s): soft rush, common rush, smooth rush
habit: densely cespitose, stout-rhizomatous, perennial herb
range: Newfoundland, s. to FL, w. to TX, n. to MN and Ontario

state frequency: common
habitat: swamps, thickets, pool margins
notes: —

Juncus filiformis L.
synonym(s): —
common name(s): thread rush
habit: filiform, cespitose to subcespitose, slender-rhizomatous, perennial herb
range: circumboreal; s. to WV, w. to UT and OR
state frequency: uncommon
habitat: bogs, wet shores and ditches
notes: —

Juncus gerardii Loisel.
synonym(s): —
common name(s): black-grass
habit: ascending to erect, cespitose, slender-rhizomatous, perennial herb
range: Newfoundland, s. to FL
state frequency: occasional
habitat: saline to brackish marshes
key to the subspecies:
    1a. Pedicels 1.0–5.0 mm long; inflorescence 2.0–11.0 cm tall; tepals 2.0–3.0 mm long ....
    ....*J. g.* var. *gerardii*
    1b. Pedicels 3.0–10.0 mm long; inflorescence 10.0–20.0 cm tall; tepals 3.3–5.0 mm long ....
    ....*J. g.* var. *pedicellatus* Fern.
notes: —

Juncus greenei Oakes & Tuckerman
synonym(s): —
common name(s): Greene's rush
habit: erect, densely cespitose, short-rhizomatous, perennial herb
range: New Brunswick and Nova Scotia, s. to NJ, w. to IA, n. to MN and Ontario
state frequency: occasional
habitat: sandy soil of roadsides and fields
notes: —

Juncus marginatus Rostk.
synonym(s): —
common name(s): grass-leaved rush
habit: slender, cespitose, short- and thick-rhizomatous, perennial herb
range: Nova Scotia and ME, s. to FL, w. to TX, n. to SD, MN, and Ontario
state frequency: uncommon
habitat: wet meadows, swales
notes: —

Juncus militaris Bigelow
synonym(s): —
common name(s): bayonet rush
habit: stout, erect, solitary, stout-rhizomatous, perennial herb
range: Newfoundland, s. to MD, w. to MI, n. to Ontario
state frequency: occasional
habitat: shallow water, wet shores
notes: —

Juncus nodosus L.

synonym(s): —
common name(s): knotted rush
habit: slender, erect, slender-rhizomatous, perennial herb
range: Newfoundland, s. to VA, w. to CA, n. to AK
state frequency: uncommon
habitat: bogs, swamps, marshes, gravelly banks, wet shores
notes: Known to hybridize with *Juncus alpinoarticulatus* and *J. nodosus*.

Juncus pelocarpus E. Mey.

synonym(s): —
common name(s): brown-fruited rush
habit: slender, erect to ascending, colonial, rhizomatous, perennial herb
range: Newfoundland, s. to MD, w. to MN, n. to Ontario
state frequency: uncommon
habitat: damp shores, boggy areas
notes: —

Juncus secundus Beauv. *ex* Poir.

synonym(s): —
common name(s): —
habit: loosely cespitose, perennial herb
range: ME, s. to GA, w. to KS, n. to Ontario
state frequency: very rare
habitat: ledges, rocky forests
notes: This species is ranked S2 by the Maine Natural Areas Program.

Juncus stygius L. ssp. *americanus* (Buch.) Hultén

synonym(s): —
common name(s): bog rush
habit: solitary to several together, slender-rhizomatous, perennial herb
range: circumboreal; s. to ME, w. to MI, MN, and Saskatchewan
state frequency: rare
habitat: circumneutral fens
notes: This species is ranked S2 by the Maine Natural Areas Program.

Juncus subtilis E. Mey.

synonym(s): —
common name(s): slender rush
habit: slender, floating or creeping, mat-forming, perennial herb
range: Greenland, s. to ME, w. to Quebec
state frequency: very rare
habitat: wet shores, shallow water
notes: This species is ranked SH by the Maine Natural Areas Program.

Juncus tenuis Willd.

synonym(s): *Juncus tenuis* Willd. var. *williamsii* Fern.
common name(s): path rush
habit: slender, cespitose, perennial herb
range: Newfoundland, s. to FL, w. to CA, n. to AK
state frequency: common
habitat: fields, roadsides, open areas
notes: —

Juncus torreyi Coville

synonym(s): —
common name(s): Torrey's rush
habit: stout, erect, rhizomatous, perennial herb
range: Ontario, s. to KY, w. to TX and CA, n. to WA and Saskatchewan; also ME, s. to NJ
state frequency: very rare
habitat: low areas, shallow water, sometimes tidal
notes: This species is ranked SH by the Maine Natural Areas Program. Known to hybridize with
　　Juncus alpinoarticulatus.

Juncus trifidus L.

synonym(s): —
common name(s): highland rush
habit: erect, densely cespitose, short-rhizomatous, perennial herb
range: Newfoundland, s. to mountains of VA and NC
state frequency: rare
habitat: alpine areas
notes: —

Juncus vaseyi Engelm.

synonym(s): *Juncus greenei* Oakes & Tuckerman var. *vaseyi* (Engelm.) Boivin
common name(s): Vasey's rush
habit: erect, densely cespitose, short-rhizomatous, perennial herb
range: Labrador, s. to ME, w. to UT, n. to British Columbia
state frequency: very rare
habitat: wetlands, ledges, shores, forests
notes: This species is ranked S1 by the Maine Natural Areas Program.

Luzula
(7 species)

KEY TO THE SPECIES
1a. Flowers occurring singly (paired) at the tips of branches
　　2a. Inflorescence an umbelliform raceme, the flowers borne mostly on primary branches;
　　　　seeds with a distinct appendage; plants flowering May–early June .... *L. acuminata*
　　2b. Inflorescence a many-branched dichasium, the flowers borne on secondary or
　　　　further-divided branches; seeds with a tuft of hairs; plants flowering mid-June–July
　　　　.. *L. parviflora*
1b. Flowers occurring in glomerules of 3–12
　　3a. Tepals white (pink-tinged); inflorescence a branched, decompound dichasium,
　　　　3.5–7.0 x 5.0–7.0 cm; glomerules mostly 3- to 6-flowered .... *L. luzuloides*
　　3b. Tepals light to dark brown; inflorescence an umbel-like raceme, 1.0–6.0 x
　　　　0.4–5.0 cm; glomerules mostly 5- to 12-flowered
　　　　4a. Leaves involute, with pointed tips; plants of exposed, alpine habitats
　　　　　　5a. Some of the glomerules nodding; bracts subtending glomerules long and
　　　　　　　　slender-pointed, projecting beyond the glomerules .... *L. spicata*
　　　　　　5b. Glomerules ascending; bracts subtending glomerules obtuse to acuminate, not
　　　　　　　　projecting beyond the glomerules .... *L. confusa*
　　　　4b. Leaves flat, with blunt, callous tips; plants of various forested to open areas of
　　　　　　lower elevation

6a. Tepals 1.5–2.5 mm long, the inner ones conspicuously shorter than the outer; styles 0.2–0.3 mm long; fruiting glomerules mostly 4.0–5.0 mm wide .... *L. pallidula*

6b. Tepals 2.5–4.5 mm long, the inner and outer ones not of markedly different lengths; styles 0.4–0.7 mm long; fruiting glomerules mostly 6.0–9.0 mm wide .... *L. multiflora*

Luzula acuminata Raf.

synonym(s): *Juncoides saltuense* (Fern.) Heller
common name(s): hairy woodrush
habit: loosely cespitose to clustered, perennial herb
range: Newfoundland, s. to VA, w. to IA, n. to Saskatchewan
state frequency: occasional
habitat: moist woods and open areas
notes: —

Luzula confusa Lindeberg

synonym(s): *Juncoides hyperboreum* (R. Br.) Sheldon
common name(s): alpine woodrush
habit: cespitose, perennial herb
range: circumboreal; s. to ME, w. to Alberta
state frequency: very rare
habitat: alpine areas
notes: This species is ranked S1 by the Maine Natural Areas Program.

Luzula luzuloides (Lam.) Dandy & Wilmott

synonym(s): *Juncoides nemorosum* (Pollard) Kuntze
common name(s): forest woodrush
habit: slender, loosely cespitose, perennial herb
range: Eurasia; naturalized to ME, s. to PA, w. to MN, n. to Ontario
state frequency: very rare
habitat: open woods, roadsides, lawns
notes: —

Luzula multiflora (Ehrh.) Lej.

synonym(s): <*L. m.* ssp. *frigida*> = *Luzula multiflora* (Ehrh.) Lej. var. *fusconigra* Celak.; <*L. m.* ssp. *multiflora*> = *Luzula multiflora* (Ehrh.) Lej. var. *acadiensis* Fern.
common name(s): common woodrush
habit: erect, densely cespitose, perennial herb
range: Eurasia; Newfoundland, s. to DE and mountains of VA, w. to MO, n. to MN
state frequency: occasional
habitat: open woods, fields, meadows
key to the subspecies:

1a. Tepals yellow-brown to brown near the center, all pointed at the apex; capsules brown .... *L. m.* ssp. *multiflora*

1b. Tepals dark brown near the center, the outer pointed, the inner truncate-mucronate; capsules dark brown to nearly black .... *L. m.* ssp. *frigida* (Buch.) Krecz.
notes: —

Luzula pallidula Kirsch.

synonym(s): *Luzula campestris* (L.) DC. *ex* Lam. & DC. var. *pallescens* (Wahlenb.) Wahlenb., *Luzula pallescens* (Wahlenb.) Bess.
common name(s): —

habit: erect, cespitose, perennial herb
range: Eurasia; introduced to Newfoundland, s. to ME, w. to NY, n. to Saskatchewan
state frequency: very rare
habitat: woods
notes: —

Luzula parviflora (Ehrh.) Desv. ssp. *melanocarpa* (Michx.) Hämet-Ahti
synonym(s): *Luzula melanocarpa* (Michx.) Tolm.
common name(s): small-flowered woodrush
habit: clustered, short-rhizomatous, perennial herb
range: circumboreal; s. to MA, w. to WY and CA
state frequency: uncommon
habitat: moist to wet thickets, ravines, slopes, and open woods
notes: —

Luzula spicata (L.) DC.
synonym(s): *Juncoides spicatum* (L.) Kuntze
common name(s): nodding woodrush, spiked woodrush
habit: densely cespitose, perennial herb
range: circumboreal; s. to ME, w. to AZ and CA
state frequency: very rare
habitat: gravel shores, slopes, and meadows of alpine areas
notes: This species is ranked S1 by the Maine Natural Areas Program.

CYPERACEAE
(14 genera, 214 species)

KEY TO THE GENERA

1a. Flowers unisexual, lacking a perianth; carpellate flowers subtended by an enclosing, sac-like scale [called a perigynium], the stigmas protruding through an orifice at the top ....
.... *Carex*

1b. Flowers bisexual, with or without a perianth; carpellate flowers subtended by flat scales
    2a. Spikes 1 per culm; involucral bracts resembling enlarged scales
        3a. Perianth composed of numerous, white, capillary bristles that greatly exceed the floral scales; spike 2.5–5.0 cm wide, with 10–15 basal, sterile scales ....
.... *Eriophorum*
        3b. Perianth composed of 3–6 brown bristles that are shorter than the floral scales (white and longer than floral scales in *Trichophorum alpinum*); spike much narrower, with up to 3 basal, sterile scales
            4a. Achenes not crowned by a tubercle; at least the upper leaf sheaths blade-bearing .... *Trichophorum*
            4b. Achenes crowned by a persistent tubercle, usually of different color and texture; leaf sheaths lacking blades .... *Eleocharis*
    2b. Spikes usually 2 or more per culm; involucral bracts elongate, not resembling enlarged scales
        5a. Scales of spikes distichous [*i.e.*, arranged in 2 ranks]
            6a. Inflorescence terminal; flowers lacking a perianth; style deciduous from the achene .... *Cyperus*

6b. Inflorescence partly or wholly axillary; flowers with a perianth composed of bristles; style persistent on the achene as a beak .... *Dulichium*
5b. Scales of spikes spirally arranged [*i.e.*, arranged in 3 or more ranks]
    7a. Achenes crowned with an evident tubercle as broad as the achene summit and clearly differentiated from it; perianth composed of bristles; only the 1–5 most apical scales subtending flowers, the lower ones empty .... *Rhynchospora*
    7b. Achenes with minute tubercles or lacking them altogether; perianth and flower number various
        8a. Perianth composed of numerous, capillary bristles that greatly exceed the floral scales .... *Eriophorum*
        8b. Perianth absent or of scales or bristles, when composed of bristles, the bristles up to 7 and shorter than to exceeding the floral scales
            9a. Principal involucral bract erect, of similar cross-section as the culm and seeming to be a continuation of it, giving the appearance of a lateral inflorescence
                10a. Achenes 0.5–0.7 mm long; spikelets 2.0–6.0 mm long; flowers subtended by a single, diminutive scale; floral scales 1.0–2.0 mm long .... *Lipocarpha*
                10b. Achenes 1.3–4.0 mm long; spikelets 5.0–15.0 mm long; flowers subtended by bristles, these sometimes caducous; floral scales 2.0–6.0 mm long .... *Schoenoplectus*
            9b. Principal involucral bract erect to spreading, flat or setaceous, not seeming a continuation of the stem, the inflorescence appearing terminal
                11a. Only the upper 2 scales of the spikelet subtending flowers, the ultimate flower bisexual, the penultimate functionally unisexual, the ovary aborting; leaves terete, at least in the apical portion .... *Cladium*
                11b. Normally most or all the scales of the spikelet subtending bisexual and fertile flowers; leaves flat, conduplicate or involute
                    12a. Plants annual, with soft stem bases and fibrous roots; leaves up to 4.0 mm wide
                        13a. Stems capillary; leaves up to 0.5 mm wide; apex of achene with a minute tubercle .... *Bulbostylis*
                        13b. Stems flat; larger leaves wider than 1.0 mm; apex of achene with a swollen style base .... *Fimbristylis*
                    12b. Plants perennial, with hardened stem bases and rhizomes; leaves 3.0–20.0 mm wide
                        14a. Scales pubescent on the abaxial surface; rhizomes with tuberous thickenings; spikelets 1.0–4.0 cm long; achenes 2.5–5.0 mm long; leaves without ligules ... *Bolboschoenus*
                        14b. Scales glabrous on the abaxial surface; rhizomes without tubers; spikelets 0.2–1.0 cm long; achenes up to 2.0 mm; leaves with ligules .... *Scirpus*

Bolboschoenus
(3 species)

KEY TO THE SPECIES

1a. Styles trifid; achenes distinctly trigonous, gradually tapering to the apex; perianth bristles as long as and persistent on the achene; inflorescence with some elongate rays .... ***B. fluviatilis***

1b. Styles bifid or both bifid and trifid; achenes more compressed, abruptly tapering to the apex; perianth bristles either absent or weakly attached to and often shorter than the achene; inflorescence usually congested, sometimes with 1 or 2 elongate rays

    2a. Scales opaque, red-brown, with awns *ca.* 0.5 mm wide at the base; anthers red; veins usually abruptly diverging at the summit of the leaf sheath opposite the blade, without a prominent veinless area .... ***B. robustus***

    2b. Scales translucent, light brown to red-brown, with awns *ca.* 0.25 mm wide at the base; anthers yellow; veins usually gradually diverging at the summit of the leaf sheath opposite the blade, leaving a triangular veinless area .... ***B. maritimus***

Bolboschoenus fluviatilis (Torr.) Sojak
synonym(s): *Scirpus fluviatilis* (Torr.) Gray
common name(s): river bulrush
habit: rhizomatous, perennial herb
range: New Brunswick, s. to VA, w. to CA, n. to Alberta
state frequency: uncommon
habitat: fresh to fresh tidal shores
notes: Hybrids with *Bolboschoenus robustus*, called *B.* ×*novae-angliae* (Britt.) S. G. Smith, occur in Maine. These hybrids can be detected by their intermediacy in achene shape, perianth bristle length and persistence, carpel number, openness of inflorescence, and habitat.

Bolboschoenus maritimus (L.) Palla ssp. *paludosus* (A. Nels.) T. Koyama
synonym(s): *Scirpus maritimus* L. var. *fernaldii* (Bickn.) Beetle, *Scirpus paludosus* A. Nels. var. *atlanticus* Fern.
common name(s): saltmarsh bulrush, alkali bulrush
habit: rhizomatous, perennial herb
range: New Brunswick, s. to VA
state frequency: uncommon
habitat: coastal marshes and shores
notes: Hybrids with *Bolboschoenus robustus* occur in Maine.

Bolboschoenus robustus (Pursh) Sojak
synonym(s): *Scirpus robustus* Pursh
common name(s): saltmarsh bulrush
habit: stout, rhizomatous, perennial herb
range: Nova Scotia, s. to FL, w. to TX and Mexico; also CA
state frequency: very rare
habitat: brackish to saline, coastal marshes
notes: Hybrids with *Bolboschoenus fluviatilis* (*q.v.*) and *B. maritimus* occur in Maine.

Bulbostylis
(1 species)

Bulbostylis capillaris (L.) Kunthe *ex* C. B. Clarke
synonym(s): *Bulbostylis capillaris* (L.) Kunthe *ex* C. B. Clarke var. *crebra* Fern.
common name(s): vagabond
habit: cespitose, annual herb
range: New Brunswick, s. to FL, w. to AZ, n. to MN
state frequency: occasional
habitat: dry, open, sandy or rocky places
notes: —

Carex
(150 species)

This group of sedges has conspicuous, sac-like scales, called perigynia, that enclose the carpellate flowers. Because the perigynium occupies more space than stamens, is frequently inflated, and is eventually filled with a mature achene, the carpellate spikes or carpellate portions of spikes are much thicker than the staminate spikes or portions. This makes possible distinguishing gynaecandrous from androgynous spikes. Gynaecandrous spikes most often have a basal portion of narrow, overlapping scales. These scales subtend staminate flowers. The carpellate flowers are borne in the apical portion and occupy more space, so that the scales frequently spread wider apart. Androgynous spikes show the reverse situation. The narrow portion of the spike, where the staminate flowers are found, is located at the apex of the spike. Some species of *Carex* cannot be identified without mature perigynia [*i.e.*, in fruit] and should always be collected with them until one has intimate familiarity with these sedges. Collections should include the plant bases and underground portions. The plant base will indicate whether the plant is phyllopodic [*i.e.*, the lower sheaths blade-bearing] or aphyllopodic [*i.e.*, the lower sheaths without elongate blades]. The underground portion is needed to determine presence or absence of rhizomes. At the very least, notes should be taken as to the number of stems arising from one point. Many times, cespitose species lack prominent rhizomes and those that have separate stems also have rhizomes.

KEY TO THE SPECIES
1a. Spikes entirely staminate .... **Group 1**
1b. At least some of the spikes partly or entirely carpellate
    2a. Spikes 1 per stem (rarely carpellate plants of *C. scirpoidea* with 1 or 2 additional, very small spikes); rachilla often present in perigynium .... **Group 2**
    2b. Spikes normally 2 or more per stem (except depauperate individuals); rachilla absent from perigynium
        3a. Stigmas 2; achenes lenticular to planoconvex
            4a. Spikes all similar and sessile, bearing staminate flowers at base or apex; bracteoles usually absent from the base of each spike .... **Group 3**
            4b. Spikes dissimilar and some or all usually peduncled; staminate flowers borne in separate or mixed spikes; bracteoles usually present at the base of each spike .... **Group 4**
        3b. Stigmas 3; achenes trigonous to terete
            5a. Style articulated to achene and deciduous from it at maturity; perigynia 2.5–8.3 mm long, with or without an entire to bidentate beak
                6a. Perigynia pubescent or scabrous .... **Group 5**
                6b. Perigynia glabrous

7a. Bract of lowest carpellate spike consisting of a blade only, the sheath absent or extremely short, usually not exceeding 3.0 mm ...... **Group 6**

7b. Bract of lowest carpellate spike consisting of a blade and a prolonged, closed sheath, usually longer than 4.0 mm .... **Group 7**

5b. Style not articulated to achene and persistent on it at maturity; perigynia 4.2–19.0 mm long, with a bidentate beak (entire beak in *C. oligosperma*) ......
.... **Group 8**

Group 1

1a. Plants of organic soil wetlands

2a. Stems produced singly or few together from long rhizomes, 0.5–2.0 dm tall ....
.... *Dioicae*

2b. Stems cespitose, 1.0–7.0 dm tall

3a. Spikes 1(–3) per stem; anthers 2.0–3.6 mm long (*C. exilis*) .... *Stellulatae*

3b. Spikes 3–8 per stem; anthers 0.6–2.2 mm long (*C. sterilis*) .... *Stellulatae*

1b. Plants of mineral soils or alpine and subalpine habitats

4a. Spikes 1–5 per stem; stems aphyllopodic

5a. Leaves 1.0–3.0 mm wide; plants mostly of alpine and subalpine habitats ....
.... *Scirpinae*

5b. Leaves 3.0–5.0 mm wide; plants of open, sandy soils (*C. polymorpha*) ....
.... *Paniceae*

4b. Spikes 6–25 per stem; stems phyllopodic .... *Divisae*

Group 2

1a. Stigmas 2; achenes lenticular

2a. Stems produced separately from long, thin rhizomes, 0.5–2.0 dm tall; spike unisexual or androgynous .... *Dioicae*

2b. Stems cespitose, 2.0–7.0 dm tall; spike unisexual or gynaecandrous (*C. exilis*) ....
.... *Stellulatae*

1b. Stigmas 3; achenes trigonous

3a. Spike unisexual; perigynia mostly pubescent

4a. Spike 1.0–3.0 cm long; perigynium beak 0.2 mm long .... *Scirpinae*

4b. Spike 0.4–1.0 cm long; perigynium beak 0.5–1.0 mm long (*C. umbellata*) ....
.... *Acrocystis*

3b. Spike bisexual; perigynia glabrous

5a. Spike gynaecandrous; body of the perigynia obconic to obovoid .... *Squarrosae*

5b. Spike androgynous; body of the perigynia lance-subulate to globose

6a. Perigynia 6.0–7.5 mm long, lance-subulate, spreading to reflexed; style persistent on achene .... *Orthocerates*

6b. Perigynia 2.5–5.4 mm long, lance-elliptic to globose, ascending to appressed; style deciduous from achene

7a. Lower carpellate scales foliaceous, surpassing and concealing the perigynium; leaves 3.0–6.0 mm wide .... *Phyllostachyae*

7b. Lower carpellate scales not foliaceous, shorter than to slightly exceeding the perigynium; leaves 0.7–1.2 mm wide .... *Polytrichoideae*

Group 3

1a. Spikes, at least the uppermost, gynaecandrous, the staminate flowers borne at the base of the spike or mixed throughout it

  2a. Perigynia with obvious wing-like margins, not distended to the edges by achene, not spongy at base .... ***Ovales***

  2b. Perigynia with rounded or sharp, unwinged margins, distended to the edges by achene or nearly so, at least somewhat spongy at the base

    3a. Perigynia sharp-margined, spreading to reflexed, spongy at base .... ***Stellulatae***

    3b. Perigynia round-margined, appressed to ascending, only somewhat spongy at base

      4a. Perigynia 1.5–4.0 mm long, with a beak up to 0.7 mm long, nearly filled by achene .... ***Glarosae***

      4b. Perigynia 4.0–5.5 mm long, with a beak up to 1.0–2.0 mm long, occupied in upper 0.65 by achene .... ***Deweyanae***

1b. Spikes androgynous, the staminate flowers borne at the summit of the spike

  5a. Stems produced singly from elongate rhizomes, stolons, or prostrate stems

    6a. Stems leaning or prostrate, producing upright stems from the axils of dried leaves of the previous year .... ***Chordorrhizae***

    6b. Stems upright, produced from elongate rhizomes

      7a. Spikes aggregated in a crowded inflorescence, each spike with many perigynia; perigynia definitely compressed

        8a. Perigynia sharp-margined but not wing-margined, nerveless or inconspicuously nerved on the adaxial surface; plants phyllopodic, with basal tufts of leaves .... ***Divisae***

        8b. Perigynia wing-margined, conspicuously nerved on the adaxial surface; plants aphyllopodic, without basal tufts of leaves .... ***Arenariae***

      7b. Spikes remote, each with 1–4 perigynia; perigynia terete in cross-section .... ***Dispermae***

  5b. Stems cespitose, the rhizomes none or short

    9a. Spikes mostly 1–8 per stem, occurring singly at the nodes .... ***Phaestoglochin***

    9b. Spikes numerous, usually more than 8, 2 or more found from at least the lower nodes

      10a. Perigynia flat to often convex on the adaxial surface, thick-walled, firm, and dark colored, olive-brown to black (light brown in *C. prairea*) ... ***Paniculatae***

      10b. Perigynia flat on the adaxial surface, thin-walled and light colored, yellow-brown to light brown or green

        11a. Carpellate scales with a conspicuous awn 1.0–5.0 mm long; culms firm, not wing-angled; base of perigynia rounded, inconspicuously spongy .... ***Multiflorae***

        11b. Carpellate scales without awns; culms soft, concave-sided to wing-angled; base of perigynia rounded to truncate, conspicuously spongy .... ***Vulpinae***

Group 4

1a. Perigynia white or gold-orange, elliptic in cross-section; lowest bract of carpellate spikes with a prolonged, closed sheath .... ***Bicolores***

1b. Perigynia green to brown to black, lenticular to biconvex in cross-section; lowest bract of carpellate spikes lacking a sheath

2a. Style deciduous from achene in fruit; leaf sheaths with no, or very few, septate partitions; perigynia essentially beakless .... *Phacocystis*

2b. Style persistent on achene in fruit; leaf sheaths with septate partitions; perigynia with a short, distinct beak *ca.* 0.5 mm long (*C. saxatilis*) .... *Vesicariae*

Group 5

1a. Perigynium beak terminated by prominent, stiff teeth 0.3–2.0 mm long (except *C. vestita* with short teeth, the beak eventually splitting with age); stems strongly aphyllopodic, without basal tufts of leaves

  2a. Leaf sheaths and abaxial surface of carpellate scales pubescent; teeth of perigynium beak 1.0–1.75 mm long; plants introduced, weedy (*C. hirta*) .... *Carex*

  2b. Leaf sheaths and abaxial surface of carpellate scales glabrous; teeth of perigynium beak less than 1.0 mm long; plants native .... *Paludosae*

1b. Perigynium beak entire, minutely toothed, or with an irregularly split or torn beak, lacking 2 prominent and distinct teeth; stems phyllopodic, arising from tufts of basal leaves

  3a. Perigynia with 20 or more fine nerves; basal leaves 10.0–30.0 mm wide (*C. plantaginea*) .... *Careyanae*

  3b. Perigynia with fewer nerves; basal leaves up to 8.0 mm wide

    4a. Stems, sheaths, and leaf blades glabrous or scabrous, not pubescent

      5a. Perigynium pubescent with soft hairs, with only 2 evident ribs, the other nerves obscure; plants 1.0–6.0 dm tall

        6a. Excluding the spikes borne on peduncles originating from near the base of the plant, the spikes long-pedunculate, the peduncles hidden in long, sheathing bracts; achenes distinctly trigonous, with flat or concave sides ... . *Clandestinae*

        6b. Excluding the spikes borne on peduncles originating from near the base of the plant, the spikes sessile or short-pedunculate; achenes obscurely trigonous, with rounded or convex sides

          7a. Perigynium beak less than 0.5 mm long; style base expanded into a dark ring; carpellate scales with a long, broad, scabrous awn often 1.0 mm long or longer .... *Praecoces*

          7b. Perigynium beak 0.5–1.5 mm long; style base not expanded; carpellate scales awnless or with an awn less than 1.0 mm long .... *Acrocystis*

      5b. Perigynium scabrous, with several conspicuous nerves; plants 4.0–9.0 dm tall .... *Anomalae*

    4b. Stems, sheaths, and/or leaf blades pubescent

      8a. Uppermost spike bearing some perigynia; summit of perigynium not filled by achene .... *Porocystis*

      8b. Uppermost spike entirely staminate; summit of perigynium occupied by achene .... *Halleranae*

Group 6

1a. Uppermost spike with at least some carpellate flowers [*i.e.,* perigynia present in spike]

  2a. Carpellate spikes very slender, 2.0–6.0 cm long; carpellate flowers approximate, the axis of the spike visible

3a. Lower sheaths brown, not anthocyanic (*C. prasina*) .... **Gracillimae**
3b. Lower sheaths anthocyanic, red-brown to purple
    4a. Perigynium distinctly beaked, 2-ribbed but otherwise obscurely nerved ....
    .... **Sylvaticae**
    4b. Perigynium beakless, 2-ribbed and also conspicuously nerved (*C. gracillima*)
    .... **Gracillimae**
2b. Carpellate spikes cylindric to globose, 0.5–2.0 cm long; carpellate flowers close and overlapping, concealing the axis of the spike
    5a. Leaves pubescent on the sheath and/or blade
        6a. Perigynium with a distinct, toothed beak, the beak *ca.* 0.5 times the length of the perigynium body (*C. castanea*) .... **Sylvaticae**
        6b. Perigynium beakless .... **Porocystis**
    5b. Leaves glabrous
        7a. Perigynia spreading to reflexed at maturity, with a distinct beak 0.6–1.5 mm long .... **Ceratocystis**
        7b. Perigynia ascending, beakless or with a minute beak up to 0.5 mm long
            8a. Carpellate scales clearly exceeding the perigynia; roots densely pubescent
            .... **Limosae**
            8b. Carpellate scales shorter than to equaling the perigynia (scales longer in *C. buxbaumii*); roots not densely pubescent
                9a. Uppermost spike composed about equally of staminate and carpellate flowers; leaves more or less flat, green .... **Atratae**
                9b. Uppermost spike composed mostly of staminate flowers, the carpellate flowers 1–3; leaves involute, glaucous (*C. livida*) ....
                .... **Paniceae**
1b. Uppermost spike entirely staminate [*i.e.*, perigynia absent]
    10a. Perigynia with a prominent beak; stems 0.5–1.0(–2.0) dm tall; plants with 1–3 carpellate spikes borne on peduncles originating from near the base of the plant (*C. tonsa*) .... **Acrocystis**
    10b. Perigynia beakless; stems 2.0–5.0 dm tall; plants without carpellate spikes borne on basal peduncles
        11a. Perigynia 2.1–2.8 mm long; leaves usually pubescent, green; plants of fields and woods (*C. pallescens*) .... **Porocystis**
        11b. Perigynia 3.0–5.0 mm long; leaves glabrous, glaucous; plants of fens and bogs (*C. livida*) .... **Paniceae**

Group 7

1a. Carpellate spikes spreading or drooping on slender peduncles (erect to ascending in *C. debilis* var. *strictior*)
    2a. Perigynia with 2 evident ribs, the remaining 10 or fewer nerves inconspicuous
        3a. Perigynia 2.2–3.3 mm long; carpellate spikes 0.5–1.3 cm long; leaves 1.0–3.0 mm wide .... **Capillares**
        3b. Perigynia 2.9–8.3 mm long; carpellate spikes 1.0–6.0 cm long; leaves 1.5–8.0 mm wide
            4a. Uppermost spike with carpellate flowers (*C. prasina*) .... **Gracillimae**
            4b. Uppermost spike entirely staminate-flowered

5a. Perigynium beak with distinct teeth; lower sheaths brown .... ***Longirostres***

5b. Perigynium beak with minute teeth; lower sheaths red-brown to purple ....
.... ***Sylvaticae***

2b. Perigynium with 2 evident ribs as well as 10 or more conspicuous nerves

6a. Uppermost spike with carpellate flowers (*C. gracillima*) .... ***Gracillimae***

6b. Uppermost spike entirely staminate-flowered

7a. Perigynia sharply trigonous, with flat or slightly convex faces; plants usually with a carpellate spike borne on a peduncle originating from the base of the plant .... ***Careyanae***

7b. Perigynia obscurely trigonous, with rounded or convex faces, especially in the basal portion; carpellate spikes all borne on stems, none on basal peduncles
.... ***Laxiflorae***

1b. Carpellate spikes erect or ascending, sessile or on upright peduncles

8a. Perigynia 1.0–2.0 mm long; leaves 0.5 mm wide or narrower; carpellate scales white or white-brown .... ***Albae***

8b. Perigynia 2.2 mm or longer; leaves 1.0 mm wide or wider; carpellate scales green, brown, or black

9a. Perigynia spreading to reflexed, in thick-cylindric to globose spikes ....
.... ***Ceratocystis***

9b. Perigynia spreading to usually ascending or appressed, in narrowly cylindric to linear spikes

10a. Stems 1 or a few together, produced from long rhizomes

11a. Carpellate scales brown to red-brown in fruit; perigynia often resin-dotted (*C. crawei*) .... ***Granulares***

11b. Carpellate scales dark brown to purple in fruit; perigynia not resin-dotted .... ***Paniceae***

10b. Stems cespitose, produced from dense tufts of leaves, the rhizomes none or very short

12a. Perigynia with 2 evident ribs, the remaining 10 or fewer nerves inconspicuous

13a. Lowest spike subtended by a long, sheathing bract that lacks a prolonged, flat blade; perigynium occupied to the apex by an achene
.... ***Clandestinae***

13b. Lowest spike subtended by a bract with a prolonged, flat blade; perigynium not occupied in the apical portion by an achene

14a. Carpellate spikes 6.0–10.0 mm wide; perigynia rounded to the base (*C. polymorpha*) .... ***Paniceae***

14b. Carpellate spikes 3.0–4.0 mm wide; perigynia distinctly narrowed to the base (*C. leptonervia*) .... ***Laxiflorae***

12b. Perigynia with 2 evident ribs as well as 10 or more conspicuous nerves

15a. Perigynia with impressed nerves; carpellate scales with distinct, scabrous awns often surpassing the perigynia .... ***Griseae***

15b. Perigynia with elevated nerves; carpellate scales unawned or with short awns not exceeding the perigynia

16a. Perigynia ovoid to nearly globose, the base rounded and not spongy (*C. granularis*) .... ***Granulares***

16b. Perigynia elliptic to fusiform, the base distinctly narrowed to a spongy stipe

17a. Perigynia sharply trigonous, with flat or slightly convex faces; plants usually with a carpellate spike borne on a peduncle originating from the base of the plant .... *Careyanae*

17b. Perigynia obscurely trigonous, with rounded or convex faces, especially in the basal portion; carpellate spikes all borne on stems, none on basal peduncles .... *Laxiflorae*

Group 8

1a. Carpellate scales with long, scabrous awns, the awns equal to the length of the scale; carpellate spikes ascending to drooping .... *Vesicariae*

1b. Carpellate scales awnless or with glabrous awns shorter than the scale; carpellate spikes ascending

2a. Perigynia thick-walled, coriaceous, only slightly inflated, olive- to dark brown in fruit; lower sheaths prominently fibrillose

3a. Teeth of the perigynium beak up to 1.0 mm long; leaves glabrous; perigynia 4.7–7.3 mm long (*C. lacustris*) .... *Paludosae*

3b. Teeth of the perigynium beak 1.0–3.0 mm long; leaves pubescent on the sheath and abaxial blade surface; perigynia 7.0–10.0 mm long (*C. atherodes*) ..... *Carex*

2b. Perigynia thin-walled, membranaceous, slightly to much inflated, often green or light brown in fruit; lower sheaths often not fibrillose

4a. Perigynia obconic, widest above the middle, the body truncate to abruptly tapering to a long beak .... *Squarrosae*

4b. Perigynia lance-subulate to lanceolate to ovoid, widest below the middle, rounded to gradually tapering to the beak

5a. Perigynia lance-subulate, extremely long and slender, 4.0–6.5 times as long as wide, slightly inflated .... *Folliculatae*

5b. Perigynia lanceolate to ovoid, up to 4.0 times as long as wide, much inflated

6a. Lowest bract of the carpellate spikes composed of a blade only, the sheath absent or very short; perigynia 3.0–10.0 mm long; staminate spikes often more than 1 .... *Vesicariae*

6b. Lowest bract of the carpellate spikes composed of a blade and a prolonged, closed sheath; perigynia 10.0–20.0 mm long; staminate spikes usually only 1 .... *Lupulinae*

Acrocystis

1a. Spikes sessile or short-pedunculate, all of them elevated above the base of the plant on stems, the stems leafy at least at the base

2a. Perigynium body subglobose, as wide as long

3a. Leaves 3.0–7.0 mm wide; basal sheaths only slightly fibrillose if at all; plants cespitose, without long rhizomes; staminate spikes 1.0–2.5 mm wide .... *C. communis*

3b. Leaves 1.0–3.0 mm wide; basal sheaths fibrillose; plants not cespitose, with long rhizomes; staminate spikes wider than 2.0 mm

4a. Perigynia with a beak 0.5–0.9 mm long and 0.1–0.5 times as long as the body of the perigynium; uppermost stem leaves with a well developed blade; apical portion of the sheath opposite the leaf blade shallowly concave, the cleft extending 0.2–1.6 mm below the junction with the leaf blade .... *C. pensylvanica*

    4b. Perigynia with a beak 1.0–2.0 mm long and 0.5–1.0 times as long as the body of the perigynium; uppermost stem leaves with a poorly developed blade; apical portion of the sheath opposite the leaf blade deeply concave, the cleft extending 0.4–6.9 mm below the junction with the leaf blade ...... *C. lucorum*

2b. Perigynium body ellipsoid to obovoid, distinctly longer than wide

    5a. Lowest carpellate spike short-pedunculate, separated from the upper spikes by more than 7.0 mm .... *C. novae-angliae*

    5b. Lowest carpellate spike sessile, approximate to or overlapping the upper spikes, usually less than 7.0 mm from the upper spikes

        6a. Carpellate scales *ca.* 0.5 times as long as the perigynia; achenes longer than 1.4 mm

            7a. Leaves usually surpassing the stem, dark green; carpellate scales purple at the margins; perigynia 2.0–3.0 mm long .... *C. deflexa*

            7b. Leaves usually strongly surpassed by the stem, light green; carpellate scales white at the margins; perigynia 2.7–4.1 mm long .... *C. peckii*

        6b. Carpellate scales nearly equaling to slightly exceeding the perigynia; achenes up to 1.4 mm long .... *C. albicans*

1b. Some of the spikes sessile or short-peduncled, borne on leafy stems and others [which are carpellate] borne on short to elongate peduncles originating from the base of the plant, sometimes hidden among the leaf bases

8a. Carpellate scales *ca.* half as long as the perigynia; leaf base slightly, if at all, fibrillose; lowest spike of the stems subtended by a prolonged bract that usually exceeds the uppermost spike .... *C. deflexa*

8b. Carpellate scales nearly equal to or exceeding the perigynia; leaf bases fibrillose; lowest spike of the stems subtended by a short bract that is overtopped by the uppermost spike

    9a. Perigynium 3.2–4.7 mm long, with a beak 0.9–1.7 mm long .... *C. tonsa*

    9b. Perigynium 2.2–3.3 mm long, with a beak 0.5–1.0 mm long .... *C. umbellata*

Albae
only 1 species .... *C. eburnea*

Anomalae
only 1 species .... *C. scabrata*

Arenariae
only 1 species .... *C. siccata*

Atratae
1a. Carpellate spikes on long, often arching, peduncles 1.0–4.0 cm long .... *C. atratiformis*

1b. Carpellate spikes sessile or on short peduncles less than 1.0 cm long

    2a. Stems aphyllopodic, arising 1 or few together from long, horizontal stolons; perigynia 2.7–4.3 mm long, conspicuously papillate; carpellate scales equaling to, more commonly, exceeding the perigynia; spikes in a cluster 3.0–10.0 cm long .... *C. buxbaumii*

2b. Stems phyllopodic and cespitose, arising from leafy tufts, without stolons; perigynia 2.1–3.0 mm long, not conspicuously papillate; carpellate scales up to as long as the perigynia; spikes in a congested cluster 1.0–3.0 cm long .... *C. norvegica*

Bicolores

1a. Uppermost spike usually entirely staminate; lateral spikes relatively sparsely flowered, the internodes of the mid-portion (0.5–)0.7–1.5 mm long; carpellate scales pointed, white-brown to light brown; perigynia white, turning gold-brown at maturity .... *C. aurea*

1b. Uppermost spike with carpellate flowers in the apical portion; lateral spikes densely flowered, the internodes of the mid-portion 0.2–0.7 mm long; carpellate scales blunt, brown to dark red-brown; perigynia white and remaining so in maturity .... *C. garberi*

Capillares

only 1 species .... *C. capillaris*

Carex

1a. Perigynia pubescent; style articulated to the summit of the achene and deciduous from it at maturity; plants of dry soils .... *C. hirta*

1b. Perigynia glabrous; style not articulated to the summit of the achene and persistent on it in maturity; plants of wetlands .... *C. atherodes*

Careyanae

1a. Stems with long-sheathing, bladeless or very short-bladed leaves; lower sheaths anthocyanic, red-brown to purple .... *C. plantaginea*

1b. Stems with blade-bearing leaves; lower sheaths white to light brown

2a. Leaves dimorphic—those of the flowering shoots 2.0–6.0 mm wide, those of the vegetative shoots 10.0–25.0 mm wide; upper stem leaves up to 3.0 times the length of their sheath .... *C. platyphylla*

2b. Leaves only slightly dimorphic, 1.0–12.0 mm wide, those of the flowering shoots only *ca.* 2.0–4.0 mm narrower than those of the vegetative shoots; upper stem leaves more than 3.0 times the length of their sheath

3a. Leaves of the vegetative shoots 3.0–12.0 mm wide, light green or glaucous; carpellate spikes with 1–3 empty scales or staminate flowers at base, the lower spikes 4.0–6.0 mm thick .... *C. laxiculmis*

3b. Leaves of the vegetative shoots 1.0–5.0 mm wide, green; carpellate spikes entirely of carpellate flowers, the lower spikes 3.0–4.0 mm thick .... *C. digitalis*

Ceratocystis

1a. Perigynia 2.2–3.3 mm long, straight, nearly filled by the mature achene; carpellate spikes 4.0–7.0 mm thick; ligules of the upper stem leaves very reduced or absent .... *C. viridula*

1b. Perigynia 3.2–6.2 mm long, many strongly bent, occupied in the lower half by the achene; carpellate spikes 6.0–14.0 mm thick; ligules of the upper stem leaves well developed

2a. Perigynium 3.7–6.2 mm long, with a serrate beak 1.4–2.3 mm long; carpellate scales copper-brown, of a color different from that of the perigynia; leaves 3.0–5.5 mm wide; carpellate spikes 10.0–14.0 mm thick .... *C. flava*

2b. Perigynium 3.2–4.8 mm long, with a smooth beak 1.2–1.5 mm long; carpellate scales yellow-green to yellow-brown, of a color similar to that of the perigynia; leaves 1.5–3.5 mm wide; carpellate spikes 6.0–10.0 mm thick .... *C. cryptolepis*

Chordorrhizae
only 1 species .... **C. chordorrhiza**

Clandestinae
only 1 species .... **C. pedunculata**

Deweyanae
1a. Perigynia 0.8–1.2 mm wide, 4.0–5.0 times as long as wide, conspicuously nerved
    abaxially; leaves 1.0–2.5 mm wide .... **C. bromoides**
1b. Perigynia 1.3–1.6 mm wide, 3.0–4.0 times as long as wide, nerveless or inconspicuously
    nerved abaxially; leaves 2.0–5.0 mm wide .... **C. deweyana**

Dioicae
only 1 species .... **C. gynocrates**

Dispermae
only 1 species .... **C. disperma**

Divisae
only 1 species .... **C. praegracilis**

Folliculatae
1a. Leaves mostly 1.5–4.0 mm wide; sheaths concave at the summit; staminate spikes
    0.6–1.5 cm long; carpellate scales up to 0.5 times as long as the perigynia, without awns
    .... **C. michauxiana**
1b. Leaves mostly 4.0–16.0 mm wide; sheaths convex at the summit; staminate spikes
    1.2–4.0 cm long; carpellate scales 0.5–0.9 times as long as the perigynia, usually with an
    awn .... **C. folliculata**

Glarosae
1a. Spikes 2–4, aggregated and overlapping in a dense cluster; perigynia beakless; carpellate
    scales white-hyaline .... **C. tenuiflora**
1b. Spikes 1–9, separate, the lower not overlapping and often remote; perigynia definitely,
    although shortly, beaked; carpellate scales green to brown or red-brown
    2a. Spikes 1–3, the lowest widely separated from the upper spikes (when present); the
        lowest spike subtended by a prolonged, setaceous bract 2.0–4.0 cm long; spikes
        composed of 1–5 carpellate flowers .... **C. trisperma**
    2b. Spikes 3–9, the lowest approximate to shortly separated from the upper spikes; the
        lowest spike subtended by a short, setaceous bract mostly less than 2.0 cm; spikes
        composed of 5–30 carpellate flowers
        3a. Carpellate scales red-brown, covering the perigynia; stems smooth on the angles;
            uppermost spike with a prolonged, basal, staminate portion; plants of saline
            habitats .... **C. mackenziei**
        3b. Carpellate scales gray, gray-green, or light brown, not completely covering the
            perigynia; stems scabrous on the angles; uppermost spike with a short, basal,
            staminate portion; plants not of saline habitats
            4a. Spikes composed of 5–15 carpellate flowers; perigynia ascending-spreading,
                green, becoming brown; abaxial side of the beak with a distinct slit; leaves
                1.0–2.5 mm wide, green .... **C. brunnescens**

4b. Spikes composed of 10–30 carpellate flowers; perigynia ascending-appressed, white-green, becoming white-brown; abaxial side of the beak without a slit; leaves 2.0–4.0 mm wide, glaucous .... *C. canescens*

Gracillimae

1a. Perigynia sharply angled at the 2 ribs, obscurely nerved otherwise, beaked; lower sheaths brown to green; carpellate spikes 3.0–5.0 mm thick .... *C. prasina*
1b. Perigynia bluntly angled, the angles not at the 2 ribs, with several additional, evident nerves, beakless; lower sheaths anthocyanic, red-brown to purple; carpellate spikes 1.0–3.0 mm thick .... *C. gracillima*

Granulares

1a. Leaves flat, 3.0–12.0 mm wide; staminate spike sessile or short-peduncled, not overlapping the uppermost carpellate spike; stems tufted on short rhizomes .... *C. granularis*
1b. Leaves often folded, 1.5–4.0 mm wide; staminate spike long-peduncled, overtopping the uppermost carpellate spike; stems produced singly or a few together from elongate rhizomes .... *C. crawei*

Griseae

1a. Leaves mostly 2.0–4.0 mm wide; perigynia 2.5–3.8 x 1.5–1.7 mm; staminate spikes long-peduncled; peduncles and stem above lowest spike scabrous .... *C. conoidea*
1b. Leaves mostly 4.0–8.0 mm wide; perigynia 4.0–5.3 x 2.0–2.5 mm; staminate spikes sessile or nearly so; peduncles and stem above lowest spike smooth .... *C. grisea*

Halleranae

only 1 species .... *C. hirtifolia*

Laxiflorae

1a. Lower sheaths brown; leaves of the sterile shoots 3.5–25.0 mm wide, commonly more than 5.0 mm wide
  2a. Perigynia with 2 evident ribs, the remaining nerves inconspicuous ...... *C. leptonervia*
  2b. Perigynia with 2 evident ribs and 22–34 conspicuous nerves
    3a. Beak of the perigynium abruptly bent; leaves of the sterile shoots 4.0–10.0 mm wide; summit of the stem and the bract subtending the inflorescence minutely scabrous on the angles; spikes with crowded perigynia .... *C. blanda*
    3b. Beak of the perigynium straight or slightly bent; leaves of the sterile shoots 5.0–25.0 mm wide; summit of the stem and the bract subtending the inflorescence usually smooth on the angles; spikes with approximate perigynia .... *C. laxiflora*
1b. Lower sheaths anthocyanic, red-brown to purple; leaves of the sterile shoots 1.0–5.0 mm wide .... *C. ormostachya*

Limosae

1a. Carpellate scales lanceolate, narrower and exceeding the perigynia; stems phyllopodic and cespitose .... *C. magellanica*
1b. Carpellate scales ovate to elliptic, as wide as and about equaling the length of the perigynia; stems aphyllopodic and occurring singly or a few together

2a. Stems obscurely angled and glabrous; leaves green and flat; peduncle of the uppermost staminate spike 1.0–10.0 mm long; carpellate scales wrapping around the bases of the immature perigynia, caducous, leaving the persistent perigynia .... *C. rariflora*

2b. Stems sharply angled and scabrous on the angles; leaves glaucous and involute or caniculate; peduncle of the uppermost staminate spike 5.0–30.0 mm long; carpellate scales not enclosing perigynia bases, persistent .... *C. limosa*

Longirostres
only 1 species .... *C. sprengelii*

Lupulinae
1a. Perigynia in an aggregated, nearly globose spike, with a beak 2.0–4.2 mm long; stems cespitose, without rhizomes; uppermost leaf without a sheath or with a sheath up to 1.0 cm long .... *C. intumescens*

1b. Perigynia in ovoid to cylindric spikes, with a beak 6.0–10.0 mm long; stems produced singly or a few together from long rhizomes; uppermost leaf with a sheath 1.7–10.0 cm long .... *C. lupulina*

Multiflorae
1a. Leaves equaling or exceeding the stem; perigynia pale brown, gradually tapering to a beak 0.5–0.75 times as long as the body of the perigynium; cluster of spikes commonly 7.0–10.0 cm tall .... *C. vulpinoidea*

1b. Leaves shorter than the stem; perigynia yellow-brown, abruptly tapering to a beak less than 0.5 times as long as the body; cluster of spikes commonly 4.0–7.0 cm tall .... *C. brachyglossa*

Orthocerates
only 1 species .... *C. pauciflora*

Ovales
1a. Carpellate scales nearly equaling to slightly exceeding the length of the perigynia, mostly concealing the beaks of the perigynia .... **Group 1**

1b. Carpellate scales clearly exceeded by perigynia (scales nearly reaching apex of beak in *C. brevior*), the apex of the perigynia clearly visible
    2a. Perigynia up to 2.0 mm at widest point .... **Group 2**
    2b. Perigynia wider than 2.0 mm .... **Group 3**

Group 1
1a. Leaves stiff, very glaucous, many with rounded auricles at the summit of the sheath; floral scales white or white-brown; plants of Atlantic coast, rocky or sandy beaches .... *C. silicea*

1b. Leaves softer, slightly or not at all glaucous, without auricles; floral scales white-green or brown; plants usually not coastal
    2a. Lowest bract widened and flat at the base, often twice as long as the spike it subtends; perigynia 2.0–3.0 mm wide; achene 1.7–2.0 mm wide; carpellate scales about as wide as perigynia .... *C. adusta*
    2b. Lowest bract setaceous, shorter; perigynia 0.4–1.6 mm wide; achene 0.4–1.6 mm wide; carpellate scales narrower than perigynia

3a. Spikes aggregated in a dense cluster
  4a. Distal 0.5 mm portion of the perigynium beak slender, subterete, and smooth; perigynia 1.7–3.0 mm wide, prominently winged in the apical and basal portions; apical, herbaceous portion of leaf sheath smooth, without papillae ... . *C. ovalis*
  4b. Distal 0.5 mm portion of the perigynium beak wider, flat, and serrulate; perigynia 1.0–1.7 mm wide, prominently winged only in the apical half; apical, herbaceous portion of leaf sheath minutely papillose .... *C. oronensis*
3b. Spikes separated in an elongate cluster, usually, at least the lower, remote (sometimes approximate in *C. praticola*)
  5a. Distal 0.5 mm portion of the perigynium beak slender, subterete, and smooth; perigynia 4.0–6.5 mm long .... *C. praticola*
  5b. Distal 0.5 mm portion of the perigynium beak wider, flat-margined, and serrulate; perigynia 3.1–5.0 mm long
    6a. Perigynia white-green, evidently 4- to 10-nerved on the adaxial surface, 3.1–4.2 mm long, finely granular-papillose; lower spikes with a short, basal, staminate portion; upper spikes usually crowded ......*C. argyrantha*
    6b. Perigynia light brown to brown, nerveless or inconspicuously nerved in the basal portion of the adaxial surface, 3.4–5.0 mm long, not papillose; lower spikes with a prolonged, narrow, basal, staminate portion; upper spikes usually separate .... *C. foenea*

Group 2
1a. Carpellate scales acuminate, often with a short awn
  2a. Perigynia 0.4–1.0 mm wide, 3.5–5.0 times as long as wide, tapering to a distally terete beak .... *C. crawfordii*
  2b. Perigynia 1.0–3.0 mm wide, 1.5–4.0 times as long as wide, mostly tapering to a flat beak
    3a. Apical portion of sheaths opposite the leaf blade firm and green-veined nearly to the summit; perigynia 1.7–2.5 times as long as wide; plants usually of saline habitats .... *C. hormathodes*
    3b. Apical portion of sheaths opposite the leaf blade with a prolonged, thin, hyaline area; perigynia 2.5–3.0 times as long as wide; plants not of saline habitats .... *C. scoparia*
1b. Carpellate scales obtuse to acute, without awns
  4a. Perigynia often lacking wings in the basal portion; achenes very flat, the perigynia not obviously distended over them and flat throughout; leaf sheaths with sharp wings continuous with midrib and margins of the leaf blade; leaves 3.0–7.0 mm wide; non-flowering stems with numerous leaves spaced along the upper portion
    5a. Spikes 8.0–12.0 mm long, congested and overlapping, each composed of 30 or more perigynia; perigynia appressed to ascending; apical portion of the leaf sheath opposite the blade firm and green-veined nearly to the summit, the hyaline area very short .... *C. tribuloides*
    5b. Spikes 5.0–8.0 mm long, separate and in an elongate cluster, each composed of 15–30 perigynia; perigynia ascending to recurved; apical portion of the leaf sheath opposite the blade with a prolonged, weak, hyaline area .... *C. projecta*

4b. Perigynia with wings in the basal portion; achenes thicker, the perigynia obviously distended over them; leaf sheaths with rounded edges; leaves 1.0–4.5 mm wide (up to 6.0 mm in *C. normalis*); non-flowering stems with leaves crowded in the upper portion

    6a. Body of the perigynium obovate or elliptic to orbicular, broadest at or above the middle

        7a. Lower spikes often with a prolonged, narrow, basal, staminate portion; perigynia 2.5–3.5 mm long, usually with 3 or fewer nerves on the adaxial surface; achenes 0.95–1.2 mm wide .... *C. festucacea*

        7b. Lower spikes lacking a conspicuous, basal, staminate portion; perigynia 3.0–4.5 mm long, usually with 4 to 7 nerves on the adaxial surface; achenes 0.75–1.0 mm wide .... *C. longii*

    6b. Body of the perigynium broad-lanceolate to ovate, widest below the middle

        8a. Spikes densely aggregated, the basal internodes less than 6.0 mm long

            9a. Achenes 0.6–0.8 mm wide; perigynia nerveless on the adaxial surface, 1.0–1.8 mm wide .... *C. bebbii*

            9b. Achenes 0.9 mm wide or wider; perigynia nerved or nerveless on the adaxial surface, 1.7–2.0 mm wide

                10a. Floral scales dark brown or red-brown; leaves mostly 2.0–4.0 mm wide; apical portions of the sheaths minutely papillose on the herbaceous portion, the hyaline area not prolonged above the orifice; perigynia mostly 3.5–5.0 mm long .... *C. tincta*

                10b. Floral scales brown, at least in the center; leaves mostly 3.5–6.5 mm wide; apical portion of the sheaths without minute papillae, the hyaline area evidently prolonged; perigynia mostly 3.0–4.0 mm long .... *C. normalis*

        8b. Spikes separate, forming an elongate cluster (except in late-season individuals), the proximal internodes longer than 6.0 mm

            11a. Leaves mostly 3.5–6.5 mm wide, loosely sheathing; flowering stems with 4–7 leaves; sheaths green- and white-mottled .... *C. normalis*

            11b. Leaves mostly 1.0–2.5 mm wide, tightly sheathing; flowering stems with 3–5 leaves; sheaths uniformly colored .... *C. tenera*

Group 3

1a. Leaves stiff, very glaucous, many with rounded auricles at the summit of the sheath; scales white or white-brown; plants of Atlantic coast, rocky or sandy beaches .. *C. silicea*

1b. Leaves softer, not or only slightly glaucous, without auricles; scales light green to green or white-brown to brown; plants not of coastal areas (except *C. hormathodes*)

    2a. Carpellate scales acuminate, often with a short awn .... *C. hormathodes*

    2b. Carpellate scales obtuse to acute, without awns

        3a. Perigynia obovate, widest above the middle; leaf sheaths firm and green-veined nearly to the summit, the hyaline area very short, usually less than 0.5 cm

            4a. Perigynia 2.4–3.8 mm wide, nerveless on the adaxial surface; widest leaves 3.0–6.0 mm wide; sheaths truncate and slightly prolonged at the summit .... *C. cumulata*

4b. Perigynia 1.5–2.6 mm wide, evidently nerved on the adaxial surface; widest leaves 2.0–4.0 mm wide; sheaths concave and not prolonged at the summit .... *C. longii*

3b. Perigynia lanceolate, ovate, elliptic, or orbicular, widest at or below the middle; portion of the sheath opposite the leaf blade with or without an elongate, hyaline area near the summit

5a. Perigynia 1.2–2.2 mm wide

6a. Leaves 2.5–6.5 mm wide; sheaths loose, mottled with green and white, truncate and prolonged up to 2.0 mm at the summit .... *C. normalis*

6b. Leaves 1.5–4.0 mm wide; sheaths tight, evenly colored, concave and not prolonged at the summit

7a. Carpellate scales dark brown or red-brown; flowering stems with 2–4 leaves; apical portions of the sheaths minutely papillose on the herbaceous portion .... *C. tincta*

7b. Carpellate scales white-brown to light brown; flowering stems with 3–6 leaves; apical portions of the sheaths without minute papillae .... *C. festucacea*

5b. Perigynia 2.2–4.8 mm wide

8a. Perigynia 4.2–7.0 mm long, thin and translucent except over the achene, conspicuously nerved on the adaxial surface; apical, herbaceous portion of the leaf sheath distinctly papillose .... *C. bicknellii*

8b. Perigynia 3.2–4.8 mm long, thicker and opaque (except at wing-margins in *C. merritt-fernaldii*), usually inconspicuously nerved on the adaxial surface; apical, herbaceous portion of the leaf sheath papillose or not

9a. Apical, herbaceous portion of the leaf sheath minutely papillose; perigynium wing thin, translucent, sometimes erose, often with 1 or 2 nerves .... *C. merritt-fernaldii*

9b. Apical, herbaceous portion of the leaf sheath smooth, without papillae; perigynium wing thicker, nearly opaque, regularly and finely toothed, usually unnerved .... *C. brevior*

Paludosae

1a. Perigynia pubescent; style articulated with the summit of the achene and deciduous from it at maturity; leaves 1.0–6.0 mm wide

2a. Beak of perigynium with distinct, firm teeth 0.2–0.9 mm long; staminate spikes 1–3, long-pedunculate, usually equaled or overtopped by the lowest bract

3a. Perigynia 4.0–7.0 mm long, the pubescence short and not concealing the ribs; plants of dry, sandy or rocky soil .... *C. houghtoniana*

3b. Perigynia 2.8–5.0 mm long, the pubescence dense and concealing the ribs; plants of wet shores, meadows, and bogs

4a. Leaves flat, 2.0–5.0 mm wide; stems sharply angled and scabrous on the angles; teeth of perigynium beak 0.3–0.9 mm long .... *C. pellita*

4b. Leaves folded, 1.0–2.0 mm wide when measured as folded; stems obscurely angled and smooth, except sometimes near apex; teeth of perigynium beak 0.2–0.5 mm long .... *C. lasiocarpa*

2b. Beak of perigynium with soft, hyaline, obscure teeth, the beak eventually splitting in age; staminate spikes 1–3 but often solitary, subsessile to short-pedunculate, exceeding the lowest bract .... *C. vestita*
1b. Perigynia glabrous; style persistent on the achene in maturity; leaves 6.0–15.0 mm wide .... *C. lacustris*

Paniceae

1a. Perigynium beakless or with a short beak less than 0.5 mm long
  2a. Achene widest near the middle, rounded at the base; perigynia, if beaked at all, with a straight or slightly bent beak
    3a. Leaves involute, 0.5–3.5 mm wide, coriaceous, very glaucous; perigynia erect to ascending; plants of wet meadows and fens .... *C. livida*
    3b. Leaves involute at the base, flat above, becoming revolute in age, 1.5–7.0 mm wide, herbaceous, green or slightly glaucous; perigynia ascending to spreading; plants of dry fields and openings .... *C. panicea*
  2b. Achene widest above the middle, narrowed to the base; the minute beak of the perigynium bent at the base .... *C. tetanica*
1b. Perigynium with a distinct beak 0.7–2.2 mm long
  4a. Spikes with perigynia mostly aligned in 2 sparsely flowered rows, erect; stems phyllopodic, the basal sheaths producing elongate blades; plant of wet, calcareous woods and fens .... *C. vaginata*
  4b. Spikes with perigynia aligned in 3 or more densely flowered rows, spreading to drooping; flowering stems aphyllopodic, the basal sheaths lacking elongate blades; plants of dry woods and fields .... *C. polymorpha*

Paniculatae

1a. Apical portion of sheath opposite the leaf blade pale and often dotted on the hyaline surface; perigynia spreading, not covered by their scales, olive-brown to black, with a pale, depressed stripe on the abaxial surface; spikes closely aggregated .... *C. diandra*
1b. Apical portion of sheath opposite the leaf blade yellow-brown to bronze and not dotted on the surface; perigynia appressed, covered by their scales, light brown, without a pale, depressed stripe; spikes separate .... *C. prairea*

Phacocystis

1a. Carpellate spikes sessile, erect or ascending (the lowest sometimes drooping on short peduncles up to 3.0 cm long); carpellate scales brown or red-brown to black, unawned or with a short awn in *C. recta* and *C. vacillans*
  2a. Carpellate scales awned; achenes often with an invagination on one side; plants of saline habitats
    3a. Perigynia conspicuously nerved and densely long-papillose on each surface; orifice of beak often scabrous; midrib of carpellate scales not reaching the apex, 0.1–0.35 times the width of the scale; achenes dull, weakly to somewhat invaginated .... *C. vacillans*
    3b. Perigynia inconspicuously nerved and short-papillose; orifice of beak glabrous; midrib of carpellate scales reaching the apex, 0.35–0.5 times the width of the scale; achenes lustrous, strongly invaginated .... *C. recta*
  2b. Carpellate scales blunt to pointed; achenes without an invagination; plants mostly not of saline habitats (except sometimes *C. nigra*)

4a. Longest bract usually exceeding the spikes

    5a. Perigynia obscurely nerved, spotted or suffused with red-brown or purple; leaves often glaucous, 2.5–8.0 mm wide; rhizome elongate .... *C. aquatilis*

    5b. Perigynia with a few, conspicuous nerves on each surface, not spotted; leaves green, 1.5–3.5 mm wide; rhizome very short .... *C. lenticularis*

4b. Longest bract shorter than to equaling uppermost spike

    6a. Perigynium with a bent, recurved, or twisted beak; lowest spike commonly spreading on a short peduncle; carpellate scales purple-brown to black ....
.... *C. torta*

    6b. Perigynium with a straight, planar beak; lowest spike usually ascending, sessile; carpellate scales variously colored

        7a. Floral scales purple to black; perigynia suffused with or entirely purple-brown (sometimes entirely green in *C. nigra*); stems arising from the center of the previous year's leaf tufts, surrounded by senescing leaves

            8a. Perigynia nerveless or inconspicuously nerved; leaves 2.0–6.0 mm wide; plants of exposed, alpine areas .... *C. bigelowii*

            8b. Perigynia with a few, conspicuous nerves on each surface; leaves 2.0–3.0 mm wide; plants of lower habitats, mostly near the coast ....
.... *C. nigra*

        7b. Floral scales commonly brown or red-brown (to sometimes purple-brown in *C. stricta*); perigynia green to olive to brown, sometimes spotted or suffused with red; stems arising laterally, not surrounded by the previous year's leaves

            9a. Lower sheaths prominently fibrillose; spikes appressed to the stem; carpellate scales as long as perigynia; rhizome elongate; sheaths scabrous on the surface opposite the leaf blade; flowering stems not, or only slightly, surpassing the leaves; perigynia ovate to elliptic, not inflated; carpellate spikes often staminate at the apex .... *C. stricta*

            9b. Lower sheaths not, or only slightly, fibrillose; spikes not appressed to the stem; carpellate scales longer than the perigynia; rhizome short; sheaths glabrous on the surface opposite the leaf blade; flowering stems surpassing the leaves; perigynia obovoid to nearly globose, slightly inflated; carpellate spikes rarely staminate at the apex ....
.... *C. haydenii*

1b. Carpellate spikes borne on elongate peduncles (2.0–)3.0–7.0 cm long, spreading to drooping; carpellate scales copper-brown to brown, with long, usually scabrous, awns up to 10.0 mm long

    10a. Stems 1 or a few together, from elongate rhizomes; carpellate spikes 5.0–13.0 mm wide; perigynia coriaceous, stipitate; plants of saline habitats .... *C. paleacea*

    10b. Stems cespitose, without long rhizomes; carpellate spikes 4.0–8.0 mm wide; perigynia chartaceous, sessile; plants not of saline habitats

        11a. Lower sheaths scabrous; perigynia compressed, gradually tapering from near the middle to the beak; body of carpellate scales acute to acuminate at the apex ....
.... *C. gynandra*

11b. Lower sheaths glabrous; perigynia slightly inflated, usually obovoid, abruptly tapering from above the middle to the beak; body of carpellate scales truncate to retuse .... *C. crinita*

Phaestoglochin
1a. Sheaths loosely clasping the stem, septate-nodulose, mottled green and white on the abaxial surface
  2a. Inflorescence elongate, 4.0–10.0 cm long; lower spikes remote; carpellate scales equaling (or nearly) the length of the perigynium .... *C. sparganioides*
  2b. Inflorescence congested, 2.0–4.0 cm long; lower spikes crowded; carpellate scales only half (or slightly more than half) the length of the perigynium ...... *C. cephaloidea*
1b. Sheaths tightly clasping the stem, not septate-nodulose, not mottled
  3a. Spikes aggregated in a dense cluster and concealing the stem .... *C. cephalophora*
  3b. Spikes less congested, at least the lower ones clearly separated, the stem visible between at least the lower spikes
    4a. Carpellate scales brown, red-brown, or purple; ligule much longer than broad; perigynia 4.0–5.5 mm long; plants weedy, of disturbed habitats .... *C. spicata*
    4b. Carpellate scales white, green, or light brown; ligule not longer than wide; perigynia 2.0–4.0 mm long; plants native, of woods and dry, open areas
      5a. Perigynia spongy-thickened at the base, nerveless or inconspicuously nerved at the spongy base, with slightly revolute margins
        6a. Stigmas straight or reflexed; achene occupying upper half of perigynium .... *C. radiata*
        6b. Stigmas 1- to 3-times coiled; achene occupying lower half of perigynium
          7a. Widest leaves 0.9–1.5 mm wide; base of flowering stems 0.7–1.3 mm wide .... *C. appalachica*
          7b. Widest leaves 1.8–2.6 mm wide; base of flowering stems 1.5–2.2 mm thick .... *C. rosea*
      5b. Perigynia not spongy-thickened at the base, conspicuously nerved on the abaxial surface, with plane margins .... *C. muehlenbergii*

Phyllostachyae
only 1 species .... *C. backii*

Polytrichoideae
only 1 species .... *C. leptalea*

Porocystis
1a. Uppermost spike composed entirely of staminate flowers, rarely with 1–3 carpellate flowers; fruiting in summer .... *C. pallescens*
1b. Uppermost spike with carpellate flowers in the apical portion, staminate in the basal portion; fruiting in late spring–early summer
  2a. Perigynia glabrous or sparsely pubescent; spikes 3.0–11.0 mm thick
    3a. Perigynia strongly compressed, appressed-ascending, obscurely nerved; carpellate scales awnless or essentially so; achenes 1.6–2.1 mm long .... *C. hirsutella*
    3b. Perigynia nearly terete, spreading, conspicuously nerved; carpellate scales with an awn 0.5–2.0 mm long; achenes 2.0–2.6 mm long .... *C. bushii*

2b. Perigynia densely pubescent; spikes 2.0–5.0 mm thick
    4a. Carpellate spikes linear-cylindric, 2.0–4.0 cm long, the flowers approximate only
        near the base; anthers mostly 1.5–2.5 mm long .... ***C. virescens***
    4b. Carpellate spikes nearly globose to thick-cylindric, 1.0–2.0 cm long, densely
        flowered throughout; anthers mostly 0.7–1.5 mm long .... ***C. swanii***

Praecoces
only 1 species .... ***C. caryophyllea***

Scirpinae
only 1 species .... ***C. scirpoidea***

Squarrosae
only 1 species .... ***C. typhina***

Stellulatae
1a. Spike usually 1 per stem; anthers 2.0–3.6 mm long; leaves involute .... ***C. exilis***
1b. Spikes 2 or more per stem; anthers 0.6–2.2 mm long; leaves flat or folded
    2a. Spikes 7–15 per stem, closely aggregated in an elongate cluster, each spike
        composed of 20–40 carpellate flowers .... ***C. arcta***
    2b. Spikes 2–8 per stem, approximate to remote, usually with fewer carpellate flowers
        3a. Plants subdioecious, the spikes all or nearly all of a single sex on each stem ....
            .... ***C. sterilis***
        3b. Plants monoecious, at least the uppermost spike with a lower staminate and
        upper carpellate portion, the spikes never entirely staminate
        4a. Perigynia 1.5–3.0 mm wide, up to 1.0–1.7 times as long as wide ....
            .... ***C. atlantica***
        4b. Perigynia 0.9–1.8 mm wide, mostly more than 1.7 times as long as wide
            5a. Widest leaves 2.8–5.0 mm wide .... ***C. wiegandii***
            5b. Widest leaves up to 2.7 mm wide
                6a. Perigynia 2.8–3.5 mm long, 1.8–3.2 times as long as wide, with a long
                beak 1.0–1.6 mm long and 0.5–1.0 times as long as the body of the
                perigynium .... ***C. echinata***
                6b. Perigynia 1.9–3.2 mm long, 1.1–2.0 times as long as wide, with a
                short beak 0.5–1.2 mm long and 0.25–0.5 times as long as the body
                7a. Perigynia convexly tapered in the apical portion, forming a
                    shoulder below the beak, nerveless or few-nerved adaxially, with
                    many, fine serrations on the beak .... ***C. interior***
                7b. Perigynia gradually or concavely tapered in the apical portion,
                    several-nerved adaxially, with few, coarse serrations on the beak
                    .... ***C. atlantica***

Sylvaticae
1a. Leaves pubescent; carpellate spikes thick-cylindric, 10.0–25.0 x 6.0–8.0 mm; beak of
    the perigynium *ca.* 0.5 times the length of the body of the perigynium .... ***C. castanea***
1b. Leaves glabrous or scabrous; carpellate spikes linear-cylindric, 20.0–60.0 x
    2.0–6.0 mm; beak of the perigynium very short

2a. Leaves 3.0–8.0 mm wide; achene sessile at the base of the perigynium; perigynium 3.2–4.8 mm long, distinctly trigonous .... *C. arctata*
2b. Leaves 2.0–4.0 mm wide; achene not at the base of the perigynium, raised on a stipe 0.5–1.5 mm long; perigynium 4.5–7.0 mm long, obscurely trigonous .... *C. debilis*

Vesicariae

1a. Carpellate scales with long, scabrous awns, the awns equal to the length of the scale; carpellate spikes ascending to drooping
  2a. Perigynia ascending to spreading, thin-textured, inflated, nearly terete in cross-section, yellow-green, aging yellow-brown, not or only slightly stipitate; ligules approximately as wide as long
    3a. Perigynia mostly 15–20 nerved, slightly inflated, 1.5–2.0 mm thick, the body lanceolate to lance-ovate .... *C. hystericina*
    3b. Perigynia mostly *ca.* 10-nerved, very inflated, 2.0–4.0 mm thick, the body ovoid or ovoid-globose
      4a. Carpellate spikes 14.0–20.0 mm thick; perigynium beak shorter than the body of the perigynium; leaves (2.0–)4.0–7.0 mm wide .... *C. lurida*
      4b. Carpellate spikes 8.0–13.0 mm thick; perigynium beak longer than the body; leaves 2.0–4.0 mm wide .... *C. baileyi*
  2b. Perigynia spreading to reflexed, firm, coriaceous, only scarcely inflated, definitely 2-edged, green, aging light brown, stipitate; ligule longer than wide
    5a. Teeth of the perigynium beak 0.5–1.0 mm long, straight and nearly parallel; carpellate spikes 8.0–11.0 mm thick .... *C. pseudocyperus*
    5b. Teeth of the perigynium beak 1.2–2.3 mm long, curving and divergent; carpellate spikes 13.0–17.0 mm thick .... *C. comosa*
1b. Carpellate scales awnless or with glabrous awns shorter than the scales; carpellate spikes ascending
  6a. Leaves 1.0–4.0 mm wide, involute or becoming so; perigynium beak emarginate, lacking conspicuous teeth
    7a. Stigmas 2; achenes lenticular; perigynia nerveless or inconspicuously nerved on each surface .... *C. saxatilis*
    7b. Stigmas 3; achenes trigonous; perigynia conspicuously nerved on each surface ....
    .... *C. oligosperma*
  6b. Leaves 3.0–12.0 mm wide, flat or folded (sometimes involute in *C. bullata*); perigynium beak with evident, slender teeth
    8a. Perigynia in fruit spreading to reflexed, arranged in 8–12 ranks
      9a. Perigynia horizontally spreading to reflexed; sheaths loosely clasping the stem, truncate to prolonged at the summit; bract of the lowest carpellate spike 2 or more times as long as the cluster of spikes; stems cespitose, without long rhizomes .... *C. retrorsa*
      9b. Perigynia spreading; sheaths tightly clasping the stem, concave at the summit; bract of the lowest carpellate spike 1–2 times the length of the cluster of spikes; stems solitary or few together, produced from long rhizomes
        10a. Leaves mostly 5.0–12.0 mm wide; perigynia 4.0–7.0 mm long, with a smooth beak; stems thick and spongy at the base .... *C. utriculata*

10b. Leaves 2.0–5.0 mm wide; perigynia 6.0–9.0 mm long, with a serrulate
        beak; stems slender and firm at the base .... *C. bullata*
  8b. Perigynia ascending, arranged in 6 ranks (sometimes as many as 8 ranks in *C.
      vesicaria*)
      11a. Achenes with a deep indentation on one of the angles, 3.0–4.0 mm long;
            perigynia 7.0–10.0 x 4.0–7.0 mm; bract of the lowest carpellate spike 3 or
            more times as long as the cluster of spikes .... *C. tuckermanii*
      11b. Achenes without an indentation, 1.7–2.4 mm long; perigynia 5.0–8.0 x
            2.0–3.0 mm; bract of the lowest carpellate spike up to 3 times as long as the
            cluster of spikes .... *C. vesicaria*

Vulpinae

1a. Perigynia rounded and inconspicuously spongy-thickened at the base, with a beak shorter
    than the body of the perigynium; sheaths red- or brown-dotted near the margins ....
    .... *C. alopecoidea*
1b. Perigynia truncate and conspicuously spongy-thickened at the base, with a beak as long
    or longer than the body; sheaths not dotted
    2a. Leaf sheaths thin and fragile at the margins, prolonged beyond the base of the blade,
        commonly wrinkled or corrugated; leaves not epistomic, not papillose on the adaxial
        surface; inflorescence often larger, 5.0–10.0 x 1.0–3.0 cm .... *C. stipata*
    2b. Leaf sheaths thick and firm at the margins, not prolonged beyond the base of the
        blade, only rarely wrinkled or corrugated; leaves epistomic, papillose on the adaxial
        surface; inflorescence often smaller, 3.0–6.0 x 1.0–2.0 cm .... *C. laevivaginata*

Carex adusta Boott
synonym(s): —
common name(s): swarthy sedge
habit: cespitose, perennial herb
range: Newfoundland, s. to NY, w. to MN, n. to British Columbia
state frequency: very rare
habitat: dry, open woods and gravels
notes: Member of section *Ovales*. This species is ranked S1 by the Maine Natural Areas Program.
    Often appearing after fire or other disturbance.

Carex albicans Willd. *ex* Spreng.
synonym(s): <*C. a.* var. *albicans*> = *Carex artitecta* Mackenzie, *Carex nigromarginata* Schwein.
    var. *muhlenbergii* (Gray) Gleason; <*C. a.* var. *emmonsii*> = *Carex emmonsii* Dewey *ex* Torr.,
    Carex varia Muhl. *ex* Willd. *sensu* Britton and Brown (1913)
common name(s): <*C. a.* var. *albicans*> = white-tinged sedge; <*C. a.* var. *emmonsii*> = Emmons'
    sedge
habit: erect, somewhat cespitose, short-rhizomatous, perennial herb
range: Nova Scotia, s. to FL, w. to OK, n. to WI and Ontario
state frequency: uncommon
habitat: acidic to circumneutral soils of forests and open areas
key to the subspecies:
    1a. Stems erect, 2.0–5.0 dm, clearly surpassing the leaves; staminate scales obtuse, the midrib
        inconspicuous in the distal portion; staminate spikes 8.4–11.1 mm long; plants of rich, often
        calcareous, soils .... *C. a.* var. *albicans*

1b. Stems weaker, 0.5–2.0 dm, often shorter than the leaves; staminate scales acute to acuminate, with a distally prominent midrib prolonged as a short cusp; spikes 5.0–8.5 mm long; plants of acid soils .... *C. a.* var. *emmonsii* (Dewey *ex* Torr.) J. Rettig
notes: Member of section *Acrocystis.*

Carex alopecoidea Tuckerman

synonym(s): —
common name(s): foxtail sedge
habit: stout, clustered, perennial herb
range: Quebec, s. to NJ, w. to MO, n. to Saskatchewan
state frequency: very rare
habitat: wet, often circumneutral, meadows
notes: Member of section *Vulpinae*. This species is ranked S1 by the Maine Natural Areas Program.

Carex appalachica J. Webber & P. W. Ball

synonym(s): *Carex radiata* (Wahlenb.) Dew. *sensu* Fernald (1950)
common name(s): —
habit: cespitose, erect, perennial herb
range: ME, s. to DE and mountains of NC, w. to OH, n. to Ontario
state frequency: uncommon
habitat: open woods, thickets
notes: Member of section *Phaestoglochin.*

Carex aquatilis Wahlenb.

synonym(s): <*C. a.* var. *aquatilis*> = *Carex aquatilis* Wahlenb. var. *altior* (Rydb.) Fern.; <*C. a.* var. *substricta*> = *Carex substricta* (Kükenth.) Mackenzie
common name(s): water sedge
habit: densely cespitose, rhizomatous, perennial herb
range: circumboreal; s. to NJ, w. to NM and CA
state frequency: uncommon
habitat: shallow water, wet soil
key to the subspecies:
   1a. Carpellate spikes 4.0–7.0 mm wide; carpellate scales pale, with a narrow, red-brown margin and a broad, pale midrib .... *C. a.* var. *substricta* Kükenth.
   1b. Carpellate spikes 3.0–5.0 mm wide; carpellate scales red-brown to black throughout except for the narrow, pale midrib .... *C. a.* var. *aquatilis*
notes: Member of section *Phacocystis.*

Carex arcta Boott

synonym(s): —
common name(s): northern clustered sedge
habit: densely cespitose, perennial herb
range: Quebec, s. to ME, w. to CA, n. to AK
state frequency: uncommon
habitat: wet woods, thickets, meadows, and stream banks
notes: Member of section *Stellulatae.*

Carex arctata Boott ex Hook.

synonym(s): —
common name(s): drooping wood sedge
habit: densely cespitose, perennial herb
range: Newfoundland, s. to CT, w. to OH, n. to MN and Ontario
state frequency: occasional

habitat: moist woods, borders, thickets, meadows, and clearings
notes: Member of section *Sylvaticae*.

Carex argyrantha Tuckerman

synonym(s): —
common name(s): —
habit: densely cespitose, perennial herb
range: Nova Scotia, s. to NC, w. to OH, n. to MI and Ontario
state frequency: uncommon
habitat: dry, sandy or rocky woods, openings, and banks
notes: Member of section *Ovales*. This species is minutely papillose on the apical, herbaceous portion of the leaf sheath.

Carex atherodes Spreng.

synonym(s): —
common name(s): awned sedge
habit: stout, solitary or few together, long-rhizomatous, stoloniferous, perennial herb
range: ME, s. to WV, w. to CO, n. to Yukon
state frequency: very rare
habitat: calcareous marshes, meadows, swales, shores, and shallow water
notes: Member of section *Carex*. This species is ranked SH by the Maine Natural Areas Program.

Carex atlantica Bailey

synonym(s): <*C. a.* ssp. *atlantica*> = *Carex incomperta* Bickn.; <*C. a.* ssp. *capillacea*> = *Carex howei* Mackenzie
common name(s): eastern sedge
habit: coarse, cespitose, perennial herb
range: New Brunswick, s. to FL, w. to TX, n. to MI and Ontario
state frequency: uncommon, predominantly in coastal areas
habitat: open or forested wetlands
key to the subspecies:
    1a. Widest leaves 1.5–4.0 mm wide; spikes 3–8 per stem, in a cluster 1.8–4.5 cm long; perigynia 2.3–3.8 x 1.5–3.0 mm .... *C. a.* ssp. *atlantica*
    1b. Widest leaves 0.5–1.5 mm; spikes 2–5 per stem, in a cluster 0.8–2.0 cm long; perigynia 1.9–3.0 x 1.3–2.0 mm .... *C. a.* ssp. *capillacea* (Bailey) Reznicek
notes: Member of section *Stellulatae*.

Carex atratiformis Britt.

synonym(s): —
common name(s): black sedge
habit: loosely cespitose, perennial herb
range: Labrador, s. to ME, w. to Alberta, n. to Yukon
state frequency: rare
habitat: often calcareous ledges, banks, ravines, and alpine meadows
notes: Member of section *Atratae*. This species is ranked S2 by the Maine Natural Areas Program.

Carex aurea Nutt.

synonym(s): —
common name(s): golden-fruited sedge
habit: slender, solitary or loosely cespitose, rhizomatous, perennial herb
range: Newfoundland, s. to PA, w. to CA, n. to AK
state frequency: uncommon
habitat: moist to wet meadows, shores, and springy banks
notes: Member of section *Bicolores*.

Carex backii Boott
synonym(s): —
common name(s): Back's sedge
habit: cespitose, mat-forming, perennial herb
range: Quebec, s. to NJ, w. to UT and OR, n. to British Columbia
state frequency: very rare
habitat: dry ledges and rocky or sandy woods and thickets
notes: Member of section *Phyllostachyae*. This species is ranked S1 by the Maine Natural Areas
    Program.

Carex baileyi Britt.
synonym(s): —
common name(s): Bailey's sedge
habit: cespitose, perennial herb
range: ME, s. to VA, w. to TN
state frequency: very rare
habitat: wet woods and meadows, swamps
notes: Member of section *Vesicariae*. This species is ranked S1 by the Maine Natural Areas Program.

Carex bebbii Olney ex Fern.
synonym(s): —
common name(s): —
habit: densely cespitose, perennial herb
range: Newfoundland, s. to NJ, w. to CO and OR, n. to AK
state frequency: rare
habitat: wet meadows and shores, often calcareous
notes: Member of section *Ovales*.

Carex bicknellii Britt.
synonym(s): —
common name(s): Bicknell's sedge
habit: cespitose, short-rhizomatous, perennial herb
range: ME, s. to DE, w. to NM, n. to Saskatchewan
state frequency: very rare
habitat: open woods, thickets, slopes, meadows
notes: Member of section *Ovales*. This species is ranked SH by the Maine Natural Areas Program.

Carex bigelowii Torr. ex Schwein.
synonym(s): *Carex concolor* R. Br., *Carex rigida* Goodenough var. *bigelowii* Tuckerman
common name(s): Bigelow's sedge
habit: solitary or few together, rhizomatous, perennial herb
range: circumboreal; s. to mountains of ME and NH, w. to AK
state frequency: rare
habitat: alpine meadows
notes: Member of section *Phacocystis*. This species is ranked S2 by the Maine Natural Areas Program.

Carex blanda Dewey
synonym(s): —
common name(s): woodland sedge
habit: cespitose, perennial herb
range: ME, s. to FL, w. to TX, n. to ND and Ontario
state frequency: rare
habitat: woods, thickets, meadows
notes: Member of section *Laxiflorae*.

Carex brachyglossa Mackenzie
synonym(s): *Carex annectens* (Bickn.) Bickn., *Carex vulpinoidea* Michx. var. *ambigua* F. Boott
common name(s): yellow-fruited sedge
habit: stout, clustered, aphyllopodic, perennial herb
range: ME, s. to VA, w. to IN, n. to WI
state frequency: uncommon
habitat: moist to wet, often sandy, meadows, ditches, shores, and marshes
notes: Member of section *Multiflorae*.

Carex brevior (Dewey) Mackenzie *ex* Lunell
synonym(s): —
common name(s): —
habit: densely cespitose, perennial herb
range: ME, s. to VA, w. to NM, n. to British Columbia
state frequency: rare
habitat: dry, open soil
notes: Member of section *Ovales*.

Carex bromoides Schkuhr *ex* Willd.
synonym(s): —
common name(s): brome-like sedge
habit: slender, erect to ascending, densely cespitose, perennial herb
range: New Brunswick, s. to FL, w. to Mexico, n. to MN and Ontario
state frequency: uncommon
habitat: rich, wet woods, swamps, bogs
notes: Member of section *Deweyanae*.

Carex brunnescens (Pers.) Poir.
synonym(s): —
common name(s): brownish sedge
habit: cespitose, perennial herb
range: circumboreal; s. to mountains of NC, w. to UT and WA
state frequency: occasional
habitat: bogs, wet woods, headlands, summits
key to the subspecies:
    1a. Leaves (1.0–)1.5–2.5 mm wide; spikes ellipsoid, the 2 lowest 0.1–1.0 cm apart, the terminal not, or only sub-, clavate-based; stems erect, mostly 1.5–6.0 dm tall ...... *C. b.* ssp. *brunnescens*
    1b. Leaves (0.5–)1.0–1.5 mm wide; spikes nearly globose, the 2 lowest 1.0–2.5 cm apart, the terminal usually clavate-based; stems ascending to arching, 3.0–9.0 dm ....
.... *C. b.* ssp. *sphaerostachya* (Tuckerman) Kalela
notes: Member of section *Glarosae*.

Carex bullata Schkuhr *ex* Willd.
synonym(s): —
common name(s): button sedge
habit: slender, rhizomatous, perennial herb
range: Nova Scotia, s. to GA
state frequency: rare
habitat: meadows, swamps, bogs
notes: Member of section *Vesicariae*.

Carex bushii Mackenzie

synonym(s): —
common name(s): Bush's sedge
habit: cespitose, perennial herb
range: ME, s. to VA and MS, w. to TX, n. to MI
state frequency: very rare
habitat: meadows, fields, moist, open woods
notes: Member of section *Porocystis*. This species is ranked SX by the Maine Natural Areas Program.

Carex buxbaumii Wahlenb.

synonym(s): —
common name(s): brown sedge
habit: solitary or several together, rhizomatous, stoloniferous, perennial herb
range: circumboreal; s. to NC, w. to AR, CO, and CA
state frequency: uncommon
habitat: ledges or gravel of circumneutral river shores or rarely coastal islands
notes: Member of section *Atratae*.

Carex canescens L.

synonym(s): *Carex canescens* L. var. *subloliacea* (Laestad.) Hartman *sensu* Fernald (1950)
common name(s): silvery sedge
habit: cespitose, perennial herb
range: circumboreal and bipolar; s. to VA, w. to AZ and CA
state frequency: occasional
habitat: swamps, bogs, shallow water
key to the subspecies:
    1a. Spikes, especially the lower, remote, in a cluster 6.0–15.0 cm long, the 2 lower ones
        2.0–4.0 cm apart; perigynia mostly broad-ovate; stems 3.0–9.0 dm tall ....
        .... *C. c.* ssp. *disjuncta* (Fern.) Toivonen
    1b. Spikes more approximate, in a cluster 2.0–7.0 cm long, the 2 lower ones 0.5–2.5 cm apart;
        perigynia mostly oval-ovate; stems 1.5–6.0 dm tall .... *C. c.* ssp. *canescens*
notes: Member of section *Glarosae*. Known to hybridize with *Carex tenuiflora* and *C. trisperma*.

Carex capillaris L.

synonym(s): *Carex capillaris* L. var. *major* Blytt
common name(s): hair-like sedge, capillary sedge
habit: densely cespitose, perennial herb
range: circumboreal; s. to ME and NH, w. to mountains of NM
state frequency: very rare
habitat: circumneutral shores and seepy ledges
notes: Member of section *Capillares*. This species is ranked S1 by the Maine Natural Areas Program.

Carex caryophyllea Lat.

synonym(s): —
common name(s): vernal sedge
habit: loosely cespitose, short-rhizomatous, perennial herb
range: Eurasia; naturalized to New Brunswick, s. to DC
state frequency: very rare
habitat: dry fields and roadsides
notes: Member of section *Praecoces*.

Carex castanea Wahlenb.
synonym(s): —
common name(s): chestnut sedge
habit: loosely cespitose, perennial herb
range: Newfoundland, s. to CT, w. to MN, n. to Manitoba
state frequency: rare
habitat: rich soil of forests, meadows, fens, and shores
notes: Member of section *Sylvaticae*.

Carex cephaloidea (Dewey) Dewey
synonym(s): *Carex sparganioides* Muhl. *ex* Willd. var. *cephaloidea* (Dewey) Carey
common name(s): thin-leaved sedge
habit: cespitose, perennial herb
range: New Brunswick, s. to NJ, w. to IL, n. to MN and Ontario
state frequency: uncommon
habitat: rich forests, thickets
notes: Member of section *Phaestoglochin*.

Carex cephalophora Muhl. *ex* Willd.
synonym(s): —
common name(s): oval-headed sedge
habit: densely cespitose, perennial herb
range: ME, s. to FL, w. to TX, n. to Manitoba
state frequency: rare
habitat: dry, open areas and woodlands
notes: Member of section *Phaestoglochin*.

Carex chordorrhiza Ehrh. *ex* L. f.
synonym(s): —
common name(s): creeping sedge
habit: reclining to ascending, perennial herb
range: circumboreal; s. to MA, w. to MT
state frequency: uncommon
habitat: organic soil wetlands
notes: Member of section *Chordorrhizae*.

Carex communis Bailey
synonym(s): —
common name(s): fibrous-rooted sedge
habit: densely cespitose, perennial herb
range: Quebec, s. to GA, w. to AR, n. to MN and Ontario
state frequency: occasional
habitat: clearings, forested areas
notes: Member of section *Acrocystis*.

Carex comosa Boott
synonym(s): —
common name(s): bristly sedge
habit: short-rhizomatous, perennial herb
range: New Brunswick, s. to FL, w. to LA, n. to MN and Ontario; also ID, s. and w. to CA, n. to WA
state frequency: uncommon
habitat: swamps, wet meadows, shallow water
notes: Member of section *Vesicariae*.

Carex conoidea Schkuhr *ex* Willd.
synonym(s): *Carex katahdinensis* Fern.
common name(s): field sedge
habit: densely cespitose, perennial herb
range: Newfoundland, s. to NC, w. to MO, n. to MN and Ontario
state frequency: uncommon
habitat: moist, gravelly or rocky shores, grassy areas
notes: Member of section *Griseae*.

Carex crawei Dewey
synonym(s): —
common name(s): Crawe's sedge
habit: solitary to loosely cespitose, rhizomatous, perennial herb
range: Newfoundland, s. to NJ and AL, w. to WA, n. to British Columbia
state frequency: very rare
habitat: wet, calcareous shores, meadows, and ledges
notes: Member of section *Granulares*. This species is ranked S1 by the Maine Natural Areas Program.

Carex crawfordii Fern.
synonym(s): —
common name(s): —
habit: slender, erect, cespitose, perennial herb
range: Newfoundland, s. to NJ, w. to WA, n. to AK
state frequency: occasional
habitat: wet meadows and shores, swamps
notes: Member of section *Ovales*.

Carex crinita Lam.
synonym(s): <*C. c.* var. *crinita*> = *Carex crinita* Lam. var. *minor* Boott
common name(s): drooping sedge, fringed sedge
habit: densely cespitose, perennial herb
range: Newfoundland, s. to GA, w. to TX, n. to MN and Manitoba
state frequency: occasional
habitat: wet woods and thickets, swales
key to the subspecies:
    1a. Perigynia slightly inflated, ascending-spreading, conspicuously longer than and loosely enclosing the mature achene; carpellate spikes densely flowered; carpellate scales long-awned, at least those of the basal portion of the spike .... *C. c.* var. *crinita*
    1b. Perigynia not inflated, ascending, barely longer than and tightly enclosing the mature achene; carpellate spikes less densely flowered; carpellate scales lance-attenuate .... *C. c.* var. *porteri* (Olney) Fern.
notes: Member of section *Phacocystis*.

Carex cryptolepis Mackenzie
synonym(s): *Carex flava* L. var. *fertilis* Peck *sensu* Fernald (1950)
common name(s): —
habit: cespitose, perennial herb
range: Newfoundland, s. to NJ, w. to IN, n. to MN and Ontario
state frequency: uncommon
habitat: wet, calcareous, meadows and shores
notes: Member of section *Ceratocystis*.

Carex cumulata (Bailey) Fern.
synonym(s): —
common name(s): —
habit: cespitose, perennial herb
range: New Brunswick, s. to NJ, w. to IL, n. to Manitoba
state frequency: uncommon
habitat: dry to moist, rocky or sandy soil
notes: Member of section *Ovales.*

Carex debilis Michx.
synonym(s): —
common name(s): white-edged sedge
habit: cespitose, perennial herb
range: Newfoundland, s. to FL, w. to TX, n. to MN and Ontario
state frequency: occasional
habitat: moist to wet woods, thickets, meadows, clearings, and shores
key to the subspecies:
    1a. Perigynia yellow- to red-brown, *ca.* 2.0 times the length of the scales; basal leaves 2.0–4.0 mm
       wide; carpellate spikes spreading to drooping .... *C. d.* var. *rudgei* Bailey
    1b. Perigynia green to green-brown, up to 1.5 times the length of the scales; basal leaves
       4.0–7.0 mm wide; carpellate spikes erect to spreading .... *C. d.* var. *strictior* Bailey
notes: Member of section *Sylvaticae.*

Carex deflexa Hornem.
synonym(s): —
common name(s): northern sedge
habit: loosely to densely cespitose, short-rhizomatous, perennial herb
range: Greenland, s. to MA, w. to MN, n. to AK
state frequency: uncommon
habitat: moist clearings, slopes, and open woods, swamps
notes: Member of section *Acrocystis.*

Carex deweyana Schwein.
synonym(s): —
common name(s): —
habit: erect to ascending, densely cespitose, perennial herb
range: Labrador, s. to PA, w. to Mexico and CA, n. to British Columbia
state frequency: uncommon
habitat: rich, open woods
notes: Member of section *Deweyanae.*

Carex diandra Schrank
synonym(s): —
common name(s): —
habit: loosely to densely cespitose, short-rhizomatous, perennial herb
range: circumboreal; s. to NJ, w. to CO and CA
state frequency: rare
habitat: calcareous fens, swamps, and meadows
notes: Member of section *Paniculatae.*

Carex digitalis Willd.
synonym(s): —
common name(s): slender wood sedge

habit: densely cespitose, perennial herb
range: ME, s. to FL, w. to TX, n. to WI and Ontario
state frequency: uncommon
habitat: dry woods
notes: Member of section *Careyanae*.

Carex disperma Dewey

synonym(s): —
common name(s): two-seeded sedge, soft-leaved sedge
habit: slender, loosely cespitose, rhizomatous, perennial herb
range: circumboreal; s. to NJ, w. to NM and CA
state frequency: occasional
habitat: bogs, wet woods, thickets, and clearings
notes: Member of section *Dispermae*. Sometimes confused with *Carex trisperma* (*q.v.*).

Carex eburnea Boott

synonym(s): —
common name(s): ebony sedge, bristle-leaved sedge
habit: slender, densely cespitose, rhizomatous, perennial herb
range: Newfoundland, s. to VA, w. to TX, n. to AK
state frequency: very rare
habitat: circumneutral rock of cliffs and river shore ledges
notes: Member of section *Albae*. This species is ranked S1 by the Maine Natural Areas Program.

Carex echinata Murr.

synonym(s): *Carex angustior* Mackenzie, *Carex cephalantha* (Bailey) Bickn., *Carex josselynii* (Fern.) Mackenzie *ex* Pease
common name(s): star sedge
habit: densely cespitose, perennial herb
range: circumboreal; s. to mountains of NC, w. to UT and CA
state frequency: occasional
habitat: swamps, bogs, wet meadows and ditches
notes: Member of section *Stellulatae*.

Carex exilis Dewey

synonym(s): —
common name(s): coast sedge
habit: stiff, densely cespitose, perennial herb
range: Labrador, s. to MD, w. to MN, n. to Ontario; also NC, MS, and AL
state frequency: uncommon
habitat: *Sphagnum* bogs, wet, peaty soil
notes: Member of section *Stellulatae*.

Carex festucacea Schkuhr ex Willd.

synonym(s): —
common name(s): fescue sedge
habit: slender, cespitose, perennial herb
range: ME, s. to GA, w. to TX, n. to MI and Ontario
state frequency: rare
habitat: marshes, swales, low woods
notes: Member of section *Ovales*.

Carex flava L.
synonym(s): *Carex flava* L. var. *gaspensis* Fern.
common name(s): —
habit: cespitose, perennial herb
range: Newfoundland, s. to NJ, w. to MT, n. to AK
state frequency: uncommon
habitat: fens, meadows, shores, swales, often circumneutral
notes: Member of section *Ceratocystis*. Known to hybridize with *Carex viridula*.

Carex foenea Willd.
synonym(s): *Carex aenea* Fern.
common name(s): —
habit: densely cespitose, rhizomatous, perennial herb
range: Labrador, s. to NJ, w. to AZ, n. to Yukon
state frequency: uncommon
habitat: dry soil of woodlands, slopes, and ledges
notes: Member of section *Ovales*. This species is minutely papillose on the apical, herbaceous portion of the leaf sheath.

Carex folliculata L.
synonym(s): —
common name(s): long sedge
habit: coarse, clumped, perennial herb
range: Newfoundland, s. to GA, w. and n. to WI
state frequency: uncommon
habitat: wet thickets, swampy woods, swales
notes: Member of section *Folliculatae*.

Carex garberi Fern.
synonym(s): *Carex garberi* Fern. var. *bifaria* Fern.
common name(s): Garber's sedge
habit: loosely cespitose, rhizomatous, perennial herb
range: Quebec, s. to ME, w. to British Columbia, n. to AK
state frequency: very rare
habitat: calcareous ledges and sandy or gravelly shores
notes: Member of section *Bicolores*. This species is ranked S2 by the Maine Natural Areas Program.

Carex gracillima Schwein.
synonym(s): —
common name(s): graceful sedge
habit: cespitose, perennial herb
range: Newfoundland, s. to NC, w. to MO, n. to Manitoba
state frequency: occasional
habitat: woods, thickets, meadows
notes: Member of section *Gracillimae*.

Carex granularis Muhl. *ex* Willd. var. **haleana** (Olney) Porter
synonym(s): *Carex haleana* Olney, *Carex shriveri* Britt.
common name(s): Shriver's sedge
habit: cespitose, short-rhizomatous, perennial herb
range: Quebec, s. to VA, w. to KS, n. to Saskatchewan
state frequency: rare
habitat: calcareous, wet meadows, swales, shores
notes: Member of section *Granulares*.

Carex grisea Wahlenb.
synonym(s): *Carex amphibola* Steud. var. *turgida* Fern.
common name(s): narrowleaf sedge
habit: slender, loosely or densely cespitose, perennial herb
range: New Brunswick, s. to GA, w. to TX, n. to Ontario
state frequency: rare
habitat: rich woods, moist meadows
notes: Member of section *Griseae*. This species is ranked S2 by the Maine Natural Areas Program.

Carex gynandra Schwein.
synonym(s): *Carex crinita* Lam. var. *gynandra* (Schwein.) Schwein. & Torr., *Carex crinita* Lam. var.
    simulans Fern.
common name(s): nodding sedge
habit: densely cespitose, perennial herb
range: Newfoundland, s. to mountains of NC, w. to TN, n. to MN and Ontario
state frequency: common
habitat: wet woods, clearings, and ditches
notes: Member of section *Phacocystis*.

Carex gynocrates Wormsk. *ex* Drej.
synonym(s): *Carex dioica* L. var. *gynocrates* (Wormsk. *ex* Drej.) Ostenf.
common name(s): northern bog sedge
habit: slender, rhizomatous herb
range: circumboreal; s. to PA, w. to UT
state frequency: rare
habitat: circumneutral fens
notes: Member of section *Dioicae*. This species is ranked S2/S3 by the Maine Natural Areas Program.

Carex haydenii Dewey
synonym(s): —
common name(s): Hayden's sedge
habit: slender, cespitose, perennial herb
range: New Brunswick, s. to NJ, w. to MO and NE, n. to SD and Ontario
state frequency: uncommon
habitat: swales, wet meadows, open woods, and thickets
notes: Member of section *Phacocystis*.

Carex hirsutella Mackenzie
synonym(s): *Carex complanata* Torr. & Hook. var. *hirsuta* (Willd.) Gleason
common name(s): —
habit: slender, loosely cespitose, perennial herb
range: ME, s. to mountains of SC, w. to OK, n. to MI and Ontario
state frequency: very rare
habitat: open woods, clearings, fields, meadows
notes: Member of section *Porocystis*.

Carex hirta L.
synonym(s): —
common name(s): hairy sedge, hammer sedge
habit: creeping, mat-forming, rhizomatous, perennial herb
range: Europe; naturalized to Nova Scotia, s. to DC, w. and n. to WI
state frequency: very rare
habitat: dry fields, roadsides, and waste places
notes: Member of section *Carex*.

Carex hirtifolia Mackenzie
synonym(s): —
common name(s): pubescent sedge
habit: loosely cespitose, perennial herb
range: New Brunswick, s. to MD, w. to KS, n. to MN and Ontario
state frequency: rare
habitat: rich forests and meadows
notes: Member of section *Halleranae*.

Carex hormathodes Fern.
synonym(s): *Carex straminea* Willd. *ex* Schkuhr var. *invisa* W. Boott
common name(s): marsh straw sedge
habit: densely cespitose, perennial herb
range: Newfoundland, s. to VA
state frequency: uncommon
habitat: coastal, sandy or rocky areas and edges of tidal marshes
notes: Member of section *Ovales*.

Carex houghtoniana Torr. *ex* Dewey
synonym(s): —
common name(s): Houghton's sedge
habit: creeping, stoloniferous, rhizomatous, perennial herb
range: Newfoundland, s. to ME, w. to MN, n. to Saskatchewan
state frequency: uncommon
habitat: dry, sandy or gravelly clearings, banks, and roadsides
notes: Member of section *Paludosae*.

Carex hystericina Muhl. *ex* Willd.
synonym(s): —
common name(s): porcupine sedge
habit: slender, erect, loosely cespitose, short-rhizomatous, perennial herb
range: Newfoundland, s. to VA, w. to AZ and CA, n. to British Columbia
state frequency: uncommon
habitat: swales, swamps, wet meadows, muddy shores
notes: Member of section *Vesicariae*.

Carex interior Bailey
synonym(s): —
common name(s): inland sedge
habit: erect to ascending, cespitose, perennial herb
range: Labrador, s. to VA, w. to Mexico and CA, n. to AK
state frequency: uncommon
habitat: wet soil
notes: Member of section *Stellulatae*.

Carex intumescens Rudge
synonym(s): *Carex intumescens* Rudge var. *fernaldii* Bailey
common name(s): —
habit: solitary or cespitose, perennial herb
range: Newfoundland, s. to FL, w. to TX, n. to Manitoba
state frequency: common
habitat: moist to wet, meadows and alluvial woods, swales
notes: Member of section *Lupulinae*.

Carex lacustris Willd.

synonym(s): —
common name(s): lake bank sedge
habit: stout, solitary-stemmed or somewhat clustered, rhizomatous, perennial herb
range: Quebec, s. to VA, w. to NE, n. to ID and Alberta
state frequency: uncommon
habitat: swamps, marshes, ditches, shallow water
notes: Member of section *Paludosae*.

Carex laevivaginata (Kükenth.) Mackenzie

synonym(s): —
common name(s): smooth-sheathed sedge
habit: cespitose, perennial herb
range: ME, s. to SC, w. to IL, n. to MI
state frequency: very rare
habitat: wet woods and meadows
notes: Member of section *Vulpinae*. This species is ranked SU by the Maine Natural Areas Program.

Carex lasiocarpa Ehrh. var. *americana* Fern.

synonym(s): —
common name(s): slender sedge
habit: cespitose, rhizomatous, stoloniferous, perennial herb
range: circumboreal; s. to NJ, w. to WA
state frequency: occasional
habitat: bogs, swales, shallow water
notes: Member of section *Paludosae*.

Carex laxiculmis Schwein.

synonym(s): —
common name(s): spreading sedge
habit: slender, weak, ascending to spreading, cespitose, perennial herb
range: ME, s. to GA, w. to MO, n. to WI and Ontario
state frequency: very rare
habitat: moist, rich woods
notes: Member of section *Careyanae*. This species is ranked S1 by the Maine Natural Areas Program.

Carex laxiflora Lam.

synonym(s): —
common name(s): loose-flowered sedge
habit: erect to ascending, cespitose, perennial herb
range: Nova Scotia and ME, s. to mountains of NC and GA, w. to TN, n. to WI and Ontario
state frequency: uncommon
habitat: rich woods
notes: Member of section *Laxiflorae*. Perigynia prone to distortion, making identification difficult.

Carex lenticularis Michx.

synonym(s): *Carex lenticularis* Michx. var. *blakei* Dewey
common name(s): lenticular sedge
habit: slender, erect, densely cespitose, perennial herb
range: Labrador, s. to MA, w. to CA, n. to AK
state frequency: uncommon
habitat: meadows, swales, wet, gravelly shores
notes: Member of section *Phacocystis*.

Carex leptalea Wahlenb.

synonym(s): —
common name(s): bristle-stalked sedge
habit: slender, densely cespitose, rhizomatous, perennial herb
range: Labrador, s. to NC, w. to CO and CA, n. to AK
state frequency: uncommon
habitat: wet woods, swales, fens
notes: Member of section *Polytrichoideae*.

Carex leptonervia (Fern.) Fern.

synonym(s): *Carex laxiflora* Lam. var. *leptonervia* Fern.
common name(s): two-edged sedge
habit: slender, cespitose, perennial herb
range: Labrador, s. to mountains of NC, w. to TN, n. to MN and Ontario
state frequency: occasional
habitat: moist, low woods, thickets, clearings
notes: Member of section *Laxiflorae*. Perigynia prone to distortion, making identification difficult.

Carex limosa L.

synonym(s): —
common name(s): mud sedge
habit: solitary or clustered, long-rhizomatous, perennial herb
range: circumboreal; s. to NJ, irregularly w. to CA
state frequency: uncommon
habitat: bogs, pond margins
notes: Member of section *Limosae*. Known to hybridize with *Carex paleacea*.

Carex livida (Wahlenb.) Willd. var. *radicaulis* Paine

synonym(s): *Carex livida* (Wahlenb.) Willd. var. *grayana* (Dewey) Fern.
common name(s): livid sedge
habit: solitary or clustered, long-rhizomatous, perennial herb
range: circumboreal; s. in North America to NJ, interruptedly w. to CA; also Panama and South
America
state frequency: rare
habitat: often circumneutral, wet soil, meadows, fens
notes: Member of section *Paniceae*. This species is ranked S1/S2 by the Maine Natural Areas
Program.

Carex longii Mackenzie

synonym(s): —
common name(s): —
habit: densely cespitose, perennial herb
range: Nova Scotia, s. to FL, w. to TX and Mexico, n. to MI and Ontario
state frequency: very rare
habitat: swamps, bogs, often coastal
notes: Member of section *Ovales*.

Carex lucorum Willd. *ex* Link

synonym(s): *Carex pensylvanica* Lam. var. *distans* Peck
common name(s): —
habit: cespitose, long-rhizomatous, perennial herb
range: New Brunswick, s. to NC, w. to TN, n. to MN and Ontario

state frequency: common
habitat: dry, open woods, thickets, and barrens
notes: Member of section *Acrocystis*. Along with other members of its section, *Carex lucorum* is
    among the earliest flowering sedges. It can be seen with exserted anthers in April.

Carex lupulina Muhl. *ex* Willd.
synonym(s): *Carex lupulina* Muhl. *ex* Willd. var. *pedunculata* Gray
common name(s): —
habit: stout, solitary or clustered, long-rhizomatous, perennial herb
range: New Brunswick, s. to FL, w. to TX, n. to MN and Ontario
state frequency: occasional
habitat: wet woods, meadows, and shores, swamps, marshes
notes: Member of section *Lupulinae*.

Carex lurida Wahlenb.
synonym(s): —
common name(s): sallow sedge
habit: cespitose, perennial herb
range: New Brunswick, s. to FL, w. to TX and Mexico, n. to MN and Ontario
state frequency: common
habitat: swamps, swales, wet meadows and woods
notes: Member of section *Vesicariae*.

Carex mackenziei Krecz.
synonym(s): —
common name(s): —
habit: erect, cespitose, long-rhizomatous, perennial herb
range: circumboreal; s. in the east to ME
state frequency: uncommon
habitat: saline to brackish marshes and shores
notes: Member of section *Glarosae*.

Carex magellanica Lam. ssp. *irrigua* (Wahlenb.) Hultén
synonym(s): *Carex paupercula* Michx. var. *irrigua* (Wahlenb.) Fern
common name(s): bog sedge
habit: cespitose, perennial herb
range: Labrador, s. to PA, w. to UT, n. to WA and AK
state frequency: uncommon
habitat: organic soil wetlands
notes: Member of section *Limosae*.

Carex merritt-fernaldii Mackenzie
synonym(s): —
common name(s): Fernald's sedge
habit: cespitose, perennial herb
range: Quebec, s. to CT, w. and n. to MN and Manitoba
state frequency: uncommon
habitat: dry banks, meadows, borders of woods
notes: Member of section *Ovales*.

Carex michauxiana Boeckl.
synonym(s): *Carex abacta* Bailey
common name(s): yellowish sedge

habit: erect, densely cespitose, perennial herb
range: Newfoundland, s. to MA, w. to MN, n. to Saskatchewan
state frequency: uncommon
habitat: bogs, wet meadows
notes: Member of section *Folliculatae*.

Carex muehlenbergii Schkuhr *ex* Willd.

synonym(s): —
common name(s): Muhlenberg sedge
habit: slender, stiff, cespitose, perennial herb
range: ME, s. to FL, w. to TX, n. to MN and Ontario
state frequency: very rare
habitat: dry woods and fields
notes: Member of section *Phaestoglochin*. This species is ranked SH by the Maine Natural Areas
    Program.

Carex nigra (L.) Reichard

synonym(s): *Carex goodenowii* J. Gay, *Carex nigra* (L.) Reichard var. *strictiformis* (Bailey) Fern.,
    Carex vulgaris E. Fries var. *strictiformis* Bailey
common name(s): Goodenough's sedge
habit: loosely cespitose, rhizomatous, stoloniferous, perennial herb
range: Greenland, s. to CT; also MI
state frequency: uncommon
habitat: wet meadows, edges of saltmarshes
notes: Member of section *Phacocystis*.

Carex normalis Mackenzie

synonym(s): —
common name(s): larger straw sedge
habit: slender, erect, cespitose, perennial herb
range: New Brunswick, s. to NC, w. to OK, n. to SD and Manitoba
state frequency: uncommon
habitat: open woods, thickets, swales
notes: Member of section *Ovales*.

Carex norvegica Retz. ssp. *inferalpina* (Wahlenb.) Hultén

synonym(s): *Carex media* R. Br.
common name(s): intermediate sedge
habit: slender, densely cespitose, rhizomatous, perennial herb
range: circumboreal; s. to ME, w. to UT and Manitoba
state frequency: very rare
habitat: stream banks, seepy areas
notes: Member of section *Atratae*. This species is ranked S1 by the Maine Natural Areas Program.

Carex novae-angliae Schwein.

synonym(s): —
common name(s): New England sedge
habit: slender, cespitose, perennial herb
range: Newfoundland, s. to CT, w. to WI, n. to Ontario
state frequency: uncommon
habitat: moist woods and slopes
notes: Member of section *Acrocystis*.

Carex oligosperma Michx.

synonym(s): —
common name(s): few-seeded sedge
habit: slender, rhizomatous, stoloniferous, perennial herb
range: Labrador, s. to CT, w. to IN, n. to Alberta
state frequency: uncommon
habitat: bogs, swamps, wet meadows, shallow water
notes: Member of section *Vesicariae*.

Carex ormostachya Wieg.

synonym(s): *Carex laxiflora* Lam. var. *ormostachya* (Wieg.) Gleason
common name(s): —
habit: slender, cespitose, perennial herb
range: Quebec, New Brunswick, and Nova Scotia, s. to CT, w. to WI, n. to Ontario
state frequency: uncommon
habitat: sandy or rocky, rich woods, clearings
notes: Member of section *Laxiflorae*. Perigynia prone to distortion, making identification difficult.

Carex oronensis Fern.

synonym(s): —
common name(s): Orono sedge
habit: loosely cespitose, perennial herb
range: ME
state frequency: rare; known predominantly from the Penobscot River drainage
habitat: fields, meadows, clearings
notes: Member of section *Ovales*. This species is ranked S2 by the Maine Natural Areas Program.

Carex ovalis Goodenough

synonym(s): *Carex leporina auct. non* L.
common name(s): —
habit: spreading to erect, cespitose, perennial herb
range: Eurasia; naturalized to Newfoundland, s. to MA; also PA, NC, and the Pacific northwest
state frequency: uncommon
habitat: thickets, swales, wet meadows, fields, roadsides, ditches
notes: Member of section *Ovales*.

Carex paleacea Schreb. *ex* Wahlenb.

synonym(s): —
common name(s): —
habit: stout, solitary or clustered, rhizomatous, perennial herb
range: Greenland, s. to MA
state frequency: uncommon
habitat: brackish to saline marshes and shores
notes: Member of section *Phacocystis*. Known to hybridize with *Carex limosa* and *C. recta*.

Carex pallescens L.

synonym(s): *Carex pallescens* L. var. *neogaea* Fern.
common name(s): —
habit: slender, erect, cespitose, perennial herb
range: Newfoundland, s. to NJ, w. to OH, n. to WI and Ontario
state frequency: occasional
habitat: moist woods, thickets, and meadows
notes: Member of section *Porocystis*.

Carex panicea L.

synonym(s): —
common name(s): grass-like sedge
habit: cespitose, long-rhizomatous, perennial herb
range: Eurasia; naturalized to Newfoundland, s. to NJ
state frequency: uncommon
habitat: meadows, fields, lawns
notes: Member of section *Paniceae*.

Carex pauciflora Lightf.

synonym(s): —
common name(s): —
habit: solitary or clustered, long-rhizomatous, perennial herb
range: circumboreal; s. to CT, w. to WA
state frequency: uncommon
habitat: bogs, fens
notes: Member of section *Orthocerates*.

Carex peckii Howe

synonym(s): —
common name(s): —
habit: slender, erect to ascending, loosely cespitose, perennial herb
range: Quebec, s. to NJ, w. to British Columbia, n. to AK
state frequency: uncommon
habitat: often calcareous, dry, open woods and rocky slopes
notes: Member of section *Acrocystis*.

Carex pedunculata Muhl. *ex* Willd.

synonym(s): —
common name(s): long-stalked sedge
habit: low, mat-forming, short-rhizomatous, perennial herb
range: Newfoundland, s. to DE and uplands of GA, w. to IA, n. to Saskatchewan
state frequency: uncommon
habitat: rich woods
notes: Member of section *Clandestinae*.

Carex pellita Muhl.

synonym(s): *Carex lanuginosa* Michx., *Carex lasiocarpa* Ehrh. var. *latifolia* (Boeckl.) Gilly
common name(s): woolly sedge
habit: cespitose, long-rhizomatous, perennial herb
range: Newfoundland, s. to VA, w. to AZ and CA, n. to British Columbia
state frequency: uncommon
habitat: wet meadows, ditches, and shores
notes: Member of section *Paludosae*.

Carex pensylvanica Lam.

synonym(s): —
common name(s): Pennsylvania sedge
habit: rhizomatous, perennial herb
range: Quebec and Nova Scotia, s. to GA, w. to OK, n. to Manitoba
state frequency: very rare
notes: Member of section *Acrocystis*.

Carex plantaginea Lam.
synonym(s): —
common name(s): plantain-leaved sedge
habit: cespitose, perennial herb
range: New Brunswick, s. to MD and mountains of GA, w. to IN, n. to MN and Manitoba
state frequency: uncommon
habitat: rich woods
notes: Member of section *Careyanae*. Vegetative specimens are often confused with *Carex platyphylla*, but are easily separable. *Carex plantaginea* has green leaves with anthocyanic lower sheaths. *Carex platyphylla* has glaucous leaves and white to light brown lower sheaths.

Carex platyphylla Carey
synonym(s): —
common name(s): wide-leaved sedge
habit: cespitose, perennial herb
range: ME, s. to NC, w. and n. to WI and Ontario
state frequency: rare
habitat: rich woods, rocky slopes
notes: Member of section *Careyanae*. Similar to *Carex plantaginea* (*q.v.*).

Carex polymorpha Muhl.
synonym(s): —
common name(s): variable sedge
habit: stout, erect, rhizomatous, perennial herb
range: ME, s. to MD and WV
state frequency: very rare
habitat: dry, open woods and clearings.
notes: Member of section *Paniceae*. This species is ranked S1 by the Maine Natural Areas Program. *Carex polymorpha* is highly variable in the flowers it produces. Individuals in a population may be entirely staminate or with only a few carpellate flowers in the basal portion of the lowest spike.

Carex praegracilis W. Boott
synonym(s): —
common name(s): freeway sedge
habit: rhizomatous, perennial herb
range: Manitoba, s. to Mexico, w. to CA, n. to AK; naturalized to Quebec and ME, s. to NY, w. to MO, n. to Ontario
state frequency: very rare
habitat: disturbed soil, roadsides
notes: Member of section *Divisae*.

Carex prairea Dewey *ex* Wood
synonym(s): —
common name(s): prairie sedge
habit: loosely cespitose to clustered, rhizomatous, perennial herb
range: New Brunswick, s. to NJ, w. to NE, n. to British Columbia
state frequency: very rare
habitat: circumneutral fens
notes: Member of section *Paniculatae*. This species is ranked S1 by the Maine Natural Areas Program.

Carex prasina Wahlenb.

synonym(s): —
common name(s): drooping sedge
habit: loosely cespitose, perennial herb
range: ME, s. to GA, w. to AL, n. to MI and Ontario
state frequency: rare
habitat: moist to wet stream banks and low woods
notes: Member of section *Gracillimae*.

Carex praticola Rydb.

synonym(s): —
common name(s): northern meadow sedge
habit: cespitose, aphyllopodic, perennial herb
range: Greenland, s. to ME and Quebec, w. to CA, n. to AK
state frequency: very rare
habitat: open woods, clearings, meadows
notes: Member of section *Ovales*. This species is ranked SX by the Maine Natural Areas Program. This species is minutely papillose on the apical, herbaceous portion of the leaf sheath.

Carex projecta Mackenzie

synonym(s): —
common name(s): necklace sedge
habit: cespitose, perennial herb
range: Newfoundland, s. to NJ, w. to MO, n. to MN and Manitoba
state frequency: uncommon
habitat: damp woods, thickets, and meadows, shores
notes: Member of section *Ovales*.

Carex pseudocyperus L.

synonym(s): —
common name(s): cyperus-like sedge
habit: stout, loosely cespitose, perennial herb
range: Newfoundland, s. to NJ, w. to IN, n. to ND and Saskatchewan
state frequency: uncommon
habitat: swamps, bogs, shallow water
notes: Member of section *Vesicariae*.

Carex radiata (Wahlenb.) Small

synonym(s): *Carex rosea* Schkuhr *ex* Willd. *sensu* Fernald (1950)
common name(s): —
habit: erect to ascending, cespitose, perennial herb
range: Nova Scotia, s. to NC and AL, w. to MO, n. to ND and Manitoba
state frequency: uncommon
habitat: moist woods, thickets, pond margins
notes: Member of section *Phaestoglochin*.

Carex rariflora (Wahlenb.) Sm.

synonym(s): —
common name(s): loose-flowered sedge
habit: clustered, rhizomatous, perennial herb
range: circumboreal; s. to ME, w. to AK
state frequency: very rare
habitat: bogs, pond margins
notes: Member of section *Limosae*. This species is ranked SH by the Maine Natural Areas Program.

Carex recta Boott

synonym(s): *Carex salina* Wahlenb. var. *kattegatensis* (Fries) Almquist
common name(s): saltmarsh sedge, cuspidate sedge
habit: cespitose, rhizomatous, perennial herb
range: Labrador, s. to MA
state frequency: very rare
habitat: saline to brackish marshes and shores
notes: Member of section *Phacocystis*. *Carex recta* is thought to be derived through hybridization
between *C. aquatilis* and *C. paleacea*. This species is ranked S1 by the Maine Natural Areas
Program. Reported to hybridize with *Carex paleacea* and *C. stricta*.

Carex retrorsa Schwein.

synonym(s): —
common name(s): —
habit: coarse, cespitose, short-rhizomatous, perennial herb
range: Newfoundland, s. to MD, w. to UT and OR, n. to British Columbia
state frequency: uncommon
habitat: low, swampy woods, wet meadows
notes: Member of section *Vesicariae*.

Carex rosea Schkuhr *ex* Willd.

synonym(s): *Carex convoluta* Mackenzie, *Carex sylvicola* J. Webber & P. W. Ball
common name(s): stellate sedge
habit: cespitose, perennial herb
range: Nova Scotia, s. to GA, w. to AR, n. to ND
state frequency: uncommon
habitat: rich woods, thickets
notes: Member of section *Phaestoglochin*.

Carex saxatilis L.

synonym(s): —
common name(s): russet sedge
habit: rhizomatous, perennial herb
range: circumboreal; s. to ME
state frequency: very rare
habitat: moist to wet, sandy or gravelly soil, shores
key to the subspecies:
  1a. Plants robust; stems 3.0–7.0 mm wide at the base; leaves 2.0–4.0 mm wide; staminate spikes
  (1)2 or 3 per stem; anthers 2.0–4.0 mm long; perigynia 3.0–5.0 mm long ....
  .... *C. s.* var. *rhomalea* Fern.
  1b. Plants slender; stems 2.0–4.0 mm wide at the base; leaves 1.0–2.5 mm wide; staminate spike
  1(2) per stem; anthers 1.0–2.5 mm long; perigynia 2.5–3.5 mm long ....
  .... *C. s.* var. *miliaris* (Michx.) Bailey
notes: Member of section *Vesicariae*. This species is ranked S1 by the Maine Natural Areas Program.
Known to hybridize with *Carex vesicaria*, producing *C.* ×*stenolepis* Lessing, a mostly sterile
hybrid that occurs in central Maine.

Carex scabrata Schwein.

synonym(s): —
common name(s): rough sedge
habit: cespitose, rhizomatous, perennial herb
range: Quebec, s. to MD and mountains of GA, w. to MO, n. to MI and Ontario

state frequency: uncommon
habitat: wet woods and meadows, swamps
notes: Member of section *Anomalae*.

Carex scirpoidea Michx.

synonym(s): —
common name(s): bulrush sedge
habit: dioecious, slender, erect, solitary to clustered, rhizomatous, perennial herb
range: circumboreal; s. to ME, w. to NV
state frequency: very rare
habitat: wet rocks, shores, and meadows in alpine and subalpine areas
notes: Member of section *Scirpinae*. This species is ranked S1 by the Maine Natural Areas Program.

Carex scoparia Schkuhr ex Willd.

synonym(s): —
common name(s): pointed broom sedge
habit: slender, mostly erect, cespitose, perennial herb
range: Newfoundland, s. to FL, w. to NM, n. to OR and British Columbia
state frequency: occasional
habitat: wide variety of wet to dry soils
key to the subspecies:
    1a. Carpellate scales yellow- to light brown, of nearly the same color as the perigynia; perigynia 4.0–7.0 mm long, the body lanceolate to lance-ovate .... *C. s.* var. *scoparia*
    1b. Carpellate scales dark brown to black, of a color distinctly different from that of the perigynia; perigynia 4.0–5.0 mm long, the body elliptic-ovate .... *C. s.* var. *tessellata* Fern. & Wieg.
notes: Member of section *Ovales*.

Carex siccata Dewey

synonym(s): *Carex foenea* Willd. *sensu* Fernald (1950)
common name(s): dry land sedge
habit: rhizomatous, perennial herb
range: ME, s. to NJ, w. to AZ, n. to Mackenzie
state frequency: very rare
habitat: dry, sandy soil of open areas
notes: Member of section *Arenariae*. This species is ranked SH by the Maine Natural Areas Program.

Carex silicea Olney

synonym(s): —
common name(s): seabeach sedge
habit: densely cespitose, perennial herb
range: Newfoundland, s. to MD
state frequency: rare
habitat: coastal rocks, sand, and sandy soil
notes: Member of section *Ovales*. This species is ranked S2 by the Maine Natural Areas Program. This species is minutely papillose on the apical, herbaceous portion of the leaf sheath.

Carex sparganioides Muhl. ex Willd.

synonym(s): —
common name(s): bur-reed sedge
habit: cespitose, perennial herb
range: New Brunswick, s. to VA, w. to KS, n. to MN and Ontario
state frequency: very rare
habitat: rich forests

notes: Member of section *Phaestoglochin*. This species is ranked S1 by the Maine Natural Areas Program.

Carex spicata Huds.

synonym(s): *Carex contigua* Hoppe.
common name(s): —
habit: densely cespitose, perennial herb
range: Europe; naturalized to Nova Scotia, s. to VA, w. to OH, n. to MI and Ontario
state frequency: uncommon
habitat: dry fields and roadsides
notes: Member of section *Phaestoglochin*.

Carex sprengelii Dewey *ex* Spreng.

synonym(s): —
common name(s): long-beaked sedge
habit: cespitose, short-rhizomatous, perennial herb
range: New Brunswick, s. to NJ, w. to CO, n. to British Columbia
state frequency: very rare
habitat: rocky, open woods, thickets, meadows
notes: Member of section *Longirostres*. This species is ranked S1 by the Maine Natural Areas Program.

Carex sterilis Willd.

synonym(s): *Carex elachycarpa* Fern.
common name(s): dioecious sedge
habit: dioecious, cespitose, perennial herb
range: Newfoundland, s. to VA, w. to MO, n. to Alberta
state frequency: very rare
habitat: wet, calcareous soil
notes: Member of section *Stellulatae*. This species is ranked S1 by the Maine Natural Areas Program.

Carex stipata Muhl. *ex* Willd.

synonym(s): —
common name(s): awl-fruited sedge
habit: stout, cespitose, perennial herb
range: Newfoundland, s. to FL, w. to CA, n. to AK
state frequency: occasional
habitat: wet, low areas
notes: Member of section *Vulpinae*.

Carex stricta Lam.

synonym(s): *Carex stricta* Lam. var. *strictior* (Dewey) Carey
common name(s): tussock sedge
habit: erect, cespitose, rhizomatous, perennial herb
range: Quebec and New Brunswick, s. to VA, w. to TX, n. to MN and Manitoba
state frequency: common
habitat: swales, swamps
notes: Member of section *Phacocystis*. Reported to hybridize with *Carex recta*.

Carex swanii (Fern.) Mackenzie

synonym(s): —
common name(s): —
habit: cespitose, perennial herb
range: New Brunswick, s. to NC, w. to AR, n. to WI and Ontario
state frequency: uncommon

habitat: dry woods, thickets, and fields
notes: Member of section *Porocystis*.

Carex tenera Dewey

synonym(s): —
common name(s): —
habit: slender, cespitose, perennial herb
range: Quebec, s. to NC, w. to MO, n. to MT and Alberta
state frequency: uncommon
habitat: moist woodlands, thickets, and meadows
notes: Member of section *Ovales*.

Carex tenuiflora Wahlenb.

synonym(s): —
common name(s): sparse-flowered sedge
habit: loosely cespitose, rhizomatous, perennial herb
range: circumboreal; s. to ME, w. to Alberta
state frequency: rare
habitat: bogs, wet woods and pond margins
notes: Member of section *Glarosae*. This species is ranked S2 by the Maine Natural Areas Program.
    Hybrids with *Carex trisperma*, called *C.* ×*trichina* Fern., occur in northern Maine. Also known to
    hybridize with *C. canescens*.

Carex tetanica Schkuhr

synonym(s): —
common name(s): Wood's sedge
habit: cespitose, fibrillose-based, long-rhizomatous, phyllopodic, perennial herb
range: MA, s. to VA, w. to IA, n. to MN and Alberta; also ME
state frequency: very rare
habitat: bogs, meadows, low woods
notes: Member of section *Paniceae*.

Carex tincta (Fern.) Fern.

synonym(s): —
common name(s): —
habit: cespitose, perennial herb
range: New Brunswick, s. to CT, w. to WA, n. to British Columbia
state frequency: uncommon
habitat: sandy soil of fields and open woods
notes: Member of section *Ovales*.

Carex tonsa (Fern.) Bickn.

synonym(s): <*C. t.* var. *rugosperma*> = *Carex rugosperma* Mackenzie; <*C. t.* var. *tonsa*> = *Carex*
    umbellata Schkuhr *ex* Willd. var. *tonsa* Fern.
common name(s): deep green sedge
habit: densely cespitose, perennial herb
range: Labrador and Newfoundland, s. to GA, w. to IL, n. to MN and Ontario; also British Columbia,
    Alberta, and Saskatchewan
state frequency: uncommon
habitat: dry, sandy soils of open areas, oak-pine forests, and roadsides
key to the subspecies:
    1a. Perigynia glabrous except for a few hairs near the beak; leaves coriaceous, pale green, usually
        smooth on the adaxial surface, often relatively short .... *C. t.* var. *tonsa*

1b. Perigynia pubescent on the beak and body; leaves herbaceous, bright green, scabrous to papillose on the adaxial surface, usually longer than the flowering stems .... *C. t.* var. *rugosperma* (Mackenzie) W. J. Crins

notes: Member of section *Acrocystis*.

Carex torta Boott *ex* Tuckerman
synonym(s): —
common name(s): twisted sedge
habit: erect, densely cespitose, rhizomatous, perennial herb
range: Quebec, s. to NC and mountains of GA, w. to AR, n. to MN and Ontario
state frequency: uncommon
habitat: stream banks, shallow water
notes: Member of section *Phacocystis*. One of the earlier flowering species.

Carex tribuloides Wahlenb.
synonym(s): —
common name(s): blunt broom sedge
habit: loosely cespitose, perennial herb
range: New Brunswick, s. to FL, w. to LA and OK, n. to MN
state frequency: rare
habitat: wet meadows and low woods
notes: Member of section *Ovales*.

Carex trisperma Dewey
synonym(s): —
common name(s): three-seeded sedge
habit: loosely cespitose, short-rhizomatous, perennial herb
range: Labrador, s. to mountains of NC, w. to IL, n. to MN and Saskatchewan
state frequency: occasional
habitat: bogs, wet woods, clearings
key to the subspecies:
    1a. Leaves 1.0–2.0 mm wide; perigynia 2–5 per spike, 3.3–4.0 mm long; spikes 2 or 3 per stem .... *C. t.* var. *trisperma*
    1b. Leaves 0.2–0.5 mm wide; perigynia 1–3 per spike, 2.5–3.3 mm long; spikes 1 or 2 per stem .... *C. t.* var. *billingsii* Knight
notes: Member of section *Glarosae*. Known to hybridize with *Carex tenuiflora* (*q.v.*) and *C. canescens*. Sometimes confused with *C. disperma*, which has green sheaths and terete perigynia. *Carex trisperma* has sheaths that are white-dotted and compressed perigynia.

Carex tuckermanii Dewey
synonym(s): —
common name(s): Tuckerman's sedge
habit: slender, erect to arching, loosely cespitose, short-rhizomatous, perennial herb
range: New Brunswick, s. to NJ, w. to IA, n. to MN and Ontario
state frequency: uncommon
habitat: wet meadows, pond margins, and low woods
notes: Member of section *Vesicariae*.

Carex typhina Michx.
synonym(s): —
common name(s): cat-tail sedge
habit: stout, densely cespitose, perennial herb
range: ME, s. to GA, w. to LA, n. to MN

state frequency: very rare
habitat: moist to wet woods and meadows, marshes
notes: Member of section *Squarrosae*. This species is ranked SH by the Maine Natural Areas
    Program.

Carex umbellata Schkuhr *ex* Willd.
synonym(s): *Carex abdita* Bickn.
common name(s): umbel-like sedge
habit: densely cespitose to mat-forming, perennial herb
range: Newfoundland, s. to GA, w. to TX, n. to British Columbia
state frequency: uncommon
habitat: dry, sandy, open areas
notes: Member of section *Acrocystis*.

Carex utriculata Boott
synonym(s): *Carex rostrata* Stokes var. *utriculata* (Boott) Bailey
common name(s): beaked sedge
habit: long-rhizomatous, perennial herb
range: Newfoundland, s. to VA, w. to NM and CA, n. to AK
state frequency: occasional
habitat: swamps, wet meadows and ditches, pond margins
notes: Member of section *Vesicariae*.

Carex vacillans Drejer *ex* Hartman
synonym(s): *Carex* ×*subnigra* Lepage, *Carex* ×*super-goodenoughii* (Kükenth.) Lepage
common name(s): —
habit: rhizomatous, perennial herb
range: Quebec and Nova Scotia, s. along coast to MA; also Norway, Sweden, and Finland
state frequency: very rare
habitat: saline to brackish shores and marshes
notes: Member of section *Phacocystis*. This species is though to be derived through hybridization
    between *Carex nigra* and *C. paleacea*.

Carex vaginata Tausch
synonym(s): *Carex saltuensis* Bailey
common name(s): sheathed sedge
habit: slender, clustered, long-rhizomatous, perennial herb
range: circumboreal; s. to ME, w. to British Columbia
state frequency: very rare
habitat: circumneutral fens
notes: Member of section *Paniceae*. This species is ranked S1 by the Maine Natural Areas Program.

Carex vesicaria L.
synonym(s): *Carex vesicaria* L. var. *distenta* Fries, *Carex vesicaria* L. var. *jejuna* Fern., *Carex*
    vesicaria L. var. *laurentiana* Fern., *Carex vesicaria* L. var. *monile* (Tuckerman) Fern., *Carex*
    vesicaria L. var. *raeana* (Boott) Fern.
common name(s): inflated sedge
habit: cespitose, short-rhizomatous, perennial herb
range: circumboreal; s. to DE, w. to CA
state frequency: occasional
habitat: wet meadows, stream banks, and pond shores, bogs, swamps
notes: Member of section *Vesicariae*. Known to hybridize with *Carex saxatilis* (*q.v.*).

Carex vestita Willd.

synonym(s): —
common name(s): clothed sedge, velvet sedge
habit: stiff, slender, rhizomatous, perennial herb
range: ME, s. to VA
state frequency: very rare
habitat: dry, sandy woods, clearings
notes: Member of section *Paludosae*. This species is ranked SH by the Maine Natural Areas Program.

Carex virescens Muhl. *ex* Willd.

synonym(s): —
common name(s): ribbed sedge
habit: erect to ascending, cespitose, perennial herb
range: ME, s. to GA, w. to MO, n. to MI
state frequency: uncommon
habitat: sandy woods, thickets, and clearings
notes: Member of section *Porocystis*.

Carex viridula Michx.

synonym(s): —
common name(s): —
habit: cespitose, perennial herb
range: circumboreal; s. to NJ, w. to NM and CA
state frequency: uncommon
habitat: boggy shores, springy places, often calcareous
notes: Member of section *Ceratocystis*. Known to hybridize with *Carex flava*.

Carex vulpinoidea Michx.

synonym(s): —
common name(s): fox sedge
habit: stout, cespitose, perennial herb
range: Newfoundland, s. to FL, w. to AZ, n. to British Columbia
state frequency: uncommon
habitat: marshes, wet meadows, shores, and low areas
notes: Member of section *Multiflorae*.

Carex wiegandii Mackenzie

synonym(s): —
common name(s): Wiegand's sedge
habit: cespitose, perennial herb
range: Labrador, s. to MA, w. to PA, n. to MI and Ontario
state frequency: rare
habitat: *Sphagnum* bogs, wet thickets and shores
notes: Member of section *Stellulatae*. This species is ranked S2 by the Maine Natural Areas Program.

Cladium
(1 species)

This genus is superficially similar to *Rhynchospora* in flower. *Cladium*, however, lacks perianth bristles.

Cladium mariscoides (Muhl.) Torr.

synonym(s): —
common name(s): twig-rush

habit: stiff, slender, solitary or clustered, rhizomatous herb
range: Newfoundland, s. to FL, w. to TX, n. to MN and Saskatchewan
state frequency: uncommon
habitat: fresh to brackish swamps, marshes, and shores
notes: —

Cyperus
(10 species)

KEY TO THE SPECIES

1a. Styles 2; achenes lenticular; plants annual, with soft, fibrous roots
   2a. Scales light brown, pointed at the apex; plants of tidal shores and marshes ....
   .... *C. polystachyos*
   2b. Scales red-brown to dark brown, blunt at the apex; plants of inland shores and wet
      places, only occasionally found in tidal habitats
      3a. Styles split nearly to the base, persistent in fruit; scales with more pigment in the
      distal portion .... *C. diandrus*
      3b. Style split to about the middle, often deciduous in fruit; scales with more pigment
      in the proximal portion .... *C. bipartitus*
1b. Styles 3; achenes trigonous; plants annual or perennial
   4a. Plants 5.0–20.0 cm tall; scales recurved at the tip; leaves 0.5–2.0 mm wide; flowers
      with 1 stamen .... *C. squarrosus*
   4b. Plants 10.0–70.0 cm tall; scales straight at the tip; leaves 1.0–10.0 mm wide; flowers
      with 3 stamens
      5a. Plants annual, with red, fibrous roots; scales 1.2–1.5 mm long .. *C. erythrorhizos*
      5b. Plants perennial, with brown, tuberiferous or knotty rhizomes (commonly with a
      knotty, corm-like base in *C. strigosus*)
         6a. Spikes very flat, borne on elongate axes, forming short-cylindric clusters
            7a. Scales mostly 2.3–3.0 mm long, not or scarcely keeled; plants with long,
            slender rhizomes ending in tubers; floral scales deciduous from the rachis
            .... *C. esculentus*
            7b. Scales mostly 3.0–4.5 mm long, keeled; plants without well developed
            rhizomes, the base tuberous; scales persistent on and falling with the
            rachis .... *C. strigosus*
         6b. Spikes flat to subterete, borne at the summits of short axes, forming
         hemispherical to fully spherical clusters
            8a. Achenes up to 1.0 mm long; scales with 3–5 nerves arranged near the
            center of the scale; spikelets often proliferated, forming vegetative bulbils
            .... *C. dentatus*
            8b. Achenes 1.4–2.2 mm long; scales with 7–13 nerves uniformly distributed
            over the scale; spikelets not proliferous
               9a. Scales orbicular; achenes with evidently concave sides, 0.6 times as
               wide as long; spikes mostly ascending, forming more or less
               hemispherical clusters .... *C. houghtonii*
               9b. Scales oblong-elliptic; achenes with flat to slightly concave sides, 0.5
               times as wide as long; many of the spikes horizontally spreading,
               forming subglobose clusters .... *C. lupulinus*

Cyperus bipartitus Torr.
synonym(s): *Cyperus rivularis* Kunth
common name(s): —
habit: cespitose, annual herb
range: New Brunswick, s. to GA, w. to CA, n. to MN and Ontario
state frequency: uncommon
habitat: sandy, gravelly, muddy, or peaty shores
notes: —

Cyperus dentatus Torr.
synonym(s): —
common name(s): —
habit: short-rhizomatous, perennial herb
range: Quebec, s. to NC, w. and n. to IN
state frequency: occasional
habitat: sandy, gravelly, or rocky shores
notes: —

Cyperus diandrus Torr.
synonym(s): —
common name(s): —
habit: erect to prostrate, cespitose or solitary, annual herb
range: New Brunswick, Nova Scotia, and ME, s. to SC, w. to NM, n. to ND
state frequency: uncommon
habitat: wet shores
notes: —

Cyperus erythrorhizos Muhl.
synonym(s): —
common name(s): red-root flatsedge
habit: fibrous-rooted, annual herb
range: ME, s. to FL, w. to CA, n. to ND and Ontario
state frequency: very rare; known from southern Maine
habitat: wet soil of shores and marshes
notes: This species is ranked S1 by the Maine Natural Areas Program.

Cyperus esculentus L.
synonym(s): —
common name(s): yellow nutsedge
habit: rhizomatous, perennial herb
range: New Brunswick, s. to FL, w. to TX and Mexico, n. to WA and Manitoba
state frequency: occasional
habitat: damp to wet soil, sandy shores
notes: —

Cyperus houghtonii Torr.
synonym(s): —
common name(s): Houghton's umbrella-sedge
habit: short-rhizomatous, perennial herb
range: ME, s. to VA, w. to IA, n. to MN and Manitoba
state frequency: very rare
habitat: dry, sandy soil
notes: This species is ranked SH by the Maine Natural Areas Program.

Cyperus lupulinus (Spreng.) Marcks ssp. ***macilentus*** (Fern.) Marcks
synonym(s): *Cyperus filiculmis* Vahl var. *macilentus* Fern.
common name(s): —
habit: short-rhizomatous, perennial herb
range: New Brunswick, s. to FL, w. to TX and CO, n. to SD and MN
state frequency: uncommon
habitat: dry woods and fields, rocky or sandy soil
notes: —

Cyperus polystachyos Rottb. var. ***filicinus*** (Vahl) C. B. Clarke
synonym(s): *Cyperus filicinus* Vahl, *Cyperus nuttallii* Eddy
common name(s): —
habit: cespitose, annual herb
range: ME, s. to FL, w. to LA
state frequency: uncommon; found in the southern, coastal regions of the state
habitat: brackish to saline shores and marshes
notes: —

Cyperus squarrosus L.
synonym(s): *Cyperus aristatus* Rottb., *Cyperus inflexus* Muhl.
common name(s): —
habit: slender, cespitose, annual herb
range: New Brunswick and ME, s. to FL, w. to TX and Mexico, n. to British Columbia
state frequency: very rare
habitat: moist to wet, silt or sandy soil
notes: This species is ranked SH by the Maine Natural Areas Program.

Cyperus strigosus L.
synonym(s): —
common name(s): false nutsedge
habit: rhizomatous, short-lived perennial herb
range: ME, s. to FL, w. to TX and NM, n. to WA
state frequency: uncommon
habitat: damp thickets, fields, meadows, and shores, swales
notes: —

Dulichium
(1 species)

Dulichium arundinaceum (L.) Britt.
synonym(s): —
common name(s): three-way sedge
habit: erect, rhizomatous, perennial herb
range: Newfoundland, s. to FL, w. to CA, n. to British Columbia
state frequency: occasional
habitat: swamps, marshes, margins of ponds and streams
notes: —

Eleocharis
(16 species)

The glumes in *Eleocharis* are the lowest 1–3 scales of the spike and usually distinguishable from the floral scales by their larger size, firmer texture, and less translucent nature. Collections should include underground portions, or notes should be made about stem density and presence/absence of rhizomes. Timing of collections is also important. Our species are identifiable in flower, but specimens with mature fruits are easier to assign to species because the achenes add several valuable characters. Stamen number can be assessed even in late season. The filaments, which usually persist on the achene, are smooth and relatively broad and flat compared with the narrow, barbed perianth bristles.

KEY TO THE SPECIES

1a. Stems dimorphic—the flowering stems triangular and firm, the vegetative stems [produced when in water] capillary, flaccid, and floating; spike of similar thickness as stem; scales persistent .... *E. robbinsii*
1b. Stems monomorphic, terete to compressed but not triangular; spike noticeably thicker than the stem; scales deciduous
  2a. Tubercle confluent with the summit of the achene, not of a different color or texture, appearing as a beak; styles 3
    3a. Plants small; stems up to 1.0 mm thick, 2.0–30.0 cm tall, not rooting at the tip; spikes with 2–9 flowers; rhizomes long and slender, often terminated by a thickened bud
      4a. Scales 2.5–5.0 mm long; achenes 1.9–2.6 mm long; plants of calcareous shores and fens .... *E. quinqueflora*
      4b. Scales 1.5–2.5 mm long; achenes 0.9–1.3 mm long; plants of saline habitats .... *E. parvula*
    3b. Plants larger; stems 1.0–2.0 mm thick, 40.0–100.0 cm tall, some bending and rooting at the tip; spikes with mostly 10–20 flowers; rhizomes lacking .... *E. rostellata*
  2b. Tubercle clearly differentiated from and articulated to the achene, usually of another color and texture; styles 2 or 3
    5a. Achenes with prominent longitudinal ridges; stems capillary, up to 0.25 mm thick, 3.0–12.0 cm tall .... *E. acicularis*
    5b. Achenes variously smooth to ornamented, but without prominent longitudinal ridges; stems thicker, 5.0–100.0 cm tall
      6a. Styles 2(3); achenes lenticular, smooth or inconspicuously ornamented
        7a. Plants perennial, with elongate rhizomes; tubercle narrowed or constricted at the junction with the achene
          8a. Plants usually cespitose, with soft, thin rhizomes; leaf sheath with a prolonged, scarious, white tip; anthers 0.7–1.0 mm long .. *E. olivacea*
          8b. Plants produced separately or a few to a clump on firm, thick rhizomes; leaf sheath firm at the summit, lacking a tip or blade; anthers 1.0–3.0 mm long
            9a. Spike subtended by 2 or 3 glumes; bristles 4, reaching the base of the tubercle .... *E. palustris*
            9b. Spike subtended by 1 glume; bristles caducous or 1–4 and delicate
              10a. Achenes 0.7–1.0 mm wide; scales membranaceous, dull, closely appressed, those of the lower half of the spike

　　　　1.8–3.0 mm long; plants not of saline habitats ....
　　　　... *E. erythropoda*
　　10b. Achenes 1.0–1.4 mm wide; scales firm, lustrous, loosely
　　　　ascending, those of the lower half of the spike 3.0–5.0 mm
　　　　long; plants of saline habitats .... *E. halophila*
　7b. Plants annual, with soft, fibrous roots; tubercle lacking a constriction at
　　junction with the achene
　　　11a. Perianth bristles, if present, rarely reaching the base of the tubercle;
　　　　tubercle very short, less than 0.25 times as tall as the achene; spike
　　　　narrow-cylindric, 4.0–20.0 mm tall .... *E. engelmannii*
　　　11b. Perianth bristles exceeding the tubercle (reduced or absent in tidal
　　　　forms of *E. obtusa*); tubercle taller, 0.3–0.5 times as tall as the
　　　　achene; spike ovoid to ovoid-cylindric, 2.0–12.0 mm tall
　　　　　12a. Perianth bristles shorter than the achene or absent; floral scales
　　　　　　green-brown to light brown; plants of tidal shores .... *E. obtusa*
　　　　　12b. Perianth bristles present, overtopping the achene; floral scales
　　　　　　brown to purple-brown; plants of wet soil and non-tidal shores
　　　　　　　13a. Stamens 2 per flower; tubercle less than 0.65 times as wide
　　　　　　　　as the achene, 0.30–0.48 mm wide .... *E. ovata*
　　　　　　　13b. Stamens 3 per flower; tubercle greater than 0.65 times as
　　　　　　　　wide as the achene, 0.52–0.83 mm wide .... *E. obtusa*
　6b. Styles 3; achenes trigonous to terete, conspicuously reticulate to
　　inconspicuously ornamented, or smooth in *E. intermedia*
　　　14a. Tubercle as tall as the achene or nearly so; scales broadly rounded at the
　　　　apex; stems cespitose; plants annual .... *E. tuberculosa*
　　　14b. Tubercle up to 0.5 times as tall as the achene; scales obtuse to acute at
　　　　the apex; stems cespitose or separate; plants annual or perennial
　　　　　15a. Achenes smooth or minutely marked; tubercle slender-conic, 0.3–0.5
　　　　　　times as long as the achene; sheaths pale at the base; anthers up to
　　　　　　0.5 mm long; stems cespitose; rhizomes lacking; plants annual ....
　　　　　　... *E. intermedia*
　　　　　15b. Achenes conspicuously reticulate-sculpted; tubercle wide-deltate,
　　　　　　often less than 0.3 times as long as the achene; sheaths red-purple at
　　　　　　the base; anthers 1.5–2.0 mm long; stems produced separately or few
　　　　　　to a clump from creeping rhizomes; plants perennial
　　　　　　　16a. Stems 4- to 5-angled; mature achenes olive-green, with
　　　　　　　　inconspicuous latitudinal bands .... *E. tenuis*
　　　　　　　16b. Stems 6- to 8-angled; mature achenes yellow to orange, with
　　　　　　　　prominent latitudinal bands .... *E. elliptica*

Eleocharis acicularis (L.) Roemer & J. A. Schultes
synonym(s): *Eleocharis acicularis* (L.) Roemer & J. A. Schultes var. *submersa* (Hj. Nilss.) Svens.
common name(s): needle spike-rush
habit: diminutive, cespitose, terrestrial to aquatic, perennial herb
range: circumboreal; s. in the east to FL
state frequency: occasional
habitat: wetlands and muddy shores
notes: Atypically long stems and rhizomes can be found on submersed plants.

Eleocharis elliptica Kunth
synonym(s): *Eleocharis tenuis* (Willd.) J. A. Schultes var. *borealis* (Svens.) Gleason
common name(s): —
habit: slender, rhizomatous, perennial herb
range: Newfoundland, s. to NJ, w. to IN and IL, n. to British Columbia
state frequency: occasional
habitat: various wet soils, wet or dry sands
notes: —

Eleocharis engelmannii Steud.
synonym(s): —
common name(s): Engelmann's spike-rush
habit: cespitose, annual herb
range: ME, s. to GA, w. to AZ and CA, n. to WA and Saskatchewan
state frequency: very rare
habitat: wet soil
notes: This species is ranked SH by the Maine Natural Areas Program.

Eleocharis erythropoda Steud.
synonym(s): *Eleocharis calva* Torr.
common name(s): —
habit: rhizomatous, perennial herb
range: Quebec, s. to VA, w. to AR and OK, n. to Manitoba
state frequency: uncommon
habitat: wet soil
notes: —

Eleocharis halophila (Fern. & Brack.) Fern. & Brack.
synonym(s): *Eleocharis uniglumis* (Link) J. A. Schultes var. *halophila* Fern. & Brack.
common name(s): —
habit: rhizomatous, halophytic, perennial herb
range: Newfoundland, s. to VA, w. to NY, n. to Quebec
state frequency: uncommon
habitat: saline marshes and shores
notes: —

Eleocharis intermedia J. A. Schultes
synonym(s): —
common name(s): matted spike-rush
habit: cespitose, rhizomatous, perennial herb
range: Quebec, s. to WV and TN, w. to IL, n. to and MN
state frequency: rare
habitat: wet, calcareous soil
notes: —

Eleocharis obtusa (Willd.) J. A. Schultes
synonym(s): *Eleocharis diandra* C. Wright, *Eleocharis obtusa (Willd.) J. A. Schultes* var. *jejuna* Fern., *Eleocharis obtusa* (Willd.) J. A. Schultes var. *peasei* Svens.
common name(s): blunt spike-rush
habit: cespitose, annual herb
range: New Brunswick and Nova Scotia, s. to FL, w. to TX, n. to MN; also British Columbia, s. to CA
state frequency: occasional
habitat: wet soil, sometimes intertidal
notes: —

Eleocharis olivacea Torr.

synonym(s): *Eleocharis flavescens* (Poir.) Urban var. *olivacea* (Torr.) Gleason
common name(s): bright green spike-rush
habit: cespitose, rhizomatous, perennial herb
range: Nova Scotia, s. to FL, w. to PA and OH, n. to MI
state frequency: uncommon
habitat: wet, mineral or organic soil
notes: *Eleocharis olivacea* sometimes flowers in the first year, and these collections will lack
rhizomes, causing confusion with the *E. ovata* complex.

Eleocharis ovata (Roth) Roemer & J. A. Schultes

synonym(s): *Eleocharis ovata* (Roth) Roemer & J. A. Schultes var. *heuseri* Uechtr.
common name(s): ovoid spike-rush
habit: cespitose, annual herb
range: Newfoundland, s. to CT, w. to NY and IN, n. to MN
state frequency: occasional
habitat: wet soil
notes: —

Eleocharis palustris (L.) Roemer & J. A. Schultes

synonym(s): *Eleocharis palustris* (L.) Roemer & J. A. Schultes var. *major* Sonder, *Eleocharis*
palustris (L.) Roemer & J. A. Schultes var. *vigens* Bailey, *Eleocharis smallii* Britt., *E. smallii*
Britt. var. *major* (Sonder) Seymour
common name(s): creeping spike-rush
habit: rhizomatous, terrestrial or aquatic, perennial herb
range: Labrador, s. to WV, w. to IL, IN, WY, and CA, n. to British Columbia; also Eurasia
state frequency: occasional
habitat: shores, wetlands
notes: —

Eleocharis parvula (Roemer & J. A. Schultes) Link *ex* Bluff, Nees, & Schauer

synonym(s): —
common name(s): —
habit: diminutive, rhizomatous, perennial herb
range: irregularly cosmopolitan; in the east, Newfoundland, s. to LA
state frequency: rare
habitat: saline shores and marshes
notes: —

Eleocharis quinqueflora (F. X. Hartmann) Schwarz

synonym(s): *Eleocharis pauciflora* (Lightf.) Link var. *fernaldii* Svens.
common name(s): few-flowered spike-rush
habit: rhizomatous, perennial herb
range: circumboreal; in the east, s. to NJ
state frequency: very rare
habitat: calcareous shores and fens
notes: This species is ranked S1 by the Maine Natural Areas Program.

Eleocharis robbinsii Oakes

synonym(s): —
common name(s): Robbin's spike-rush
habit: rhizomatous, terrestrial or aquatic, perennial herb
range: New Brunswick and Nova Scotia, s. to FL, w. and n. to MI and WI

state frequency: uncommon
habitat: muddy soil, shallow, still or slow-moving, water
notes: —

Eleocharis rostellata (Torr.) Torr.

synonym(s): —
common name(s): beaked spike-rush
habit: cespitose, rhizomatous, stoloniferous, halophytic, perennial herb
range: Nova Scotia, s. to FL, w. to IL, n. to WI
state frequency: very rare
habitat: saline shores and marshes
notes: This species is ranked SH by the Maine Natural Areas Program.

Eleocharis tenuis (Willd.) J. A. Schultes

synonym(s): —
common name(s): slender spike-rush
habit: slender, rhizomatous, perennial herb
range: Nova Scotia and ME, s. to SC and TN
state frequency: occasional
habitat: various wet soils, wet to dry sands
notes: —

Eleocharis tuberculosa (Michx.) Roemer & J. A. Schultes

synonym(s): —
common name(s): long-tubercled spike-rush
habit: cespitose, annual herb
range: Nova Scotia, s. to FL, w. to TX, n. to AR
state frequency: very rare
habitat: sandy shores
notes: This species is ranked S1 by the Maine Natural Areas Program.

Eriophorum
(6 species)

KEY TO THE SPECIES

1a. Spikelets solitary, erect, with 10–15 sterile scales in the basal portion, not subtended by foliaceous involucral bracts; plants cespitose .... *E. vaginatum*
1b. Spikelets 2–many, spreading or drooping on peduncles; scales all fertile; foliaceous involucral bracts present; stems produced singly or a few together from a rhizome
  2a. Involucral bract 1 (sometimes with another very reduced, scale-like bract), erect; leaves 1.0–2.0 mm wide, channeled
    3a. Uppermost leaf blade 3.0–25.0 cm long, as long or longer than its sheath; stems scabrous near the top; peduncles scabrous; involucral bract 1.5–6.0 cm long, green to red-brown at the base .... *E. tenellum*
    3b. Uppermost leaf blade 1.0–4.0 cm long, shorter than its sheath; stems smooth; peduncles pubescent; involucral bract 1.0–2.0 cm long, gray to black at the base
.... *E. gracile*
  2b. Involucral bracts 2 or 3, somewhat spreading; leaves 2.0–8.0 mm wide, flat, at least in the basal portion

4a. Scales with 3–7 nerves, red-brown near the margin; flowers with 1 stamen; perianth bristles red-brown (white) .... *E. virginicum*

4b. Scales with 1 central nerve, green or gray to black near the margin; flowers with 3 stamens; perianth bristles white

    5a. Midrib of scales inconspicuous near the apical portion; summit of the sheath darkened; anthers 2.5–5.0 mm long .... *E. angustifolium*

    5b. Midrib of scale prominent to the tip; summit of the sheath not darkened; anthers 1.0–1.3 mm long .... *E. viridicarinatum*

Eriophorum angustifolium Honckeny

synonym(s): <*E. a.* ssp. *subarcticum* var. *majus*> = *Eriophorum angustifolium* Honckeny var. *majus* F. W. Schultz

common name(s): —

habit: soft, slender, drooping, perennial herb

range: Newfoundland, s. to ME, w. to OR, n. to AK; also NM

state frequency: uncommon

habitat: organic soil wetlands

key to the subspecies:

    1a. Plants smaller, with stems 2.0–6.0 dm tall and leaves 1.5–4.0 mm wide ....
.... *E. a.* ssp. *angustifolium*

    1b. Plants larger, with stems 3.0–9.0 dm tall and leaves 4.0–8.0 mm wide ....
.... *E. a.* ssp. *subarcticum* (Vassiljev) Hultén *ex* Kartesz & Gandhi var. *majus* F. W. Schultz

notes: —

Eriophorum gracile W. D. J. Koch

synonym(s): —

common name(s): slender cotton-grass

habit: slender, drooping, rhizomatous, perennial herb

range: circumboreal; s. to DE, w. to CO and CA

state frequency: uncommon

habitat: organic soil wetlands

notes: —

Eriophorum tenellum Nutt.

synonym(s): —

common name(s): conifer cotton-grass

habit: drooping, rhizomatous, perennial herb

range: Newfoundland, s. to NJ, w. to IL, n. to MN and Saskatchewan

state frequency: uncommon

habitat: organic soil wetlands

notes: —

Eriophorum vaginatum L. var. *spissum* (Fern.) Boivin

synonym(s): *Eriophorum spissum* Fern.

common name(s): tussock cotton-grass, hare's tail

habit: densely cespitose, perennial herb

range: circumboreal; s. to NJ, w. to IN, Saskatchewan, and AK

state frequency: occasional

habitat: organic soil wetlands, alpine bogs and folists

notes: —

Eriophorum virginicum L.

synonym(s): —
common name(s): tawny cotton-grass
habit: slender, solitary-stemmed or somewhat clustered, perennial herb
range: Labrador, s. to FL, w. and n. to MN and Manitoba
state frequency: uncommon
habitat: organic soil wetlands
notes: —

Eriophorum viridicarinatum (Engelm.) Fern.

synonym(s): —
common name(s): darkscale cotton-grass
habit: slender, drooping, cespitose, rhizomatous, perennial herb
range: Newfoundland, s. to coastal NY, w. to CO, n. to WA and British Columbia
state frequency: uncommon
habitat: organic soil wetlands, often with higher pH
notes: —

Fimbristylis
(1 species)

Fimbristylis autumnalis (L.) Roemer & J. A. Schultes

synonym(s): *Fimbristylis autumnalis* (L.) Roemer & J. A. Schultes var. *mucronulata* (Michx.) Fern.
common name(s): dwarf-bulrush
habit: spreading to erect, cespitose, annual herb
range: ME, s. to GA, w. to LA, n. to MN and Ontario
state frequency: rare
habitat: moist to wet, sandy or peaty soil
notes: This species is ranked S2 by the Maine Natural Areas Program.

Lipocarpha
(1 species)

Lipocarpha micrantha (Vahl) G. Tucker

synonym(s): *Hemicarpha micrantha* (Vahl) Pax
common name(s): dwarf-bulrush
habit: cespitose, arching, annual herb
range: ME, s. to FL, w. to TX and Mexico, n. to WA
state frequency: very rare
habitat: moist, sandy soil of pond and stream margins/borders
notes: This species is ranked S1 by the Maine Natural Areas Program.

Rhynchospora
(5 species)

KEY TO THE SPECIES

1a. Plants large; leaves 3.0–20.0 mm wide; spikelets 10.0–30.0 mm long; body of achenes
    3.5–5.5 mm long, with a tubercle 13.0–22.0 mm long; style unbranched, except
    sometimes at the very tip .... *R. macrostachya*

1b. Plants smaller; leaves 0.2–3.5 mm wide; spikelets 2.0–10.0 mm long; body of achenes 1.1–2.1 mm long, with a tubercle 0.7–1.6 mm long; style deeply divided

    2a. Perianth bristles antrorsely barbed; achenes narrowed to a thick base, not obviously stipe-like .... *R. fusca*

    2b. Perianth bristles retrorsely barbed; achenes distinctly narrowed to a thin, stipe-like base

        3a. Perianth bristles 8–14 per flower, often with tiny, upward-projecting hairs at the base .... *R. alba*

        3b. Perianth bristles 6 per flower, without hairs at the base

            4a. Leaves 0.2–0.4 mm wide; achenes up to 0.5 times as wide as long, dark brown near the margins and paler with dark lines near center .... *R. capillacea*

            4b. Leaves 1.5–3.5 mm wide; achenes 0.5 times or more as wide as long, uniformly dark brown .... *R. capitellata*

Rhynchospora alba (L.) Vahl

synonym(s): —
common name(s): white beak-rush
habit: erect, cespitose, perennial herb
range: circumboreal; s. to NC, w. to CA
state frequency: common
habitat: bogs and fens
notes: —

Rhynchospora capillacea Torr.

synonym(s): —
common name(s): horned beak-rush, capillary beak-rush
habit: cespitose, perennial herb
range: Newfoundland, s. to VA, w. to MO, n. to Saskatchewan
state frequency: very rare
habitat: calcareous swamps, fens, and shores
notes: This species is ranked S1 by the Maine Natural Areas Program.

Rhynchospora capitellata (Michx.) Vahl

synonym(s): —
common name(s): —
habit: erect, cespitose, perennial herb
range: New Brunswick, s. to FL, w. to TX, n. to WI and Ontario
state frequency: uncommon
habitat: bogs, damp shores, ledges
notes: —

Rhynchospora fusca (L.) Ait. f.

synonym(s): —
common name(s): brown beak-rush
habit: rhizomatous, stoloniferous, perennial herb
range: Newfoundland, s. to DE and MD, w. to MI, n. to Saskatchewan
state frequency: uncommon
habitat: bogs, marshes, springy areas, wet sands
notes: —

Rhynchospora macrostachya Torr. *ex* Gray
synonym(s): —
common name(s): tall beak-rush, horned-rush
habit: coarse, erect, annual or short-lived perennial herb
range: ME, s. to FL, w. to TX and KS, n. to MI
state frequency: very rare
habitat: swamps, wet sand
notes: This species is ranked SH by the Maine Natural Areas Program.

Schoenoplectus
(7 species)

KEY TO THE SPECIES

1a. Plants submerged aquatics with flaccid leaves; spikelets solitary on an erect, emergent
   stem .... *S. subterminalis*
1b. Plants rarely submerged, the leaves stiffer; spikelets 2 or more
  2a. Spikelets 1–15, in a congested inflorescence without elongate rays; stems
    0.5–3.0 mm in diameter at the midpoint; plants up to 1.0 m tall
    3a. Plants annual, tufted; culms terete to obscurely trigonous; anthers 0.4–0.6 mm
      long
      4a. Scales with a distinct midvein, lacking a green midstrip; achenes noticeably
        pitted, 1.75–2.0 mm long, rounded-obovate; principal involucral bract usually
        divergent .... *S. purshianus*
      4b. Scales with an obscure midvein, possessing a broad green midstrip; achenes
        nearly smooth, 1.5–1.8 mm long, cuneate-obovate; principal involucral bract
        usually erect .... *S. smithii*
    3b. Plants perennial, rhizomatous; culms sharply trigonous; anthers 1.5–3.0 mm long
      5a. Leaves short, less than half the length of the stem; styles bifid; achenes
        planoconvex, mucronate; scales notched, several-nerved .... *S. pungens*
      5b. Leaves longer, more than half as long as the stem; styles trifid; achenes
        compressed-trigonous, distinctly beaked; scales entire, with only a single,
        central nerve .... *S. torreyi*
  2b. Spikelets commonly more than 15; inflorescence only rarely without visible
    branches; stems 2.0–10.0 mm in diameter at the midpoint; plants 1.0–3.0 m tall
    6a. Stems firm, with 10–12 aerenchymal lacunae in cross-section at 0.65 of the plant
      height; lacunae mostly 0.3–0.5 mm in diameter; spikelets mostly in clusters of
      3–7; midrib of scale commonly exserted as a contorted awn 0.5–2.0 mm long;
      floral scales gray, mostly 3.5–4.0 mm long; achenes 2.0–2.8 mm long; style beak
      mostly 0.2–0.5 mm long .... *S. acutus*
    6b. Stems softer, with 2–4 aerenchymal lacunae in cross-section at 0.65 of the plant
      height; lacunae mostly 1.0–2.5 mm in diameter; spikelets solitary or in clusters of
      mostly 2 or 3; the mostly straight awns 0.2–0.8 mm long; floral scales red-brown
      or brown, mostly 2.5–3.2 mm long; achenes 1.7–2.3 mm long; style beak mostly
      0.1–0.3 mm long .... *S. tabernaemontanii*

Schoenoplectus acutus (Bigel.) A. & D. Löve
synonym(s): *Scirpus acutus* Muhl. *ex* Bigelow
common name(s): hardstem bulrush, great bulrush

habit: stout, rhizomatous, perennial herb
range: Newfoundland, s. to NJ, w. to TX and CA, n. to AK
state frequency: uncommon
habitat: marshes, shores of ponds and streams
notes: Hybrids with Schoenoplectus tabernaemontanii occur in Maine. This hybrid can be identified by its intermediate morphology and high sterility (in both pollen and fruit production).

Schoenoplectus pungens (Vahl) Palla
synonym(s): *Scirpus americanus* Pers.
common name(s): common three-square, chair-maker's-rush
habit: rhizomatous, perennial herb
range: Newfoundland, s. to FL, w. to CA, n. to AK
state frequency: occasional
habitat: fresh to saline shores and marshes
notes: Hybrids with *Schoenoplectus americanus* (Pers.) Volk *ex* Schinz & R. Keller occur in Maine. This suggests *S. americanus* may be found in Maine, considering it is documented from Nova Scotia and Massachusetts.

Schoenoplectus purshianus (Fern.) M. T. Strong
synonym(s): *Scirpus juncoides* Roxb. var. *williamsii* (Fern.) T. Koyama, *Scirpus purshianus* Fern., *Scirpus smithii* Gray var. *williamsii* (Fern.) Beetle
common name(s): weak-stalk bulrush
habit: cespitose, annual herb
range: ME, s. to GA, w. to AL, n. to MN
state frequency: uncommon
habitat: bogs and open, acid swamps
notes: —

Schoenoplectus smithii (Gray) Soják
synonym(s): *Scirpus smithii* Gray
common name(s): bluntscale bulrush, Smith's bulrush
habit: cespitose, annual herb
range: New Brunswick, s. to VA, w. to IL, n. to MN
state frequency: uncommon
habitat: shores, often tidal
notes: Similar to *Schoenoplectus purshianus* and often confused with it in spite of differences in morphology and habitat. Three distinct forms occur in our area . "Var. smithii" is characterized by very reduced or completely absent perianth bristles. "Var. levisetus" has 1–3(–4) weakly retrorse-barbed perianth bristles of various lengths, but never exceeding the length of the achene. "Var. setosus" has 4–6 strongly retrorse-barbed bristles equaling to exceeding the length of the achene. These forms are taxa that warrant recognition, but, as the quotation marks are meant to illustrate, the proper combinations, under the genus *Schoenoplectus*, have not yet been made.

Schoenoplectus subterminalis (Torr.) Soják
synonym(s): *Scirpus subterminalis* Torr.
common name(s): water bulrush, swaying bulrush
habit: rhizomatous, aquatic, perennial herb
range: Newfoundland, s. to GA, w. to CA, n. to AK
state frequency: uncommon
habitat: shallow, still or slowly moving water, bogs
notes: —

Schoenoplectus tabernaemontanii (Gmel.) Palla
synonym(s): *Scirpus validus* Vahl
common name(s): softstem bulrush, great bulrush
habit: rhizomatous, perennial herb
range: Newfoundland, s. to South America, w. to CA, n. to AK; Eurasia, Australia, Africa, Pacific
    islands
state frequency: occasional
habitat: fresh to brackish marshes, shores of ponds and streams
notes: Hybrids with *Schoenoplectus heterochaetus* (Chase) Soják, called *S.* ×*steinmetzii* (Fern.) S. G.
    Smith & A. E. Schuyler, occur in central Maine and can be distinguished from *S.*
    tabernaemontanii, which they most closely resemble, by their pedicellate spikelets, 5–7
    aerenchymal lacunae in diameter at 0.65 of the stem height, and mostly sterile spikelets with 0–2
    mature achenes. *Schoenoplectus tabernaemontanii* also hybridizes with *S. acutus*. This hybrid is
    commonly found in mixed populations.

Schoenoplectus torreyi (Olney) Palla
synonym(s): *Scirpus torreyi* Olney
common name(s): Torrey's bulrush, Torrey three-square
habit: rhizomatous, perennial herb
range: New Brunswick, s. to NJ and mountains of VA, w. to MO and NE, n. to Manitoba
state frequency: uncommon
habitat: fresh to brackish marshes and shores
notes: —

Scirpus
(9 species)

Hybrids are common when closely related species occur at the same site. In addition to morphological
intermediacy, hybrids can be recognized by pollen and fruit abortion.

KEY TO THE SPECIES
1a. Perianth bristles retrorsely barbed, straight or nearly so, at most 1.5 times the length of
    the achene
    2a. Stigmas mostly 2; achenes mostly lenticular; lower sheaths anthocyanic ....
    .... **S. microcarpus**
    2b. Stigmas mostly 3; achenes mostly trigonous; lower sheaths green, or anthocyanic in
    S. expansus
        3a. Perianth bristles barbed nearly to the base with sharp, stout, thick-walled teeth,
            often caducous; rays of the inflorescence antrorsely scabrous over entire length;
            lower sheaths red or red-brown .... **S. expansus**
        3b. Perianth bristles barbed only on the apical half with dull, slender teeth, persistent;
            rays of the inflorescence smooth, scabrous only near the apical portion, if at all;
            lower sheaths green
            4a. Lower sheaths and blades inconspicuously septate-nodulose, the blades
                2.0–10.0 mm wide; spikelets 2.0–3.5 mm long; scales black; longer perianth
                bristles shorter than to equaling the length of the achene; achenes 0.8–1.1 mm
                long .... **S. hattorianus**
            4b. Lower sheaths and blades conspicuously septate-nodulose, the blades
                2.0–18.0 mm wide; spikelets mostly 3.5–5.0 mm long; scales brown; longer

> perianth bristles exceeding the length of the achene; achenes 1.0–1.3 mm
> long .... *S. atrovirens*
1b. Perianth bristles smooth, bent and curled, greatly exceeding the achene when extended
>   5a. Mature perianth bristles mostly contained within the scales; achenes 1.0–1.2 mm
>   long; midrib of scale conspicuous and green; spikelets 3.0–13.0 mm long ....
>   .... *S. pendulus*
>   5b. Mature perianth bristles evidently exceeding the scales; achenes 0.6–1.0 mm long;
>   midrib of scale inconspicuous; spikelets mostly 3.0–6.0 mm long
>   6a. Achenes red or red-brown; scales 2.0–3.0 mm long; base of the involucral bract
>   glutinous; plants with long, stout rhizomes .... *S. longii*
>   6b. Achenes white to yellow-gray; scales 1.0–2.0 mm long; base of the involucral
>   bract not glutinous; plants cespitose from very short rhizomes
>   7a. Spikelets mostly pedicellate; achenes maturing June–July; floral scales
>   1.4–1.8 mm long
>   8a. Involucels and the base of involucral bracts black; achenes maturing late
>   June and early July; stems 1.0–4.0 mm thick below the inflorescence;
>   leaves 2.0–5.0 mm wide .... *S. atrocinctus*
>   8b. Involucels and the base of involucral bracts light brown to brown;
>   achenes maturing in mid- to late July; stems 2.0–5.0 mm thick below the
>   inflorescence; leaves 3.0–10.0 mm wide .... *S. pedicellatus*
>   7b. Spikelets glomerulate, in clusters of 2 or more; achenes maturing in August
>   and September; floral scales mostly shorter than 1.5 mm .... *S. cyperinus*

Scirpus atrocinctus Fern.
synonym(s): —
common name(s): black-girdled wool-grass
habit: slender, cespitose, short-rhizomatous, perennial herb
range: Newfoundland, s. to NJ, w. to IA and WA, n. to British Columbia
state frequency: occasional
habitat: meadows, swamps
notes: Hybrids with *Scirpus cyperinus* and *S. hattorianus*, called *S.* ×*peckii* Britt., occur in Maine. The
   former hybrid has glomerulate spikelets, but flowers and fruits earlier than *S. cyperinus*. The latter
   hybrid can be identified by its perianth bristles of intermediate length and serration.

Scirpus atrovirens Willd.
synonym(s): —
common name(s): black bulrush
habit: cespitose, short-rhizomatous, perennial herb
range: Newfoundland, s. to GA, w. to MO, n. to Saskatchewan
state frequency: uncommon
habitat: wetlands, shores
notes: Most closely resembles *Scirpus hattorianus* and is often confused with it. *Scirpus atrovirens* is
   more often found in lower, wetter areas, whereas *S. hattorianus* is more often found in moist,
   more upland sites. These two species are known to hybridize.

Scirpus cyperinus (L.) Kunth
synonym(s): *Scirpus cyperinus* (L.) Kunth var. *pelius* Fern.
common name(s): wool-grass
habit: densely cespitose, short-rhizomatous, perennial herb
range: Newfoundland, s. to NC, w. to OK, n. to MN and Ontario

state frequency: occasional
habitat: bogs, marshes, swamps, wet meadows
notes: This species has a variable scale and bract color, but tends to be black in the north, resembling
   Scirpus atrocinctus. Known to hybridize with *S. atrocinctus* (*q.v.*), *S. longii*, and *S. pedicellatus*.

Scirpus expansus Fern.

synonym(s): —
common name(s): —
habit: long-rhizomatous, perennial herb
range: ME, s. to GA, w. to OH, n. to MI
state frequency: uncommon
habitat: swamps, meadows, low thickets, margins of streams and rills
notes: Superficially similar to and known to hybridize with *Scirpus microcarpus*. *Scirpus expansus* is
   normally more robust, with wider leaves and conspicuously septate sheaths. Hybrids between these
   species show a mixture of bicarpellate and tricarpellate flowers in the same spikelet and a high
   percentage of immature achenes.

Scirpus hattorianus Makino

synonym(s): *Scirpus atrovirens* Willd. var. *georgianus* (Harper) Fern. *sensu* Fernald (1950) in part
common name(s): —
habit: cespitose, short-rhizomatous, perennial herb
range: Newfoundland, s. to GA, w. to AR, n. to MN and Ontario
state frequency: occasional
habitat: moist soil of meadows, marshes, and ditches
notes: Similar to *Scirpus atrovirens* (*q.v.*) and known to hybridize with it. In areas where both species
   are found, *S. hattorianus* is more likely to be found in more upland settings such as ditches and
   meadows, whereas *S. atrovirens* is more likely to be found in lower areas, such as marshes and
   swamps.

Scirpus longii Fern.

synonym(s): —
common name(s): Long's bulrush
habit: long-rhizomatous, perennial herb
range: Nova Scotia, s. to NJ; also NC
state frequency: very rare
habitat: meadows, swamps, marshes, near the coast but freshwater
notes: This species is ranked S1 by the Maine Natural Areas Program. Known to hybridize with
   Scirpus cyperinus. *Scirpus longii* rarely flowers, requiring drought or fire to initiate stem
   production.

Scirpus microcarpus J. & K. Presl

synonym(s): *Scirpus rubrotinctus* Fern.
common name(s): —
habit: coarse, solitary or clustered, rhizomatous, perennial herb
range: Newfoundland, s. to WV, w. to CA, n. to AK
state frequency: occasional
habitat: marshes, wet meadows and low thickets
notes: Superficially similar to *Scirpus expansus* (*q.v.*) and known to hybridize with it.

Scirpus pedicellatus Fern.

synonym(s): *Scirpus pedicellatus* Fern. var. *pullus* Fern.
common name(s): —
habit: cespitose, short-rhizomatous, perennial herb

range: Quebec, s. to NJ, w. to IA, n. to MN and Ontario
state frequency: uncommon
habitat: swales, wet thickets, meadows, and shores
notes: Known to hybridize with *Scirpus cyperinus*. Similar to *S. atrocinctus* but more robust, with lighter bracts and scales, and flowers later.

Scirpus pendulus Muhl.
synonym(s): *Scirpus lineatus auct. non* Michx.
common name(s): pendulous bulrush
habit: cespitose, short-rhizomatous, perennial herb
range: ME, s. to FL, w. to Mexico and NM, n. to MN and SD
state frequency: very rare
habitat: wet woods, thickets, and meadows, swales
notes: This species is ranked S1 by the Maine Natural Areas Program.

Trichophorum
(3 species)

KEY TO THE SPECIES

1a. Stems triangular in cross-section, scabrous on the angles
    2a. Perianth bristles 6 from each flower, white, 1.0–3.0 cm long at maturity, clearly surpassing the scales .... *T. alpinum*
    2b. Perianth bristles 3–6, brown, about 0.2 cm long, hidden by the scales ..... *T. clintonii*
1b. Stems subterete, glabrous .... *T. cespitosum*

Trichophorum alpinum (L.) Pers.
synonym(s): *Scirpus hudsonianus* (Michx.) Fern.
common name(s): —
habit: clustered, short-rhizomatous, perennial herb
range: circumboreal; s. to CT, w. to MT and British Columbia
state frequency: uncommon
habitat: bogs, fens, springy meadows, wet gravel
notes: —

Trichophorum cespitosum (L.) Hartman
synonym(s): *Scirpus cespitosus* L. var. *callosus* Bigelow, *Scirpus cespitosus* L. var. *delicatulus* Fern.
common name(s): tufted club-rush, deer's hair
habit: densely cespitose, short-rhizomatous, perennial herb
range: circumboreal; s. to ME, w. to OR; disjunct to mountains of NC and TN
state frequency: uncommon
habitat: bogs, alpine areas
notes: —

Trichophorum clintonii (Gray) S. G. Smith
synonym(s): *Scirpus clintonii* Gray, *Scirpus planifolius* Muhl. var. *brevifolius* Torr.
common name(s): Clinton's bulrush, Clinton's club-rush
habit: densely cespitose, short-rhizomatous, perennial herb
range: Quebec, s. to ME, w. to MN, n. to Alberta
state frequency: rare
habitat: dry to wet ledges, gravel, or rocky shores
notes: This species is ranked S2 by the Maine Natural Areas Program.

POACEAE
(69 genera, 209 species)

KEY TO THE GENERA

1a. Spikelets appearing spiny, either enclosed in a bur-like involucre or with enlarged, spiny glumes .... **Group 1**
1b. Spikelets not concealed by a spiny bur, although pubescence or bristles may be present
  2a. Spikelets with 2 or more evident florets; florets bisexual, unisexual, or sterile
    3a. Inflorescence an open or dense panicle
      4a. Spikelets dimorphic with the 2 types paired—the lower spikelets sterile, flabellate, consisting of 2 glumes and several narrow, acuminate lemmas, concealing the upper, fertile spikelets; fertile spikelets 2- to 4-flowered; inflorescence a spike-like panicle .... **Group 2**
      4b. Spikelets of a single plant monomorphic; inflorescence various
        5a. Glumes equaling or exceeding the body of the most apical lemma, more or less concealing the florets .... **Group 3**
        5b. Glumes shorter than the body of the most apical lemma, the florets clearly visible .... **Group 4**
    3b. Inflorescence a collection of 1 or more spikes or secund racemes .... **Group 5**
  2b. Spikelets with only 1 evident floret (sometimes with a second floret in *Sporobolus*), the remaining floret, if present, unisexual, sterile, or rudimentary
    6a. Glumes and lemmas compressed, appearing tightly folded and somewhat to strongly keeled
      7a. Inflorescence composed of 1 or more spikes or racemes, the spikes and racemes secund .... **Group 6**
      7b. Inflorescence an open or dense panicle (falsely appearing as a spike in *Alopecurus, Lycurus, Phleum,* and *Polypogon,* but the inflorescence spirally arranged .... **Group 7**
    6b. Glumes and lemmas flat, arched, or rounded
      8a. Inflorescence composed of a single, terminal spike, raceme, or spike-like panicle, the branches, when present, scarcely discernible .... **Group 8**
      8b. Inflorescence a panicle, or composed of 2 or more spikes or racemes, or the plant with terminal as well as axillary racemes
        9a. Inflorescence composed of 2 or more spikes or racemes disposed on an elongate axis, or the plant with both terminal and axillary racemes .... **Group 9**
        9b. Inflorescence a panicle .... **Group 10**

Group 1
1a. Bur formed from concrescent branchlets; inflorescence a dense panicle .... *Cenchrus*
1b. Bur formed from enlarged glumes; inflorescence a collection of racemes .... *Tragus*

Group 2
only 1 genus .... *Cynosurus*

Group 3
1a. Spikelets 1.9–2.6 cm long; glumes 5- to 11-veined .... *Avena*

1b. Spikelets up to 1.3 cm long; glumes 1- to 5-veined
2a. Spikelets each with 1 bisexual, fertile floret, the remaining 1 or 2 florets unisexual [staminate] or sterile
    3a. Apex of the lemma dissected into numerous, spreading awns .... ***Pappophorum***
    3b. Apex of the lemma without awns or with a single awn
        4a. Spikelets each with 1 bisexual floret and either 2 unisexual or 2 sterile florets; foliage mildly fragrant
            5a. Inflorescence an open panicle; glumes subequal in length; spikelets each with 1 bisexual and 2 unisexual florets .... ***Hierochloe***
            5b. Inflorescence a densely congested panicle; glumes distinctly unequal in length; spikelets each with 1 bisexual and 2 sterile florets .... ***Anthoxanthum***
        4b. Spikelets each with 1 bisexual floret and 1 unisexual [staminate] floret; foliage not fragrant
            6a. Spikelets 6.5–10.0 mm long; awns straight or bent; upper floret unisexual .... ***Arrhenatherum***
            6b. Spikelets 4.0–5.0 mm long; awns reflexed or hooked; lower floret bisexual .... ***Holcus***
2b. Spikelets with 2 or more bisexual florets
    7a. Each spikelet with, or mostly with, 2 florets
        8a. Lemmas awnless or with a very tiny, short point
            9a. Glumes dissimilar in shape—the first more or less linear, the second obovate, dilated upwards .... ***Sphenopholis***
            9b. Glumes similar in shape, both tapering from base to apex
                10a. Sheaths pubescent; rachilla glabrous; lemma untoothed at the apex ... . ***Koeleria***
                10b. Sheaths glabrous; rachilla pubescent; lemma 2-toothed at the apex ... . ***Trisetum***
        8b. Lemmas (at least 1 of them in the spikelet) with a conspicuous awn
            11a. Lemmas with an awn attached above the middle; disarticulation below the glumes, the falling spikelet leaving the pedicel behind .... ***Trisetum***
            11b. Lemmas with an awn attached below the middle; disarticulation above the glumes, the falling floret leaving the empty glumes behind
                12a. Glumes equaling to slightly exceeding the lemmas; anthers 1.3–3.0 mm long; stems with prominent tufts of basal leaves .... ***Deschampsia***
                12b. Glumes approximately twice as long as the lemmas; anthers 0.4–0.8 mm long; stems without prominent tufts of basal leaves .... ***Vahlodea***
    7b. Each spikelet with 3 or more florets
        13a. Lemmas with a conspicuous, bent awn arising from between 2 apical teeth; ligule a band of hairs .... ***Danthonia***
        13b. Lemmas awnless, without apical teeth; ligule of scarious tissue ...... ***Koeleria***

Group 4

1a. Plants robust, 1.5–4.0 m tall; inflorescence appearing plumose due to a long-pubescent rachilla .... ***Phragmites***

1b. Plants smaller, mostly less than 1.5 m tall; inflorescence not appearing plumose

  2a. Lemmas with 3 nerves [veins]

    3a. Callus and/or nerves of the lemma pubescent

      4a. Lemmas 2-lobed at the apex, awned from between the lobes, without arachnoid hairs on the callus; mature stems disarticulating at the nodes, the fragments including the leaf sheath and enclosed cleistogamous inflorescence; ligule a band of hairs .... ***Triplasis***

      4b. Lemmas entire at the apex, without awns, arachnoid-pubescent on the callus; mature stem not disarticulating at the nodes; ligule membranaceous .... ***Poa***

    3b. Callus and nerves of the lemma glabrous

      5a. Lemmas keeled, longer than the paleas; rachilla typically not disarticulating, the floral scales falling from it .... ***Eragrostis***

      5b. Lemmas rounded, shorter than the paleas; rachilla disarticulating ...... ***Molinia***

  2b. Lemmas with 5 or more nerves (the nerves inconspicuous, though still numbering 5 or more, in *Deschampsia*, *Puccinellia*, and *Sphenopholis*)

    6a. Panicle composed of dense, somewhat secund, glomerules of spikelets on few, stiff branches .... ***Dactylis***

    6b. Panicle composed of pedicellate spikelets on spirally arranged branches

      7a. Plants dioecious [the flowers of a given plant unisexual and of one type]; plants of saline habitats .... ***Distichlis***

      7b. Plants synoecious [the flowers of a given plant bisexual}; plants of various habitats

        8a. At least the basal half of the sheath fused [with united edges]

          9a. Lemmas 2-lobed at the apex, mostly awned, with arching veins that converge near the apex

            10a. Callus of the lemma pubescent; styles borne close together from the summit of the ovary; caryopsis free from the palea .... ***Schizachne***

            10b. Callus of the lemma glabrous; styles borne below the summit of the ovary and separated; caryopsis adnate to the palea ..... ***Bromus***

          9b. Lemmas entire at the apex, without awns, with parallel veins that do not converge near the apex .... ***Glyceria***

        8b. Leaf sheath with distinct, overlapping edges except at the very base

          11a. Glumes dissimilar in shape—the first more or less linear, the second obovate, dilated upwards .... ***Sphenopholis***

          11b. Glumes similar in shape, both with parallel sides or tapering from base to tip

            12a. Lemmas awned, the awn attached below the middle of the keel .... ***Deschampsia***

            12b. Lemmas without awns or with awns attached to the tip of the lemma

              13a. Lemmas blunt at the apex, with parallel veins that do not converge near the apex

14a. Lemmas prominently veined; plants of freshwater habitats
.... ***Torreyochloa***
14b. Lemmas inconspicuously veined; plants of saline habitats
.... ***Puccinellia***
13b. Lemmas pointed or awned at the apex, with arching veins
that converge near the apex
15a. Lemmas prominently keeled, often with arachnoid
pubescence on the callus, without awns; leaf tips cucullate
.... ***Poa***
15b. Lemmas rounded, or keeled only at the apex, with a
glabrous callus, usually awned; leaf tips flat
16a. Plants perennial; flowers open-pollinated
[chasmogamous]; anthers 3 per floret, 0.8–4.0 mm
long
17a. Body of the lemma 5.5–9.0 mm long; apical half
of the lemma with a prominent, scarious margin *ca.*
0.5 wide; leaves 3.0–10.0 mm wide .... ***Lolium***
17b. Body of the lemma 2.3–6.1 mm long; lemmas
with firm margins or very narrow, inconspicuous,
scarious margins; leaves 0.2–3.0 mm wide (up to
8.0 mm in *Festuca subverticillata*) .... ***Festuca***
16b. Plants annual; flowers remaining closed
[cleistogamous]; anthers usually 1 per floret, up to
1.0 mm long .... ***Vulpia***

Group 5

1a. Spikelets dimorphic, the 2 types paired—the lower sterile, flabellate, consisting of 2
glumes and several, narrow, acuminate lemmas, concealing the upper, fertile, 2- to 4-
flowered spikelet; inflorescence a spike-like panicle .... ***Cynosurus***
1b. Spikelets monomorphic; inflorescence various
2a. Inflorescence a solitary spike
3a. Spikelets secund along the rachis; lemmas with 3 excurrent veins or awns ....
.... ***Bouteloua***
3b. Spikelets on alternating sides of the rachis; lemmas awnless or with a single awn
4a. Spikelets with the narrow side toward the rachis; only the glume away from
the rachis developed, the other obsolete .... ***Lolium***
4b. Spikelets with the wide side toward the rachis; both glumes developed
5a. Spikelets 2 or more at each node of the rachis, although only 1 may be
fertile
6a. Spikelets consistently 3 at each node, the outer pair usually pedicellate
and unisexual [staminate] or sterile .... ***Hordeum***
6b. Spikelets 2–4 at each node, commonly 2, all bisexual and fertile
7a. Plants mostly self-pollinating, with small anthers 1.0–3.0 mm
long, cespitose, with short or no rhizomes; lemmas mostly awned
.... ***Elymus***

7b. Plants mostly outcrossing, with large anthers mostly 4.0–9.0 mm long, not cespitose, with elongate, stout rhizomes; lemmas mostly without awns .... *Leymus*

5b. Spikelets mostly 1 per node

8a. Plants annual, short-persisting crops; glumes 1- to 3-veined

9a. Glumes with 1 vein, linear-subulate; keel of lemmas ciliate .... *Secale*

9b. Glumes with 3 veins, ovate; keel of lemmas glabrous ..... *Triticum*

8b. Plants perennial, native or introduced; glumes 5- to 7-veined .... *Elymus*

2b. Inflorescence composed of 2 or more spikes or racemes

10a. Spikes or racemes secund, not appearing plumose; plants commonly less than 1.5 m tall

11a. Spikelets 2- to 12-flowered, the florets all bisexual and fertile

12a. Inflorescence with an elongate central axis on which the racemes are borne; second glume 1-veined; neither the sheaths nor the stems compressed .... *Leptochloa*

12b. Inflorescence lacking an elongate central axis, all the spikes branching from nearly the same point; second glume 3- or 5-veined; both stems and sheaths compressed .... *Eleusine*

11b. Spikelets 2- or 3-flowered, only 1 of the florets bisexual and fertile, the other(s) sterile or unisexual [staminate]

13a. Inflorescence with an elongate central axis

14a. Glumes 3-veined, obovate, inflated, their tips crossing; lemmas acuminate .... *Beckmannia*

14b. Glumes 1-veined, linear to lanceolate, not inflated, their tips separate; lemmas 3-awned .... *Bouteloua*

13b. Inflorescence lacking an elongate central axis, all florets branching from nearly the same point .... *Chloris*

10b. Racemes not secund, appearing plumose due to a ring of silky hairs subtending the glumes; plants robust, 1.5–2.0 m tall .... *Miscanthus*

Group 6

1a. Glumes obovate, inflated, their tips crossing; spikelets nearly orbicular from side view .... *Beckmannia*

1b. Glumes linear to lanceolate, not inflated, their tips separate; spikelets lanceolate to ovate or obovate to obtriangular from side view

2a. Spikelets each with 1 bisexual, fertile floret; plants native, of marshes and shores [fresh or saline] .... *Spartina*

2b. Spikelets each with 1 bisexual, fertile floret and 1 or 2 additional, sterile florets; plants introduced, of dry woods, sandy soils, and lawns

3a. Inflorescence consisting of several spikes branching from nearly the same point, appearing whorled; lemmas with a single awn .... *Chloris*

3b. Inflorescence consisting of a single spike or 2 or more spikes disposed on an elongate axis, not branching from the same point; lemmas with 3 excurrent veins or awns .... *Bouteloua*

Group 7

1a. Inflorescence a dense, cylindrical, spike-like panicle, the branches virtually absent

    2a. Glumes without awns, connate at the base; paleas wanting .... *Alopecurus*

    2b. Glumes with awns, free at the base; paleas present, equaling the length of the lemmas or nearly so

        3a. Glumes exceeding the length of the floret and concealing it, each glume with a single awn

            4a. Awn of glumes 3.0–8.0 mm long; lemmas with awns; glumes 2-lobed at the apex, the awn emerging from between the lobes .... *Polypogon*

            4b. Awn of glumes 0.7–3.0 mm long; lemmas without awns; glumes entire at the apex, the awn terminal .... *Phleum*

        3b. Glumes shorter than and not concealing the floret, the first glume with 2(3) awns, the second with a single awn .... *Lycurus*

1b. Inflorescence an open or dense panicle, the branches identifiable

    5a. Lemmas indurate and lustrous at maturity, of much firmer texture than the glumes

        6a. Lemmas awnless; spikelets with 1 fertile floret and at its base 2 sterile, rudimentary florets represented by pubescent scales .... *Phalaris*

        6b. Lemmas with awns; spikelets without any sterile florets

            7a. Lemmas pubescent only on the callus, with 3 persistent awns; callus of lemmas slender and acute; plants annual .... *Aristida*

            7b. Lemmas pubescent over the abaxial surface, with a single awn articulated and eventually deciduous from the summit of the lemma; callus of lemmas short and oblique; plants perennial

                8a. Lemmas dark brown to nearly black; leaves without ligules, the uppermost with flat blades usually longer than 12.0 cm .... *Piptatherum*

                8b. Lemmas yellow-green, pale green, gray, or pale brown; leaves with ligules, though very small in *Oryzopsis asperifolia*, the uppermost with flat or involute blades usually shorter than 12.0 cm .... *Oryzopsis*

    5b. Lemmas soft and dull at maturity, not of conspicuously firmer texture than the glumes

        9a. Glumes virtually obsolete, reduced to minute cupules; leaves and sheaths prominently scabrous .... *Leersia*

        9b. Glumes present and evident; leaves and sheaths glabrous to slightly scabrous

            10a. Floret stipitate, with 1 stamen; spikelets articulated below the glumes and falling intact leaving the pedicel behind .... *Cinna*

            10b. Floret sessile, mostly with 3 stamens; spikelets articulated above the glumes, the florets falling and leaving the empty glumes behind

                11a. Spikelets 10.0–15.0 mm long; plants robust, to 1.0 m tall, of Atlantic coast sands .... *Ammophila*

                11b. Spikelets 1.2–8.0 mm long; plants of various statures and habitats, but not regularly of coastal sands

                    12a. Glumes shorter than the lemmas and/or the lemmas awned from the apex; caryopsis tightly enclosed by the lemma .... *Muhlenbergia*

                    12b. At least 1 of the glumes equaling the length of the floret; awn of the lemma, if present, originating from below the apex on the keel of the lemma; caryopsis loosely enclosed by the lemma

13a. Lemma evidently, though sometimes shortly, pubescent on the callus; palea 2-nerved, membranaceous, 0.5–1.0 times as long as the lemma; rachilla prolonged beyond the upper floret as a minute bristle

14a. Awn straight, inserted just below the apex and between the terminal teeth of the lemma, 5.0–12.0 mm long; pubescence of the callus of the lemma *ca.* 0.2 times as long as the lemma; plants annual .... ***Apera***

14b. Awn often bent, definitely originating from below the apex of the lemma, up to 3.0 mm long; pubescence of the callus of the lemma 0.25–1.0 times as long as the lemma; plants perennial .... ***Calamagrostis***

13b. Lemma glabrous on the callus (pubescent in *Agrostis elliottiana*); palea 2-nerved or, more often, nerveless, delicate, up to 0.65 times as long as the lemma; rachilla not prolonged .... ***Agrostis***

Group 8

1a. Inflorescence appearing bristly or silky due to long-awned glumes, subtending bristles, or a false involucre of sterile, awn-like lemmas; lemmas awnless or with a single awn

2a. Inflorescence composed of spirally arranged spikelets; spikelets 1.6–3.5 mm long

3a. Spikelets subtended by 1–12 bristles; glumes awnless, with 3 or more veins .... ***Setaria***

3b. Spikelets not subtended by bristles; glumes awned, 1-veined .... ***Polypogon***

2b. Inflorescence composed of spikelets that are positioned on alternate sides of the rachis; spikelets, including awns, 6.0–150.0 mm long .... ***Hordeum***

1b. Inflorescence not appearing bristly or silky; lemmas with 3 excurrent veins or awns .... ***Bouteloua***

Group 9

1a. Inflorescence composed of 2 or more spikes or racemes, lacking an elongate axis, branching from nearly the same point

2a. Spikelets in pairs, both pedicelled, one short- and the other long-pedicelled; glumes subtended by a ring of partially concealing hairs .... ***Miscanthus***

2b. Spikelets sessile or, if conspicuously paired, then one of the spikelets sessile; glumes not subtended by a ring of long hairs

3a. Spikelets all fertile, subsessile, borne on and falling free from a continuous, non-pubescent rachis; plants to 0.6 m tall .... ***Digitaria***

3b. Spikelets both fertile [staminate] and non-fertile, the non-fertile pedicellate, borne on and falling with sections of an articulated, pubescent rachis; plants commonly exceeding 1.0 m tall .... ***Andropogon***

1b. Inflorescence composed of 2 or more spikes or racemes disposed on an elongate axis, or the racemes solitary, several per plant, both terminal and axillary

4a. Spikelets all sessile or subsessile, borne on a continuous rachis; rachis and the virtually absent pedicels glabrous

5a. Lemmas with 3 excurrent veins or awns; glumes 1-nerved .... ***Bouteloua***

5b. Lemmas awnless or with a single awn; first glume 3-nerved; second glume 5-nerved .... ***Echinochloa***

4b. One or both of the paired spikelets pedicellate (in *Sorghastrum*, only a pedicel present to represent the second spikelet of a pair) on a continuous or, more commonly, articulated rachis; rachis and/or pedicels pubescent or the glumes subtended by a ring of long hairs

   6a. Both spikelets of a pair pedicelled, one short- and the other long-pedicelled; glumes subtended by a ring of partially concealing hairs .... ***Miscanthus***

   6b. One of the paired spikelets sessile; glumes not subtended by a ring of long hairs

      7a. Inflorescence a single raceme; rachis composed of many articulated segments that prominently dehisce .... ***Schizachyrium***

      7b. Inflorescence consisting of many racemes in a panicle-like arrangement; rachis composed of 2 or 3 tardily separating segments

         8a. Pedicelled floret evident, staminate; plants cultivated or escaped weeds .... ***Sorghum***

         8b. Pedicelled floret rudimentary, represented by the pubescent pedicel; plants native .... ***Sorghastrum***

Group 10

1a. Spikelets in conspicuous pairs (or trios at the tips of the branches) on an articulated rachis—one spikelet sessile and fertile, the other pedicelled and staminate [evident] or rudimentary [represented by a pubescent pedicel]

   2a. Pedicelled floret evident, staminate; plants cultivated or escaped weeds ..... ***Sorghum***

   2b. Pedicelled floret rudimentary, represented by the pubescent pedicel; plants native .... ***Sorghastrum***

1b. Spikelets neither in conspicuous pairs nor on an articulated rachis

   3a. Plants monoecious, with staminate flowers below and carpellate flowers above; plants robust, aquatic annuals .... ***Zizania***

   3b. Plants synoecious or polygamous; plants of various stature and habit, but not aquatic

      4a. Lemmas without awns

         5a. Spikelets at maturity with 1 pair of lemmas and paleas indurate and lustrous, of much firmer texture than the glumes

            6a. Spikelet with 1 bisexual, fertile floret, articulated above the glumes, the floret falling leaving the empty glumes behind .... ***Milium***

            6b. Spikelet with 1 bisexual, fertile floret and 1 inconspicuous, sterile or staminate floret, articulated below the glumes, the entire spikelet falling and leaving the pedicel behind .... ***Panicum***

         5b. Lemmas soft or firm at maturity, dull, not conspicuously indurate

            7a. Lemmas 1-veined; fruit a utricle, gelatinizing when wet, with an easily separated pericarp .... ***Sporobolus***

            7b. Lemmas 3-veined; fruit a caryopsis, not gelatinizing when wet, the pericarp adnate to the ovule .... ***Muhlenbergia***

      4b. Lemmas with awns

         8a. Lemmas with 3 awns at the apex, indurate; plants annual .... ***Aristida***

         8b. Lemmas with a single awn from the apex, indurate or softer; plants perennial

9a. Lemmas indurate and lustrous, conspicuously firmer than the glumes, pilose over the abaxial surface; awn of lemma articulated and eventually deciduous

   10a. Lemmas dark brown to nearly black; leaves without ligules, the uppermost with flat blades usually longer than 12.0 cm .. ***Piptatherum***

   10b. Lemmas yellow-green, pale green, gray, or pale brown; leaves with ligules, though very small in *Oryzopsis asperifolia*, the uppermost with flat or involute blades usually shorter than 12.0 cm ..... ***Oryzopsis***

9b. Lemmas firm, but not indurate, dull, glabrous or scabrous to pubescent on the nerves or callus (minutely pubescent between the nerves in some *Muhlenbergia*); awn of lemma not articulated and persistent

   11a. First glume diminutive or obsolete; rachilla prolonged, located near the abaxial surface of the palea and appearing as a minute bristle; plants with rhizomes .... ***Brachyelytrum***

   11b. First glume present and evident (except in *Muhlenbergia schreberi*, which lacks rhizomes); rachilla not prolonged; plants with or without rhizomes .... ***Muhlenbergia***

Agrostis
(10 species)

KEY TO THE SPECIES

1a. Palea 2-nerved, 0.5–0.65 times as long as the lemma; plants mostly with rhizomes and/or stolons

   2a. Ligules 0.5–2.0 mm long, usually wider than long; branches of panicle bearing flowers only in the distal half; leaves 1.0–5.0 mm wide .... ***A. capillaris***

   2b. Ligules 2.0–6.0 mm long, usually longer than wide; branches of panicle bearing flowers in both distal and proximal halves; leaves 1.0–9.0 mm wide

      3a. Plants with rhizomes and erect stems; leaves 3.0–9.0 mm wide; inflorescence 10.0–20.0 x 6.0–20.0 cm, usually red-purple at anthesis .... ***A. gigantea***

      3b. Plants without rhizomes, though often stoloniferous, with basally decumbent stems; leaves 1.0–5.0 mm wide; inflorescence up to 10.0 x 8.0 cm, usually yellow-brown at anthesis .... ***A. stolonifera***

1b. Palea unnerved and minute, less than 0.5 times as long as the lemma; plants cespitose, without prolonged rhizomes or stolons (sometimes stoloniferous in *A. canina*)

   4a. Plants annual; lemmas usually with a delicate, flexuous awn 6.0–10.0 mm long; callus of the lemma pubescent .... ***A. elliottiana***

   4b. Plants perennial; lemmas with or without awns, the awns, if present, stouter and up to 6.0 mm long; callus of the lemma glabrous

      5a. Lemmas awnless or with a very short awn produced from near the tip

         6a. Branches of panicle forking near or below middle, smooth or sparsely scabrous; leaves 2.0–6.0 mm wide .... ***A. perennans***

         6b. Branches of panicle forking beyond the middle, conspicuously scabrous; leaves 1.0–3.0 mm wide

            7a. Anthers globose, 0.2–0.3 mm long; pedicels 0.3–2.0 mm long; glumes 1.2–2.4 mm long; plants flowering March–July .... ***A. hyemalis***

       7b. Anthers oblong, 0.4–0.8 mm long; pedicels 0.5–5.0 mm long; glumes
           2.0–3.0 mm long; plants flowering June–November

           8a. Fully expanded panicle diffuse, 0.5–0.65 of the height of the plant;
              branches of the panicle gibbous at the base; lemma 0.5–0.65 times as
              long as the glume .... *A. scabra*

           8b. Fully expanded panicle less diffuse, 0.15–0.35 of the height of the
              plant; branches of the panicle not or only slightly gibbous at the base;
              lemma *ca.* 0.65 times as long as the glume .... *A. geminata*

    5b. Lemmas with an awn 1.3–6.0 mm long, the awn produced from near the middle
       of the lemma

       9a. Branches of panicle scabrous, bearing flowers only in the distal half ....
       .... *A. geminata*

       9b. Branches of panicle glabrous or scabrous, bearing flowers in both distal and
          proximal halves

          10a. Spikelets 1.9–2.5 mm long; anthers 1.0–1.8 mm long; branches of
             panicle scabrous; plants of elevation less than 1000 m .... *A. canina*

          10b. Spikelets 2.0–4.0 mm long; anthers 0.5–0.7 mm long; branches of
             panicle smooth; plants of alpine habitats, elevation exceeding 1000 m ......
             .... *A. mertensii*

Agrostis canina L.

synonym(s): —
common name(s): velvet bentgrass, brown bentgrass
habit: cespitose, stoloniferous, perennial herb
range: Newfoundland and Quebec, s. to DE, w. to TN, n. to MN
state frequency: uncommon
habitat: fields, meadows
notes: —

Agrostis capillaris L.

synonym(s): *Agrostis tenuis* Sibthorp
common name(s): Rhode Island bentgrass
habit: cespitose to matted, perennial herb
range: Europe; introduced to Labrador, s. to NC, w. to MO, n. to ND; also OR
state frequency: common
habitat: pastures, roadsides
notes: —

Agrostis elliottiana J. A. Schultes

synonym(s): —
common name(s): Elliott's bentgrass
habit: cespitose, annual herb
range: MD, s. to FL, w. to TX, n. to KS; introduced to ME and MA
state frequency: very rare
habitat: fields, pastures
notes: Introduced to southern Maine in grass seed.

Agrostis geminata Trin.

synonym(s): *Agrostis hyemalis* (Walt.) B. S. P. var. *geminata* (Trin.) A. S. Hitchc.
common name(s): —
habit: cespitose, perennial herb

range: Labrador, s. to ME, w. to CO and CA, n. to AK
state frequency: rare
habitat: shores, barrens, rocky or peaty areas
notes: —

Agrostis gigantea Roth
synonym(s): *Agrostis alba auct. non* L.
common name(s): redtop, black bentgrass
habit: rhizomatous, perennial herb
range: Europe; introduced throughout the United States
state frequency: common
habitat: shores, moist areas
notes: —

Agrostis hyemalis (Walt.) B. S. P.
synonym(s): —
common name(s): ticklegrass, hairgrass
habit: slender, cespitose, perennial herb
range: ME, s. to FL, w. to TX, n. to MN
state frequency: rare
habitat: open areas, bogs, thin woods
notes: —

Agrostis mertensii Trin.
synonym(s): *Agrostis borealis* Hartman, *Agrostis borealis* Hartman var. *americana* (Scribn.) Fern.
common name(s): boreal bentgrass
habit: cespitose, perennial herb
range: circumboreal; s. to mountains of New England and NY; also WV
state frequency: rare
habitat: rocky soil, moist, peaty areas
notes: This species is ranked S2 by the Maine Natural Areas Program.

Agrostis perennans (Walt.) Tuckerman
synonym(s): *Agrostis perennans* (Walt.) Tuckerman var. *aestivalis* Vasey
common name(s): autumn bentgrass, upland bentgrass
habit: cespitose, perennial herb
range: Quebec and Nova Scotia, s. to FL, w. to TX, n. to MN
state frequency: common
habitat: dry, rocky soil and open woods
notes: —

Agrostis scabra Willd.
synonym(s): *Agrostis hyemalis* (Walt.) B. S. P. var. *scabra* (Willd.) Blomq.
common name(s): ticklegrass, hairgrass, fly-away grass
habit: cespitose, perennial herb
range: Labrador, s. to GA, w. to Mexico and CA, n. to AK
state frequency: common
habitat: wet or dry, open soil
notes: —

Agrostis stolonifera L.
synonym(s): *Agrostis alba* L. var. *palustris* (Huds.) Pers.
common name(s): creeping bentgrass

habit: usually stoloniferous, perennial herb
range: Europe; naturalized to Newfoundland, s. to SC, interruptedly w. to CA, n. to AK
state frequency: common
habitat: damp shores
notes: —

Alopecurus
(4 species)

KEY TO THE SPECIES

1a. Glumes 4.0–6.0 mm long, acute to acuminate and firm at the apex; anthers 2.2–3.5 mm long
  2a. Glumes connate at the base for 0.25 of their length, ciliate on the keel with long hairs, the hairs on the distal half of the keel 1.0–1.5 mm long .... *A. pratensis*
  2b. Glumes connate at the base for 0.35 to 0.5 of their length, short-ciliate on the proximal half of the keel, scabrous on the distal half .... *A. myosuroides*
1b. Glumes 2.0–3.2 mm long, blunt and scarious at the apex; anthers 0.5–2.0 mm long
  3a. Awn inserted on the lemma just below the midpoint, equaling the glumes or exserted from the spikelet up to 1.0 mm; anthers 0.5–1.0 mm long .... *A. aequalis*
  3b. Awn inserted on the lemma halfway between the midpoint and base, exserted from the spikelet 1.5–3.0 mm; anthers 1.3–2.0 mm long .... *A. geniculatus*

Alopecurus aequalis Sobol.
synonym(s): —
common name(s): short-awn foxtail
habit: cespitose, annual or short-lived perennial herb
range: circumboreal; s. in North America to PA, w. to NM and CA
state frequency: uncommon
habitat: muddy shores, ditches, shallow water
notes: —

Alopecurus geniculatus L.
synonym(s): —
common name(s): marsh foxtail, water foxtail
habit: matted, perennial herb
range: Eurasia; introduced to Newfoundland, s. to VA, interruptedly w. to AZ and CA, n. to British Columbia
state frequency: uncommon
habitat: ditches, mud, shallow water
notes: —

Alopecurus myosuroides Huds.
synonym(s): —
common name(s): slender foxtail
habit: cespitose, annual herb
range: Europe; introduced to ME, s. to NC, w. to TX, n. to KS; also OR and WA
state frequency: very rare
habitat: fields, waste places
notes: —

Alopecurus pratensis L.
synonym(s): —
common name(s): meadow foxtail
habit: rhizomatous, perennial herb
range: Eurasia; introduced to Newfoundland, s. to GA, w. to KS, n. to AK
state frequency: uncommon
habitat: moist fields, pastures, meadows, open areas
notes: —

Ammophila
(1 species)

Ammophila breviligulata Fern.
synonym(s): —
common name(s): beach grass
habit: rhizomatous, perennial herb
range: Labrador, s. along the coast to NC; also Lake Champlain and Great Lakes
state frequency: uncommon; locally abundant on sandy seashores
habitat: sand shores and dunes of the Atlantic coast
notes: —

Andropogon
(1 species)

Andropogon gerardii Vitman
synonym(s): —
common name(s): big bluestem
habit: cespitose, rhizomatous, perennial herb
range: Quebec, s. to FL, w. to AZ, n. to Saskatchewan
state frequency: uncommon
habitat: dry or moist, open areas
notes: —

Anthoxanthum
(2 species)

Maine's members of this genus have gold-brown, pubescent, sterile lemmas that enclose the brown, lustrous, fertile lemma.

KEY TO THE SPECIES
1a. Plants perennial; leaves 2.0–6.0 mm wide; awn of the lower, sterile lemma straight; spikelets 7.0–9.0 mm long .... *A. odoratum*
1b. Plants annual; leaves mostly less than 2.0 mm wide; awn of both sterile lemmas bent; spikelets 5.0–7.0 mm long .... *A. aristatum*

Anthoxanthum aristatum Boiss.
synonym(s): *Anthoxanthum puelii* Lecoq & Lamotte
common name(s): —
habit: branched, basally geniculate, cespitose, annual herb
range: Europe; introduced in eastern North America to ME, s. to FL, w. to AR, n. to IA
state frequency: very rare

habitat: waste places
notes: —

Anthoxanthum odoratum L.
synonym(s): —
common name(s): sweet vernal grass
habit: cespitose, perennial herb
range: Eurasia; introduced in eastern North America to Greenland, s. to GA, w. to LA, n. to British Columbia
state frequency: common
habitat: fields, roadsides, waste places
notes: —

Apera
(2 species)

KEY TO THE SPECIES
1a. Leaves 3.0–5.0 mm wide; branches of the panicle ascending to spreading at anthesis, bearing flowers in the distal half .... *A. spica-venti*
1b. Leaves 1.0–3.0 mm wide; branches of the panicle contracted, erect at anthesis, bearing flowers in both distal and proximal halves .... *A. interrupta*

Apera interrupta (L.) Beauv.
synonym(s): *Agrostis interrupta* L.
common name(s): —
habit: cespitose, annual herb
range: Europe; introduced to several, widely scattered locations in North America
state frequency: very rare
habitat: waste places
notes: —

Apera spica-venti (L.) Beauv.
synonym(s): *Agrostis spica-venti* L.
common name(s): —
habit: cespitose, annual herb
range: Europe; introduced to ME, s. to DE, w. to MO, n. to Ontario
state frequency: rare
habitat: roadsides, waste places
notes: —

Aristida
(3 species)

KEY TO THE SPECIES
1a. Longest glume 9.5–30.0 mm long; central awn 11.0–70.0 mm long; lateral awns 4.0–13.0 mm long; glumes distinctly unequal in length
  2a. Central awn 40.0–70.0 mm long and, when dry, curved but not coiled at the base; first glume 3- to 5-veined .... *A. oligantha*
  2b. Central awn 11.0–19.0 mm long and, when dry, coiled 0.5–3.0 times at the base; first glume 1-veined .... *A. basiramea*

1b. Longest glume up to 10.5 mm long; central awn 3.0–10.0 mm long; lateral awns
    0.7–3.3 mm long; glumes subequal in length .... *A. dichotoma*

Aristida basiramea Engelm. *ex* Vasey
synonym(s): —
common name(s): forktip three-awn, forktip triple-awn, branching needle-grass
habit: cespitose, annual herb
range: ME, s. to FL, w. to CO, n. to ND
state frequency: very rare
habitat: dry, sandy soil
notes: This species is ranked S1 by the Maine Natural Areas Program.

Aristida dichotoma Michx.
synonym(s): —
common name(s): poverty grass, churchmouse three-awn
habit: cespitose, annual herb
range: ME, s. to FL, w. to TX, n. to ND
state frequency: rare
habitat: dry, sterile soil
notes: —

Aristida oligantha Michx.
synonym(s): —
common name(s): prairie three-awn
habit: cespitose, annual herb
range: MA, s. to FL, w. to TX, n. to SD and MI; adventive to northeastern North America
state frequency: very rare
habitat: dry, open, sterile ground
notes: —

Arrhenatherum
(1 species)

Arrhenatherum elatius (L.) J. & K. Presl
synonym(s): —
common name(s): tall oatgrass
habit: cespitose, short-lived perennial herb
range: Europe; introduced to Newfoundland, s. to GA, w. to CA, n. to British Columbia
state frequency: uncommon
habitat: moist soil of fields, roadsides, and waste places
notes: —

Avena
(2 species)

Our species of *Avena* are adventive annuals, more common in areas of extensive cultivation.

KEY TO THE SPECIES
1a. Lemmas pubescent with appressed, brown hairs; usually the 2 basalmost florets with
    awned lemmas; awns bent .... *A. fatua*
1b. Lemmas glabrous or scabrous; usually only the basalmost floret with an awned lemma;
    awns, when present, straight .... *A. sativa*

Avena fatua L.

synonym(s): —
common name(s): wild oats
habit: stout, erect, cespitose, annual herb
range: Europe; naturalized to Newfoundland, s. to PA, w. to Mexico and NM, n. to British Columbia
state frequency: rare
habitat: fields, cultivated ground, waste places
notes: Known to hybridize with *Avena sativa*.

Avena sativa L.

synonym(s): —
common name(s): oats
habit: stout, erect, cespitose, annual herb
range: Eurasia; escaped from cultivation
state frequency: uncommon
habitat: fields, roadsides, waste places
notes: Known to hybridize with *Avena fatua*.

Beckmannia
(1 species)

Beckmannia syzigachne (Steud.) Fern.

synonym(s): —
common name(s): American sloughgrass
habit: solitary or cespitose, annual herb
range: Asia; naturalized to western Quebec, s. to NY, w. to KS, NM, and CA, n. to AK; occasionally adventive in our area
state frequency: very rare
habitat: marshes, wet ground
notes: —

Bouteloua
(3 species)

Valuable forage grasses from the western portion of the continent.

KEY TO THE SPECIES
1a. Inflorescence usually with 7–12 spikes; disarticulation below the glumes, the entire
    spikelet falling intact; leaves mostly 2.0–3.0 mm wide .... *B. radicosa*
1b. Inflorescence with 1–3(–6) spikes; disarticulation above the glumes, the florets falling
    individually without the glumes; leaves mostly 1.0–2.0 mm wide
    2a. Plants annual; inflorescence usually a solitary spike 1.5–2.5 cm long .... *B. simplex*
    2b. Plants perennial; inflorescence usually composed of 2 spikes 2.5–5.0 cm long ....
    .... *B. gracilis*

Bouteloua gracilis (Willd. *ex* Kunth) Lag. *ex* Griffiths
synonym(s): *Chondrosum gracile* Willd. *ex* Kunth
common name(s): blue grama
habit: slender, erect, cespitose, short-rhizomatous, perennial herb
range: Manitoba, s. to IL, MO, and TX, w. to CA, n. to British Columbia; adventive in the east

state frequency: very rare
habitat: dry, sandy areas
notes: —

Bouteloua radicosa (Fourn.) Griffiths
synonym(s): *Atheropogon radicosus* Fourn.
common name(s): purple grama
habit: rhizomatous, cespitose, perennial herb
range: CA, s. to Mexico; adventive in ME
state frequency: very rare
habitat: waste places
notes: —

Bouteloua simplex Lag.
synonym(s): *Chondrosum procumbens* Desv. *ex* Beauv.
common name(s): mat grama
habit: cespitose, prostrate to ascending, annual herb
range: UT and CO, s. to Mexico; adventive in our area
state frequency: very rare
habitat: waste places
notes: —

Brachyelytrum
(1 species)

Brachyelytrum septentrionale (Babel) G. Tucker
synonym(s): *Brachyelytrum erectum* (Schreb. *ex* Spreng.) Beauv. var. *septentrionale* Babel
common name(s): —
habit: erect, short-rhizomatous, perennial herb
range: Newfoundland, s. to NJ, w. to OH, n. to MN
state frequency: occasional
habitat: forests
notes: —

Bromus
(14 species)

KEY TO THE SPECIES
1a. Lobes of the lemma apex 0.1–0.5 mm long; plants perennial
    2a. Lower glume 1-nerved; upper glume 3-nerved
        3a. Plants with long rhizomes; branches of the panicle ascending; anthers
            4.0–6.0 mm long
            4a. Spikelets longer than their pedicels; leaves 2.0–3.0 mm wide; awns
                4.0–7.0 mm long .... ***B. erectus***
            4b. Spikelets shorter than their pedicels; leaves 5.0–15.0 mm wide; awns none or
                up to 4.0(5.0) mm long .... ***B. inermis***
        3b. Plants without rhizomes; branches of the panicle spreading to drooping; anthers
            0.9–5.0 mm long
            5a. Lemma glabrous or, more commonly, pubescent along the margins; anthers
                0.9–1.7 mm long

6a. Rachilla visible in mature, intact spikelet; glumes and lemmas strongly folded; paleas linear, closely embraced by lemmas .... *B. ciliatus*

6b. Rachilla concealed in mature, intact spikelet; glumes and lemmas flat or arched; paleas oblong, weakly embraced by lemmas .... *B. canadensis*

5b. Lemma sparsely to densely pubescent over the abaxial surface; anthers 1.5–5.0 mm long

    7a. Sheaths auriculate, densely pubescent at the apex, overlapping and concealing the nodes; leaves 8–20 per stem; anthers 1.5–2.2 mm long ......
    .... *B. latiglumis*

    7b. Sheaths lacking both auricles and a prominently pubescent apex, at least the upper sheaths not overlapping and exposing the nodes; leaves 4–6 per stem; anthers 2.5–5.0 mm long .... *B. pubescens*

2b. Lower glume 3- or 5-nerved; upper glume 5- to 9-nerved

    8a. Lemmas compressed and strongly keeled; ligules 1.0–4.0 mm long; first and second glumes 7.0–11.0 and 9.0–13.0 mm long, respectively .... *B. marginatus*

    8b. Lemmas convex, not strongly compressed; ligules up to 1.0 mm long; first and second glumes 6.0–7.0 and 7.0–9.0 mm long, respectively .... *B. kalmii*

1b. Lobes of the lemma apex 0.6–3.0 mm long; plants annual or biennial

    9a. First glume 1-veined; second glume 3-veined; lemmas narrow from a side view, 1.0–1.5 mm deep .... *B. tectorum*

    9b. First glume 3- or 5-veined; second glume 5- or 7-veined; lemmas wider from a side view, 1.2–2.2 mm deep

        10a. Pedicels mostly shorter than their spikelets; inflorescence compact, erect, 3.0–10.0 cm tall; lemmas prominently plicate-nerved .... *B. hordeaceus*

        10b. Pedicels as long as or longer than their associated spikelets; inflorescence open, ascending to drooping or narrow and erect in *B. racemosus*, 4.0–20.0 cm tall; lemmas merely nerved

            11a. Margin of the lemma involute at maturity, tightly enclosing the caryopsis and exposing the rachilla

                12a. Palea about as long as its associated lemma; awns of lemma 0.0–6.6 mm long; branches of panicle mostly ascending .... *B. secalinus*

                12b. Palea 1.0–2.5 mm shorter than its associated lemma; awns of lemma 7.0–12.0 mm long; branches of panicle widely spreading .... *B. japonicus*

            11b. Margin of the lemma not tightly involute, overlapping the caryopsis and concealing the rachilla

                13a. Branches of the panicle erect, straight or nearly so, usually 2 per node ....
                .... *B. racemosus*

                13b. Branches of the panicle ascending to, more commonly, spreading or even drooping, flexuous or arching, 2–6 per node

                    14a. Awns straight or nearly so and erect to ascending, less than 2.0 times as long as the lowest lemma of the spikelet; branches of the panicle stiff; palea up to 1.5 mm shorter than its associated lemma ....
                    .... *B. commutatus*

                    14b. Awns flexuous to recurved, more than 2.0 times as long as the lowest lemma of the spikelet; branches of panicle lax and flexuous; palea 1.0–2.5 mm shorter than its associated lemma .... *B. japonicus*

Bromus canadensis Michx.

synonym(s): *Bromus dudleyi* Fern.
common name(s): Dudley's bromegrass
habit: solitary or cespitose, perennial herb
range: Newfoundland, s. to CT and mountains of WV, w. to MT, n. to British Columbia
state frequency: uncommon
habitat: moist to wet meadows, thickets, and shores
notes: —

Bromus ciliatus L.

synonym(s): *Bromus ciliatus* L. var. *intonsus* Fern., *Bromus ramosus* Huds.
common name(s): fringed bromegrass
habit: usually cespitose, perennial herb
range: Labrador, s. to mountains of NC, w. to TX, n. to AK
state frequency: uncommon
habitat: moist forests, thickets, slopes, and shores
notes: —

Bromus commutatus Schrad.

synonym(s): —
common name(s): hairy chess
habit: annual herb
range: Europe; introduced to Quebec, s. to LA, w. to KS, n. to British Columbia
state frequency: very rare
habitat: dry roadsides, waste places
notes: —

Bromus erectus Huds.

synonym(s): —
common name(s): —
habit: cespitose, long-rhizomatous, perennial herb
range: Europe; introduced to ME, s. to DC, w. to WI, n. to Ontario
state frequency: rare
habitat: fields, roadsides, wet places
notes: —

Bromus hordeaceus L.

synonym(s): *Bromus mollis* L. *sensu* Fernald (1950)
common name(s): soft chess
habit: annual or biennial herb
range: Europe; introduced to Newfoundland, s. to NC, w. to CA, n. to British Columbia
state frequency: rare
habitat: roadsides, fields, waste places
notes: —

Bromus inermis Leyss.

synonym(s): <*B. i.* ssp. *pumpellianus* var. *arcticus*> = *Bromus inermis* Leyss. var. *aristatus* Schur *ex* Scribn. & Merr.
common name(s): smooth bromegrass, awnless bromegrass, Hungarian brome
habit: mostly solitary, stoloniferous, long-rhizomatous, perennial herb
range: Europe; escaped from cultivation to Newfoundland, s. to VA, w. to NM, n. to British Columbia
state frequency: common
habitat: fields, roadsides

key to the subspecies:
  1a. Lemmas glabrous or scabrous, with awns 0.0–2.5 mm long; nodes and leaf blades mostly
      glabrous; ligules 0.5–1.0 mm long .... *B. i.* ssp. *inermis*
  1b. Lemmas pubescent near the margins, with awns 1.5–4.0(–5.0) mm long; nodes and leaf blades
      pubescent; ligules 1.0–2.5 mm long ....
      ......*B. i.* ssp. *pumpellianus* (Scribn.) Wagnon var. *arcticus* (Shear *ex* Scribn. & Merr.) Wagnon
notes: The 2 subspecies of *Bromus inermis* found in Maine hybridize with each other.

Bromus japonicus Thunb. ex Murr.
synonym(s): —
common name(s): Japanese chess
habit: annual herb
range: Europe; introduced to ME, s. to GA, w. to CA, n. to Alberta
state frequency: very rare
habitat: roadsides, waste places
notes: —

Bromus kalmii Gray
synonym(s): —
common name(s): wild chess
habit: solitary or clustered, perennial herb
range: ME, s. to MD, w. to SD, n. to Manitoba
state frequency: very rare
habitat: dry to moist woods, thickets, and banks, often calcareous and/or sandy
notes: This species is ranked SH by the Maine Natural Areas Program.

Bromus latiglumis (Shear) A. S. Hitchc.
synonym(s): *Bromus altissimus* Pursh
common name(s): —
habit: solitary or cespitose, perennial herb
range: New Brunswick, s. to NC, w. to NM, n. to MT
state frequency: uncommon
habitat: moist thickets and woods
notes: This species is similar to and sometimes difficult to separate from *Bromus pubescens*. Where
      the 2 species grow together, *B. latiglumis* flowers 20–30 days later than *B. pubescens*.

Bromus marginatus Nees ex Steud.
synonym(s): *Ceratochloa marginata* (Nees *ex* Steud.) W. A. Weber
common name(s): —
habit: cespitose, perennial herb
range: western North America; naturalized or adventive in the east
state frequency: very rare
habitat: waste places
notes: —

Bromus pubescens Muhl. ex Willd.
synonym(s): *Bromus purgans auct. non* L.
common name(s): hairy wood bromegrass, Canada brome
habit: solitary or clustered, perennial herb
range: Quebec and ME, s. to FL, w. to AZ, n. to Alberta
state frequency: very rare
habitat: rich woods, thickets, rocky slopes
notes: This species is ranked SH by the Maine Natural Areas Program.

Bromus racemosus L.

synonym(s): —
common name(s): —
habit: ascending, annual herb
range: Eurasia; naturalized to Newfoundland, s. to NC, w. to KS, n. to MN
state frequency: occasional
habitat: roadsides, waste places
notes: —

Bromus secalinus L.

synonym(s): —
common name(s): cheat, chess
habit: annual herb
range: Europe; naturalized throughout much of the United States
state frequency: uncommon
habitat: fields, roadsides, waste places
notes: —

Bromus tectorum L.

synonym(s): —
common name(s): Junegrass, downy chess
habit: annual herb
range: Europe; introduced to Quebec, s. to VA and MS, w. to CA, n. to British Columbia
state frequency: very rare
habitat: roadsides, waste places
notes: —

Calamagrostis

(4 species)

A genus complicated by polyploidy and apomixis.

KEY TO THE SPECIES

1a. Awn attached at or above the middle of the lemma, neither twisted nor conspicuously
   bent (the awn twisted and bent in a form of *C. stricta* var. *inexpansa*); callus of the
   lemmas densely pubescent with hairs 0.5–1.0 times the length of the lemma
   2a. Awn attached above the middle of the lemma; rachilla vestige glabrous except near
      the apex; spikelets mostly 6.0–8.0 mm long .... *C. cinnoides*
   2b. Awn attached near the middle of the lemma; rachilla vestige pubescent throughout its
      length; spikelets mostly 2.0–6.0 mm long
      3a. Inflorescence loose and open, with spreading to ascending branches; rachilla
         vestige 0.1–0.3 mm long; callus of the lemma pubescent with hairs about
         equaling the length of the lemma .... *C. canadensis*
      3b. Inflorescence dense, with appressed branches (except at anthesis); rachilla
         vestige 0.6–1.0(–1.5) mm long; callus of the lemma pubescent with hairs
         0.5–0.75(–1.0) times the length of the lemma .... *C. stricta*
1b. Awn attached in the basal portion of the lemma [about 1.0 mm from the base of the
   lemma], twisted near the base and bent near the middle; callus of the lemmas sparsely
   pubescent with hairs *ca.* 0.2–0.25 times the length of the lemma .... *C. pickeringii*

Calamagrostis canadensis (Michx.) Beauv.

synonym(s): *Calamagrostis canadensis* (Michx.) Beauv. var. *robusta* Vasey
common name(s): bluejoint, Canada bluejoint
habit: rhizomatous, perennial herb
range: Greenland, s. to NC, w. to CA, n. to AK
state frequency: common
habitat: wet meadows, swamps, and bogs
notes: —

Calamagrostis cinnoides (Muhl.) Bart.

synonym(s): —
common name(s): small reedgrass
habit: stout, solitary or clustered, erect to ascending, rhizomatous, perennial herb
range: Nova Scotia and ME, s. to GA, w. to AL, n. to KY
state frequency: rare
habitat: wet woods, swamps
notes: This species is ranked SH by the Maine Natural Areas Program.

Calamagrostis pickeringii Gray

synonym(s): *Calamagrostis pickeringii* Gray var. *debilis* (Kearney) Fern. & Wieg.
common name(s): Pickering's bluejoint, Pickering's reed bentgrass
habit: mostly solitary, slender-rhizomatous, perennial herb
range: Newfoundland, s. to MA, w. to NY, n. to Ontario; also NJ
state frequency: very rare
habitat: bogs, wet shores and open woods
notes: This species is ranked S1 by the Maine Natural Areas Program.

Calamagrostis stricta (Timm) Koel.

synonym(s): <*C. s.* ssp. *inexpansa*> = *Calamagrostis fernaldii* Louis-Marie, *Calamagrostis inexpansa* Gray var. *novae-angliae* Stebbins; <*C. s.* ssp. *stricta*> = *Calamagrostis neglecta* (Ehrh.) P. G. Gaertn., B. Mey., & Scherb.
common name(s): <*C. s.* ssp. *inexpansa*> = New England northern reedgrass; <*C. s.* ssp. *stricta*> = neglected reedgrass
habit: solitary or cespitose, rhizomatous, perennial herb
range: circumboreal; s. in North America to WV, w. to AZ and CA
state frequency: very rare
habitat: wet meadows, marshes, damp woods
key to the subspecies:
　1a. Anthers usually fully formed and containing pollen; leaves 2.0–4.0 mm wide, thin, often involute, scabrous only on the margins and near the apex, if at all; ligules mostly 1.0–3.0 mm long, with an entire apex; glumes 2.0–4.5 mm long, thin, hyaline near the margins; callus of the lemma pubescent with hairs 0.5–0.75 times the length of the lemma; panicle 5.0–12.0 cm tall, relatively less dense .... *C. s.* ssp. *stricta*
　1b. Anthers often degenerate and lacking pollen; leaves 3.0–6.0 mm wide, stiffer, usually flat, strongly scabrous; ligules mostly 3.0–6.0 mm long, with an erose apex; glumes 3.0–6.0 mm long, thicker, usually indurate and opaque; callus of the lemma pubescent with hairs 0.65–1.0 times the length of the lemma; panicle 6.0–20.0 cm tall, relatively more dense ....
　.. *C. s.* ssp. *inexpansa* (Gray) C. W. Greene
notes: Both subspecies are ranked S1 by the Maine Natural Areas Program.

Cenchrus
(2 species)

Spikelets are subtended by bristle- or spine-like branchlets that have coalesced into an enclosing bur.

KEY TO THE SPECIES
1a. Each bur with 45–75 terete spines .... *C. longispinus*
1b. Each bur with 8–40 basally flattened spines .... *C. tribuloides*

Cenchrus longispinus (Hack.) Fern.
synonym(s): —
common name(s): common sandbur
habit: spreading to ascending, annual herb
range: ME, s. to FL, w. to CA, n. to OR
state frequency: very rare
habitat: river banks, beaches, disturbed areas
notes: This species is ranked S1 by the Maine Natural Areas Program.

Cenchrus tribuloides L.
synonym(s): *Cenchrus pauciflorus* Benth.
common name(s): dune sandbur, sandspur
habit: stout, branched, decumbent to trailing, annual herb
range: NY, s. to FL, w. to TX; also ME
state frequency: very rare
habitat: coastal sands
notes: —

Chloris
(3 species)

Spikelets of this genus usually contain 2 or 3 florets. The lowest floret of a spikelet is fertile, possessing both stamens and carpels. The uppermost 1 or 2 florets are reduced and empty.

KEY TO THE SPECIES
1a. Lowest rudimentary floret acute at the apex; plants robust, commonly 10.0–15.0 dm tall,
   producing long, stout stolons .... *C. gayana*
1b. Rudimentary florets broad-truncate at the apex; plants smaller, 2.0–6.0(–10.0) dm tall,
   not stoloniferous
   2a. Lemmas conspicuously long-ciliate on the nerves and margin; plants annual; leaves
      flat, 2.0–6.0 mm wide .... *C. virgata*
   2b. Lemmas short-ciliate or glabrous on the nerves and margin; plants perennial; leaves
      conduplicate, 1.0–2.0 mm wide [when folded] .... *C. cucullata*

Chloris cucullata Bisch.
synonym(s): *Chloris verticillata* Nutt.
common name(s): windmill grass, windmill fingergrass
habit: cespitose, erect to basally decumbent, perennial herb
range: MO, s. to LA, w. to AZ, n. to CO; adventive in the east
state frequency: very rare
habitat: dry soil of waste places
notes: —

Chloris gayana Kunth
synonym(s): —
common name(s): Rhodes grass
habit: stoloniferous, perennial herb
range: Africa; escaped from cultivation n. to MA; also ME
state frequency: very rare
habitat: fields, waste places
notes: —

Chloris virgata Sw.
synonym(s): —
common name(s): feather finger-grass
habit: spreading to erect, annual herb
range: tropical America; introduced sparingly to the eastern United States
state frequency: very rare
habitat: old fields, waste places
notes: —

Cinna
(2 species)

KEY TO THE SPECIES

1a. Second [upper] glume 1-veined, 2.6–4.1 mm long, with wide, scarious margins, each
scarious margin constituting nearly 0.25 times the total width of the glume; anthers
0.5–0.8 mm long; branches of the panicle spreading to drooping; lemma with an awn up
to 1.5 mm long; leaves 6.0–25.0 cm long; palea with 1 nerve or with 2 closely spaced
nerves .... **C. latifolia**
1b. Second glume 3-veined, 4.1–6.6 mm long, with narrow, scarious margins; anthers
0.8–1.8 mm long; branches of the panicle ascending to spreading; lemma with an awn up
to 0.5 mm long; leaves 15.0–40.0 cm long; palea with 1 nerve .... **C. arundinacea**

Cinna arundinacea L.
synonym(s): —
common name(s): common woodreed
habit: tall, erect, solitary or clustered, perennial herb
range: ME and New Brunswick, s. to GA, w. to TX, n. to MN and Ontario
state frequency: uncommon
habitat: moist, commonly alluvial, forests
notes: —

Cinna latifolia (Trev. ex Goepp.) Griseb.
synonym(s): —
common name(s): drooping woodreed
habit: tall, erect, solitary or clustered, perennial herb
range: circumboreal; s. to PA and mountains of NC, w. to CA
state frequency: occasional
habitat: moist woods, thickets, clearings
notes: —

Cynosurus
(1 species)

Cynosurus cristatus L.
synonym(s): —
common name(s): crested dog's-tail grass
habit: cespitose, perennial herb
range: Europe; introduced to Newfoundland, s. to NC, w. to OH, n. to Ontario
state frequency: very rare
habitat: fields, roadsides, waste places
notes: —

Dactylis
(1 species)

This genus has fused, united edges of leaf sheaths.

Dactylis glomerata L.
synonym(s): *Dactylis glomerata* L. var. *ciliata* Peterm., *Dactylis glomerata* L. var. *detonsa* Fries
common name(s): orchard grass
habit: cespitose, perennial herb
range: Europe; introduced to Newfoundland, s. to AL, w. to CA, n. to British Columbia
state frequency: common
habitat: moist fields, roadsides, waste places
notes: —

Danthonia
(2 species)

Our species produce enlarged, 1- or 2-flowered, cleistogamous spikelets in the lower sheaths. The ligule is a band of hairs.

KEY TO THE SPECIES
1a. Apical teeth of the lemma 0.8–1.8 mm long, triangular-acuminate; panicle 2.0–5.0 cm tall, crowded, the very short branches with 1 or sometimes 2 spikelets; leaves commonly involute, 1.0–2.0 mm wide; basal leaves rarely over 12.0 cm long; stems commonly terete, although sometimes trigonous .... *D. spicata*
1b. Apical teeth of the lemma 2.0–4.5 mm long, aristate or setaceous; panicle 5.0–10.0 cm tall, open, the branches longer and often with 2 or 3 spikelets; leaves flat, 2.0–4.0 mm wide; basal leaves up to 20.0 cm long; stems trigonous or compressed ...... *D. compressa*

Danthonia compressa Austin *ex* Peck
synonym(s): *Danthonia allenii* Austin
common name(s): wild oatgrass
habit: slender, densely cespitose, perennial herb
range: Nova Scotia and ME, s. to VA and mountains of NC and GA, w. to OH, n. to Ontario
state frequency: uncommon
habitat: old roadbeds, trail edges, and other forest openings
notes: Often grows in more wooded, less open habitats than *Danthonia spicata*.

Danthonia spicata (L.) Beauv. *ex* Roemer & J. A. Schultes
synonym(s): *Danthonia spicata* (L.) Beauv. *ex* Roemer & J. A. Schultes var. *pinetorum* Piper
common name(s): poverty oatgrass
habit: erect, densely cespitose, perennial herb
range: Newfoundland, s. to FL, w. to NM, n. to British Columbia
state frequency: common
habitat: open, sandy or stony soil
notes: This species is apomictic and often has abortive anthers. It often grows in more open habitats
  that *Danthonia compressa.*

Deschampsia
(4 species)

Grasses with shining, silver to purple spikelets and obscurely 5-veined lemmas.

KEY TO THE SPECIES
1a. Plants annual; leaves few, short, the blades less than 0.35 times the height of the stem .....
.... ***D. danthonioides***
1b. Plants perennial; leaves relatively more numerous, longer, 0.35–0.5 times the length of
  the stem
    2a. Panicle very narrow, with appressed branches; both glumes with 3 nerves ....
    .... ***D. elongata***
    2b. Panicle open, with ascending to spreading branches; first, and sometimes also the
      second, glume with 1 nerve
        3a. Leaves involute-margined, 1.0–2.0 mm wide; ligule 1.0–2.5 mm long; lemmas
          scabrous, bearing a conspicuously bent awn that evidently exceeds the glumes;
          palea not bifid at the apex .... ***D. flexuosa***
        3b. Leaves flat or conduplicate, 1.0–5.0 mm wide; ligule usually 3.0–12.0 mm long;
          lemmas glabrous, bearing a straightish awn that is shorter than to slightly longer
          than the glumes; palea bifid at the apex .... ***D. cespitosa***

Deschampsia cespitosa (L.) Beauv.
synonym(s): *Deschampsia cespitosa* (L.) Beauv. ssp. *glauca* (Hartman) Hartman, *Deschampsia
  cespitosa* (L.) Beauv. ssp. *parviflora* (Thuill.) Jarmolenko & Soó
common name(s): tufted hairgrass
habit: slender, densely cespitose, perennial herb
range: circumboreal; s. to NC, w. to AZ and CA
state frequency: uncommon
habitat: shores, thickets, fields, roadsides, lawns
notes: This species has both native and introduced forms occurring in Maine.

Deschampsia danthonioides (Trin.) Munro
synonym(s): —
common name(s): annual hairgrass
habit: slender, erect, annual herb
range: MT, s. to TX, w. to Mexico and CA, n. to AK; introduced to ME and OH
state frequency: very rare
habitat: fields, roadsides, open areas
notes: Introduced in grass seed.

Deschampsia elongata (Hook.) Munro
synonym(s): —
common name(s): slender hairgrass
habit: erect, cespitose, perennial herb
range: WY, s. to AZ and Mexico, w. to CA, n. to AK; also Chile; adventive in ME
state frequency: very rare
habitat: open areas
notes: —

Deschampsia flexuosa (L.) Trin.
synonym(s): —
common name(s): common hairgrass, wavy hairgrass
habit: slender, erect, densely cespitose, perennial herb
range: circumboreal; s. to GA, w. to OK; also Mexico
state frequency: occasional
habitat: rocky forests, cliffs, open, subalpine areas
notes: —

Digitaria
(2 species)

KEY TO THE SPECIES

1a. Spikelets 1.7–2.3 mm long, pubescent with capitate hairs; fertile lemma dark brown to
    purple-black; first glume obsolete; second glume 0.75–1.0 times as long as the fertile
    lemma; leaves glabrous (sometimes apical portion of sheath pustulose-pubescent) ....
    .... ***D. ischaemum***
1b. Spikelets 2.4–3.2 mm long; pubescent with hairs that taper to the apex; fertile lemma
    gray-brown; first glume present; second glume 0.35–0.6 times as long as the fertile
    lemma; leaves pustulose-pubescent .... ***D. sanguinalis***

Digitaria ischaemum (Schreb.) Muhl.
synonym(s): —
common name(s): smooth crabgrass, small crabgrass
habit: prostrate to ascending, branched, loosely cespitose, annual herb
range: Eurasia; naturalized to New Brunswick and Nova Scotia, s. to FL, w. to TX, n. to OR
state frequency: occasional
habitat: cultivated areas, waste places
notes: Known to hybridize with *Digitaria sanguinalis*. Frequently misidentified as *D. sanguinalis*, *D. ischaemum* is more common than realized.

Digitaria sanguinalis (L.) Scop.
synonym(s): —
common name(s): northern crabgrass
habit: prostrate to ascending, branched, annual herb
range: Europe; naturalized throughout the United States
state frequency: occasional
habitat: fields, cultivated areas, lawns, waste places
notes: Known to hybridize with *Digitaria ischaemum*.

Distichlis
(1 species)

Distichlis spicata (L.) Greene
synonym(s): —
common name(s): —
habit: spreading to ascending, colonial, rhizomatous, perennial herb
range: New Brunswick and Nova Scotia, s. to FL, w. to TX; also Pacific coast
state frequency: uncommon
habitat: saltmarshes
notes: —

Echinochloa
(3 species)

KEY TO THE SPECIES
1a. Fertile lemma abruptly narrowed to an acuminate, firm, and persistent apex; second
    glume and sterile lemma usually with stout, pustulose-based hairs .... *E. muricata*
1b. Fertile lemma obtuse to broadly acute at the thin, membranaceous, and withering apex
    [the apex separable by a line of minute hairs]; second glume and sterile lemma lacking
    pustulose-based hairs (except sometimes a few near the margin)
    2a. Spikelets gray-purple, awnless or nearly so, crowded onto appressed-ascending
        branches; plants mostly 0.7–1.5 m tall; leaves mostly 15.0–30.0 mm wide ....
        .... *E. frumentacea*
    2b. Spikelets green or purple-tinged, commonly with awns, set on relatively fewer-
        flowered, erect to spreading branches; plants mostly 0.3–0.7 m tall; leaves mostly
        5.0–15.0 mm wide .... *E. crus-galli*

Echinochloa crus-galli (L.) Beauv.
synonym(s): —
common name(s): barnyard grass
habit: stout, branched or simple, decumbent-based, erect, annual herb
range: Eurasia; naturalized throughout North America
state frequency: occasional
habitat: cultivated ground, waste places
notes: —

Echinochloa frumentacea Link
synonym(s): *Echinochloa crus-galli* (L.) Beauv. var. *frumentacea* (Roxb.) W. Wight
common name(s): Japanese-millet, billion dollar grass
habit: coarse, annual herb
range: Asia; escaped from cultivation
state frequency: uncommon
habitat: fields, roadsides, waste places
notes: —

Echinochloa muricata (Beauv.) Fern.
synonym(s): <*E. m.* var. *microstachya*> = *Echinochloa muricata* (Beauv.) Fern. var. *wiegandii*
    Fassett, *Echinochloa pungens* (Poir.) Rydb. var. *wiegandii* Fassett; <*E. m.* var. *muricata*> =
    Echinochloa pungens (Poir.) Rydb.
common name(s): barnyard grass

habit: erect to decumbent, branched, annual herb
range: Quebec and ME, s. to FL, w. to northern Mexico and CA, n. to WA and Alberta
state frequency: uncommon
habitat: disturbed, sometimes muddy, ground
key to the subspecies:
    1a. Spikelets, excluding the awns, 3.3–4.5 x 1.8–2.2 mm; awns lacking or, more commonly,
      6.0–25.0 mm long; anthers 0.7–0.9 mm long .... *E. m.* var. *muricata*
    1b. Spikelets, excluding the awns, 2.5–3.4 x 1.4–1.8 mm; awns lacking or sometimes up to
      6.0 mm, rarely longer; anthers 0.3–0.7 mm long .... *E. m.* var. *microstachya* Wieg.
notes: —

Eleusine
(1 species)

Eleusine indica (L.) Gaertn.
synonym(s): —
common name(s): yard grass, wire grass
habit: coarse, spreading to ascending, cespitose, annual herb
range: tropics of the eastern hemisphere; now pantropical and naturalized in the United States to ME
    and Quebec, s. to FL, w. to CA, n. to OR
state frequency: very rare
habitat: gardens, lawns, waste places
notes: —

Elymus
(8 species)

The number of leaves on a stem is an important identification character. Be sure to count the lower,
shriveled leaves as part of the total leaf number.

KEY TO THE SPECIES

1a. Spikelets spreading; glumes obsolete or represented by bristles .... ***E. hystrix***
1b. Spikelets ascending to appressed-ascending; glumes present, 0.4 mm wide or wider
    2a. Spikelets mostly solitary at each node; glumes 5.0–12.0 mm long
        3a. Anthers 3.0–7.0 mm long [plants outcrossing]; plants rhizomatous; spikelets
        disarticulating below the glumes
            4a. Apical portion of the stem hollow at anthesis; leaf blades soft, lax, finely
            veined, flat; internodes of the rachis thin, semi-circular in cross-section;
            plants typically of inland areas .... ***E. repens***
            4b. Apical portion of the stem solid at anthesis; leaf blades stiff, coarsely veined,
            often involute; internodes of the rachis thick, quadrangular; plants of coastal
            areas .... ***E. pycnanthus***
        3b. Anthers 1.0–2.0 mm long [plants self-pollinating]; plants cespitose, without
        rhizomes; spikelets disarticulating above the glumes .... ***E. trachycaulus***
    2b. Spikelets 2–4 at each node, commonly 2; glumes 10.0–30.0 mm long
        5a. Awns arching at maturity; paleas 8.5–15.0 mm long; glumes thin or compressed
        at the base
            6a. Leaves thin, flat, 15.0–25.0 mm wide, numbering mostly 9–15 per stem;
            spikes conspicuously drooping from near the base .... ***E. wiegandii***

6b. Leaves firm, flat or involute apically, 3.0–15.0 mm wide, numbering mostly 5–9 per stem; spikes erect to arching .... ***E. canadensis***
5b. Awns straight or nearly so; paleas 6.5–8.5 mm long; glumes terete at the base
7a. Glumes flattened apically, 0.8–2.0 mm wide, bowed out at the base and exposing the florets; palea nearly equaling the body of the lemma ....
.... ***E. virginicus***
7b. Glumes setaceous, only slightly broadened apically, 0.2–1.0 mm wide, straight or slightly bowed at the base; palea much shorter than the body of the lemma .... ***E. riparius***

Elymus canadensis L.
synonym(s): —
common name(s): Canada wild rye, broad-leaved wild rye
habit: coarse, erect, cespitose, perennial herb
range: New Brunswick and Nova Scotia, s. to NC, w. to TX and CA, n. to AK
state frequency: uncommon
habitat: sandy, gravelly, or rocky soil of fields, meadows, and stream banks
notes: Known to hybridize with *Hordeum jubatum*.

Elymus hystrix L.
synonym(s): *Elymus hystrix* L. var. *bigeloviana* (Fern.) Bowden, *Hystrix patula* Moench, *Hystrix patula* Moench var. *bigeloviana* (Fern.) Deam
common name(s): bottlebrush grass
habit: solitary, erect, cespitose, perennial herb
range: New Brunswick and ME, s. to GA, w. to OK, n. to ND
state frequency: rare
habitat: rich, low woods, clearings
notes: This species is ranked S2 by the Maine Natural Areas Program.

Elymus pycnanthus (Godr.) Melderis
synonym(s): *Agropyron pungens* (Pers.) Roemer & J. A. Schultes, *Agropyron pycnanthus* (Godr.) Godr. & Gren., *Elytrigia pungens* (Pers.) Tutin
common name(s): saltmarsh wheatgrass
habit: long-rhizomatous, perennial herb
range: Nova Scotia and ME, s. to MA
state frequency: uncommon
habitat: borders of saltmarshes
notes: —

Elymus repens (L.) Gould
synonym(s): *Agropyron repens* (L.) Beauv., *Agropyron repens* (L.) Beauv. var. *subulatum* (Schreb.) Roemer & J. A. Schultes, *Elytrigia repens* (L.) Desv. *ex* B. D. Jackson
common name(s): witch grass, quick grass, quack grass
habit: long-rhizomatous, perennial herb
range: Eurasia; naturalized in North America to Newfoundland, s. to NC, w. to MO, n. to AK
state frequency: common
habitat: sandy or gravelly shores, waste places
notes: It is thought that some forms of *Elytrigia repens* may be native to the Atlantic coast states and provinces.

Elymus riparius Wieg.

synonym(s): —
common name(s): stream bank wild rye
habit: cespitose, perennial herb
range: ME, s. to FL, w. to AR, n. to WI
state frequency: rare
habitat: moist woods, thickets, and stream banks
notes: —

Elymus trachycaulus (Link) Gould *ex* Shinners

synonym(s): *Agropyron trachycaulum* (Link) Malte *ex* H. F. Lewis var. *glaucum* (Pease & Moore)
     Malte; *Agropyron trachycaulum* (Link) Malte *ex* H. F. Lewis var. *majus* (Vasey) Fern., *Agropyron*
     trachycaulum (Link) Malte *ex* H. F. Lewis var. *novae-angliae* (Scribn.) Fern., *Agropyron*
     trachycaulum (Link) Malte *ex* H. F. Lewis ssp. *subsecundus* (Link) A. & D. Löve
common name(s): slender wheatgrass
habit: cespitose, perennial herb
range: Newfoundland, s. to WV and KY, w. to AZ and CA, n. to AK
state frequency: uncommon
habitat: rocky or gravelly soils of various habitats
notes: Known to hybridize with *Hordeum jubatum*.

Elymus virginicus L.

synonym(s): <*E. v.* var. *virginicus*> = *Elymus virginicus* L. var. *glabriflorus* (Vasey) Bush
common name(s): Virginia wild rye
habit: cespitose, perennial herb
range: Newfoundland, s. to FL, w. to NM, n. to British Columbia
state frequency: occasional
habitat: moist, rich woods, thickets, meadows, and shores
key to the subspecies:
     1a. Plants of brackish marshes and moist, coastal sands, with long-exserted, smaller spikes, slender,
          glaucous leaves, and glabrous floral scales .... *E. v.* var. *halophilus* (Bickn.) Wieg.
     1b. Plants of inland, non-saline soils, with various morphologies .... *E. v.* var. *virginicus*
notes: Known to hybridize with *Hordeum jubatum*.

Elymus wiegandii Fern.

synonym(s): *Elymus canadensis* L. var. *wiegandii* (Fern.) Bowden
common name(s): broad-leaved wild rye
habit: coarse, erect, cespitose, perennial herb
range: Quebec, s. to PA
state frequency: uncommon
habitat: alluvial soil of flood plains and river banks
notes: —

Eragrostis
(8 species)

Our species of *Eragrostis* have gray to purple spikelets and small anthers [up to 0.5 mm long].

KEY TO THE SPECIES
1a. Plants extensively creeping by stolons, rooting at the nodes, forming mats; inflorescence
     2.0–8.0 cm tall .... *E. hypnoides*

1b. Plants without stolons, the stems erect or ascending, sometimes decumbent at the base, not rooting at the nodes; inflorescence 5.0–40.0 cm tall
    2a. Plants perennial, with firm bases, sometimes with rhizomes; spikelets 5.0–7.0 mm long, 7- to 11-flowered .... *E. spectabilis*
    2b. Plants annual, with soft bases, without rhizomes; spikelets various
        3a. Leaves with elevated glands or glandular pits along the margin of the leaves; fresh plants with a strong odor
            4a. Lemmas 1.5–2.0 mm long, lacking glandular pits along the keel, loosely imbricate in the spikelet, their bases visible; sheaths pubescent on the margin; spikelets 4.0–11.0 x 1.5–2.0 mm, 5- to 12(20)-flowered .... *E. minor*
            4b. Lemmas 2.0–2.8 mm long, with glandular pits along the keel, tightly imbricate in the spikelet, their bases concealed; sheaths not pubescent on the margin; spikelets 4.0–25.0 x 2.0–3.0 mm, 10- to 40-flowered .... *E. cilianensis*
        3b. Leaves without warty-glands along the margin; plants without a strong odor
            5a. Panicle large and diffuse, often 0.75–1.0 of the entire height of the plant; spikelets 2- to 5-flowered .... *E. capillaris*
            5b. Panicle smaller, denser, up to 0.5 of the entire height of the plant; spikelets 5- to 15-flowered
                6a. Caryopsis with a furrow extending its entire length; leaves mostly 3.0–10.0 mm wide; lower lemmas 1.8–2.5 mm long .... *E. mexicana*
                6b. Caryopsis without a longitudinal furrow; leaves 1.0–4.0 mm wide; lower lemmas 1.0–2.2 mm long
                    7a. First [lower] glume 0.3–0.8 mm long, less than half as long as the lowest lemma; branches of the panicle usually whorled at 1 of the 2 lower nodes, bearing spikelets in the distal 0.5–0.65 of the branch; paleas mostly deciduous with the lemmas and not persistent on the rachilla; lemmas with inconspicuous, lateral nerves .... *E. pilosa*
                    7b. First glume 0.5–1.1 mm long, half or more as long as the lowest lemma; branches of the panicle usually solitary or paired at the nodes, bearing spikelets throughout the length of the branch; paleas persistent on the rachilla after the lemmas have fallen; lemmas with conspicuous, lateral nerves .... *E. pectinacea*

Eragrostis capillaris (L.) Nees
synonym(s): —
common name(s): lacegrass
habit: erect to ascending, cespitose, annual herb
range: ME, s. to GA, w. to TX, n. to WI
state frequency: very rare
habitat: dry, sandy or rocky soil, open woods
notes: This species is ranked S1 by the Maine Natural Areas Program. Like *Eragrostis mexicana*, this species has a furrowed caryopsis.

Eragrostis cilianensis (All.) Lut. *ex* Janchen
synonym(s): *Eragrostis megastachya* (Koel.) Link
common name(s): stink grass
habit: ascending to depressed, densely cespitose, annual herb

range: Europe; naturalized throughout much of North America
state frequency: uncommon
habitat: roadsides, waste places
notes: Plants with malodorous herbage and superficially similar to *Eragrostis minor* but larger, with more flowers, and with more glands along the leaves, lemmas, sheaths, and panicle branches.

Eragrostis hypnoides (Lam.) B. S. P.

synonym(s): —
common name(s): —
habit: creeping, mat-forming, annual herb
range: ME, s. to FL, w. to Mexico, n. to WA
state frequency: very rare
habitat: mud flats, gravelly or sandy shores
notes: This species is ranked SH by the Maine Natural Areas Program.

Eragrostis mexicana (Hornem.) Link

synonym(s): *Eragrostis neomexicana* Vasey
common name(s): Mexican lovegrass
habit: annual herb
range: southwestern United States and Mexico; rarely adventive in the northeast
state frequency: very rare
habitat: roadsides, waste places
notes: —

Eragrostis minor Host

synonym(s): *Eragrostis poaeoides* Beauv. *ex* Roemer & J. A. Schultes
common name(s): little lovegrass
habit: slender, cespitose, annual herb
range: Europe; naturalized to New Brunswick and Nova Scotia, s. to FL, w. to CA, n. to WI and Ontario
state frequency: rare
habitat: gardens, roadsides, waste places
notes: Similar to *Eragrostis cilianensis* (*q.v.*).

Eragrostis pectinacea (Michx.) Nees *ex* Steud.

synonym(s): —
common name(s): Canada lovegrass, tufted lovegrass
habit: slender, erect to ascending, cespitose, annual herb
range: New Brunswick and Nova Scotia, s. to FL, w. to Mexico and CA, n. to British Columbia
state frequency: uncommon
habitat: shores, ditches, gardens, roadsides, waste places
notes: —

Eragrostis pilosa (L.) Beauv.

synonym(s): *Eragrostis multicaulis* Steud.
common name(s): India lovegrass
habit: basally decumbent, erect to ascending, annual herb
range: pantropical, n. to ME, w. to WI and CO
state frequency: uncommon
habitat: fields, roadsides, waste places
notes: —

Eragrostis spectabilis (Pursh) Steud.

synonym(s): *Eragrostis pectinacea* (Michx.) Nees *ex* Steud. var. *spectabilis* Gray, *Eragrostis spectabilis* (Pursh) Steud. var. *sparsihirsuta* Farw.
common name(s): purple lovegrass, tumblegrass
habit: erect to ascending, short-rhizomatous, cespitose, perennial herb
range: ME, s. to FL, w. to TX, n. to ND
state frequency: rare
habitat: dry fields and open woods
notes: Like *Eragrostis capillaris*, this species has a large and diffuse panicle occupying much of the height of the plant.

Festuca
(8 species)

KEY TO THE SPECIES
1a. Leaves flat, 4.0–10.0 mm wide; basal leaves not densely tufted; apex of ovary pubescent
.... *F. subverticillata*
1b. Leaves usually involute or conduplicate (flat in *F. heteromalla*), 0.2–2.0 mm wide; basal leaves conspicuously tufted; apex of ovary glabrous
    2a. Most of the flowers proliferous, replaced by vegetative bulbils; lemmas awnless or with very short awns; anthers rarely present; plants of alpine habitats ..... *F. prolifera*
    2b. Flowers fertile, not proliferous; lemmas with or without awns; anthers present; plants not of alpine habitats
        3a. Lemmas awnless or with short awns up to 0.6 mm, 2.3–4.0 mm long; anthers 1.5–2.2 mm long .... *F. filiformis*
        3b. Lemmas with awns 1.0–3.0 mm long, 3.0–6.1 mm long; anthers 1.8–4.5 mm long
            4a. Basal sheaths white-brown to light brown, persistent, not soon shredding into fibers, the margins closed for less than 0.75 of their length; rhizomes absent
                5a. Spikelets mostly 5.0–7.0 mm long; lemmas 3.0–3.5 mm long; anthers 2.0–2.5 mm long; leaves 0.4–0.6 mm wide, with 1–3 ribs .... *F. ovina*
                5b. Spikelets mostly 7.0–10.0 mm long; lemmas 3.8–5.5 mm long; anthers 2.3–3.5 mm long; leaves 0.6–1.2 mm wide, with 1 rib .... *F. trachyphylla*
            4b. Basal sheaths brown or red-brown, soon shredding into many fibers, the margins closed for more than 0.75 of their length; rhizomes often present
                6a. Leaves, both basal and cauline, flat, 0.8–2.0 mm wide; anthers 3.5–4.3 mm long .... *F. heteromalla*
                6b. Leaves tightly involute or conduplicate, 0.3–1.0 mm wide; anthers 1.8–3.5 mm long
                    7a. Plants without creeping rhizomes, the shoots erect, densely cespitose; anthers 1.8–2.2 mm long .... *F. nigrescens*
                    7b. Plants with creeping rhizomes, the shoots decumbent, loosely cespitose; anthers 2.4–3.5 mm long .... *F. rubra*

Festuca filiformis Pourret

synonym(s): *Festuca capillata* Lam., *Festuca ovina* L. var. *capillata* (Lam.) Alef., *Festuca tenuifolia* Sibthorp
common name(s): fine-leaved sheep fescue

habit: densely cespitose, perennial herb
range: Europe; naturalized to Newfoundland, s. to NC, w. to IL, n. to MN
state frequency: uncommon
habitat: lawns, waste places
notes: —

Festuca heteromalla Pourret

synonym(s): *Festuca diffusa* Dumort., *Festuca rubra* L. var. *multiflora* (Hoffmann) Aschers. & Graebn.
common name(s): flatleaf red fescue
habit: loosely cespitose, rhizomatous, perennial herb
range: Eurasia; naturalized to Greenland, s. to MA
state frequency: very rare
habitat: open areas
notes: —

Festuca nigrescens Lam.

synonym(s): *Festuca rubra* L. var. *commutata* Gaudin, *Festuca rubra* L. ssp. *fallax* Thuill. in part
common name(s): chewings fescue
habit: erect, densely cespitose, perennial herb
range: Europe; naturalized to Newfoundland, s. to NC, w. and n. to MI
state frequency: very rare
habitat: open areas
notes: —

Festuca ovina L.

synonym(s): —
common name(s): sheep fescue
habit: erect, densely cespitose, perennial herb
range: Eurasia; naturalized to ME, s. to NJ, w. to WV, n. to MI
state frequency: uncommon
habitat: dry, open areas and rocky slopes
notes: —

Festuca prolifera (Piper) Fern.

synonym(s): *Festuca rubra* L. ssp. *arctica* (Hack.) Govor. in part, *Festuca rubra* L. var. *prolifera* Piper
common name(s): proliferous fescue
habit: loosely cespitose, sometimes rhizomatous, perennial herb
range: Newfoundland, s. to ME and NH
state frequency: very rare
habitat: alpine areas
notes: This species is ranked S1 by the Maine Natural Areas Program.

Festuca rubra L.

synonym(s): *Festuca rubra* L. var. *juncea* (Hack.) Richter
common name(s): red fescue
habit: decumbent, loosely cespitose, rhizomatous, perennial herb
range: Eurasia; Greenland, s. to GA, w. to CA, n. to AK
state frequency: common
habitat: open areas
notes: —

Festuca subverticillata (Pers.) Alexeev
synonym(s): *Festuca obtusa* Biehler
common name(s): nodding fescue
habit: erect, solitary or few together, perennial herb
range: ME and Nova Scotia, s. to FL, w. to TX, n. to Manitoba
state frequency: rare
habitat: moist woods
notes: —

Festuca trachyphylla (Hack.) Krajina
synonym(s): *Festuca ovina* L. var. *duriuscula* (L.) W. D. J. Koch
common name(s): hard fescue
habit: cespitose, perennial herb
range: Europe; naturalized to Newfoundland, s. to PA, w. and n. to MN
state frequency: very rare
habitat: dry, open soil
notes: —

Glyceria
(8 species)

Often confused with 2 other genera: *Puccinellia* and *Torreyochloa*. *Glyceria* is characterized by united sheath edges, usually 7 prominent veins on the lemma, upper glume 1-nerved, rhizome present, and freshwater habitats. *Puccinellia* is identified by overlapping sheath edges, 5 inconspicuous veins on the lemma, upper glume 3-nerved, rhizomes absent, and saline habitats. The somewhat intermediate *Torreyochloa* combines characters from each genus—sheaths overlapping, 5 conspicuous veins on the lemma, upper glume 3-nerved, rhizome absent, and freshwater habitats.

KEY TO THE SPECIES
1a. Spikelets 10.0–40.0 mm long, with 7–16 florets; internodes of the rachilla 1.0–4.0 mm
    long; stamens 3 per floret; sheaths conspicuously compressed
    2a. Spikelets 20.0–40.0 mm long; lemmas 7.0–8.5 mm long, acute; paleas exceeding
        lemmas by 1.5–3.0 mm .... *G. acutiflora*
    2b. Spikelets 10.0–20.0 mm long; lemmas 3.1–5.3 mm long, blunt; paleas equaling or
        exceeding lemmas by no more than 1.0 mm
        3a. Lemmas glabrous between the veins, 3.1–3.9 mm long; anthers 0.6–1.0 mm long;
            caryopsis 1.0–1.2 mm long .... *G. borealis*
        3b. Lemmas scabrous on, as well as between, the veins, 3.7–5.3 mm long; anthers
            1.0–1.7 mm long; caryopsis 1.5–2.0 mm long .... *G. septentrionalis*
1b. Spikelets 3.0–10.0 mm long, with 3–10 florets; internodes of the rachilla shorter than
    1.0 mm; stamens 2 per floret (except *G. grandis* with 3 stamens); sheaths not
    conspicuously compressed
    4a. Panicle dense, with strictly ascending branches; ligules shorter than 1.0 mm
        5a. Spikelets *ca.* 4.0 mm long; inflorescence very narrow, linear-cylindric, with 1–3
            branches at each node; internodes of the inflorescence 2.0–7.0 cm long; new
            cauline leaves numbering 3–6 per stem .... *G. melicaria*
        5b. Spikelets 4.0–7.0 mm long; inflorescence wider, ovoid to oblong, with 3–8
            branches at each node; internodes of the inflorescence up to 2.5 cm long,

commonly less than 2.0 cm; new cauline leaves numbering 7–9 per stem ....
.... *G. obtusa*
4b. Panicle open, the branches ascending to drooping; ligules 1.0–6.0 mm long
    6a. Spikelets 3.0–5.0 mm wide; veins of the lemma visible but not raised; palea less
        than 2.0 times as long as wide .... *G. canadensis*
    6b. Spikelets 1.5–2.5 mm wide; veins of the lemma conspicuous and raised; palea
        3.0–5.0 times as long as wide
        7a. Spikelets 2.0–4.5 mm long; glumes obtuse, the first and second 0.5–1.0 mm
            and 0.8–1.3 mm long, respectively; lemmas firm, caducous; stamens 2 per
            floret; new cauline leaves 5–10 per stem .... *G. striata*
        7b. Spikelets 4.0–6.5 mm long; glumes acute, the first and second 1.2–1.9 mm
            and 1.5–2.4 mm long, respectively; lemmas membranaceous; stamens mostly
            3 per floret; new cauline leaves 3–6 per stem .... *G. grandis*

Glyceria acutiflora Torr.
synonym(s): *Panicularia acutiflora* (Torr.) Kuntze
common name(s): sharp-scaled mannagrass
habit: slender, decumbent, perennial herb
range: ME, s. to VA, w. to MO, n. to MI
state frequency: very rare
habitat: shallow water, wet soil, pond margins
notes: This species is ranked S1 by the Maine Natural Areas Program. Known to hybridize with
    Glyceria septentrionalis.

Glyceria borealis (Nash) Batchelder
synonym(s): *Panicularia borealis* Nash
common name(s): northern mannagrass, small floating mannagrass
habit: decumbent, perennial herb
range: Newfoundland, s. to NJ, w. to AZ and CA, n. to AK
state frequency: occasional
habitat: shallow water, wet soil
notes: —

Glyceria canadensis (Michx.) Trin.
synonym(s): *Panicularia canadensis* (Michx.) Kuntze
common name(s): rattlesnake mannagrass, rattlesnake grass
habit: erect, solitary to cespitose, perennial herb
range: Newfoundland, s. to VA and mountains of TN, w. to IL, n. to MN and Ontario
state frequency: occasional
habitat: damp to wet woods, thickets, meadows, and shores, swamps
notes: Hybrids with *Glyceria striata*, called *G.* ×*laxa* (Scribn.) Scribn., occur in Maine. Similar to *G.*
    canadensis, these hybrids possess wide spikelets and relatively wide paleas. However, *G.* ×*laxa*
    has obovate paleas not, or only scarcely, exceeded by the 2.3–2.9 mm long lemmas, whereas *G.*
    canadensis has orbicular-obovate paleas exceeded 0.5–1.0 mm by the 2.9–4.0 mm long lemmas.

Glyceria grandis S. Wats.
synonym(s): *Panicularia grandis* (S. Wats.) Nash
common name(s): American mannagrass, reed meadowgrass
habit: stout, erect, clustered, perennial herb
range: Newfoundland, s. to VA, w. to NM, n. to OR and AK
state frequency: uncommon

habitat: wet meadows and brooksides, swamps, marshes, shallow water
notes: —

Glyceria melicaria (Michx.) F. T. Hubbard
synonym(s): *Panicularia melicaria* (Michx.) A. S. Hitchc., *Panicularia torreyana* (Spreng.) Merrill
common name(s): northeastern mannagrass
habit: slender, erect, solitary or few together, perennial herb
range: New Brunswick and Nova Scotia, s. to MD and mountains of NC, w. to OH, n. to Quebec
state frequency: uncommon
habitat: forested wetlands
notes: —

Glyceria obtusa (Muhl.) Trin.
synonym(s): *Panicularia obtusa* (Muhl.) Kuntze
common name(s): coastal mannagrass
habit: basally decumbent, stiff, erect, perennial herb
range: New Brunswick and Nova Scotia, s. to NC
state frequency: rare
habitat: wet woods, swamps, bogs, shallow water
notes: —

Glyceria septentrionalis A. S. Hitchc.
synonym(s): *Panicularia septentrionalis* (A. S. Hitchc.) Bickn.
common name(s): eastern mannagrass, floating mannagrass
habit: soft, decumbent, perennial herb
range: ME, s. to GA, w. to TX, n. to MN and Ontario
state frequency: very rare
habitat: wet woods and ditches, swamps, shallow water
notes: Known to hybridize with *Glyceria acutiflora*.

Glyceria striata (Lam.) A. S. Hitchc.
synonym(s): *Glyceria nervata* (Willd.) Trin., *Glyceria stricta* (Scribn.) Fern., *Panicularia nervata*
(Willd.) Kuntze, *Panicularia striata* (Lam.) A. S. Hitchc.
common name(s): fowl mannagrass, fowl meadowgrass
habit: slender, erect, cespitose, perennial herb
range: Newfoundland, s. to FL, w. to CA, n. to British Columbia
state frequency: uncommon
habitat: wet woods, pastures, and stream margins, marshes, swamps
notes: Hybridizes with *Glyceria canadensis (q.v.)*.

Hierochloe
(2 species)

Spikelets of this genus have 3 florets. The lower 2 are unisexual [staminate] and often of a form or texture different from the uppermost, bisexual floret.

KEY TO THE SPECIES
1a. Panicle compact, 3.0–5.0 cm tall; both lemmas of the staminate florets with awns; glumes 6.0–8.0 mm long; plants of alpine areas at elevation exceeding 1000 m .... *H. alpina*
1b. Panicle more open, 5.0–10.0 cm tall; none of the lemmas with awns; glumes 4.0–6.0 mm long; plants not of alpine habitats .... *H. odorata*

Hierochloe alpina (Sw. *ex* Willd.) Roemer & J. A. Schultes

synonym(s): —
common name(s): —
habit: cespitose, short-rhizomatous, perennial herb
range: circumboreal; s. in the east to mountains of ME, NH, VT, and NY
state frequency: very rare
habitat: open, alpine areas
notes: This species is ranked S1 by the Maine Natural Areas Program.

Hierochloe odorata (L.) Beauv.

synonym(s): —
common name(s): sweet grass
habit: slender-rhizomatous, perennial herb
range: circumboreal; s. to NJ, w. to AZ
state frequency: uncommon
habitat: meadows, shores, swales, bog margins
notes: —

Holcus
(1 species)

Holcus lanatus L.

synonym(s): —
common name(s): common velvet grass
habit: erect, cespitose, perennial herb
range: Europe; naturalized to Newfoundland, s. to GA; also British Columbia, s. to CA
state frequency: rare
habitat: meadows, fields, roadsides
notes: —

Hordeum
(5 species)

Grasses with clusters of 3 spikelets borne on alternating sides of a rachis. The 2 lateral spikelets are usually sterile, the central spikelet fertile. The glumes are commonly elongate and resemble bristles.

KEY TO THE SPECIES
1a. Anthers 2.0–2.5 mm long; leaves 5.0–15.0 mm wide; rachis continuous and not
disarticulating .... *H. vulgare*
1b. Anthers 0.5–1.8 mm long; leaves 1.5–5.0(–7.0) mm wide; rachis disarticulating
   2a. Plants perennial; awns very slender
      3a. Spikes arching or nodding; glumes 2.5–15.0 cm long; fertile lemmas with an awn
      1.0–6.0 cm long .... *H. jubatum*
      3b. Spikes erect; glumes 0.7–2.0 cm long; fertile lemmas with an awn
      0.5–1.0(–2.0) cm long .... *H. brachyantherum*
   2b. Plants annual; awns stout
      4a. Glumes of the central spikelet of each triad with ciliate margins; some leaf
      sheaths with auriculate summits; body of fertile lemmas exceeding 12.0 mm long
      .... *H. murinum*

4b. Glumes of the central spikelet of each triad with scabrous margins; summit of the leaf sheath without auricles; body of fertile lemmas 5.0–7.0 mm long ....
.... ***H. pusillum***

Hordeum brachyantherum Nevski

synonym(s): *Hordeum nodosum* L.
common name(s): meadow barley
habit: cespitose, perennial herb
range: MT, s. to NM, w. to CA, n. to AK; adventive locally in our area
state frequency: very rare
habitat: meadows, shores, saltmarshes, bottomlands
notes: —

Hordeum jubatum L.

synonym(s): —
common name(s): foxtail barley, squirrel-tail grass
habit: cespitose, perennial herb
range: Newfoundland, s. to VA, w. to TX, Mexico, and CA, n. to AK
state frequency: uncommon
habitat: fields, meadows, roadsides
notes: Known to hybridize with *Elymus canadensis, E. trachycaulus,* and *E. virginicus.*

Hordeum murinum L. ssp. *leporinum* (Link) Arcang.

synonym(s): —
common name(s): wall barley, way barley
habit: annual herb
range: Europe; naturalized locally in the northeast
state frequency: very rare
habitat: roadsides, waste places
notes: —

Hordeum pusillum Nutt.

synonym(s): —
common name(s): little barley
habit: erect, annual herb
range: DE, s. to FL, w. to Mexico and CA, n. to WA; adventive locally in the east
state frequency: very rare
habitat: pastures, roadsides, waste places, borders of marshes
notes: —

Hordeum vulgare L.

synonym(s): *Hordeum vulgare* L. var. *trifurcatum* (Schlecht.) Alef.
common name(s): barley
habit: annual herb
range: Eurasia; escaped from cultivation locally
state frequency: uncommon
habitat: roadsides, waste places
notes: Two cultivated forms appear—those with all 3 spikelets of a triad sessile and fertile, called 6-row barley, and those with only the central spikelet of a triad fertile, the lateral spikelets sterile and pedicelled, called 2-row barley.

Koeleria
(1 species)

Koeleria macrantha (Ledeb.) J. A. Schultes
synonym(s): *Koeleria cristata* (L.) Pers.
common name(s): —
habit: erect, cespitose, perennial herb
range: Quebec, s. to DE, w. to TX and Mexico, n. to British Columbia; adventive locally in the
    northeast
state frequency: very rare
habitat: sandy soil of open areas
notes: —

Leersia
(2 species)

Our species are weak-stemmed grasses with scabrous leaves, rooting at the lower nodes, of moist to wet
soils.

KEY TO THE SPECIES
1a. Branches of the panicle 2 or more at some of the nodes; florets with 3 stamens; lemmas
    4.0–7.5 mm long; principal leaves 15.0–30.0 cm long .... ***L. oryzoides***
1b. Branches of the panicle 1 per node; florets with 1 or 2 stamens; lemmas 2.9–4.1 mm
    long; principal leaves 5.0–20.0 cm long .... ***L. virginica***

Leersia oryzoides (L.) Sw.
synonym(s): —
common name(s): rice cut-grass
habit: decumbent, long-rhizomatous, perennial herb
range: New Brunswick and Nova Scotia, s. to FL, w. to AZ and CA, n. to British Columbia
state frequency: occasional
habitat: wet meadows, ditches, shores, swamps
notes: —

Leersia virginica Willd.
synonym(s): *Leersia virginica* Willd. var. *ovata* (Poir.) Fern.
common name(s): white grass
habit: slender, decumbent, long-rhizomatous, perennial herb
range: New Brunswick and ME, s. to GA, w. to TX, n. to SD and Ontario
state frequency: very rare
habitat: moist to wet woods and thickets
notes: —

Leptochloa
(1 species)

Leptochloa uninervia (J. Presl) A. S. Hitchc. & Chase
synonym(s): —
common name(s): —
habit: annual herb

range: southern and western United States, Mexico, and South America; adventive to ME, MA, and
 NJ
state frequency: very rare
habitat: ditches, moist areas
notes: —

Leymus
(1 species)

Leymus mollis (Trin.) Hara
synonym(s): *Elymus arenarius* L. var. *villosus* E. Mey.
common name(s): American dunegrass, sea lymegrass
habit: stout, erect, long-rhizomatous, perennial herb
range: Greenland, s. to MA; also AK, s. to CA
state frequency: uncommon; frequent on some coastal shores
habitat: Atlantic coast dunes and sandy beaches
notes: —

Lolium
(5 species)

KEY TO THE SPECIES
1a. Inflorescence a panicle; both glumes well developed
    2a. Lemmas 5.5–7.0 mm long; auricles of leaves glabrous; basal sheaths brown to red-
        brown, soon shredding into many fibers; internodes of the rachilla glabrous or nearly
        so .... *L. pratense*
    2b. Lemmas 7.0–9.0 mm long; auricles of leaves pubescent; basal sheaths white-brown
        to light brown, persistent, not soon shredding into fibers; internodes of the rachilla
        antrorsely scabrous .... *L. arundinaceum*
1b. Inflorescence a spike, the spikelets on alternating sides of the rachis; only 1 glume
    developed
    3a. Plants commonly perennial; glumes 4.0–12.0 mm long; upper lemmas exceeding the
        body of the glumes
        4a. Rachis of the inflorescence rough on the surface opposite the spikelet; spikelets
            with 11–22 florets; apex of the glume not reaching the tip of the lowest lemma on
            the same side [observe in the upper spikelets]; lemmas of the upper florets with a
            conspicuous awn; leaves up to 8.0 mm wide .... *L. multiflorum*
        4b. Rachis of the inflorescence smooth on the surface opposite the spikelet, rough
            only on the angles; spikelets with mostly 5–10 florets; apex of the glume more or
            less extending to the tip of the lowest lemma on the same side [observe in the
            upper spikelets]; lemmas without awns; leaves 2.0–5.0 mm wide .... *L. perenne*
    3b. Plants annual; glumes (12.0–)15.0–25.0 mm long; upper lemmas exceeded by the
        body of the glumes .... *L. temulentum*

Lolium arundinaceum (Schreb.) Darbyshire
synonym(s): *Festuca arundinacea* Schreb., *Festuca elatior* L. var. *arundinacea* (Schreb.) C. F. H.
 Wimmer
common name(s): reed fescue, Alta fescue
habit: erect, cespitose, perennial herb

range: Europe; irregularly introduced to North America
state frequency: uncommon
habitat: meadows, roadsides
notes: Known to hybridize with *Lolium pratense*.

Lolium multiflorum Lam.

synonym(s): *Lolium multiflorum* Lam. var. *diminutum* Mutel, *Lolium perenne* L. var. *aristatum*
   Willd.
common name(s): Italian rye grass
habit: slender, annual or perennial herb
range: Europe; escaped from cultivation in our area
state frequency: rare
habitat: fields, roadsides
notes: —

Lolium perenne L.

synonym(s): —
common name(s): ryegrass, common darnel
habit: slender, short-lived perennial herb
range: Europe; escaped from cultivation in our area
state frequency: uncommon
habitat: fields, roadsides, waste places
notes: —

Lolium pratense (Huds.) Darbyshire

synonym(s): *Festuca elatior* L., *Festuca pratensis* Huds.
common name(s): meadow fescue, tall fescue
habit: erect, loosely cespitose, perennial herb
range: Europe; introduced to temperate and boreal North America
state frequency: occasional
habitat: fields, meadows
notes: Known to hybridize with *Lolium arundinaceum*.

Lolium temulentum L.

synonym(s): —
common name(s): bearded darnel, poison darnel
habit: solitary-stemmed or somewhat clustered, annual herb
range: Europe; naturalized to New Brunswick and ME, s. to CT, w. to KS, n. to MN
state frequency: very rare
habitat: grainfields, waste places
notes: —

Lycurus
(1 species)

Lycurus phleoides Kunth

synonym(s): —
common name(s): wolftail
habit: erect or decumbent-based, cespitose, perennial herb
range: southwestern United States and Mexico; adventive in ME
state frequency: very rare
habitat: dry, sandy soil
notes: Originally introduced in wool waste.

Milium
(1 species)

Milium effusum L.
synonym(s): —
common name(s): spreading millet grass
habit: slender, erect, perennial herb
range: Newfoundland, s. to MD, w. to IL, n. to MN and Ontario
state frequency: uncommon
habitat: rich woods, thickets
notes: —

Miscanthus
(1 species)

Miscanthus sacchariflorus (Maxim.) Franch.
synonym(s): —
common name(s): Amur silvergrass
habit: long- and stout-rhizomatous, colonial, perennial herb
range: East Asia; escaped from cultivation locally
state frequency: very rare
habitat: roadsides, waste places
notes: —

Molinia
(1 species)

Molinia caerulea (L.) Moench
synonym(s): —
common name(s): moorgrass
habit: densely cespitose, perennial herb
range: Europe; naturalized to ME, s. to PA, w. to WI, n. to Ontario
state frequency: very rare
habitat: fields, meadows, roadsides
notes: —

Muhlenbergia
(9 species)

Glume measurements in the key include the apical awns.

KEY TO THE SPECIES
1a. Panicle open and diffuse, 4.0 cm wide or wider; spikelets on long, thin pedicels; callus of
the lemma glabrous .... **M. uniflora**
1b. Panicle slender and contracted, up to 2.5 cm wide; spikelets subsessile or short-
pedicelled; callus of the lemma pubescent (glabrous in *M. richardsonis*)
2a. Callus of the lemma glabrous; leaves 1.0–2.0 mm wide, involute; plants cespitose,
without rhizomes .... **M. richardsonis**
2b. Callus of the lemma pubescent; leaves 1.0–7.0 mm wide, normally flat; plants
rhizomatous (except *M. schreberi* with sprawling, rooting stems)

3a. Glumes 4.5–8.0 mm long, prolonged at the apex into long, slender awns, exceeding the awnless lemma
    4a. Internodes of the stem pubescent and dull; anthers 0.8–1.5 mm long; ligules 0.2–0.6 mm long; mature caryopsis 1.0–1.6 mm long; margin of the lemma pubescent .... *M. glomerata*
    4b. Internodes of the stem glabrous, except sometimes at the very summit, lustrous; anthers 0.6–0.8 mm long; ligules 0.6–1.5 mm long; mature caryopsis 1.4–2.3 mm long; margin of the lemma glabrous .... *M. racemosa*
3b. Glumes 0.1–3.0 mm long, pointed and awned, shorter than to equaling the body of the awned or awnless lemma
    5a. Glumes minute and veinless, the larger only 0.1–0.3 mm long; plants with sprawling stems that root at the lower nodes, without rhizomes .... *M. schreberi*
    5b. Glumes larger, 1- or 3-veined; plants not rooting at the nodes, with rhizomes
        6a. Internodes of the stem glabrous
            7a. Lemma 1.8–2.5 mm long, the margin sigmoid-curved; glumes 1.3–2.5 mm long; scales of rhizome tightly appressed ... *M. sobolifera*
            7b. Lemma 2.9–3.6 mm long, the margin straight; glumes 2.0–5.0 mm long; scales of rhizome loosely imbricate .... *M. frondosa*
        6b. Internodes of the stem pubescent
            8a. Ligule 1.0–2.5 mm long; panicle loose; spikelets short-pedicelled; lemma white or white-green .... *M. sylvatica*
            8b. Ligule 0.5–1.0 mm long; panicle dense; spikelets subsessile; lemma green or purple .... *M. mexicana*

Muhlenbergia frondosa (Poir.) Fern.
synonym(s): —
common name(s): wirestem muhly
habit: ascending to decumbent, thick-rhizomatous, perennial herb
range: New Brunswick and ME, s. to GA, w. to TX, n. to ND and Ontario
state frequency: uncommon
habitat: open woods, thickets, fields, gardens
notes: —

Muhlenbergia glomerata (Willd.) Trin.
synonym(s): *Muhlenbergia setosa* (Biehler) Trin.
common name(s): marsh muhly
habit: erect, rhizomatous, perennial herb
range: Newfoundland, s. to NC, w. to NV, n. to British Columbia
state frequency: uncommon
habitat: wet meadows and shores, marshes, bogs
notes: —

Muhlenbergia mexicana (L.) Trin.
synonym(s): —
common name(s): wirestem muhly
habit: erect to ascending, rhizomatous, perennial herb
range: coastal Quebec, New Brunswick, and Nova Scotia, s. to NC, w. to CA, n. to British Columbia
state frequency: occasional
habitat: thickets, shores, open areas
notes: —

Muhlenbergia racemosa (Michx.) B. S. P.

synonym(s): —
common name(s): —
habit: ascending, slender-rhizomatous, perennial herb
range: IL, s. to OK, w. to AZ, n. to WA and Alberta; adventive in our area
state frequency: uncommon
habitat: woods, roadsides, disturbed areas
notes: —

Muhlenbergia richardsonis (Trin.) Rydb.

synonym(s): —
common name(s): mat muhly, soft-leaf muhly
habit: ascending, clustered, perennial herb
range: coastal Quebec and New Brunswick, s. to ME, w. to NM, n. to WA and Yukon
state frequency: rare
habitat: damp thickets, wet, gravelly shores
notes: This species is ranked S1 by the Maine Natural Areas Program.

Muhlenbergia schreberi J. F. Gmel.

synonym(s): —
common name(s): nimblewill
habit: slender, erect to spreading, perennial herb
range: ME, s. to FL, w. to Mexico, n. to NE and MN
state frequency: very rare
habitat: woods, thickets, roadsides, lawns, gardens
notes: —

Muhlenbergia sobolifera (Muhl. *ex* Willd.) Trin.

synonym(s): —
common name(s): cliff muhly
habit: erect, solitary-stemmed or somewhat clustered, rhizomatous, perennial herb
range: ME, s. to VA, w. to TX, n. to NE and WI
state frequency: very rare
habitat: rocky or gravelly woods, ledges
notes: This species is ranked SH by the Maine Natural Areas Program.

Muhlenbergia sylvatica Torr. *ex* Gray

synonym(s): —
common name(s): forest muhly
habit: solitary-stemmed or somewhat clustered, erect to spreading, rhizomatous, perennial herb
range: ME, s. to NC, w. to TX, n. to MN and Ontario
state frequency: uncommon
habitat: rocky woods, thickets, ledges, stream banks
notes: —

Muhlenbergia uniflora (Muhl.) Fern.

synonym(s): —
common name(s): —
habit: slender, cespitose, perennial herb
range: Newfoundland, s. to NJ, w. to WI, n. to Ontario
state frequency: occasional
habitat: moist to wet fields and meadows, shores, swales, bogs
notes: —

Oryzopsis
(3 species)

KEY TO THE SPECIES

1a. Leaves flat or nearly so, 4.0–15.0 mm wide, the uppermost bladeless or with a tiny blade less than 1.0 cm long; glumes 6.0–9.0 mm long; ligules 0.2–0.7 mm long ....
.... ***O. asperifolia***

1b. Leaves normally involute, 1.0–2.0 mm wide, the uppermost with a well developed blade longer than 1.0 cm; glumes 3.5–4.8 mm long; ligules 1.5–3.0 mm long

2a. Awn of lemma 6.0–11.0 mm long, often bent or twisted in the basal half; glumes glabrous .... ***O. canadensis***

2b. Awn of lemma 0.5–2.0 mm long, sometimes absent, straight or slightly bent; glumes minutely scabrous in the apical portion .... ***O. pungens***

Oryzopsis asperifolia Michx.
synonym(s): —
common name(s): rough-leaved ricegrass
habit: loosely cespitose, perennial herb
range: Newfoundland, s. to coastal NY, w. to NM, n. to British Columbia
state frequency: occasional
habitat: woods, thickets
notes: This species has erect, sterile stems and wide-spreading to prostrate, fertile stems.

Oryzopsis canadensis (Poir.) Torr.
synonym(s): *Stipa canadensis* Poir.
common name(s): Canada mountain ricegrass
habit: loosely cespitose, erect, perennial herb
range: Newfoundland, s. to New Brunswick and ME, w. to MN, n. to Alberta; also WV
state frequency: rare
habitat: woods, slopes
notes: This species is ranked S? by the Maine Natural Areas Program.

Oryzopsis pungens (Torr. *ex* Spreng.) A. S. Hitchc.
synonym(s): —
common name(s): —
habit: slender, erect, densely cespitose, perennial herb
range: Newfoundland, s. to NJ, w. to CO, n. to British Columbia
state frequency: uncommon
habitat: rocky or sandy woods
notes: —

Panicum
(16 species)

Many species of the subgenus *Dichanthelium* [1b in the key] flower twice during the growing season. In early to mid-summer, the terminal inflorescence is produced. The flowers of this inflorescence are open-pollinated. In late season to early fall, the axillary inflorescences are produced. The flowers of these later inflorescences are self-pollinated, and often partly enclosed in the leaf sheaths. Late season forms of these grasses are often more branched and leaf and spikelet measurements may not fit those stated in the key.

KEY TO THE SPECIES

1a. Basal and cauline leaves similar in shape and length; panicles of early and late season similar, their branches not forked at the base
  2a. Plants perennial, either rhizomatous or with a firm caudex; stems normally unbranched in the lower portion
    3a. Spikelets secund along the smaller branches of the panicle, 1.8–3.0 mm long; stems and sheaths compressed; ligule membranaceous .... *P. rigidulum*
    3b. Spikelets not secund, 2.2–5.6 mm long; stems and sheaths terete or only weakly compressed; ligule a band of silky hairs .... *P. virgatum*
  2b. Plants annual, with fibrous roots; stems commonly branched near the base
    4a. Leaf sheaths glabrous; first glume rounded to obtuse at the apex, 0.2–0.3 times as long as the spikelet .... *P. dichotomiflorum*
    4b. Leaf sheaths pustulose-pubescent; first glume obtuse to acuminate, 0.35–0.5 times as long as the spikelet
      5a. Spikelets 4.5–6.0 mm long; panicle arching to nodding; caryopsis *ca.* 2.0 mm wide .... *P. miliaceum*
      5b. Spikelets 1.6–3.5 mm long; panicle erect; caryopsis less than 1.0 mm wide
        6a. Mature caryopsis darkly pigmented, 2.0(–3.0) times as long as wide; margin of the lemma slightly inrolled over the palea, concealing 0.35–0.50 the width of the palea; spikelets ovoid, 1.6–2.4 mm long; panicle usually less than 0.5 of the plant height .... *P. philadelphicum*
        6b. Mature caryopsis light brown, 3.0–4.0 times as long as wide; margin of the lemma conspicuously inrolled over the palea, concealing 0.50–0.65 the width of the palea; spikelets lanceolate, 1.8–3.5 mm long; panicle commonly 0.5 or more of the plant height .... *P. capillare*
1b. Basal and cauline leaves dissimilar (except *Panicum depauperatum* and *P. linearifolium*), the basal shorter and broader than the more elongate cauline leaves; panicles of the early season exserted, bearing chasmogamous flowers, their branches often forked at the base; panicles of the late season reduced, bearing cleistogamous flowers, partly or wholly enclosed in leaf sheaths
  7a. Leaves crowded, firm, elongate, 2.0–5.0 mm wide, the longer *ca.* 20.0 times as long as wide, the basal and cauline similar
    8a. Second glume and sterile lemma rounded to subacute at the apex, equaling the length of the fertile lemma; inflorescence usually extended much beyond the leaves, with pedicels 0.8–1.8 cm long, the terminal panicles often with more than 20 spikelets; spikelets 1.7–3.1 mm long .... *P. linearifolium*
    8b. Second glume and sterile lemma pointed at the apex, extending 0.5–1.5 mm beyond the fertile lemma; inflorescence usually only somewhat extended beyond

the leaves, with pedicels 3.0–8.0 cm long, the terminal panicles with fewer than 20 spikelets; spikelets 2.7–4.1 mm long .... *P. depauperatum*
7b. Leaves often not crowded, usually softer, broader, 2.0–40.0 mm wide, the longer usually much less than 20.0 times as long as wide, the basal and cauline dissimilar
   9a. Pubescence of some of the sheaths and internodes of 2 types—minute hairs 0.1–0.4 mm long and longer, appressed hairs to 1.0 mm long ..... *P. acuminatum*
   9b. Pubescence of sheaths and internodes various, but not of 2 types
      10a. Ligule a band of hairs 2.0–5.0 mm long, conspicuously protruding from the sheath .... *P. acuminatum*
      10b. Ligule none or a band of hairs up to 2.0 mm long, not projecting above the sheath
         11a. Leaves cordate- or auriculate-clasping at the base
            12a. Spikelets subglobose, 1.3–1.9 mm long .... *P. sphaerocarpon*
            12b. Spikelets distinctly longer than wide, 1.9–5.2 mm long
               13a. Spikelets 2.5–3.7 mm long; primary veins of leaf, including midvein, thicker and raised above the finer, secondary veins
                  14a. Some or all of the sheaths pubescent with pustulose-based hairs; axillary panicles often concealed within the sheaths; lower floret sterile .... *P. clandestinum*
                  14b. Sheaths glabrous or pubescent, the hairs, if present, not pustulose-based; axillary panicles often at least partly exserted from the sheaths; lower floret staminate .... *P. latifolium*
               13b. Spikelets 1.9–2.3 mm long; primary veins of leaf, excluding midvein, scarcely differentiated from, and of similar size, as the secondary veins .... *P. boreale*
         11b. Leaves narrowed or rounded at the base
            15a. Spikelets 1.5–2.9 mm long
               16a. Uppermost leaf of the stem ascending or erect; branches of panicle ascending; spikelets pubescent .... *P. boreale*
               16b. Uppermost leaf of the stem spreading; branches of panicle spreading; spikelets glabrous or sparsely pubescent
                  17a. Spikelets 1.5–2.9 mm long; second glume and sterile lemma slightly shorter than the mature caryopsis; late season phase highly branched .... *P. dichotomum*
                  17b. Spikelets 2.7–4.0 mm long; second glume and sterile lemma equaling the mature caryopsis; late season phase sparingly branched .... *P. oligosanthes*
            15b. Spikelets 3.0–4.0 mm long
               18a. Cauline leaves ascending, the primary 10.0–15.0 x 1.0–2.0 cm; panicle narrow, with strictly erect branches; first glume triangular-ovate, *ca.* 0.5 times as long as the second glume ....
               .... *P. xanthophysum*
               18b. Cauline leaves spreading, the primary 6.0–12.0 x 0.7–1.2 cm; panicle ovoid, with spreading to ascending branches; first glume broad-ovate, *ca.* 0.4 times as long as the second glume ....
               .... *P. oligosanthes*

Panicum acuminatum Sw.

synonym(s): <*P. a.* ssp. *columbianum*> = *Dichanthelium sabulorum* (Lam.) Gould & C. A. Clark var. *thinium* (A. S. Hitchc. & Chase) Gould & C. A. Clark, *Panicum columbianum* Scribn. var. *thinium* A. S. Hitchc. & Chase; <*P. a.* var. *fasciculatum*> = *Dichanthelium acuminatum* (Sw.) Gould & C. A. Clark var. *fasciculatum* (Torr.) Freckmann, *Panicum huachucae* Ashe, *Panicum lanuginosum* Ell. var. *fasciculatum* (Torr.) Fern., *Panicum lanuginosum* Ell. var. *tennesseense* (Ashe) Gleason , *Panicum subvillosum* Ashe, *Panicum tennesseense* Ashe; <*P. a.* ssp. *implicatum*> = *Panicum lanuginosum* Ell. var. *implicatum* (Scribn.) Fern.; <*P. a.* var. *lindheimeri*> = *Dichanthelium acuminatum* (Sw.) Gould & C. A. Clark var. *lindheimeri* (Nash) Gould & C. A. Clark, *Panicum lanuginosum* Ell. var. *lindheimeri* (Nash) Fern., *Panicum lanuginosum* Ell. var. *septentrionale* (Fern.) Fern.; <*P. a.* ssp. *spretum*> = *Dichanthelium spretum* (J. A. Schultes) Freckmann, *Panicum spretum* J. A. Schultes

common name(s): woolly panic grass
habit: slender, erect to ascending, perennial herb
range: ME, s. to FL, w. to CA, n. to British Columbia
state frequency: common
habitat: dry to wet, sandy, or sometimes peaty, soils
key to the subspecies:
  1a. Sheaths and stems pubescent
      2a. Sheaths and stems densely and minutely pubescent with hairs 0.1–0.4 mm long, and usually with some longer, appressed hairs to 1.0 mm long .... *P. a.* ssp. *columbianum* (Scribn.) Freckmann & Lelong
      2b. Sheaths and stems pustulose-pilose with long, mostly ascending to spreading, hairs
         3a. Leaves usually wider than 6.0 mm, spreading to ascending, short-pilose or nearly glabrous on the adaxial surface; spikelets 1.5–2.0 mm long .... *P. a.* ssp. *fasciculatum* (Torr.) Freckmann & Lelong
         3b. Leaves usually narrower than 6.0 mm, erect to ascending, long-pilose on the adaxial surface; spikelets 1.3–1.6 mm long .... *P. a.* ssp. *implicatum* (Scribn.) Freckmann & Lelong
  1b. Sheaths and stems glabrous or sparsely pubescent
      4a. Panicle broad, with spreading branches, up to 2.0 times as tall as wide; leaf blades often yellow-green, with conspicuous long cilia near the base; spikelets mostly obovoid .... *P. a.* ssp. *lindheimeri* (Nash) Freckmann & Lelong
      4b. Panicle narrow, with ascending branches, 3.0–4.0 times as tall as wide; leaf blades green or tinged with purple, with few or no long cilia near the base; spikelets mostly ellipsoid .... *P. a.* ssp. *spretum* (J. A. Schultes) Freckmann & Lelong
notes: —

Panicum boreale Nash

synonym(s): *Dichanthelium boreale* (Nash) Freckmann
common name(s): northern panic grass
habit: slender, erect to ascending, cespitose, perennial herb
range: Newfoundland, s. to NJ, w. to IN, n. to MN and Ontario
state frequency: occasional
habitat: thickets, fields, meadows, shores, river- and stream banks
notes: —

Panicum capillare L.

synonym(s): *Panicum capillare* L. var. *occidentale* Rydb.
common name(s): witch grass
habit: stout, erect to ascending, basally branched, annual herb
range: New Brunswick and Nova Scotia, s. to FL, w. to TX, n. to MT and Manitoba

state frequency: occasional
habitat: fields, gardens
notes: —

Panicum clandestinum L.
synonym(s): *Dichanthelium clandestinum* (L.) Gould
common name(s): deer tongue grass
habit: stout, erect, solitary or clustered, perennial herb
range: Nova Scotia and ME, s. to FL, w. to TX, n. to MI
state frequency: occasional
habitat: borders of woods, thickets, shores
notes: —

Panicum depauperatum Muhl.
synonym(s): *Dichanthelium depauperatum* (Muhl.) Gould, *Panicum depauperatum* Muhl. var.
    psilophyllum Fern.
common name(s): starved panic grass
habit: slender, erect, densely cespitose, perennial herb
range: New Brunswick and Nova Scotia, s. to GA, w. to TX, n. to MN and Manitoba
state frequency: uncommon
habitat: open woods, shores
notes: —

Panicum dichotomiflorum Michx.
synonym(s): —
common name(s): fall panicum
habit: erect to decumbent, annual herb
range: New Brunswick and Nova Scotia, s. to FL, w. to TX, n. to SD and Ontario; introduced to
    western United States and West Indies
state frequency: rare
habitat: cultivated ground, shores, waste places
notes: —

Panicum dichotomum L.
synonym(s): *Dichanthelium dichotomum* (L.) Gould
common name(s): —
habit: slender, erect, branched, cespitose, perennial herb
range: New Brunswick and ME, s. to FL, w. to TX, n. to MI and Ontario
state frequency: very rare
habitat: thin woods, thickets, open areas
notes: Known to hybridize with *Panicum latifolium*.

Panicum latifolium L.
synonym(s): *Dichanthelium latifolium* (L.) Gould & C. A. Clark
common name(s): broad-leaved panic grass
habit: slender, erect, clustered, perennial herb
range: ME, s. to NC, w. to KS, n. to MN
state frequency: uncommon
habitat: open woods, thickets
notes: Known to hybridize with *Panicum dichotomum*.

Panicum linearifolium Scribn. *ex* Nash

synonym(s): *Dichanthelium linearifolium* (Scribn. *ex* Nash) Gould, *Panicum linearifolium* Scribn. *ex* Nash var. *werneri* (Scribn.) Fern.
common name(s): linear-leafed panic grass
habit: slender, erect to ascending, clustered, perennial herb
range: New Brunswick and Nova Scotia, s. to GA, w. to TX, n. to MN and Manitoba
state frequency: uncommon
habitat: open woods, banks, roadsides
notes: —

Panicum miliaceum L.

synonym(s): —
common name(s): broomcorn millet, proso millet
habit: stout, annual herb
range: Eurasia; adventive in our area
state frequency: uncommon
habitat: roadsides, waste places
notes: —

Panicum oligosanthes (J. A. Schultes) Gould var. *scribnerianum* (Nash) Gould

synonym(s): *Dichanthelium oligosanthes* (J. A. Schultes) Gould var. *scribnerianum* (Nash) Gould, *Panicum scribnerianum* Nash
common name(s): —
habit: erect to ascending, loosely cespitose, perennial herb
range: ME, s. to FL, w. to CA, n. to British Columbia
state frequency: very rare
habitat: open woods, open areas
notes: —

Panicum philadelphicum Bernh. *ex* Trin.

synonym(s): *Panicum tuckermanii* Fern.
common name(s): Philadelphia witch grass
habit: slender, basally branched, ascending to erect, rarely decumbent, annual herb
range: New Brunswick and Nova Scotia, s. to GA, w. to TX, n. to MN
state frequency: uncommon
habitat: open woods, fields
notes: —

Panicum rigidulum Bosc *ex* Nees

synonym(s): *Panicum agrostoides* Spreng.
common name(s): redtop panicum
habit: densely cespitose, perennial herb
range: ME, s. to FL, w. to TX, n. to MI
state frequency: uncommon
habitat: wet soil
notes: —

Panicum sphaerocarpon Ell.

synonym(s): *Dichanthelium sphaerocarpon* (Ell.) Gould
common name(s): round-fruited panic grass
habit: ascending to spreading, cespitose, perennial herb
range: MA, s. to FL, w. to TX and Mexico, n. to MI and Ontario; also ME

state frequency: very rare
habitat: thin woods
notes: —

Panicum virgatum L. var. *spissum* Linder
synonym(s): —
common name(s): switchgrass
habit: stout, erect, solitary or cespitose, rhizomatous, perennial herb
range: Nova Scotia and ME, s. to PA
state frequency: uncommon
habitat: open woods, fresh to brackish shores and marshes
notes: —

Panicum xanthophysum Gray
synonym(s): *Dichanthelium xanthophysum* (Gray) Freckmann
common name(s): pale panic grass
habit: loosely tufted, erect to ascending, glabrous, perennial herb
range: Quebec and ME, s. to WV, w. to MN, n. to Manitoba
state frequency: occasional
habitat: dry, sandy soil
notes: —

Pappophorum
(1 species)

Pappophorum vaginatum Buckl.
synonym(s): *Pappophorum mucronulatum auct. non* Nees
common name(s): pappusgrass
habit: erect, cespitose, perennial herb
range: southwestern United States, s. to South America; naturalized in Maine
state frequency: very rare
habitat: waste places
notes: This species was introduced in wool ballast. It has spikelets with 1 or 2 fertile florets and 2 or 3 sterile ones.

Phalaris
(2 species)

Spikelets with 3 florets—2 sterile, reduced, and usually pubescent florets on either side of a fertile, larger, indurate, and shining floret.

KEY TO THE SPECIES
1a. Plants perennial, with creeping rhizomes; keel of glumes not winged; panicle mostly
    7.0–25.0 cm tall, some branching visible; glumes 4.0–6.5 mm long; sterile lemmas
    1.0–2.0 mm long .... *P. arundinacea*
1b. Plants annual, without rhizomes; keel of glumes winged; panicle 1.5–4.0 cm tall, very
    dense; glumes 7.0–10.0 mm long; sterile lemmas 2.5–4.5 mm long .... *P. canariensis*

Phalaris arundinacea L.
synonym(s): —
common name(s): reed canarygrass
habit: stout, colonial, rhizomatous, perennial herb

range: circumboreal; s. to NC, w. to AZ and CA
state frequency: common
habitat: meadows, shores, stream banks, marshes
notes: —

Phalaris canariensis L.
synonym(s): —
common name(s): canarygrass
habit: erect, annual herb
range: Europe; adventive throughout much of North America
state frequency: uncommon
habitat: roadsides, waste places
notes: —

Phleum
(2 species)

KEY TO THE SPECIES
1a. Glumes with an awn 1.5–3.0 mm long; stems not bulbous-thickened at the base, glabrous near the apex; inflorescence normally ellipsoid to ellipsoid-cylindric, 1.0–5.0 cm tall .... *P. alpinum*
1b. Glumes with an awn 0.7–1.5 mm long; stems bulbous-thickened at the base, scabrous near the apex; inflorescence normally cylindric, 5.0–10.0(–22.0) cm tall ...... *P. pratense*

Phleum alpinum L.
synonym(s): *Phleum commutatum* Gaudin var. *americanum* (Fourn.) Hultén
common name(s): mountain timothy
habit: decumbent to creeping, cespitose, perennial herb
range: circumboreal; s. in the east to mountains of New England
state frequency: very rare
habitat: wet meadows, damp shores, bogs
notes: This species is ranked S1 by the Maine Natural Areas Program.

Phleum pratense L.
synonym(s): —
common name(s): timothy
habit: erect, solitary or clumped, perennial herb
range: Europe; naturalized throughout most of the United States and southern Canada
state frequency: common
habitat: fields, roadsides
notes: —

Phragmites
(1 species)

Phragmites australis (Cav.) Trin. *ex* Steud.
synonym(s): *Phragmites communis* Trin.
common name(s): common reed
habit: tall, stout, erect, long-rhizomatous, perennial herb
range: cosmopolitan
state frequency: occasional
habitat: wetlands, shores, ditches
notes: —

Piptatherum
(1 species)

Piptatherum racemosum (Sm.) Eat.
synonym(s): *Oryzopsis racemosa* (Sm.) Ricker *ex* A. S. Hitchc.
common name(s): blackseed ricegrass
habit: erect, loosely cespitose, rhizomatous, perennial herb
range: ME, s. to DE and VA, w. to MO, n. to ND and Ontario
state frequency: uncommon
habitat: woods
notes: —

Poa
(10 species)

KEY TO THE SPECIES

1a. Plants annual, with soft stem bases, sometimes rooting at the nodes; callus of the lemma without arachnoid hairs .... *P. annua*
1b. Plants perennial, with firm stem bases and persistent, dried leaves from the previous season, not rooting at the nodes; callus of the lemma various
   2a. Plants with elongate, slender rhizomes
      3a. Stems strongly compressed, with 3–7 leaves; panicle branches mostly 2 per node; only 3 veins of the lemma prominent, the intermediate 2 veins inconspicuous .... *P. compressa*
      3b. Stems terete to slightly compressed, with 2–4 leaves; panicle branches 2–5 per node; all 5 veins of the lemma prominent .... *P. pratensis*
   2b. Plants without rhizomes
      4a. Callus of the lemma with arachnoid hairs
         5a. Marginal veins of the lemma pubescent, at least toward the base; leaves 1.0–2.0 mm wide (up to 4.0 in *P. palustris*)
            6a. Lower branches of the panicle 1–3 per node; plants of alpine habitats, elevation exceeding 1000 m .... *P. fernaldiana*
            6b. Lower branches of the panicle 5 per node; plants not of alpine habitats, elevation less than 1000 m
               7a. Ligule 2.0–5.0 mm long; anthers 0.8–1.2 mm long; rachilla glabrous; stems stout, 5.0–15.0 dm tall .... *P. palustris*
               7b. Ligule 0.2–1.5 mm long; anthers 1.2–1.6 mm long; rachilla minutely pubescent; stems slender, 4.0–8.0 dm tall .... *P. nemoralis*
         5b. Marginal veins of the lemma glabrous; leaves 2.0–5.0 mm wide
            8a. Branches of the panicle 4–8 per node; lemmas pubescent or scabrous on the keel
               9a. Sheaths usually scabrous; ligule 2.5–7.0 mm long; all 5 veins of the lemma prominent .... *P. trivialis*
               9b. Sheaths glabrous; ligule 0.7–3.0 mm long; only 3 veins of the lemma prominent, the intermediate 2 veins inconspicuous .... *P. alsodes*
            8b. Branches of the panicle mostly 2 per node; lemmas glabrous on the keel .... *P. saltuensis*

4b. Callus of the lemma without arachnoid hairs
    10a. Inflorescence at anthesis tall and slender, 3.0–5.0 times as tall as wide; ligule 1.0–2.0 mm long; anthers 1.0–2.2 mm long; branches of the panicle 2–5 per node .... *P. glauca*
    10b. Inflorescence at anthesis broad and open, up to 2.0 times as tall as wide; ligule 2.5–3.5 mm long; anthers 0.7–0.9 mm long; branches of the panicle mostly 2 per node .... *P. fernaldiana*

Poa alsodes Gray

synonym(s): —
common name(s): —
habit: slender, loosely cespitose, perennial herb
range: New Brunswick and Nova Scotia, s. to DE and mountains of NC
state frequency: uncommon
habitat: moist woods, thickets
notes: The lemmas of this species have only 3 distinct veins.

Poa annua L.

synonym(s): —
common name(s): speargrass, low speargrass, annual bluegrass
habit: decumbent to ascending, seldom erect, cespitose, annual herb
range: Eurasia; naturalized throughout most of North America
state frequency: common
habitat: fields, cultivated ground, lawns, waste places
notes: The lemmas of this species have 5 distinct veins.

Poa compressa L.

synonym(s): —
common name(s): Canada bluegrass
habit: erect, slender- and long-rhizomatous, perennial herb
range: Europe; naturalized to Newfoundland, s. to GA, w. to CA, n. to AK
state frequency: occasional
habitat: dry soil
notes: The lemmas of this species have only 3 distinct veins.

Poa fernaldiana Nannf.

synonym(s): —
common name(s): wavy bluegrass
habit: slender, matted to cespitose, perennial herb
range: Newfoundland, s. to mountains of ME, NH, VT, and NY
state frequency: very rare
habitat: alpine areas
notes: This species is ranked S1 by the Maine Natural Areas Program. The lemmas of this species have only 3 distinct veins.

Poa glauca Vahl

synonym(s): —
common name(s): —
habit: erect, clustered, perennial herb
range: circumboreal; s. to ME, w. to UT and WA
state frequency: very rare
habitat: dry, rocky or gravelly areas, often calcareous
notes: The lemmas of this species have only 3 distinct veins.

Poa nemoralis L.

synonym(s): —
common name(s): wood bluegrass
habit: slender, cespitose, perennial herb
range: Europe; naturalized to Newfoundland, s. to CT, w. to NE, n. to AK
state frequency: occasional
habitat: open woods, thickets, moist meadows, shores, roadsides
notes: The lemmas of this species have only 3 distinct veins.

Poa palustris L.

synonym(s): *Poa triflora* Gilib.
common name(s): fowl meadowgrass
habit: stout, loosely cespitose, perennial herb
range: circumboreal; s. to VA, w. to NM
state frequency: occasional
habitat: damp to wet thickets, meadows, and shores
notes: The lemmas of this species have only 3 distinct veins.

Poa pratensis L.

synonym(s): <*P. p.* ssp. *angustifolia*> = *Poa angustifolia* L.
common name(s): Kentucky bluegrass
habit: cespitose, rhizomatous, perennial herb
range: Eurasia; introduced and naturalized throughout most of North America
state frequency: common
habitat: fields, meadows, shores, lawns
key to the subspecies:
    1a. Cauline leaves 2.0–6.0 mm wide; stems 2.0–3.0 mm wide at the base, equal in width to the basal leaves .... *P. p.* ssp. *pratensis*
    1b. Cauline leaves 1.0–2.0 mm wide; stems 1.0–2.0 mm wide at the base, wider than the basal leaves .... *P. p.* ssp. *angustifolia* (L.) Arcang.
notes: The lemmas of this species have 5 distinct veins.

Poa saltuensis Fern. & Wieg.

synonym(s): *Poa languida* A. S. Hitchc., *Poa saltuensis* Fern. & Wieg. var. *microlepis* Fern. & Wieg.
common name(s): —
habit: slender, cespitose, perennial herb
range: Newfoundland, s. to MD and mountains of VA, w. to MN, n. to Ontario
state frequency: uncommon
habitat: open woods, thickets, clearings
notes: The lemmas of this species have 5 distinct veins.

Poa trivialis L.

synonym(s): —
common name(s): rough bluegrass, rough-stalked meadowgrass
habit: basally decumbent, erect, cespitose, perennial herb
range: Europe; naturalized to Newfoundland, s. to GA, w. to KS, n. to MN and Ontario
state frequency: uncommon
habitat: moist woods, thickets, brooksides, meadows, roadsides
notes: The lemmas of this species have 5 distinct veins.

Polypogon
(1 species)

Polypogon monspeliensis (L.) Desf.
synonym(s): —
common name(s): rabbitfoot grass
habit: basally decumbent, ascending to erect, annual herb
range: Europe; naturalized to ME, s. to GA, w. to CA, n. to AK
state frequency: very rare/rare
habitat: waste places
notes: —

Puccinellia
(5 species)

Sometimes confused at the generic level with *Glyceria* (*q.v.*) and *Torreyochloa*.

KEY TO THE SPECIES

1a. Anthers 1.5–2.0 mm long; first glume 2.0–3.5 mm long; second glume 2.8–4.0 mm long; spikelets with 5–11 florets .... *P. maritima*
1b. Anthers 0.3–1.2 mm long; first glume 0.7–2.7 mm long; second glume 1.5–4.0 mm long, commonly less than 3.0 mm; spikelets with 3–9 florets
  2a. Inflorescence contracted, even the lower branches ascending; spikelets borne in the apical as well as the basal portion of branches; lemma firm at the tip, with an excurrent midrib; anthers 0.4–0.8 mm long .... *P. fasciculata*
  2b. Inflorescence often open, at least the lower branches spreading or divergent; spikelets borne mostly in the apical portion of branches; lemma hyaline and weak at the apex, the midrib evanescent, not reaching the tip; anthers 0.7–1.5 mm long
    3a. Branches of the panicle usually glabrous or slightly scabrous; distal margin of the lemma entire or with a few scattered cilia or scabrules; caryopsis 1.3–2.1 mm long .... *P. tenella*
    3b. Branches of the panicle prominently scabrous; distal margin of the lemma prominently scabrous-ciliolate; caryopsis 1.0–1.6 mm long
      4a. Lemmas broad-obtuse to truncate at the apex, the lowest spikelet 1.7–2.5 mm long; inflorescence open, the lower branches spreading or reflexed; spikelets 3.0–5.0 mm long .... *P. distans*
      4b. Lemmas acute to narrow-obtuse at the apex, the lowest of the spikelets 2.1–3.5 mm long; inflorescence more upright, the branches commonly ascending; spikelets 4.0–8.0 mm long .... *P. nuttalliana*

Puccinellia distans (Jacq.) Parl.
synonym(s): —
common name(s): European alkali-grass
habit: basally decumbent, cespitose, perennial herb
range: Europe; naturalized to New Brunswick and Nova Scotia, s. to DE, w. to WI, n. to Ontario
state frequency: very rare
habitat: roadsides, waste places, usually coastal
notes: —

Puccinellia fasciculata (Torr.) Bickn.

synonym(s): —
common name(s): —
habit: coarse, cespitose, perennial herb
range: New Brunswick and Nova Scotia, s. to VA; also UT, NE, and AZ
state frequency: very rare
habitat: saltmarshes, sandy shores
notes: —

Puccinellia maritima (Huds.) Parl.

synonym(s): —
common name(s): seaside alkali-grass
habit: coarse, erect, cespitose, perennial herb
range: Quebec and Nova Scotia, s. to PA
state frequency: rare
habitat: saltmarshes, shores
notes: —

Puccinellia nuttalliana (J. A. Schultes) A. S. Hitchc.

synonym(s): —
common name(s): Nuttall's alkali-grass
habit: slender, erect, cespitose, perennial herb
range: WI, s. to TX, w. to NM and CA, n. to British Columbia; adventive locally in the northeast
state frequency: very rare
habitat: saltmarshes, coastal sands
notes: —

Puccinellia tenella (Lange) Holmb. ssp. *alascana* (Scribn. & Merr.) Tzvelev

synonym(s): *Puccinellia paupercula* (Holmb.) Fern. & Weatherby var. *alascana* (Scribn. & Merr.)
    Fern. & Weatherby, *Puccinellia pumila* (Vasey) A. S. Hitchc. *sensu* Gleason and Cronquist (1991)
common name(s): —
habit: densely cespitose or matted, perennial herb
range: Labrador, s. to CT; also AK, s. to British Columbia
state frequency: uncommon
habitat: saline shores
notes: —

Schizachne
(1 species)

Schizachne purpurascens (Torr.) Swallen

synonym(s): *Melica striata* (Michx.) A. S. Hitchc.
common name(s): —
habit: loosely cespitose, perennial herb
range: Newfoundland, s. to PA and KY, w. to Mexico and NM, n. to AK
state frequency: uncommon
habitat: dry, sandy or rocky thickets and woods
notes: —

Schizachyrium
(1 species)

Schizachyrium scoparium (Michx.) Nash

synonym(s): *Andropogon scoparius* Michx. var. *ducis* Fern. & Grisc., *Andropogon scoparius* Michx.
  var. *frequens* F. T. Hubbard, *Andropogon scoparius* Michx. var. *neomexicanus* (Nash) A. S.
  Hitchc., *Andropogon scoparius* Michx. var. *septentrionalis* Fern. & Grisc., *Schizachyrium*
  scoparium (Michx.) Nash ssp. *neomexicanum* (Nash) A. S. Hitchc.
common name(s): little bluestem, broom beardgrass
habit: loosely or densely cespitose, perennial herb
range: New Brunswick and Nova Scotia, s. to FL, w. to TX and Mexico, n. to MT
state frequency: common
habitat: sandy soil, gravelly or rocky shores
notes: —

Secale
(1 species)

Secale cereale L.

synonym(s): —
common name(s): rye
habit: robust, annual herb
range: Europe; escaped from cultivation in our area
state frequency: uncommon
habitat: roadsides, disturbed areas
notes: —

Setaria
(5 species)

Spikelets of *Setaria* possess 2 florets—the apical floret fertile and bisexual, the basal floret sterile and either
staminate or neutral.

KEY TO THE SPECIES

1a. Each spikelet subtended by 4–12 bristles; spikelets 3.0–3.5 mm long .... **S. pumila**
1b. Each spikelet subtended by 1–3 bristles (up to 6 in *S. faberi*); spikelets 1.6–3.0 mm long
  2a. Axis of the panicle scabrous; branches of the panicle verticillate; inflorescence
    0.5–1.2 cm thick; leaves usually spreading; bristles of the inflorescence usually
    retrorsely scabrous .... **S. verticillata**
  2b. Axis of the panicle pubescent but not scabrous; branches of the panicle not
    verticillate; inflorescence 0.7–3.5 cm thick; leaves usually ascending; bristles of the
    inflorescence antrorsely scabrous
    3a. Fertile lemma transversely rugulose; disarticulation below the glumes, the entire
      spikelet, including the caryopsis, falling intact; mature caryopsis green
      4a. Leaves pubescent and scabrous on the adaxial surface; spikelets 2.5–3.0 mm
        long; inflorescence nodding from near the base, 2.0–3.0 cm wide; second
        glume 0.65–0.75 times as long as the spikelet .... **S. faberi**

4b. Leaves scabrous on the adaxial surface; spikelets 1.6–2.5 mm long; inflorescence erect or nodding from near the apex, 1.0–2.3 cm wide; second glume nearly as long as the spikelet .... *S. viridis*

3b. Fertile lemma smooth; disarticulation above the glumes, the caryopsis falling free from the more persistent glumes; mature caryopsis yellow to red or brown to black .... *S. italica*

Setaria faberi Herrm.
synonym(s): —
common name(s): nodding foxtail grass, giant foxtail grass
habit: annual herb
range: Asia; naturalized to ME, s. to NC, w. to MO, n. to NE
state frequency: rare
habitat: fields, roadsides, waste places
notes: —

Setaria italica (L.) Beauv.
synonym(s): —
common name(s): foxtail-millet, German-millet, Hungarian-millet
habit: erect, annual herb
range: Eurasia; escaped from cultivation in our area
state frequency: uncommon
habitat: roadsides, waste places
notes: This is essentially a cultivated form of *Setaria viridis*. Some forms may be very robust with coarse, stout stems and very large panicles.

Setaria pumila (Poir.) Roemer & J. A. Schultes
synonym(s): *Setaria glauca* (L.) Beauv.
common name(s): foxtail, yellow foxtail grass
habit: erect, cespitose, annual herb
range: Europe; naturalized throughout most of North America
state frequency: occasional
habitat: cultivated ground, roadsides, waste places
notes: —

Setaria verticillata (L.) Beauv.
synonym(s): —
common name(s): bur foxtail grass
habit: weak, often basally forked, annual herb
range: Eurasia; naturalized to ME, s. to VA and AL, w. to TX, n. to ND
state frequency: very rare
habitat: gardens, stream banks, roadsides, waste places
notes: —

Setaria viridis (L.) Beauv.
synonym(s): *Setaria viridis* (L.) Beauv. var. *weinmannii* (Roemer & J. A. Schultes) Borbás
common name(s): green foxtail grass, green bristlegrass
habit: basally branched, annual herb
range: Eurasia; naturalized in our area
state frequency: uncommon
habitat: gardens, fields, roadsides, waste places
notes: —

Sorghastrum
(1 species)

Sorghastrum nutans (L.) Nash
synonym(s): —
common name(s): Indian grass
habit: simple, cespitose, short-rhizomatous, perennial herb
range: ME, s. to FL, w. to Mexico and AZ, n. to WY and Manitoba
state frequency: very rare
habitat: open woods, borders of woods, fields
notes: This species is ranked S1 by the Maine Natural Areas Program.

Sorghum
(1 species)

Sorghum bicolor (L.) Moench
synonym(s): *Sorghum vulgare* Pers.
common name(s): sorghum
habit: coarse, annual herb
range: Eurasia; escaped from cultivation locally
state frequency: very rare
habitat: waste places
notes: —

Spartina
(3 species)

Marsh and shore plants with strongly flattened spikelets arranged in 2 secund rows.

KEY TO THE SPECIES

1a. Fresh leaves flat, 30.0–80.0 x 0.5–1.5 cm; inflorescence commonly composed of 5–60 spikes; lemmas 7.0–10.0 mm long; stems 5.0–25.0 mm thick at the base; rhizome 4.0–20.0 mm in diameter
    2a. Margins and apex of leaves scabrous; glumes scabrous on the keel; rhizomes firm; rachis of individual spikes prolonged but not surpassing the uppermost spikelets; second [upper] glume with an awn 3.0–10.0 mm long .... *S. pectinata*
    2b. Margins and apex of leaves glabrous; glumes glabrous to somewhat scabrous on the keel; rhizomes soft; rachis of individual spikes prolonged and usually conspicuously exceeding the uppermost spikelets; second glume lacking an awn .... *S. alterniflora*
1b. Fresh leaves involute, 4.0–30.0 x 0.1–0.4 cm; inflorescence composed of 1–9 spikes; lemmas 5.0–7.0 mm long; stems 1.0–6.0 mm thick at the base; rhizome 1.0–6.0 mm in diameter .... *S. patens*

Spartina alterniflora Loisel.
synonym(s): *Spartina alterniflora* Loisel. var. *pilosa* (Merr.) Fern.
common name(s): smooth cordgrass, saltwater cordgrass
habit: coarse, long-rhizomatous, perennial herb
range: Newfoundland, s. to FL, w. to TX; introduced to WA; Atlantic coast of Europe
state frequency: common
habitat: saltmarshes, coastal shores
notes: The stems of this species provide a rank, sulphur scent when bruised.

Spartina patens (Ait.) Muhl.

synonym(s): —
common name(s): saltmeadow cordgrass
habit: slender, stiff, slender-rhizomatous, perennial herb
range: Newfoundland, s. to VA, w. and n. to MI and Ontario
state frequency: common
habitat: saltmarshes, coastal beaches and shores, saline areas
notes: Hybrids with *Spartina pectinata*, called *S.* ×*caespitosa* A. A. Eat., occur in Maine and are
   relatively intermediate between the 2 parents in many morphological characters. However, unlike
   the parents, their stems are normally tufted, and rhizomes are commonly lacking or poorly
   developed.

Spartina pectinata Link

synonym(s): *Spartina pectinata* Link var. *suttiei* (Farw.) Fern.
common name(s): fresh water cordgrass
habit: stout, erect, long-rhizomatous, perennial herb
range: Newfoundland, s. to NC, w. to TX and NM, n. to WA and Alberta
state frequency: occasional
habitat: marshes, shores, swamps, wet gravel
notes: Hybrids with *Spartina patens* (*q.v.*) occur in Maine.

Sphenopholis
(2 species)

Grasses with 2-flowered spikelets bearing a prolonged rachilla behind the upper palea.

KEY TO THE SPECIES

1a. Second [upper] glume acute at the apex; spikelets 3.0–4.2 mm long; lowest rachilla
   segment 0.8–1.0 mm long; inflorescence looser and somewhat open; anthers *ca.* 0.5 mm
   long .... ***S. intermedia***
1b. Second glume broadly-rounded to truncate at the apex; spikelets 2.5–3.0 mm long;
   lowest rachilla segment 0.5–0.7 mm long; inflorescence dense; anthers 0.5–0.8 mm long
   .... ***S. obtusata***

Sphenopholis intermedia (Rydb.) Rydb.

synonym(s): *Sphenopholis obtusata* (Michx.) Scribn. var. *major* (Torr.) K. S. Erdman
common name(s): —
habit: solitary or cespitose, annual or short-lived perennial herb
range: Newfoundland, s. to FL, w. to AZ, n. to AK
state frequency: uncommon
habitat: moist meadows, shores, stream banks, and slopes
notes: —

Sphenopholis obtusata (Michx.) Scribn.

synonym(s): *Sphenopholis obtusata* (Michx.) Scribn. var. *lobata* (Trin.) Scribn.
common name(s): wedgegrass, prairie wedgegrass
habit: solitary or cespitose, annual or short-lived perennial herb
range: ME, s. to FL, w. to CA, n. to British Columbia
state frequency: very rare
habitat: moist meadows, stream banks, and shores
notes: This species is ranked SH by the Maine Natural Areas Program.

Sporobolus
(6 species)

Unique among our grasses, the members of *Sporobolus* have a loose pericarp that can be removed from the seed when moistened [the fruit is a utricle].

KEY TO THE SPECIES

1a. Panicle open, with spreading to ascending branches; plants perennial
    2a. Plants with rhizomes; stems 15.0–40.0 cm tall; leaves mostly less than 5.0 cm long; panicle mostly 3.0–8.0 cm tall .... *S. nealleyi*
    2b. Plants without rhizomes; stems 30.0–100.0 cm tall; leaves 6.0–20.0 cm long; panicle 10.0–20.0 cm tall .... *S. cryptandrus*
1b. Panicle strict, with erect to appressed branches; plants annual or perennial
    3a. Panicle 10.0–50.0 cm tall; plants perennial
        4a. Lemmas 2.0–2.5 mm long; leaves mostly 6.0–20.0 cm tall, much shorter than the length of the stem .... *S. contractus*
        4b. Lemmas (3.5–)4.0–6.5 mm long; leaves elongate, nearly as long as the stem ....
        .... *S. asper*
    3b. Panicle 1.0–5.0 cm tall; plants annual
        5a. Lemmas 3.0–5.0 mm long, sparsely pubescent, apex sharply acute; first [lower] glume 2.8–4.1 mm long; caryopsis permanently enclosed in the lemma and palea
        .... *S. vaginiflorus*
        5b. Lemmas 1.5–3.0 mm long, glabrous, apex acute; first glume 1.5–2.4 mm long; caryopsis falling free from the lemma and palea [freeing the seed when moistened] .... *S. neglectus*

Sporobolus asper (Beauv.) Kunth
synonym(s): —
common name(s): tall dropseed, longleaf dropseed
habit: stout, erect, cespitose, perennial herb
range: New Brunswick and ME, s. to NC and AL, w. to NM, n. to WA
state frequency: very rare
habitat: dry and/or sandy soil
notes: This species is ranked S1 by the Maine Natural Areas Program.

Sporobolus contractus A. S. Hitchc.
synonym(s): *Sporobolus cryptandrus* (Torr.) Gray var. *strictus* Scribn.
common name(s): spike dropseed
habit: stout, erect to spreading, cespitose, perennial herb
range: MI, s. to TX, w. to CA. n. to WA; adventive in NY and ME
state frequency: very rare
habitat: dry, sandy soil
notes: —

Sporobolus cryptandrus (Torr.) Gray
synonym(s): —
common name(s): sand dropseed
habit: stout, erect to spreading, cespitose, perennial herb
range: ME and NH, s. to NC, w. to TX and Mexico, n. to WA
state frequency: very rare
habitat: dry, sandy soil
notes: —

Sporobolus nealleyi Vasey
synonym(s): —
common name(s): Nealley dropseed
habit: slender, erect, rhizomatous herb
range: TX, NV, NM, and AZ; adventive to ME
state frequency: very rare
habitat: sandy soil
notes: —

Sporobolus neglectus Nash
synonym(s): —
common name(s): small dropseed
habit: erect to spreading, cespitose, annual herb
range: ME, s. to VA and LA, w. to TX, n. to ND and Ontario; also AZ and WA
state frequency: very rare
habitat: dry, sterile or sandy soil, often calcareous
notes: This species is ranked SH by the Maine Natural Areas Program.

Sporobolus vaginiflorus (Torr. *ex* Gray) Wood
synonym(s): *Sporobolus vaginiflorus* (Torr. *ex* Gray) Wood var. *inaequalis* Fern.
common name(s): poverty grass
habit: erect to spreading, cespitose, annual herb
range: New Brunswick and Nova Scotia, s. to GA, w. to AZ, n. to ND
state frequency: uncommon
habitat: dry, sterile or sandy soil
notes: —

Torreyochloa
(1 species)

Sometimes confused at the generic level with *Glyceria* (*q.v.*) and *Puccinellia*.

Torreyochloa pallida (Torr.) Church
synonym(s): <*T. p.* var. *fernaldii*> = *Glyceria fernaldii* (A. S. Hitchc.) St. John; <*T. p.* var. *pallida*> =
   Glyceria pallida (Torr.) Trin., *Panicularia pallida* (Torr.) Kuntze
common name(s): Fernald's manna-grass
habit: slender, decumbent, trailing, matted, perennial herb
range: Newfoundland, s. to NC, w. to MO, n. to MN
state frequency: uncommon
habitat: swamps, shallow water
key to the subspecies:
   1a. Spikelets 5.0–7.0 mm long; lemmas 2.5–3.5 mm long; larger leaves mostly 4.0–10.0 mm
      wide, bearing ligules 5.0–8.0 mm long; anthers definitely longer than wide, 0.6–1.5 mm long;
      caryopsis *ca.* 1.5 mm long .... *T. p.* var. *pallida*
   1b. Spikelets 3.0–5.0 mm long; lemmas 2.0–2.8 mm long; larger leaves 1.0–3.5 mm wide, bearing
      ligules 2.0–4.0 mm long; anthers nearly globose, 0.2–0.5 mm long; caryopsis *ca.* 0.8 mm long
      .... *T. p.* var. *fernaldii* (A. S. Hitchc.) Dore *ex* Koyama & Kawano
notes: —

Tragus
(2 species)

The florets of this genus are enclosed inside a bur formed from enlarged second glumes that bear 6 rows of stout, hooked prickles [1 row on each side of the 3 nerves].

KEY TO THE SPECIES

1a. Second glume 3.6–4.5 mm long; spikelets distinctly pedicelled .... *T. racemosus*
1b. Second glume 2.3–3.0 mm long; spikelets subsessile .... *T. berteronianus*

Tragus berteronianus J. A. Schultes
synonym(s): —
common name(s): —
habit: spreading, basally branched, annual herb
range: TX, w. to AZ, s. to Argentina and warm regions of the old world; adventive to MA and ME
state frequency: very rare
habitat: dry, open ground
notes: Found originally in wool waste in Maine.

Tragus racemosus (L.) All.
synonym(s): *Nazia racemosa* (L.) Kuntze
common name(s): prickle grass
habit: annual herb
range: old world; adventive to ME, interruptedly s. to NC, w. to TX, NM, and AZ
state frequency: very rare
habitat: waste places
notes: Introduced to the United States in ballast.

Triplasis
(1 species)

Triplasis purpurea (Walt.) Chapman
synonym(s): —
common name(s): purple sandgrass
habit: slender, spreading to ascending, cespitose, annual herb
range: ME and NH, s. to FL, w. to TX, n. to MN and Ontario
state frequency: very rare
habitat: dry sand, usually coastal
notes: —

Trisetum
(2 species)

KEY TO THE SPECIES

1a. Inflorescence a dense, spike-like panicle 3.0–10.0 cm tall; lemmas tipped by 2 apical teeth 1.3–2.0 mm long, at least the uppermost lemma awned; sheaths pubescent; rachilla pubescent with hairs *ca.* 0.5 mm long .... *T. spicatum*
1b. Inflorescence an open and flexuous panicle 8.0–20.0 cm tall; lemmas entire at the apex, without awns; sheaths usually glabrous; rachilla pubescent with hairs *ca.* 1.3–2.0 mm long .... *T. melicoides*

Trisetum melicoides (Michx.) Vasey *ex* Scribn.
synonym(s): *Graphephorum melicoides* (Michx.) Desv., *Trisetum melicoides* (Michx.) Vasey *ex*
    Scribn. var. *majus* (Gray) A. S. Hitchc.
common name(s): graphephorum
habit: slender, cespitose or solitary, perennial herb
range: Newfoundland, s. to New Brunswick and ME, w. to WI, n. to Ontario
state frequency: very rare
habitat: rocky or gravelly shores and banks
notes: This species is ranked S1 by the Maine Natural Areas Program.

Trisetum spicatum (L.) Richter
synonym(s): *Aira spicata* L., *Trisetum spicatum* (L.) Richter var. *molle* (Kunth) Beal, *Trisetum*
    spicatum (L.) Richter var. *pilosiglume* Fern., *Trisetum triflorum* (Bigelow) A. & D. Löve
common name(s): narrow false oat
habit: erect, cespitose or solitary, perennial herb
range: circumboreal; s. in the east to NC
state frequency: uncommon
habitat: rocky soil, ledges, river shore outcrops
notes: —

Triticum
(1 species)

Triticum aestivum L.
synonym(s): —
common name(s): common wheat
habit: annual or winter-annual herb
range: Eurasia; escaped from cultivation in our area
state frequency: uncommon
habitat: fields, roadsides
notes: —

Vahlodea
(1 species)

Vahlodea atropurpurea (Wahlenb.) Fries *ex* Hartman
synonym(s): *Deschampsia atropurpurea* (Wahlenb.) Scheele
common name(s): mountain hairgrass
habit: erect, solitary-stemmed or cespitose, perennial herb
range: circumboreal; s. to mountains of ME and NH, w. to CO
state frequency: very rare
habitat: bogs, meadows, and wet rocks of alpine areas
notes: This species is ranked SH by the Maine Natural Areas Program.

Vulpia
(2 species)

Annuals with cleistogamous flowers, morphologically similar to *Festuca*.

KEY TO THE SPECIES

1a. Spikelets 5- to 13-flowered; first [lower] glume 1.7–4.5 mm long, more than half as long as the second glume; lowest lemma with an awn 0.3–6.0(–9.0) mm long; caryopsis 1.7–3.3 mm long .... *V. octoflora*

1b. Spikelets 3- to 7-flowered; first glume 0.5–2.5 mm long, up to half as long as the second glume; lowest lemma with an awn 7.5–22.0 mm long; caryopsis 3.0–4.5 mm long ....
.... *V. myuros*

Vulpia myuros (L.) K. C. Gmel.

synonym(s): *Festuca megalura* Nutt., *Festuca myuros* L., *Vulpia megalura* (Nutt.) Rydb.
common name(s): rat-tail fescue
habit: erect to ascending, solitary or cespitose, annual herb
range: Europe; naturalized to ME, s. to FL, w. to TX, n. to WI; also WA, s. to CA and AZ
state frequency: very rare
habitat: dry fields, waste places
notes: —

Vulpia octoflora (Walt.) Rydb. var. *glauca* (Nutt.) Fern.

synonym(s): *Festuca octoflora* Walt. var. *glauca* (Nutt.) Fern., *Vulpia octoflora* (Walt.) Rydb. var. *tenella* (Willd.) Fern.
common name(s): six weeks fescue
habit: slender, decumbent to erect, often cespitose, annual herb
range: ME, s. to FL, w. to CA, n. to British Columbia
state frequency: very rare
habitat: dry, sterile soil
notes: —

Zizania
(2 species)

KEY TO THE SPECIES

1a. Leaves 8.0–50.0 mm wide; ligules 10.0–15.0 mm long; carpellate lemmas scabrous on and between the nerves; abortive lemmas less than 1.5 mm wide; plants commonly 2.0–3.0 m tall ....*Z. aquatica*

1b. Leaves 4.0–15.0 mm wide; ligules 3.0–8.0 mm long; carpellate lemmas scabrous only on the nerves, glabrous between them; abortive lemmas 1.5–2.0 mm wide; plants 0.7–1.5 m tall .... *Z. palustris*

Zizania aquatica L.

synonym(s): —
common name(s): annual wildrice
habit: robust, simple or basally branched, annual herb
range: ME, s. to FL, w. and n. to WI and Ontario
state frequency: uncommon
habitat: quiet water, shores
notes: —

Zizania palustris L.

synonym(s): *Zizania aquatica* L. var. *angustifolia* A. S. Hitchc.
common name(s): narrow-leaved wildrice
habit: robust, simple or basally branched, annual herb

range: New Brunswick and Nova Scotia, s. to PA, w. to IA, n. to MN
state frequency: uncommon
habitat: quiet water
notes: —

NELUMBONACEAE
(1 genus, 1 species)

Nelumbo
(1 species)

Nelumbo lutea Willd.
synonym(s): —
common name(s): American lotus-lily, water chinkapin, yellow nelumbo
habit: rhizomatous, aquatic, perennial herb
range: ME, s. to FL, w. to TX, n. to MN and Ontario; also ME
state frequency: very rare
habitat: ponds, quiet streams, estuaries
notes: Populations of *Nelumbo lutea* in the northeastern part of its range are thought to have been
   introductions.

PLATANACEAE
(1 genus, 1 species)

Platanus
(1 species)

Platanus occidentalis L.
synonym(s): —
common name(s): sycamore, buttonwood
habit: tree
range: ME, s. to FL, w. to TX, n. to Ontario
state frequency: very rare
habitat: moist or wet, rich, alluvial soil
notes: This species is ranked SX by the Maine Natural Areas Program.

PAPAVERACEAE
(8 genera, 11 species)

KEY TO THE GENERA
1a. Perianth zygomorphic, spurred or saccate at the base; plants with a watery latex; petals
   not crumpled in bud; stamens 6 per flower, the filaments diadelphous; leaves finely
   divided
   2a. The 2 outer petals dissimilar, only 1 of which is spurred; flowers borne laterally on
      the pedicels

3a. Ovary linear-cylindric; flowers 10.0–17.0 mm long; fruit dehiscent, 25.0–50.0 mm long, many-seeded, with a persistent style; seed with an aril .... *Corydalis*

3b. Ovary subglobose; flowers 6.0–9.0 mm long; fruit indehiscent, subglobose, 2.5–3.0 mm long, 1-seeded; style deciduous; seed without an aril .... *Fumaria*

2b. The 2 outer petals similar and spurred; flowers borne on pendulous pedicels

4a. Plants not climbing; corolla connate only at the base, deciduous .... *Dicentra*

4b. Plants climbing by means of tendril-like, reduced leaflets and their petiolules; corolla connate, except at the very apex, persistent, becoming spongy, enclosing the capsule .... *Adlumia*

1b. Perianth actinomorphic, without a nectary spur; plants with a milky or colored latex; petals often crumpled in bud [therefore, wrinkled on expansion]; stamens *ca.* 16–30 per flower, the filaments distinct; leaves lobed or once-divided

5a. Leaves 1, basal, the blade orbicular in outline; corolla with 8–12 petals; latex red .... *Sanguinaria*

5b. Leaves several to many, borne on a stem, the blade not orbicular in outline; corolla with 0 or 4 petals; latex yellow or white

6a. Petals absent or present and yellow; stigmas 2-lobed; capsule with 2–4 valves that dehisce to the base; latex yellow

7a. Corolla present, of 4 yellow petals; sepals green; leaves pinnately divided, the leaflets again lobed; flowers few, in an umbel .... *Chelidonium*

7b. Corolla absent; sepals whitish; leaves pinnately lobed; flowers numerous, in a large panicle .... *Macleaya*

6b. Petals variously colored, but never yellow; stigmas forming a 5- to 18-lobed disk; capsule dehiscing by 5–18 subapical pores .... *Papaver*

Adlumia
(1 species)

Adlumia fungosa (Ait.) Greene *ex* B. S. P.
synonym(s): —
common name(s): Allegheny vine, climbing fumitory, mountain fringe
habit: biennial vine
range: Quebec, s. to NC, w. and n. to MN and Ontario
state frequency: very rare
habitat: moist woods, wooded slopes
notes: This species is ranked S1 by the Maine Natural Areas Program.

Chelidonium
(1 species)

This genus has a yellow latex.

Chelidonium majus L.
synonym(s): —
common name(s): celandine
habit: biennial herb

range: Eurasia; naturalized to Quebec and Nova Scotia, s. to NC, w. to MO, n. to Ontario; also scattered locations in the northwest United States
state frequency: occasional
habitat: damp soil of roadsides and waste places
notes: This species has a yellow to orange-yellow latex.

Corydalis
(1 species)

Corydalis sempervirens (L.) Pers.
synonym(s): *Capnoides sempervirens* (L.) Borkh.
common name(s): pale corydalis, tall corydalis
habit: erect, biennial herb
range: Newfoundland, s. to PA and mountains of GA, w. to MT, n. to AK
state frequency: uncommon
habitat: rocky woods, cliffs, talus
notes: —

Dicentra
(3 species)

KEY TO THE SPECIES

1a. Leaves finely divided, the ultimate segments linear to narrow-oblong, less than 5.0 mm wide; flowers borne on a scape; corolla white or white with yellow; seeds very obscurely reticulate-patterned

    2a. Spurs of the corolla divergent, subacute at the apex, mostly 2.0–3.0 mm long; bulblets of the rhizome white or pink; apical crest of the inner 2 petals inconspicuous; leaves typically yellow-green to green; flowers infragrant .... ***D. cucullaria***

    2b. Spurs of the corolla straight, rounded at the apex, 0.5–1.0 mm long; bulblets of the rhizome yellow; apical crest of the inner 2 petals conspicuous and projecting; leaves typically blue-green; flowers fragrant .... ***D. canadensis***

1b. Leaves divided, the ultimate segments lanceolate or oblanceolate to elliptic or wider, more than 5.0 mm wide; flowers borne on a leafy stem; corolla largely pink; seeds reticulate-patterned .... ***D. spectabilis***

Dicentra canadensis (Goldie) Walp.
synonym(s): *Bicuculla canadensis* (Goldie) Millsp.
common name(s): squirrel-corn, turkey-corn
habit: spring ephemeral, scapose, perennial herb
range: Quebec and Nova Scotia, s. to NC, w. to MO, n. to MN
state frequency: very rare
habitat: rich, often rocky, hardwood forests
notes: This species is ranked S1 by the Maine Natural Areas Program.

Dicentra cucullaria (L.) Bernh.
synonym(s): *Bicuculla cucullaria* (L.) Millsp.
common name(s): Dutchman's breeches
habit: spring ephemeral, scapose, perennial herb
range: Quebec and Nova Scotia, s. to GA, w. to KS, n. to ND

state frequency: occasional
habitat: rich woods
notes: This species is often confused with the much rarer *Dicentra canadensis*. *Dicentra cucullaria* flowers 7–10 days earlier than *D. canadensis* when both are present at the same site.

Dicentra spectabilis (L.) Lem.
synonym(s): —
common name(s): bleeding heart
habit: rhizomatous, perennial herb
range: perhaps native to Asia; cultivated nearly worldwide
state frequency: very rare
habitat: cultivated areas
notes: —

Fumaria
(1 species)

Fumaria officinalis L.
synonym(s): —
common name(s): fumitory, earth smoke
habit: branched, annual herb
range: Europe; naturalized in our area
state frequency: uncommon
habitat: waste places
notes: —

Macleaya
(1 species)

Macleaya cordata (Willd.) R. Br.
synonym(s): —
common name(s): plume-poppy
habit: erect, perennial herb
range: Asia; escaped from cultivation in the northeast
state frequency: very rare
habitat: fields, roadsides, waste places
notes: —

Papaver
(2 species)

Our species in this genus have a white latex.

KEY TO THE SPECIES

1a. Leaves of the stem cordate-clasping; stems glabrous; flower buds 2.0–4.0 cm long; capsule 2.5–6.0 cm long .... *P. somniferum*
1b. Leaves of the stem not clasping; stems bristly; flower buds 0.5–2.0 cm long; capsule 1.0–2.0 cm long .... *P. rhoeas*

Papaver rhoeas L.

synonym(s): —
common name(s): corn poppy, Shirley poppy, field poppy
habit: annual or rarely biennial herb
range: Eurasia and n. Africa; escaped from cultivation to Nova Scotia, s. to VA, w. to MO, n. to ND
state frequency: rare
habitat: fields, roadsides, disturbed soil
notes: —

Papaver somniferum L.

synonym(s): —
common name(s): opium poppy, common poppy
habit: annual herb
range: Europe; escaped from cultivation locally throughout North America
state frequency: rare
habitat: fields, roadsides, waste places
notes: —

Sanguinaria
(1 species)

Sanguinaria canadensis L.

synonym(s): *Sanguinaria canadensis* L. var. *rotundifolia* (Greene) Fedde
common name(s): bloodroot
habit: rhizomatous, perennial herb
range: Quebec and New Brunswick, s. to FL, w. to OK, n. to Manitoba
state frequency: uncommon
habitat: rich forests, rocky slopes, and alluvial flats
notes: —

BERBERIDACEAE
(3 genera, 4 species)

KEY TO THE GENERA

1a. Plants woody, with spines; perianth yellow .... *Berberis*
1b. Plants herbaceous, without spines; perianth white to pink, yellow-green, or green-purple
    2a. Leaves lobed, the basal ones peltate; inflorescence a solitary flower; perianth white or pink; anthers dehiscing by longitudinal slits; ovary maturing as a berry .... *Podophyllum*
    2b. Leaves compound, none peltate; inflorescence a cyme; perianth yellow-green or green-purple; anthers dehiscing by uplifting valves; ovary ruptured by the 2 enlarging seeds, the stalked, blue seeds ripening fully exposed .... *Caulophyllum*

Berberis
(2 species)

Perianth composed of 6 petaloid sepals and 6 petals with 2 glands near the base.

KEY TO THE SPECIES

1a. Margins of leaves entire; spines simple; inflorescence a cluster of 1–4 flowers; berries rather dry .... *B. thunbergii*

1b. Margins of leaves spinulose-toothed; many of the spines compound; inflorescence a raceme of 10–25 flowers; berries fleshy .... *B. vulgaris*

Berberis thunbergii DC.

synonym(s): —
common name(s): Japanese barberry
habit: spiny shrub
range: Asia; escaped from cultivation to Nova Scotia, s. to NC, w. to MO, n. to MI
state frequency: common
habitat: thickets, roadsides, open areas
notes: Hybrids with *Berberis vulgaris*, called *B.* ×*ottawensis* Schneid., occur in Maine. *Berberis thunbergii* is more common than *B. vulgaris* due to attempts to eradicate the latter, the alternate host of a wheat rust.

Berberis vulgaris L.

synonym(s): —
common name(s): common barberry, European barberry
habit: spiny shrub
range: Nova Scotia, s. to DE, w. to MO, n. to MN
state frequency: uncommon
habitat: thickets, pastures
notes: Hybrids with *Berberis thunbergii* (*q.v.*) occur in Maine. *Berberis vulgaris* is the alternate host of a wheat rust.

Caulophyllum
(1 species)

Caulophyllum thalictroides (L.) Michx.

synonym(s): —
common name(s): blue cohosh, papoose-root
habit: erect, perennial herb
range: New Brunswick and Nova Scotia, s. to SC, w. to MO, n. to Manitoba
state frequency: uncommon
habitat: moist, rich forests, slopes, and alluvial flats
notes: —

Podophyllum
(1 species)

Podophyllum peltatum L.

synonym(s): —
common name(s): May-apple, mandrake
habit: erect, perennial herb
range: Quebec, s. to FL, w. to TX, n. to MN and Ontario
state frequency: rare
habitat: moist, usually open, rich woods

notes: Introduced to southern Maine by the Shaker community. Developmental irregularities sometimes occur in this species. Populations or individuals have been observed with 2–8 carpels, red fruits, aerial stems, or with 0, 1, or 3 leaves.

RANUNCULACEAE
(12 genera, 39 species)

KEY TO THE GENERA

1a. Shrubs with yellow roots and sparsely pubescent follicles .... *Xanthorhiza*
1b. Herbs without yellow roots (except *Coptis*), with achenes, berries, or glabrous follicles
   2a. Fruit a follicle or berry; ovaries with 2 or more ovules
      3a. Flowers zygomorphic
         4a. Upper sepal arched or hooded; petals not spurred .... *Aconitum*
         4b. Upper sepal spurred; upper petal spurred .... *Delphinium*
      3b. Flowers actinomorphic
         5a. Leaves simple; perianth composed of 5–9 yellow sepals .... *Caltha*
         5b. Leaves compound or dissected; perianth otherwise
            6a. Sepals petaloid, not caducous; ultimate segments of leaves crenately toothed or crenately lobed
               7a. Petals with long spurs; leaves cauline, 2–3 times compound .... *Aquilegia*
               7b. Petals without long spurs; leaves all basal, once-compound ..... *Coptis*
            6b. Sepals inconspicuous and caducous; ultimate segments of leaves sharply cleft and toothed
               8a. Racemes compound; stigma borne on a style; fruit a follicle; carpels 1–8 .... *Cimicifuga*
               8b. Racemes simple; stigma sessile or subsessile; fruit a berry; carpels 1 .... *Actaea*
   2b. Fruit an achene; ovaries with 1 ovule
      9a. Petals present, yellow or both white and yellow; sepals (3–)5, sepaloid .... *Ranunculus*
      9b. Petals absent or inconspicuous; sepals 4–20, petaloid or sepaloid in some species of *Thalictrum*
         10a. Plants trailing or climbing, vines; sepals valvate in bud; style plumose .... *Clematis*
         10b. Plants not vines, never climbing; sepals imbricate in bud; style not plumose
            11a. Flowers in a panicle; leaves alternate .... *Thalictrum*
            11b. Flowers solitary; leaves whorled [sometimes appearing basal] .... *Anemone*

Aconitum
(1 species)

Aconitum napellus L.
synonym(s): —
common name(s): garden monkshood, garden aconite, garden wolfsbane
habit: erect, perennial herb
range: Europe; escaped from cultivation to Newfoundland, s. to NY, w. and n. to Ontario
state frequency: rare
habitat: roadsides, fields
notes: This species has blue or purple flowers.

Actaea
(2 species)

KEY TO THE SPECIES
1a. Pedicels of infructescence slender, 0.3–0.7 mm in diameter; berries red (white); stigma
    narrower than ovary; leaflets commonly pubescent on the abaxial [lower] surface ....
    .... *A. rubra*
1b. Pedicels of infructescence thick, 1.0–2.5 mm in diameter; berries white (red); stigma
    wider than ovary; leaflets glabrous or nearly so on the abaxial surface .... *A. pachypoda*

Actaea pachypoda Ell.
synonym(s): *Actaea alba* (L.) P. Mill. *sensu* Gleason and Cronquist (1991)
common name(s): white baneberry, doll's eyes
habit: erect, perennial herb
range: Nova Scotia, s. to GA, w. to OK, n. to Manitoba
state frequency: occasional
habitat: rich woods, thickets
notes: —

Actaea rubra (Ait.) Willd.
synonym(s): *Actaea rubra* (Ait.) Willd. forma *neglecta* (Gillman) Robins.
common name(s): red baneberry, snakeberry
habit: erect, perennial herb
range: circumboreal; s. to NJ, w. to AZ
state frequency: occasional
habitat: rich woods, thickets
notes: —

Anemone
(7 species)

KEY TO THE SPECIES
1a. Leaves simple and of 2 types—the basal 3(5)-lobed, the cauline [subtending the flower]
    lanceolate; stem leaves closely subtending the flower
    2a. Apex of leaf lobes rounded to blunt; sinuses indented, at most, to middle of leaf;
        bracts rounded at the apex; achenes 1.0–1.4 mm wide .... *A. americana*
    2b. Apex of leaf lobes acute to acuminate; sinuses indented beyond middle of leaf; bracts
        acute to obtuse at the apex; achenes 1.4–1.8 mm wide .... *A. acutiloba*

1b. Leaves compound or nearly so, often with more lobes or segments, the basal and cauline ones similar; stem leaves remotely subtending the flower

   3a. Carpels and achenes glabrous or thinly pubescent, not woolly; achenes in capitate to spherical clusters; plants with rhizomes

      4a. Leaves subtending flowers distinctly petioled; stems simple, with 1 flower; sepals pink to white; achenes sharp-margined at most; above-ground stems produced directly from rhizomes .... *A. quinquefolia*

      4b. Leaves subtending flowers sessile or nearly so; stems branched, with 1–6 flowers; sepals white; achenes broadly wing-margined; above-ground stems produced from caudices on rhizomes .... *A. canadensis*

   3b. Carpels and achenes densely pubescent, the achenes with long, woolly hairs; achenes in oblong to cylindric clusters; plants commonly without rhizomes

      5a. Leaves dissected into numerous, narrow segments 1.5–3.0(–4.3) mm wide, those subtending flowers sessile or nearly so; styles deciduous from achenes .... *A. multifida*

      5b. Leaves divided into 3–5 broad segments (4.0)6.0 mm wide or wider, those subtending flowers distinctly petioled; styles persistent on achenes

         6a. Flowers subtended by (3–)5–9 leaves; peduncles naked; cluster of achenes 2.0–4.5 cm tall, 2.0–5.0 times as tall as broad; styles crimson; beak of the achene 0.3–1.0 mm long .... *A. cylindrica*

         6b. Flowers subtended by 2–3(–5) leaves; peduncles with 2 additional subtending leaves; cluster of achenes 1.5–3.0 cm tall, mostly 1.0–1.5 times as tall as broad; styles pale or crimson-tipped; beak of the achene 1.0–1.5 mm long .... *A. virginiana*

Anemone acutiloba (de Candolle) G. Lawson
synonym(s): *Hepatica acutiloba* DC., *Hepatica nobilis* P. Mill. var. *acuta* (Pursh) Steyermark
common name(s): sharp-leaved hepatica
habit: spring-flowering, rhizomatous, perennial herb
range: Quebec and ME, s. to GA, w. to AR, n. to MN
state frequency: very rare
habitat: rich, often rocky, forests
notes: This species is ranked SX by the Maine Natural Areas Program.

Anemone americana (de Candolle) H. Hara
synonym(s): *Hepatica americana* (DC.) Ker-Gawl., *Hepatica nobilis* P. Mill. var. *obtusa* (Pursh) Steyermark
common name(s): round-lobed hepatica
habit: spring-flowering, rhizomatous, perennial herb
range: Nova Scotia, s. to FL, w. to AR, n. to Manitoba
state frequency: uncommon
habitat: rich, often rocky, forests
notes: —

Anemone canadensis L.
synonym(s): —
common name(s): Canada anemone
habit: rhizomatous, caudex-bearing, perennial herb
range: Quebec and Nova Scotia, s. to MD, w. to NM, n. to British Columbia
state frequency: occasional

habitat: damp meadows, sandy or gravelly shores
notes: —

Anemone cylindrica Gray

synonym(s): —
common name(s): long-headed anemone
habit: caudex-bearing, perennial herb
range: ME, s. to NJ, w. to AZ, n. to British Columbia
state frequency: uncommon
habitat: dry, open woods and prairies
notes: —

Anemone multifida Poir.

synonym(s): —
common name(s): cut-leaved anemone
habit: caudex-bearing, perennial herb
range: Newfoundland, s. to ME, w. to SD, n. to AK; also NM and CA
state frequency: very rare
habitat: calcareous shores and rocky areas
notes: This species is ranked S1 by the Maine Natural Areas Program.

Anemone quinquefolia L.

synonym(s): —
common name(s): wood anemone
habit: erect, slender-rhizomatous, perennial herb
range: Quebec, s. to NC and mountains of GA, w. to IA, n. to Manitoba
state frequency: occasional
habitat: rich forests, rocky banks, streamsides
notes: —

Anemone virginiana L.

synonym(s): <*A. v.* var. *alba*> = *Anemone riparia* Fern., *Anemone virginiana* L. var. *riparia* (Fern.)
Boivin
common name(s): thimbleweed, pasque flower, prairie smoke
habit: caudex-bearing, perennial herb
range: Quebec, s. to GA, w. to AR and KS, n. to ND
state frequency: uncommon
habitat: dry or open woods, rocky places
key to the subspecies:
  1a. Clusters of fruits 1.1–1.4 cm wide; achenes with divergent styles; anthers 1.2–1.6 mm long;
  leaf segments usually cordate to truncate at the base .... *A. v.* var. *virginiana*
  1b. Clusters of fruits 0.8–1.0 cm wide; achenes with spreading-ascending to ascending styles;
  anthers 0.7–1.2 mm long; leaf segments usually cuneate to truncate at the base ....
  .... *A. v.* var. *alba* (Oakes) Wood
notes: —

Aquilegia
(2 species)

KEY TO THE SPECIES
1a. Flowers red and yellow; spurs straight to arching; follicle with a persistent style beak
1.5–2.0 cm long .... *A. canadensis*

1b. Flowers purple to white or pink; spurs strongly curved; follicle with a persistent style
    beak 0.5–1.0 cm long .... *A. vulgaris*

Aquilegia canadensis L.
synonym(s): —
common name(s): wild columbine, Canada columbine
habit: perennial herb
range: Quebec and Nova Scotia, s. to FL, w. to TX, n. to Saskatchewan
state frequency: occasional
habitat: rich or rocky forests, cliffs
notes: —

Aquilegia vulgaris L.
synonym(s): —
common name(s): garden columbine, European columbine
habit: perennial herb
range: Eurasia; escaped from cultivation to Newfoundland, s. to PA, w. to IA, n. to Ontario
state frequency: rare
habitat: fields, roadsides, borders of woods
notes: —

Caltha
(1 species)

Caltha palustris L.
synonym(s): —
common name(s): marsh-marigold, cowslip
habit: hollow-stemmed, perennial herb
range: circumboreal; s. to SC, w. to OR
state frequency: uncommon
habitat: swamps, streams, ditches, shallow water
notes: —

Cimicifuga
(1 species)

Cimicifuga racemosa (L.) Nutt.
synonym(s): —
common name(s): black snakeroot, black cohosh
habit: perennial herb
range: MA, s. to GA, w. to MO, n. to Ontario; escaped from cultivation in ME
state frequency: very rare
habitat: moist or dry, rich woods
notes: Often the outermost set of stamens are transformed into petaloid, 2-horned staminodes.

Clematis
(2 species)

KEY TO THE SPECIES
1a. Flowers [and hence, the fruit cluster] single in axil of leaves; sepals 3.0–5.0 cm long,
    purple; plants mostly of rich, rocky forests .... *C. occidentalis*

1b. Flowers [and hence, fruit clusters] in many-flowered panicles; sepals 1.0–1.5 cm long, white; plants of open areas, borders, roadsides, and disturbed places .... *C. virginiana*

Clematis occidentalis (Hornem.) DC.
synonym(s): *Atragene americana* Sims, *Clematis verticillaris* DC.
common name(s): purple virgin's bower, purple clematis
habit: trailing to climbing, woody-based perennial
range: Quebec, s. to DE and mountains of NC, w. to IA, n. to Manitoba; also in mountains of NC
state frequency: rare
habitat: rich, rocky forests and slopes
notes: This species is ranked S1/S2 by the Maine Natural Areas Program.

Clematis virginiana L.
synonym(s): —
common name(s): virgin's bower
habit: climbing, woody-based perennial
range: Quebec and Nova Scotia, s. to FL, w. to KS, n. to Manitoba
state frequency: occasional
habitat: moist thickets, borders of woods
notes: —

Coptis
(1 species)

Coptis trifolia (L.) Salisb.
synonym(s): *Coptis groenlandica* (Oeder) Fern., *Coptis trifolia* (L.) Salisb. var. *groenlandica* (Oeder) Fassett
common name(s): goldthread
habit: rhizomatous, perennial herb
range: Greenland, s. to NC, w. to IA, n. to British Columbia
state frequency: common
habitat: mossy woods, swamps, bogs
notes: —

Delphinium
(1 species)

Delphinium exaltatum Ait.
synonym(s): —
common name(s): tall larkspur
habit: perennial herb
range: PA, s. to NC, w. to MO, n. to OH; escaped from cultivation
state frequency: very rare
habitat: gardens, abandoned fields and lots
notes: —

Ranunculus
(16 species)

KEY TO THE SPECIES

1a. Submersed leaves finely divided into narrow segments less than 1.0 mm wide, the emersed leaves, if present, with broader segments; plants aquatic

  2a. Petals white with yellow bases; leaf segments capillary; achenes coarsely wrinkled on the surface, with a beak 0.2–0.5 mm long .... *R. aquatilis*

  2b. Petals yellow; leaf segments flat; achenes smooth on the surface, with a beak 0.5–1.5 mm long

    3a. Petals 7.0–12.0 mm long; achenes 1.8–2.2 mm, the lower portion thickened and spongy; achene beak 1.0–1.8 mm long; styles 0.8–1.2 mm long ...... *R. flabellaris*

    3b. Petals 3.0–7.0 mm long; achenes 1.0–1.6 mm long, not thickened in the lower half; achene beak 0.4–0.8 mm long; styles 0.2–0.4 mm long .... *R. gmelinii*

1b. Leaves simple or divided with broad segments; plants aquatic or terrestrial

  4a. Sepals 3(4); flowering stems with 1(2) basal leaves; achenes with a prominent, corky appendage prolonged distally to the seed .... *R. lapponicus*

  4b. Sepals 5; flowers borne on leafy stems, at least arising from a tuft of leaves; achenes not prolonged beyond the seed body

    5a. Leaves simple

      6a. Leaf blades cordate to reniform; plants of tidal shores; achenes with longitudinal nerves .... *R. cymbalaria*

      6b. Leaf blades linear to lanceolate; plants mostly not of tidal shores; achenes without longitudinal nerves

        7a. Leaf blades 0.2–0.7 cm wide; plants with filiform, arching stolons; sepals 2.0–4.0 mm long; achenes essentially smooth .... *R. flammula*

        7b. Leaf blades 1.0–3.0 cm wide; plants lacking stolons, rooting at the base; sepals 5.0–7.0 mm long; achenes reticulate .... *R. ambigens*

    5b. Leaves at the base of the stem and/or on the stem evidently lobed to dissected

      8a. Leaves dimorphic—the basal ones simple, with orbicular to reniform blades, the cauline ones deeply divided .... *R. abortivus*

      8b. Leaves monomorphic, the basal and cauline ones more or less similar except perhaps in size or deepness of the sinuses

        9a. At least the larger leaves with definite, unwinged petiolules [therefore, these leaves compound]

          10a. Petals 2.0–5.0 mm long; anthers up to 1.0 mm long .... *R. pensylvanicus*

          10b. Petals 5.0–15.0 mm long; anthers 1.2 mm or longer

            11a. Sepals reflexed along a defined, transverse fold 2.0–3.0 mm above the base; stems bulbous-thickened at base .... *R. bulbosus*

            11b. Sepals spreading (sometimes reflexed from the very base in age); stems not thickened at the base

              12a. Leaf blades more or less pinnately divided, longer than wide; some roots tuberous-thickened; stems erect to ascending, not rooting at the nodes .... *R. fascicularis*

12b. Leaf blades more or less palmately divided, about as wide as long; roots all slender; stems decumbent to creeping, sometimes rooting at the nodes

 13a. Style elongate with a terminal, deciduous stigma; achene beak 1.5–3.0 mm long .... ***R. hispidus***

 13b. Style short, curved, the persistent stigma spread out over upper side of style; achene beak 0.7–1.4 mm long .... ***R. repens***

9b. Leaves with flattened, winged, sometimes ill defined petiolules [therefore, the leaves lobed but not divided]

 14a. Stems floating or prostrate and rooting at the nodes; achenes plump, without a sharp or wing-like keel

  15a. Petals 7.0–12.0 mm long; achenes 1.8–2.2 mm, the lower portion thickened and spongy; achene beak 1.0–1.8 mm long; styles 0.8–1.2 mm long .... ***R. flabellaris***

  15b. Petals 3.0–7.0 mm long; achenes 1.0–1.6 mm long, not thickened in the lower half; achene beak 0.4–0.8 mm long; styles 0.2–0.4 mm long .... ***R. gmelinii***

 14b. Stems erect, rooting only at the base; achenes flattened, with a sharp or wing-like keel between the faces

  16a. Petals 7.0–15.0 mm long, exceeding the length of the sepals; receptacle glabrous; achenes 2.0–3.0 x 1.8–2.4 mm .... ***R. acris***

  16b. Petals 2.0–5.0 mm long, not exceeding the length of the sepals; receptacle pubescent (sometimes glabrous in *R. scleratus*); achenes 1.0–2.2 x 0.8–1.8 mm

   17a. Anthers up to 1.0 mm long; style absent; achenes in a cylindric cluster, with a very short, straight beak *ca.* 0.1 mm long; stems not thickened at the base .... ***R. scleratus***

   17b. Anthers longer than 1.0 mm; style present; achenes in a spherical cluster, with a hooked beak *ca.* 1.0 mm long; stems bulbous-thickened at the base .... ***R. recurvatus***

Ranunculus abortivus L.

synonym(s): *Ranunculus abortivus* L. var. *acrolasius* Fern., *Ranunculus abortivus* L. var. *eucyclus* Fern.
common name(s): small-flowered crowfoot, kidneyleaf buttercup
habit: erect, branched, perennial herb
range: Labrador, s. to FL, w. to TX, n. to AK
state frequency: uncommon
habitat: moist or dry woods, clearings
notes: —

Ranunculus acris L.

synonym(s): *Ranunculus acris* L. var. *latisectus* G. Beck
common name(s): common buttercup, meadow buttercup, tall buttercup
habit: erect, perennial herb
range: Labrador, s. to VA, w. to KS, n. to MN and Ontario
state frequency: common
habitat: fields, roadsides
notes: —

Ranunculus ambigens S. Wats.

synonym(s): —
common name(s): creeping spearwort, water-plantain spearwort
habit: stoloniferous, perennial herb
range: ME, s. to GA, w. to LA, n. to MN and Ontario
state frequency: very rare
habitat: swamps, ditches, shores
notes: This species is ranked SH by the Maine Natural Areas Program.

Ranunculus aquatilis L. var. *diffusus* Withering

synonym(s): *Batrachium trichophyllum* (Chaix) F. W. Schultz, *Ranunculus longirostris* Godr.,
    Ranunculus subrigidus W. Drew, *Ranunculus trichophyllus* Chaix, *Ranunculus trichophyllus*
    Chaix var. *calvescens* W. Drew
common name(s): white water crowfoot
habit: submerged, aquatic herb
range: Greenland, s. to South America, w. to CA, n. to AK; also Europe and Australia
state frequency: occasional
habitat: shallow, still or slow-moving water
notes: —

Ranunculus bulbosus L.

synonym(s): *Ranunculus bulbosus* L. var. *dissectus* Barbey
common name(s): bulbous buttercup
habit: perennial herb
range: Newfoundland, s. to GA, w. to LA, n. to Ontario
state frequency: uncommon
habitat: fields, roadsides
notes: —

Ranunculus cymbalaria Pursh

synonym(s): —
common name(s): seaside crowfoot
habit: stoloniferous, leafy-tufted, perennial herb
range: Greenland and Newfoundland, s. to NJ, w. to NM, n. to AK
state frequency: uncommon
habitat: brackish to saline marshes and shores
notes: —

Ranunculus fascicularis Muhl. *ex* Bigelow

synonym(s): —
common name(s): early buttercup, thick-root buttercup, early crowfoot
habit: ascending to erect, short-rhizomatous, perennial herb
range: ME, s. to GA, w. to TX, n. to MN and Ontario
state frequency: very rare
habitat: thin soils, dry woods, open areas
notes: This species is ranked S1 by the Maine Natural Areas Program.

Ranunculus flabellaris Raf.

synonym(s): *Ranunculus delphiniifolius* Torr. *ex* Eat.
common name(s): yellow water crowfoot, yellow water buttercup
habit: aquatic, perennial herb
range: New Brunswick, s. to NC, w. to CA, n. to British Columbia
state frequency: uncommon

habitat: quiet waters, muddy shores
notes: —

Ranunculus flammula L.

synonym(s): <*R. f.* var. *reptans*> = *Ranunculus flammula* L. var. *filiformis* (Michx.) Hook., *Ranunculus reptans* L.; <*R. f.* var. *ovalis*> = *Ranunculus reptans* L. var. *ovalis* (Bigelow) Torr. & Gray
common name(s): creeping spearwort
habit: creeping to erect, perennial herb
range: Greenland, s. to PA, w. to CA, n. to AK; also Eurasia
state frequency: occasional
habitat: wet shores, sometimes intertidal
key to the subspecies:
    1a. Leaf blades filiform to linear, 0.4–1.5 mm wide .... *R. f.* var. *reptans* (L.) E. Meyer
    1b. Leaf blades lanceolate to oval, 1.5–7.0 mm wide .... *R. f.* var. *ovalis* (Bigelow) L. Benson
notes: The arching, green, filiform stolons serve as a useful vegetative character.

Ranunculus gmelinii DC.

synonym(s): *Ranunculus gmelinii* DC. var. *hookeri* (D. Don) L. Benson, *Ranunculus gmelinii* DC. var. *purshii* (Richards.) Hara
common name(s): small yellow water crowfoot
habit: aquatic, perennial herb
range: Newfoundland, s. to ME, w. to OR, n. to AK
state frequency: very rare; known only from northern Maine
habitat: wet meadows, springy places, mud
notes: This species is ranked S1 by the Maine Natural Areas Program.

Ranunculus hispidus Michx. var. caricetorum (Greene) T. Duncan

synonym(s): *Ranunculus septentrionalis* Poir. var. *caricetorum* (Greene) Fern.
common name(s): swamp buttercup, northern crowfoot
habit: decumbent, fibrous-rooted, short-rhizomatous, perennial herb
range: Newfoundland, s. to VA, w. to MO, n. to Manitoba
state frequency: uncommon
habitat: swamps, marshes, ditches, stream shores
notes: —

Ranunculus lapponicus L.

synonym(s): —
common name(s): Lappland buttercup
habit: rhizomatous, perennial herb
range: circumboreal; s. to ME, w. to British Columbia
state frequency: very rare; known only from northern Maine
habitat: wet woods, swamps
notes: This species is ranked S1/S2 by the Maine Natural Areas Program.

Ranunculus pensylvanicus L. f.

synonym(s): —
common name(s): bristly crowfoot
habit: erect to ascending, simple or branched, annual or perennial herb
range: Labrador, s. to DE, w. to OR, n. to British Columbia
state frequency: uncommon
habitat: wet meadows, ditches
notes: Similar to *Ranunculus bulbosus*, this species has reflexed sepals that bend at a transverse fold above the base of the sepal.

Ranunculus recurvatus Poir.

synonym(s): —
common name(s): hooked crowfoot
habit: perennial herb
range: Eurasia, Australia, Pacific islands; naturalized to Greenland, irregularly s. to South America, w. to TX, n. to Ontario; also AK, s. to CA and UT
state frequency: uncommon
habitat: shores, marshes, meadows, roadsides
notes: —

Ranunculus repens L.

synonym(s): *Ranunculus repens* L. var. *degeneratus* Schur, *Ranunculus repens* L. var. *erectus* DC., *Ranunculus repens* L. var. *glabratus* DC., *Ranunculus repens* L. var. *linearilobus* DC., *Ranunculus repens* L. var. *pleniflorus* Fern., *Ranunculus repens* L. var. *villosus* Lamotte
common name(s): creeping buttercup
habit: stoloniferous, perennial herb
range: Europe; naturalized to Labrador, s. to NC, w. to MO, n. to MI and Ontario
state frequency: uncommon
habitat: fields, meadows, roadsides, lawns
notes: —

Ranunculus sceleratus L.

synonym(s): —
common name(s): cursed crowfoot
habit: erect, annual or short-lived, often weedy, perennial herb
range: Newfoundland, s. to FL, w. to TX, n. to Alberta; also British Columbia, s. to CA; also Europe
state frequency: very rare
habitat: wetlands, often in disturbed soils
notes: This species may be naturalized from Europe, but its origin is uncertain.

Thalictrum
(4 species)

Leaf morphology is sometimes variable on a single plant. Therefore, positive identification requires the use of both reproductive and vegetative characters.

KEY TO THE SPECIES

1a. Leaves of the stem opposite or whorled; sepals 10.0–15.0 mm long, petaloid, deciduous; plants 1.0–3.0 dm tall .... *T. thalictroides*
1b. Leaves alternate; sepals up to 5.0 mm long, often sepaloid and/or caducous; plants 3.0–30.0 dm tall
    2a. Leaflets with 4 or more lobes [usually appearing as 3 lobes and each lobe with 1–3 tooth-like sublobes]; anthers 2.0–4.5 mm long; filaments slender, much narrower than the anthers
        3a. Lowest branch of the inflorescence subtended by a long-petioled leaf, the petiole 3.0–6.0 cm long; achenes straight; plants arising from a stout caudex, without rhizomes, flowering in May .... *T. dioicum*
        3b. Lowest branch of the inflorescence subtended by a short-petioled leaf, the petiole less than 3.0 cm long; achenes inwardly curved; plants with rhizomes, flowering in mid-June–early July .... *T. confine*

2b. Leaflets mostly with 3 lobes, each lobe entire and without additional sublobes; anthers 0.5–1.5(–2.1) mm long; filaments usually dilated distally, nearly as wide as the anthers, constricted at the apex just below the anther .... *T. pubescens*

Thalictrum confine Fern.
synonym(s): *Thalictrum venulosum* Trel. var. *confine* (Fern.) Boivin
common name(s): northern meadow-rue
habit: rhizomatous, perennial herb
range: Quebec, s. to ME, VT, and NY, w. and n. to Ontario
state frequency: very rare
habitat: alluvial or rocky river shores, talus
notes: This species is ranked SH by the Maine Natural Areas Program.

Thalictrum dioicum L.
synonym(s): —
common name(s): early meadow-rue
habit: erect, fibrous-rooted, perennial herb
range: Quebec and ME, s. to GA, w. to MO, n. to Manitoba
state frequency: uncommon
habitat: rich, rocky woods and slopes
notes: —

Thalictrum pubescens Pursh
synonym(s): *Thalictrum polygamum* Muhl. *ex* Spreng., *Thalictrum polygamum* Muhl. *ex* Spreng. var. *hebecarpum* Fern.
common name(s): tall meadow-rue, king-of-the-meadow
habit: erect, fibrous-rooted, perennial herb
range: Newfoundland, s. to GA, w. to TN, n. to Ontario
state frequency: common
habitat: moist soil of fields, ditches, and swamps
notes: —

Thalictrum thalictroides (L.) Eames & Boivin
synonym(s): *Anemonella thalictroides* (L.) Spach
common name(s): rue-anemone
habit: tuberous-rooted, perennial herb
range: ME, s. to FL, w. to MS, AR, and OK, n. to MN
state frequency: very rare
habitat: woods
notes: This species is ranked SX by the Maine Natural Areas Program.

Xanthorhiza
(1 species)

Xanthorhiza simplicissima Marsh.
synonym(s): *Zanthorhiza apiifolia* L'Hér.
common name(s): yellowroot, shrub yellowroot
habit: low shrub
range: VA, s. to GA, w. to MS, n. to TN; escaped from cultivation in the northeast United States
state frequency: very rare
habitat: moist woods, thickets, stream banks
notes: —

VITACEAE
(2 genera, 6 species)

KEY TO THE GENERA

1a. Petals connate in the apical portion, deciduous as a single unit; bark becoming loose and exfoliating on older stems; pith brown; berry juicy; leaves simple; tendrils without expanded disks .... *Vitis*
1b. Petals distinct; bark close, not exfoliating; pith white; berry with thin flesh; leaves compound or simple; tendrils with or without expanded disks .... *Parthenocissus*

Parthenocissus
(3 species)

KEY TO THE SPECIES

1a. Leaves simple or sometimes compound with 3 leaflets; plants introduced, escaped from cultivation, normally found growing on buildings .... *P. tricuspidata*
1b. Leaves compound, mostly with 5 leaflets; plants native, normally growing on the ground or on other plants
  2a. Tendrils without expanded disks; inflorescence mostly 10- to 60-flowered, without a central axis, normally with 2 subequal branches; berry 8.0–10.0 mm long .... *P. vitacea*
  2b. Tendrils with expanded disks; inflorescence mostly 25- to 200-flowered, with a central axis; berry 5.0–7.0 mm long .... *P. quinquefolia*

Parthenocissus quinquefolia (L.) Planch.
synonym(s): —
common name(s): Virginia creeper
habit: climbing liana
range: Quebec, s. to FL, w. to TX and Mexico, n. to MN and Ontario; naturalized to Newfoundland, New Brunswick, and Nova Scotia
state frequency: occasional
habitat: rocky banks, moist woods
·notes: —

Parthenocissus tricuspidata (Sieb. & Zucc.) Planch.
synonym(s): —
common name(s): Boston-ivy
habit: climbing liana
range: Asia; escaped from cultivation locally
state frequency: uncommon
habitat: sides of brick and stone buildings
notes: —

Parthenocissus vitacea (Knerr) A. S. Hitchc.
synonym(s): *Parthenocissus inserta* (Kerner) Fritsch *sensu* Gleason and Cronquist (1991)
common name(s): grape woodbine
habit: trailing to climbing liana
range: Quebec, s. to NJ, w. to AZ, n. to Manitoba
state frequency: occasional
habitat: moist woods, thickets, and banks
notes: —

Vitis

(3 species)

KEY TO THE SPECIES

1a. Tendrils and/or panicles produced from 3–7 successive nodes; berries 1.0–2.5 cm wide; abaxial surface of the leaves persistently pubescent with rusty (later turning gray), tomentose hairs; axis of the inflorescence tomentose .... *V. labrusca*

1b. Tendrils and/or panicles produced from no more than 2 successive nodes; berries 0.5–1.2 cm wide; abaxial surface of the leaves glabrate to sparsely pubescent at maturity; axis of the inflorescence glabrous to sparsely pubescent

2a. Leaves, when young, pubescent on the abaxial surface between the veins with arachnoid hairs that lay parallel with the leaf surface, some spreading hairs may also be found on the veins; leaves commonly with teeth shorter than 5.0 mm; abaxial surface of leaves glaucous .... *V. aestivalis*

2b. Leaves pubescent on the abaxial surface with short, straight, spreading hairs, nearly glabrous at maturity except in the axils of veins; leaves with large, prominent teeth 5.0–20.0 mm long; abaxial surface of leaves green .... *V. riparia*

Vitis aestivalis Michx. var. *bicolor* Deam

synonym(s): *Vitis aestivalis* Michx. var. *argentifolia* (Munson) Fern.
common name(s): summer grape, pigeon grape
habit: climbing liana
range: ME, s. to GA, w. to TX, n. to MN and Ontario
state frequency: very rare
habitat: moist or dry, rocky areas, woods, thickets
notes: This species is ranked S1 by the Maine Natural Areas Program. Known to hybridize with *Vitis labrusca* (*q.v.*).

Vitis labrusca L.

synonym(s): —
common name(s): fox grape
habit: climbing liana
range: ME, s. to SC, w. to KY, n. to MI and Ontario
state frequency: occasional
habitat: woods, thickets, roadsides
notes: The hybrid with *Vitis aestivalis*, called *V. ×labruscana* Bailey is a cultivar that is sometimes planted in Maine. *Vitis labrusca* also hybridizes with *V. riparia*, yielding *V. ×novae-angliae* Fern., a relatively common hybrid in Maine that can be identified by its intermediate morphology.

Vitis riparia Michx.

synonym(s): —
common name(s): riverbank grape, frost grape
habit: climbing liana
range: Quebec, New Brunswick, and Nova Scotia, s. to VA, w. to TX, n. to MT and Manitoba
state frequency: occasional
habitat: river banks, thickets, moist woods
notes: Known to hybridize with *Vitis labrusca* (*q.v.*).

VISCACEAE
(1 genus, 1 species)

Arceuthobium
(1 species)

Arceuthobium pusillum Peck
synonym(s): —
common name(s): dwarf mistletoe, eastern dwarf mistletoe
habit: hemiparasitic shrublet
range: Newfoundland, s. to NJ, w. and n. to MI and Saskatchewan
state frequency: uncommon
habitat: branches of *Picea mariana*, *P. glauca*, sometimes *Larix laricina* and *Pinus strobus*
notes: —

SANTALACEAE
(2 genera, 2 species)

KEY TO THE GENERA
1a. Flowers borne in axillary cymes, lacking a hypanthium, usually in groups of 2–4, the
    lateral flowers functionally staminate and deciduous, the central flower bisexual; tepals
    green-purple; plants of bogs, coniferous forests, and alpine areas .... *Geocaulon*
1b. Flowers borne in terminal, panicle-like cymes, with a hypanthium, all the flowers
    bisexual; tepals white; plants of dry, sterile, often acidic, soils .... *Comandra*

Comandra
(1 species)

Comandra umbellata (L.) Nutt.
synonym(s): *Comandra richardsiana* Fern.
common name(s): bastard toadflax
habit: low, erect, rhizomatous, perennial herb
range: Newfoundland, s. to GA, w. to NM, n. to British Columbia
state frequency: uncommon
habitat: dry, sandy areas
notes: —

Geocaulon
(1 species)

Geocaulon lividum (Richards.) Fern.
synonym(s): —
common name(s): northern comandra
habit: erect, rhizomatous, perennial herb
range: Labrador, s. to ME, w. to MN, n. to AK
state frequency: rare
habitat: *Sphagnum* bogs, coniferous woods, alpine areas
notes: This species is ranked S2 by the Maine Natural Areas Program.

HAMAMELIDACEAE
(1 genus, 1 species)

Hamamelis
(1 species)

Shrubs with naked, scurfy-pubescent, scalpel-shaped buds.

Hamamelis virginiana L.
synonym(s): *Hamamelis virginiana* L. var. *parvifolia* Nutt.
common name(s): witch-hazel
habit: shrub to small tree
range: Quebec and Nova Scotia, s. to FL, w. to TX, n. to MN
state frequency: uncommon
habitat: moist or dry woods
notes: —

PENTHORACEAE
(1 genus, 1 species)

Penthorum
(1 species)

Penthorum sedoides L.
synonym(s): —
common name(s): ditch-stonecrop
habit: basally decumbent, simple or branched, stoloniferous, perennial herb
range: New Brunswick, s. to FL, w. to TX, n. to MN and Manitoba
state frequency: uncommon
habitat: marshes, muddy or gravelly shores, ditches
notes: —

HALORAGACEAE
(2 genera, 8 species)

KEY TO THE GENERA
1a. Perianth 3-merous; flowers with 3 stamens and 3 carpels; leaves alternate, the emersed
ones foliaceous .... *Proserpinaca*
1b. Perianth 4-merous; flowers with 4 or 8 stamens and 4 carpels; leaves opposite or
alternate, the emersed ones reduced and bract-like .... *Myriophyllum*

Myriophyllum
(6 species)

Water-milfoils are monoecious plants. The flowers in most species are borne on an emersed spike. The
carpellate flowers are found in the lower portion, and the staminate flowers are found in the upper portion

of the spike. Some of the species perennate by a turion. These are vegetatively plastic species showing a broad array of leaf lengths and divisions based on water depth.

KEY TO THE SPECIES

1a. Leaves, when present, and bracts entire, represented by small, scale-like bumps on the
    erect stems .... *M. tenellum*
1b. Leaves pinnately divided; bracts entire or, more commonly, toothed or divided
    2a. Foliage leaves regularly whorled; staminate flowers commonly with 8 stamens
        1.2–2.0 mm long; flowers in emersed spikes
        3a. Flowers and their subtending bracts alternate (sometimes the lowest opposite);
            leaves commonly with 3–7 pairs of narrow segments; spikes mostly 2.0–5.0 cm
            tall; fruits 1.5–2.0 mm long .... *M. alterniflorum*
        3b. Flowers and their subtending bracts whorled; leaves commonly with 6–13 pairs
            of narrow segments; spikes mostly 4.0–15.0 cm tall; fruits 2.0–3.0 mm long
            4a. Staminate flowers subtended by pinnately lobed bracts; carpellate bracts
                (1.0–)2.0 or more times as long as the flowers or fruits; stems only slightly
                whitening in drying .... *M. verticillatum*
            4b. Staminate flowers subtended by entire or toothed bracts; carpellate bracts
                about equaling the length of the flowers or fruits; stems whitening in drying ...
                .... *M. sibiricum*
    2b. Foliage leaves alternate, subopposite, or irregularly subverticillate; staminate flowers
        commonly with 4 stamens 0.4–0.6 mm long; flowers in the axils of submersed leaves
        or in emersed spikes
        5a. Mericarps with 2 tuberculate, longitudinal ridges on the outside [back] surface;
            plants producing terminal turions; fruit 2.0–2.5 mm long .... *M. farwellii*
        5b. Mericarps rounded on the outside surface, without ridges or projections; plants
            not producing turions; fruit 0.7–1.2 mm long .... *M. humile*

Myriophyllum alterniflorum DC.
synonym(s): —
common name(s): slender water-milfoil
habit: very slender, forked, aquatic, perennial herb
range: Greenland, s. to CT, w. and n. to British Columbia
state frequency: uncommon
habitat: shallow water of lakes, ponds, streams, and rivers
notes: —

Myriophyllum farwellii Morong
synonym(s): —
common name(s): Farwell's water-milfoil
habit: submerged, aquatic, perennial herb
range: Quebec, s. to ME, w. to MN, n. to Ontario
state frequency: uncommon
habitat: ponds, slow streams
notes: —

Myriophyllum humile (Raf.) Morong
synonym(s): —
common name(s): low water-milfoil
habit: terrestrial to submerged, aquatic, perennial herb

range: New Brunswick, s. to MD, w. to IL, n. to MN and Ontario
state frequency: rare
habitat: sandy, peaty, or muddy shores, ponds
notes: This species, like other water-milfoils, is highly variable vegetatively. It is typically found as a subterrestrial plant with pinnately lobed leaves. When entirely submerged, *Myriophyllum humile* may have finely divided leaves and elongate stems. It sometimes also produces an emersed spike instead of the typical submersed inflorescences.

Myriophyllum sibiricum Komarov
synonym(s): *Myriophyllum exalbescens* Fern.
common name(s): common water-milfoil, parrot feather
habit: simple or forked, emergent, aquatic, perennial herb
range: Labrador, s. to VA, w. to NM and CA, n. to AK
state frequency: rare
habitat: ponds, quiet streams, pools
notes: —

Myriophyllum tenellum Bigelow
synonym(s): —
common name(s): leafless water-milfoil
habit: creeping, usually simple, aquatic, perennial herb
range: Newfoundland, s. to NJ, w. to MN, n. to Ontario
state frequency: uncommon
habitat: sandy, muddy, peaty, or gravelly margins of ponds, lakes, and pools
notes: —

Myriophyllum verticillatum L.
synonym(s): *Myriophyllum verticillatum* L. var. *pectinatum* Wallr.
common name(s): comb water-milfoil
habit: simple or branched, emergent, aquatic, perennial herb
range: circumboreal; s. to MD, w. to TX and CA
state frequency: uncommon
habitat: shallow, quiet water of lakes, ponds, and streams
notes: —

Proserpinaca
(2 species)

KEY TO THE SPECIES
1a. Flowers subtended by serrate leaves; leaves 1.5–6.0 cm long, the submersed ones mostly with 8–14 pairs of divisions .... *P. palustris*
1b. Flowers subtended by pinnately lobed or divided leaves; leaves 1.0–3.0 cm long, mostly with 4–9 pairs of divisions .... *P. pectinata*

Proserpinaca palustris L. var. *crebra* Fern. & Grisc.
synonym(s): —
common name(s): common mermaidweed
habit: decumbent, emergent, aquatic, perennial herb
range: New Brunswick, s. to FL, w. to TX, n. to MN and Ontario
state frequency: uncommon
habitat: shallow water, wet shores
notes: —

Proserpinaca pectinata Lam.
synonym(s): —
common name(s): comb-leaved mermaidweed
habit: repent, emergent, aquatic, perennial herb
range: Nova Scotia, s. to FL, w. to TX
state frequency: rare
habitat: coastal bogs
notes: This species is ranked S3 by the Maine Natural Areas Program.

GROSSULARIACEAE
(1 genus, 10 species)

Ribes
(10 species)

KEY TO THE SPECIES

1a. Inflorescence a raceme; pedicels articulated at the summit, the fruits falling from them; twigs unarmed (armed in *R. lacustre*)

  2a. Ovary and fruit pubescent with bristly, glandular hairs

    3a. Stems with prickles and spines; racemes spreading to drooping; berries purple or black; petals broad-flabellate to orbicular .... ***R. lacustre***

    3b. Stems unarmed; racemes ascending to erect; berries dark red; petals cuneate-obovate .... ***R. glandulosum***

  2b. Ovary and fruit glabrous or with sessile, resinous glands

    4a. Ovary, fruit, buds, twig apices, and leaves (at least on the abaxial surface) resin-dotted; berries black; hypanthium above the ovary campanulate

      5a. Pedicels 0.0–2.0 mm long, much exceeded by the conspicuous, lanceolate bracts; sepals glabrous or sparsely pubescent on the abaxial surface .... ***R. americanum***

      5b. Pedicels 2.0–8.0 mm long, much longer than the minute, ovate bracts; sepals pubescent on the abaxial surface .... ***R. nigrum***

    4b. Ovary, fruit, buds, twig apices, and leaves without resin glands; berries red; hypanthium above the ovary saucer-shaped

      6a. Axis of raceme and pedicels glandular; anther sacs adjacent to one another; lateral leaf lobes commonly directed forward; terminal leaf lobe deltate; sepals green-purple; stems straggling to ascending .... ***R. triste***

      6b. Axis of raceme and pedicels without glands; anther sacs widely separated by a broad connective; lateral leaf lobes spreading; terminal leaf lobe ovate; sepals yellow to green; stems ascending to erect .... ***R. rubrum***

1b. Inflorescence a solitary flower or a corymb-like cluster of 2–5 flowers; pedicels not articulated at the summit, the fruits remaining on them; twigs usually armed with slender spines, prickles, and/or bristles, at least at the nodes

  7a. Berry covered with prickles at maturity; pedicels elongate, 5.0–16.0 mm long; tubular portion of the hypanthium above the ovary longer than the calyx lobes .... ***R. cynosbati***

7b. Berry glabrous, glandular, or bristly, but without prickles; pedicels short, up to 6.0 mm long; tubular portion of the hypanthium above the ovary equaling or shorter than the calyx lobes

    8a. Hypanthium, sepals, and fruit pubescent; hypanthium above the ovary campanulate to cupulate; sepals round-obovate .... *R. uva-crispa*

    8b. Hypanthium, sepals, and fruit glabrous (rarely the hypanthium and fruit pubescent in *R. hirtellum*); hypanthium above the ovary obconic to cylindric; sepals oblong to broad-oblong

        9a. Stamens *ca.* 2.0 times as long as the petals and approximately equaling the sepals; leaves without glands; internodal prickles often absent on the upper half of the stem .... *R. hirtellum*

        9b. Stamens *ca.* 1.0 times as long as the petals and shorter than the sepals; leaves glandular on the abaxial surface; internodal prickles commonly present on the upper half of the stem .... *R. oxyacanthoides*

Ribes americanum P. Mill.

synonym(s): —
common name(s): eastern black currant, wild black currant
habit: erect, unarmed shrub
range: New Brunswick, s. to DE, w. to NM, n. to Alberta
state frequency: uncommon
habitat: rich, moist woods and thickets
notes: —

Ribes cynosbati L.

synonym(s): —
common name(s): dogberry, prickly gooseberry
habit: prickly shrub
range: ME, s. to GA, w. to AR, n. to ND and Manitoba
state frequency: rare
habitat: moist, open or rocky woods
notes: —

Ribes glandulosum Grauer

synonym(s): *Ribes prostratum* L'Hér.
common name(s): skunk currant
habit: spreading to ascending shrub
range: Newfoundland, s. to ME and mountains of NC, w. to OH, n. to AK
state frequency: occasional
habitat: wet woods
notes: —

Ribes hirtellum Michx.

synonym(s): *Ribes hirtellum* Michx. var. *calcicola* (Fern.) Fern., *Ribes hirtellum* Michx. var. *saxosum* (Hook.) Fern.
common name(s): bristly gooseberry
habit: sparsely prickly to unarmed shrub
range: Labrador, s. to PA, w. to NE, n. to British Columbia
state frequency: occasional
habitat: rocky woods and openings, often near the coast
notes: —

Ribes lacustre (Pers.) Poir.
synonym(s): —
common name(s): spiny swamp currant, bristly black currant
habit: prickly shrub
range: Labrador, s. to PA and mountains of TN, w. to CA, n. to AK
state frequency: uncommon
habitat: cold, wet woods, swamps
notes: —

Ribes nigrum L.
synonym(s): —
common name(s): black currant, garden black currant
habit: erect, unarmed shrub
range: Eurasia; escaped from cultivation occasionally in the northeast
state frequency: rare
habitat: thickets, old gardens, cultivated land
notes: —

Ribes oxyacanthoides L.
synonym(s): —
common name(s): northern gooseberry
habit: prickly shrub
range: Hudson Bay, s. to MI, w. to NV, n. to AK; also ME
state frequency: rare
habitat: moist woods
notes: —

Ribes rubrum L.
synonym(s): *Ribes sativum* Syme, *Ribes vulgare* Lam.
common name(s): garden red currant, garden currant, red currant
habit: ascending to erect, unarmed shrub
range: Eurasia; escaped from cultivation
state frequency: uncommon
habitat: open woods, thickets
notes: —

Ribes triste Pallas
synonym(s): —
common name(s): swamp red currant, wild red currant
habit: straggling to ascending, unarmed shrub
range: Newfoundland, s. to NJ and WV, w. to OR, n. to AK
state frequency: uncommon; more common in northern Maine
habitat: cool, wet woods, bogs
notes: —

Ribes uva-crispa L. var. sativum DC.
synonym(s): *Ribes grossularia* L.
common name(s): European gooseberry, garden gooseberry
habit: prickly shrub
range: Europe; escaped from cultivation locally
state frequency: very rare
habitat: thickets, roadsides
notes: —

SAXIFRAGACEAE
(4 genera, 7 species)

KEY TO THE GENERA

1a. Perianth 4-merous; petals absent; at least the lower leaves opposite; plants low, decumbent herbs with forking stems .... *Chrysosplenium*
1b. Perianth 5-merous; petals present; leaves alternate or all basal; plants herbaceous, of various habits, the stems normally simple to the inflorescence
  2a. Petals conspicuously pectinately fringed .... *Mitella*
  2b. Petals entire
    3a. Ovary 2-locular, the placentation axile; leaf blades oblanceolate to obovate or lanceolate to ovate, simple, narrowed at the base .... *Saxifraga*
    3b. Ovary 1-locular, the placentation parietal; leaf blades broad-ovate to suborbicular, shallowly 3- to 5-lobed, cordate at the base .... *Tiarella*

Chrysosplenium
(1 species)

Chrysosplenium americanum Schwein. *ex* Hook.
synonym(s): —
common name(s): water carpet, golden-saxifrage
habit: low, creeping herb
range: Quebec and Nova Scotia, s. to VA and uplands of GA, w. to IA, n. to MN and Saskatchewan
state frequency: occasional
habitat: springy places, muddy soil
notes: —

Mitella
(1 species)

Mitella nuda L.
synonym(s): —
common name(s): naked miterwort
habit: rhizomatous, perennial herb
range: Labrador, s. to PA, w. to OH, n. to MT
state frequency: occasional
habitat: bogs, wet, mossy woods
notes: This species is vegetatively similar to, and sometimes grows with, *Rubus dalibarda*. However, *Mitella nuda* has leaves that have double-crenate or obscurely lobed margins and are lighter green. *Rubus dalibarda* has leaves that have evenly crenate margins and are dark green.

Saxifraga
(4 species)

KEY TO THE SPECIES

1a. Larger leaves 10.0–20.0 cm long; stems 3.0–10.0 dm tall; petals green-white or purple; plants of wet, low ground .... *S. pensylvanica*

1b. Leaves 1.0–5.0 cm long; stems 0.5–4.0 dm tall; petals white; plants of rocky woods, ledges, and alpine areas

    2a. Each tooth of the leaf with a conspicuous, white, lime-encrusted pore; stems with alternate leaves .... ***S. paniculata***

    2b. Teeth of leaves without lime-encrusted pores; stems scapose, bearing only reduced bracts

        3a. Inflorescence usually with a single, terminal flower, the lower flowers modified into vegetative bulbils; ovary superior, the hypanthium not adnate to the carpels; calyx lobes reflexed in fruit; plants of alpine habitats .... ***S. foliolosa***

        3b. Inflorescence bearing flowers only; ovary partly inferior, the hypanthium adnate to the base of the carpels; calyx lobes ascending or spreading in fruit; plants of non-alpine habitats .... ***S. virginiensis***

Saxifraga foliolosa R. Br.

synonym(s): *Hydatica foliolosa* (R. Br.) Small, *Saxifraga stellaris* L. var. *comosa* Poir.
common name(s): naked bulbil saxifrage
habit: perennial herb
range: circumboreal; s. in the east to ME
state frequency: very rare
habitat: mossy, alpine rocks
notes: This species is ranked S1 by the Maine Natural Areas Program.

Saxifraga paniculata P. Mill.

synonym(s): *Saxifraga aizoon* Jacq. var. *neogaea* Butters, *Chondrosea aizoon* (Jacq.) Haw.
common name(s): white alpine saxifrage
habit: perennial herb
range: circumboreal; s. to ME, w. to MN and Saskatchewan
state frequency: very rare
habitat: exposed, calcareous gravel and rocks
notes: This species is ranked S1 by the Maine Natural Areas Program.

Saxifraga pensylvanica L.

synonym(s): *Micranthes pensylvanica* (L.) Haw.
common name(s): swamp saxifrage
habit: erect, thick-rhizomatous, perennial herb
range: ME, s. to VA, w. to MO, n. to MN
state frequency: rare
habitat: wet meadows, bogs, swamps
notes: This species is ranked S2 by the Maine Natural Areas Program.

Saxifraga virginiensis Michx.

synonym(s): *Micranthes virginiensis* (Michx.) Small
common name(s): early saxifrage
habit: perennial herb
range: New Brunswick, s. to GA, w. to OK, n. to MN and Manitoba
state frequency: occasional
habitat: shaded, rocky ledges
notes: —

Tiarella
(1 species)

Tiarella cordifolia L.
synonym(s): —
common name(s): foam flower, false miterwort
habit: erect, rhizomatous, perennial herb
range: New Brunswick and Nova Scotia, s. to GA, w. to AL, n. to WI and Ontario
state frequency: occasional
habitat: rich, or sometimes moist, forests
notes: —

CRASSULACEAE
(3 genera, 10 species)

KEY TO THE GENERA

1a. Plants annual, aquatic or subaquatic; flowers solitary in the axils of the leaves; perianth
    3- or 4-merous; stamens 3 or 4 per flower; leaves connate around the stem ...... ***Crassula***
1b. Plants perennial, terrestrial; flowers usually borne in cymes; perianth 4- to 16-merous;
    stamens usually 8–32 per flower; leaves not connate around the stem
    2a. Perianth mostly 12- to 16-merous; stamens mostly 24–32 per flower; leaves in
        crowded, basal rosettes; plants monocarpic .... ***Sempervivum***
    2b. Perianth mostly 4- or 5-merous; stamens mostly 8 or 10 per flower; leaves borne on
        a stem; plants not monocarpic .... ***Sedum***

Crassula
(1 species)

Crassula aquatica (L.) Schoenl.
synonym(s): *Tillaea aquatica* L.
common name(s): pygmyweed
habit: branched, mat-forming, annual herb
range: Newfoundland, s. to MD, w. to UT, n. to AK; also LA and TX
state frequency: rare
habitat: muddy, often tidal, shores and pool margins
notes: This species is ranked S2 by the Maine Natural Areas Program.

Sedum
(8 species)

KEY TO THE SPECIES

1a. Plants dioecious or polygamous; staminate plants with yellow flowers; carpellate plants
    with yellow to purple flowers; follicles erect and connivent in fruit .... ***S. rosea***
1b. Plants synoecious, with white, yellow, or pink to purple flowers; follicles spreading in
    fruit

2a. Petals yellow or white-yellow; leaves of the flowering stems and branches terete or subterete in cross-section, linear-cylindric to ovoid

    3a. Leaves whorled [3 at each node]; sepals 3.5–5.0 mm long .... ***S. sarmentosum***

    3b. Leaves alternate [1 at each node]; sepals usually shorter or longer

        4a. Sepals 2.0–3.0 mm long; petals yellow; leaves 2.0–6.0 mm long .... ***S. acre***

        4b. Sepals 5.0–6.0 mm long; petals pale yellow or white-yellow; leaves 5.0–20.0 mm long .... ***S. ochroleucum***

2b. Petals white to pink to purple; leaves definitely flat, distally widened, elliptic or oblong to obovate

    5a. Leaves of the flowering stems 1.0–3.0 cm long; vegetative stems depressed, mat-forming; flowering stems upright, 1.0–2.0 dm tall

        6a. Inflorescence branched at the base into 2–4 secund cymes; leaves entire, those of the vegetative stems usually whorled; perianth mostly 4-merous .... ***S. ternatum***

        6b. Inflorescence repeatedly branched and not secund; leaves coarsely crenate in the apical portion, those of the vegetative stems usually opposite; perianth mostly 5-merous .... ***S. spurium***

    5b. Leaves 3.0–8.0 cm long; stems tufted, erect to ascending, 1.5–6.0 dm tall

        7a. Petals pink to red; leaves reduced in size from the base to the apex of the stem; sepals exceeding 5.0 mm long [mean = 6.0 mm]; flowers fertile .... ***S. telephium***

        7b. Petals pink-white; leaves only slightly reduced in size near the apex of the stem; sepals less than 5.0 mm long [mean = 2.5 mm]; flowers often sterile, lacking some or all of the stamens and carpels .... ***S. alboroseum***

Sedum acre L.

synonym(s): —
common name(s): mossy stonecrop, golden carpet
habit: creeping, mat-forming, evergreen, perennial herb
range: Eurasia; escaped from cultivation to Quebec, s. to FL, w. to IL, n. to WA and British Columbia
state frequency: uncommon
habitat: dry, open areas
notes: —

Sedum alboroseum Baker

synonym(s): *Sedum ×erythrostictum* Miq. *sensu* Gleason and Cronquist (1991)
common name(s): garden orpine
habit: spreading to ascending, perennial herb
range: Asia; escaped from cultivation to Newfoundland, s. to VA
state frequency: very rare
habitat: cultivated areas
notes: —

Sedum ochroleucum Chaix

synonym(s): *Sedum anopetalum* DC.
common name(s): —
habit: creeping, evergreen, perennial herb
range: Europe; escaped from cultivation locally
state frequency: very rare
habitat: waste places, fields
notes: —

Sedum rosea (L.) Scop.

synonym(s): *Rhodiola rosea* L.
common name(s): —
habit: erect, dioecious, perennial herb
range: circumboreal; s. to ME, w. to NM and CA; also NC
state frequency: rare
habitat: coastal cliffs, ledges, and shores
notes: —

Sedum sarmentosum Bunge

synonym(s): —
common name(s): stringy stonecrop
habit: slender, creeping, perennial herb
range: Asia; escaped from cultivation in the northeast
state frequency: very rare
habitat: roadsides, waste places, open woods
notes: —

Sedum spurium Bieb.

synonym(s): —
common name(s): —
habit: creeping, perennial herb
range: Eurasia; escaped from cultivation to Newfoundland, s. to PA
state frequency: rare
habitat: roadsides, banks, fields
notes: —

Sedum telephium L.

synonym(s): *Sedum purpureum* (L.) J. A. Schultes, *Sedum triphyllum* (Haw.) S. F. Gray
common name(s): live-forever, garden orpine
habit: coarse, fleshy, erect, perennial herb
range: Europe; escaped from cultivation to Newfoundland, s. to MD, w. to IN, n. to WI and Ontario;
    also Manitoba and British Columbia
state frequency: uncommon
habitat: roadsides, banks, open woods
notes: —

Sedum ternatum Michx.

synonym(s): —
common name(s): mountain stonecrop
habit: creeping, perennial herb
range: NJ, s. to GA, w. to AR, n. to IA; escaped from cultivation to ME, s. to CT, w. to NY
state frequency: very rare
habitat: damp rocks, woods, banks, and streamsides
notes: —

Sempervivum
(1 species)

Sempervivum tectorum L.

synonym(s): —
common name(s): hen and chickens

habit: fleshy, mat-forming, perennial herb
range: Europe; escaped from cultivation in ME and MA
state frequency: very rare
habitat: areas of cultivation
notes: —

DROSERACEAE
(1 genus, 4 species)

Drosera
(4 species)

A characteristic genus of insectivorous plants with leaves circinate in bud, like the ferns.

KEY TO THE SPECIES FOR USE WITH VEGETATIVE MATERIAL
1a. Leaf blades suborbicular, wider than long; petioles glandular-pubescent ....
.... ***D. rotundifolia***
1b. Leaf blades obovate to linear, longer than wide; petioles glabrous (sometimes sparsely
glandular-pubescent in *D. anglica*)
   2a. Stipules adnate to the petiole for much of their length; leaf blades 15.0–50.0 x
      1.5–4.0 mm
      3a. Leaf blades obovate to elongate-spatulate, 3.0–4.0 mm wide .... ***D. anglica***
      3b. Leaf blades linear, 1.5–3.0 mm wide .... ***D. linearis***
   2b. Stipules adnate for only *ca.* 1.0 mm, free for most of their length; leaf blades
      8.0–20.0 x 4.0–5.0 mm .... ***D. intermedia***

KEY TO THE SPECIES FOR USE WITH REPRODUCTIVE MATERIAL
1a. Seeds fusiform, 1.0–1.5 mm long, the testa loose and extending beyond the ovule at each
end
   2a. Seeds black, striate-alveolate .... ***D. anglica***
   2b. Seeds light brown, striate .... ***D. rotundifolia***
1b. Seeds narrow-ovoid, 0.5–1.0 mm long, the testa close and not extending beyond the ends
of the ovule
   3a. Seeds red-brown, papillose .... ***D. intermedia***
   3b. Seeds black, pitted .... ***D. linearis***

Drosera anglica Huds.
synonym(s): *Drosera longifolia* L.
common name(s): English sundew
habit: low, insectivorous, perennial herb
range: circumboreal; s. to ME, w. to CA
state frequency: very rare; known only from northern Maine
habitat: circumneutral and basic fens
notes: This species is ranked S1 by the Maine Natural Areas Program. Resembling a large *Drosera
intermedia*, *D. anglica* has strictly erect peduncles and petioles that are stipitate-glandular near the
apex. *Drosera intermedia* has basally geniculate peduncles and glabrous petioles. *Drosera anglica*
is thought to be derived through hybridization between *D. linearis* and *D. rotundifolia*.

Drosera intermedia Hayne

synonym(s): —
common name(s): spatulate-leaved sundew
habit: low, insectivorous, perennial herb
range: circumboreal; s. to FL, w. to TX and British Columbia
state frequency: occasional
habitat: wet places, shallow water
notes: Superficially similar to a small *Drosera anglica* (*q.v.*).

Drosera linearis Goldie

synonym(s): —
common name(s): slender-leaved sundew
habit: low, insectivorous, perennial herb
range: Newfoundland, s. to ME, w. to MT, n. to British Columbia
state frequency: very rare; known only from northern Maine
habitat: circumneutral and basic fens
notes: This species is ranked S1 by the Maine Natural Areas Program.

Drosera rotundifolia L.

synonym(s): —
common name(s): round-leaved sundew
habit: low, insectivorous, perennial herb
range: circumboreal; s. to FL, w. to CA
state frequency: occasional
habitat: wet, organic soil of bogs, shores, and ledges
key to the subspecies:

    1a. Inflorescence usually composed of 2–25 approximate, but not capitate-crowded, flowers; sepals green; petals white to pink; carpels normally developed .... *D. r.* var. *rotundifolia*
    1b. Inflorescence usually with 1–few crowded flowers in a capitate cluster; sepals red; petals green or red; carpels (and sometimes other parts of the flower) modified into tiny, stipitate-glandular leaves .... *D. r.* var. *comosa* Fern.

notes: —

PLUMBAGINACEAE
(1 genus, 1 species)

Limonium
(1 species)

Attractive seaside plants with a cluster of basal leaves and a leafless scape bearing numerous purple flowers.

Limonium carolinianum (Walt.) Britt.

synonym(s): *Limonium nashii* Small
common name(s): sea lavender, marsh-rosemary
habit: acaulescent, perennial herb
range: Labrador, s. to FL, w. to TX
state frequency: occasional
habitat: Atlantic coast shores
notes: —

POLYGONACEAE
(7 genera, 47 species)

KEY TO THE GENERA

1a. Blade of basal leaves 20.0–40.0 x 20.0–40.0 cm, cordate-ovate to deltate, with broad, stout petioles; achenes distinctly winged; stamens commonly 9 per flower .... **Rheum**
1b. Blade of basal or lower leaves not as above, commonly shorter and narrower, with slimmer petioles; achenes not winged; stamens commonly 8 or fewer per flower
  2a. Tepals 6, sepaloid or herbaceous, in 2 series of 3—the outer series spreading to reflexed, the inner series erect and distinctly enlarged in fruit .... **Rumex**
  2b. Tepals 5 (rarely 4 or 6), often petaloid, in a single series, all erect, equal or subequal in size in fruit
    3a. Flowers borne singly on pedicels that are articulated near the base and recurved in fruit; inflorescence a terminal raceme; leaves 5.0–20.0 x up to 1.0 mm, articulated to the summit of the ocrea, caducous, revolute, firm .... **Polygonella**
    3b. Flowers borne in various types of inflorescences, the pedicels articulated at or near the summit and not recurved in fruit (except *Polygonum douglasii*); leaves wider, articulated to the ocrea or not, not caducous, with various margins and textures
      4a. Mature achenes conspicuously surpassing the persistent but shriveled perianth; plants annual, with deltate or deltate-hastate leaf blades ....
      .... **Fagopyrum**
      4b. Mature achenes nearly or completely enclosed by the somewhat accrescent perianth; plants annual or perennial, with variously shaped leaf blades
        5a. Three tepals of the perianth prominently keeled or winged, especially in fruit; stems trailing to climbing or robust and subwoody, 1.0–5.0 m long or tall .... **Fallopia**
        5b. None of the tepals prominently keeled or winged, even in fruit; stems neither trailing nor climbing, not robust and subwoody, to 1.5(2.5) m tall
          6a. Flowers few to a cluster in the axils of leaves; ocreae 2-lobed, eventually lacerate; leaves articulated at the base of the petiole; filaments, at least the innermost, dilated; plants annual .... **Polygonum**
          6b. Flowers in terminal (and sometimes also axillary) racemes; ocreae various but not 2-lobed; leaves not articulated at the base of the petiole; filaments linear; plants annual or perennial .... **Persicaria**

Fagopyrum
(2 species)

KEY TO THE SPECIES

1a. Tepals green, mostly 3.0–4.0 mm long; achenes 5.0–7.0 mm long; inflorescences clustered near the apex of the plant .... **F. esculentum**
1b. Tepals white or cream, mostly 2.0–3.0 mm long; achenes 4.0–5.5 mm long; inflorescences scattered along the stem, not clustered near the apex of the plant ....
.... **F. tataricum**

Fagopyrum esculentum Moench
synonym(s): *Fagopyrum sagittatum* Gilib.
common name(s): buckwheat
habit: annual herb
range: Eurasia; escaped from cultivation locally
state frequency: uncommon
habitat: fields, roadsides, waste places
notes: —

Fagopyrum tataricum (L.) Gaertn.
synonym(s): —
common name(s): India-wheat
habit: annual herb
range: Eurasia; escaped from cultivation in the northeast
state frequency: uncommon
habitat: fields, waste places
notes: —

Fallopia
(5 species)

KEY TO THE SPECIES
1a. Stems trailing or climbing, slender, solid; stigmas capitate; perianth usually not
    accrescent
    2a. Ocreae reflexed-bristly at the base .... ***F. cilinodis***
    2b. Ocreae glabrous
        3a. Perianth strongly surpassing the mature achenes; 3 of the tepals conspicuously
            winged; achenes lustrous .... ***F. scandens***
        3b. Perianth scarcely surpassing the mature achenes; 3 of the tepals keeled, but not
            winged; achenes usually dull .... ***F. convolvulus***
1b. Stems erect, stout, with hollow internodes; stigmas fimbriate; perianth accrescent
    4a. Leaf blades broad-cuneate to truncate at the base, abruptly tapering to the apex,
        5.0–15.0 x 2.0–10.0 cm; flowers functionally unisexual .... ***F. japonica***
    4b. Leaf blades cordate at the base, gradually tapering to the apex, 10.0–30.0 x
        7.0–20.0 cm; flowers bisexual .... ***F. sachalinensis***

Fallopia cilinodis (Michx.) Holub
synonym(s): *Polygonum cilinode* Michx., *Tiniaria cilinodis* (Michx.) Small
common name(s): fringed bindweed
habit: trailing and twining to erect, perennial herb
range: Newfoundland, s. to mountains of NC, w. to TN, n. to Saskatchewan
state frequency: occasional
habitat: dry woods and thickets
notes: —

Fallopia convolvulus (L.) A. Löve
synonym(s): *Polygonum convolvulus* L., *Polygonum convolvulus* L. var. *subalatum* Lej. & Court.,
    Tiniaria convolvulus (L.) Webb & Moq.
common name(s): black bindweed
habit: trailing, twining, annual herb

range: Europe; naturalized to ME, s. to PA, w. and n. to MI
state frequency: uncommon
habitat: roadsides, waste places
notes: —

Fallopia japonica (Houtt.) Decraene
synonym(s): *Pleuropterus zuccarinii* (Small) Small, *Polygonum cuspidatum* Sieb. & Zucc.
common name(s): Japanese knotweed, bamboo
habit: tall, stout, branched, perennial herb
range: Asia; escaped from cultivation to Newfoundland, s. to MD, w. to MN, n. to Ontario
state frequency: common
habitat: waste places, old gardens
notes: —

Fallopia sachalinensis (F. S. Petro.) Decraene
synonym(s): *Polygonum sachalinense* F. Schmidt *ex* Maxim., *Reynoutria sachalinensis* (F. Schmidt
    ex Maxim.) Nakai
common name(s): giant knotweed
habit: tall, stout, branched, perennial herb
range: Asia; escaped from cultivation to ME, s. to MD
state frequency: rare
habitat: waste places
notes: —

Fallopia scandens (L.) Holub
synonym(s): *Polygonum scandens* L., *Tiniaria scandens* (L.) Small
common name(s): false buckwheat, climbing false buckwheat
habit: twining, perennial herb
range: Quebec, s. to FL, w. to TX, n. to ND and Manitoba
state frequency: uncommon
habitat: moist thickets and woods, roadsides
notes: —

Persicaria
(14 species)

The underground portion of the stem is vital for assessing whether plants are annual or perennial, an
important separation in the following key.

KEY TO THE SPECIES
1a. Stems armed with reflexed prickles; leaf blades sagittate or hastate at the base
    (sometimes the uppermost without basal lobes)
    2a. Leaf blades lanceolate to elliptic, 0.5–3.1 cm wide, sagittate at the base, the basal
        lobes directed backward; styles trifid, 1.0–1.5 mm long; achenes trigonous ....
        .... *P. sagittata*
    2b. Leaf blades deltate, 1.0–15.0 cm wide, hastate at the base, the basal lobes divergent;
        styles bifid, 0.5–0.6 mm long; achenes biconvex .... *P. arifolia*
1b. Stems unarmed; leaf blades cuneate to cordate, lacking distinct basal lobes
    3a. Leaves chiefly basal; stems simple, with a single inflorescence; plants perennial

4a. Raceme 3.0–6.0 x 0.5–1.0 cm, bearing sterile flowers in the apical half and
    bulbils in the basal half; leaves 2.0–10.0 cm long, narrowed at the base; plants
    native, of alpine habitats, elevation exceeding 1000 m .... *P. vivipara*
4b. Raceme 5.0–9.0 x 1.0–2.0 cm, bearing only fertile flowers; leaves 6.0–20.0 cm
    long, rounded to cordate at the base; plants escaped from cultivation, not of
    alpine habitats .... *P. bistorta*
3b. Leaves chiefly cauline; stems usually branched, with several inflorescences (except
    P. amphibia with 1 or 2 inflorescences); plants annual or perennial
    5a. Plants perennial with rhizomes or stolons
        6a. Racemes terminal, solitary or paired, densely flowered, 10.0–20.0 mm wide;
            tepals deep pink to red .... *P. amphibia*
        6b. Racemes terminal and often also axillary, usually remotely flowered at the
            base, mostly 2.0–5.0 mm wide; tepals pink, white, and/or green
            7a. Tepals spotted with glandular dots; inflorescence 3.0–15.0 cm long,
                sparsely flowered, with many of the bracteoles non-overlapping ....
                .... *P. punctata*
            7b. Tepals not spotted with glandular dots; inflorescence 3.0–8.0 cm long,
                more densely flowered, the bracteoles overlapping except at the very base
                .... *P. hydropiperoides*
    5b. Plants annual from fibrous roots and/or a taproot
        8a. Sheathing stipules [ocreae] ciliate with a prominent ring of bristles around the
            summit
            9a. Tepals conspicuously spotted with glandular dots
                10a. Perianth consisting of 4(5) tepals; inflorescences terminating the
                    main stem and often axillary, commonly with overlapping bracteoles
                    [ocreolae], interrupted by small leaves; achenes not lustrous ....
                    .... *P. hydropiper*
                10b. Perianth consisting of 5 tepals; inflorescences terminating the main
                    stem or branches, commonly with non-overlapping bracteoles (at least
                    in the basal portion), without small leaves; achenes lustrous ....
                    .... *P. punctata*
            9b. Tepals not spotted with glandular dots
                11a. Leaf blades ovate, 3.0–16.0 cm wide; sheathing stipules [ocreae]
                  horizontally divergent at the summit creating a conspicuous flange ......
                  .... *P. orientalis*
                11b. Leaf blades narrow-lanceolate to elliptic or oblanceolate, 1.0–3.0 cm
                  wide; sheathing stipules without a horizontally spreading summit
                    12a. Upper portion of the stem and peduncles stipitate-glandular;
                      inflorescence 3.0–10.0 cm long, lax or drooping .... *P. careyi*
                    12b. Upper portion of the stem and peduncles without stipitate glands;
                      inflorescence 1.0–4.5 cm long, commonly erect to ascending
                        13a. Inflorescence densely flowered near the base, 7.0–12.0 mm
                        wide; bracteoles of the flowers [ocreolae] with glabrous or
                        short-ciliate (less than 1.0 mm) summits; leaves often with
                        purple blotches; achenes commonly lenticular ..... *P. maculosa*

13b. Inflorescence often remotely flowered near the base, 3.0–6.0 mm wide; bracteoles of the flowers with long-ciliate (2.0–3.5 mm) summits; leaves without purple blotches; achenes trigonous .... *P. caespitosa*
8b. Sheathing stipules glabrous or lacerate into segments, lacking a prominent ring of cilia around the summit
14a. Perianth consisting of 4(5) tepals; tepals conspicuously 3-nerved, each nerve forking near the end into 2 recurved branches .... *P. lapathifolia*
14b. Perianth consisting of 5 tepals; tepals obscurely nerved ....
.... *P. pensylvanica*

Persicaria amphibia (L.) S. F. Gray
synonym(s): <*P. a.* var. *emersa*> = *Polygonum amphibium* L., *Polygonum coccineum* Muhl. *ex* Willd.; <*P. a.* var. *stipulacea*> = *Polygonum amphibium* L. var. *stipulaceum* Coleman
common name(s): water smartweed
habit: rhizomatous, aquatic, amphibious, or terrestrial, perennial herb
range: Eurasia and southern Africa; Newfoundland, s. to NJ, w. to UT, n. to AK
state frequency: occasional
habitat: shallow water, wet shores and meadows
key to the subspecies:
1a. Inflorescence up to 4.0 cm long; leaves more or less elliptic, usually glabrous, acute at the apex, thinner, and floating; sheathing stipules lacking an outward-flanging summit; plants aquatic ....
.... *P. a.* var. *stipulacea* (Coleman) Hara
1b. Inflorescence 4.0–15.0 cm long; leaves more or less lanceolate, often pubescent, acuminate at the apex, thicker, not floating; sheathing stipules usually with a horizontal flange at the summit; plants terrestrial or strongly emergent .... *P. a.* var. *emersa* (Michx.) Hickman
notes: *Persicaria amphibia* var. *amphibia* is native to Eurasia and is naturalized to North America.

Persicaria arifolia (L.) Haroldson
synonym(s): *Polygonum arifolium* L. var. *pubescens* (Keller) Fern., *Tracaulon arifolium* (L.) Raf.
common name(s): halberd-leaved tearthumb
habit: reclining to erect, annual herb
range: New Brunswick, s. to GA, w. to MO, n. to MN and Ontario
state frequency: uncommon
habitat: wet meadows, swamps, marshes
notes: —

Persicaria bistorta (L.) Sampaio
synonym(s): *Polygonum bistorta* L.
common name(s): bistort
habit: rhizomatous, perennial herb
range: Eurasia; escaped from cultivation to Nova Scotia, s. to MA
state frequency: very rare
habitat: thickets, fields, meadows, cultivated ground
notes: —

Persicaria caespitosa (Blume) Nakai
synonym(s): *Polygonum cespitosum* Blume var. *longisetum* (de Bruyn) A. N. Steward
common name(s): —
habit: simple or branched, prostrate to ascending, annual herb
range: Asia; naturalized to New Brunswick, s. to FL, w. and n. to British Columbia

state frequency: very rare
habitat: roadsides, waste places, damp areas
notes: —

Persicaria careyi (Olney) Greene
synonym(s): *Polygonum careyi* Olney
common name(s): Carey's smartweed
habit: erect to ascending, branched or simple, annual herb
range: New Brunswick, s. to FL, w. and n. to IN and MN
state frequency: uncommon
habitat: moist to wet fields and roadsides
notes: —

Persicaria hydropiper (L.) Opiz
synonym(s): *Polygonum hydropiper* L.
common name(s): water-pepper, common smartweed
habit: simple or branched, erect to spreading, annual herb
range: Europe; naturalized to Newfoundland, s. to AL, w. to CA, n. to British Columbia
state frequency: occasional
habitat: damp to wet soil, open areas
notes: —

Persicaria hydropiperoides (Michx.) Small
synonym(s): *Polygonum hydropiperoides* Michx.
common name(s): false water-pepper, mild water-pepper, mild smartweed
habit: decumbent to erect, rhizomatous, perennial herb
range: Quebec, s. to FL, w. to TX and Mexico, n. to AK
state frequency: uncommon
habitat: wet, muddy soil, shallow water
notes: —

Persicaria lapathifolia (L.) S. F. Gray
synonym(s): *Persicaria lapathifolia* L. ssp. *salicifolia* Sibthorp, *Polygonum lapathifolium* L.,
Polygonum lapathifolium L. var. *salicifolium* Sibthorp, *Polygonum scabrum* Moench
common name(s): dock-leaved smartweed, willow-weed
habit: erect to depressed, usually branched, annual herb
range: Greenland, s. to FL, w. to CA, n. to British Columbia
state frequency: occasional
habitat: gravelly shores, damp clearings, cultivated fields
notes: —

Persicaria maculosa S. F. Gray
synonym(s): *Persicaria persicaria* (L.) Small, *Persicaria vulgaris* Webb & Moq., *Polygonum*
dubium Stein, *Polygonum persicaria* L., *Polygonum persicaria* L. var. *ruderale* (Salisb.) Meisn.,
Polygonum puritanorum Fern.
common name(s): lady's thumb
habit: erect to ascending, simple or branched, annual herb
range: Europe; naturalized to Greenland, s. to FL, w. to CA, n. to AK
state frequency: common
habitat: wet, sandy or gravelly shores, swamps
notes: —

Persicaria orientalis (L.) Spach

synonym(s): *Polygonum orientale* L.
common name(s): prince's feather
habit: tall, erect, branched, annual herb
range: Eurasia; escaped from cultivation in our area
state frequency: rare
habitat: waste places
notes: —

Persicaria pensylvanica (L.) G. Maza

synonym(s): *Polygonum pensylvanicum* L. var. *laevigatum* Fern.
common name(s): pinkweed, Pennsylvania smartweed
habit: ascending to erect, branched, annual herb
range: New Brunswick, s. to FL, w. to TX and CO, n. to ND
state frequency: occasional
habitat: damp shores, fields, and waste places
notes: —

Persicaria punctata (Ell.) Small

synonym(s): *Polygonum punctatum* Ell., *Polygonum punctatum* Ell. var. *leptostachyum* (Meisn.)
    Small, *Polygonum punctatum* Ell. var. *parvum* Victorin & Rouss.
common name(s): dotted smartweed, water smartweed
habit: erect to ascending, simple or branched, annual or perennial herb
range: New Brunswick and Nova Scotia, s. to FL, w. to CA, n. to British Columbia
state frequency: uncommon
habitat: open swamps, shallow water, wet, sandy or muddy shores
notes: —

Persicaria sagittata (L.) H. Gross

synonym(s): *Polygonum sagittatum* L., *Tracaulon sagittatum* (L.) Small
common name(s): arrow-leaved tearthumb
habit: reclining to erect, branched, annual herb
range: Newfoundland, s. to FL, w. to TX, n. to Saskatchewan
state frequency: occasional
habitat: wet meadows and ditches, low ground
notes: —

Persicaria vivipara (L.) S. F. Gray

synonym(s): *Bistorta vivipara* (L.) S. F. Gray, *Polygonum viviparum* L.
common name(s): alpine bistort, serpent-grass
habit: erect, simple, perennial herb
range: circumboreal; s. to mountains of New England, w. to MN
state frequency: very rare
habitat: wet, mossy rocks and seeps in alpine areas
notes: This species is ranked S1 by the Maine Natural Areas Program.

Polygonella
(1 species)

Polygonella articulata (L.) Meisn.

synonym(s): —
common name(s): jointweed

habit: erect, branched, annual herb
range: New Brunswick, s. to NC, w. to IA, n. to MN and Ontario
state frequency: uncommon
habitat: dry sands, roadsides
notes: —

Polygonum
(11 species)

Plants in this genus have fascicles of flowers produced in the axils of the leaves. If the leaves subtending the flowers are subequal in size to the ordinary, foliage leaves, the plants are termed homophyllous. If the subtending leaves are smaller than the foliage leaves, the plants are termed heterophyllous. Most of our species have green tepals with a white or pink margin. Only *Polygonum erectum* and forms of *P. ramosissimum* have tepals that are yellow-green to green at the margin. Achene texture requires 30× magnification for accurate assessment. Achenes produced late in the season may be of a different size, texture, and color from those produced during the summer. The fruiting key utilizes summer achenes.

KEY TO THE SPECIES FOR USE WITH FLOWERING MATERIAL
1a. Plants heterophyllous
    2a. Pedicels abruptly recurved, the flowers nodding .... *P. douglasii*
    2b. Pedicels straight or nearly so, the flowers erect
        3a. Leaves pleated with 2 longitudinal folds, minutely serrate on the margin ....
        .... *P. tenue*
        3b. Leaves without folds, entire or obscurely erose on the margin
            4a. Three of the tepals cucullate, longer than and partially concealing the other 2 flat ones
                5a. Leaves linear to lanceolate or oblanceolate, mostly 1.0–5.0 mm wide, 4.0–12.0 times as long as wide
                    6a. Pedicels 2.0–3.5 mm long; leaf apex obtuse to acuminate; plants strongly heterophyllous .... *P. ramosissimum*
                    6b. Pedicels less than 2.0 mm long; leaf apex rounded to obtuse; plants weakly heterophyllous .... *P. prolificum*
                5b. Leaves elliptic to ovate, 10.0–30.0 mm wide, 2.0–3.0 times as long as wide .... *P. erectum*
            4b. All the tepals flat or merely keeled, subequal .... *P. aviculare*
1b. Plants homophyllous
    7a. Three of the tepals cucullate, longer than and partially concealing the other 2 flat ones
        8a. Tepals connate for 0.65 times their length, only the tips of the tepals distinct; plants of inland habitats, only rarely coastal .... *P. achoreum*
        8b. Tepals connate for less than 0.5 times their length; plants of coastal habitats, only rarely inland
            9a. Leaves oblong or elliptic-oblong to oblanceolate, 1.0–20.0 mm wide, mostly 2.0–4.0 times as long as wide
                10a. Tepals 2.0–3.0 mm long; plants prostrate and mat-forming ....
                .... *P. buxiforme*
                10b. Tepals 3.0–5.0 mm long; plants prostrate to ascending .... *P. fowleri*

9b. Leaves linear-oblong to oblanceolate 1.0–5.0 mm wide, 4.0–12.0 times as
    long as wide .... *P. prolificum*
7b. All the tepals flat or merely keeled, subequal
   11a. Leaves oval-oblong to broad-elliptic, mostly 2.0–5.0 times as long as wide;
      plants prostrate .... *P. arenastrum*
   11b. Leaves linear-lanceolate to linear-oblong, mostly 5.0–9.0 times as long as wide;
      plants ascending .... *P. neglectum*

KEY TO THE SPECIES FOR USE WITH FRUITING MATERIAL

1a. Stems and branches sharply quadrangular, erect; fruits in the axils of tiny leaves, giving
   the appearance of a terminal infructescence
   2a. Pedicels straight, the fruits erect; leaves pleated with 2 longitudinal folds, minutely
      serrate on the margin .... *P. tenue*
   2b. Pedicels abruptly recurved, the fruits nodding; leaves without longitudinal folds, with
      entire margins .... *P. douglasii*
1b. Stems and branches terete or 8- to 16-angled, prostrate to ascending; fruits in the axils of
   normal or slightly reduced leaves
   3a. Achenes papillose or striate-papillose
      4a. Achenes uniformly papillose, yellow-green to tan .... *P. achoreum*
      4b. Achenes striate-papillose, dark brown to black
         5a. Plants heterophyllous
            6a. Leaves lanceolate to ovate-lanceolate, each with an acute apex, gray-
               green, deciduous; fruiting perianth not strongly constricted above the
               achene .... *P. aviculare*
            6b. Leaves elliptic to ovate, each with an obtuse apex, green or, more
               commonly, yellow-green, persistent; fruiting perianth strongly constricted
               above the achene .... *P. erectum*
         5b. Plants homophyllous
            7a. Three of the tepals cucullate, longer than and partially concealing the
               other 2 flat ones; plants of coastal habitats, only rarely inland ....
               .... *P. buxiforme*
            7b. All the tepals flat, subequal; plants of inland habitats, only rarely coastal
               8a. Leaves oval-oblong to broad-elliptic, 2.0–5.0 times as long as wide;
                  plants prostrate .... *P. arenastrum*
               8b. Leaves linear-lanceolate to linear-oblong, mostly 5.0–9.0 times as
                  long as wide; plants ascending .... *P. neglectum*
   3b. Achenes smooth, irregularly roughened or granular-roughened, but not papillose
      9a. Achenes abruptly short-beaked, uniformly granular-roughened .... *P. fowleri*
      9b. Achenes gradually tapering to the tip, smooth or irregularly roughened
         10a. Flowers evidently exserted from their bracteoles [ocreolae], on pedicels
            2.0–3.5 mm long; plants heterophyllous .... *P. ramosissimum*
        10b. Flowers barely or not exserted from their bracteoles, on pedicels less than
            2.0 mm long; plants homophyllous or nearly so .... *P. prolificum*

Polygonum achoreum Blake

synonym(s): —
common name(s): leathery knotweed

habit: erect to prostrate, branched, annual herb
range: Newfoundland, s. to NY, w. to MO and ID, n. to AK
state frequency: rare
habitat: waste places, roadsides
notes: —

Polygonum arenastrum Bor.

synonym(s): —
common name(s): dooryard knotweed
habit: prostrate, often mat-forming, branched, annual herb
range: Europe; naturalized throughout much of North America
state frequency: occasional
habitat: dooryards, sidewalks
notes: *Polygonum arenastrum* may have olive-green, smooth achenes in late season. Similar to *P. aviculare*, and sometimes confused with it. In addition to leaf morphology, tepal and achene size can be useful for separating these species. *Polygonum arenastrum* has tepals 1.8–2.6 mm long and achenes 1.5–2.3 mm long. *Polygonum aviculare* has tepals 2.5–4.0 mm long and achenes 2.2–3.2 mm long.

Polygonum aviculare L.

synonym(s): *Polygonum aviculare* L. var. *vegetum* Ledeb.
common name(s): prostrate knotweed
habit: prostrate to ascending, branched, annual herb
range: Europe; naturalized throughout North America
state frequency: occasional
habitat: roadsides, waste places, cultivated ground
notes: Similar to *Polygonum arenastrum* (*q.v.*).

Polygonum buxiforme Small

synonym(s): *Polygonum aviculare* L. var. *littorale* (Link) Mert.
common name(s): —
habit: prostrate, mat-forming, branched, annual herb
range: Newfoundland, s. to SC, w. to OK, n. to AK
state frequency: uncommon
habitat: saltmarshes, sand dunes, alkaline soil
notes: —

Polygonum douglasii Greene

synonym(s): —
common name(s): Douglas' knotweed
habit: erect to ascending, branched, annual herb
range: ME, s. to NY, w. to NM and CA, n. to British Columbia
state frequency: very rare
habitat: rocky or gravelly slopes and open areas
notes: This species is ranked S1 by the Maine Natural Areas Program.

Polygonum erectum L.

synonym(s): —
common name(s): —
habit: stout, erect to prostrate, branched, annual herb
range: ME, s. to GA, w. to KS, n. to IA and WI
state frequency: uncommon
habitat: waste places
notes: —

Polygonum fowleri B. L. Robins.
synonym(s): *Polygonum allocarpum* Blake
common name(s): Fowler's knotweed
habit: prostrate to ascending, somewhat branched, annual herb
range: Labrador, s. to ME; also AK, s. to WA
state frequency: uncommon
habitat: sea beaches, saltmarshes
notes: —

Polygonum neglectum Bess.
synonym(s): *Polygonum aviculare* L. var. *angustissimum* Meisn., *Polygonum rurivagum* Jord. *ex*
    Boreau
common name(s): —
habit: prostrate to ascending, branched, annual herb
range: Eurasia; naturalized throughout much of North America
state frequency: very rare
habitat: roadsides, waste places
notes: —

Polygonum prolificum (Small) B. L. Robins.
synonym(s): *Polygonum ramosissimum* Michx. var. *prolificum* Small
common name(s): —
habit: erect to ascending, sometimes prostrate, branched, annual herb
range: Nova Scotia and ME, s. to VA, w. to TX, n. to MN and Saskatchewan; also WA, s. to NM
state frequency: rare
habitat: saline or brackish shores and marshes
notes: —

Polygonum ramosissimum Michx.
synonym(s): *Polygonum atlanticum* (B. L. Robins.) Bickn., *Polygonum exsertum* Small *sensu* Fernald
    (1950)
common name(s): bushy knotweed
habit: erect to ascending, branched, annual herb
range: New Brunswick, s. to DE, w. to NM, n. to WA and British Columbia
state frequency: uncommon
habitat: brackish or saline marshes, sandy soils
notes: Summer achenes are trigonous, 2.0–3.5 mm long. Those produced late in the season are
    lenticular, 4.0–6.5 mm long.

Polygonum tenue Michx.
synonym(s): —
common name(s): slender knotweed
habit: erect to ascending, branched, annual herb
range: ME, s. to GA, w. to TX, n. to MN
state frequency: very rare
habitat: dry, open, often acid, soil
notes: This species is ranked SH by the Maine Natural Areas Program.

Rheum
(1 species)

Rheum rhaponticum L.
synonym(s): —
common name(s): rhubarb
habit: coarse, perennial herb
range: Asia; escaped from cultivation
state frequency: uncommon
habitat: old homesites, cultivated areas
notes: —

Rumex
(13 species)

In *Rumex*, the inner whorl of tepals [at maturity called valves] often enlarge in fruit. The midrib of the inner tepals in some species swells into an evident protuberance called a tubercle. Hybrids occur in the genus and generally possess an intermediate morphology.

KEY TO THE SPECIES
1a. Leaves sagittate or hastate, acid tasting; plants dioecious
    2a. Leaves predominantly hastate; plants arising from long, slender rhizomes; carpellate perianth closely enclosing the achenes at maturity, neither greatly exceeding the achenes nor prominently reticulate-veiny, lacking tubercles on the midribs; staminate perianth 1.5–2.0 mm long; achenes 1.0–1.5 mm long .... *R. acetosella*
    2b. Leaves predominantly sagittate; plants arising from a stout root; carpellate perianth [the 3 outer tepals] greatly exceeding the achenes, reticulate-veiny, developing a tubercle on the midrib of 1 of the 3 inner tepals; staminate perianth 2.0–3.0 mm long; achenes 2.0–2.5 mm long .... *R. acetosa*
1b. Leaves without basal lobes, mostly not acid tasting; plants synoecious, monoecious, or polygamo-monoecious
    3a. Inner tepals [valves] with prominent teeth or spines on the margins
        4a. All 3 of the inner tepals with an enlarged tubercle on the midrib; margins of the inner tepals with slender spines; plants fibrous-rooted annuals .... *R. maritimus*
        4b. Only 1 of the 3 inner tepals with an enlarged tubercle on the midrib; margins of the inner tepals with short teeth; plants taprooted perennials .... *R. obtusifolius*
    3b. Inner tepals with entire margins
        5a. Stems with lateral branches and/or axillary leaf tufts; leaves usually pale-green or glaucous, with flat, entire margins
            6a. Inner tepals with a broad tubercle, the tubercle more than half as wide and nearly as long as its associated inner tepal; plants mostly of coastal habitats .... *R. pallidus*
            6b. Inner tepals with a narrow tubercle, the tubercle less than half as wide and much shorter than its associated tepal; plants mostly of inland habitats
                7a. Inner tepals approximately broad-ovate, each with an obtuse apex, usually only 1 or 2 of the 3 inner tepals forming a tubercle; leaf blades lanceolate to, more commonly, lance-ovate .... *R. altissimus*

7b. Inner tepals approximately deltate, each with an acute apex, all 3 of the inner tepals forming a tubercle; leaf blades narrow-lanceolate .... *R. salicifolius*

5b. Stems lacking both lateral branches and prominent axillary leaf tufts; leaves dark green to red-green, with crenate, undulate, and/or crisped margins

8a. Plants with rhizomes; leaf blades round-ovate to reniform, as long as wide, or nearly so; stems without a basal rosette of leaves; inner tepals without tubercles .... *R. alpinus*

8b. Plants with taproots, lacking rhizomes; leaf blades oblanceolate to lanceolate, more than 2.0 times as long as wide; stems with a basal rosette; inner tepals with or without tubercles

9a. Inner tepals reniform, wider than long, lacking well developed tubercles .... *R. longifolius*

9b. Inner tepals ovate to ovate-orbicular, longer than wide, with or without tubercles

10a. Inner tepals lacking well developed tubercles .... *R. aquaticus*

10b. At least 1 of the inner tepals with a well developed tubercle

11a. Lower margin of the tubercle located above the base of the inner tepal; pedicels without a distinctly swollen node at the articulation point; plants of wetlands and shores .... *R. orbiculatus*

11b. Lower margin of the tubercle located at or below the base of the inner tepal; pedicels with a distinctly swollen node at the articulation point; plants mostly of disturbed habitats

12a. Only 1 of the inner tepals with a well developed tubercle, the tubercle less than 0.5 times as long as its associated inner tepal; inner tepals 6.0–9.0 mm long .... *R. patientia*

12b. One to 3 of the inner tepals with a well developed tubercle, the tubercle more than 0.5 times as long as its associated inner tepal; inner tepals 4.0–5.0 mm long .... *R. crispus*

Rumex acetosa L.
synonym(s): —
common name(s): garden sorrel, green sorrel
habit: simple, dioecious, perennial herb
range: Eurasia; naturalized to Labrador, s. to PA, w. to British Columbia, n. to AK
state frequency: uncommon
habitat: fields, meadows, roadsides
notes: —

Rumex acetosella L.
synonym(s): —
common name(s): sheep sorrel, common sorrel, red sorrel
habit: dioecious, long-rhizomatous, perennial herb
range: Eurasia; naturalized throughout North America
state frequency: common
habitat: dry fields, lawns, and waste places
notes: —

Rumex alpinus L.

synonym(s): —
common name(s): —
habit: stout, rhizomatous, perennial herb
range: Europe; naturalized to Nova Scotia and ME, w. to VT
state frequency: very rare
habitat: fields, meadows
notes: —

Rumex altissimus Wood

synonym(s): —
common name(s): pale dock
habit: tall, stout, often branched, perennial herb
range: ME, s. to GA, w. to AZ, n. to MN
state frequency: very rare
habitat: swamps, rich, wet, alluvial soil
notes: —

Rumex aquaticus L. var. *fenestratus* (Greene) Dorn

synonym(s): *Rumex fenestratus* Greene, *Rumex occidentalis* S. Wats.
common name(s): western dock
habit: tall, stout, perennial herb
range: Labrador, s. to ME, w. to CA, n. to AK
state frequency: very rare
habitat: wet soil of swales and shores
notes: —

Rumex crispus L.

synonym(s): —
common name(s): curly dock, yellow dock
habit: stout, taprooted, perennial herb
range: Europe; naturalized throughout much of North America
state frequency: common
habitat: roadsides, fields, cultivated areas, waste places
notes: Hybrids with *Rumex obtusifolius* occur in Maine. Also known to hybridize with *R. patientia*.

Rumex longifolius DC.

synonym(s): *Rumex domesticus* Hartman
common name(s): yard dock, long-leaved dock
habit: erect, coarse, perennial herb
range: Europe; naturalized to Greenland, s. to CT, w. to British Columbia, n. to AK
state frequency: uncommon
habitat: fields, roadsides, waste places
notes: —

Rumex maritimus L.

synonym(s): *Rumex maritimus* L. var. *fueginus* (Phil.) Dusen
common name(s): golden dock
habit: hollow-stemmed, branched, annual herb
range: Quebec, s. to coastal NY, w. to AZ and CA, n. to British Columbia
state frequency: rare
habitat: saline to brackish marshes and shores
notes: —

Rumex obtusifolius L.

synonym(s): —
common name(s): bitter dock, blunt-leaved dock, red-veined dock
habit: stout, erect, simple, perennial herb
range: Europe; naturalized to Quebec, s. to FL, w. to AZ, n. to British Columbia
state frequency: occasional
habitat: moist, shaded roadsides and waste places
notes: Hybrids with *Rumex crispus* occur in Maine. Also known to hybridize with *R. patientia*.

Rumex orbiculatus Gray

synonym(s): *Rumex britannica* L. *sensu* Britton and Brown (1913)
common name(s): water dock, great water dock
habit: tall, stout, simple, perennial herb
range: Newfoundland, s. to NJ, w. to NE, n. to ND
state frequency: uncommon
habitat: wet meadows, swamps, shores, shallow water
notes: —

Rumex pallidus Bigelow

synonym(s): —
common name(s): seabeach dock, white dock
habit: ascending to decumbent, branched, perennial herb
range: Newfoundland, s. to coastal NY, w. and n. to Quebec
state frequency: uncommon
habitat: saline to brackish marshes, swamps, and rocky and sandy areas
notes: —

Rumex patientia L.

synonym(s): —
common name(s): patience dock
habit: stout, simple, perennial herb
range: Eurasia; naturalized to ME, s. to FL, w. to OK, n. to MN
state frequency: rare
habitat: rich thickets, roadsides, waste places
notes: Known to hybridize with *Rumex crispus* and *R. obtusifolius*.

Rumex salicifolius Weinm. var. mexicanus (Meisn.) C. L. Hitchc.

synonym(s): *Rumex mexicanus* Meisn., *Rumex salicifolius* Weinm. var. *triangulivalvis* (Danser) C. L. Hitchc., *Rumex triangulivalvis* (Danser) Rech. f.
common name(s): narrow-leaved dock
habit: erect, branched, perennial herb
range: Newfoundland, s. to PA, w. to MO and Mexico, n. to British Columbia and AK
state frequency: uncommon
habitat: moist, often brackish or saline, soil
notes: —

PHYTOLACCACEAE
(1 genus, 1 species)

Phytolacca
(1 species)

Phytolacca americana L.
synonym(s): *Phytolacca decandra* L.
common name(s): pokeweed
habit: perennial herb
range: Quebec, s. to FL, w. to TX, n. to MN and Ontario
state frequency: uncommon
habitat: fields, roadsides, clearings
notes: —

MOLLUGINACEAE
(1 genus, 1 species)

Mollugo
(1 species)

Mollugo verticillata L.
synonym(s): —
common name(s): carpetweed
habit: prostrate, mat-forming, annual herb
range: tropical America; naturalized to Quebec, s. to FL, w. to Mexico, n. to Ontario
state frequency: occasional
habitat: moist, sandy soil of cultivated ground, river banks, and roadsides
notes: —

NYCTAGINACEAE
(1 genus, 2 species)

Mirabilis
(2 species)

Flowers lacking a corolla, but with a sepaloid involucre and a petaloid calyx, thereby appearing as though the flowers had sepals and petals.

KEY TO THE SPECIES
1a. Leaf blades linear to lance-ovate, 1.0–2.0 cm wide, tapering to the base; stems pubescent; axes of the inflorescence glandular-pubescent .... *M. hirsuta*
1b. Leaf blades ovate to deltate-ovate, 2.5–7.5 cm wide, truncate to cordate at the base; stems glabrous or nearly so; axes of the inflorescence without glands .... *M. nyctaginea*

Mirabilis hirsuta (Pursh) MacM.
synonym(s): *Allionia hirsuta* Pursh, *Oxybaphus hirsutus* (Pursh) Sweet
common name(s): hairy umbrellawort
habit: erect to decumbent, perennial herb
range: WI, s. to TX, w. to AZ, n. to Saskatchewan; adventive in eastern United States
state frequency: very rare
habitat: dry, open areas
notes: —

Mirabilis nyctaginea (Michx.) MacM.
synonym(s): *Allionia nyctaginea* Michx., *Oxybaphus nyctagineus* (Michx.) Sweet
common name(s): heart-leaved umbrellawort, wild four-o'clock
habit: perennial herb
range: IL, s. to AL, w. to TX, n. to MT and Manitoba; adventive in eastern United States
state frequency: very rare
habitat: dry areas, waste places
notes: —

PORTULACACEAE
(3 genera, 3 species)

KEY TO THE GENERA
1a. Leaves alternate; flowers with yellow petals, 6–10 stamens, and 4–6 stigmas; plants
  weeds, of disturbed ground; fruit a pyxis; ovary partly inferior .... *Portulaca*
1b. Leaves opposite; flowers with white or pink petals, 5 stamens, and 3 stigmas; plants
  native; fruit a capsule; ovary superior
  2a. Plants perennial from a corm, flowering mid-April–early June; leaves 3.0–10.0 cm
    long; petal 9.0–15.0 mm long; capsule with 3–6 seeds .... *Claytonia*
  2b. Plants annual, flowering mid-June–early July; leaves 1.5–3.0 cm long; petals *ca.*
    1.5 mm long; capsule with 1–3 seeds .... *Montia*

Claytonia
(1 species)

Claytonia caroliniana Michx.
synonym(s): —
common name(s): Carolina spring beauty
habit: corm-bearing, spring-ephemeral, perennial herb
range: Newfoundland, s. to mountains of NC and GA, w. to IL, n. to MN
state frequency: uncommon
habitat: rich, cool woods
notes: —

Montia
(1 species)

Montia fontana L.
synonym(s): *Montia lamprosperma* Cham.
common name(s): blinks

habit: annual herb
range: circumboreal, s. along ocean coasts to ME and AK
state frequency: rare
habitat: springy soil, shores
notes: This species is ranked S2 by the Maine Natural Areas Program. It appears to senesce early, rotting away by late summer and leaving only the lustrous, black seeds.

Portulaca
(1 species)

Portulaca oleracea L.
synonym(s): —
common name(s): common purslane, pusley
habit: prostrate, branched, mat-forming, annual herb
range: Eurasia; naturalized throughout much of the world
state frequency: uncommon
habitat: cultivated ground, waste places
notes: —

AMARANTHACEAE
(10 genera, 40 species)

KEY TO THE GENERA

1a. Sepals usually dry and scarious; filaments distinct or, more commonly, connate near the base; ovary with 3 stigmas .... *Amaranthus*
1b. Sepals membranaceous to herbaceous (hyaline in *Axyris*); filaments distinct; ovary with 2 (–5) stigmas
   2a. Stems jointed; leaves opposite and scale-like; flowers sunken into the fleshy stem
      3a. Plants perennial with woody rhizomes; all 3 flowers of each cluster inserted at the same level .... *Sarcocornia*
      3b. Plants annual, without rhizomes; the middle flower of each cluster conspicuously elevated above the 2 lateral flowers .... *Salicornia*
   2b. Stems not jointed; leaves alternate or the lower often opposite in *Atriplex*, but not scale-like; flowers not sunken into a fleshy stem
      4a. Leaves stellate-pubescent on the abaxial surface; sepals hyaline; fruit obovate, with a bilobed wing at the apex .... *Axyris*
      4b. Leaves glabrous or pubescent, but not stellate-pubescent; sepals usually pigmented; fruit orbicular to obovate to oblong, without a bilobed wing at the apex
         5a. Plants monoecious; fruit concealed by a pair of accrescent bracteoles .... *Atriplex*
         5b. Plants synoecious or sometimes polygamous [with bisexual and unisexual flowers]; fruit visible or concealed by a persistent calyx, the bracteoles not accrescent
            6a. Leaf blades linear to narrow-lanceolate, sessile, entire
               7a. Leaves tipped with spines; fruits globose to ovoid .... *Salsola*
               7b. Leaves not tipped with spines, at most pointed; fruits lenticular

8a. Leaves terete to planoconvex in cross-section; flowers borne in groups of 3 in the axils of the leaves .... **Suaeda**

8b. Leaves flat, at least in the basal portion; flowers borne variously, but not consistently in groups of 3

    9a. Calyx minute at anthesis, *ca.* 0.3 mm long; sepals of fruiting plants transversely winged on the abaxial surface .... **Kochia**

    9b. Calyx larger and evident at anthesis; sepals of fruiting plants rounded to keeled, but not winged .... **Chenopodium**

6b. Leaf blades lanceolate to ovate, at least the lower ones petioled, entire or toothed

    10a. Calyx composed of 3–5 sepals, enclosing and often somewhat concealing the fruit at maturity; androecium composed of 1–5 stamens, commonly 5 .... **Chenopodium**

    10b. Calyx composed of 1 sepal, not enclosing the fruit at maturity; androecium composed of 1 stamen .... **Monolepis**

Amaranthus

(7 species)

KEY TO THE SPECIES

1a. Plants dioecious; carpellate flowers lacking a calyx; fruit an indehiscent utricle or the utricle irregularly rupturing in *A. tuberculatus*

    2a. Utricle 2.5–4.0 mm long, fleshy, not dehiscent; seeds 2.0–3.0 mm long; bracts of staminate flowers up to 1.0 mm long, the midrib weakly excurrent; leaf blades linear to lanceolate; plants of saline wetlands and shores .... **A. cannabinus**

    2b. Utricle 1.5–2.0 mm long, membranaceous, sometimes irregularly dehiscing; seeds 0.8–1.0 mm long; bracts of staminate flowers *ca.* 1.0–1.5 mm long, the midrib evidently excurrent; leaf blades lanceolate to ovate; plants of fresh water wetlands and shores .... **A. tuberculatus**

1b. Plants monoecious; flowers with a calyx of 3–5 sepals, the carpellate sepals persistent in fruit; fruit a pyxis

    3a. All or most of the flowers in small, axillary clusters; blades of principal leaves 0.5–8.0 cm long; staminate flowers each with 3 stamens .... **A. albus**

    3b. Flowers in evident, terminal inflorescences, axillary clusters may also be present; blades of principal leaves 5.0–30.0 cm long; staminate flowers each with 3–5 stamens

        4a. Pyxis exceeding the bracts and sepals; inflorescence large, showy, with red to purple bracts .... **A. cruentus**

        4b. Pyxis exceeded by the bracts and longer sepals; inflorescence relatively smaller, dull, with green to brown bracts (the bracts sometimes red-tinged in *A. hybridus*)

            5a. Sepals of the carpellate flowers obtuse to emarginate at the apex, outwardly curved in the apical portion; lateral spikes 1.0–5.0 cm tall; carpellate calyx 2.0–3.5 mm long (up to 4.0 mm in fruit) .... **A. retroflexus**

            5b. Sepals of the carpellate flowers acute at the apex, not outwardly curved; lateral spikes 1.0–30.0 cm long; carpellate calyx 1.5–3.5 mm long

6a. Branches of the style spreading; staminate flowers with 3–5 sepals and stamens; calyx of carpellate flowers 2.0–3.5 mm long; inflorescence stiff, simple or with a few, long branches .... *A. powellii*

6b. Branches of the style erect; staminate flowers with 5 sepals and stamens; calyx of carpellate flowers 1.5–2.0 mm long; inflorescence lax, branched with many lateral branches .... *A. hybridus*

Amaranthus albus L.

synonym(s): *Amaranthus graecizans auct. non* L.
common name(s): tumbleweed
habit: monoecious, erect to ascending, branched, annual herb
range: central North America; adventive to eastern North America, Eurasia, Africa, and South America
state frequency: uncommon
habitat: waste places, fields
notes: —

Amaranthus cannabinus (L.) Sauer

synonym(s): *Acnida cannabina* L.
common name(s): saltmarsh-hemp
habit: dioecious, erect, annual herb
range: ME, s. to FL
state frequency: rare
habitat: saline to brackish marshes
notes: —

Amaranthus cruentus L.

synonym(s): *Amaranthus hybridus* L. var. *cruentus* (L.) Moq.
common name(s): red amaranth, purple amaranth
habit: erect, monoecious, annual herb
range: Asia; escaped from cultivation
state frequency: very rare
habitat: roadsides, waste places
notes: —

Amaranthus hybridus L.

synonym(s): —
common name(s): green amaranth, wild-beet, smooth pigweed
habit: erect, monoecious, branched, annual herb
range: tropical America; nearly cosmopolitan weed, naturalized in North America to ME, s. to FL, w. to Mexico, n. to Manitoba
state frequency: rare
habitat: waste places, fields
notes: —

Amaranthus powellii S. Wats.

synonym(s): —
common name(s): —
habit: erect, monoecious, branched, annual herb
range: Nova Scotia, s. to PA, w. to TX and Mexico, n. to British Columbia
state frequency: very rare
habitat: waste places, open areas
notes: —

Amaranthus retroflexus L.

synonym(s): —
common name(s): redroot, green amaranth, pigweed
habit: erect, stout, monoecious, usually branched, annual herb
range: tropical America; naturalized to much of the world's temperate regions, in North America to
    New Brunswick, s. to FL, w. to Mexico, n. to Northwest Territories
state frequency: uncommon
habitat: cultivated ground, waste places
notes: —

Amaranthus tuberculatus (Moq.) Sauer

synonym(s): *Acnida altissima* (Riddell) Moq. *ex* Standl.
common name(s): —
habit: prostrate to erect, dioecious, annual herb
range: ME, s. to NJ, w. to CO, n. to ND and Ontario
state frequency: very rare
habitat: margins of fresh water, often disturbed
notes: —

Atriplex
(5 species)

In this genus, the lower leaves of the primary plant axis are crucial for shape assessment. Unfortunately, they are occasionally deciduous before maturation of the bracteoles. Leaves of the upper part of the stem are often of a different shape and are not diagnostic among the species. All our species, except *Atriplex acadiensis* and *A. patula*, have spongy-thickened bracteoles. These appear as a thickened, brown or silver-brown region on the inner surface of the bracteole. Fruits of this group are dimorphic—the larger brown and the smaller black.

KEY TO THE SPECIES
1a. Inflorescence with leafy bracts throughout its length; fruits usually brown, black ones
    rare or absent .... *A. glabriuscula*
1b. Inflorescence with leafy bracts only at its base or not at all; both brown and black fruits
    present
    2a. Blades of lower leaves triangular-hastate, each with a more or less truncate base;
        bracteoles somewhat spongy-thickened near the base .... *A. prostrata*
    2b. Blades of lower leaves linear to ovate, with or without basal lobes, narrowed to
        cuneate to rounded at the base; bracteoles spongy-thickened or not
        3a. Blades of lower leaves linear to lanceolate, usually without basal lobes;
            bracteoles spongy-thickened at their base .... *A. littoralis*
        3b. Blades of lower leaves lanceolate to rhombic, commonly with basal lobes;
            bracteoles not spongy-thickened at their base
            4a. Bracteoles rhombic, with commonly pointed, lateral margins that are united to
                near the middle; blades of lower leaves usually with well developed basal
                lobes; plants of symmetric appearance with opposing branches of equal
                length; inland species of disturbed ground (or sometimes of disturbed, coastal
                shores), usually lacking red coloration .... *A. patula*
            4b. Bracteoles ovate, with rounded, lateral margins that are united only near the
                base; blades of lower leaves with poorly developed basal lobes; plants of

asymmetric appearance with opposing branches of unequal length; coastal halophytes, frequently with red coloration on the stems and margins of leaves and bracteoles .... *A. acadiensis*

Atriplex acadiensis Taschereau
synonym(s): —
common name(s): —
habit: erect to decumbent, annual herb
range: Quebec, New Brunswick, Nova Scotia, and ME
state frequency: rare
habitat: saltmarshes, coastal shores
notes: —

Atriplex glabriuscula Edmondston
synonym(s): —
common name(s): —
habit: prostrate to erect, annual herb
range: Greenland and Newfoundland, s. to RI
state frequency: uncommon
habitat: ocean beaches
notes: —

Atriplex littoralis L.
synonym(s): *Atriplex patula* L. var. *littoralis* (L.) Gray
common name(s): —
habit: erect, annual herb
range: New Brunswick and Nova Scotia, s. to NJ, w. to IN, n. to Ontario
state frequency: uncommon
habitat: Atlantic coast shores
notes: —

Atriplex patula L.
synonym(s): —
common name(s): spearscale
habit: erect to decumbent, annual herb
range: Eurasia; introduced to Newfoundland, s. to coastal NY, w. to MO, n. to British Columbia
state frequency: uncommon
habitat: disturbed soils, both inland and coastal
notes: —

Atriplex prostrata Bouchér *ex* DC.
synonym(s): *Atriplex hastata* L. *sensu* Gleason and Cronquist (1991), *Atriplex patula* L. var. *hastata* Gray *sensu* Gleason and Cronquist (1991), *Atriplex triangularis* Willd.
common name(s): —
habit: erect to prostrate, annual herb
range: Eurasia; introduced to Newfoundland, s. to SC, w. to MO, n. to British Columbia
state frequency: occasional
habitat: saline to brackish soils
notes: —

Axyris
(1 species)

Axyris amaranthoides L.
synonym(s): —
common name(s): Russian pigweed
habit: erect, branched, annual herb
range: Asia; naturalized in our area
state frequency: very rare
habitat: agricultural areas
notes: Introduced to Maine in feed grains.

Chenopodium
(17 species)

The fruits in *Chenopodium* are essential for identification of certain species complexes. They are of 2 types—an achene and a utricle. The pericarp in the achene is tightly adhered. In the utricle, however, it is loosely attached and can be rubbed off or flaked away with a probe.

KEY TO THE SPECIES
1a. Leaves either stipitate-glandular or resinous-glandular
    2a. Leaves stipitate-glandular, pinnately lobed, the blades often less than 3.0 cm long; inflorescence an elongate thyrse; achenes dull black
        3a. Sepals corniculate, provided with yellow, resinous glands; apex of leaf lobes acute .... **C. graveolens**
        3b. Sepals not corniculate, glandular-pubescent; apex of leaf lobes obtuse .... **C. botrys**
    2b. Leaves resinous-glandular, blades of the lower ones serrate and often longer than 6.0 cm, the upper entire; inflorescence composed of small glomerules of flowers aggregated in elongate spikes; achenes lustrous, dark brown .... **C. ambrosioides**
1b. Leaves not glandular, either glabrous or farinose
    4a. All or most fruits arranged vertically in the fruiting calyx; calyx of 3–5 sepals
        5a. Achenes 1.5–2.0 mm in diameter; styles persistent on the fruit, 0.8–1.5 mm long; calyx of 4 or 5 sepals; plants perennial .... **C. bonus-henricus**
        5b. Achenes 0.5–1.0 mm in diameter; styles inconspicuous in fruit, less than 0.8 mm long; calyx of 3 sepals; plants annual
            6a. Leaves conspicuously farinose on the abaxial surface; glomerules of flowers 1.8–2.5 mm wide .... **C. glaucum**
            6b. Leaves green or sparsely farinose on the abaxial surface; glomerules of flowers 2.0–12.0 mm wide
                7a. Glomerules of flowers 5.0–12.0 mm wide, not subtended by bracts; plants not halophytic .... **C. capitatum**
                7b. Glomerules of flowers 2.0–5.0 mm wide, subtended by bracts; plants halophytic
                    8a. Mature achene 0.5–0.6 mm in diameter; sepals herbaceous; flowers in leafy-bracted, spike-like inflorescences; plants erect, up to 8.0 dm tall; blades of lower leaves toothed, up to 15.0 cm long .... **C. rubrum**

8b. Mature achene 0.8–1.0 mm in diameter; sepals fleshy; flowers in axillary glomerules; plants spreading-ascending, 0.3–2.5 dm tall; lower leaves entire or nearly so, the blade 0.7–3.0 cm long .... *C. humile*

4b. All or most fruits arranged horizontally in the fruiting calyx; calyx of 5 sepals

9a. Leaves, axes of the inflorescence, and abaxial surface of sepals glabrous and green (sometimes sparsely farinose in *C. berlandieri* and *C. murale*); sepals rounded on the abaxial surface, at maturity conforming to the fruit, the calyx appearing circular

10a. Pericarp loosely adhered to and easily removed from mature fruit [fruit a utricle] .... *C. berlandieri*

10b. Pericarp tightly adhered to and difficult to remove from mature fruit [fruit an achene]

11a. Leaves entire, or rarely with a single tooth on each margin; stems quadrangular; sepals apiculate at the apex .... *C. polyspermum*

11b. Leaves prominently toothed; stems terete to angled, but not quadrangular; sepals not apiculate at the apex

12a. Leaves dull on the adaxial surface; flowers developing asynchronously [flowers and fruits may be present in the same glomerule]; mature achenes 1.3–1.9 x 0.9–1.0 mm .... *C. simplex*

12b. Leaves lustrous on the adaxial surface; flowers developing synchronously [all the flowers in a glomerule of similar stage]; mature achenes 0.8–1.5 x 0.5–0.8 mm

13a. Achenes rounded on the margin, lustrous; anthers 0.2–0.3 mm long .... *C. urbicum*

13b. Achenes keeled on the margin, dull; anthers 0.4–0.5 mm long .... *C. murale*

9b. Leaves, axes of inflorescence, and abaxial surface of sepals farinose; sepals keeled, cucullate, or corniculate on the abaxial surface, not conforming to the fruit, the calyx appearing pentagonal or star-shaped

14a. Blade of principal leaves linear to lanceolate, mostly 1-nerved, 2.0–4.0 mm wide .... *C. leptophyllum*

14b. Blade of principal leaves lanceolate to broad-ovate, rhombic, or deltate, with well developed, secondary veins, 5.0–80.0 mm wide

15a. Pericarp loosely adhered to and easily removed from mature fruit

16a. Leaf blades ovate to rhombic-ovate; utricles mostly 0.8–1.0 mm wide; stems 1.0–2.5(–5.0) dm tall .... *C. incanum*

16b. Leaf blades narrow-ovate; utricles 1.0–1.3 mm wide; stems 2.0–10.0 dm tall .... *C. foggii*

15b. Pericarp tightly adhered to and difficult to remove from mature fruit

17a. Achenes mostly 1.3–2.3 mm wide, with a conspicuously cellular-reticulate pericarp .... *C. berlandieri*

17b. Achenes mostly 1.0–1.5 mm wide, with a smooth or obscurely mottled pericarp .... *C. album*

Chenopodium album L.

synonym(s): *Chenopodium lanceolatum* Muhl. *ex* Willd.
common name(s): lamb's quarters, pigweed
habit: erect, annual herb
range: Europe; introduced throughout much of North America
state frequency: occasional
habitat: cultivated ground, waste places
notes: Known to hybridize with *Chenopodium berlandieri*.

Chenopodium ambrosioides L.

synonym(s): —
common name(s): Mexican-tea, wormseed
habit: erect, annual or short-lived perennial herb
range: tropical America; naturalized to ME, s. to FL, w. to TX, n. to WI and Ontario
state frequency: very rare
habitat: waste places
notes: Plants with aromatic herbage.

Chenopodium berlandieri Moq.

synonym(s): <*C. b.* var. *boscianum*> = *Chenopodium boscianum* Moq., *Chenopodium standleyanum* Aellen *sensu* Gleason and Cronquist (1991); <*C. b.* var. *bushianum*> = *Chenopodium bushianum* Aellen, *Chenopodium paganum auct. non* Reichenb.; <*C. b.* var. *macrocalycium*> = *Chenopodium macrocalycium* Aellen
common name(s): pitseed goosefoot, Bosc's goosefoot
habit: annual herb
range: Quebec, s. to Mexico, w. and n. to AK
state frequency: very rare
habitat: disturbed soil of fields, roadsides, and agricultural areas
key to the subspecies:
  1a. Leaves, axes of inflorescence, and abaxial surface of sepals glabrous and green; sepals rounded on abaxial surface, at maturity conforming to the fruit, the calyx appearing circular; pericarp loosely adhered to and easily removed from mature fruit .... *C. b.* var. *boscianum* (Moq.) H. A. Wahl
  1b. Leaves, axes of inflorescence, and abaxial surface of sepals farinose; sepals keeled on the abaxial surface, not conforming to the fruit, the calyx appearing pentagonal; pericarp tightly adhered to and difficult to remove from mature fruit
    2a. Achenes mostly 1.3–1.7 mm wide; blade of larger leaves 5.0–6.0 cm long; stems rarely taller than 5.0 dm; plants of coastal areas and sea beaches .... *C. b.* var. *macrocalycium* (Aellen) Cronq.
    2b. Achenes mostly 1.5–2.0 mm wide; blade of larger leaves often longer, up to 15.0 cm long; stems mostly 7.0–15.0 dm tall; plants of cultivated ground and disturbed soil .... *C. b.* var. bushianum (Aellen) Cronq.
notes: This species is ranked S1 by the Maine Natural Areas Program. Known to hybridize with *Chenopodium album*.

Chenopodium bonus-henricus L.

synonym(s): —
common name(s): good King Henry
habit: erect to ascending, perennial herb
range: Europe; naturalized locally to Quebec, s. to coastal NY, w. to IA, n. to Ontario
state frequency: very rare
habitat: waste places, roadsides
notes: Plants with aromatic herbage.

Chenopodium botrys L.

synonym(s): —
common name(s): Jerusalem-oak, feather-geranium
habit: erect, annual herb
range: Eurasia; naturalized to Quebec, s. to VA, w. to MO, n. to WA
state frequency: uncommon
habitat: waste places, cultivated ground, roadsides
notes: Plants with aromatic herbage.

Chenopodium capitatum (L.) Aschers.

synonym(s): *Blitum capitatum* L.
common name(s): strawberry-blight, Indian paint
habit: erect to ascending, annual herb
range: circumboreal; s. to NJ, w. to IN, MN, and AK
state frequency: rare
habitat: clearings, waste places, often burned areas
notes: —

Chenopodium foggii H. A. Wahl

synonym(s): —
common name(s): Fogg's goosefoot
habit: slender, erect, annual herb
range: Quebec, s. to VA, w. to IL, n. to MN
state frequency: very rare
habitat: woodlands
notes: —

Chenopodium glaucum L.

synonym(s): —
common name(s): oak-leaved goosefoot
habit: prostrate to erect, annual herb
range: Europe; naturalized to Quebec, s. to VA, w. to MO and NE, n. to Saskatchewan
state frequency: uncommon
habitat: waste places, cultivated land, roadsides
notes: —

Chenopodium graveolens Willd.

synonym(s): —
common name(s): fetid pigweed
habit: erect, annual herb
range: southwest United States, s. to South America; naturalized in the northeast
state frequency: very rare
habitat: cultivated ground, waste places
notes: Plants with aromatic herbage.

Chenopodium humile Hook.

synonym(s): *Chenopodium rubrum* L. var. *humile* (Hook.) S. Wats.
common name(s): —
habit: annual herb
range: New Brunswick and Nova Scotia, s. to ME, w. to CA, n. to British Columbia
state frequency: very rare
habitat: saline or brackish soil
notes: —

Chenopodium incanum (S. Wats.) Heller
synonym(s): —
common name(s): —
habit: branched, farinose, annual herb
range: western United States; adventive to ME and MO
state frequency: very rare
habitat: waste places
notes: —

Chenopodium leptophyllum (Moq.) Nutt. *ex* S. Wats.
synonym(s): —
common name(s): narrow-leaved goosefoot
habit: erect, annual herb
range: ME, s. to VA, w. to TX and Mexico, n. to Alberta
state frequency: uncommon
habitat: dry soil of disturbed areas
notes: —

Chenopodium murale L.
synonym(s): —
common name(s): nettle-leaved goosefoot
habit: erect, annual herb
range: Europe; introduced to Quebec and New Brunswick, s. to FL, w. to Mexico, n. to British
    Columbia
state frequency: rare
habitat: waste places
notes: —

Chenopodium polyspermum L.
synonym(s): —
common name(s): many-seeded goosefoot
habit: erect to ascending, annual herb
range: Eurasia; introduced to Quebec, s. to MD and TN, w. to IA, n. to Saskatchewan
state frequency: rare
habitat: waste places, cultivated ground
notes: —

Chenopodium rubrum L.
synonym(s): —
common name(s): coast-blite, alkali-blite, red goosefoot
habit: erect, annual herb
range: Newfoundland, s. to NJ, w. to AZ, n. to Yukon
state frequency: very rare
habitat: saltmarshes, saline or brackish soil
notes: This species is ranked S1 by the Maine Natural Areas Program.

Chenopodium simplex (Torr.) Raf.
synonym(s): *Chenopodium gigantospermum* Aellen, *Chenopodium hybridum* L. var.
    gigantospermum (Aellen) Rouleau
common name(s): maple-leaved goosefoot
habit: erect, annual herb
range: Quebec, s. to VA, w. to CA, n. to Yukon
state frequency: uncommon

habitat: rocky woods, disturbed areas
notes: —

Chenopodium urbicum L.
synonym(s): —
common name(s): city goosefoot, upright goosefoot
habit: erect, annual herb
range: Europe; introduced to New Brunswick, s. to MD, w. to MO, n. to British Columbia
state frequency: very rare
habitat: waste places
notes: —

Kochia
(1 species)

Kochia scoparia (L.) Schrad.
synonym(s): *Bassia scoparia* (L.) A. J. Scott
common name(s): summer-cypress
habit: erect, annual herb
range: Eurasia; escaped from cultivation to New Brunswick and Nova Scotia, s. to CT
state frequency: rare
habitat: roadsides, waste places
notes: —

Monolepis
(1 species)

Monolepis nuttalliana (J. A. Schultes) Greene
synonym(s): —
common name(s): povertyweed
habit: annual herb
range: Manitoba, s. to MO and TX, w. to CA; adventive e. to New England
state frequency: very rare
habitat: dry, alkaline soil
notes: —

Salicornia
(3 species)

Succulent, halophytic herbs best identified using fresh material. Most characters are absent or distorted in dried herbarium specimens due to shrinking of the stem. Reconstituting stems in warm water may help restore original dimensions.

KEY TO THE SPECIES
1a. Leaf and scale apex acute to acuminate, with a prominent mucro; inflorescence
    4.5–6.0 mm thick, notably wider than the stem; flowers all concealed by bracts ....
    .... **S. bigelovii**
1b. Leaf and scale apex rounded to acute, without a mucro; inflorescence 1.5–4.0 mm thick,
    usually of similar thickness as the stem; central flower exceeding the bract and visible

2a. Inflorescences cylindric to long-tapering, the terminal with (5–)7–23(–25) fertile segments; scarious margin of leaves 0.3–0.4 mm wide; flowers with exserted stamens; fertile segments cylindric .... **S. depressa**

2b. Inflorescences swollen and rounded near apex, the terminal with (3–)5–10(–14) fertile segments; scarious margin of leaves less than 0.3 mm wide; flowers not exserting stamens, or sometimes, but then after dehiscence; fertile segments widened in the apical portion .... **S. maritima**

Salicornia bigelovii Torr.
synonym(s): —
common name(s): dwarf glasswort, dwarf saltwort
habit: erect, annual herb
range: Nova Scotia, s. along coast to TX and Mexico; also CA
state frequency: very rare
habitat: saltmarshes
notes: —

Salicornia depressa Standl.
synonym(s): *Salicornia europaea* L. *sensu* most American authors, *Salicornia virginica* L.
common name(s): samphire
habit: erect to prostrate, simple or branched, succulent, annual herb
range: Quebec, s. along coast to FL; also w. to IL and WI
state frequency: common
habitat: saltmarshes, saline areas
notes: —

Salicornia maritima Wolff & Jefferies
synonym(s): *Salicornia europaea* L. var. *prostrata* (Pallas) Fern., *Salicornia prostrata* Pallas
common name(s): samphire
habit: erect to prostrate, simple or branched, succulent, annual herb
range: Newfoundland, s. to New Brunswick and ME
state frequency: very rare
habitat: saltmarshes, saline areas
notes: —

Salsola
(1 species)

Salsola kali L.
synonym(s): <*S. k.* ssp. *tragus*> = *Salsola kali* L. ssp. *tenuifolia* Tausch
common name(s): common saltwort
habit: prostrate to ascending, annual herb
range: Newfoundland, s. along coast to LA
state frequency: uncommon
habitat: ocean beaches
key to the subspecies:

1a. Sepals firm, with a midvein prolonged into a stiff, sharp point; plants of Atlantic coast shores .... *S. k.* ssp. *kali*

1b. Sepals soft at anthesis, with an obscure midvein, blunt; plants of inland areas .... *S. k.* ssp. *tragus* (L.) Aellen

notes: —

Sarcocornia
(1 species)

Sarcocornia perennis (P. Mill.) A. J. Scott
synonym(s): *Arthrocnemum perenne* (P. Mill.) Moss, *Salicornia ambigua* Michx., *Salicornia perennis* P. Mill., *Salicornia virginica* L. *sensu* Gleason and Cronquist (1991)
common name(s): —
habit: low, rhizomatous shrub
range: ME, s. to TX; also AK, s. to CA
state frequency: very rare; known only from southern Maine
habitat: Atlantic coast shores
notes: This species is ranked S1 by the Maine Natural Areas Program.

Suaeda
(3 species)

KEY TO THE SPECIES
1a. Sepals corniculate or keeled on the abaxial surface; utricle 1.0–1.5 mm long
   2a. Sepals unequal at maturity, 1 or 2 of them corniculate, the others keeled; anthers 0.3–0.4 mm long .... **S. calceoliformis**
   2b. Sepals equally keeled at maturity; anthers *ca.* 0.2 mm long .... **S. linearis**
1b. Sepals usually rounded on the abaxial surface; utricle 1.0–2.0 mm long .... **S. maritima**

Suaeda calceoliformis (Hook.) Moq.
synonym(s): *Dondia depressa auct. non* (Pursh) Britt., *Suaeda americana* (Pers.) Fern.
common name(s): northeastern sea-blite
habit: decumbent to erect, annual herb
range: Quebec, s. along coast to NJ
state frequency: very rare
habitat: saltmarshes and coastal strands
notes: This species is ranked SH by the Maine Natural Areas Program.

Suaeda linearis (Ell.) Moq.
synonym(s): *Dondia linearis* (Ell.) Heller
common name(s): southern sea-blite
habit: erect to ascending, annual herb
range: ME, s. along coast to TX
state frequency: uncommon
habitat: saltmarshes, sandy, coastal areas
notes: —

Suaeda maritima (L.) Dumort.
synonym(s): <*S. m.* ssp. *maritima*> = *Dondia maritima* (L.) Druce; <*S. m.* ssp. *richii*> = *Suaeda richii* Fern.
common name(s): white sea-blite
habit: prostrate to erect, annual herb
range: Eurasia; naturalized to Quebec, s. along coast to LA
state frequency: occasional
habitat: saltmarshes

key to the subspecies:

1a. Utricle 1.5–2.0 mm wide; leaves mostly 1.0–3.0 cm long, commonly glaucous .... *S. m.* ssp. *maritima*

1b. Utricle 1.0–1.5 mm wide; leaves mostly 0.3–1.5 cm long, usually green .... *S. m.* ssp. *richii* (Fern.) Bassett & C. W. Crompton

notes: *Suaeda maritima* ssp. *richii* is ranked S1 by the Maine Natural Areas Program.

CARYOPHYLLACEAE
(19 genera, 53 species)

KEY TO THE GENERA

1a. Sepals distinct or essentially so [appearing connate in *Scleranthus*, but the tube actually a hypanthium]; ovary sessile; petals without a prominent, narrow, basal portion [*i.e.*, lacking a claw]

  2a. Leaves with conspicuous, scarious or hyaline stipules

    3a. Flowers without petals; ovules and seeds 1 per flower; ovary maturing as a utricle

      4a. Stipules with entire margins, longer than 4.0 mm; calyx 3.5–4.0 mm long; flowers more or less concealed by silver, hyaline bracts .... ***Paronychia***

      4b. Stipules with fringed margins, 0.5–1.0 mm long; calyx 0.5–1.0 mm long; flowers not concealed by bracts .... ***Herniaria***

    3b. Flowers with white or pink to red-purple petals; ovules and seeds many per flower; ovary maturing as a capsule

      5a. Leaves clustered at the nodes in 2 sets of 6–8, appearing whorled, with small stipules mostly less than 1.0 mm long; flowers normally with 5 styles; capsules normally with 5 valves .... ***Spergula***

      5b. Leaves opposite (with axillary fascicles in *Spergularia rubra*), with stipules 1.0–5.0 mm long; flowers normally with 3 styles; capsules normally with 3 valves .... ***Spergularia***

  2b. Leaves without stipules

    6a. Flowers perigynous [with a cupulate hypanthium]; fruit a 1-seeded utricle; perianth sepaloid, composed only of sepals that have a narrow, scarious border ... . ***Scleranthus***

    6b. Flowers hypogynous or essentially so; fruit a many-seeded capsule; perianth in part petaloid, composed of both sepals and petals (perianth sepaloid, the petals absent in *Sagina procumbens* and rarely absent in *Stellaria alsine*)

      7a. Perianth monochlamydeous, only the sepals present

        8a. Flowers with 3 styles; capsule dehiscing by *ca.* 6 valves; leaves linear-oblong to elliptic, 1.5–6.0(–10.0) mm wide .... ***Stellaria***

        8b. Flowers with 4(5) styles; capsule dehiscing by 4(5) valves; leaves linear-subulate, up to 1.0 mm wide .... ***Sagina***

      7b. Perianth dichlamydeous, both sepals and petals present

        9a. Petals deeply notched, in some *Stellaria* so deeply divided as to appear as 2 distinct, but proximate, petals (very rarely the petals absent)

          10a. Capsule cylindric, opening by *ca.* 10 apical teeth; flowers with mostly 5 styles .... ***Cerastium***

10b. Capsule ovoid to ellipsoid, the *ca.* 6 valves dehiscing to near the middle; flowers with mostly 3 styles .... ***Stellaria***

9b. Petals entire, at most the apex retuse or erose

11a. Plants fleshy; seeds 3.0–5.0 mm long; petals and stamens inserted on a conspicuous, 10-lobed disk .... ***Honckenya***

11b. Plants not fleshy; seeds 0.4–1.4 mm long; staminal disk inconspicuous

12a. Flowers with 4 or 5 styles; capsule dehiscing by 4 or 5 valves .... ***Sagina***

12b. Flowers with 3 styles; capsule dehiscing by 3 or 6 valves

13a. Seeds 1.0–1.4 mm long, strophiolate; plants rhizomatous perennials, the stems not tufted .... ***Moehringia***

13b. Seeds 0.4–0.8 mm long, not strophiolate; plants annuals or perennials, the stems tufted

14a. Capsule dehiscing by 6 valves; leaves ovate, 3.0–5.0 mm wide .... ***Arenaria***

14b. Capsule dehiscing by 3 valves; leaves linear to linear-oblanceolate, 0.3–0.8 mm wide .... ***Minuartia***

1b. Sepals connate in the basal portion; ovary stipitate; petals with a prominent, narrow, basal portion [*i.e.*, possessing a claw]

15a. Flowers with 2 or 3 styles; capsules dehiscing by 4 or 6 valves

16a. Calyx subtended by 1–3 pairs of scarious or foliaceous bracts, (20-)25- to 40-nerved; seeds disciform .... ***Dianthus***

16b. Calyx not subtended by bracts, 5- to 30-nerved; seeds orbicular to reniform

17a. Flowers with 3 styles; capsules dehiscing by 6 valves .... ***Silene***

17b. Flowers usually with 2 styles; capsules dehiscing by 4 valves

18a. Adaxial surface of petals with a subulate appendage at the junction of the claw and blade; calyx with 20, often inconspicuous, nerves ..... ***Saponaria***

18b. Petals without appendages; calyx with 5 nerves

19a. Petals 18.0–22.0 mm long; seeds globose; calyx nerved with stout ribs, these becoming wing angles in fruit, not at all scarious .... ***Vaccaria***

19b. Petals 2.0–15.0 mm long; seeds subreniform; calyx merely nerved, without prominent wing angles in fruit, evidently scarious between its green nerves .... ***Gypsophila***

15b. Flowers with (4)5 styles or 1 style in staminate individuals of *Silene dioica* and *S. latifolia*; capsules dehiscing by 5 or 10 valves (or dehiscing by 8 valves in 4-styled flowers)

20a. Calyx lobes 0.25–0.9 cm long, much shorter than the connate portion; petals with auricles and/or appendages at the junction of the claw and blade (except in some *Silene*)

21a. Calyx both glandular-pubescent and inflated in fruit; plants dioecious .... ***Silene***

21b. Calyx neither glandular-pubescent nor inflated in fruit; plants synoecious .... ***Lychnis***

20b. Calyx lobes 2.0–4.0 cm long, much longer than the connate portion; petals lacking both auricles and appendages at the junction of the claw and blade ....
.... *Agrostemma*

Agrostemma
(1 species)

Agrostemma githago L.
synonym(s): —
common name(s): purple cockle
habit: erect, annual herb
range: Europe; widely naturalized, especially in the northeastern United States
state frequency: uncommon
habitat: grainfields, roadsides, waste places
notes: —

Arenaria
(1 species)

Arenaria serpyllifolia L.
synonym(s): <*A. s.* ssp. *leptoclados*> = *Arenaria serpyllifolia* L. var. *tenuior* Mert. & Koch
common name(s): thyme-leaved sandwort
habit: annual herb
range: Eurasia; naturalized to Quebec, s. to FL, w. to MO, n. to British Columbia
state frequency: uncommon
habitat: sterile, sandy or rocky soils of fields and roadsides
key to the subspecies:
    1a. Leaves commonly ovate, 4.3–8.5 mm long; calyx 3.0–4.0 mm long and 2.0–3.0 mm wide at the base in fruit; capsule ovoid, tapering in the apical portion; seeds *ca.* 0.6 mm long ....
    .... *A. s.* ssp. *serpyllifolia*
    1b. Leaves lanceolate to lance-ovate, 3.0–5.3 mm long; calyx 2.0–3.0 mm long and 1.2–2.0 mm wide at the base in fruit; capsule nearly cylindric, not conspicuously tapering in the apical portion; seeds 0.4–0.5 mm long .... *A. s.* ssp. *leptoclados* (Reichenb.) Nyman
notes: —

Cerastium
(3 species)

KEY TO THE SPECIES
1a. Petals 2.0–3.0 times the length of the sepals; leaves mostly 2.0–7.0 cm long
    2a. Stems, leaves, and sepals glabrous to villous .... *C. arvense*
    2b. Stems, leaves, and sepals densely tomentose .... *C. tomentosum*
1b. Petals approximately as long as the sepals; leaves 1.0–2.0 cm long .... *C. fontanum*

Cerastium arvense L.
synonym(s): —
common name(s): field chickweed
habit: perennial herb
range: Labrador, s. to GA, w. to CA, n. to AK
state frequency: uncommon

habitat: fields, meadows, rocky areas

notes: Our native chickweed, this species has leafy, marcescent, basal branches that bear axillary leaf tufts.

Cerastium fontanum Baumg. ssp. *vulgare* (Hartman) Greuter & Burdet
synonym(s): *Cerastium vulgatum* L.
common name(s): common mouse-ear chickweed
habit: short-lived, perennial herb
range: Eurasia; naturalized throughout North America
state frequency: occasional
habitat: lawns, fields, roadsides, cultivated ground
notes: —

Cerastium tomentosum L.
synonym(s): —
common name(s): snow-in-summer
habit: depressed, mat-forming, perennial herb
range: Eurasia; escaped from cultivation locally
state frequency: rare
habitat: disturbed soil, vacant lots, cultivated areas
notes: —

Dianthus
(4 species)

KEY TO THE SPECIES

1a. Plants with scattered, solitary flowers borne on slender pedicels 1.0–4.0 cm long
  2a. Basal leaves oblanceolate, 1.5–3.0 cm long; blade of petals 5.0–10.0 mm long, the apex toothed; calyx equaling the length of the fruit; bracts 0.5 times as long as the calyx .... *D. deltoides*
  2b. Basal leaves linear, 2.0–8.0 cm long; blade of petals 12.0–18.0 mm long, fringed-cleft to near the middle; calyx shorter than the length of the fruit; bracts 0.25–0.35 times as long as the calyx .... *D. plumarius*
1b. Plants with closely crowded, sessile or short-pedicellate flowers borne in terminal cymes
  3a. Plants perennial; leaves of the stem lanceolate or oblanceolate, 1.0–3.0 cm wide; stems normally simple, glabrous (or sometimes scabrous); calyx glabrous, about 40-nerved; petals red to white .... *D. barbatus*
  3b. Plants annual or biennial; leaves of the stem linear to lanceolate, 0.2–0.8 cm wide; stems normally branched, strigose; calyx pubescent, 20- to 25-nerved; petals pink to red, dotted with white .... *D. armeria*

Dianthus armeria L.
synonym(s): —
common name(s): Deptford pink
habit: annual or biennial herb
range: Europe; naturalized to Quebec, s. to FL, w. to AR, n. to British Columbia
state frequency: rare
habitat: dry fields, roadsides
notes: —

Dianthus barbatus L.
synonym(s): —
common name(s): sweet William
habit: stout, erect, simple, densely tufted, perennial herb
range: Eurasia; naturalized to Quebec, s. to DE, w. and n. to ND
state frequency: rare
habitat: roadsides
notes: —

Dianthus deltoides L.
synonym(s): —
common name(s): maiden pink
habit: low, perennial herb
range: Europe; naturalized to New Brunswick, s. to NJ, w. to IL, n. to MI
state frequency: uncommon
habitat: roadsides, fields
notes: —

Dianthus plumarius L.
synonym(s): —
common name(s): grass pink, garden pink
habit: decumbent to erect, perennial herb
range: Europe; escaped from cultivation to ME, w. to NE
state frequency: rare
habitat: fields, roadsides
notes: —

Gypsophila
(3 species)

KEY TO THE SPECIES

1a. Slender annuals with green stems 0.5–4.0 dm tall; leaves linear, 1.0–2.0 mm wide;
    capsule oblong .... *G. muralis*
1b. Stout annuals or perennials with glaucous stems 2.0–10.0 dm tall; leaves lanceolate,
    2.5–15.0 mm wide; capsule spherical
    2a. Plants perennial; longer pedicels 0.5–1.2 cm long; calyx 1.5–3.0 mm long, about as
        long as the petals .... *G. paniculata*
    2b. Plants usually annual; longer pedicels 1.2–3.5 cm long; calyx 3.0–5.0 mm long,
        0.35–0.5 times as long as the petals .... *G. elegans*

Gypsophila elegans Bieb.
synonym(s): —
common name(s): —
habit: annual or rarely short-lived perennial herb
range: Eurasia; escaped from cultivation in our area
state frequency: very rare
habitat: roadsides, waste places
notes: —

Gypsophila muralis L.
synonym(s): —

common name(s): cushion baby's breath
habit: slender, branched, annual herb
range: Eurasia; naturalized to ME, s. to NJ, w. to IN, n. to MN and Ontario
state frequency: rare
habitat: roadsides, fields
notes: —

Gypsophila paniculata L.
synonym(s): —
common name(s): baby's breath
habit: branched, perennial herb
range: Eurasia; escaped from cultivation to New Brunswick, s. to FL, w. to NE, n. to British Columbia
state frequency: very rare
habitat: roadsides, waste places
notes: —

Herniaria
(1 species)

Herniaria glabra L.
synonym(s): —
common name(s): —
habit: branched, mat-forming, annual or perennial herb
range: Europe; introduced to our area
state frequency: very rare
habitat: waste places
notes: —

Honckenya
(1 species)

Honckenya peploides (L.) Ehrh. ssp. *robusta* (Fern.) Hultén
synonym(s): *Arenaria peploides* L. var. *robusta* Fern.
common name(s): sea-beach sandwort
habit: fleshy, rhizomatous, perennial herb
range: circumboreal; s. in the east to VA
state frequency: uncommon
habitat: sands of sea beaches and dunes
notes: —

Lychnis
(4 species)

KEY TO THE SPECIES
1a. Plants densely tomentose; calyx lobes twisted; capsule dehiscent by 5 teeth; inflorescence few-flowered .... *L. coronaria*
1b. Plants glabrous to pubescent, but not tomentose; calyx lobes planar, not twisted; capsule dehiscent by 5 or 10 teeth; inflorescence with more flowers
  2a. Stems pubescent with long hairs, with mostly 10–20 pairs of lanceolate to ovate leaves 2.0–5.0 cm wide .... *L. chalcedonica*

2b. Stems glabrous or pubescent with short hairs, with mostly 2–5 pairs of linear to lanceolate leaves up to 0.8–1.5 cm wide
    3a. Petals cleft into 4 lobes; capsule unilocular; stems not viscid .... *L. flos-cuculi*
    3b. Petals entire or retuse at the apex; capsule 5-locular in the basal half and unilocular in the apical half; stems viscid at the nodes and near the apex .... *L. viscaria*

Lychnis chalcedonica L.

synonym(s): —
common name(s): scarlet lychnis, Maltese cross
habit: erect, perennial herb
range: Asia; escaped from cultivation to New Brunswick, s. to CT, w. and n. to MN
state frequency: uncommon
habitat: thickets, open woods, roadsides
notes: —

Lychnis coronaria (L.) Desr.

synonym(s): —
common name(s): mullein-pink
habit: stout, rarely branched, perennial herb
range: Europe; escaped from cultivation to ME, s. to DE, w. to IN, n. to Ontario
state frequency: very rare
habitat: woods, roadsides, waste places
notes: —

Lychnis flos-cuculi L.

synonym(s): —
common name(s): ragged robin
habit: erect, perennial herb
range: Europe; escaped from cultivation to New Brunswick and ME, s. and w. PA, n. to Quebec
state frequency: occasional
habitat: fields, meadows, swales
notes: —

Lychnis viscaria L.

synonym(s): *Viscaria vulgaris* Bernh.
common name(s): German catchfly
habit: erect, perennial herb
range: Eurasia; escaped from cultivation to ME, s. and w. to OH
state frequency: very rare
habitat: roadsides, thickets
notes: —

Minuartia
(3 species)

KEY TO THE SPECIES
1a. Sepals acute at the apex, strongly 3-nerved; leaves with 3 nerves; stems glabrous or glandular-pubescent .... *M. rubella*
1b. Sepals obtuse at the apex, faintly 1-nerved; leaves with 1 nerve; stems glabrous

2a. Stems 7.0–27.0 cm tall, lacking a dense, basal tuft of leaves; upper leaves 8.0–30.0 mm long; cymes with 3–7 flowers; pedicels 1.2–4.5 cm long in fruit; sepals 4.0–5.5 mm long; petals 6.0–10.0 mm long; plants not alpine, usually of summits less than 1000 m .... **M. glabra**

2b. Stems 2.0–15.0 cm tall, with a dense, basal tuft of leaves; upper leaves 2.0–9.0 mm long; cymes with 9–15 flowers; pedicels 0.6–2.3 cm long in fruit; sepals 3.0–4.0 mm long; petals 5.0–6.0(–8.0) mm long; plants alpine, usually of summits more than 1000 m .... **M. groenlandica**

Minuartia glabra (Michx.) Mattf.

synonym(s): *Arenaria groenlandica* (Retz.) Spreng. var. *glabra* (Michx.) Fern.
common name(s): smooth sandwort
habit: low, erect to ascending, annual herb
range: ME, s. to SC and TN
state frequency: rare
habitat: gravel of exposed areas
notes: This species is ranked S1/S2 by the Maine Natural Areas Program.

Minuartia groenlandica (Retz.) Ostenf.

synonym(s): *Arenaria groenlandica* (Retz.) Spreng.
common name(s): mountain sandwort
habit: low, tufted, annual or perennial herb
range: Greenland, s. to mountains of ME, w. to NY; also mountains of NC
state frequency: rare
habitat: gravel of exposed areas
notes: This species is ranked S3 by the Maine Natural Areas Program.

Minuartia rubella (Wahlenb.) Hiern

synonym(s): *Arenaria rubella* (Wahlenb.) Sm.
common name(s): arctic sandwort
habit: tufted to mat-forming, decumbent to ascending, annual or short-lived perennial herb
range: circumboreal; s. to ME, w. to NM, AZ, and CA
state frequency: very rare
habitat: moist gravel or rock
notes: This species is ranked S1 by the Maine Natural Areas Program.

Moehringia
(1 species)

Moehringia lateriflora (L.) Fenzl

synonym(s): *Arenaria lateriflora* L.
common name(s): grove sandwort
habit: low, rhizomatous, perennial herb
range: circumboreal; s. to MD, w. to CA
state frequency: uncommon
habitat: open woods, meadows, shores
notes: The strophiole is a pale, spongy appendage that is visible on the ovules as well as the seeds.

Paronychia
(1 species)

Paronychia argyrocoma (Michx.) Nutt.
synonym(s): *Paronychia argyrocoma* (Michx.) Nutt. var. *albimontana* Fern.
common name(s): silver whitlowwort, silverling
habit: low, tufted, perennial herb
range: ME, s. along the Appalachian Mountains to GA
state frequency: very rare
habitat: bare, rocky slopes
notes: This species is ranked S1/S2 by the Maine Natural Areas Program.

Sagina
(2 species)

KEY TO THE SPECIES
1a. Petals about 2.0 times as long as the sepals; axils of the upper leaves with short, sterile shoots replacing the flowers; seeds black, subreniform to subglobose, tuberculate in lines; flowers usually 5-merous, each with (5)10 stamens .... **S. nodosa**
1b. Petals absent or about as long as the sepals; sterile shoots absent; seeds brown, obliquely triangular, sulcate along the 2 dorsal ridges, minutely roughened elsewhere; flowers usually 4-merous, each with 4(5) stamens .... **S. procumbens**

Sagina nodosa (L.) Fenzl
synonym(s): <*S. n.* ssp. *nodosa*> = *Sagina nodosa* (L.) Fenzl var. *pubescens* (Bess.) Mert. & Koch
common name(s): knotted pearlwort
habit: decumbent to erect, tufted, perennial herb
range: circumboreal; s. to MA, w. to Alberta
state frequency: uncommon
habitat: moist, rocky, gravelly, or peaty soil
key to the subspecies:
    1a. Stems, pedicels, base of calyx, and sometimes the leaf margins glandular-pubescent ....
    .... *S. n.* ssp. *nodosa*
    1b. Plants glabrous, or sometimes the pedicels and base of the calyx glandular-pubescent ....
    .... *S. n.* ssp. *borealis* Crow
notes: *Sagina nodosa* ssp. *borealis* is our native phase of this plant; the glandular-pubescent subspecies is naturalized from Europe.

Sagina procumbens L.
synonym(s): —
common name(s): birdseye
habit: prostrate to ascending, mat-forming, branched, perennial herb
range: circumboreal; s. to MD, w. to KS and AK
state frequency: occasional
habitat: moist, rocky soil
notes: —

Saponaria
(1 species)

Saponaria officinalis L.
synonym(s): —
common name(s): soapwort, bouncing Bet
habit: stout, erect, rhizomatous, perennial herb
range: Eurasia; naturalized to much of North America
state frequency: uncommon
habitat: roadsides, waste places
notes: —

Scleranthus
(1 species)

Scleranthus annuus L.
synonym(s): —
common name(s): annual knawel
habit: low, branched, spreading, annual or biennial herb
range: Eurasia; naturalized to New Brunswick, s. to FL, w. to MS, n. to British Columbia
state frequency: rare
habitat: roadsides, waste places, fields
notes: —

Silene
(13 species)

Annual, biennial, and perennial habits are very useful for separating certain species of *Silene*. Annual and biennial plants tend to have weak or soft stem bases, whereas perennials usually have firm stem bases. This information is available in the species notes. Noting habit on herbarium specimens is helpful when the plant bases are not included in the collections. Flowers may be open during the day or at night. Night-flowering species can be determined by inrolled petals and relatively closed flowers during the daytime. Petals in the genus are often divided into a conspicuous, narrow, basal portion [the limb] and an expanded, apical portion [the claw]. At the junction of the limb and claw can sometimes be found a pair of small, lateral lobes called auricles. Also at the junction can be found a pair of appendages. These appendages are located on the adaxial surface, and can form a small, crown-like structure when the petals are spread open.

KEY TO THE SPECIES
1a. Plants matted, pulvinate, alpine, perennial, 0.3–0.6 dm tall; flowers solitary at the tips of
    branches; petals purple .... *S. acaulis*
1b. Plants upright, simple or branched, mostly weedy of non-alpine habitats, annual to
    perennial, 1.0–15.0 dm tall; flowers axillary or, more commonly, in terminal cymes;
    petals white to pink to red (purple)
    2a. Plants dioecious (sometimes polygamous); carpellate flowers with 5 styles; staminate
        flowers with a simple, undivided style; capsules dehiscing by 10 valves
        3a. Flowers with white petals, fragrant, opening in the evening; valves of the capsule
            erect or slightly recurved after dehiscence .... *S. latifolia*
        3b. Flowers with red petals, inodorous, opening during the day; valves of the capsule
            recurved after dehiscence .... *S. dioica*
    2b. Plants synoecious; flowers with 3(4) styles; capsules dehiscing by 6(8) valves
        4a. Inflorescence a monochasial cyme, appearing as a secund raceme

5a. Axis of the inflorescence commonly forking; petals lobed to below the middle; stems coarse, 3.0–8.0(–15.0) dm tall .... *S. dichotoma*

5b. Axis of the inflorescence usually simple; petals retuse or lobed nearly to the middle; stems shorter, 1.0–4.5 dm tall

    6a. Calyx 6.0–9.0 mm long in fruit, its lobes acuminate at the apex; petals entire to retuse at the apex; flowers erect to ascending .... *S. gallica*

    6b. Calyx 12.0–18.0 mm long in fruit, its lobes obtuse at the apex; petals lobed nearly to the middle; flowers divergent or nodding .... *S. pendula*

4b. Inflorescence of solitary flowers from the axils or a cyme (the cyme sometimes panicle-like)

7a. Flowers solitary on a peduncle from the axils of the upper leaves ...... *S. nivea*

7b. Flowers in a terminal cyme

    8a. Calyx glabrous (or sometimes minutely pubescent near the base)

        9a. Stems glutinous below the nodes; calyx fitting rather tightly over the capsule, 10-nerved

            10a. Ovary on a carpophore 7.0–8.0 mm long; appendages of petals 2.0–3.0 mm long; calyx 13.0–17.0 mm long; inflorescence tending to be compact with short pedicels .... *S. armeria*

            10b. Ovary on a carpophore 1.0 mm long; appendages of petals minute or absent; calyx 4.0–10.0 mm long; inflorescence tending to be open with long, slender pedicels .... *S. antirrhina*

        9b. Stems glabrous; calyx inflated, fitting loosely over the body of the capsule, 20-nerved

            11a. Calyx with 20 obscure veins of equal length connected by many, smaller, anastomosing veins, about 2.0 cm long in fruit .... *S. vulgaris*

            11b. Calyx with 10 long veins and 10 short veins, with few or no anastomosing veins, 1.0–1.2 cm long in fruit .... *S. csereii*

    8b. Calyx pubescent, especially on the nerves

        12a. Calyx *ca*. 1.5 cm long in flower, becoming 2.5–3.0 cm long in fruit; petals pink on the adaxial surface, yellow on the abaxial surface, opening at night, with auricles 1.0–1.5 mm long; leaves lanceolate to lance-ovate, 2.0–4.0 cm wide .... *S. noctiflora*

        12b. Calyx 0.7–1.0 cm long; petals white or pink, open during the day, with inconspicuous auricles; leaves narrow-lanceolate to oblanceolate, 0.5–1.5 cm wide .... *S. nutans*

Silene acaulis (L.) Jacq. var. ***exscapa*** (All.) DC.
synonym(s): —
common name(s): moss campion
habit: pulvinate-cespitose, perennial herb
range: circumboreal; s. to ME and NH, w. to MT and WA
state frequency: very rare
habitat: gravelly, rocky barrens of alpine areas
notes: This species is ranked SX by the Maine Natural Areas Program.

Silene antirrhina L.

synonym(s): —
common name(s): sleepy catchfly
habit: erect to ascending, branched, annual or biennial herb
range: Quebec, s. to FL, w. to Mexico, n. to British Columbia
state frequency: uncommon
habitat: dry woods, fields, sandy waste places
notes: The flowers in the species sometimes lack petals.

Silene armeria L.

synonym(s): —
common name(s): sweet William catchfly
habit: erect, branched, annual herb
range: Europe; escaped from cultivation to Quebec, s. to FL, w. to IA, n. to MN
state frequency: uncommon
habitat: roadsides, waste places
notes: —

Silene csereii Baumg.

synonym(s): —
common name(s): —
habit: erect, biennial herb
range: Europe; naturalized to ME, s. to NJ, w. to MO, n. to MT
state frequency: rare
habitat: roadsides, waste places
notes: —

Silene dichotoma Ehrh.

synonym(s): —
common name(s): forking catchfly
habit: erect, annual or biennial herb
range: Eurasia; naturalized to much of North America
state frequency: uncommon
habitat: fields, waste places
notes: —

Silene dioica (L.) Clairville

synonym(s): *Lychnis dioica* L.
common name(s): red campion, red cockle
habit: erect, branched, perennial herb
range: Eurasia; naturalized to Newfoundland, s. to DE, w. to MO, n. to Ontario
state frequency: uncommon
habitat: waste places, roadsides
notes: Known to hybridize with *Silene latifolia*.

Silene gallica L.

synonym(s): *Silene anglica* L.
common name(s): small-flowered catchfly
habit: erect, annual or biennial herb
range: Eurasia; naturalized to New Brunswick, s. to FL, w. to MO, n. to British Columbia
state frequency: rare
habitat: waste places, roadsides
notes: —

Silene latifolia Poir. ssp. **alba** (P. Mill.) Greuter & Burdet
synonym(s): *Lychnis alba* P. Mill.
common name(s): white campion, white cockle
habit: branched, annual, biennial, or short-lived perennial herb
range: Europe; naturalized to Quebec, s. to NC, w. to UT and CA, n. to British Columbia
state frequency: occasional
habitat: roadsides, fields, waste places
notes: Known to hybridize with *Silene dioica*.

Silene nivea (Nutt.) Muhl. *ex* Otth
synonym(s): *Silene alba* Muhl., *Cucubalus niveus* Nutt.
common name(s): white campion, snowy campion
habit: ascending, rhizomatous, perennial herb
range: NJ, s. to VA, w. to MO, n. to SD and MN; naturalized to New Brunswick and ME
state frequency: rare
habitat: rich woods
notes: —

Silene noctiflora L.
synonym(s): —
common name(s): sticky cockle, night-flowering catchfly
habit: annual herb
range: Europe; naturalized throughout most of the United States and southern Canada
state frequency: occasional
habitat: waste places
notes: —

Silene nutans L.
synonym(s): —
common name(s): —
habit: erect, perennial herb
range: Europe; escaped from cultivation to ME, s. to NY, w. to OH
state frequency: very rare
habitat: roadsides
notes: —

Silene pendula L.
synonym(s): —
common name(s): nodding catchfly
habit: erect, annual or biennial herb
range: Europe; escaped from cultivation in ME
state frequency: very rare
habitat: roadsides, fields
notes: —

Silene vulgaris (Moench) Garcke
synonym(s): *Cucubalus latifolius* P. Mill, *Silene cucubalus* Wibel, *Silene inflata* Sm.
common name(s): bladder campion
habit: robust, depressed to ascending, perennial herb
range: Europe; naturalized to Newfoundland, s. to VA, w. to CO and OR, n. to British Columbia
state frequency: common
habitat: roadsides, fields, shores, waste places
notes: —

Spergula
(1 species)

Spergula arvense L.
synonym(s): —
common name(s): corn-spurrey
habit: simple to branched, annual herb
range: Europe; naturalized to Newfoundland, s. to FL, w. to CA, n. to AK
state frequency: uncommon
habitat: cultivated ground, waste places
notes: —

Spergularia
(3 species)

KEY TO THE SPECIES
1a. Flowers with 6–10 stamens; seeds 0.4–0.6 mm long; plants with fascicles of leaves in
the axils; leaves scarcely fleshy; plants not, or only casually, halophytic .... *S. rubra*
1b. Flowers with 2–5 stamens; seeds 0.6–1.4 mm long; plants without fascicles of leaves in
the axils; leaves fleshy; plants halophytic
 2a. Leaves blunt at the apex; seeds 0.8–1.4 mm long, minutely reticulate; stipules
 1.0–2.8 mm long; sepals glabrous .... *S. canadensis*
 2b. Leaves short-mucronate at the apex; seeds 0.6–0.8 mm long, glandular-papillate;
 stipules 2.0–4.0 mm long; sepals glabrous or glandular-pubescent .... *S. salina*

Spergularia canadensis (Pers.) G. Don
synonym(s): *Tissa canadensis* (Pers.) Britt.
common name(s): northern sand spurrey
habit: prostrate to decumbent, branched, annual herb
range: Labrador, s. to NY; also AK, s. to CA
state frequency: uncommon
habitat: muddy or sandy soil of estuaries and other tidal areas
notes: —

Spergularia rubra (L.) J. & K. Presl
synonym(s): *Tissa rubra* (L.) Britt.
common name(s): roadside sand spurrey
habit: prostrate to ascending, simple or branched, annual or short-lived perennial herb
range: Europe; naturalized to Newfoundland, s. to VA, w. to AL, n. to MI, WI, and Ontario; also NM,
CA, and British Columbia
state frequency: uncommon
habitat: dry, sandy or gravelly soil
notes: —

Spergularia salina J. & K. Presl
synonym(s): *Spergularia marina* (L.) Griseb., *Spergularia marina* (L.) Griseb. var. *leiosperma*
(Kindb.) Guerke, *Tissa marina* (L.) Britt.
common name(s): saltmarsh sand spurrey
habit: fleshy, erect to prostrate, simple or branched, annual herb
range: Eurasia; naturalized to Newfoundland, s. to FL, w. to TX; also British Columbia; occasionally
inland along salted highways

state frequency: uncommon
habitat: saltmarshes, estuaries
notes: —

Stellaria
(8 species)

KEY TO THE SPECIES

1a. Leaves with distinct petioles, the blade ovate; stems terete, without angles, pubescent in 1 or 2 lines; sepals pubescent on the abaxial surface .... *S. media*

1b. Leaves sessile or subsessile, the blade linear to ovate; stems commonly 4-angled and glabrous; sepals glabrous, or sometimes ciliate on the margin

  2a. Flowers solitary from the axils of leaves or in terminal cymes and subtended by foliaceous bracts (the foliaceous bracts sometimes with scarious margins in *S. borealis*)

    3a. Leaves 0.3–1.0(–1.5) x 0.15–0.5 cm, fleshy, many of the axils bearing sterile tufts or branchlets; capsule shorter than to equaling the length of the sepals; plants halophytic .... *S. humifusa*

    3b. Leaves 0.7–6.0 x 0.2–0.8 cm, not fleshy, usually without sterile tufts or branchlets; capsule much longer than the sepals; plants not halophytic

      4a. Leaves lance-elliptic or lance-ovate to ovate, 0.7–2.5 cm long ....
.... *S. calycantha*

      4b. Leaves lance-linear to lanceolate, 2.5–6.0 cm long .... *S. borealis*

  2b. Flowers in terminal cymes, subtended by conspicuously reduced, scarious bracts

    5a. Sepals 2.5–3.5 mm long; petals shorter than the sepals or absent; seeds 0.5–0.7 mm long .... *S. alsine*

    5b. Sepals 3.5–8.0 mm long; petals as long as or longer than the sepals; seeds 0.7–1.6 mm long

      6a. Inflorescence open, with spreading to divaricate pedicels; seeds 0.7–1.2 mm long; leaves commonly ciliate at the base; flowers 0.9–1.3 cm wide

        7a. Seeds coarsely rugose-tuberculate; sepals conspicuously 3-veined, 4.5–5.5 mm long; leaves widest below the middle; inflorescence many-flowered .... *S. graminea*

        7b. Seeds obscurely sculpted, nearly smooth; sepals weakly 3-veined, 3.5–4.5 mm long; leaves commonly widest at or above the middle; inflorescence with relatively fewer flowers .... *S. longifolia*

      6b. Inflorescence closer, with erect pedicels; seeds 1.4–1.6 mm long; leaves glabrous; flowers 1.5–2.2 cm wide .... *S. palustris*

Stellaria alsine Grimm
synonym(s): *Alsine uliginosa* (Murr.) Britt.
common name(s): bog stitchwort
habit: decumbent to ascending, annual or biennial herb
range: Newfoundland, s. to MD, w. to WV, n. to NY and Quebec
state frequency: uncommon
habitat: cold, wet marshes and springy areas
notes: —

Stellaria borealis Bigelow
synonym(s): *Alsine borealis* (Bigelow) Britt., *Stellaria calycantha* (Ledeb.) Bong. var. *floribunda* (Fern.) Fern., *Stellaria calycantha* (Ledeb.) Bong. var. *isophylla* (Fern.) Fern.
common name(s): northern stitchwort
habit: branched, rhizomatous, perennial herb
range: Greenland, s. to WV, w. to AZ, n. to AK
state frequency: uncommon
habitat: moist, shaded soils
notes: —

Stellaria calycantha (Ledeb.) Bong.
synonym(s): *Alsine calycantha* (Ledeb.) Rydb.
common name(s): alpine stitchwort, chickweed
habit: stoloniferous, perennial herb
range: Newfoundland, s. to MA, w. to CA, n. to AK
state frequency: uncommon
habitat: seepy areas
notes: —

Stellaria graminea L.
synonym(s): *Alsine graminea* (L.) Britt.
common name(s): common stitchwort
habit: weak-stemmed, perennial herb
range: Europe; naturalized to Newfoundland, s. to SC, w. to KS, n. to MN and Ontario
state frequency: occasional
habitat: fields, roadsides, grassy areas
notes: —

Stellaria humifusa Rottb.
synonym(s): *Alsine humifusa* (Rottb.) Britt.
common name(s): saltmarsh stitchwort
habit: fleshy, ascending to trailing, perennial herb
range: circumboreal; s. to ME; also s. along the west coast to OR
state frequency: rare
habitat: saltmarshes, saline or brackish shores
notes: —

Stellaria longifolia Muhl. *ex* Willd.
synonym(s): *Alsine longifolia* (Muhl. *ex* Willd.) Britt.
common name(s): long-leaved stitchwort
habit: weak-stemmed, perennial herb
range: circumboreal; s. to VA, w. to AZ and CA
state frequency: uncommon
habitat: damp shores and meadows
notes: —

Stellaria media (L.) Vill.
synonym(s): *Alsine media* L.
common name(s): common chickweed
habit: weak-stemmed, branched, annual herb
range: Eurasia; naturalized throughout North America
state frequency: occasional
habitat: lawns, cultivated ground, waste places
notes: —

Stellaria palustris (Murr.) Retz.
synonym(s): *Alsine glauca* (With.) Britt.
common name(s): meadow starwort
habit: slender, erect, rhizomatous, perennial herb
range: Europe; naturalized locally to Quebec and ME
state frequency: very rare
habitat: shores, grassy areas
notes: —

Vaccaria
(1 species)

Vaccaria hispanica (P. Mill.) Rauschert
synonym(s): *Saponaria vaccaria* L.
common name(s): cow herb, cow-cockle
habit: branched, taprooted, annual herb
range: Europe; naturalized throughout much of North America
state frequency: rare
habitat: waste places, cultivated ground
notes: —

PODOSTEMACEAE
(1 genus, 1 species)

Podostemum
(1 species)

Podostemum ceratophyllum Michx.
synonym(s): —
common name(s): threadfoot
habit: submerged, aquatic, perennial herb
range: Quebec and New Brunswick, s. to GA, w. to LA and OK, n. to Ontario
state frequency: very rare
habitat: fast-flowing, rocky or ledgey streams
notes: This species is ranked S1/S2 by the Maine Natural Areas Program.

GERANIACEAE
(2 genera, 11 species)

KEY TO THE GENERA
1a. Leaves pinnately lobed or pinnately divided; stamens 10 per flower, but the outer series
of filaments sterile and lacking anthers; seeds smooth; carpel beaks separating from the
style column, dehiscing from apex to base, the free portions of the carpel beaks
becoming spirally coiled; carpel beaks pubescent on the adaxial [inner] surface ....
.... ***Erodium***

1b. Leaves palmately lobed or palmately divided; stamens 10 per flower, all fertile (except *Geranium pusillum* with 3–5 sterile filaments); seeds reticulate or striate (smooth in *G. pusillum*); carpel beaks remaining partially adnate to the style column, dehiscing from base to apex, the free portions of the carpel beaks merely recurving; carpel beaks glabrous on the adaxial surface .... ***Geranium***

Erodium
(3 species)

Ovary and fruit are similar to *Geranium* (*q.v.*) except as noted in the key.

KEY TO THE SPECIES

1a. Leaves pinnately divided; style column mostly 2.5–4.5 cm long
    2a. Leaflets sessile, prominently lobed, the sinuses extending half to nearly the entire distance to the leaflet midrib; anther-bearing [fertile] filaments without basal teeth ....
    .... ***E. cicutarium***
    2b. Leaflets petiolulate, toothed to lobed, the sinuses rarely extending more than half way to the leaflet midrib; anther-bearing filaments with 2 teeth at the base ....
    .... ***E. moschatum***
1b. Leaves pinnately lobed, the leaf segments not completely divided to the midrib or the midrib with a prominent wing; style column mostly 5.5–11.3 cm long .... ***E. botrys***

Erodium botrys (Cav.) Bertol.
synonym(s): —
common name(s): bristly storksbill
habit: prostrate to erect, taprooted, annual or biennial herb
range: Europe; escaped from cultivation
state frequency: very rare
habitat: areas of cultivation
notes: —

Erodium cicutarium (L.) L'Hér. *ex* Ait.
synonym(s): —
common name(s): alfileria, redstem filaree
habit: simple or branched, annual or biennial herb
range: Europe; naturalized to New Brunswick and Nova Scotia, s. to VA, w. to TX and Mexico, n. to MI
state frequency: rare
habitat: fields, roadsides, waste places
notes: —

Erodium moschatum (L.) L'Hér. *ex* Ait.
synonym(s): —
common name(s): musk storksbill, whitestem filaree
habit: simple or branched, annual or biennial herb
range: Europe; naturalized to ME, s. to DE
state frequency: very rare
habitat: waste places
notes: —

Geranium
(8 species)

The gynoecium of *Geranium* is composed of 5 connate carpels that form a compound ovary. Each carpel is made up of a basal, thickened portion, called the carpel body, that contains the ovules and an elongate, apical portion, called the carpel beak, that covers and is adnate to the style column. Above the carpel beaks project the styles, which form a style beak in fruit. The ovary matures as a schizocarp, the 5 1-seeded mericarps separating from each other, attached to the style column by their associated carpel beaks.

KEY TO THE SPECIES

1a. Leaves palmately compound, the terminal segment petiolulate; carpel bodies separating from the carpel beaks but remaining attached to the style column for a period of time by 2 subapical filaments .... *G. robertianum*
1b. Leaves palmately lobed, the leaf segments confluent; carpel bodies remaining attached to the carpel beaks without subapical filaments
    2a. Petals 12.0–20.0 mm long; anthers 2.0–3.0 mm long; carpel beaks 5.0–10.0 mm long; plants perennial from a stout rhizome
        3a. Stems with a single pair of leaves, the other leaves basal; petals entire or very slightly retuse at the apex, blue-purple or pink (white)
            4a. Sepals, pedicels, and carpel beaks glandular-hirsute; pedicels deflexed after anthesis; petals blue-purple .... *G. pratense*
            4b. Sepals, pedicels, and carpel beaks pubescent but not glandular; pedicels suberect after anthesis; petals pink (white) .... *G. maculatum*
        3b. Stems with several leaves; petals retuse at the apex, deep red (white) ....
        .... *G. sanguineum*
    2b. Petals 2.5–10.0 mm long; anthers 0.5–0.8 mm long; carpel beaks 0.0–6.0 mm long; plants annual or biennial, commonly from a taproot
        5a. Three to 5 of the 5 outer stamens sterile and lacking anthers; seeds smooth; petals 2.5–4.5 mm long .... *G. pusillum*
        5b. All of the stamens fertile and bearing anthers; seeds reticulate or striate, sometimes obscurely so; petals 5.0–10.0 mm long
            6a. Sepals acute at the apex, sometimes with a short mucro; carpel bodies glabrous; seeds striated .... *G. molle*
            6b. Sepals acuminate at the apex or tipped with an awn 0.7–3.0 mm long; carpel bodies at least sparsely hirsute; seeds reticulate
                7a. Pedicels up to 2.0 times as long as the sepals; style beak 1.0–2.0 mm long; inflorescence rather dense .... *G. carolinianum*
                7b. Pedicels elongate, more than 2.0 times as long as the sepals; style beak 4.0–5.0 mm long; inflorescence less dense .... *G. bicknellii*

Geranium bicknellii Britt.
synonym(s): —
common name(s): Bicknell's wild geranium, northern geranium
habit: simple or branched, annual or biennial herb
range: Newfoundland, s. to CT, w. to IA, n. to SD
state frequency: uncommon
habitat: open woods, fields, clearings, disturbed areas
notes: —

Geranium carolinianum L.

synonym(s): —
common name(s): Carolina cranesbill
habit: simple or branched, annual herb
range: ME, s. to FL, w. to TX and CA, n. to British Columbia
state frequency: rare
habitat: dry fields, waste places, and rocky woods
key to the subspecies:
  1a. Upper internodes short, creating a compact inflorescence composed of umbel-like clusters; pubescence *ca.* 1.0 mm long .... *G. c.* var. *confertiflorum* Fern.
  1b. Upper internodes longer, the inflorescence open; pubescence *ca.* 0.5 mm long .... *G. c.* var. *carolinianum*
notes: —

Geranium maculatum L.

synonym(s): —
common name(s): wild geranium, wild cranesbill, spotted cranesbill
habit: erect, rhizomatous, perennial herb
range: ME, s. to GA, w. to KS, n. to Manitoba
state frequency: uncommon
habitat: woods, thickets, meadows
notes: —

Geranium molle L.

synonym(s): —
common name(s): dovesfoot
habit: spreading to ascending, basally branched, annual herb
range: Eurasia; naturalized to our area
state frequency: rare
habitat: lawns, waste places
notes: —

Geranium pratense L.

synonym(s): —
common name(s): meadow geranium, meadow cranesbill
habit: erect, rhizomatous, perennial herb
range: Eurasia; escaped from cultivation to Labrador, s. to CT, w. and n. to Manitoba
state frequency: uncommon
habitat: fields, roadsides
notes: —

Geranium pusillum L.

synonym(s): —
common name(s): small geranium, small-flowered cranesbill
habit: slender, spreading to ascending, branched, annual or biennial herb
range: Europe; naturalized to MA and CT, s. to NC, w. to AR and OR, n. to British Columbia; also ME
state frequency: very rare
habitat: fields, roadsides, waste places
notes: —

Geranium robertianum L.
synonym(s): *Robertiella robertiana* (L.) Hanks
common name(s): herb Robert
habit: branched, annual or biennial herb
range: Eurasia; naturalized to Newfoundland, s. to MD, w. to IL, n. to Manitoba
state frequency: uncommon
habitat: rich, rocky woods, gravelly shores
notes: —

Geranium sanguineum L.
synonym(s): —
common name(s): blood-red cranesbill
habit: leafy-stemmed, perennial herb
range: Europe; escaped from cultivation to locally
state frequency: very rare
habitat: areas of cultivation
notes: —

OXALIDACEAE
(1 genus, 4 species)

Oxalis
(4 species)

Character of the pubescence is important for identifying some of the weedy species. Proper assessment requires high magnification, 20× or greater.

KEY TO THE SPECIES
1a. Petals white with pink to purple veins (or rarely entirely pink to purple); plants lacking a leafy stem; peduncles with 1 flower .... *O. montana*
1b. Petals yellow; plants with leafy stems; peduncles with 1–9 flowers
    2a. Stipules absent; pubescence of the stem pointed at the apex, septate; plants with rhizomes .... *O. stricta*
    2b. Stipules present, though sometimes small; pubescence of the stem blunt at the apex, without septa; plants without rhizomes, sometimes stoloniferous
        3a. Stems trailing and rooting at the nodes; seeds brown; the 5 longer filaments glabrous .... *O. corniculata*
        3b. Stems erect to decumbent, but neither trailing nor rooting; seeds brown with white ridges; the 5 longer filaments pubescent .... *O. dillenii*

Oxalis corniculata L.
synonym(s): *Xanthoxalis corniculata* (L.) Small
common name(s): creeping yellow wood-sorrel
habit: trailing, perennial herb
range: tropics; naturalized to Newfoundland, s. to FL, w. to TX, n. to ND
state frequency: uncommon
habitat: areas of cultivation, fields, roadsides
notes: —

Oxalis dillenii Jacq. ssp. *filipes* (Small) Eiten

synonym(s): *Oxalis filipes* Small, *Oxalis florida* Salisb., *Xanthoxalis brittoniae* (Small) Small, *Xanthoxalis filipes* (Small) Small
common name(s): southern yellow wood-sorrel
habit: erect to decumbent, tufted, perennial herb
range: New Brunswick, s. to FL, w. to MO, n. to IA; now a cosmopolitan weed
state frequency: uncommon
habitat: sandy fields, thickets, and borders of woods
notes: —

Oxalis montana Raf.

synonym(s): *Oxalis acetosella auct. non* L.
common name(s): common wood-sorrel, northern wood-sorrel
habit: creeping, rhizomatous, perennial herb
range: Newfoundland, s. to mountains of NC, w. to TN, n. to MN and Saskatchewan
state frequency: occasional
habitat: rich, moist woods
notes: This species produces cleistogamous flowers late in the season on short, recurved peduncles.

Oxalis stricta L.

synonym(s): *Oxalis europaea* Jord., *Xanthoxalis bushii* Small, *Xanthoxalis stricta* (L.) Small
common name(s): common yellow wood-sorrel
habit: prostrate to erect, long-rhizomatous, perennial herb
range: New Brunswick, s. to FL, w. to Mexico, n. to British Columbia
state frequency: common
habitat: dry fields, roadsides, and open areas
notes: The 5 longer filaments in this species are glabrous.

PARNASSIACEAE
(1 genus, 1 species)

Parnassia
(1 species)

Parnassia glauca Raf.

synonym(s): *Parnassia americana* Muhl.
common name(s): grass-of-parnassus
habit: erect, scapose, perennial herb
range: Newfoundland, s. to NJ, w. to SD, n. to Saskatchewan
state frequency: rare
habitat: calcareous fens
notes: This species is ranked S3 by the Maine Natural Areas Program.

CELASTRACEAE
(2 genera, 4 species)

KEY TO THE GENERA

1a. Leaves alternate; twigs terete; perianth 5-merous; fruit an orange or orange-yellow capsule; some of the flowers unisexual; plants woody vines .... *Celastrus*

1b. Leaves opposite; twigs quadrangular; perianth 4-merous; fruit a pink or purple capsule; flowers bisexual; plants upright shrubs .... *Euonymus*

Celastrus
(2 species)

KEY TO THE SPECIES

1a. Leaf blades elliptic to ovate, serrate, about 2.0 times as long as wide; inflorescence a many-flowered, terminal panicle .... *C. scandens*

1b. Leaf blades broad-ovate to suborbicular, crenate, usually less than 2.0 times as long as wide; inflorescence a few-flowered, axillary cyme .... *C. orbiculata*

Celastrus orbiculata Thunb.
synonym(s): —
common name(s): oriental bittersweet
habit: liana
range: Asia; escaped from cultivation to ME, s. to VA
state frequency: uncommon
habitat: thickets, open woods, roadsides
notes: —

Celastrus scandens L.
synonym(s): —
common name(s): American bittersweet, climbing bittersweet, waxwork
habit: liana
range: New Brunswick and ME, s. to GA, w. to TX, n. to WY and Saskatchewan
state frequency: uncommon
habitat: woods, thickets, river banks, roadsides
notes: —

Euonymus
(2 species)

KEY TO THE SPECIES

1a. Leaves glabrous on the abaxial surface; capsule pink; aril of the seed orange; petals green-white; cymes with usually 2–5 flowers .... *E. europaea*

1b. Leaves pubescent on the abaxial surface; capsule purple; aril of the seed red; petals brown-purple; cymes with usually 7–15 flowers .... *E. atropurpurea*

Euonymus atropurpurea Jacq.
synonym(s): —
common name(s): burning bush, wahoo
habit: large, erect shrub to small tree
range: NY, s. to FL, w. to TX, n. to ND and Ontario; escaped from cultivation in ME

state frequency: very rare
habitat: rich, moist woods and thickets
notes: —

Euonymus europaea L.
synonym(s): —
common name(s): European spindle-tree
habit: shrub
range: Europe; escaped from cultivation in our area
state frequency: very rare
habitat: hedgerows, fields, roadsides, waste places
notes: —

MYRICACEAE
(2 genera, 3 species)

KEY TO THE GENERA
1a. Leaves pinnately lobed, with stipules; bracts exceeding the fruit, persistent, forming an
    enclosing bur .... *Comptonia*
1b. Leaves without lobes, entire or toothed, without stipules; bracts shorter than to equaling
    the fruit, not forming a bur .... *Myrica*

Comptonia
(1 species)

Distinctive shrub with aromatic, marcescent leaves.

Comptonia peregrina (L.) Coult.
synonym(s): *Myrica aspleniifolia* L., *Myrica peregrina* (L.) Kuntze
common name(s): sweetfern
habit: colonial, aromatic shrub
range: Nova Scotia and New Brunswick, s. to VA and mountains of GA, w. to IL, n. to MN and
    Manitoba
state frequency: common
habitat: dry, sandy or gravelly, open areas
notes: Sweetfern has marcescent leaves, making winter identification easier.

Myrica
(2 species)

Shrubs with aromatic foliage and resin-dotted twigs.

KEY TO THE SPECIES
1a. Carpellate bracts 2, accrescent, clasping the achene in fruit; achene flattened, not
    concealed by a thick layer of wax; staminate aments borne at the apex of the previous
    year's twig; leaf blades serrate in the distal third; winter buds dark brown, ovoid, acute at
    the apex; preformed staminate aments present .... *M. gale*
1b. Carpellate bracts 4–6, small and inconspicuous, even in fruit; achene subglobose,
    covered with a thick layer of blue-white wax; staminate aments borne below the apex of

the previous year's twig; leaf blades serrate in the distal half; winter buds red, subglobose, obtuse at the apex; preformed staminate aments absent ...... *M. pensylvanica*

Myrica gale L.

synonym(s): *Gale palustris* Chev., *Myrica gale* L. var. *subglabra* (Chev.) Fern.
common name(s): sweet gale
habit: branched, resinous, aromatic shrub
range: circumboreal; s. to NJ and mountains of NC, w. to OR
state frequency: occasional
habitat: swamps, shores
notes: —

Myrica pensylvanica Loisel.

synonym(s): *Cerothamnus pensylvanica* (Mirbel) Moldenke
common name(s): bayberry, northern bayberry
habit: aromatic, colonial shrub
range: Newfoundland, s. to NC, w. to OH, n. to Ontario
state frequency: uncommon
habitat: shores, edges, woodlands, often coastal areas
notes: —

JUGLANDACEAE
(2 genera, 3 species)

KEY TO THE GENERA

1a. Leaves with 11–19 leaflets; pith chambered; nut with an indehiscent husk; middle lateral leaflets the largest; each staminate ament borne separately; staminate flowers with 8–40 stamens and glabrous anthers .... *Juglans*

1b. Leaves with 5–9(–11) leaflets; pith continuous; nut with a 4-valved, dehiscent husk; terminal leaflet the largest; staminate aments borne in clusters of 3; staminate flowers with 3–10 stamens and pilose anthers .... *Carya*

Carya
(2 species)

To facilitate identification of herbarium material, every effort should be made to collect winter buds and fruit, in addition to the leaves.

KEY TO THE SPECIES

1a. Winter buds orange-yellow, with 4–6 valvate scales; husk prominently keeled at the sutures, dehiscing to near the middle; leaves commonly with 7–9 leaflets; seeds bitter tasting .... *C. cordiformis*

1b. Winter buds gray, with 10–12 imbricate scales; husk lacking prominent keels at the sutures, dehiscing to near the base; leaves commonly with 5 leaflets; seeds sweet tasting
.... *C. ovata*

Carya cordiformis (Wangenh.) K. Koch
synonym(s): —
common name(s): bitternut hickory, pignut

habit: tree
range: Quebec and ME, s. to FL, w. to TX, n. to NE and MN
state frequency: very rare
habitat: wet to dry woods, stream banks, swamps
notes: This species is ranked S1 by the Maine Natural Areas Program.

Carya ovata (P. Mill.) K. Koch
synonym(s): —
common name(s): shagbark hickory
habit: tree
range: Quebec and ME, s. to GA, w. to TX, n. to MN
state frequency: uncommon
habitat: rocky woods and bottomlands
notes: —

Juglans
(1 species)

Although commonly planted, trees of this genus are uncommon as native plants in Maine forests.

Juglans cinerea L.
synonym(s): —
common name(s): butternut, white walnut
habit: tree
range: New Brunswick, s. to GA, w. to AR, n. to ND
state frequency: uncommon
habitat: moist, rich woods
notes: Ellipsoidal fruits and an evident, pubescent ridge on the distal edge of the leaf scar distinguish *Juglans cinerea* from the cultivated species *J. nigra*, which has globose fruits and lacks a pubescent ridge distal to the leaf scar.

BETULACEAE
(5 genera, 14 species)

KEY TO THE GENERA FOR USE WITH VEGETATIVE MATERIAL

1a. Bark smooth, light gray, with slightly raised, rounded, longitudinal ridges; buds quadrangular in cross-section .... ***Carpinus***
1b. Bark smooth or exfoliating, variously colored but not light gray, without sinewy ridges; buds terete in cross-section
    2a. Bark gray-brown and furrowed, exfoliating in vertical strips; bud scales conspicuously longitudinally striate .... ***Ostrya***
    2b. Bark white, yellow, or brown, smooth or exfoliating in horizontal strips or in patches; bud scales not, or only inconspicuously, striate
        3a. Phyllotaxis 1/3; pith often triangular; buds stalked except in *A. viridis* .... ***Alnus***
        3b. Phyllotaxis 2/5; pith terete; buds not stalked
            4a. Plants without conspicuous short shoots; buds with 4 or more visible bud scales; bark not exfoliating .... ***Corylus***

4b. Plants with conspicuous, often 2-leaved short shoots; buds of the long shoots usually with 3 visible bud scales; bark in many species exfoliating in horizontal strips or in patches .... ***Betula***

KEY TO THE GENERA FOR USE WITH REPRODUCTIVE MATERIAL

1a. Fruit a nut, subtended and often concealed by the greatly accrescent involucre; scales of the carpellate ament deciduous; anthers pubescent; staminate flowers lacking a calyx

    2a. Nut 1.0–1.5 cm wide, closely invested by the husk-like involucre; carpellate aments compact, ovoid .... ***Corylus***

    2b. Nut *ca.* 0.5 cm wide, loosely invested or merely subtended by a bladder-like or leaf-like involucre; carpellate aments slender and longer

        3a. Involucre a closed, inflated sac; preformed staminate aments present .... ***Ostrya***

        3b. Involucre an open, leaf-like bract; preformed staminate aments absent .... ***Carpinus***

1b. Fruit a samara, not subtended by an involucre; scales of the carpellate ament persistent; anthers glabrous; staminate flowers with a minute calyx

    4a. Carpellate scales thin, deciduous with the fruit; staminate flowers with 4 monothecal half-anthers .... ***Betula***

    4b. Carpellate scales thick, woody, long-persistent after the falling fruit; staminate flowers with 4 dithecal anthers .... ***Alnus***

Alnus
(3 species)

KEY TO THE SPECIES

1a. Winter buds sessile, with 3–6 scales of unequal sizes, acute to acuminate at the apex, glabrous and glutinous; samara with a thin wing; only the staminate aments emerging from naked buds .... ***A. viridis***

1b. Winter buds stalked, with 2 or 3 scales of equal size, rounded to acute at the apex, scurfy-pubescent; samara merely thin-edged; aments emerging from naked buds

    2a. Leaf blades coarsely double-serrate, elliptic to ovate, with conspicuous cross-veins on the abaxial surface; bark with evident, white lenticels; axis of the carpellate inflorescence arching, without abrupt bends .... ***A. incana***

    2b. Leaf blades finely and regularly serrulate, elliptic to obovate, with delicate or inconspicuous cross-veins on the abaxial surface; bark with fewer, darker lenticels; axis of the young, carpellate inflorescence with 1 or more abrupt bends .... ***A. serrulata***

Alnus incana (L.) Moench ssp. ***rugosa*** (Du Roi) Clausen
synonym(s): *Alnus rugosa* (Du Roi) Spreng., *Alnus rugosa* (Du Roi) Spreng. var. *americana* (Regel) Fern.
common name(s): speckled alder
habit: shrub to small tree
range: circumboreal; s. in the east to MD
state frequency: common
habitat: wet soil of swamps and stream margins
notes: Known to hybridize with *Alnus serrulata*.

Alnus serrulata (Ait.) Willd.
synonym(s): *Alnus serrulata* (Ait.) Willd. var. *subelliptica* Fern.
common name(s): smooth alder, common alder
habit: shrub to small tree
range: Quebec and Nova Scotia, s. to FL, w. to TX, n. to MO and IL
state frequency: uncommon
habitat: stream margins, swamps, wet woods
notes: Known to hybridize with *Alnus incana* ssp. *rugosa*.

Alnus viridis (Vill.) Lam. & DC. ssp. *crispa* (Ait.) Turrill
synonym(s): *Alnus crispa* (Ait.) Pursh
common name(s): green alder, mountain alder
habit: shrub
range: circumboreal; s. in the east to MA and NY
state frequency: uncommon
habitat: bogs, rocky shores, mountains
notes: —

Betula
(7 species)

Branchlets in *Betula* are of 2 different types—long shoots and short shoots. Long shoots are elongate, with remote leaf scars. The short shoots are conspicuous, spur-like knobs with crowded leaf scars. Lateral vein counts used in the following key are for the larger leaves.

KEY TO THE SPECIES
1a. Leaf blades 0.5–5.0 cm long, with 2–4 pairs of lateral veins, obovate to orbicular; leaf apex obtuse to rounded; shrubs to 3.0 m tall
    2a. Plants of fens and bogs; leaf blades mostly 2.5–5.0 cm long; twigs with few, scattered resin glands; scales of fruiting ament with widely divergent lateral lobes ... ***B. pumila***
    2b. Plants of exposed mountain summits; leaf blades 0.5–3.0 cm long; twigs conspicuously warty, with numerous resin glands; scales of fruiting ament with the lateral lobes upturned .... ***B. glandulosa***
1b. Leaf blades 3.0–10.0 cm long, with 5 or more pairs of lateral veins, ovate, ovate-oblong to deltate; leaf apex acute to long-acuminate; trees to 30.0 m tall, or shorter in severe habitats
    3a. Larger leaves often with 12–18 pairs of lateral veins; fruiting aments 1.0–3.5 cm broad; twigs with a wintergreen odor; body of samara wider than individual wings
        4a. Bark smooth, not exfoliating, brown; scales of fruiting aments glabrous, 5.0–7.0 mm long; leaves finely and regularly serrate, generally 6 or more teeth per cm; twigs often glabrous .... ***B. lenta***
        4b. Bark exfoliating, yellow to silver; scales of fruiting aments pubescent and/or ciliate, 6.0–13.0 mm long; leaves more coarsely, and often irregularly, serrate, generally fewer than 6 teeth per cm; twigs often pubescent ...... ***B. alleghaniensis***
    3b. Larger leaves with 5–12 pairs of lateral veins; fruiting aments 0.5–1.0 cm broad; twigs without a wintergreen odor; body of samara narrower than individual wings
        5a. Leaf blades cordate at the base, with 9–12 pairs of lateral veins; scales of fruiting aments with upturned, lateral lobes .... ***B. cordifolia***

5b. Leaf blades cuneate to truncate at the base, with 5–9 pairs of lateral veins; scales of fruiting aments with divergent, lateral lobes

    6a. Leaf blades deltate, mostly with 5–7 pairs of lateral veins; leaf apex long-acuminate; staminate aments borne singly; twigs thin, with numerous, small, warty resin glands .... *B. populifolia*

    6b. Leaf blades ovate to elliptic, mostly with 7–9 pairs of lateral veins; leaf apex acute; staminate aments borne in clusters of 2–4; twigs thicker, lacking resin glands .... *B. papyrifera*

Betula alleghaniensis Britt.

synonym(s): *Betula alleghaniensis* Britt. var. *macrolepis* (Fern.) Brayshaw, *Betula lutea* Michx. f.
common name(s): yellow birch
habit: tree
range: Newfoundland, s. to DE and mountains of GA, w. to IA, n. to British Columbia
state frequency: common
habitat: moist, rich woods
notes: Known to hybridize with *Betula papyrifera* and *B. pumila*.

Betula cordifolia Regel

synonym(s): *Betula papyrifera* Marsh. var. *cordifolia* (Regel) Fern.
common name(s): mountain paper birch
habit: small tree
range: Labrador, s. to NC, w. to IA, n. to Ontario
state frequency: occasional
habitat: rocky, upland and subalpine forests, exposed summits
notes: Hybrids with *Betula glandulosa* (*q.v.*) and *B. populifolia*, called *B.* ×*caerulea* Blanch., occur in Maine.

Betula glandulosa Michx.

synonym(s): —
common name(s): dwarf birch, tundra dwarf birch
habit: depressed, matted to erect shrub
range: arctic and boreal North America, s. in the east to ME
state frequency: very rare
habitat: rocky summits and gullies of alpine areas
notes: This species is ranked S1 by the Maine Natural Areas Program. Hybrids with *Betula cordifolia*, called *B.* ×*minor* (Tuckerman) Fern., occur in central Maine. This hybrid is ranked S1 by the Maine Natural Areas Program. Also known to hybridize with *B. pumila*. This hybrid has not been reported from Maine and is unlikely due to the extreme habitat separation of the parents.

Betula lenta L.

synonym(s): —
common name(s): sweet birch, cherry birch
habit: tree
range: Quebec, s. to MD and mountains of GA, w. to OH, n. to Ontario
state frequency: uncommon; more common in the southern portion of the state
habitat: moist, rich woods
notes: —

Betula papyrifera Marsh.

synonym(s): *Betula papyrifera* Marsh. var. *commutata* (Regel) Fern., *Betula papyrifera* Marsh. var. *macrostachya* Fern., *Betula papyrifera* Marsh. var. *pensilis* Fern.

common name(s): paper birch, white birch
habit: tree
range: Newfoundland, s. to VA, w. to OR, n. to Yukon AK
state frequency: common
habitat: forests, rocky slopes
notes: Known to hybridize with *Betula alleghaniensis* and *B. pumila*.

Betula populifolia Marsh.
synonym(s): —
common name(s): gray birch
habit: small tree
range: Quebec and Nova Scotia, s. to DE, w. to IN, n. to Ontario
state frequency: common
habitat: dry to wet soil of old fields and upland woods
notes: Hybrids with *B. cordifolia*, called *B. ×caerulea* Blanch., occur in Maine.

Betula pumila L.
synonym(s): —
common name(s): swamp birch, low birch
habit: prostrate to erect shrub
range: Newfoundland, s. to MD, w. to IN, n. to Ontario
state frequency: rare
habitat: bogs, wooded swamps
notes: This species is ranked S3 by the Maine Natural Areas Program. Known to hybridize with *Betula alleghaniensis*, *B. glandulosa*, and *B. papyrifera*.

Carpinus
(1 species)

Carpinus caroliniana Walt. ssp. *virginiana* (Marsh.) Furlow
synonym(s): —
common name(s): American hornbeam, blue-beech, musclewood
habit: tall shrub to small tree
range: Nova Scotia, s. to FL, w. to TX, n. to MN
state frequency: uncommon
habitat: moist, rich woods, swamps
notes: The leaves are dotted with dark glands, especially on the abaxial surface.

Corylus
(2 species)

KEY TO THE SPECIES

1a. Husk-like involucre 4.0–7.0 cm long, with a pronounced, tubular beak, bristly with slender spicules; twigs without red, stipitate glands; staminate aments sessile; apex of buds acute .... *C. cornuta*
1b. Husk-like involucre 1.5–3.0 cm long, without a tubular beak, pubescent but not bristly; twigs with red, stipitate glands; staminate aments on a short, woody peduncle; apex of buds rounded .... *C. americana*

Corylus americana Walt.
synonym(s): —
common name(s): American hazelnut
habit: shrub
range: ME, s. to GA, w. to OK, n. to Saskatchewan
state frequency: uncommon
habitat: thickets, woods
notes: —

Corylus cornuta Marsh.
synonym(s): —
common name(s): beaked hazelnut
habit: shrub
range: Newfoundland, s. to NJ and mountains of GA, w. to CO, n. to British Columbia
state frequency: common
habitat: moist woods, borders of woods, thickets, clearings
notes: —

Ostrya
(1 species)

Ostrya virginiana (P. Mill.) K. Koch
synonym(s): *Ostrya virginiana* (P. Mill.) K. Koch var. *lasia* Fern.
common name(s): eastern hophornbeam
habit: tall shrub to tree
range: Nova Scotia, s. to FL, w. to TX, n. to Manitoba
state frequency: occasional
habitat: dry or moist, rich woods
notes: —

FAGACEAE
(3 genera, 11 species)

KEY TO THE GENERA

1a. Nuts triangular in cross-section; pith terete or obscurely angled in cross-section; winter buds slender-fusiform, 8.0–19.0 mm long; bark smooth and light gray, even in age staminate flowers in dense, spherical clusters .... ***Fagus***

1b. Nuts terete or compressed; pith with 5 prominent points; winter buds ovoid to globose, up to 10.0 mm long; bark gray to brown, furrowed in age; staminate flowers in slender aments

    2a. Carpellate flowers 2–4 per involucre; involucre spiny, enclosing 1–3 nuts at maturity; buds not clustered toward distal end of twig .... ***Castanea***

    2b. Carpellate flowers 1 per involucre; involucre not spiny, enclosing 1 nut at maturity; buds clustered toward distal end of twig .... ***Quercus***

Castanea
(1 species)

A distinctive genus of Maine members of the Fagaceae for its flowers appearing after the leaves and being, to some extent, entomophilous.

Castanea dentata (Marsh.) Borkh.
synonym(s): —
common name(s): American chestnut
habit: tree
range: ME, s. to KY and AL, w. to IL, n. to MI
state frequency: rare
habitat: dry, rocky or gravelly, acid soil
notes: This species is ranked S2/S3 by the Maine Natural Areas Program. *Castanea dentata* has been
    nearly exterminated by the chestnut blight and rarely reaches reproductive maturity. Its leaves are
    morphologically similar to *Fagus grandifolia* (*q.v.*).

Fagus
(1 species)

Fagus grandifolia Ehrh.
synonym(s): —
common name(s): American beech
habit: tree
range: Nova Scotia, s. to FL, w. to TX, n. to WI
state frequency: common
habitat: upland soils, sometimes rich
notes: The leaves of this species are somewhat similar to *Castanea dentata*. *Fagus grandifolia* has
ovate (obovate) to narrow-ovate leaves borne on petioles 4.0–12.0 mm long. *Castanea dentata* has
narrow-obovate to oblanceolate leaves borne on petioles mostly 10.0–30.0 mm long.

Quercus
(9 species)

The leaves of oaks, as well as many species, possess different dimensions depending on their exposure to sunlight. Leaves from the lower, more shaded portion of the plant will have shallower sinuses and less pubescence than leaves from the exposed portion of the plant. The number of seasons required for fruit maturation can be used to separate our oaks into 2 groups. Species with nuts maturing in the first year show a single size class of fruits on the current year's growth. Species with nuts maturing in the second year will show small, carpellate flowers on the current year's growth and larger, developing on the previous year's growth. Hybridization is common in mixed populations.

KEY TO THE SPECIES FOR USE WITH VEGETATIVE MATERIAL
1a. Leaves with bristle-tipped lobes or teeth; twigs and buds dark brown; bark usually dark
    and not flaky
  2a. Leaves gray to brown and stellate-pubescent on the abaxial surface
    3a. Leaves irregularly lobed with deltate to ovate lobes; buds rounded at the apex;
        twigs pubescent; shrubs or small trees to 6.0 m .... *Q. ilicifolia*
    3b. Leaves regularly lobed with oblong lobes; buds pointed at the apex; twigs
        glabrous; trees to 40.0 m .... *Q. velutina*

2b. Leaves green and glabrous on the abaxial surface (some hairs may be present in the axils of veins)

    4a. Buds conspicuously 4- to 5-angled, pubescent on the abaxial surface, 7.0–10.0 mm long .... *Q. velutina*

    4b. Buds terete to inconspicuously angled, glabrous or ciliate, 3.0–7.0 mm long

        5a. Leaves moderately sinused, the distance between the midvein and the tip of the lobe 3.0 times the distance between the midvein and the base of the adjacent sinuses; buds sharply pointed .... *Q. rubra*

        5b. Leaves deeply sinused, the distance between the midvein and the tip of the lobe 4.0–7.0 times the distance between the midvein and the base of the adjacent sinuses; buds merely pointed

            6a. Terminal buds 4.0–7.0 mm long, pubescent in the distal half ....
            .... *Q. coccinea*

            6b. Terminal buds 3.0–5.0 mm long, glabrous or with a few hairs near the apex .... *Q. palustris*

1b. Leaves with rounded to acute, but not bristle-tipped, lobes or teeth; twigs and buds light-, yellow-, gray-, or red-brown or purple; bark light gray and flaky to dark and furrowed

    7a. Leaves with 6–15 pairs of rounded to obtuse teeth and shallow sinuses, the distance between the midvein and the tip of the lobe up to 1.6 times as long as the distance between the midvein and the base of the adjacent sinuses

        8a. Leaves with 10–15 regular teeth, pubescent with minute, appressed, stellate hairs and tufts of simple, spreading hairs along the veins; plants of upland, rocky woods .... *Q. montana*

        8b. Leaves with 6–10 irregular teeth, pubescent with both appressed and erect, stellate hairs; plants of wet, low areas .... *Q. bicolor*

    7b. Leaves with 3–7 pairs of oblong to deltate lobes, with deep sinuses, the distance between the midvein and the tip of some of the lobes exceeding 1.5 times the distance between the midvein and the base of the adjacent sinuses

        9a. Leaves regularly lobed, glabrous at maturity; twigs glabrous, without corky ridges; plants mostly of well drained soils .... *Q. alba*

        9b. Leaves irregularly lobed, usually with a deeper central sinus, persistently stellate-pubescent; twigs glabrous to pubescent, often with corky ridges; plants of poorly drained soils .... *Q. macrocarpa*

KEY TO THE SPECIES FOR USE WITH REPRODUCTIVE MATERIAL

1a. Fruit maturing in the second year; inner surface of shell tomentose; bracts that comprise the carpellate involucre membranaceous; staminate flowers with 4–6 stamens; styles elongate

    2a. Marginal bracts of the carpellate involucre loose and projecting, forming a fringe around the nut .... *Q. velutina*

    2b. Marginal bracts of the carpellate involucre covered and concealed, not forming a fringe

        3a. Nut 8.0–11.0 mm wide, pubescent on the outer surface; inner surface of involucre pubescent .... *Q. ilicifolia*

3b. Nut 9.0–21.0 mm wide, glabrous on the outer surface; inner surface of involucre commonly glabrous or with a ring of hairs around the scar only

    4a. Carpellate involucre concealing 0.35–0.5 of the nut; nut with 1 or more rings of small pits near the apex .... *Q. coccinea*

    4b. Carpellate involucre concealing 0.2–0.35 of the nut; nut without a ring of pits near the apex

        5a. Nut 15.0–30.0 mm long; carpellate involucre 5.0–12.0 x 18.0–30.0 mm .... *Q. rubra*

        5b. Nut 10.0–16.0 mm long; carpellate involucre 3.0–6.0 x 9.5–16.0 mm wide .... *Q. palustris*

1b. Fruit maturing in the first year; inner surface of shell glabrous; bracts that comprise the carpellate involucre woody or corky at the base; staminate flowers with 6–8 stamens; styles very short, the stigmas sessile or nearly so

    6a. Marginal bracts of the carpellate involucre acuminate to caudate at the apex, forming a fringe around the nut

        7a. Fruit with a peduncle 2.0–7.0 cm long; carpellate involucre concealing 0.35–0.5 of the nut .... *Q. bicolor*

        7b. Fruit sessile or short-pedunculate; carpellate involucre concealing 0.35–0.65 or more of the nut .... *Q. macrocarpa*

    6b. Marginal bracts of the carpellate involucre relatively blunter, not forming a fringe around the nut

        8a. Carpellate involucre concealing 0.25–0.35 of the nut, the marginal scales distinct .... *Q. alba*

        8b. Carpellate involucre concealing 0.35–0.5 of the nut, the marginal scales concrescent .... *Q. montana*

Quercus alba L.

synonym(s): —
common name(s): white oak
habit: tree
range: ME, s. to FL, w. to TX, n. to MI and MN
state frequency: uncommon
habitat: dry, upland woods
notes: Known to hybridize with *Quercus bicolor*, *Q. macrocarpa*, and *Q. montana*.

Quercus bicolor Willd.

synonym(s): —
common name(s): swamp white oak
habit: tree
range: Quebec and ME, s. to NC, w. to AR, n. to MN
state frequency: very rare
habitat: floodplain forests and other low, wet areas
notes: This species is ranked S1 by the Maine Natural Areas Program. Known to hybridize with *Quercus alba* and *Q. macrocarpa*.

Quercus coccinea Muenchh.

synonym(s): —
common name(s): scarlet oak
habit: tree

range: ME, s. to GA, w. to OK, n. to MI and Ontario
state frequency: very rare
habitat: dry soil
notes: This species is ranked S1 by the Maine Natural Areas Program. Known to hybridize with
     Quercus ilicifolia, *Q. palustris*, *Q. rubra*, and *Q. velutina*.

Quercus ilicifolia Wangenh.
synonym(s): —
common name(s): bear oak, scrub oak
habit: shrub to small tree
range: ME, s. to NC, w. and n. to OH
state frequency: uncommon; found in southern and western Maine
habitat: rocky or sandy soil
notes: Hybrids with *Quercus velutina*, called *Q. ×rehderi* Trel., occur in Maine. Also known to
     hybridize with *Q. coccinea* and *Q. rubra*.

Quercus macrocarpa Michx.
synonym(s): —
common name(s): bur oak
habit: tree
range: Quebec and New Brunswick, s. to VA, w. to TX, n. to Ontario and Manitoba
state frequency: uncommon
habitat: floodplain forests and other moist soils
notes: Known to hybridize with *Quercus alba* and *Q. bicolor*.

Quercus montana Willd.
synonym(s): *Quercus prinus* L. *sensu* Gleason and Cronquist (1991)
common name(s): mountain chestnut oak, rock chestnut oak
habit: tree
range: ME, s. to GA, w. to MS, n. to IL and MI
state frequency: very rare; known only from extreme southern Maine
habitat: upland, rocky woods
notes: This species is ranked S1 by the Maine Natural Areas Program. Known to hybridize with
     Quercus alba.

Quercus palustris Muenchh.
synonym(s): —
common name(s): pin oak
habit: tree
range: MA and VT, s. to NC, w. to OK and KS, n. to IA and MI; escaped from cultivation in ME
state frequency: uncommon
habitat: low, wet soil; often planted in upland areas
notes: Known to hybridize with *Quercus coccinea*, *Q. rubra*, and *Q. velutina*.

Quercus rubra L.
synonym(s): *Quercus borealis* Michx. f., *Quercus borealis* Michx. f. var. *maxima* (Marsh.) Ashe,
     Quercus maxima (Marsh.) Ashe, *Quercus rubra* L. var. *ambigua* (Gray) Fern., *Quercus rubra* L.
     var. *borealis* (Michx. f.) Farw.
common name(s): northern red oak
habit: tree
range: New Brunswick and Quebec, s. to GA, w. to OK, n. to MN
state frequency: common
habitat: dry, upland woods

notes: Hybrids with the *Quercus velutina*, called *Q.* ×*hawkinsiae* Sudworth, occur in Maine. Also known to hybridize with *Q. coccinea*, *Q. ilicifolia*, and *Q. palustris*.

Quercus velutina Lam.
synonym(s): —
common name(s): black oak
habit: tree
range: ME, s. to FL, w. to TX, n. to MN
state frequency: uncommon; found in southern and western Maine
habitat: dry, upland soils
notes: Hybrids with *Quercus ilicifolia* (*q.v.*) and *Q. rubra* (*q.v.*) occur in Maine. Also known to hybridize with *Q. coccinea*.

CUCURBITACEAE
(4 genera, 4 species)

KEY TO THE GENERA
1a. Ovary and fruit evidently prickly; petals shorter than 1.0 cm
    2a. Fruit inflated, 4-seeded, 3.0–5.0 cm long, solitary, dehiscent by 2 apical pores; staminate corolla 6-merous; herbage glabrous .... *Echinocystis*
    2b. Fruit not inflated, 1-seeded, 1.3–1.5 cm long, in clusters of 3–10, indehiscent; staminate corolla 5-merous; herbage pubescent .... *Sicyos*
1b. Ovary and fruit smooth or with small prickles in longitudinal stripes; petals 1.0 cm long or longer
    3a. Leaves with deep sinuses, the principal ones less than 10.0 cm wide; petals *ca.* 1.0–1.5 cm long; anthers distinct .... *Citrullus*
    3b. Leaves with shallow sinuses, the principal ones wider than 15.0 cm; petals *ca.* 5.0–9.0 cm long; anthers connivent .... *Cucurbita*

Citrullus
(1 species)

Citrullus lanatus (Thunb.) Matsumura & Nakai
synonym(s): *Citrullus citrullus* (L.) Karst., *Citrullus vulgaris* Schrad.
common name(s): watermelon
habit: trailing vine
range: tropical Africa; extensively planted and sometimes escaping
state frequency: very rare
habitat: compost piles, disturbed ground
notes: —

Cucurbita
(1 species)

Cucurbita maxima Duchesne
synonym(s): —
common name(s): squash
habit: trailing vine

range: Asia; extensively planted and sometimes escaping
state frequency: very rare
habitat: compost piles, disturbed ground
notes: —

Echinocystis
(1 species)

Echinocystis lobata (Michx.) Torr. & Gray
synonym(s): *Micrampelis lobata* (Michx.) Greene
common name(s): wild cucumber, prickly cucumber, balsam-apple
habit: high-climbing, monoecious, annual herb
range: New Brunswick, s. to GA, w. to AZ, n. to British Columbia
state frequency: uncommon
habitat: moist thickets, along streams
notes: —

Sicyos
(1 species)

Sicyos angulatus L.
synonym(s): —
common name(s): bur-cucumber
habit: high-climbing, vining, monoecious, annual herb
range: ME, s. to FL, w. to AZ, n. to MN
state frequency: rare
habitat: damp soil, river banks
notes: —

ROSACEAE
(24 genera, 115 species)

KEY TO THE GENERA
1a. Leaves simple, although sometimes lobed
    2a. Plants trees or shrubs, commonly taller than 1.0 m
        3a. Gynoecium syncarpous, composed of 2–5 carpels; fruit a pome or drupe
            4a. Flowers with 1 style; fruit a drupe; ovary superior .... ***Prunus***
            4b. Flowers with 2–5 styles; fruit a pome; ovary inferior, or at least partly so in fruit
                5a. Inflorescence a raceme (except in *Amelanchier bartramiana*); ovary and fruit 10-locular .... ***Amelanchier***
                5b. Inflorescence a cyme, umbel, or cluster, in any case, lacking an elongate, main axis; ovary and fruit 2- to 5-locular
                    6a. Carpels hard and bony, difficult to open to expose the seeds; styles distinct; plants usually armed with evident thorns; leaves often lobed .... ***Crataegus***

6b. Carpels papery or cartilaginous, easily opened to expose the seeds;
styles connate at the base (except in *Malus sylvestris*); plants unarmed
(sometimes with spinescent short shoots in *Malus sylvestris*); leaves
not lobed

7a. Large shrubs or small trees to 10.0 m; leaf midrib without a row of
glands on the adaxial surface; petals 10.0–20.0 mm long; fruit
0.8–12.0 cm in diameter, usually sweet at maturity .... ***Malus***

7b. Small shrubs to 2.0 m; leaf midrib with a row of glands on the
adaxial surface; petals 4.0–6.0 mm long; fruit 0.4–1.0 cm in
diameter, bitter at maturity .... ***Photinia***

3b. Gynoecium apocarpous, composed of 3–many, separate carpels; fruit a follicle or
aggregate of drupes

8a. Fruit dry and dehiscent; petals up to 0.5 cm long; leaf blades 3.0–7.0 cm long

9a. Leaves simple, without stipules; carpels not inflated in fruit .... ***Spiraea***

9b. Leaves usually 3-lobed, with stipules [or stipule scars]; carpels
prominently inflated in fruit .... ***Physocarpus***

8b. Fruit somewhat fleshy, indehiscent; petals 1.5–2.5 cm long; leaf blades
10.0–20.0 cm long .... ***Rubus***

2b. Plants herbs, less than 1.0 m tall

10a. Stamens 4; petals absent; bractlets present and alternating with the sepals;
gynoecium with 1(–3) ovaries, permanently enclosed by the hypanthium ....
.... ***Alchemilla***

10b. Stamens many; petals present (absent in apetalous flowers of *Rubus dalibarda*);
bractlets absent; gynoecium with 5–many ovaries, not enclosed by the
hypanthium .... ***Rubus***

1b. Leaves compound

11a. Leaves pinnately compound with 3 leaflets or palmately compound

12a. Flowers with 2–6 carpels; petals yellow; stems scapose .... ***Waldsteinia***

12b. Flowers with many carpels; petal color various; stems with or without leaves

13a. Styles elongate, jointed near or above the middle, the basal portion with a
hooked tip, becoming indurate and persistent in fruit, the apical portion
deciduous; lower and upper stem leaves distinctly different in size, shape, and
division .... ***Geum***

13b. Styles short, not jointed near middle, deciduous in fruit; lower and upper
stem leaves similar, except in size

14a. Flowers without bractlets alternating with the sepals [*i.e.*, the sepals
appear to number 5 per flower]; carpels maturing as small drupes that are
coherent in fruit; plants often armed with bristles or prickles .... ***Rubus***

14b. Flowers with bractlets that resemble the sepals and alternate with them
[causing the sepals appear to number (8)10 per flower]; carpels maturing
as achenes; plants unarmed

15a. Receptacle enlarged in fruit and becoming fleshy; petals white or
tinged with pink; bractlets subequal in size to the sepals ...... ***Fragaria***

15b. Receptacle neither enlarged nor fleshy in fruit; petals yellow or
white; bractlets in most species narrower and/or shorter than the
sepals

16a. Style originating from the side of the carpel [*i.e.*, lateral]; petals white; carpels pubescent [even in fruit]; leaves with 3 leaflets that are toothed at the very apex .... ***Sibbaldiopsis***

16b. Style originating from or very near the apex of the carpel [*i.e.*, terminal]; petals yellow; carpels glabrous; leaves either with more leaflets or toothed nearly the length of the margin, or both .... ***Potentilla***

11b. Leaves once- or twice-pinnately compound, always with more than 3 leaflets

17a. Plants trees or shrubs

18a. Flowers with 2–5 carpels; inflorescence many-flowered

19a. Flowers with 2–4 carpels, partly epigynous; fruit a pome .... ***Sorbus***

19b. Flowers with 5 carpels, perigynous; fruit a follicle .... ***Sorbaria***

18b. Flowers with many carpels; inflorescence with 1–many flowers

20a. Plants unarmed; leaflets entire; fruit and hypanthium not fleshy .... ***Pentaphylloides***

20b. Plants usually armed with bristles or prickles; leaflets toothed; fruit fleshy or the fruit enclosed in a fleshy hypanthium

21a. Fruit an aggregate of achenes, surrounded and concealed by a fleshy hypanthium; stipules adnate to the petiole for at least half their length; plants perennial .... ***Rosa***

21b. Fruit an aggregate of drupes, not surrounded by a hypanthium and therefore clearly visible; stipules free from the petiole; plants biennial—producing vegetative stems the first year and reproductive the second .... ***Rubus***

17b. Plants herbaceous (sometimes woody at the very base)

22a. Leaves twice-pinnately compound; fruit dehiscent .... ***Aruncus***

22b. Leaves once-pinnately compound; fruit indehiscent

23a. Flowers with 4 petaloid sepals, 0 petals, and 0–2 carpels .... ***Sanguisorba***

23b. Flowers with 5 sepaloid sepals, 5 petals, and 2–many carpels

24a. Flowers with 2 carpels; hypanthium armed with hooked bristles; inflorescence an elongate raceme .... ***Agrimonia***

24b. Flowers with 5–many carpels; hypanthium not armed; inflorescence a solitary flower, corymb, or cyme

25a. Styles elongate, jointed near or above the middle, the basal portion with a hooked tip, becoming indurate and persistent in fruit, the apical portion deciduous; lower and upper stem leaves distinctly different in size, shape, and division .... ***Geum***

25b. Styles short, not jointed near the middle, deciduous in fruit; lower and upper stem leaves similar, except in size

26a. Flowers with bractlets that resemble the sepals and alternate with them [*i.e.*, the sepals appear to number 10 per flower]; petals yellow, white, or red-purple; carpels many per flower, arranged in a cluster

27a. Styles originating from or very near the apex of the carpel [*i.e.*, terminal]; petals yellow .... ***Potentilla***

27b. Styles originating from the side of the carpel [*i.e.*, lateral]; petals white, red-purple, or yellow

    28a. Petals white or yellow-white; stamens 25–30; plants glandular-pubescent with brown hairs; styles thickened near the middle .... ***Potentilla***

    28b. Petals red-purple or yellow; stamens 5–20; plants glandular or not; styles not thickened near the middle

        29a. Petals red-purple; leaves with 5–7 leaflets; peduncles usually with more than 1 flower; plants without stolons .... ***Comarum***

        29b. Petals yellow; leaves with 7–25 leaflets; peduncles with 1 flower; plants with stolons ....
.... ***Argentina***

26b. Flowers without bractlets alternating with the sepals [*i.e.*, the sepals appear to number 5 per flower]; petals white or pink; carpels 5–15 per flower, arranged in a single whorl ....
.... ***Filipendula***

Agrimonia
(2 species)

Agrimonia pubescens Wallr. is reported to occur in Maine by Gleason and Cronquist (1991) but voucher specimens are unknown.

KEY TO THE SPECIES

1a. Axis of the inflorescence conspicuously glandular; hooked bristles of the hypanthium spreading, the outer sometimes even reflexed .... ***A. gryposepala***

1b. Axis of the inflorescence without glands; hooked bristles of the hypanthium ascending and connivent .... ***A. striata***

Agrimonia gryposepala Wallr.
synonym(s): —
common name(s): common agrimony
habit: stout, fibrous-rooted, perennial herb
range: New Brunswick, s. to mountains of NC, w. to KS, n. to MT; also British Columbia, s. to NM
state frequency: occasional
habitat: moist to dry, open woods, borders, and thickets
notes: —

Agrimonia striata Michx.
synonym(s): —
common name(s): roadside agrimony
habit: stout, fibrous-rooted, perennial herb
range: Newfoundland, s. to NJ, w. to AZ, n. to British Columbia
state frequency: occasional
habitat: dry to moist woods, borders, and thickets
notes: —

Alchemilla
(1 species)

Alchemilla xanthochlora Rothm.
synonym(s): *Alchemilla pratensis auct. non* F. W. Schmidt, *Alchemilla vulgaris auct. non* L.
common name(s): lady's mantle
habit: stout, rhizomatous, perennial herb
range: Nova Scotia, s. and w. to NY
state frequency: very rare
habitat: thickets, fields, roadsides
notes: —

Amelanchier
(7 species)

This genus probably hybridizes more than any other group in Maine and essentially every combination of parents is possible [however, these combinations are not included in the species notes]. Hybrids are often difficult to separate from other species due to their fertility [apomixis is prevalent]. Plants are most easily identified during anthesis, which is the only time petal length can be assessed. One useful character is leaf pubescence at anthesis. Those species that are densely tomentose when flowering will still have remnant pubescence on the leaves later in the season. Species that have nearly glabrous leaves when flowering will be completely devoid of hairs later in the season. Flowering time can be a useful character after one gains some familiarity with this group. For example, *Amelanchier ×neglecta* flowers before either of its parents at the same site and *A. stolonifera* flowers later than other species at the same site. Small petals are sometimes produced on a branch of a species that normally bears larger petals, a phenomenon called micropetaly. Not all material will be possible to identify, as hybrids can be disjunct from parents, and there are yet-undescribed morphologies in Maine.

KEY TO THE SPECIES
1a. Flowers borne in fascicles of 1–4; leaves imbricate in bud, at maturity with petioles
    2.0–10.0 mm long, cuneate at the base; bark often brown; ovary summit conical and
    pubescent; small, colonial shrub .... *A. bartramiana*
1b. Flowers borne in racemes; leaves conduplicate in bud, at maturity with petioles
    10.0–30.0 mm long, rounded to cordate at the base; bark often gray; ovary summit
    rounded, pubescent or glabrous; size and habit various
    2a. Petals 3.0–4.0 mm long, sometimes pollen-bearing; plants colonial, with many close,
        often straight and upright stems to 2.0 m tall; ovary summit mostly pubescent in
        Maine plants .... *A. nantucketensis*
    2b. Petals 5.0–22.0 mm long, not bearing pollen; plants colonial or not; ovary summit
        various
        3a. Summit of the ovary pubescent [visible in flower and fruit]; sepals reflexed from
            near the middle in fruit
            4a. Margin of leaf blade with 0–5 teeth per cm, often entire near the base; teeth
                less than twice as many as veins; veins of leaf prominent, nearly straight,
                forking 1 or more times, extending into the teeth or not .... *A. sanguinea*
            4b. Margin of leaf blade with 6–10 teeth per cm; teeth more than twice as many
                as veins; veins of leaf less prominent, curved, forking 3 or more times,
                anastomosing and evanescent near the margin, not extending into the teeth .....
                .... *A. stolonifera*
        3b. Summit of the ovary glabrous; sepals variously arranged

5a. Petals 5.0–12.0 mm long; racemes straight, erect or ascending; sepals ascending or spreading in fruit; leaf blade usually rounded at the base; colonial shrubs .... *A. canadensis*

5b. Petals 12.0–22.0 mm long; racemes lax or arching; sepals recurved in fruit; leaf blade usually cordate at the base; one- or few-stemmed shrubs or trees

6a. Leaves at anthesis half or more developed, nearly glabrous, tinged with red; pedicels 2.5–5.0 cm in fruit; pomes fleshy and sweet .... *A. laevis*

6b. Leaves at anthesis less than half developed; densely tomentose on the abaxial surface, green; pedicels up to 2.0 cm in fruit; pomes drier and bland tasting .... *A. arborea*

Amelanchier arborea (Michx. f.) Fern.

synonym(s): —
common name(s): downy shadbush, downy serviceberry
habit: shrub to tree
range: New Brunswick, s. to FL, w. to OK, n. to MN and Ontario
state frequency: uncommon
habitat: dry or rocky, upland woods, fields
notes: Known to hybridize frequently with *Amelanchier bartramiana*, *A. canadensis*, and *A. laevis*. Sometimes with pubescence on the ovary summit according to Gleason and Cronquist (1991).

Amelanchier bartramiana (Tausch) M. Roemer

synonym(s): —
common name(s): mountain shadbush, mountain serviceberry
habit: colonial shrub
range: Labrador, s. to mountains of WV, w. and n. to Ontario
state frequency: occasional
habitat: moist woods, thickets, and stream banks, swamps, mountain summits
notes: Hybrids with *Amelanchier laevis*, called *A.* ×*neglecta* Egglest. *ex* G. N. Jones, occur in Maine.

Amelanchier canadensis (L.) Medik.

synonym(s): —
common name(s): eastern serviceberry
habit: colonial shrub
range: Newfoundland, s. to GA, w. to MS, n. to NY and Quebec
state frequency: uncommon
habitat: swamps, moist thickets and low woods, shores
notes: Known to hybridize frequently with *Amelanchier arborea*.

Amelanchier laevis Wieg.

synonym(s): —
common name(s): smooth shadbush
habit: shrub to medium-sized tree
range: Newfoundland, s. to mountains of GA, w. to IA, n. to MN and Ontario
state frequency: common
habitat: thickets, forests, fields
notes: Hybrids with *Amelanchier bartramiana* (*q.v.*), *A. canadensis*, called *A.* ×*intermedia* Spach, and *A. sanguinea* (*q.v.*) occur in Maine. The red-purple-tinged leaves at anthesis, a character shared only with *A. bartramiana*, serve as a useful landmark for determining the role of *A. laevis* in hybridization events.

Amelanchier nantucketensis Bickn.

synonym(s): *Amelanchier stolonifera* Wieg. forma *micropetala* (Robins.) Rehd.
common name(s): Nantucket shadbush
habit: shrub
range: Nova Scotia, s. along coastal plain to MD
state frequency: rare
habitat: dry, open areas
notes: This species is ranked S2 by the Maine Natural Areas Program.

Amelanchier sanguinea (Pursh) DC.

synonym(s): <*A. s.* var. *gaspensis*> = *Amelanchier gaspensis* (Wieg.) Fern. & Weatherby
common name(s): Gaspé shadbush, New England serviceberry
habit: solitary-stemmed to colonial shrub
range: Quebec, s. to NJ, w. to MN, n. to Ontario
state frequency: uncommon
habitat: ledges, rocky river shores, forests, slopes
key to the subspecies:

    1a. Stems solitary or few together, 1.0–3.0 m tall; sepals recurved; veins of the leaf extending into the teeth .... *A. s.* var. *sanguinea*
    1b. Stems colonial, many together, up to 1.0 m tall; sepals ascending; veins of the leaf often anastomosing before the margin .... *A. s.* var. *gaspensis* Wieg.

notes: *Amelanchier sanguinea* var. *gaspensis* is ranked S2 by the Maine Natural Areas Program. Hybrids with *A. laevis*, called *A.* ×*interior* Nielson, occur in Maine and can be recognized by its intermediate morphology [short-acuminate leaf apex, sepals reflexed from near the base, leaves slightly red-tinged, with nearly straight, few-forked veins].

Amelanchier stolonifera Wieg.

synonym(s): —
common name(s): thicket shadbush
habit: low, colonial shrub
range: Newfoundland, s. to VA, w. and n. to Ontario
state frequency: occasional
habitat: dry, open, sandy or rocky areas
notes: —

Argentina
(2 species)

KEY TO THE SPECIES

1a. Leaves densely sericeous over the abaxial surface with lustrous hairs that overlie and conceal the tomentum; stolons, peduncles, and petioles usually villous; achenes as thick as wide, with a deep groove on the abaxial ridge .... *A. anserina*
1b. Leaves sericeous on the veins of the abaxial side, tomentose on the remainder of the surface; stolons, peduncles, and petioles glabrous to sparsely villous; achenes flattened, not as thick as wide, rounded or weakly grooved on the abaxial ridge .... *A. egedii*

Argentina anserina (L.) Rydb.

synonym(s): *Potentilla anserina* L.
common name(s): silverweed
habit: stoloniferous, perennial herb
range: circumboreal; s. to CT and NY, w. to NM and CA

state frequency: uncommon
habitat: moist or wet, gravelly or sandy shores, banks, and open places
notes: —

Argentina egedii (Wormsk.) Rydb. ssp. ***groenlandica*** (Tratt.) A. Löve
synonym(s): *Potentilla egedii* Wormsk. var. *groenlandica* (Tratt.) Polunin; *Potentilla pacifica* T. J.
Howell *sensu* Gleason and Cronquist (1991)
common name(s): coastal silverweed
habit: stoloniferous, perennial herb
range: circumboreal; s. in eastern North America to coastal NY
state frequency: occasional
habitat: Atlantic coast and tidal shores
notes: —

Aruncus
(1 species)

Aruncus dioicus (Walt.) Fern.
synonym(s): —
common name(s): goat's beard
habit: tall, erect, short-rhizomatous, perennial herb
range: PA, s. to GA, w. to AR, n. to IA; escaped from cultivation in ME
state frequency: rare
habitat: rich woods and ravines
notes: —

Comarum
(1 species)

Comarum palustre L.
synonym(s): *Potentilla palustris* (L.) Scop., *Potentilla palustris* (L.) Scop. var. *villosa* (Pers.) Lehm.
common name(s): marsh-potentilla
habit: coarse, stout, long-rhizomatous, perennial herb
range: circumboreal; s. to NJ, w. to CA
state frequency: uncommon
habitat: margins of bogs, swamps, and marshes, stream banks
notes: —

Crataegus
(22 species)

Contrary to its reputation, *Crataegus* contains some species are readily identifiable. However, a few of the taxa are separated by one or few characters, so sequential collections from flowering and fruiting specimens are required to make positive identifications. Floral leaves are those found on the flowering [and later, fruiting] branches. Vegetative leaves are found on the non-flowering branches. Be sure these distinctions are followed as differences exists between these 2 types of leaves. Stamen number is used extensively to separate species and can often be assessed in fruit by observing persistent filaments and/or filament scars. Anther color is also useful and should be recorded when the anthers are fresh, although the color is often still recognizable when they are dry. Hybrids are reported to be common and are generally intermediate in morphology.

KEY TO THE SPECIES FOR USE WITH FLOWERING MATERIAL

1a. Veins of the leaves running to the sinuses as well as the points of the lobes; flowers 9.0–13.0 mm wide; thorns under 5.0 cm long; plants escaped from cultivation

  2a. Leaf blades 3-lobed in general outline, the apex sharply acute, the base cordate, 2.0–5.0 cm wide; anthers yellow; styles 3–5; thorns up to 4.5(5.5) cm long .... *C. phaenopyrum*

  2b. Leaf blades 3- to 7-lobed, the apex rounded to acute, the base cuneate or truncate, often less than 2.0 cm wide; anthers pink; styles usually 1; thorns up to 2.5 cm long, usually much shorter .... *C. monogyna*

1b. Veins of the leaves running only to the points of the lobes or to the larger teeth; flowers 8.0–26.0 mm wide; thorns various; plants mostly native

  3a. Floral leaf blades acute at the base [*i.e.*, the margins at the base of the blade forming an angle of less than 90°]

    4a. Floral leaf blades unlobed or sometimes obscurely lobed at the apex, mostly obovate (rarely elliptic), coriaceous (except in the shade), lustrous, mostly 1.5–3.0 times as long as wide; veins on the adaxial surface not deeply impressed and appearing faint; petioles often shorter than 1.0 cm; styles 1 or 2(–5) .... *C. crus-galli*

    4b. Floral leaf blades usually lobed, mostly ovate to elliptic (to oblong-obovate in *C. succulenta*), usually broadest near or below the middle, coriaceous or thinner, lustrous or dull above, mostly 1.0–1.5 times as long as wide; veins on the adaxial surface distinct; petioles longer than 1.0 cm; styles 1–5

      5a. Styles 2 or 3; leaves coriaceous and lustrous above, with obscure lobes; calyx lobes deeply glandular-serrate; inflorescence not copiously bracteate; twigs and thorns chestnut- to dark brown; bracts of leaf opening conspicuous and red

        6a. Anthers 5–10, yellow, 1.2–1.5 mm long; thorns 7.0–9.0(–12.0) cm long; leaf blades mostly 2.5–5.0 cm long .... *C. macracantha*

        6b. Anthers 15–20, red, 0.5–0.7 mm long; thorns 3.0–5.0(–8.0) cm long; leaf blades mostly 6.0–8.0 mm long .... *C. succulenta*

      5b. Styles 2–5, often 3 or more; leaves not lustrous, usually distinctly lobed; sepals glandular-serrate or not; inflorescence sometimes copiously bracteate; twigs lighter colored; bracts not red

        7a. Leaf blades rounded to obtuse or acute at the apex, relatively more orbicular [isodiametric] in outline (blades elliptic to elliptic-ovate in *C. brunetiana*), often with glandular petioles; inflorescence glabrous or often villous, sometimes copiously glandular-bracteate; stamens 5–10

          8a. Anthers white to yellow; sepals glandular-serrate and often deeply so; flowers 1.2–2.0 cm wide; inflorescence glabrous or villous; mature leaf blades 2.5–5.0(–6.0) cm long

            9a. Floral leaf blades elliptic to elliptic-ovate, definitely longer than wide .... *C. brunetiana*

            9b. Floral leaf blades oval or rhombic to orbicular, nearly isodiametric

              10a. Floral leaves with uniform, sharp, spreading lobes; inflorescence glabrous to villous; sepals relatively more deeply serrate .... *C. chrysocarpa*

10b. Floral leaves with shallow and rounded lobes that are developed only above the middle of the blade; inflorescence glabrous; sepals shallowly serrate, if at all .... *C. dodgei*

8b. Anthers pink to purple; sepals entire; flowers 2.0–2.3 cm wide; inflorescence villous; mature leaf blades usually 6.0–8.0 cm long .... *C. jonesiae*

7b. Leaf blades usually abruptly short-acuminate at the apex, ovate to elliptic, usually with eglandular petioles; inflorescence glabrous or glabrate in common forms, not copiously bracteate; stamens 5–20

11a. Leaf blades scabrous to glabrous above, mostly elliptic to oblong-obovate, widest at or above the middle; stamens 5–10; sepals glandular-serrate .... *C. scabrida*

11b. Leaf blades more or less glabrous above, mostly ovate, widest below the middle; stamens 15–20; sepals entire or weakly glandular-serrate .... *C. brainerdii*

3b. Floral, as well as vegetative, leaf blades obtuse, broadly rounded, truncate, or cordate at the base [*i.e.*, the margins at the base of the blade forming an angle greater than 90°]

12a. Inflorescence, calyx, and leaves (at least along main veins) tomentose to villous-tomentose; filaments much shorter than the petals; anthers white to yellow; sepals glandular-serrate; styles (3–)5; inflorescence sometimes conspicuously bracteate .... *C. submollis*

12b. Inflorescence, calyx, and leaves villous or strigose or glabrous; filaments nearly as long as the petals; anthers commonly pink to purple; sepals glandular-serrate or not; styles (2–)3–5; inflorescence copiously bracteate or not

13a. Sepals usually deeply glandular-serrate; leaf blades larger, usually longer than 5.0 cm

14a. Flowers large, 2.2–2.6 cm wide, with 15–20 stamens and 5 styles; leaf blades truncate to subcordate at the base, glabrous from very early, the margins usually crisped .... *C. coccinioides*

14b. Flowers smaller, 1.3–2.0 cm wide, with 5–10 stamens and 3–5 styles; leaf blades cuneate to truncate at the base, usually pubescent on 1 or both surfaces, the margins not crisped

15a. Hypanthium and inflorescence villous; leaves often slightly convex ... . *C. pringlei*

15b. Hypanthium and inflorescence sparsely villous to glabrous; leaves planar .... *C. pedicellata*

13b. Sepals entire or weakly glandular-serrate; leaf blades usually smaller, 5.0 cm long or less

16a. Leaves usually glabrous from early on; inflorescence glabrous, sometimes copiously bracteate; stamens usually 15–20 .... *C. pruinosa*

16b. Leaves, at least the younger ones, strigose on the adaxial surface; inflorescence glabrous or villous, usually not copiously bracteate; stamens 5–20, commonly 10 or fewer

17a. Inflorescence glabrous

18a. Leaves acute at the apex, with spreading lateral lobes

19a. Thorns stout, curved; flowers 1.3–1.5 cm wide, in anthesis
earlier .... *C. iracunda*
19b. Thorns finer, nearly straight; flowers 1.4–1.8 cm wide, in
anthesis later .... *C. compta*
18b. Leaves acuminate at the apex, with recurved lateral lobes
20a. Stamens 5–10 per flower .... *C. macrosperma*
20b. Stamens 15–20 per flower .... *C. schuettei*
17b. Inflorescence villous
21a. Leaves herbaceous, with 4 or 5 pairs of lobes; flowers
1.3–1.5 cm wide; stamens 5–10 .... *C. lemingtonensis*
21b. Leaves dark green and coriaceous, with 5 or 6 pairs of lobes;
flowers 1.5–1.8 cm wide; stamens 15–20 .... *C. flabellata*

KEY TO THE SPECIES FOR USE WITH MATURE FRUITS
1a. Veins of the leaves running to the sinuses as well as the points of the lobes; pome
5.0–8.0 mm wide
2a. Leaf blades 3-lobed in general outline, the apex sharply acute, the base cordate,
2.0–5.0 cm wide; calyx deciduous in fruit; nutlets 3–5; thorns up to 4.5(5.5) cm long
.... *C. phaenopyrum*
2b. Leaf blades 3- to 7-lobed, the apex rounded to acute, the base cuneate to truncate,
often less than 2.0 cm wide; calyx persistent in fruit; nutlets 1; thorns up to 2.5 cm
long, often much shorter .... *C. monogyna*
1b. Veins of the leaves running only to the points of the lobes or to the larger teeth; pome
6.0–15.0 mm wide
3a. Nutlets of the pome with pits or depressions on the inner faces
4a. Nutlets 2 or 3, with a definite depression occupying most of each half of the inner
face; leaf apex acute to obtuse; twigs and thorns chestnut- to dark brown; thorns
2.5–9.5 cm long; inflorescence often villous; sepals deeply glandular-serrate
5a. Thorns 3.0–5.0(–8.0) cm long; leaf blades mostly 6.0–8.0 mm long ....
.... *C. succulenta*
5b. Thorns 7.0–9.0(–12.0) cm long; leaf blades mostly 3.5–5.0 cm long ....
.... *C. macracantha*
4b. Nutlets 2–5, commonly 3, with a shallow and irregular depression on each half of
the inner face; leaf apex narrowly acute to short-acuminate; twigs and thorns
brown to gray-brown; thorns 3.5–4.5(–5.5) cm long; inflorescence glabrous or
glabrate; sepals entire or serrate
6a. Leaf blades scabrous to glabrous above, mostly elliptic to oblong-obovate,
widest at or above the middle .... *C. scabrida*
6b. Leaf blades more or less glabrous above, mostly ovate, widest below the
middle .... *C. brainerdii*
3b. Nutlets smooth on the inner faces, without depressions
7a. Floral leaf blades acute at the base [*i.e.*, the margins at the base of the blade
forming an angle of less than 90°]
8a. Floral leaf blades unlobed or sometimes obscurely lobed at the apex, mostly
obovate (rarely elliptic), coriaceous (except in the shade), lustrous, 1.5–3.0
times as long as wide; nutlets 1 or 2(–5) .... *C. crus-galli*

8b. Floral leaf blades usually lobed, mostly ovate to elliptic, usually broadest near or below the middle, thinner, dull above, 1.0–1.5 times as long as wide; nutlets 2–5

  9a. Leaf blades rounded to obtuse or acute at the apex, more orbicular in outline (blades elliptic to elliptic-ovate in *C. brunetiana*), often with glandular petioles; infructescence glabrous or often villous

    10a. Lobes of the floral leaves shallow and rounded, developed only above the middle of the leaf blade; infructescence glabrous .... ***C. dodgei***

    10b. Lobes of the floral leaves uniform, deeper, sharp, and spreading; infructescence glabrous to villous

      11a. Floral leaf blades elliptic to elliptic-ovate, definitely longer than wide .... ***C. brunetiana***

      11b. Floral leaf blades oval or rhombic to orbicular, nearly isodiametric

        12a. Pome dark red or green-red, 0.8–1.2 cm in diameter; sepals serrate; leaf blades 3.5–5.0(–6.0) cm long ...... ***C. chrysocarpa***

        12b. Pome bright red, 1.0–1.3 cm in diameter; sepals entire; leaf blades mostly 6.0–8.0 cm long .... ***C. jonesiae***

  9b. Leaf blades usually abruptly short-acuminate at the apex, ovate to elliptic, usually borne on eglandular petioles; infructescence glabrous in common forms

    13a. Leaf blades scabrous to glabrous above, mostly elliptic to oblong-obovate, widest at or above the middle .... ***C. scabrida***

    13b. Leaf blades more or less glabrous above, mostly ovate, widest below the middle .... ***C. brainerdii***

7b. Floral, as well as vegetative, leaf blades obtuse, broadly rounded, truncate, or cordate at the base [*i.e.*, the margins at the base of the blade forming an angle greater than 90°]

  14a. Infructescence, calyx, and leaves (at least along main veins) tomentose to villous-tomentose; pome pubescent, at least at the ends; sepals glandular-serrate; pome red, sometimes with pale dots, becoming mellow tasting; nutlets (3–)5 .... ***C. submollis***

  14b. Infructescence, calyx, and leaves villous or strigose or glabrous; pome usually glabrous; sepals glandular-serrate or entire, slender; pome color and taste various; nutlets (2–)3–5

    15a. Pome usually 1.0 cm long or longer; sepals usually deeply glandular-serrate and prominent in fruit; leaf blades larger, often exceeding 5.0 cm long

      16a. Nutlets 2–4(–5); leaf blades cuneate to truncate at the base, the margins not crisped; plants native

        17a. Infructescence villous; pome sometimes sparsely villous, fleshy, dull red; leaf blades orbicular-ovate, often slightly convex; sepals narrow, mostly less than 1.5 mm wide at the base .... ***C. pringlei***

        17b. Infructescence glabrous or nearly so; pome glabrous, with thin, dry flesh, bright red; leaves narrower, more ovate, planar; sepals

larger, especially in fruit, mostly 1.5–2.0 mm wide at the base ....
.... *C. pedicellata*
16b. Nutlets 5; leaf blades truncate to subcordate at the base, the margins
crisped; plants escaped from cultivation .... *C. coccinioides*
15b. Pome usually 1.0 cm long or shorter; sepals usually entire or weakly
glandular-serrate; leaf blades smaller, often less than 5.0 cm long
18a. Leaves usually glabrous; infructescence glabrous; fruiting calyx
prominent and elevated; pome conspicuously glaucous .... *C. pruinosa*
18b. Leaves often strigose on the adaxial surface at maturity;
infructescence glabrous or villous; fruiting calyx usually small and
closely sessile, in some species deciduous from the fruit; pome not
glaucous
19a. Infructescence glabrous
20a. Pome subglobose to globose
21a. Pome 0.8–1.0(–1.2) cm in diameter; nutlets evidently
ridged and grooved on the outer surface; usually at least
some vegetative leaf blades more or less deltate ....
.... *C. iracunda*
21b. Pome 1.0–1.3(–1.5) cm in diameter; nutlets smooth or
slightly ridged on the outer surface; vegetative leaf blades
ovate to oval .... *C. schuettei*
20b. Pome obloid to obovoid
22a. Pome dull red to green-red at maturity, not fleshy; leaves
acute at the apex, with spreading lateral lobes ... *C. compta*
22b. Pome usually bright red and succulent at maturity; leaves
acuminate at the apex, with recurved lateral lobes ....
.... *C. macrosperma*
19b. Infructescence with at least some hairs remaining
23a. Pome 0.7–1.0 cm wide; leaves green, herbaceous, with 4 or 5
pairs of lobes; nutlets usually 3 .... *C. lemingtonensis*
23b. Pome 1.0–1.2 cm wide; leaves dark green and coriaceous,
with 5 or 6 pairs of lobes; nutlets 3–5 .... *C. flabellata*

Crataegus brainerdii Sarg.

synonym(s): —
common name(s): Brainerd hawthorn
habit: shrub to small tree
range: New Brunswick, s. to mountains of NC, w. and n. to MI and Ontario
state frequency: rare
habitat: thickets, pastures
notes: Thought to hybridize with *Crataegus chrysocarpa*, producing *C.* ×*ideae* Sarg. The taxonomy,
however, is unclear, and the identity of this taxon is uncertain.

Crataegus brunetiana Sarg.

synonym(s): *Crataegus brunetiana* Sarg. var. *fernaldi* (Sarg.) Palmer
common name(s): —
habit: shrub or small tree

range: Newfoundland, s. to ME, w. to MN, n. to Ontario
state frequency: uncommon
habitat: edges, rocky banks
notes: —

Crataegus chrysocarpa Ashe
synonym(s): —
common name(s): fireberry hawthorn
habit: shrub to small tree
range: Newfoundland, s. to VA, w. to NM, n. to Manitoba
state frequency: uncommon
habitat: thickets, rocky streamsides, old fields
key to the subspecies:
    1a. Inflorescence villous .... *C. c.* var. *chrysocarpa*
    1b. Inflorescence glabrous .... *C. c.* var. *phoenicia* Palmer
notes: Thought to hybridize with *Crataegus brainerdii* (*q.v.*).

Crataegus coccinioides Ashe
synonym(s): —
common name(s): Kansas hawthorn
habit: small tree to shrub
range: IL, s. to AR, w. to OK, n. to KS; escaped from cultivation in ME
state frequency: very rare
habitat: thickets, roadsides
notes: —

Crataegus compta Sarg.
synonym(s): *Crataegus levis* Sarg.
common name(s): —
habit: shrub to small tree
range: ME, s. to PA, w. to MI, n. to Ontario
state frequency: rare
habitat: streamside thickets, open woods
notes: —

Crataegus crus-galli L.
synonym(s): —
common name(s): cockspur thorn
habit: small tree to large shrub
range: ME, s. to FL, w. to TX, n. to MN
state frequency: uncommon
habitat: thickets, dry, rocky, open areas
notes: —

Crataegus dodgei Ashe
synonym(s): —
common name(s): —
habit: tall shrub to small tree
range: ME, s. to PA, w. and n. to WI
state frequency: very rare
habitat: thickets, borders of woods
notes: —

Crataegus flabellata (Spach) Kirchn. var. ***grayana*** (Egglest.) Palmer
synonym(s): *Crataegus grayana* Egglest.
common name(s): fanleaf hawthorn
habit: tall shrub to small tree
range: Quebec, s. to GA, w. to LA, n. to MN
state frequency: uncommon
habitat: thickets, open, rocky woods
notes: —

Crataegus iracunda Beadle
synonym(s): *Crataegus brumalis* Ashe
common name(s): —
habit: shrub or small tree
range: ME, s. and w. to KY, n. to MI
state frequency: very rare
habitat: edges, thickets
notes: —

Crataegus jonesiae Sarg.
synonym(s): —
common name(s): —
habit: small tree to shrub
range: New Brunswick, s. to ME, w. and n. to Quebec
state frequency: rare
habitat: open woods, stream banks
notes: —

Crataegus lemingtonensis Sarg.
synonym(s): —
common name(s): —
habit: tall shrub
range: ME, s. to MA, w. to VT
state frequency: uncommon
habitat: fields, early successional forests
notes: —

Crataegus macracantha Lodd. *ex* Loud.
synonym(s): *Crataegus succulenta* Schrad. *ex* Link var. *macracantha* (Lodd.) Egglest.
common name(s): —
habit: tall shrub to small tree
range: New Brunswick, s. to RI, w. to CO, n. to Manitoba
state frequency: occasional
habitat: early successional forests, edges of fields
notes: Flowering a little later than the similar *Crataegus succulenta* when both are found at the same site.

Crataegus macrosperma Ashe
synonym(s): *Crataegus macrosperma* Ashe var. *acutiloba* (Sarg.) Egglest., *Crataegus macrosperma* Ashe var. *pentandra* (Sarg.) Egglest., *Crataegus randiana* Sarg.
common name(s): —
habit: small tree to large shrub
range: Newfoundland, s. to mountains of NC, w. to IL, n. to WI
state frequency: common

habitat: woods, thickets

notes: Hybrids with *Crataegus pedicellata*, called *C. ×fretalis* Sarg., occur in Maine. These hybrids have glandular-serrate sepals and relatively large leaves, similar to *C. pedicellata*, but the pomes are small, like *C. macrosperma*.

Crataegus monogyna Jacq.

synonym(s): —
common name(s): English hawthorn, one-seed hawthorn
habit: small tree to large shrub
range: Europe; escaped from cultivation in the northeast
state frequency: uncommon
habitat: borders of woods, roadsides
notes: —

Crataegus pedicellata Sarg.

synonym(s): —
common name(s): scarlet hawthorn
habit: small tree to large shrub
range: ME, s. to CT, w. to IL, n. to Ontario
state frequency: rare
habitat: thickets, stream banks
notes: Hybrids with *Crataegus macrosperma* (*q.v.*) occur in Maine.

Crataegus phaenopyrum (L. f.) Medik.

synonym(s): —
common name(s): Washington thorn
habit: small tree
range: PA, s. to FL, w. to MO, n. to IL; escaped from cultivation occasionally in the northeast
state frequency: uncommon
habitat: thickets, open woods
notes: —

Crataegus pringlei Sarg.

synonym(s): —
common name(s): —
habit: small tree
range: ME, s. to CT, w. to IL, n. to Ontario
state frequency: very rare
habitat: thickets, stream borders
notes: —

Crataegus pruinosa (Wendl. f.) K. Koch

synonym(s): —
common name(s): frosted hawthorn
habit: small tree to large shrub
range: Newfoundland, s. to NC, w. to OK, n. to WI
state frequency: very rare
habitat: thickets, rocky areas
notes: —

Crataegus scabrida Sarg.

synonym(s): <*C. s.* var. *asperifolia*> = *Crataegus asperifolia* Sarg., *Crataegus brainerdii* Sarg. var. *asperifolia* (Sarg.) Egglest.

common name(s): —
habit: shrub
range: Nova Scotia, s. to ME, w. to NY, n. to Ontario
state frequency: rare
habitat: thickets, pastures
key to the subspecies:
   1a. Floral leaves mostly oval, with shallow or obscure sinuses, subcoriaceous ....
   .... *C. s.* var. *egglestonii* (Sarg.) Kruschke
   1b. Floral leaves to oblong-ovate or obovate, with evident sinuses, firm-herbaceous
      2a. Inflorescence glabrous .... *C. s.* var. *scabrida*
      2b. Inflorescence villous .... *C. s.* var. *asperifolia* (Sarg.) Kruschke
notes: *Crataegus scabrida* var. *asperifolia* is often smaller, with sharper toothed leaves and stouter
    thorns than the typical variety.

Crataegus schuettei Ashe var. *basilica* (Beadle) Phipps
synonym(s): *Crataegus basilica* Beadle
common name(s): —
habit: shrub to small tree
range: ME, s. to NC, w. and n. to Ontario
state frequency: very rare
habitat: fields, edges
notes: —

Crataegus submollis Sarg.
synonym(s): —
common name(s): Champlain thorn
habit: shrub to small tree
range: New Brunswick, s. to CT, w. and n. to Ontario
state frequency: uncommon
habitat: early successional forests, edges of fields
notes: —

Crataegus succulenta Schrad. *ex* Link
synonym(s): —
common name(s): fleshy hawthorn
habit: small tree to large shrub
range: southeast Canada, s. to PA, w. to IA, n. to Ontario
state frequency: rare
habitat: dry or rocky ground of borders of woods, thickets, and pastures
notes: —

Filipendula
(3 species)

KEY TO THE SPECIES
1a. Leaves more divided, with 8–25 pairs of well developed lateral leaflets; carpels
    pubescent; corolla usually with 6 petals .... *F. vulgaris*
1b. Leaves less divided, with 2–5 pairs of well developed lateral leaflets; carpels glabrous;
    corolla usually with 5 petals
    2a. Petals pink; lateral leaflets with 3–5 lobes; fruit straight .... *F. rubra*

2b. Petals white; lateral leaflets usually without lobes, merely coarsely toothed; fruit twisted .... *F. ulmaria*

Filipendula rubra (Hill) B. L. Robins.

synonym(s): —
common name(s): queen of the prairie
habit: rhizomatous, perennial herb
range: NY, s. to GA, w. to IA, n. to MN; occasionally escapes from cultivation in our area
state frequency: uncommon
habitat: meadows, low woods
notes: —

Filipendula ulmaria (L.) Maxim.

synonym(s): —
common name(s): queen of the meadow
habit: rhizomatous, perennial herb
range: Europe; escaped from cultivation mainly in the northeastern United States and Canada
state frequency: uncommon
habitat: roadsides, waste places, abandoned homesites
key to the subspecies:
  1a. Leaflets tomentose on the abaxial surface .... *F. u.* ssp. *ulmaria*
  1b. Leaflets sparsely pubescent with straightish hairs on the abaxial surface ....
    .... *F. u.* ssp. *denudata* (J. & K. Presl) Hayek
notes: —

Filipendula vulgaris Moench

synonym(s): *Filipendula hexapetala* Gilib. *ex* Maxim.
common name(s): dropwort
habit: rhizomatous, perennial herb
range: Eurasia; escaped from cultivation locally
state frequency: very rare
habitat: roadsides, waste places, abandoned homesites
notes: —

Fragaria
(2 species)

KEY TO THE SPECIES

1a. Terminal tooth of leaflets commonly less than half as wide as adjacent teeth and surpassed by them; petals 7.0–10.0 mm long; achenes embedded in the surface of the fruiting receptacle; inflorescence resembling a corymb .... *F. virginiana*
1b. Terminal tooth of leaflets commonly more than half as wide as adjacent teeth and surpassing them; petals 5.0–7.0 mm long; achenes not embedded in the surface of the fruiting receptacle; inflorescence resembling a raceme or panicle .... *F. vesca*

Fragaria vesca L.

synonym(s): —
common name(s): woodland strawberry, thin-leaved wild strawberry
habit: stoloniferous, perennial herb
range: Newfoundland, s. to VA, w. to NE, n. to Manitoba
state frequency: uncommon

habitat: rocky woods and open areas

key to the subspecies:

    1a. Pedicels and petioles pubescent with spreading hairs; leaflets firm, dark green ... *F. v.* ssp. *vesca*

    1b. Pedicels and petioles nearly glabrous to pubescent with appressed hairs; leaflets thin, pale green .... *F. v.* ssp. *americana* (Porter) Staudt

notes: —

Fragaria virginiana Duchesne

synonym(s): <*F. v.* ssp. *glauca*> = *Fragaria virginiana* Duchesne var. *terrae-novae* (Rydb.) Fern. & Wieg.

common name(s): wild strawberry, thick-leaved wild strawberry

habit: stoloniferous, perennial herb

range: Newfoundland, s. to GA, w. to OK, n. to Alberta

state frequency: common

habitat: fields, open areas

key to the subspecies:

    1a. Pedicels and petioles villous or hirsute with spreading hairs .... *F. v.* ssp. *virginiana*

    1b. Pedicels and petioles strigose to glabrous .... *F. v.* ssp. *glauca* (S. Wats.) Staudt

notes: *Fragaria* ×*ananassa* Duchesne, the hybrid of *F. chiloensis* (L.) P. Mill. and *F. virginiana*, is the cultivated strawberry, and sometimes escapes from cultivation in Maine. It can be identified by its larger petals 1.0–1.5 cm long and fruits longer than 1.5 cm long.

Geum
(5 species)

KEY TO THE SPECIES FOR USE WITH FLOWERING MATERIAL

1a. Sepals petaloid, purple or red-purple, ascending to erect at anthesis; petals erect to ascending; flowers somewhat to strongly nodding (erect, however, in fruit) .... **G. rivale**

1b. Sepals sepaloid, green, reflexed at anthesis; petals spreading; flowers erect

    2a. Petals white to yellow-white

        3a. Petals 3.0–5.0 x 1.0–2.0 mm, much smaller than the sepals; pedicels pubescent with both minute and long, coarse hairs, the longer hairs densely covering the surface .... **G. laciniatum**

        3b. Petals 5.0–9.0 x 1.0–4.5 mm, nearly as long as the sepals; pedicels densely and closely pubescent with minute hairs (longer hairs are sometimes present as well, but these are fewer and scattered) .... **G. canadense**

    2b. Petals yellow to orange-yellow

        4a. Terminal segment of the basal leaves suborbicular or reniform in outline, truncate or cordate at the base, much larger than the lateral segments; basal segment of the style minutely glandular; petals 5.0–10.0 x 5.0–9.0 mm .... **G. macrophyllum**

        4b. Terminal segment of the basal leaves oblanceolate to obovate, narrowed at the base, essentially similar to the lateral segments; basal segment of the style not glandular; petals 4.0–7.0 x 3.0–6.0 mm .... **G. aleppicum**

KEY TO THE SPECIES FOR USE WITH FRUITING MATERIAL

1a. Receptacle glabrous or minutely pubescent, the carpel scars visible

    2a. Basal segment of the style brown, not glandular; sepals lanceolate to lance-ovate, 4.0–10.0 mm long; pedicels thick, mostly more than 1.0 mm wide ...... **G. laciniatum**

2b. Basal segment of the style usually purple, glandular; sepals broad-triangular, 2.5–5.0 mm long; pedicels slender, mostly less than 1.0 mm wide .... ***G. macrophyllum***

1b. Receptacle hirsute, usually densely so, the carpel scars partly concealed by the hairs

3a. Basal segment of the style pubescent in the lower half; apical segment of the style pubescent with long hairs *ca.* 0.75–1.0 mm long .... ***G. rivale***

3b. Basal segment of the style glabrous or with some hairs at the very base near the apex of the achene; apical segment of the style pubescent with minute hairs *ca.* 0.1–0.5 mm long (a few longer hairs may be present)

4a. Basal leaves simple to divided with 3 lobes or segments; the spherical fruiting cluster composed of 30–160 achenes; pedicels commonly less than 1.0 mm wide .... ***G. canadense***

4b. Basal leaves simple to pinnately divided with 5–9 principal segments, often with interspersed smaller segments; the elongate-spherical fruiting cluster composed of mostly more than 200 achenes; pedicels commonly more than 1.0 mm wide .... ***G. aleppicum***

Geum aleppicum Jacq.

synonym(s): *Geum aleppicum* Jacq. var. *strictum* (Ait.) Fern.
common name(s): yellow avens
habit: stout, perennial herb
range: Quebec, s. to NJ, w. to Mexico and CA, n. to AK and Yukon
state frequency: occasional
habitat: thickets, wet meadows, swamps
notes: —

Geum canadense Jacq.

synonym(s): *Geum camporum* Rydb.
common name(s): white avens
habit: slender, perennial herb
range: New Brunswick, s. to NC, w. to TX, n. to ND
state frequency: occasional
habitat: woods, thickets, fields, roadsides
key to the subspecies:

1a. Gynoecium composed of 30–60 carpels, the carpels broad-ovate or obovate and 2.5–3.0 mm long in fruit; leaves membranaceous, those of the stem often glabrous on the adaxial surface; peduncles and calyx lobes glandular or not .... *G. c.* var. *canadense*

1b. Gynoecium composed of 60–160 carpels, the carpels narrow-obovate and 3.0–4.0 mm long in fruit; leaves firm, usually pubescent on the adaxial surface; peduncles and calyx lobes eglandular .... *G. c.* var. *camporum* (Rydb.) Fern. & Weatherby

notes: —

Geum laciniatum Murr.

synonym(s): —
common name(s): rough avens
habit: coarse, perennial herb
range: New Brunswick, s. to NC, w. to KS, n. to MN and Ontario
state frequency: uncommon
habitat: damp thickets, meadows, stream banks, and roadsides

key to the subspecies:
    1a. Carpels and achenes glabrous .... *G. l.* var. *laciniatum*
    1b. Carpels and achenes pubescent near the apex .... *G. l.* var. *trichocarpum* Fern.
notes: —

Geum macrophyllum Willd.

synonym(s): —
common name(s): big-leaved avens
habit: fibrous-rooted, perennial herb
range: Labrador, s. to ME, w. to CA, n. to AK
state frequency: uncommon
habitat: rich woods, damp thickets and meadows, rocky ledges
notes: —

Geum rivale L.

synonym(s): —
common name(s): water avens, purple avens
habit: short-rhizomatous, perennial herb
range: Labrador, s. to NJ, w. to NM and CA, n. to British Columbia
state frequency: uncommon
habitat: wet meadows and shores, bogs, swamps
notes: —

Malus
(3 species)

KEY TO THE SPECIES
1a. Adaxial surface of leaves and petioles and abaxial surface of sepals sparsely to densely
    tomentose; leaves crenate-serrate; pome 3.0–12.0 cm in diameter .... *M. sylvestris*
1b. Adaxial surface of leaves and petioles and abaxial surface of sepals glabrous or nearly
    so; leaves serrate; pome less than 3.0 cm in diameter
    2a. Twigs pubescent; sepals persistent in fruit; pome *ca.* 2.0 cm in diameter ....
    .... *M. prunifolia*
    2b. Twigs glabrous; sepals deciduous in fruit; pome *ca.* 1.0 cm in diameter .. *M. baccata*

Malus baccata (L.) Borkh.

synonym(s): *Pyrus baccata* L.
common name(s): Siberian crab
habit: small tree
range: Asia; escaped from cultivation
state frequency: very rare
habitat: thickets, clearings, early successional forests
notes: —

Malus prunifolia (Willd.) Borkh.

synonym(s): *Pyrus prunifolia* Willd.
common name(s): Chinese crab, Chinese apple
habit: small tree
range: Asia; escaped from cultivation
state frequency: very rare
habitat: thickets, roadsides
notes: —

Malus sylvestris P. Mill.
synonym(s): *Pyrus malus* L.
common name(s): apple
habit: small tree
range: Eurasia; escaped from cultivation throughout our area
state frequency: occasional
habitat: early successional forests, borders, clearings, roadsides
notes: —

Pentaphylloides
(1 species)

Pentaphylloides floribunda (Pursh) A. Löve
synonym(s): *Dasiphora fruticosa* (L.) Rydb., *Potentilla floribunda* Pursh, *Potentilla fruticosa* L.
sensu Gleason and Cronquist (1991)
common name(s): shrubby-cinquefoil
habit: much branched shrub
range: circumboreal; s. to NJ, w. to AZ and CA
state frequency: uncommon
habitat: circumneutral meadows, fens, and ledges
notes: Seemingly more common due to frequent planting near establishments.

Photinia
(1 species)

Photinia melanocarpa (Michx.) Robertson & Phipps
synonym(s): *Aronia melanocarpa* (Michx.) Ell., *Pyrus melanocarpa* (Michx.) Willd.
common name(s): black chokeberry
habit: rhizomatous, colonial shrub
range: Labrador, s. to GA, w. to AL, n. to MN and Ontario
state frequency: occasional
habitat: wet woods and thickets, bogs, swamps
notes: Hybrids with *Photinia pyrifolia* (Lam.) Robertson & Phipps, called *P.* ×*floribunda* (Lindl.)
Robertson & Phipps are common in Maine and can be identified by pubescent abaxial leaf surfaces
and inflorescences and purple pomes. *Photinia melanocarpa* has glabrous abaxial leaf surfaces
and inflorescences and black pomes. Hybrids with *Sorbus americana* and *S. aucuparia* also occur
in Maine.

Physocarpus
(1 species)

Physocarpus opulifolius (L.) Maxim.
synonym(s): *Opulaster opulifolius* (L.) Kuntze
common name(s): ninebark
habit: medium to tall shrub
range: Quebec, s. to SC, w. to AR, n. to WI; escaped from cultivation in the northeast
state frequency: uncommon
habitat: often rocky shores and thickets
notes: —

Potentilla
(9 species)

KEY TO THE SPECIES

1a. Flowers solitary, on a naked peduncle; stems (at least later in the season) prostrate and trailing; style slender, not thickened near the middle or base

  2a. Many or all of the flowers with 4 sepals and petals; leaves glabrous or glabrate; upper leaves with 3 or 4 leaflets; bractlets lanceolate; stipules narrow-ovate .... *P. anglica*

  2b. Flowers with 5 sepals and petals; leaves pubescent on the adaxial surface; upper leaves usually with 5 leaflets; bractlets linear-lanceolate; stipules lanceolate

    3a. Plants shorter, 5.0–15.0 cm at anthesis; flowers from the axil of the first well developed cauline leaf; rhizome 0.5–2.0 cm long, 4.0–8.0 mm wide .... *P. canadensis*

    3b. Plants taller, 20.0–40.0 cm at anthesis; flowers usually from the axil of the second well developed cauline leaf; rhizome longer, up to 8.0 cm long, 5.0–20.0 mm wide .... *P. simplex*

1b. Flowers few to many, arranged in a cyme; stems erect or prostrate; style thickened near the middle or base

  4a. Styles originating from the side of the carpel, thickened near the base; petals white or yellow-white; flowers with 25–30 stamens; plants glandular pubescent with brown hairs .... *P. arguta*

  4b. Styles originating from near the apex of the carpel, thickened at or near the base; petals yellow; flowers with *ca.* 20(–25) stamens; plants not glandular

    5a. Leaves sparsely to, more commonly, densely tomentose on the adaxial surface

      6a. Leaflets definitely palmately arranged, with 2–4 tooth-like lobes; styles not glandular [although papillose] at the base .... *P. argentea*

      6b. Leaflets pinnately arranged or the upper 3 confluent and appearing subpalmately arranged, deeply cleft, with more than 4 segments; styles glandular at the base .... *P. pensylvanica*

    5b. Leaves glabrous or pubescent with straight hairs on the adaxial surface

      7a. Stamens 5–10(–15); petals much shorter than the sepals; achenes smooth .... *P. rivalis*

      7b. Stamens 15–30; petals subequal to or longer than the sepals; achenes usually striate

        8a. Main stem leaves with 3 leaflets; anthers *ca.* 0.3 mm long; flowers at anthesis nearly 1.0 cm wide .... *P. norvegica*

        8b. Main stem leaves with 5–7 leaflets; anthers 1.0–1.5 mm long; flowers at anthesis 1.5–2.5 cm wide .... *P. recta*

Potentilla anglica Laicharding
synonym(s): *Potentilla procumbens* Sibthorp
common name(s): wood cinquefoil, trailing cinquefoil
habit: branched, trailing, perennial herb
range: Europe; introduced to Nova Scotia, s. to VA, w. and n. to OH
state frequency: very rare
habitat: open slopes
notes: Known in Maine only from Metinic Island, Knox County.

Potentilla argentea L.

synonym(s): —
common name(s): silvery cinquefoil
habit: depressed to ascending, branched, perennial herb
range: Eurasia; naturalized to Newfoundland, s. to MD, w. to IL, n. to ND
state frequency: common
habitat: dry, open areas and waste places
notes: —

Potentilla arguta Pursh

synonym(s): *Drymocallis agrimonioides* (Pursh) Rydb.
common name(s): tall cinquefoil, glandular cinquefoil
habit: erect, glandular, perennial herb
range: Quebec and New Brunswick, s. to MD, w. to CO, n. to British Columbia
state frequency: uncommon
habitat: dry woods and fields
notes: —

Potentilla canadensis L.

synonym(s): *Potentilla pumila* Poir.
common name(s): running cinquefoil
habit: slender, short-rhizomatous, trailing to ascending, perennial herb
range: Nova Scotia, s. to GA, w. to OH, n. to Ontario
state frequency: uncommon
habitat: dry woods and fields
notes: —

Potentilla norvegica L.

synonym(s): *Potentilla monspeliensis* L. *sensu* Britton and Brown (1913)
common name(s): rough cinquefoil, strawberry-weed
habit: stout, branched, erect to ascending, annual to short-lived perennial herb
range: circumboreal; s. to SC, w. to CA
state frequency: occasional
habitat: thickets, fields, roadsides, waste places
notes: —

Potentilla pensylvanica L. var. *pectinata* (Raf.) Boivin

synonym(s): *Potentilla pectinata* Raf., *Potentilla pensylvanica* L. var. *bipinnatifida* (Dougl.) Torr. &
   Gray
common name(s): cinquefoil
habit: densely cespitose, decumbent to erect, perennial herb
range: Newfoundland, s. to ME, w. to MN, n. to Alberta
state frequency: uncommon
habitat: open, sandy or gravelly soils
notes: —

Potentilla recta L.

synonym(s): —
common name(s): rough-fruited cinquefoil
habit: erect, perennial herb
range: Europe; naturalized to Newfoundland, s. to VA, w. to KS, n. to MN and Ontario
state frequency: uncommon
habitat: dry fields, roadsides, and waste places
notes: —

Potentilla rivalis Nutt. var. ***millegrana*** (Engelm. *ex* Lehm.) S. Wats.
synonym(s): *Potentilla millegrana* Engelm. *ex* Lehm.
common name(s): brook cinquefoil
habit: slender, ascending to erect, branched, annual or biennial herb
range: MN, s. to IL, w. to Mexico and CA, n. to British Columbia; naturalized sparingly in the east
state frequency: very rare
habitat: stream banks, waste places
notes: —

Potentilla simplex Michx.
synonym(s): *Potentilla simplex* Michx. var. *calvescens* Fern.
common name(s): old-field cinquefoil
habit: short-rhizomatous, erect to arching, perennial herb
range: Newfoundland, s. to GA, w. to TX, n. to MN and Ontario
state frequency: common
habitat: woods, thickets, fields, roadsides, lawns
notes: —

Prunus
(6 species)

Identifiable vegetatively by the bitter-almond fragrance of the twigs and a pair(s) of glands at the summit of the petiole.

KEY TO THE SPECIES FOR USE WITH VEGETATIVE MATERIAL
1a. Leaves with 15 or more pairs of lateral veins [some of which are inconspicuous], commonly pubescent along midrib on abaxial surface with red-brown hairs (white in early season) .... *P. serotina*
1b. Leaves with fewer than 15 pairs of lateral veins, glabrous or pubescent
    2a. Depressed or upright shrubs to 1.0(3.0) m tall; leaf blades remotely toothed, sometimes even subentire, especially in the basal third .... *P. pumila*
    2b. Straggling shrubs or upright shrubs or trees, 1.0–20.0 m tall; leaf blades toothed throughout the margin
        3a. Straggling shrubs 1.0–2.5 m tall, of sandy or rocky coastal areas; leaves pubescent on the abaxial surface; twigs with a pseudoterminal bud .. *P. maritima*
        3b. Upright shrubs or trees to 20.0 m tall, of various habitats, usually inland; leaves glabrous or pubescent; twigs with a terminal or pseudoterminal bud
            4a. Leaf margin with coarse, irregular, rounded teeth, often doubly serrate; leaf blades variously shaped
                5a. Leaf blades obovate to nearly orbicular, usually pubescent on the abaxial surface; twigs sometimes spinescent, with pseudoterminal buds .. *P. nigra*
                5b. Leaf blades lanceolate to narrow-ovate, usually glabrous; twigs lacking spines, with a terminal bud .... *P. pensylvanica*
            4b. Leaf margin with slender, sharp teeth; leaf blades oblong to obovate ....
.... *P. virginiana*

KEY TO THE SPECIES FOR USE WITH REPRODUCTIVE MATERIAL
1a. Inflorescence an elongate raceme, the axis longer than the pedicels, with 20 or more flowers

2a. Sepals broad-triangular to semicircular, conspicuously glandular-erose, promptly deciduous .... *P. virginiana*
2b. Sepals triangular, entire or inconspicuously glandular-erose, persistent in fruit .... *P. serotina*
1b. Inflorescence a cluster or very short raceme, the axis (if present) much shorter than the pedicels, with 2–5 flowers
   3a. Sepals pubescent on the adaxial surface; seeds compressed, with 2 edges
      4a. Sepals with glands along the margin; petals 10.0–15.0 mm long; drupes 2.0–3.0 cm long .... *P. nigra*
      4b. Sepals without glands along the margin; petals 5.0–6.0 mm long; drupes *ca.* 1.5 cm long .... *P. maritima*
   3b. Sepals glabrous on the adaxial surface; seeds subglobose
      5a. Sepals glandular-serrate; mature drupe black .... *P. pumila*
      5b. Sepals without glands; mature drupe red .... *P. pensylvanica*

Prunus maritima Marsh.

synonym(s): —
common name(s): beach plum
habit: low, straggling to ascending shrub
range: ME, s. to DE and MD
state frequency: very rare
habitat: sandy or rocky soil near the coast
notes: This species is ranked S1/S2 by the Maine Natural Areas Program.

Prunus nigra Ait.

synonym(s): —
common name(s): Canada plum
habit: small tree to large shrub
range: Canada; escaped from cultivation to New Brunswick and Nova Scotia, s. to VA and mountains of GA, w. to IA, n. to Manitoba
state frequency: uncommon
habitat: moist woods, thickets, and stream banks
notes: —

Prunus pensylvanica L. f.

synonym(s): —
common name(s): pin cherry, fire cherry, bird cherry
habit: tall shrub to small tree
range: Labrador, s. to VA and mountains of NC, w. to CO, n. to British Columbia
state frequency: common
habitat: clearings, burned areas, woods
notes: —

Prunus pumila L.

synonym(s): <*P. p.* var. *depressa*> = *Prunus depressa* Pursh; <*P. p.* var. *susquehanae*> = *Prunus pumila* L. var. *cuneata* (Raf.) Bailey, *Prunus susquehanae* hort. *ex* Willd.
common name(s): sand cherry
habit: prostrate to erect, low shrub
range: Quebec and New Brunswick, s. to NC, w. to IL, n. to Manitoba
state frequency: uncommon
habitat: rocky woods, gravelly shores, dunes

key to the subspecies:
  1a. Leaves narrow-oblanceolate, 0.5–1.5 cm wide, 3.0–6.0 times as long as wide, scarcely
    glaucous on the abaxial surface; plants depressed and mat-forming, with short, erect branches
    commonly 1.0–6.0 dm tall; seeds 6.0–8.0 mm long .... *P. p.* var. *depressa* (Pursh) Gleason
  1b. Leaves oblong to oblong-obovate, mostly 1.5–3.0 cm wide, 2.0–3.0 times as long as wide,
    glaucous on the abaxial surface; plants ascending to erect, commonly 3.0–10.0 dm tall; seeds
    5.0–6.0 mm long .... *P. p.* var. *susquehanae* (hort. *ex* Willd.) Jaeger
notes: —

Prunus serotina Ehrh.
synonym(s): —
common name(s): black cherry, wild black cherry
habit: tree
range: New Brunswick, s. to FL, w. to Mexico and AZ, n. to ND and Ontario
state frequency: common
habitat: woods, borders, thickets, roadsides, waste places
notes: —

Prunus virginiana L.
synonym(s): —
common name(s): choke cherry
habit: tall shrub to small tree
range: Newfoundland, s. to NC, w. to NM and CA, n. to British Columbia
state frequency: common
habitat: thickets, borders of woods and swamps, roadsides, shores
notes: —

Rosa
(15 species)

A genus, complicated by variable morphologies and hybridization. Prickle densities differ between the apical and basal portions of the stem, the basal portion having more prickles than the apical portion. When assessing prickle shape and densities, use the newer growth in the apical half of the plant. Those species with stipitate glands on the hypanthium will often lack some or all of the glands when in late fruit. However, the deciduous gland stipes leave a tiny scar on the surface of the hypanthium. This allows the presence/absence of stipitate glands to be assessed even in late season collections.

KEY TO THE SPECIES
  1a. Styles connate, exserted 1.5–3.0 mm from the orifice of the hypanthium, equaling the
    stamens; stems climbing or scrambling .... ***R. multiflora***
  1b. Styles distinct, not or only slightly exserted, shorter than the stamens; stems erect or
    arched
    2a. Flowers solitary (2 or 3) at the tips of branches; peduncle borne from the axil of a
      leaf but without additional bracts
      3a. Leaflets 7–11 per leaf, 0.5–2.0 x 0.5–1.0 cm; petals 1.0–2.5 cm long, white or
        yellow-white .... ***R. spinosissima***
      3b. Leaflets 3–7 per leaf, 2.0–6.0 x 2.0–3.0 cm; petals 2.5–4.5 cm long, pink ....
        .... ***R. gallica***
    2b. Flowers solitary or 2–6 and arranged in a corymb; peduncle bracteate

4a. Orifice of the hypanthium *ca.* 1.0 mm wide, the styles slightly exserted beyond it; sepals pinnatifid; achenes lining the inner wall of the hypanthium as well as the receptacle; some or all of the prickles stout, broad-based, curving

  5a. Leaves serrate with non-gland-tipped teeth, lacking stipitate glands .... ***R. canina***

  5b. Leaves doubly serrate with gland-tipped teeth, stipitate-glandular on one or both surfaces

    6a. Styles pubescent; sepals subpersistent; leaflets evidently glandular on the abaxial or on both surfaces, very aromatic .... ***R. eglanteria***

    6b. Styles glabrous; sepals promptly deciduous; leaflets not glandular on the abaxial surface, weakly aromatic .... ***R. micrantha***

4b. Orifice of the hypanthium 2.0–4.0 mm wide, the styles not exserted beyond it and the stigmas closing off the opening; sepals entire or with 1–4 linear appendages; achenes confined to the receptacle; prickles variously stout to slender

  7a. Branchlets, young prickles, and older prickle bases densely tomentose; leaves strongly rugose-veiny; petals mostly 3.0–5.0 cm long .... ***R. rugosa***

  7b. Branchlets, young prickles, and older prickle bases glabrous or sparsely pubescent; leaves not strongly rugose-veiny; petals mostly 2.0–3.0 cm long

    8a. Pedicel and hypanthium stipitate-glandular; sepals spreading or reflexed after anthesis, promptly deciduous in fruit

      9a. Leaflets with fine teeth, the teeth near the center *ca.* 0.5 mm long; plants of wetlands

        10a. Stems with a pair of stout, broad-based, often curved prickles at most of the nodes, these prickles much larger than the few, slender ones of the internodes; stipules with linear margins, scarcely widened distally .... ***R. palustris***

        10b. Stems with numerous internodal prickles that are slender, small-based, and straight, similar to the prickles found at the nodes; stipules without parallel margins, widened distally .... ***R. nitida***

      9b. Leaflets with coarse teeth, the teeth near the center *ca.* 1.0 mm long; plants of moist or dry soils

        11a. Stems with a pair of stout, broad-based, often curved prickles at most of the nodes, these prickles much larger than the few, slender ones of the internodes; stipules without parallel margins, widened distally .... ***R. virginiana***

        11b. Stems with numerous, internodal prickles that are slender, small-based, and straight, similar to the prickles found at the nodes; stipules with linear margins, scarcely widened distally .... ***R. carolina***

    8b. Pedicel and hypanthium glabrous (pedicels sometimes stipitate-glandular in *R. acicularis*); sepals erect or connivent after anthesis (reflexed in *R. blanda* var. *glabra*), persistent in fruit

      12a. Stems with a pair of stout, broad-based, often curved prickles at most of the nodes, these prickles much larger than the few, slender ones of the internodes .... ***R. cinnamomea***

12b. Stems without prickles or with few to numerous prickles, the prickles of the nodes undifferentiated from those of the internodes, all the prickles of nearly similar dimensions

    13a. Leaflets serrulate with mostly 9–11 teeth per cm in the apical half of the leaflet, tinged with red .... **R. rubrifolia**

    13b. Leaflets coarsely or doubly serrate with mostly 4–8 teeth per cm in the apical half of the leaflet, green

        14a. Stems with numerous prickles, even near the summit; usually the upper stipules, floral bracts, petioles, and/or leaf rachises stipitate-glandular, at least when young .... **R. acicularis**

        14b. Stems with no prickles, or a few that are confined to the base of the stem; upper stipules, floral bracts, petioles, and leaf rachises without stipitate glands (sometimes with sessile glands at the tips of the teeth of the stipules and/or floral bracts) .... **R. blanda**

Rosa acicularis Lindl.

synonym(s): <*R. a.* ssp. *sayi*> = *Rosa acicularis* Lindl var. *bourgeauiana* (Crépin) Crépin, *Rosa sayi* Schwein. *ex* Keating

common name(s): prickly rose, bristly rose

habit: colonial shrub

range: New Brunswick, s. to WV, w. to NM, n. to AK; also Eurasia

state frequency: rare

habitat: woods, thickets, rocky slopes

key to the subspecies:

    1a. Fruit ellipsoid to slender-pyriform, narrowed to one or both ends, the persistent calyx elevated on a tube-like process .... *R. a.* ssp. *acicularis*

    1b. Fruit subglobose, rounded at both ends, the persistent calyx sessile or slightly elevated .... *R. a.* ssp. *sayi* (Schwein.) W. H. Lewis

notes: Reported to hybridize with *Rosa carolina*. Our native plants are *R. acicularis* ssp. *sayi*.

Rosa blanda Ait.

synonym(s): <*R. b.* var. *glabra*> = *Rosa johannensis* Fern.

common name(s): smooth rose, meadow rose

habit: colonial shrub

range: New Brunswick, s. to PA, w. to MO and NE, n. to Manitoba

state frequency: uncommon

habitat: rocky slopes, shores, meadows

key to the subspecies:

    1a. Sepals reflexed in fruit; petiole and leaf rachis glabrous or promptly so; abaxial surface of leaflets glabrous, or pubescent on the veins only .... *R. b.* var. *glabra* Crépin

    1b. Sepals erect in fruit; petiole and leaf rachis pubescent (glabrous); abaxial surface of leaflets pubescent (glabrous) .... *R. b.* var. *blanda*

notes: *Rosa blanda* var. *blanda* hybridizes with *R. b.* var. *glabra* and with *R. palustris* in Maine.

Rosa canina L.

synonym(s): —

common name(s): dog rose

habit: coarse, arching shrub

range: Europe; escaped from cultivation to New Brunswick, s. to VA, w. to TN, n. to Ontario; also British Columbia

state frequency: very rare
habitat: fields, banks, thickets
notes: —

Rosa carolina L.

synonym(s): <*R. c.* var. *carolina*> = *Rosa carolina* L. var. *grandiflora* (Baker) Rehd.
common name(s): pasture rose
habit: low, slender, little-branched, colonial shrub
range: New Brunswick, s. to FL, w. to TX, n. to NE, MN, and Ontario
state frequency: occasional
habitat: dry, sandy or rocky, open areas
key to the subspecies:
    1a. Leaf rachis and abaxial surface of leaflets glabrous or nearly so (sometimes with stipitate glands on only the leaf rachis) .... *R. c.* var. *carolina*
    1b. Leaf rachis and abaxial surface of leaflets pubescent
        2a. Calyx stipitate-glandular .... *R. c.* var. *setigera* Crépin
        2b. Calyx without stipitate glands .... *R. c.* var. *villosa* (Best) Rehd.
notes: Hybrids with *Rosa palustris* occur in Maine. Also known to hybridize with *Rosa acicularis* and *R. virginiana*.

Rosa cinnamomea L.

synonym(s): *Rosa majalis* Herrm. *sensu* Gleason and Cronquist (1991)
common name(s): cinnamon rose
habit: dense, colonial shrub
range: Eurasia; escaped from cultivation
state frequency: very rare
habitat: roadsides, fields
notes: —

Rosa eglanteria L.

synonym(s): *Rosa rubiginosa* L.
common name(s): sweetbrier, eglantine
habit: coarse, erect shrub
range: Europe; escaped from cultivation
state frequency: uncommon
habitat: thickets, fields, roadsides, waste places
notes: —

Rosa gallica L.

synonym(s): —
common name(s): French rose
habit: erect, colonial shrub
range: Europe; escaped from cultivation in the northeast
state frequency: rare
habitat: thickets, roadsides
notes: The orifice of the hypanthium is 2.0–4.0 mm wide in this species.

Rosa micrantha Borrer *ex* Sm.

synonym(s): —
common name(s): sweetbrier, eglantine
habit: coarse shrub
range: Europe; escaped from cultivation to Quebec, s. to NC, w. to TX, n. to WI and Ontario
state frequency: very rare

habitat: thickets, fields, roadsides, waste places
notes: —

Rosa multiflora Thunb. *ex* Murr.
synonym(s): —
common name(s): multiflora rose
habit: trailing to arching or climbing, colonial shrub
range: Asia; escaped from cultivation
state frequency: uncommon
habitat: roadsides, borders, clearings
notes: —

Rosa nitida Willd.
synonym(s): —
common name(s): bristly rose, New England rose
habit: slender, rhizomatous, colonial shrub
range: Newfoundland, s. to CT
state frequency: uncommon
habitat: wet thickets, shores, and stream margins, bogs
notes: Hybrids with *Rosa palustris* and *R. virginiana* occur in Maine.

Rosa palustris Marsh.
synonym(s): —
common name(s): swamp rose
habit: rhizomatous shrub
range: New Brunswick, s. to FL, w. to AR, n. to MN and Ontario
state frequency: uncommon
habitat: wet thickets, stream banks, and shores, swamps, marshes
notes: Hybrids with *Rosa blanda*, *R. carolina*, and *R. nitida* occur in Maine.

Rosa rubrifolia Vill.
synonym(s): —
common name(s): red-leaved rose
habit: armed shrub
range: Europe; escaped from cultivation
state frequency: very rare
habitat: areas of cultivation
notes: —

Rosa rugosa Thunb.
synonym(s): —
common name(s): rugosa rose, salt spray rose, Japanese rose
habit: coarse, dense shrub
range: Asia; naturalized to Newfoundland, s. to NJ, w. to MN, n. to Ontario
state frequency: uncommon
habitat: roadsides, seaside thickets
notes: —

Rosa spinosissima L.
synonym(s): *Rosa pimpinellifolia* L.
common name(s): Scotch rose, burnet rose
habit: erect, colonial shrub
range: Eurasia; escaped from cultivation

state frequency: rare
habitat: areas of cultivation, old homesites
notes: The orifice of the hypanthium is 2.0–4.0 mm wide in this species.

Rosa virginiana P. Mill.
synonym(s): <*R. v.* var. *lamprophylla*> = *Rosa nanella* Rydb.
common name(s): Virginia rose
habit: stout, branched shrub
range: Newfoundland, s. to VA and upland NC, w. to AL and MO, n. to Ontario
state frequency: occasional; often frequent in coastal areas
habitat: fields, roadsides, shores, wet or dry thickets
key to the subspecies:
    1a. Plants compact, 1.5–4.0 dm tall; leaflets 0.8–2.0 x 0.4–1.3 cm .. *R. v.* var. *lamprophylla* Rehd.
    1b. Plants more open, up to 20.0 dm tall; leaflets 2.0–6.0 x 1.0–3.0 cm .... *R. v.* var. *virginiana*
notes: Hybrids with *Rosa nitida* occur in Maine. Also known to hybridize with *R. carolina*.

Rubus
(20 species)

Many of the species of *Rubus* are biennial, producing a vegetative stem the first year [called a primocane] and producing a flowering/fruiting stem the second year [the floricane]. Both are usually present in each population, and it is essential that both are collected. Leaf dimensions and stem habit can differ between the 2 types of stems. For example, the weight of snow and other vegetation can force an upright primocane to appear as a low, arching stem the second year. Equally important in the collections are notes describing the primocane stem habits [trailing, prostrate, doming, arching, or erect], as this is often difficult to distinguish from herbarium specimens. Armature morphology is used extensively for *Rubus* identification. The following explanations should be consulted: hairs - soft and pliable, very thin; bristles - stiff, brittle when dry, without expanded bases; small prickles - slender, only slightly expanded at the base, usually bending or breaking before tearing the flesh; large prickles - stout, expanded at the base, usually capable of penetrating clothing and tearing the flesh. Hybridization is thought to be extensive in the subgenus *Rubus* [5b. in the key], and nearly every combination is possible. *Rubus jaysmithii* Bailey [under the name *Rubus tetricus* Bailey] is reported to occur in Maine by Fernald (1950), but voucher specimens are unknown.

1a. Principal leaves simple; plants unarmed
    2a. Leaves 10.0–20.0 cm wide; sepals conspicuously glandular; petals 1.5–2.5 cm long, rose-purple (white); stems 1.0–2.0 m tall, woody .... ***R. odoratus***
    2b. Leaves 3.0–9.0 cm wide; sepals without glands; petals 0.4–1.8 cm long or absent, white; stems up to 0.3 m tall, herbaceous
        3a. Carpels maturing as a dry, achene-like drupe; petals 4.0–8.0 mm long or absent; leaves unlobed, the margin crenate .... ***R. dalibarda***
        3b. Carpels maturing as a white-yellow, fleshy drupe; petals 10.0–18.0 mm long; leaves shallowly 5- to 7-lobed as well as toothed on the margin ....
        .... ***R. chamaemorus***
1b. Principal leaves compound; plants usually armed (unarmed in *R. pubescens*)
    4a. Stems herbaceous, not clearly differentiated into primocanes and floricanes, unarmed; stipules oblanceolate to obovate; fruit separating from the receptacle only with difficulty .... ***R. pubescens***

4b. Stems woody, clearly differentiated into primocanes and floricanes (except *R. illecebrosus* with annual, above-ground stems), usually armed with bristles or prickles; stipules linear or setaceous; fruit separating from the receptacle or not

    5a. Leaves with 3–9 pinnately arranged leaflets (sometimes palmately arranged in *R. occidentalis*); fruit red or black, separating from the dry receptacle, the receptacle remaining behind on the pedicel (separating from the pedicel with the fleshy receptacle in *R. illecebrosus*)

        6a. Petals large, surpassing the sepals; floricane leaves with 5–9 leaflets; leaves glabrous on the abaxial surface; fruit red, not separating from the receptacle .... ***R. illecebrosus***

        6b. Petals smaller, equaled or surpassed by the sepals; floricane leaves with 3 leaflets; leaves densely tomentose with gray hairs on the abaxial surface; fruit red or black, separating from the receptacle

            7a. Fruit normally red, the individual drupes not separated by bands of tomentose pubescence; pedicels with straight, slender prickles and glandular hairs; primocanes not, or only slightly, glaucous, usually erect to arching, not rooting at the tip .... ***R. idaeus***

            7b. Fruit normally black, the individual drupes separated by bands of tomentose pubescence; pedicels with curved, stout prickles only; primocanes very glaucous, erect to mounding, sometimes rooting at the tip .... ***R. occidentalis***

    5b. Leaves with 3 pinnately arranged or 5 palmately arranged leaflets; fruit black, separating from the pedicel with the fleshy receptacle

        8a. Primocanes doming, prostrate, or trailing, rooting, or with the opportunity to root at the tip (in some species the depressed forms of the primocane not developing until midsummer when the apex of the stems finally reaches the ground)

            9a. Primocanes armed with hairs, bristles, or slender, small-based prickles; petals 5.0–12.0 mm long; leaves evergreen, lustrous, with 3 leaflets .... ***R. hispidus***

            9b. Primocanes armed, at least in part, by stout, broad-based prickles; petals 10.0–25.0 mm long; leaves usually deciduous, duller, with 3 or 5 leaflets

                10a. Inflorescence with 1–3(–5) flowers; pedicels erect, up to 6.0 cm long

                    11a. Stems very slender, 1.0–2.0 mm in diameter; central leaflet of floricane leaves oblanceolate to obovate, narrow- to broad-cuneate at the base, abruptly short-acuminate at the apex .... ***R. enslenii***

                    11b. Stems thicker, 1.5–4.0 mm in diameter; central leaflet of the floricane leaves more or less ovate, broad-rounded to cordate at the base, acute to acuminate at the apex .... ***R. flagellaris***

                10b. Inflorescence with 1–12 flowers; pedicels ascending to spreading, the central pedicels *ca.* 0.5 cm long

                    12a. Leaves glabrous or nearly so on the abaxial surface .... ***R. recurvicaulis***

                    12b. Leaves pubescent on the abaxial surface, usually noticeable to the touch .... ***R. arenicola***

8b. Primocanes arching or erect, not rooting at the tip (sometimes doming in *R. vermontanus*)

    13a. Primocanes armed with hairs, bristles, or slender, small-based prickles; stems 0.3–1.0 m tall

        14a. Leaves pubescent on the abaxial surface, usually noticeable to the touch; primocanes armed, at least in part, by slender, small-based prickles .... ***R. semisetosus***

        14b. Leaves glabrous or nearly so on the abaxial surface; primocane armature various

            15a. Primocanes armed with hairs or stiff bristles only, the hairs/bristles numbering 60–500 per cm .... ***R. setosus***

            15b. Primocanes armed, at least in part, by slender, small-based prickles, numbering 1–50 per cm .... ***R. vermontanus***

    13b. Primocanes armed, at least in part, with stout, broad-based prickles; stems 0.5–3.0 m tall

        16a. Inflorescence, and often the primocane, pubescent with stipitate glands, elongate and many-flowered, usually 8.0–20.0 cm long .... ***R. allegheniensis***

        16b. Plants essentially without stipitate glands; inflorescence commonly shorter, often partly included in the leaves, 4.0–17.0 cm long

            17a. Leaves glabrous or nearly so on the abaxial surface (sometimes with a few hairs on the main veins, but these hardly noticeable to the touch); inflorescence glabrous or nearly so

                18a. Prickles of the stem few, numbering 0–1 per cm; inflorescence much exceeding the foliage, leafy only at the base .... ***R. canadensis***

                18b. Prickles of the stem more numerous, numbering 1–10 per cm; inflorescence extending not much beyond the foliage, leafy-bracted throughout .... ***R. elegantulus***

            17b. Leaves pubescent on the abaxial surface, usually noticeable to the touch; inflorescence usually pubescent

                19a. Central leaflet of the primocane leaves elliptic to obovate, cuneate to subcordate at the base; most of the pedicels subtended by small, stipule-like bracts; leaves herbaceous; prickles mostly 2.0–4.0 mm long, numbering 0–2 per cm .... ***R. pensilvanicus***

                19b. Central leaflet of the primocane leaves ovate, broad-rounded to cordate at the base; many of the pedicels subtended by foliaceous bracts; leaves tending to be more chartaceous; prickles mostly 4.0–6.0 mm long, numbering 0–6 per cm .... ***R. frondosus***

Rubus allegheniensis Porter

synonym(s): *Rubus allegheniensis* Porter var. *gravesii* Fern., *Rubus allegheniensis* Porter var. *neoscoticus* (Fern.) Bailey
common name(s): common blackberry
habit: erect to arching, stout, biennial shrub

range: Quebec and Nova Scotia, s. to NC, w. to TN, n. to MN
state frequency: common
habitat: wide variety of habitats, mostly early successional, open areas
notes: Hybrids with *Rubus canadensis* occur in Maine. Also known to hybridize with *R. elegantulus*,
   R. hispidus, *R. pensilvanicus*, and *R. setosus*.

Rubus arenicola Blanch.

synonym(s): —
common name(s): sand-dwelling dewberry
habit: trailing, biennial shrub
range: Nova Scotia and ME, s. to MA and RI, w. to NY
state frequency: very rare
habitat: dry, open soil
notes: This species is ranked SH by the Maine Natural Areas Program.

Rubus canadensis L.

synonym(s): —
common name(s): Canada blackberry, smooth blackberry
habit: erect to arching, nearly unarmed, biennial shrub
range: Newfoundland, s. to GA, w. to TN, n. to MN and Ontario
state frequency: uncommon
habitat: forests, clearings, often boreal and commonly at the edge of wet areas
notes: Hybrids with *Rubus allegheniensis* occur in Maine.

Rubus chamaemorus L.

synonym(s): —
common name(s): baked apple-berry, cloudberry
habit: low, unarmed, dioecious, rhizomatous, perennial herb
range: circumboreal; s. to ME and NH
state frequency: rare
habitat: coastal and alpine bogs, mountain slopes
notes: —

Rubus dalibarda L.

synonym(s): *Dalibarda repens* L., *Rubus repens* (L.) Kuntze
common name(s): dewdrop
habit: slender, stoloniferous, unarmed, perennial herb
range: Quebec and Nova Scotia, s. to NC, w. to OH, n. to MI and MN
state frequency: uncommon
habitat: moist woods, forested wetlands
notes: Plants with 2 types of flowers—petaliferous flowers that usually do not bear fruit and apetalous
   flowers that bear fruit. Vegetatively similar to *Mitella nuda* (*q.v.*).

Rubus elegantulus Blanch.

synonym(s): *Rubus amicalis* Blanch., *Rubus multiformis* Blanch.
common name(s): —
habit: erect to arching, armed, biennial shrub
range: Newfoundland, s. to NJ, w. to PA
state frequency: uncommon
habitat: forests, open areas, wetlands
notes: Known to hybridize in Maine with *Rubus allegheniensis* and *R. vermontanus*. Closely
   resembling *R. vermontanus*, from which it can be separated by the following, additional characters:
   Rubus elegantulus has a central petiolule mostly 2.5–4.0 cm long, an inflorescence lacking

stipitate glands, and stems 0.5–3.0 m tall; *R. vermontanus* has a central petiolule mostly 1.4–2.0 cm long, an inflorescence commonly with stipitate glands, and stems 0.3–1.0 m tall.

Rubus enslenii Tratt.
synonym(s): —
common name(s): southern dewberry
habit: slender, prostrate to low-arching, biennial shrub
range: southeastern United States, n. to southern ME
state frequency: rare
habitat: oak-hickory woods, mixed forests, rock outcrops
notes: Sometimes difficult to distinguish from depauperate specimens of *Rubus flagellaris*.

Rubus flagellaris Willd.
synonym(s): —
common name(s): northern dewberry
habit: prostrate to mounding, biennial shrub
range: Quebec, s. to GA, w. to AR, n. to MN
state frequency: uncommon
habitat: open, sometimes rocky or ledgey, areas
notes: Hybrids with *Rubus hispidus* occur in Maine.

Rubus frondosus Bigelow
synonym(s): *Rubus bellobatus* Bailey, *Rubus multispinus* Blanch., *Rubus recurvans* Blanch.
common name(s): —
habit: arching, biennial shrub
range: ME, s. to VA, w. to IN
state frequency: uncommon; predominantly in coastal areas
habitat: open fields, roadsides, clearings
notes: —

Rubus hispidus L.
synonym(s): *Rubus hispidus* L. var. *obovalis* (Michx.) Fern.
common name(s): swamp dewberry
habit: mounding to, more often, trailing, weakly armed, biennial shrub
range: Nova Scotia, s. to NC, w. and n. to WI and Quebec
state frequency: occasional
habitat: open to forested areas, frequently of mineral or organic soil wetlands
notes: Hybrids with *Rubus allegheniensis*, *R. flagellaris*, and *R. setosus* occur in Maine.

Rubus idaeus L.
synonym(s): <*R. i.* ssp. *idaeus*> = *Rubus idaeus* L. var. *heterolasius* Fern.; <*R. i.* ssp. *strigosus*> = *Rubus idaeus* L. var. *canadensis* Richards., *Rubus strigosus* Michx.
common name(s): red raspberry
habit: erect to arching, bristly, biennial shrub
range: Europe and east Asia; naturalized to Newfoundland, s. to PA, w. to AR, n. to AK
state frequency: common
habitat: fields, roadsides, forests
key to the subspecies:
  1a. Inflorescence pubescent, at least in part, with stipitate glands ....
.... *R. i.* ssp. *strigosus* (Michx.) Focke
  1b. Inflorescence without stipitate glands .... *R. i.* ssp. *idaeus*
notes: —

Rubus illecebrosus Focke
synonym(s): —
common name(s): strawberry raspberry
habit: slender, rhizomatous, perennial shrub
range: Japan; escaped from cultivation to Nova Scotia, s. to CT, w. to NY
state frequency: very rare
habitat: roadsides, fields, old home sites
notes: —

Rubus occidentalis L.
synonym(s): —
common name(s): black raspberry
habit: arching to mounding, tip-rooting, biennial shrub
range: Quebec, s. to GA, w. to AR and CO, n. to ND
state frequency: uncommon
habitat: fields, woods, thickets
notes: —

Rubus odoratus L.
synonym(s): —
common name(s): flowering raspberry, purple flowering raspberry
habit: branched, unarmed, biennial shrub
range: Nova Scotia, s. to NC, w. to TN, n. to MI
state frequency: uncommon
habitat: forests, borders
notes: —

Rubus pensilvanicus Poir.
synonym(s): *Rubus amnicola* Blanch., *Rubus floricomus* Blanch., *Rubus pergratus* Blanch.
common name(s): Pennsylvania blackberry
habit: erect to ascending, biennial shrub
range: Newfoundland, s. to VA, w. to TN, n. to MN and Ontario
state frequency: uncommon
habitat: open areas, borders of forests
notes: Hybrids with *Rubus allegheniensis* and *R. setosus* occur in Maine.

Rubus pubescens Raf.
synonym(s): —
common name(s): dwarf raspberry
habit: unarmed, stoloniferous, perennial herb
range: Labrador, s. to WV, w. to CO, n. to Yukon
state frequency: occasional
habitat: usually moist to wet forests, thickets, and rocky shores
notes: —

Rubus recurvicaulis Blanch.
synonym(s): *Rubus plicatifolius* Blanch.
common name(s): Blanchard's dewberry
habit: prostrate to trailing, biennial shrub
range: Nova Scotia and Quebec, s. to WV, w. to IN, n. to MN
state frequency: uncommon
habitat: open, often disturbed or sandy, soils, sometimes of ledgey areas
notes: Hybrids with *Rubus setosus* and *R. vermontanus* occur in Maine.

Rubus semisetosus Blanch.

synonym(s): *Rubus ortivus* Bailey, *Rubus perinvisus* Bailey, *Rubus permixtus* Blanch.
common name(s): —
habit: slender, erect to arching, biennial shrub
range: ME, s. to NY and PA
state frequency: rare
habitat: dry, open areas
notes: —

Rubus setosus Bigelow

synonym(s): *Rubus lawrencei* Bailey, *Rubus notatus* Bailey
common name(s): bristly blackberry
habit: slender, erect to arching, bristly, biennial shrub
range: New Brunswick and ME, s. to VA, w. to IL, n. to WI
state frequency: uncommon
habitat: open areas, thickets, sometimes wetlands
notes: Hybrids with *Rubus allegheniensis, R. hispidus, R. pensilvanicus,* and *R. recurvicaulis* occur in Maine.

Rubus vermontanus Blanch.

synonym(s): *Rubus junceus* Blanch., *Rubus tardatus* Blanch.
common name(s): Vermont blackberry
habit: erect to arching, sometimes mounding, biennial shrub
range: Newfoundland, s. to PA, w. to WI and MN, n. to Ontario
state frequency: uncommon
habitat: open areas, thickets
notes: Known to hybridize with *Rubus elegantulus* and *R. recurvicaulis.* Sometimes the stem apex reaches the ground and roots. This can lead to confusion with *Rubus hispidus.* However, *R. vermontanus* is marked by a doming stem habit, non-lustrous leaflets mostly 6.0–12.0 cm long, and petals 10.0–15.0 mm long. *Rubus hispidus* usually possesses a trailing stem habit, lustrous leaflets 3.5–7.0 cm long, and petals 5.0–12.0 mm long.

Sanguisorba
(3 species)

Herbs with bisexual or unisexual flowers.

KEY TO THE SPECIES
1a. Leaflets mostly 0.5–2.0 cm long; flowers with 12 or more stamens and 2 carpels .... **S. minor**
1b. Leaflets mostly 3.0–8.0 cm long; flowers with 4 stamens and 1 carpel
    2a. Sepals white, much shorter than the stamens; spikes 3.0–12.0 cm tall; filaments distally widened .... **S. canadensis**
    2b. Sepals purple, equal to or slightly longer than the stamens; spikes 1.0–3.0 cm tall; filaments slender throughout .... **S. officinalis**

Sanguisorba canadensis L.

synonym(s): —
common name(s): Canadian burnet, Canada burnet, American burnet
habit: erect, simple to little-branched, thick-rhizomatous, perennial herb
range: Labrador, s. to NJ and mountains of NC and GA, w. to IL, n. to Manitoba

state frequency: very rare
habitat: bogs, peaty areas, wet meadows
notes: This species is ranked S1 by the Maine Natural Areas Program.

Sanguisorba minor Scop.

synonym(s): —
common name(s): garden burnet, salad burnet
habit: rhizomatous, perennial herb
range: Eurasia; naturalized to Nova Scotia, s. to VA, w. to TN, n. to Ontario
state frequency: very rare
habitat: fields, roadsides, waste places
notes: —

Sanguisorba officinalis L.

synonym(s): —
common name(s): great burnet, burnet-bloodwort
habit: erect, simple to little-branched, thick-rhizomatous, perennial herb
range: Eurasia; escaped from cultivation to ME, MN, and portions of western North America
state frequency: very rare
habitat: low fields, roadside thickets
notes: —

Sibbaldiopsis
(1 species)

Sibbaldiopsis tridentata (Ait.) Rydb.

synonym(s): *Potentilla tridentata* Ait.
common name(s): three-toothed-cinquefoil
habit: creeping, woody-based, perennial herb
range: Greenland, s. to mountains of GA, w. to IA, n. to ND and Ontario
state frequency: uncommon
habitat: rocky or gravelly soil
notes: —

Sorbaria
(1 species)

Sorbaria sorbifolia (L.) A. Braun

synonym(s): *Schizonotus sorbifolius* (L.) Lindl.
common name(s): false spiraea
habit: erect shrub
range: Asia; escaped from cultivation to New Brunswick, s. to PA, w. to IN, n. to MN and Alberta
state frequency: rare
habitat: roadsides, waste places
notes: —

Sorbus
(3 species)

KEY TO THE SPECIES

1a. Twigs, inflorescence, adaxial surface of leaflets, and hypanthium villous; bud scales villous, not glutinous .... *S. aucuparia*
1b. Twigs, inflorescence, adaxial surface of leaflets, and hypanthium glabrous or nearly so; bud scales glabrous or ciliate, glutinous
    2a. Leaflets acute to short-acuminate, 2.0–3.5 times as long as wide; bud scales ciliate; petals 4.0–5.0 mm long; pome 7.0–10.0 mm wide .... *S. decora*
    2b. Leaflets acuminate, 3.0–5.0 times as long as wide; bud scales glabrous or sparsely ciliate; petals 3.0–4.0 mm long; pome 4.0–7.0 mm wide .... *S. americana*

Sorbus americana Marsh.
synonym(s): *Pyrus americana* (Marsh.) DC.
common name(s): American mountain-ash
habit: small tree to shrub
range: Newfoundland, s. to MD and mountains of GA, w. to IL, n. to MN and Ontario
state frequency: uncommon
habitat: damp woods
notes: —

Sorbus aucuparia L.
synonym(s): *Pyrus aucuparia* (L.) Gaertn.
common name(s): European mountain-ash
habit: small tree
range: Europe; escaped from cultivation to Newfoundland, s. to MA, w. to IA, n. to AK
state frequency: uncommon
habitat: borders of woods, roadsides
notes: —

Sorbus decora (Sarg.) Schneid.
synonym(s): *Pyrus decora* (Sarg.) Hyl.
common name(s): showy mountain-ash
habit: small tree to shrub
range: Greenland, s. to CT, w. to IA, n. to MN and Manitoba
state frequency: occasional
habitat: borders of woods, rocky slopes, shores
notes: —

Spiraea
(4 species)

KEY TO THE SPECIES

1a. Inflorescence a compound corymb, mostly wider than tall; leaves 8.0–15.0 cm long .... *S. japonica*
1b. Inflorescence a panicle, mostly taller than wide; leaves 3.0–7.0 cm long
    2a. Twigs and abaxial surface of leaves tomentose; sepals reflexed after anthesis; follicles pubescent; flowers 3.0–4.0 mm wide, usually with pink petals .... *S. tomentosa*

2b. Twigs and abaxial surface of leaves glabrous or nearly so; sepals spreading after anthesis; follicles glabrous; flowers 4.0–7.0 mm wide, usually with white or pink-tinged petals

    3a. Inflorescence open, 5.0–30.0 cm tall, with elongate lower branches that usually exceed the subtending leaf; plants of low, upland, and subalpine areas ..... *S. alba*

    3b. Inflorescence dense, 1.0–10.0 cm tall, with very short lower branches, the lateral inflorescences not exceeding the subtending leaves; plants of alpine areas ....
.... *S. septentrionalis*

Spiraea alba Du Roi var. *latifolia* (Ait.) Dippel
synonym(s): *Spiraea latifolia* (Ait.) Borkh.
common name(s): meadowsweet
habit: erect, simple to rarely branched, shrub
range: Newfoundland, s. to NC, w. and n. to MI
state frequency: common
habitat: fields, shores, open areas
notes: —

Spiraea japonica L. f.
synonym(s): —
common name(s): Japanese spiraea
habit: shrub
range: Asia; escaped from cultivation to ME, s. and w. to TN, n. to IN
state frequency: very rare
habitat: thickets, roadsides
notes: —

Spiraea septentrionalis (Fern.) A. & D. Löve
synonym(s): *Spiraea latifolia* (Ait.) Borkh. var. *septentrionalis* Fern.
common name(s): northern meadowsweet
habit: low shrub
range: Labrador, s. to ME and mountains of VA, w. and n. to MI
state frequency: very rare
habitat: alpine areas
notes: This species is ranked SH by the Maine Natural Areas Program.

Spiraea tomentosa L.
synonym(s): —
common name(s): hardhack, steeple-bush
habit: simple to rarely branched shrub
range: New Brunswick, s. to NC, w. to AR, n. to Ontario
state frequency: occasional
habitat: wet meadows, fields, roadsides, and shores
notes: —

Waldsteinia
(1 species)

Waldsteinia fragarioides (Michx.) Tratt.
synonym(s): —
common name(s): barren-strawberry
habit: low, perennial herb

range: New Brunswick, s. to ME, w. to IN, n. to MN and Ontario; also mountains of GA and AL
state frequency: very rare
habitat: thickets, woods, fields
notes: This species is ranked S1 by the Maine Natural Areas Program.

ELAEAGNACEAE
(2 genera, 3 species)

KEY TO THE GENERA

1a. Leaves alternate; thorns commonly present; flowers with 8 stamens; plants synoecious or polygamous [with some staminate flowers] shrubs or small trees, escaped from cultivation .... ***Elaeagnus***

1b. Leaves opposite; thorns absent; flowers with 4 stamens; plants dioecious shrubs, native, of calcareous substrates .... ***Shepherdia***

Elaeagnus
(2 species)

Leaf shape is variable in this genus and unreliable as a character for identification.

KEY TO THE SPECIES

1a. Leaves with silver, lepidote scales on the adaxial surface; twigs brown; drupe silver to yellow, 8.0–10.0 mm long .... ***E. angustifolia***

1b. Leaves green and glabrate on the adaxial surface; twigs silver; drupe red, 6.0–8.0 mm long .... ***E. umbellata***

Elaeagnus angustifolia L.
synonym(s): —
common name(s): Russian-olive, oleaster
habit: usually thorny shrub to small tree
range: Eurasia; escaped from cultivation to much of the United States
state frequency: uncommon
habitat: roadsides, field edges
notes: —

Elaeagnus umbellata Thunb.
synonym(s): —
common name(s): —
habit: shrub or small tree
range: Asia; introduced in the east
state frequency: rare
habitat: roadsides, field edges
notes: —

Shepherdia
(1 species)

Shepherdia canadensis (L.) Nutt.
synonym(s): —
common name(s): buffaloberry, Canada buffaloberry, soapberry, rabbitberry
habit: dioecious shrub
range: Newfoundland, s. to ME, w. to NM, n. to AK
state frequency: very rare
habitat: calcareous ledges and banks
notes: This species is ranked S1 by the Maine Natural Areas Program.

RHAMNACEAE
(3 genera, 4 species)

KEY TO THE GENERA
1a. Leaves with 3 prominent nerves from the base; petals with an evident claw; fruit a dry,
    capsule-like drupe; inflorescence a terminal umbel composed of numerous flowers ....
    .... ***Ceanothus***
1b. Leaves pinnately veined; petals absent or present and with a short claw; fruit a fleshy
    drupe; inflorescence axillary, either a solitary flower or a few-flowered umbel
    2a. Leaves with entire margins; winter buds naked; style undivided; flowers bisexual ......
        .... ***Frangula***
    2b. Leaves with serrate margins; winter buds covered by scales; style divided for part of
        its length; flowers functionally unisexual .... ***Rhamnus***

Ceanothus
(1 species)

Ceanothus americanus L.
synonym(s): —
common name(s): New Jersey-tea
habit: low shrub
range: Quebec, s. to FL, w. to TX, n. to MN and Ontario
state frequency: very rare
habitat: upland woods, gravelly, rocky areas
notes: This species is ranked S1 by the Maine Natural Areas Program.

Frangula
(1 species)

Frangula alnus P. Mill.
synonym(s): *Rhamnus frangula* L.
common name(s): alder-buckthorn, European alder-buckthorn
habit: shrub to small tree
range: Europe; escaped from cultivation to Quebec and Nova Scotia, s. to NJ, w. to KY, n. to MN and
    Ontario
state frequency: occasional

habitat: wet soil
notes: This species is an aggressively invasive exotic.

Rhamnus
(2 species)

KEY TO THE SPECIES
1a. Flowers with 5 sepals, 0 petals, and 5 stamens; leaves alternate; branches unarmed; plants 0.15–1.0 m tall; drupe with 3 pyrenes .... ***R. alnifolia***
1b. Flowers with 4 sepals, petals, and stamens; leaves subopposite; branches often tipped by a spine-like process; plants up to 6.0 m tall; drupe with (3)4 pyrenes .... ***R. cathartica***

Rhamnus alnifolia L'Hér.
synonym(s): —
common name(s): alder-leaved buckthorn, American alder-buckthorn
habit: low shrub
range: Newfoundland, s. to NJ, w. to NE, n. to British Columbia; also TN, WY, and CA
state frequency: uncommon
habitat: swamps, bogs, low woods
notes: —

Rhamnus cathartica L.
synonym(s): —
common name(s): common buckthorn
habit: shrub to small tree
range: Eurasia; naturalized to Quebec and Nova Scotia, s. to VA, w. to MO, n. to MN and Saskatchewan
state frequency: occasional
habitat: fields, edges, roadsides
notes: —

CANNABACEAE
(2 genera, 3 species)

KEY TO THE GENERA
1a. Leaves lobed; stems vining, retrorsely spinulose; petioles with forked hairs; plants annual or perennial .... ***Humulus***
1b. Leaves divided; stems erect, not spinulose; petioles without forked hairs; plants annual ...
.... ***Cannabis***

Cannabis
(1 species)

Cannabis sativa L.
synonym(s): —
common name(s): hemp, Indian hemp, marijuana
habit: erect, annual herb
range: Asia; escaped from cultivation throughout most of North America

state frequency: rare
habitat: waste places, vacant fields
notes: —

Humulus
(2 species)

KEY TO THE SPECIES

1a. Plants annual; larger leaves commonly 5- to 9-lobed, with very narrow sinuses, without glandular dots; carpellate spikes with spinulose-ciliate bracts that are narrow and do not completely conceal the fruits; anthers without glands .... *H. japonicus*
1b. Plants perennial; larger leaves commonly 3- to 5-lobed, with broad and open sinuses, glandular-dotted on the abaxial surface; carpellate spikes with entire-margined bracts that are broad and conceal the fruits; anthers glandular .... *H. lupulus*

Humulus japonicus Sieb. & Zucc.
synonym(s): —
common name(s): Japanese hops
habit: annual vine
range: eastern Asia; escaped from cultivation to ME, s. to GA, w. to MO, n. to ND
state frequency: very rare
habitat: roadsides, waste places
notes: —

Humulus lupulus L.
synonym(s): —
common name(s): common hops
habit: rhizomatous, perennial vine
range: New Brunswick and Nova Scotia, s. to VA, w. to CA, n. to British Columbia
state frequency: uncommon
habitat: moist soil of waste places
key to the subspecies:
   1a. Abaxial leaf midrib with fewer than 20 hairs per cm; abaxial leaf surface with fewer than 25 glands per 10.0 mm^2 .... *H. l.* var. *lupulus*
   1b. Abaxial leaf midrib with 20–100 hairs per cm; abaxial leaf surface with more than 25 glands per 10.0 mm^2 .... *H. l.* var. *lupuloides* E. Small
notes: —

ULMACEAE
(1 genus, 3 species)

Ulmus
(3 species)

KEY TO THE SPECIES

1a. Some of the leaves with 3 acuminate lobes at the apex; sepals connate for *ca.* 0.5 of their length; samara pubescent only on the central vein of the wing; bark smooth (furrowed with age) .... *U. glabra*

1b. None of the leaves with additional lobes at the apex; sepals connate for most of their length; samara pubescent or ciliate, but not pubescent on the central vein of the wing; bark furrowed

    2a. Flowers sessile or subsessile; samaras suborbicular, not fringed on the margin, pubescent over the seed; twigs scabrous-pubescent; winter buds densely covered with red-brown hairs; inner bark very mucilaginous .... *U. rubra*

    2b. Flowers long-pedicelled, drooping; samaras elliptical, fringed on the margin, glabrous over the seed; twigs glabrous or pubescent; winter buds glabrous or sparsely pubescent; inner bark not mucilaginous .... *U. americana*

Ulmus americana L.

synonym(s): —
common name(s): American elm, white elm
habit: tree
range: Quebec and Nova Scotia, s. to FL, w. to TX, n. to Saskatchewan
state frequency: common
habitat: rich, moist, often lowland, soils
notes: —

Ulmus glabra Huds.

synonym(s): —
common name(s): Scotch elm, wych elm
habit: often multiple-trunked tree
range: Eurasia; introduced to ME, s. to RI, w. and n. to NY and VT
state frequency: rare
habitat: margins of lawns and fields, waste places
notes: —

Ulmus rubra Muhl.

synonym(s): *Ulmus fulva* Michx.
common name(s): slippery elm, red elm
habit: tree
range: Quebec and ME, s. to FL, w. to TX, n. to ND and Ontario
state frequency: very rare
habitat: rich, moist soils of alluvial plains and river banks
notes: This species is ranked SH by the Maine Natural Areas Program.

URTICACEAE
(6 genera, 7 species)

KEY TO THE GENERA

1a. Plants shrubs or trees, with a milky latex; styles deeply divided; fruit a fleshy multiple .....
.... *Morus*

1b. Plants herbaceous, with a watery latex; styles undivided; fruit simple, an achene

    2a. Leaves opposite

        3a. Plants with conspicuous, stinging hairs; carpellate calyx of 4 distinct sepals ....
.... *Urtica*

        3b. Plants without stinging hairs (or sometimes with few and minute, stinging hairs); carpellate calyx of 3 sepals or 4 connate sepals

4a. Inflorescence a spike; carpellate calyx connate at the base, 4-lobed, accrescent, enclosing the achene in fruit; leaves with rounded cystoliths .... *Boehmeria*

4b. Inflorescence a panicle; carpellate calyx distinct or nearly so, 3-lobed, not enclosing the achene in fruit; leaves with linear cystoliths .... *Pilea*

2b. Leaves alternate

5a. Plants with conspicuous, stinging hairs; leaves serrate; carpellate calyx distinct or nearly so, only 2 of the 4 sepals persistent in fruit, not enclosing the achene in fruit; staminate flowers each with 5 sepals and stamens .... *Laportea*

5b. Plants without stinging hairs (or sometimes with few and minute, stinging hairs); leaves entire; carpellate calyx connate at the base, 4-lobed, accrescent, enclosing the achene in fruit; staminate flowers each with 4 sepals and stamens .... *Parietaria*

Boehmeria
(1 species)

Boehmeria cylindrica (L.) Sw.
synonym(s): *Boehmeria drummondiana* Weddell
common name(s): false nettle, bog-hemp
habit: perennial herb
range: Quebec, s. to FL, w. to NM, n. to MN and Ontario
state frequency: uncommon
habitat: moist or wet ground
notes: —

Laportea
(1 species)

Laportea canadensis (L.) Weddell
synonym(s): *Urticastrum divaricatum* (L.) Kuntze
common name(s): wood-nettle
habit: rhizomatous, perennial herb
range: Nova Scotia, s. to FL, w. to OK, n. to Manitoba
state frequency: uncommon
habitat: moist to wet, rich woods, stream banks
notes: —

Morus
(1 species)

The fruit of *Morus* is a multiple, composed of individual achenes enclosed by a fleshy calyx.

Morus alba L.
synonym(s): —
common name(s): white mulberry
habit: tree or shrub
range: eastern Asia; introduced to Europe and eastern United States from ME, s. to FL, w. to TX, n. to ND
state frequency: very rare

habitat: fields, forest margins, thickets
notes: The fruits of *Morus alba* are black, purple, or white when mature.

Parietaria
(1 species)

Parietaria pensylvanica Muhl. *ex* Willd.
synonym(s): —
common name(s): pellitory, Pennsylvania pellitory
habit: ascending to erect, annual herb
range: ME, s. to FL, w. to Mexico and CA, n. to British Columbia
state frequency: very rare
habitat: dry woods, ledges, waste places
notes: This species is ranked SX by the Maine Natural Areas Program.

Pilea
(1 species)

Carpellate flowers in this genus possess 3 staminodes positioned opposite the sepals. The staminodes are under tension and forceably eject the mature achenes.

Pilea pumila (L.) Gray
synonym(s): —
common name(s): clearweed, richweed
habit: annual herb
range: New Brunswick and Nova Scotia, s. to FL, w. to TX, n. to SD and MN
state frequency: uncommon
habitat: moist, shaded forests and stream shores
notes: —

Urtica
(2 species)

The stinging hairs in *Urtica* can be identified by their bulbous or cylindrical base and stiff, translucent apex.

KEY TO THE SPECIES
1a. Stipules 5.0–15.0 mm long, erect; staminate and carpellate flowers usually in separate inflorescences; inflorescence exceeding the length of the petioles of the subtending leaves; plants perennial .... *U. dioica*
1b. Stipules 1.0–3.0 mm long, spreading to deflexed; staminate and carpellate flowers usually in the same inflorescence; inflorescence often shorter than the length of the petioles of the subtending leaves; plants annual .... *U. urens*

Urtica dioica L.
synonym(s): <*U. d.* ssp. *gracilis*> = *Urtica gracilis* Ait., *Urtica procera* Muhl. *ex* Willd., *Urtica viridis* Rydb.
common name(s): stinging nettle
habit: erect, rhizomatous, perennial herb
range: Eurasia; naturalized to Newfoundland and Nova Scotia, s. to VA, w. to IL, n. to Manitoba
state frequency: occasional
habitat: waste places, roadsides

key to the subspecies:
  1a. Plants typically dioecious; leaves with stinging hairs on both surfaces, cordate at the base, with
       coarse teeth 5.0–6.0 mm tall; stem with stiff bristles 0.75–2.0 mm long .... *U. d.* ssp. *dioica*
  1b. Plants typically monoecious; leaves with stinging hairs usually on the abaxial surface only,
       rounded to subcordate at the base, with smaller teeth 2.0–3.5 mm tall; stem glabrous or
       pubescent with shorter, softer hairs; bristles lacking or very sparse ....
       .... *U. d.* ssp. *gracilis* (Ait.) Seland.
notes: —

Urtica urens L.
synonym(s): —
common name(s): burning nettle, dog nettle
habit: simple or branched, taprooted, annual herb
range: Europe; naturalized to Newfoundland, s. to PA and WV, w. to CA, n. to British Columbia
state frequency: very rare
habitat: waste places
notes: —

POLYGALACEAE
(1 genus, 6 species)

Polygala
(6 species)

Milkworts have strongly zygomorphic flowers. The 3 outer sepals are small, the 2 inner sepals, called
wings, arranged laterally, are large and often petaloid. The corolla is composed of 3 petals, all of which are
basally connate. The lower petal, called the keel petal, differs from the upper 2 in being keel-shaped and
ornamented with a fringe-like crest.

KEY TO THE SPECIES
  1a. Leaves few, clustered near the tip of the stem, the blades elliptic or oval; flowers
       13.0–19.0 mm long, with 6 stamens; plants perennial from creeping rhizomes ....
       .... *P. paucifolia*
  1b. Leaves usually many and spaced throughout the stem, the blades linear to narrow-elliptic
       or obovate; flowers 3.0–6.0 mm long, with 8 stamens; plants annual, biennial, or
       perennial, but without creeping rhizomes
     2a. Chasmogamous [petaliferous] flowers borne in loose racemes, on spreading or
          recurving pedicels 1.5–3.5 mm long; cleistogamous flowers produced after the
          chasmogamous flowers, borne in prostrate or subterranean, secund racemes ....
          .... *P. polygama*
     2b. Chasmogamous flowers borne in relatively more dense, spike-like racemes, the
          pedicels up to 1.5 mm long; cleistogamous flowers not produced
        3a. Plants perennial, with clustered stems arising from a hard, knotty root; leaves
             3.0–8.0 x 0.3–1.5(–3.0) cm, irregularly serrulate .... *P. senega*
        3b. Plants annual, the solitary stems arising from a taproot; leaves 1.0–4.0 x
             0.1–0.5 cm, entire
           4a. Leaves 2–5 at each of the lower nodes (sometimes the upper nodes with only
                1 leaf); wings 0.9–4.0 mm long, about as long as the keel petal

5a. Inflorescence slender, conic or cylindric-conic, tapering to the tip, 2.0–4.5 mm wide; wings 0.9–2.6 mm long, rounded to obtuse at the apex; peduncles 0.5–8.0 cm long .... *P. verticillata*

5b. Inflorescence cylindric, rounded at the apex, 7.0–20.0 mm wide; wings 2.5–4.0 mm long, acuminate to awn-tipped at the apex; peduncles 0.0–1.0 cm long .... *P. cruciata*

4b. Leaves 1 at each node; wings 4.8–6.3 mm long, about twice as long as the keel petal .... *P. sanguinea*

Polygala cruciata L. var. *aquilonia* Fern. & Schub.

synonym(s): —
common name(s): marsh milkwort, drum heads
habit: erect, simple or branched, annual herb
range: ME, s. to VA and mountains of AL, w. to IL, n. to MN
state frequency: very rare
habitat: damp to wet, sandy or peaty soil, marshes
notes: This species is ranked SH by the Maine Natural Areas Program. The flowers are red-purple or green-purple.

Polygala paucifolia Willd.

synonym(s): —
common name(s): bird on the wing, fringed polygala, flowering-wintergreen
habit: colonial, rhizomatous, stoloniferous, perennial herb
range: Quebec, s. to mountains of GA, w. to IL, n. to MN and Saskatchewan
state frequency: occasional
habitat: moist woods
notes: The flowers in this species are red-purple to white.

Polygala polygama Walt. var. *obtusata* Chod.

synonym(s): —
common name(s): bitter milkwort
habit: clustered, decumbent, biennial herb
range: ME, s. to FL, w. to TX, n. to MN and Ontario
state frequency: uncommon
habitat: dry, sandy soil
notes: Flowers red-purple to white.

Polygala sanguinea L.

synonym(s): *Polygala viridescens* L.
common name(s): blood milkwort
habit: erect, simple or branched, annual herb
range: New Brunswick and Nova Scotia, s. to SC, w. to OK, n. to MN and Ontario
state frequency: occasional
habitat: open woods, fields, meadows, roadsides, clearings, especially after a fire
notes: The flowers in this species are red-purple, white, or green.

Polygala senega L.

synonym(s): —
common name(s): Seneca snakeroot
habit: clustered, simple, perennial herb
range: New Brunswick and ME, s. to GA, w. to AR, n. to Alberta
state frequency: very rare

habitat: dry to moist, calcareous woods and shores

notes: This species is ranked S1 by the Maine Natural Areas Program. Its flowers are white.

Polygala verticillata L.

synonym(s): —

common name(s): whorled milkwort

habit: slender, erect, branched, annual herb

range: ME, s. to FL, w. to TX, n. to MI and Ontario

state frequency: uncommon

habitat: moist to dry, sterile soil

key to the subspecies:

1a. Most or all of the nodes with whorled leaves; flower-bearing portion of the raceme 0.5–2.0 cm long, usually continuous; wings shorter than the capsule

    2a. Plants 1.5–4.0 dm tall; branches ascending; peduncles 2.0–7.0 cm long; seeds hirsute; pedicels 0.5–1.0 mm long .... *P. v.* var. *verticillata*

    2b. Plants 0.5–2.0 dm tall; branches mostly spreading; peduncles 0.5–4.0 cm long; seeds finely pubescent; pedicels 0.1–0.3 mm long .... *P. v.* var. *isocycla* Fern.

1b. Only the 1–3 lower nodes with whorled leaves, the remaining upper nodes with alternate leaves; flower-bearing portion of the raceme 1.0–5.0 cm long, often interrupted in the lower portion; wings about equaling the capsule .... *P. v.* var. *ambigua* (Nutt.) Wood

notes: The flowers in this species are white, green-white, green, or rarely pink.

FABACEAE

(26 genera, 59 species)

KEY TO THE GENERA

1a. Leaves simple; plants woody; androecium with monadelphous filaments and dimorphic anthers .... *Genista*

1b. Leaves compound (sometimes the upper 1-foliate in *Cytisus*); plants herbaceous or woody; androecium various

    2a. Leaves pinnately compound with more than 3 leaflets

        3a. Plants definitely woody

            4a. Leaves with minute, glandular dots; corolla with 1 petal [the banner petal] .... *Amorpha*

            4b. Leaves without glandular dots; corolla with 5 petals [the 2 lower fused to form the keel and thus the corolla appearing as 4 petals]

                5a. Plants lianas; petals blue; legume velvety-pubescent, without thin, sharp margins .... *Wisteria*

                5b. Plants trees or shrubs; petals white or pink to red-purple (1 species with a yellow spot on the banner petal); legume glabrous, hispid, or bristly, at least 1 of the margins thin .... *Robinia*

        3b. Plants herbaceous

            6a. Inflorescence a dense umbel; some or all of the filaments distally widened

                7a. Flowers yellow (later becoming orange and marked with red); leaves appearing to have 5 leaflets, the basal 2 [actually stipules] separated from the apical 3; fruit a legume .... *Lotus*

7b. Flowers pink and white; leaves with 11–25 evenly spaced leaflets; fruit a loment .... *Coronilla*

6b. Inflorescence a raceme or of axillary flowers; all of the filaments slender, not dilated near the apex

8a. Leaves with minute glandular scales or dots; petals pale yellow; legumes uncinate-spiny .... *Glycyrrhiza*

8b. Leaves without glandular scales or dots; petals white, or pink to red to purple; legumes not spiny

9a. Leaves with an even number of leaflets, the terminal leaflet modified into a tendril (or absent in *Vicia faba*)

10a. Styles slightly widened and flattened in the apical portion, pubescent along the distal, adaxial [inner] surface; wing petals essentially free from the keel petals; leaves with 2–12 leaflets 1.5–8.0 cm long; stems winged in some species .... *Lathyrus*

10b. Styles filiform, pubescent in a ring around the apex, or on the abaxial [outer] surface, or glabrous; wing petals cohering to the keel petals; leaves with 4–20 leaflets 0.5–3.5 cm long (4.0–10.0 cm in *Vicia faba*); stems angled or ridged, but not winged .... *Vicia*

9b. Leaves with an odd number of leaflets, the terminal leaflet present

11a. Leaves with 5–7 leaflets; corolla purple-brown; plants trailing, with tuberous rhizomes .... *Apios*

11b. Leaves with 7–31 leaflets; corolla purple or white; plants upright to decumbent, but not trailing, without tuberous rhizomes

12a. Fruit a loment, glabrous; raceme secund, the floriferous portion mostly 5.0–10.0 cm tall at anthesis .... *Hedysarum*

12b. Fruit a legume, pubescent; raceme spirally arranged in flower, the floriferous portion mostly 2.0–4.0 cm tall at anthesis

13a. Keel obtusely to acutely pointed; plants with short, but evident, stems .... *Astragalus*

13b. Keel abruptly short-pointed; plants acaulescent or nearly so .... *Oxytropis*

2b. Leaves palmately compound or with 3 or fewer leaflets

14a. Plants woody

15a. Twigs 4- or 5-angled; inflorescence composed of solitary or paired flowers in the axils of the upper leaves; legume pubescent only on the margin .... *Cytisus*

15b. Twigs terete; inflorescence a pendulous raceme 15.0–25.0 cm long; legume pubescent over the surface .... *Laburnum*

14b. Plants herbaceous

16a. Leaves palmately compound with 7–17 leaflets; androecium with monadelphous filaments and dimorphic anthers —one type of anther subglobose and versatile, the other linear and basifixed .... *Lupinus*

16b. Leaves pinnately or palmately compound with 3 or fewer leaflets;
andraecium with distinct or diadelphous filaments (monadelphous in some
Desmodium) and monomorphic anthers

17a. Leaflets toothed, at least minutely in the apical portion

18a. Inflorescence 3.0–20.0 cm tall, usually 4.0–15.0 times as tall as
wide; stipules entire, 1-nerved, the free portion 8.0 or more times as
long as wide .... ***Melilotus***

18b. Inflorescence 0.3–10.0 cm tall, up to 3.0 times as tall as wide;
stipules entire or toothed, 2- or 3-veined, the free portion commonly
less than 8.0 times as long as wide

19a. Leaflets sessile or with petiolules of similar length ...... ***Trifolium***

19b. Terminal leaflet on a petiolule distinctly longer than the
petiolules of the lateral leaflets

20a. Calyx zygomorphic, the basal, connate portion glabrous;
corolla marcescent, investing the fruit; stipules entire; legume
straight .... ***Trifolium***

20b. Calyx actinomorphic or nearly so, the basal, connate portion
pubescent; corolla deciduous in fruit; stipules toothed, at least
near the base; legume curved to coiled .... ***Medicago***

17b. Leaflets entire (sometimes with 1 or 2 lobes in *Strophostyles*)

21a. Leaves with 2 lateral leaflets, the terminal leaflet modified into a
tendril .... ***Lathyrus***

21b. Leaves with 3 leaflets—2 lateral leaflets and a terminal leaflet

22a. Stamens distinct; corolla yellow

23a. Legume 5.0–7.0 cm long; ovary sessile or very short-stipitate;
leaflets 4.0–8.0 cm long .... ***Thermopsis***

23b. Legume 0.8–1.5 cm long; ovary stipitate, the stipe
5.0–10.0 mm long in fruit; leaflets 0.6–1.8 cm long ... ***Baptisia***

22b. Stamens diadelphous, 9 stamens connate into a group with 1
separate stamen (monadelphous in some *Desmodium*); corolla
pink to purple, white, or yellow-white

24a. Leaflets without stipels; ovary with 1 ovule, maturing as a 1-
seeded, indehiscent fruit .... ***Lespedeza***

24b. Leaflets usually with stipels; ovary with 2 or more ovules,
maturing as a fruit with 2 or more seeds (*Amphicarpaea* also
producing more or less apetalous flowers from near the base of
the plant with 1 ovule)

25a. Plants erect; fruit a loment .... ***Desmodium***

25b. Plants trailing, twining, or climbing; fruit a legume

26a. Bracts of pedicels obtuse to round at the apex; flowers
without subtending bracteoles; plants with dimorphic
flowers—petaliferous flowers with 10 diadelphous
stamens and maturing as a 3- or 4-seeded aerial fruit,
and more or less apetalous flowers with fewer than 10
distinct stamens and maturing as a 1-seeded, often
subterranean, fruit .... ***Amphicarpaea***

26b. Bracts of pedicels acute at the apex; flowers with
closely subtending bracteoles that overlap the calyx;
plants with monomorphic, petaliferous flowers
27a. Plants often with some leaflets possessing a lateral
lobe on one or both sides; calyx with 4 lobes [due
to fusion of 2 lobes]; keel petals arched inward;
inflorescence a capitate spike; seeds 5.0–10.0 mm
long .... ***Strophostyles***
27b. Plants with simple leaflets; calyx with 5 lobes;
keel petals coiled inward; inflorescence an elongate
raceme; seeds about 2.0 mm long .... ***Phaseolus***

Amorpha
(1 species)

Amorpha fruticosa L.
synonym(s): —
common name(s): false indigo, indigo-bush
habit: shrub
range: PA, s. to FL, w. to KS, n. to MI and Ontario; escaped from cultivation in the northeast
state frequency: rare
habitat: thickets, stream banks
notes: —

Amphicarpaea
(1 species)

Amphicarpaea bracteata (L.) Fern.
synonym(s): *Amphicarpaea bracteata* (L.) Fern. var. *comosa* (L.) Fern., *Amphicarpaea pitcheri*
(Torr. & Gray) Kuntze, *Falcata comosa* (L.) Kuntze
common name(s): hog-peanut
habit: low, twining, annual herb
range: New Brunswick, s. to FL, w. to TX, n. to MT and Manitoba
state frequency: occasional
habitat: damp woods, thickets, and shores
notes: —

Apios
(1 species)

Apios americana Medik.
synonym(s): *Glycine apios* L.
common name(s): common groundnut
habit: twining, climbing, rhizomatous, perennial herb
range: New Brunswick, s. to FL, w. to TX and CO, n. to SD and MN
state frequency: occasional
habitat: moist woods, thickets, and shores
notes: —

Astragalus
(3 species)

KEY TO THE SPECIES

1a. Legume deeply grooved on the lower surface, falcate; stems borne separately on a branched, subterranean caudex, decumbent, forming mats .... *A. alpinus*
1b. Legume not, or only shallowly, grooved on the lower surface, straight or slightly falcate; stems clustered, borne on an erect caudex at the summit of a taproot
  2a. Legume 13.0–25.0 mm long, on a stipe 2.0–5.0 mm long; flowers 10.0–12.0 mm long .... *A. robbinsii*
  2b. Legume 6.0–8.0 mm long, sessile; flowers 6.0–8.0 mm long .... *A. eucosmus*

Astragalus alpinus L. var. *brunetianus* Fern.
synonym(s): —
common name(s): alpine milk-vetch, Brunet's milk-vetch
habit: creeping, decumbent, branched, perennial herb
range: circumboreal; s. to ME, w. to CO
state frequency: rare
habitat: gravelly river banks and shores
notes: This species is ranked S2 by the Maine Natural Areas Program.

Astragalus eucosmus B. L. Robins.
synonym(s): —
common name(s): elegant milk-vetch
habit: ascending, perennial herb
range: Labrador, s. to ME, w. to CO, n. to AK
state frequency: very rare
habitat: calcareous ledges and gravelly shores
notes: This species is ranked SX by the Maine Natural Areas Program.

Astragalus robbinsii (Oakes) Gray var. *minor* (Hook.) Barneby
synonym(s): *Astragalus blakei* Egglest.
common name(s): Robbins' milk-vetch
habit: clustered, perennial herb
range: Labrador, s. to ME, w. to VT, n. to Quebec
state frequency: very rare
habitat: calcareous ledges and slopes
notes: This species is ranked SX by the Maine Natural Areas Program.

Baptisia
(1 species)

Baptisia tinctoria (L.) R. Br. *ex* Ait. f.
synonym(s): *Baptisia tinctoria* (L.) R. Br. *ex* Ait. f. var. *crebra* Fern.
common name(s): wild-indigo, false indigo
habit: slender, branched, perennial herb
range: ME, s. to GA, w. to LA, n. to MN and Ontario
state frequency: very rare
habitat: dry, open woods and clearings
notes: This species is ranked S1 by the Maine Natural Areas Program.

Coronilla
(1 species)

Coronilla varia L.
synonym(s): —
common name(s): crown-vetch
habit: ascending, perennial herb
range: Europe; naturalized to New Brunswick, s. to VA, w. to MO, n. to British Columbia
state frequency: uncommon
habitat: roadsides, waste places
notes: —

Cytisus
(1 species)

Cytisus scoparius (L.) Link
synonym(s): —
common name(s): Scotch broom
habit: much branched shrub
range: Europe; naturalized in the east to Nova Scotia, s. to GA
state frequency: very rare
habitat: dry, sandy soil of open woods and roadsides
notes: —

Desmodium
(5 species)

KEY TO THE SPECIES
1a. Flowering stems leafless, the inflorescence arising from the ground; pedicels
    10.0–20.0 mm long; stipe 10.0–18.0 mm long .... **D. nudiflorum**
1b. Flowering stems with leaves, the inflorescence from leaf axils or the apex of the stem;
    pedicels 2.0–11.0 mm long; stipe 2.0–12.0 mm long
    2a. Stamens monadelphous; free portion of the calyx [the lobes] less than 0.5 times the
        length of the connate portion [the tube]; loment borne on a stipe exceeding the length
        of the persistent stamens, glabrous on the upper suture; terminal leaflet ovate-
        orbicular .... **D. glutinosum**
    2b. Stamens diadelphous; free portion of the calyx more than 0.5 times the length of the
        connate portion; loment borne on a stipe that is shorter than the persistent stamens,
        uncinate-pubescent on the upper suture; terminal leaflet lanceolate to ovate or
        oblong-ovate
        3a. Loment segment [article] nearly semicircular, gradually rounded on the lower
            margin; inflorescence with conspicuous lance-ovate bracts, the larger bracts
            4.0–10.0 mm long; flowers 10.0–13.0 mm long .... **D. canadense**
        3b. Loment segment nearly triangular, abruptly curved on the lower margin;
            inflorescence with inconspicuous, smaller bracts; flowers 6.0–10.0 mm long
            4a. Leaves lanceolate to oblong, 3.0–8.0 times as long as wide, 1.0–2.3 cm wide;
                stems glabrous or uncinate-puberulent .... **D. paniculatum**

4b. Leaves elliptic-ovate to ovate, 1.5–3.0 times as long as wide, 1.3–3.8 cm
　　wide; stems uncinate-puberulent and pilose .... **D. perplexum**

Desmodium canadense (L.) DC.

synonym(s): *Meibomia canadensis* (L.) Kuntze
common name(s): Canadian tick-trefoil
habit: erect, branched, perennial herb
range: New Brunswick, s. to VA, w. to OK, n. to Alberta
state frequency: occasional
habitat: open woods, thickets, shores
notes: —

Desmodium glutinosum (Muhl. *ex* Willd.) Wood

synonym(s): *Desmodium acuminatum* (Michx.) DC., *Meibomia acuminata* (Michx.) Blake
common name(s): cluster-leaf tick-trefoil
habit: erect, simple, perennial herb
range: New Brunswick, s. to FL, w. to TX and Mexico, n. to MN and Ontario
state frequency: occasional
habitat: rocky woods
notes: —

Desmodium nudiflorum (L.) DC.

synonym(s): *Meibomia nudiflora* (L.) Kuntze
common name(s): naked tick-trefoil
habit: slender, ascending to erect, simple, perennial herb
range: ME, s. to FL, w. to TX, n. to MN
state frequency: uncommon
habitat: forests
notes: —

Desmodium paniculatum (L.) DC.

synonym(s): *Meibomia paniculata* (L.) Kuntze
common name(s): —
habit: slender, erect, branched, perennial herb
range: ME, s. to FL, w. to TX and NE, n. to MI and Ontario
state frequency: uncommon
habitat: dry woods, borders, and clearings
notes: —

Desmodium perplexum Schub.

synonym(s): *Desmodium dillenii* Darl., *Desmodium glabellum* (Michx.) DC. *sensu* Gleason and
　　Cronquist (1991), *Meibomia dillenii* (Darl.) Kuntze
common name(s): —
habit: slender to stout, erect to spreading, perennial herb
range: ME, s. to FL, w. to TX, n. to WI
state frequency: uncommon
habitat: sandy woods
notes: —

Genista
(1 species)

Genista tinctoria L.
synonym(s): —
common name(s): dyer's greenweed
habit: slender, erect shrub
range: Europe; escaped from cultivation to ME, s. to VA, w. and n. to MI
state frequency: rare
habitat: dry, sterile soil, sometimes on beach dunes
notes: —

Glycyrrhiza
(1 species)

Glycyrrhiza lepidota Pursh
synonym(s): —
common name(s): wild licorice
habit: tall, colonial, perennial herb
range: MN, s. to TX, w. to CA, n. to WA and Alberta; introduced in the east
state frequency: very rare
habitat: meadows, shores, disturbed soil
notes: —

Hedysarum
(1 species)

Hedysarum alpinum L. var. **americanum** Michx.
synonym(s): —
common name(s): hedysarum, alpine sweetbroom
habit: erect to arching, perennial herb
range: Newfoundland, s. to MA, w. to WY, n. to AK
state frequency: very rare; known from northern Maine
habitat: calcareous ledges and river banks
notes: This species is ranked S3 by the Maine Natural Areas Program.

Laburnum
(1 hybrid)

Laburnum ×watereri Dipp.
synonym(s): *Laburnum vossii* Hart.
common name(s): goldenchain tree, water laburnum
habit: shrub or small tree
range: southern Europe; escaped from cultivation locally
state frequency: rare
habitat: roadsides, edges, often near establishments
notes: This plant is the hybrid of *Laburnum alpinum* Griseb. and *L. anagyroides* Medik.

Lathyrus
(6 species)

KEY TO THE SPECIES

1a. Leaves with a single pair of leaflets; plants introduced or escaped from cultivation
    2a. Stems conspicuously winged; flowers 1.5–4.0 cm long
        3a. Plants annual, without rhizomes; inflorescence with 1–3 flowers; flowers fragrant
        .... ***L. odoratus***
        3b. Plants perennial, with rhizomes; inflorescence with 4–10 flowers; flowers
        odorless .... ***L. latifolius***
    2b. Stems 2-angled, but not winged; flowers 1.2–1.8 cm long
        4a. Petals yellow; stipules 1.0–3.0 cm long .... ***L. pratensis***
        4b. Petals red-purple; stipules 0.5–1.2 cm long .... ***L. tuberosus***
1b. Leaves with 2–6 pairs of leaflets; plants native
    5a. Stipules ovate-hastate, 1.0–2.5 cm wide, attached at the base and therefore with 2
    basal lobes; leaflets fleshy; legumes mostly 8.5–10.5 mm wide; plants of seashores ...
    .... ***L. japonicus***
    5b. Stipules semi-sagittate, 0.1–1.0 cm wide, attached at the side and therefore with 1
    basal lobe; leaflets scarcely fleshy; legumes mostly less than 8.5 mm wide; plants of
    moist to wet soils of inland areas .... ***L. palustris***

Lathyrus japonicus Willd.
synonym(s): <*L. j.* var. *maritimus*> = *Lathyrus japonicus* Willd. var. *glaber* (Ser.) Fern., *Lathyrus maritimus* (L.) Bigelow
common name(s): beach-pea, seaside-pea
habit: stout, fleshy, decumbent to suberect, perennial herb
range: circumboreal; s. to NJ; also on shores of Lake Champlain and Great Lakes
state frequency: uncommon; abundant in some coastal areas
habitat: sea beaches, gravelly shores
key to the subspecies:
    1a. Plants glabrous; stems commonly 1.0–3.0 dm long and 0.5–2.5 mm thick in dried specimens;
    tendrils commonly simple .... *L. j.* var. *maritimus* (L.) Kartesz & Gandhi
    1b. Plants pubescent on at least the calyx, pedicels, and abaxial leaf surface; stems commonly
    2.0–15.0 dm long and 2.0–5.0 mm thick in dried specimens; tendrils commonly forked ....
    .... *L. j.* var. *pellitus* Fern.
notes: —

Lathyrus latifolius L.
synonym(s): —
common name(s): everlasting-pea
habit: trailing to climbing, rhizomatous, perennial herb
range: Europe; escaped from cultivation to ME, s. to VA, w. to KS
state frequency: uncommon
habitat: roadsides, thickets, waste places
notes: —

Lathyrus odoratus L.
synonym(s): —
common name(s): sweet-pea
habit: annual herb

range: Europe; escaped from cultivation
state frequency: very rare
habitat: waste places
notes: —

Lathyrus palustris L.

synonym(s): *Lathyrus palustris* L. var. *linearifolius* Ser., *Lathyrus palustris* L. var. *macranthus* (White) Fern., *Lathyrus palustris* L. var. *myrtifolius* (Muhl. *ex* Willd.) Gray, *Lathyrus palustris* L. var. *pilosus* (Cham.) Ledeb.
common name(s): vetchling, marsh-pea
habit: slender-rhizomatous, climbing, perennial herb
range: circumboreal; s. to NJ, w. to CA
state frequency: uncommon
habitat: damp to wet woods, thickets, meadows, and shores, swamps
notes: —

Lathyrus pratensis L.

synonym(s): —
common name(s): yellow vetchling, meadow-pea
habit: slender, rhizomatous, perennial herb
range: Eurasia; naturalized to Newfoundland, s. to CT, w. to IL, n. to MI and Ontario
state frequency: uncommon
habitat: damp slopes, meadows, shores, and roadsides, waste places
notes: —

Lathyrus tuberosus L.

synonym(s): —
common name(s): tuberous vetchling
habit: slender, rhizomatous, perennial herb
range: Europe; naturalized to ME, w. to WI, n. to Ontario
state frequency: rare
habitat: fields, meadows, roadsides
notes: —

Lespedeza
(3 species)

Species of *Lespedeza* have 2 types of flowers—those that bear petals [which are chasmogamous] and those that lack petals [which are cleistogamous]. The apetalous flowers may have a slightly differently shaped fruit than the petaliferous flowers [the fruit from the apetalous flowers will have a blunter apex]. Hybridization is common within this genus.

KEY TO THE SPECIES

1a. Corolla purple; apetalous flowers usually present, numerous, borne in subsessile, axillary clusters; fruit much longer than the calyx .... *L. intermedia*
1b. Corolla yellow-white, often with a purple spot near the base of the banner petal; apetalous flowers absent or few and intermixed with the petaliferous ones; fruit shorter than to equaling the length of the calyx
    2a. Peduncles usually shorter than the subtending leaves; inflorescence subglobose to short-ovoid; calyx lobes of the petaliferous flowers 6.0–10.0 mm long; fruit shorter than the calyx .... *L. capitata*

2b. Peduncles usually longer than the subtending leaves; inflorescence short-cylindric;
   calyx lobes of the petaliferous flowers 3.0–7.5 mm long; fruit about equaling the
   length of the calyx .... *L. hirta*

Lespedeza capitata Michx.

synonym(s): *Lespedeza capitata* Michx. var. *velutina* (Bickn.) Fern., *Lespedeza capitata* Michx. var.
   vulgaris Torr. & Gray
common name(s): bush-clover
habit: stiff, stout, erect, simple or branched, perennial herb
range: ME, s. to FL, w. to TX, n. to NE and MN
state frequency: uncommon
habitat: dry, open soil of borders, sand dunes
notes: Hybrids with *Lespedeza hirta*, called *L.* ×*longifolia* DC., occur in Maine.

Lespedeza hirta (L.) Hornem.

synonym(s): —
common name(s): hairy lespedeza
habit: stout, erect to ascending, perennial herb
range: ME, s. to GA, w. to TX, n. to Ontario
state frequency: very rare
habitat: dry soil
notes: This species is ranked SH by the Maine Natural Areas Program. Known to hybridize with both
   Lespedeza capitata (*q.v.*) and *L. intermedia* (*q.v.*).

Lespedeza intermedia (S. Wats.) Britt.

synonym(s): —
common name(s): wand lespedeza
habit: erect to ascending, branched, perennial herb
range: ME, s. to FL, w. to TX, n. to WI and Ontario
state frequency: uncommon
habitat: dry, open woods and thickets
notes: Hybrids with *Lespedeza hirta*, called *L.* ×*nuttallii* Darl., occur in Maine and are ranked SH by
   the Maine Natural Areas Program.

Lotus
(1 species)

Lotus corniculatus L.

synonym(s): —
common name(s): birdsfoot-trefoil
habit: prostrate to suberect, much branched, perennial herb
range: Europe; naturalized to Newfoundland, s. to VA, w. to OH, n. to MN
state frequency: occasional
habitat: fields, meadows, roadsides, waste places
notes: —

Lupinus
(2 species)

KEY TO THE SPECIES

1a. Leaves with 7–11 leaflets 1.5–5.0 cm long; stems usually 2.0–6.0 dm tall; racemes 7.0–30.0 cm tall .... *L. perennis*
1b. Leaves with 11–17 leaflets 6.0–13.0 cm long; stems usually 6.0–12.0 dm tall; racemes 30.0–70.0 cm tall .... *L. polyphyllus*

Lupinus perennis L.
synonym(s): —
common name(s): wild lupine, sundial lupine
habit: erect, perennial herb
range: ME, s. to FL, w. and n. to IN and MN
state frequency: very rare
habitat: dry, open woods and clearings
notes: This species is ranked SX by the Maine Natural Areas Program.

Lupinus polyphyllus Lindl.
synonym(s): —
common name(s): lupine
habit: tall, stout, erect, perennial herb
range: western North America; naturalized in northeastern United States and adjacent Canada
state frequency: occasional
habitat: dry roadsides, banks, and fields
notes: —

Medicago
(4 species)

KEY TO THE SPECIES

1a. Flowers 6.0–12.0 mm long, blue-purple; plants perennial .... *M. sativa*
1b. Flowers 2.0–5.0 mm long, yellow; plants annual or biennial
  2a. Legume reniform-curved or slightly spiraled near the apex, without spines; inflorescence with 10–50 flowers .... *M. lupulina*
  2b. Legume 2.0- to 7.0-times spiral-coiled, with spines; inflorescence with 2–8 flowers
    3a. Leaflets broad-obcordate, as wide or wider than long, commonly with a dark blotch or mottles on the adaxial surface; legume 6.0–9.0 mm in diameter .... *M. arabica*
    3b. Leaflets triangular-obovate, about 0.65 times as wide as long, without dark mottles; legume 4.0–6.0 mm in diameter .... *M. polymorpha*

Medicago arabica (L.) Huds.
synonym(s): —
common name(s): spotted medick
habit: depressed to spreading, annual herb
range: Eurasia; naturalized to New Brunswick, s. to FL, w. to TX and Mexico, n. to WA
state frequency: very rare
habitat: waste places
notes: —

Medicago lupulina L.

synonym(s): *Medicago lupulina* L. var. *glandulosa* Neilr.
common name(s): black medick, nonesuch
habit: prostrate to ascending, branched, annual or biennial herb
range: Eurasia; naturalized to temperate North America
state frequency: uncommon
habitat: roadsides, waste places
notes: —

Medicago polymorpha L.

synonym(s): *Medicago hispida* Gaertn.
common name(s): smooth bur-clover
habit: prostrate to ascending, branched, annual herb
range: Europe; naturalized to MI, s. to FL, w. to TX and Mexico, n. to MT and British Columbia; also
    New Brunswick, s. to MA
state frequency: very rare
habitat: waste places
notes: —

Medicago sativa L.

synonym(s): —
common name(s): alfalfa
habit: slender, ascending, perennial herb
range: Eurasia; naturalized to temperate North America
state frequency: uncommon
habitat: fields, roadsides, waste places
notes: —

Melilotus
(2 species)

KEY TO THE SPECIES
1a. Flowers yellow or white, 3.5–7.0 mm long; pedicels decurved, 1.0–2.0 mm long; legume
    2.5–5.0 mm long; plants 0.5–3.0 m tall .... *M. officinalis*
1b. Flowers yellow, 2.0–3.0 mm long; pedicels ascending, 0.5–0.8 mm long; legume
    1.5–3.0 mm long; plants 0.2–0.6 m tall .... *M. indicus*

Melilotus indicus (L.) All.

synonym(s): —
common name(s): sweet-clover
habit: ascending to erect, annual herb
range: Eurasia; cosmopolitan weed, adventive in our area
state frequency: very rare
habitat: waste places
notes: —

Melilotus officinalis (L.) Lam.

synonym(s): *Melilotus albus* Medik.
common name(s): white sweet-clover, yellow sweet-clover
habit: ascending to erect, biennial or annual herb
range: Eurasia; naturalized to Newfoundland, s. to FL, w. to Mexico, n. to British Columbia
state frequency: occasional

habitat: roadsides, waste places
notes: —

Oxytropis
(1 species)

Oxytropis campestris (L.) DC. var. ***johannensis*** Fern.
synonym(s): *Oxytropis johannensis* (Fern.) Heller
common name(s): St. John River oxytrope
habit: perennial herb
range: circumboreal; s. to ME, w. to CO and OR
state frequency: very rare
habitat: calcareous rocks and gravelly shores
notes: This species is ranked S1/S2 by the Maine Natural Areas Program. Flowers purple.

Phaseolus
(1 species)

Phaseolus polystachios (L.) B. S. P. is reported to occur in Maine by Gleason and Cronquist (1991), but voucher specimens are unknown.

Phaseolus vulgaris L.
synonym(s): *Phaseolus vulgaris* L. var. *humilis* Alef.
common name(s): kidney bean
habit: twining, perennial herb
range: Europe; escaped from cultivation
state frequency: very rare
habitat: fields, areas of cultivation
notes: —

Robinia
(3 species)

KEY TO THE SPECIES
1a. Corolla white; trees to 25.0 m tall; ovary and legume glabrous; twigs glabrous or
sparsely pubescent .... ***R. pseudoacacia***
1b. Corolla pink to red-purple; shrubs to 5.0 m tall; ovary and legume hispid; twigs
glandular, the glands sessile or stipitate
2a. Leaves with 7–13 leaflets; twigs bristly-glandular .... ***R. hispida***
2b. Leaves with 13–25 leaflets; twigs glandular with sessile or subsessile glands ....
.... ***R. viscosa***

Robinia hispida L.
synonym(s): —
common name(s): bristly locust, mossy locust, rose-acacia
habit: rhizomatous shrub
range: VA, s. to GA, w. to AL, n. to KY; escaped from cultivation to ME, w. to MN
state frequency: uncommon
habitat: woods, thickets, slopes
notes: Flowers pink-purple to rose.

Robinia pseudoacacia L.
synonym(s): —
common name(s): black locust, false acacia
habit: tree to large shrub
range: PA, s. to GA, w. to OK, n. to IN; escaped from cultivation as far north as Ontario, Quebec, and New Brunswick
state frequency: uncommon
habitat: woods, thickets
notes: —

Robinia viscosa Vent.
synonym(s): —
common name(s): clammy locust, rose-acacia
habit: large shrub to tree
range: PA, s. to GA, w. to AL, n. to WV; escaped from cultivation in the northeast
state frequency: uncommon
habitat: dry woods, roadsides
notes: —

Strophostyles
(1 species)

Strophostyles helvula (L.) Ell.
synonym(s): —
common name(s): annual woolly-bean
habit: twining to trailing, branched, annual herb
range: ME, s. to FL, w. to TX, n. to SD, MN, and Ontario
state frequency: very rare
habitat: dry, sandy soil, often in coastal areas
notes: —

Thermopsis
(1 species)

Thermopsis mollis (Michx.) M. A. Curtis *ex* Gray
synonym(s): —
common name(s): bush-pea, piedmont buckbean
habit: erect, rhizomatous, perennial herb
range: VA, s. to GA, w. to AL, n. to TN; naturalized in New England
state frequency: rare
habitat: roadsides, fields, open woods
notes: Flowers yellow.

Trifolium
(9 species)

KEY TO THE SPECIES
1a. Flowers yellow, turning brown after anthesis

2a. Central leaflet sessile; flowers 5.0–7.0 mm long; inflorescence 1.0–2.0 cm tall; stipules about as long as the petioles; style nearly equaling the length of the legume ... . *T. aureum*

2b. Central leaflet borne on a petiolule; flowers 2.5–5.0 mm long; inflorescence 0.5–1.5 cm tall; stipules shorter than the petioles; style shorter than the legume

    3a. Petiolule of the central leaflet up to 1.0 mm long; inflorescence usually with 5–15 flowers; flowers 2.5–3.5 mm long; banner petal inconspicuously veined .... *T. dubium*

    3b. Petiolule of the central leaflet 1.0–3.0 mm long; inflorescence usually with 20–30 flowers; flowers 3.5–5.0 mm long; banner petal with 10 conspicuous veins .... *T. campestre*

1b. Flowers white to pink to purple

    4a. Flowers pedicellate, the pedicels commonly exceeding 2.0 mm; petals white or white and pink

        5a. Stems creeping along the ground, rooting at the nodes; petals usually concolored; stipules adnate to the petiole in the basal portion, then connate for a distance, forming a tube that surrounds the stem .... *T. repens*

        5b. Stems ascending, not rooting at the nodes; petals bicolored; stipules adnate to the petiole in the basal portion, then distinct, the tips completely free and not forming a tube .... *T. hybridum*

    4b. Flowers sessile or subsessile; petals white to pink to purple

        6a. Flowers 4.0–7.0 mm long; calyx lobes surpassing and partly concealing the petals .... *T. arvense*

        6b. Flowers 10.0–20.0 mm long; calyx lobes shorter than the petals

            7a. Inflorescence ovoid to cylindric, on peduncles usually exceeding 3.0 cm; each flower subtended by a bracteole .... *T. incarnatum*

            7b. Inflorescence globose to round-ovoid, sessile or on peduncles up to 3.0 cm long; flowers lacking bracteoles

                8a. Free portion of the stipule broad-triangular, shorter than the adnate portion; leaflets oval to obovate, frequently with dark mottles on the adaxial surface .... *T. pratense*

                8b. Free portion of the stipule lanceolate, longer than the adnate portion; leaflets elliptic to oblong, usually without dark mottles .... *T. medium*

Trifolium arvense L.

synonym(s): —
common name(s): rabbit-foot clover
habit: erect, branched, annual herb
range: Eurasia and Africa; naturalized to much of the United States and southern Canada
state frequency: occasional
habitat: sandy soil of roadsides and waste places
notes: This species is similar to *Trifolium incarnatum* in many ways, including the relatively slender inflorescence and bracteolate flowers.

Trifolium aureum Pollich

synonym(s): *Trifolium agrarium* L.
common name(s): yellow clover, palmate hop clover
habit: ascending to erect, branched, annual or biennial herb

range: Eurasia; naturalized to Newfoundland, s. to SC, w. to AR, n. to British Columbia
state frequency: occasional
habitat: fields, roadsides, waste places
notes: —

Trifolium campestre Schreb.
synonym(s): *Trifolium procumbens* L.
common name(s): low hop clover, pinnate hop clover
habit: depressed to ascending, branched, annual herb
range: Eurasia and northern Africa; naturalized to much of North America
state frequency: occasional
habitat: roadsides, waste places
notes: —

Trifolium dubium Sibthorp
synonym(s): —
common name(s): little hop clover
habit: depressed to ascending, branched, annual herb
range: Europe; naturalized to New Brunswick, s. to FL, w. to TX, n. to British Columbia
state frequency: uncommon
habitat: fields, roadsides, waste places
notes: —

Trifolium hybridum L.
synonym(s): *Trifolium hybridum* L. var. *elegans* (Savi) Boiss.
common name(s): alsike clover
habit: ascending to erect, perennial herb
range: Eurasia; naturalized throughout most of temperate North America
state frequency: occasional
habitat: clearings, fields, roadsides
notes: —

Trifolium incarnatum L.
synonym(s): *Trifolium incarnatum* L. var. *elatius* Gibelli & Belli
common name(s): crimson clover
habit: erect, annual herb
range: Europe; introduced to our area
state frequency: uncommon
habitat: fields, roadsides, waste places
notes: —

Trifolium medium L.
synonym(s): —
common name(s): zigzag clover
habit: ascending, biennial or perennial herb
range: Europe; naturalized to New Brunswick, s. to MA, w. and n. to Quebec
state frequency: rare
habitat: borders of woods, slopes, fields, roadsides, along railroads
notes: —

Trifolium pratense L.
synonym(s): *Trifolium pratense* L. var. *sativum* (P. Mill.) Schreb.
common name(s): red clover

habit: ascending to erect, biennial or short-lived perennial herb
range: Europe; naturalized to much of temperate North America
state frequency: common
habitat: fields, clearings, roadsides, lawns
notes: —

Trifolium repens L.

synonym(s): —
common name(s): white clover
habit: creeping, perennial herb
range: Eurasia; naturalized to most of temperate North America
state frequency: occasional
habitat: open woods, roadsides, lawns
notes: —

Vicia
(7 species)

KEY TO THE SPECIES

1a. Flowers solitary or in inflorescences of 2–50, borne on long peduncles; style pubescent around the tip or glabrous

  2a. Flowers 3.0–7.0 mm long, solitary or in racemes of 2–8; corollas white to light purple; legume 0.6–1.3 cm long

    3a. Legume hirsute, usually with 2 seeds, obliquely tapering from the sutures and pointed at the tip; lobes of the calyx equal or nearly so in length; leaves usually with 12–16 leaflets .... *V. hirsuta*

    3b. Legume glabrous, usually with 4 seeds, equally rounded from the sutures and blunt at the tip; lobes of the calyx distinctly unequal in length; leaves usually with 4–10 leaflets .... *V. tetrasperma*

  2b. Flowers 12.0–27.0 mm long, in racemes of 10–50; corollas blue, white and blue, or rarely entirely white; legume 2.0–3.5 cm long

    4a. Plants perennial; calyx not gibbous or saccate on the upper side, the pedicel appearing to attach to the basal part of the flower; upper calyx lobes broad-triangular; limb of the banner petal about as long as the claw .... *V. cracca*

    4b. Plants annual or biennial; calyx gibbous or saccate on the upper side, the pedicel appearing to attach to the underside of the flower; upper calyx lobes lanceolate to linear-triangular; limb of the banner petal about half as long as the claw .... *V. villosa*

1b. Flowers solitary or paired, sessile or nearly so, borne in the axils of leaves; style pubescent on the abaxial [outer] surface near the apex

  5a. Terminal leaflet absent, not modified into a tendril; legume 8.0–20.0 cm long, with 2–4 seeds; leaves with 4 or 6 leaflets .... *V. faba*

  5b. Terminal leaflet of at least the upper leaves modified into a tendril; legume 2.0–7.0 cm long, with 3–12 seeds; leaves with 6–18 leaflets

    6a. Calyx actinomorphic or nearly so, the lobes more than half as long as the tube; flowers 1.2–3.0 cm long; plants annual .... *V. sativa*

6b. Calyx zygomorphic, the upper lobes less than half as long as the tube, the lower lobes about half as long as the tube; flowers 0.8–1.5 cm long; plants perennial .... *V. sepium*

Vicia cracca L.

synonym(s): —
common name(s): cow vetch, tufted vetch, bird vetch
habit: trailing to climbing, perennial herb
range: Eurasia; naturalized to Newfoundland, s. to VA, w. to IL, n. to British Columbia
state frequency: common
habitat: fields, meadows, roadsides, thickets, shores
notes: —

Vicia faba L.

synonym(s): —
common name(s): broad-bean
habit: stout, erect, annual herb
range: Europe; escaped from cultivation as far north as Newfoundland
state frequency: very rare
habitat: cultivated ground
notes: This species has white flowers with purple veins, the wing petals each with a purple spot.

Vicia hirsuta (L.) S. F. Gray

synonym(s): —
common name(s): tiny vetch, hairy vetch
habit: slender, decumbent to climbing, annual herb
range: Europe; naturalized to much of the United States and southern Canada
state frequency: very rare
habitat: fields, roadsides, waste places
notes: —

Vicia sativa L.

synonym(s): <*V. s.* ssp. *nigra*> = *Vicia angustifolia* L., *Vicia angustifolia* L. var. *segetalis* (Thuill.) W. D. J. Koch, *Vicia angustifolia* L. var. *uncinata* (Desv.) Rouy
common name(s): spring vetch, common vetch, narrow-leaved vetch
habit: slender, erect to ascending, simple or branched, annual herb
range: Europe; escaped from cultivation as far north as Newfoundland
state frequency: uncommon
habitat: roadsides, waste places
key to the subspecies:
  1a. Flowers 1.2–1.8 cm long; legume dark brown to black at maturity, not or only obscurely constricted between the seeds; mature seeds *ca.* 3.0 mm wide; leaves with 6–10 linear to narrow-elliptic leaflets; corolla uniformly purple .... *V. s.* ssp. *nigra* (L.) Ehrh.
  1b. Flowers 1.8–3.0 cm long; legume light brown at maturity, constricted between the seeds; mature seeds *ca.* 5.0 mm wide; leaves with 8–16 oblong to elliptic leaflets; corolla purple with light purple wing petals .... *V. s.* ssp. *sativa*
notes: —

Vicia sepium L.

synonym(s): —
common name(s): hedge vetch
habit: erect to climbing, stoloniferous, perennial herb
range: Europe; naturalized to Newfoundland, s. to ME, w. and n. to Ontario

state frequency: rare
habitat: old fields, roadsides
notes: Flowers blue.

Vicia tetrasperma (L.) Schreb.

synonym(s): *Vicia tetrasperma* (L.) Schreb. var. *tenuissima* Druce
common name(s): four-seeded vetch
habit: slender, decumbent to climbing, branched, annual herb
range: Eurasia; naturalized to Newfoundland, s. to FL, w. to TX, n. to WI and Ontario
state frequency: uncommon
habitat: fields, roadsides, waste places
notes: —

Vicia villosa Roth

synonym(s): <*V. v.* ssp. *varia*> = *Vicia dasycarpa* Ten.
common name(s): hairy vetch, winter vetch, woolly-pod vetch
habit: trailing to climbing, annual or biennial herb
range: Europe; naturalized to ME, s. to GA, w. to CA, n. to MT
state frequency: uncommon
habitat: fields, roadsides, waste places
key to the subspecies:
    1a. Raceme pubescent with long, spreading hairs, commonly 10- to 40-flowered; lower lobes of the calyx 2.0–5.0 mm long .... *V. v.* ssp. *villosa*
    1b. Raceme pubescent with short, ascending or appressed hairs, commonly 5- to 15-flowered; lower lobes of the calyx 1.0–2.0 mm long .... *V. v.* ssp. *varia* (Host) Corb.
notes: —

Wisteria
(1 species)

Wisteria floribunda (Willd.) DC.

synonym(s): —
common name(s): Japanese wisteria
habit: twining liana
range: Asia; escaped from cultivation to ME, s. to LA
state frequency: rare
habitat: old homesites, open woods
notes: —

ELATINACEAE
(1 genus, 3 species)

Elatine
(3 species)

Small, inconspicuous, aquatic herbs, best identified in fruit using quality optics to identify the seed coat.

KEY TO THE SPECIES
1a. Placentation basal-axile, therefore, the ovules [and seeds] all set at approximately the same level in the ovary [or fruit]; seeds marked with transversely elliptic, round-ended

areoles, the ends of the areoles more or less meeting the other ends [not dovetailed with], therefore, the longitudinal ridges of the seed appearing straight and distinct; leaves 0.7–5.0 mm long; perianth 2-merous; flowers with 2 stamens .... ***E. minima***

1b. Placentation axile, therefore, the ovules set at different levels in the ovary; seeds marked with transversely elongate, 6-sided, angular-ended areoles, the ends of the areoles alternating and dovetailed with areole ends from the adjacent rows, therefore, the longitudinal ridges of the seed broken and less conspicuous; leaves 2.8–8.0(–15.0) mm long; perianth 2- or 3-merous; flowers with 2 or 3 stamens

    2a. Leaves obovate to broad-spatulate, usually rounded at the apex, the larger 1.5–5.0 mm wide; seeds borne from the lower half of the ovary's central axis, ascending .... ***E. americana***

    2b. Leaves linear to lanceolate to narrow-spatulate, truncate or retuse at the apex, 0.5–3.0 mm wide; seeds borne throughout the length of the ovary's central axis, divergent .... ***E. triandra***

Elatine americana (Pursh) Arn.

synonym(s): —
common name(s): waterwort
habit: prostrate, aquatic, annual herb
range: Quebec and New Brunswick, s. to VA; also MO and OK
state frequency: uncommon
habitat: muddy shores, often tidal
notes: —

Elatine minima (Nutt.) Fisch. & C. A. Mey.

synonym(s): —
common name(s): waterwort
habit: creeping, aquatic, annual herb
range: Newfoundland, s. to VA, w. and n. to MN and Ontario
state frequency: uncommon
habitat: sandy or peaty soils of shores and pond margins
notes: —

Elatine triandra Schkuhr

synonym(s): —
common name(s): longstem waterwort
habit: creeping, aquatic, annual herb
range: Eurasia; also WI, s. to TX and Mexico, w. to CA, n. to WA and Alberta; introduced to ME
state frequency: very rare
habitat: shallow water, shores
notes: This species is ranked SH by the Maine Natural Areas Program.

LINACEAE
(2 genera, 5 species)

KEY TO THE GENERA

1a. Flowers with 4 sepals, petals, stamens, and styles; sepals with 3 apical lobes; leaves 2.0–3.0 mm long; plants diminutive annuals, 0.3–1.0 dm tall .... ***Radiola***

1b. Flowers with 5 sepals, petals, stamens, and styles; sepals without apical lobes, though sometimes ciliate or toothed along the margin; leaves 3.0–35.0 mm long; plants annual or perennial, 1.0–10.0 dm tall .... *Linum*

Linum
(4 species)

KEY TO THE SPECIES

1a. Petals blue, 10.0–23.0 mm long; sepals without glands; capsules 5.0–10.0 mm wide
    2a. Leaves with 3 nerves; sepals 7.0–9.0 mm long at maturity, the inner ciliolate; plants annual .... *L. usitatissimum*
    2b. Leaves with 1 nerve, sometimes with 2 additional, obscure nerves; sepals 5.0–7.0 mm long at maturity, entire; plants perennial .... *L. perenne*
1b. Petals white or yellow, 4.0–8.0 mm long; at least the inner sepals glandular-ciliate or glandular-toothed; capsules 2.0–2.5 mm wide
    3a. Petals white; lower pedicels 10.0–25.0 mm long; staminodes present; plants annual .... *L. catharticum*
    3b. Petals yellow; pedicels up to 5.0 mm long; staminodes absent; plants perennial .... *L. medium*

Linum catharticum L.
synonym(s): *Cathartolinum catharticum* (L.) Small
common name(s): white flax, fairy flax
habit: annual or winter-annual herb
range: Europe; naturalized to Newfoundland, s. to NJ, w. to PA, n. to MI and Ontario
state frequency: uncommon
habitat: disturbed sites, old fields, ditches, sometimes of wet soil
notes: —

Linum medium (Planch.) Britt. var. *texanum* (Planch.) Fern.
synonym(s): *Cathartolinum curtissii* (Small) Small
common name(s): common yellow flax
habit: erect, perennial herb
range: ME, s. to FL, w. to TX, n. to IL, MI, and Ontario
state frequency: very rare
habitat: sandy soil of woods, openings, and beaches
notes: —

Linum perenne L.
synonym(s): —
common name(s): wild blue flax
habit: perennial herb
range: Europe; adventive locally
state frequency: very rare
habitat: disturbed areas
notes: —

Linum usitatissimum L.
synonym(s): —
common name(s): common flax
habit: erect, annual herb

range: Europe; escaped from cultivation occasionally
state frequency: uncommon
habitat: waste places, fields, roadsides
notes: —

Radiola
(1 species)

Radiola linoides Roth
synonym(s): *Millegrana radiola* (L.) Druce
common name(s): —
habit: diminutive, annual herb
range: Eurasia and northern Africa; naturalized to Nova Scotia and ME
state frequency: very rare
habitat: roadsides, fields
notes: —

SALICACEAE
(2 genera, 30 species)

KEY TO THE GENERA
1a. Bud scales more than 1; aments drooping; flowers on a cup-like disk, each staminate
     flower with 5–80 stamens .... ***Populus***
1b. Bud scales 1; aments usually erect to spreading; flowers with 1–4 basal glands, each
     staminate flower with 1–8 stamens .... ***Salix***

Populus
(4 species)

Disturbance of one of the mature stems can cause sprouting in some species. These sprouts often have
abnormally large leaves that are not reliable for identification purposes.

KEY TO THE SPECIES FOR USE WITH MATURE, VEGETATIVE MATERIAL
1a. Petioles flattened; leaves without orange resin on the abaxial surface
     2a. Leaf blades ovate to orbicular to nearly reniform; plants native
         3a. Leaves glabrous, those of the short shoots with more than 15 small teeth, the teeth
             usually not more than 1.0 mm long .... *P. tremuloides*
         3b. Leaves tomentose, at least while young, those of the short shoots with fewer than
             15 large teeth, the teeth usually more than 1.5 mm long .... *P. grandidentata*
     2b. Leaf blades deltate; plants introduced .... *P. deltoides*
1b. Petioles terete; leaves often streaked with orange resin on the abaxial surface ....
.... *P. balsamifera*

KEY TO THE SPECIES FOR USE WITH FLOWERING MATERIAL
1a. Buds 1.5–2.5 cm long, resinous, fragrant; bracts of aments deeply lacerate; staminate
     flowers with 18–80 stamens

    2a. Branchlets of vigorous sprouts often quadrangular; winter buds weakly fragrant; terminal buds with 6 or 7 scales; plants introduced, more common around civilization .... *P. deltoides*

    2b. Branchlets of vigorous sprouts terete; winter buds strongly fragrant; terminal buds with 5 scales; plants native .... *P. balsamifera*

1b. Buds less than 1.0 cm long, usually dry and non-fragrant; bracts of aments toothed; staminate flowers with 5–12 stamens

    3a. Twigs and buds glabrous; bracts of aments with 3–5 teeth .... *P. tremuloides*

    3b. Twigs and buds pubescent; bracts of aments with 5–7 teeth .... *P. grandidentata*

Populus balsamifera L.

synonym(s): *Populus balsamifera* L. var. *subcordata* Hyl.
common name(s): balsam poplar
habit: tree
range: Labrador, s. to PA, w. to OR, n. to AK
state frequency: occasional
habitat: river banks, low, wet areas
notes: —

Populus deltoides Bartr. *ex* Marsh.

synonym(s): —
common name(s): cottonwood
habit: tree
range: Quebec, s. to FL, w. to TX, n. to Manitoba; planted in Maine
state frequency: uncommon
habitat: river banks, fields, roadsides
notes: —

Populus grandidentata Michx.

synonym(s): —
common name(s): big-toothed aspen
habit: tree
range: Nova Scotia, s. to NC, w. to MO, n. to Manitoba
state frequency: occasional
habitat: dry, upland woods
notes: —

Populus tremuloides Michx.

synonym(s): *Populus tremuloides* Michx. var. *magnifica* Victorin
common name(s): quaking aspen, trembling aspen
habit: tree
range: Labrador, s. to VA, w. to Mexico and CA, n. to AK
state frequency: common
habitat: disturbed, upland areas
notes: —

Salix
(26 species)

Willows are capable of tremendous variation in pubescence depending on the time of year and habitat. Young plants are more pubescent and gradually lose hairs throughout the season. Plants from open habitats usually have more hairs than those from shaded habitats. An important character is branch brittleness, which must be assessed in the field on several branches. Note whether branches break cleanly at the junction of different year's growth or if they bend and tear, ripping from the branch. Anther color is a character used for staminate material. The colors used in the key are for mature anthers, as the expanding anthers are often purple. Both the aments and the flowering branchlets, the latter often incorrectly termed peduncles, elongate throughout the flowering and fruiting season. Hybrids are common between certain species and usually can be detected by intermediate morphology. A few species are difficult to identify while flowering and may require sequential collections. Additionally, collecting leaves that fell the previous year may add characters helpful in identification. *Salix pellita* is always carpellate in our area and therefore is not included in the key for use with staminate reproductive material.

KEY TO THE SPECIES FOR USE WITH CARPELLATE REPRODUCTIVE MATERIAL

1a. Carpels and capsules pubescent
    2a. At least some of the carpellate aments opposite or subopposite .... **Group 1**
    2b. All carpellate aments alternate
        3a. Floral bracts of aments yellow .... **Group 2**
        3b. Floral bracts of aments brown to black, at least at the apex
            4a. Aments precocious [expanding before the leaves] .... **Group 3**
            4b. Aments coaetaneous or serotinous [expanding with or after the leaves,
                respectively] .... **Group 4**
1b. Carpels and capsules glabrous
    5a. Flowers and capsules in whorls about the rachis of the ament .... **Group 5**
    5b. Flowers and capsules spirally arranged
        6a. Floral bracts of aments yellow to white-brown, deciduous before maturation of
            the capsule .... **Group 6**
        6b. Floral bracts of aments light brown to black, persistent until dehiscence of the
            capsule .... **Group 7**

Group 1
only 1 species .... *S. purpurea*

Group 2
1a. Floral bracts deciduous before maturation of fruit; branchlets of previous year usually
    glabrous; stipes 0.5–1.5 mm long; styles 0.1–0.2 mm long .... *S. interior*
1b. Floral bracts persistent in fruit; branchlets of previous year pubescent; stipes
    3.0–6.0 mm long; styles 0.1–0.4 mm long .... *S. bebbiana*

Group 3
1a. Carpels and capsules sessile or subsessile, the stipe (if present) usually shorter than
    0.5 mm; styles 0.7–1.5 mm long
    2a. Plants of exposed, alpine areas, elevation exceeding 1000 m .... *S. planifolia*
    2b. Plants not of alpine areas, elevation less than 1000 m
        3a. Branchlets and carpels densely tomentose; shrubs to 1.5 m, of calcareous fens .....
        .... *S. candida*
        3b. Branchlets tomentose or glabrous; carpels sericeous; shrubs or small trees to
            15.0 m, not of fens
            4a. Aments sessile or on leafy, flowering branchlets; styles mostly 0.3–1.2 mm
                long; capsules 4.0–5.0 mm long; plants native .... *S. pellita*
            4b. Aments mostly on leafless, flowering branchlets; styles mostly 1.0–2.0 mm
                long; capsules 6.0–8.0 mm long; plants introduced .... *S. viminalis*

1b. Carpels and capsules stipitate, the stipe 0.5–6.0 mm long; styles 0.1–0.7 mm long
    5a. Carpels and capsules short-beaked or blunt; capsules 3.0–5.0 mm long ..... *S. sericea*
    5b. Carpels and capsules long-beaked; capsules 5.0–12.0 mm long
        6a. Stipes 3.0–6.0 mm long; floral bracts yellow-brown, sometimes red at the tip ......
        .... *S. bebbiana*
        6b. Stipes 1.0–3.0 mm long; bracts dark brown to black, at least at the apex
            7a. Aments 1.0–3.5 cm long; stipes 1.0–2.0 mm long .... *S. humilis*
            7b. Aments 3.0–8.0 cm long; stipes 2.0–3.0 mm long
                8a. Styles 0.3–1.0 mm long; capsules 6.5–11.0 mm long .... *S. discolor*
                8b. Styles 0.2–0.5 mm long; capsules 5.0–6.5 mm long .... *S. cinerea*

Group 4

1a. Plants of exposed, alpine areas, elevation usually exceeding 1000 m
    2a. Capsules 6.5–8.0 mm long; styles 1.0–1.5 mm long; stipes 0.5–1.0 mm long ....
    .... *S. arctophila*
    2b. Capsules 2.0–4.0 mm long; styles *ca.* 0.5 mm long; stipes 1.0–2.0 mm long ....
    .... *S. argyrocarpa*
1b. Plants not of alpine areas, elevation less than 1000 m
    3a. Styles 0.7–2.0 mm long; carpels and capsules sessile or subsessile, the stipes (if
        present) usually less than 0.5 mm long
        4a. Carpels and capsules tomentose; styles often red during anthesis; plants native,
            shrubs to 1.5 m, of calcareous fens .... *S. candida*
        4b. Carpels and capsules short-sericeous; styles brown during anthesis; plants
            introduced, shrubs or trees to 15.0 m .... *S. viminalis*
    3b. Styles 0.1–0.5 mm long; carpels and capsules stipitate, the stipes 0.5–6.0 mm long
        5a. Floral bracts yellow-brown, sometimes red at the apex; capsules 7.0–10.0 mm
            long .... *S. bebbiana*
        5b. Floral bracts brown to black, at least at the apex; capsules 4.0–8.0 mm long
            6a. Branchlets brittle at the base, often velutinous; capsules 3.0–5.0 mm long;
                stipes 0.5–1.5 mm long; aments 2.0–4.5 cm long; expanding leaves without
                red-brown hairs .... *S. sericea*
            6b. Branchlets flexible at the base, often glabrous; capsules 4.0–8.0 mm long;
                stipes 2.5–5.0 mm long; aments 1.0–3.5 cm long; expanding leaves
                sometimes with red-brown hairs .... *S. petiolaris*

Group 5
only 1 species .... *S. nigra*

Group 6
1a. Emerging leaves glaucous beneath; introduced trees to 25.0 m
    2a. Emerging leaves long-sericeous on the abaxial surface; styles 0.2–0.4 mm long;
        stipes *ca.* 0.2 mm long .... *S. alba*
    2b. Emerging leaves glabrous, or nearly so, on the abaxial surface; styles 0.4–1.0 mm
        long; stipes *ca.* 0.8 mm long .... *S. fragilis*
1b. Emerging leaves green beneath, without bloom; shrubs or small trees to 6.0 m
    3a. Aments occurring 1–3 together on very short, lateral branchlets; styles obsolete;
        expanding leaves remotely denticulate .... *S. interior*

3b. Aments occurring singly; styles 0.5–1.0 mm long; expanding leaves closely serrulate
    4a. Stipes 3.0–5.0 times as long as the upper nectary gland; nectary glands
        0.2–0.5 mm long; plants native .... *S. lucida*
    4b. Stipes 1.3–2.0 times as long as the upper nectary gland; nectary glands
        0.4–0.8 mm long; plants introduced .... *S. pentandra*

Group 7

1a. Dwarf shrubs, rarely exceeding 15.0 cm, of exposed, alpine areas, elevation usually
    exceeding 1000 m
    2a. Styles 0.7–1.5 mm long; leaves glaucous on the abaxial surface .... *S. uva-ursi*
    2b. Styles *ca.* 0.5 mm long; leaves green on the abaxial surface, without bloom ....
        .... *S. herbacea*
1b. Upright shrubs, to 4.0 m, not of alpine areas, elevation less than 1000 m
    3a. Expanding leaves entire; styles 0.1–0.3 mm long .... *S. pedicellaris*
    3b. Expanding leaves serrulate; styles 0.3–1.5 mm long
        4a. Expanding leaves glabrous; twigs lustrous and glabrous; foliage and buds
            strongly balsam-fragrant when dry; stipes 2.5–3.5 mm long .... *S. pyrifolia*
        4b. Expanding leaves pubescent; twigs duller, glabrous or pubescent; foliage and
            buds without a balsam fragrance; stipes 0.5–3.0 mm long
            5a. Flowering branchlets 0.3–1.0 cm long; styles *ca.* 0.5 mm long ....
                .... *S. eriocephala*
            5b. Flowering branchlets 0.5–3.0 cm long; styles 0.7–1.5 mm long
                6a. Stipes 1.0–3.0 mm long; expanding leaves glaucous on the abaxial
                    surface .... *S. myricoides*
                6b. Stipes 0.5–1.0 mm long; expanding leaves green on the abaxial surface ...
                    .... *S. cordata*

KEY TO THE SPECIES FOR USE WITH STAMINATE REPRODUCTIVE MATERIAL
1a. Flowers with 2 nectary glands at the base of each floral bract
    2a. Flowers each with 3 or more stamens .... **Group 1**
    2b. Flowers each with 1 or 2 stamens .... **Group 2**
1b. Flowers with 1 nectary gland at the base of each floral bract
    3a. Filaments pubescent (though sometimes sparsely), at least near the base ..... **Group 3**
    3b. Filaments glabrous throughout
        4a. Aments precocious [expanding before the leaves] .... **Group 4**
        4b. Aments coaetaneous or serotinous [expanding with or after the leaves,
            respectively] .... **Group 5**

Group 1

1a. Flowers borne in whorls about the rachis of the ament; emerging leaves lacking glands at
    the summit of the petiole .... *S. nigra*
1b. Flowers spirally arranged about the rachis of the ament; emerging leaves with glands at
    the summit of the petiole
    2a. Emerging leaves with stipules and often with red-brown hairs; plants native ....
        .... *S. lucida*
    2b. Emerging leaves exstipulate and without red-brown hairs; plants introduced ....
        .... *S. pentandra*

Group 2

1a. Flowers each with 1 stamen, or appearing so due to fusion of the filaments

    2a. At least some of the aments opposite or subopposite; shrubs introduced, erect, 1.0–2.4 m tall .... *S. purpurea*

    2b. Aments all alternate; shrubs depressed, alpine, rarely exceeding 0.15 m ....
.... *S. uva-ursi*

1b. Flowers each with 2 stamens

    3a. Floral bracts of aments yellow or white-brown; filaments pilose, at least in the basal portion; plants upright shrubs or trees, 1.5–25.0 m tall, of various, non-alpine habitats

        4a. Petioles of emerging leaves eglandular; filaments densely pilose in the basal half; plants native, shrubs to 3.0 m, of northern Maine gravel shores .... *S. interior*

        4b. Petioles of emerging leaves glandular at the summit; filaments pilose only at the base; plants introduced, trees to 25.0 m, more common around civilization

            5a. Aments 5.0–6.0 cm long; expanding leaves sparsely sericeous or glabrous .....
.... *S. fragilis*

            5b. Aments 3.0–3.5 cm long; expanding leaves densely sericeous .... *S. alba*

    3b. Floral bracts of aments brown to black, at least near the apex; filaments glabrous throughout; plants shrubs, depressed or upright, to 1.7 m (mostly much shorter), of exposed, alpine areas

        6a. Aments 5- to 12-flowered, on branches with 2–4 leaves .... *S. herbacea*

        6b. Aments composed of more flowers, on many-leaved branches

            7a. Expanding leaves with lustrous, silver hairs; flowering branchlets 0.5–1.2 cm long .... *S. argyrocarpa*

            7b. Expanding leaves very soon glabrous; flowering branchlets 1.0–4.0 cm long
.... *S. arctophila*

Group 3

1a. Aments coaetaneous [expanding with the leaves]

    2a. Floral bracts yellow-brown, sometimes red at the apex; branchlets and buds red or red-brown .... *S. bebbiana*

    2b. Floral bracts dark brown; rarely both the branchlets and buds red

        3a. Branchlets brittle at the base, often velutinous; expanding leaves without red-brown hairs; aments 1.0–3.5 cm long .... *S. sericea*

        3b. Branchlets flexible at the base, often glabrous; expanding leaves often with red-brown hairs; aments 1.0–2.0 cm long .... *S. petiolaris*

1b. Aments precocious [expanding before the leaves]

    4a. Staminate aments 2.0–5.0 cm long

        5a. Decorticated wood of 1–4 year old branches with prominent, slender ridges longer than 2.0 cm .... *S. cinerea*

        5b. Decorticated wood of 1–4 year old branches smooth or with irregular, scattered, short ridges shorter than 0.5 cm

            6a. Aments 1.5–2.5 cm thick; pubescence of twigs and buds, if present, often partly composed of red-brown hairs; dry anthers 0.5–1.0 mm long ....
.... *S. discolor*

6b. Aments 0.5–1.3 cm thick; pubescence of twigs and buds not red-brown; dry anthers 0.4–0.6 mm long .... *S. sericea*
4b. Staminate aments 0.5–2.0 cm long
    7a. Branchlets brittle at the base; anthers commonly yellow; pubescence of twigs and buds not red-brown; plants of wet banks and shores .... *S. sericea*
    7b. Branchlets flexible at the base; anthers commonly red to purple; pubescence of twigs and bud often partly composed of red-brown hairs; plants of dry fields and open woods .... *S. humilis*

Group 4

1a. Twigs glabrous and lustrous; plants of exposed, alpine areas, elevation usually exceeding 1000 m .... *S. planifolia*
1b. Twigs usually velutinous or tomentose, rarely glabrate; plants not of alpine areas, elevation less than 1000 m
    2a. Aments at anthesis on leafy, flowering branchlets 0.5–2.0 cm long; plants of calcareous fens .... *S. candida*
    2b. Aments sessile or subsessile at anthesis, the flowering branchlets (if present) usually less than 0.5 cm; plants not of calcareous fens
        3a. Aments 0.5–2.0 cm long; shrubs native, 1.0–3.0 m tall, of dry, open woods and fields .... *S. humilis*
        3b. Aments 2.0–5.0 cm long; shrubs or trees introduced, to 15.0 m, more common around civilization
            4a. Decorticated wood of 1–4 year old branches with prominent, slender ridges longer than 2.0 cm .... *S. cinerea*
            4b. Decorticated wood of 1–4 year old branches smooth .... *S. viminalis*

Group 5

1a. Branchlets velutinous to tomentose
    2a. Pubescence of branchlets white; plants of calcareous fens .... *S. candida*
    2b. Pubescence of branchlets gray; plants mostly of shores and banks
        3a. Floral bracts yellow-brown, sometimes red at the apex; branchlets and buds red or red-brown .... *S. bebbiana*
        3b. Floral bracts dark brown; rarely both the branchlets and buds red
            4a. Anthers 0.4–0.7 mm long when fresh, subglobose .... *S. cordata*
            4b. Anthers 0.6–1.1 mm long when fresh, elongate
                5a. Flowers of aments maturing from base to apex; floral bracts mostly 1.0–1.5 mm long .... *S. eriocephala*
                5b. Flowers of aments maturing from apex to base; floral bracts mostly 1.5–2.0 mm long, pubescent with nearly straight hairs .... *S. myricoides*
1b. Branchlets glabrous or glabrate
    6a. Floral bracts yellow-brown or yellow-green
        7a. Expanding leaves with entire margins; twigs and buds without a balsam fragrance; low shrubs to 1.0 m tall .... *S. pedicellaris*
        7b. Expanding leaves with serrulate margins; twigs and buds strongly balsam-fragrant when dry; shrubs to 5.0 m tall .... *S. pyrifolia*
    6b. Floral bracts dark brown, at least at the apex

8a. Flowers of aments maturing from base to apex; pubescence of bracts crimped or curled .... *S. eriocephala*

8b. Flowers of aments maturing from apex to base; pubescence of bracts nearly straight .... *S. myricoides*

KEY TO THE SPECIES FOR USE WITH MATURE, VEGETATIVE MATERIAL

1a. Plants found in exposed, alpine areas, elevation usually exceeding 1000 m .... **Group 1**

1b. Plants not found in alpine areas, elevation less 1000 m

    2a. At least some leaves opposite or subopposite, with more or less oblanceolate blades, glaucous below, often purple-tinged .... **Group 2**

    2b. Leaves always alternate; shape and color various

        3a. Leaves whitened or grayed beneath by at least a thin bloom or hairs

            4a. Leaves with entire, undulate, or crenate margins, the teeth, if present, rounded at the apex and often irregular .... **Group 3**

            4b. Leaves dentate, serrate, or serrulate, the teeth sharply pointed

                5a. Foliage and buds strongly balsam-fragrant when dry; mature vegetative parts glabrous; buds and twigs lustrous; buds bright red; leaves membranaceous .... **Group 4**

                5b. Foliage and buds without balsam fragrance; mature vegetative parts glabrous or pubescent; buds and twigs lustrous or dull; buds variously colored; leaves mostly firmer

                    6a. Mature leaves noticeably pubescent .... **Group 5**

                    6b. Mature leaves glabrous or glabrate .... **Group 6**

        3b. Leaves green beneath, not whitened, at most merely paler .... **Group 7**

Group 1

1a. Leaf blades green beneath, orbicular, often indented at the tip .... *S. herbacea*

1b. Leaf blades whitened beneath by hairs or bloom, longer than wide, usually pointed at the apex

    2a. Foliage and buds strongly balsam-fragrant when dry; leaves membranaceous, thin; twigs and buds lustrous; buds bright red; mature vegetative parts glabrous .... *S. pyrifolia*

    2b. Plants without balsam fragrance; leaves mostly firmer; twigs and buds not reddish; mature vegetation glabrous or pubescent

        3a. Leaf blades glandular-serrate to glandular-crenate, 0.5–2.5 cm long, commonly less than 2.0 cm; plants depressed, not exceeding 15.0 cm in height, the branches often horizontal .... *S. uva-ursi*

        3b. Leaf blades entire or crenate but not glandular, 2.0–8.0 cm long; plants trailing to ascending, usually exceeding 15.0 cm in height

            4a. Plants up to 2.5 m high; twigs often glaucous later in the season; leaves membranaceous .... *S. planifolia*

            4b. Plants much shorter; twigs not glaucous; leaves firm

                5a. Blade of later leaves lanceolate to narrowly elliptic, strongly whitened below by silky hairs or bloom; buds yellow-brown, often red-glandular-dotted, less than 3.5 mm long; petioles yellowish .... *S. argyrocarpa*

5b. Blade of later leaves elliptic to obovate, not strongly whitened beneath; buds dark purple, eglandular, up to 6.0 mm long; petioles reddish .... ***S. arctophila***

Group 2
only 1 species .... ***S. purpurea***

Group 3
1a. Plants very soon glabrous; blade of mature leaves entire, elliptic to obovate, up to 6.0 cm long, often revolute; low shrub to 1.0 m, of bogs and peaty shores .... ***S. pedicellaris***
1b. Leaf blades with some form of dentition or pubescence or both, usually longer; plants generally taller, of various habitats
　2a. Leaves, when fresh, with more or less revolute margins
　　3a. Leaf blades linear-lanceolate to elongate-linear, 5.0–17.0 cm long and rarely as much as 1.0 cm wide, commonly less, tapering to the tip from below the middle; plants escaped from cultivation .... ***S. viminalis***
　　3b. Leaf blades mostly shorter and wider, tapering to the tip from near or above the middle; plants native or escaped
　　　4a. Red-brown hairs often present on the leaves, twigs, and/or buds; leaf blades oblanceolate to elliptic or obovate; plants of dry to mesic soils
　　　　5a. Decorticated wood of 1–4 year old branches smooth; leaf blades 0.3–2.3(–3.5) cm wide; twigs densely villous to villous-tomentose .... ***S. humilis***
　　　　5b. Decorticated wood of 1–4 year old branches with prominent, slender ridges longer than 2.0 cm; leaf blades 2.2–5.2 cm wide; twigs sparsely to moderately villous .... ***S. cinerea***
　　　4b. Herbage lacking red-brown hairs; leaf blades commonly narrower; plants of shores and organic soil wetlands
　　　　6a. Twigs glabrous or becoming glabrate, glaucous later in the season; stipules absent; leaves glabrate on the adaxial surface and sericeo-tomentose and lustrous on the abaxial surface, becoming glabrate later in the season; plants shrubs to small trees, 3.0–5.0 m tall .... ***S. pellita***
　　　　6b. Twigs tomentose; stipules often present; leaves often sparsely tomentose on the adaxial surface and persistently tomentose on the abaxial surface with dull, white, woolly hairs; plants small shrubs, to 1.5 m tall .... ***S. candida***
　2b. Leaves with flat margins
　　7a. Leaves usually impressed-veiny above and rugose-veiny below, reticulate, often persistently pubescent; branchlets and buds uniformly red; branchlets persistently pubescent .... ***S. bebbiana***
　　7b. Plants not combining all of the above characters, especially lacking the combination of pubescent, red branchlets and red buds .... ***S. discolor***

Group 4
only 1 species .... ***S. pyrifolia***

Group 5

1a. Leaf blades cordate to rounded at the base; stipules large and persistent; branchlets flexible at the base .... **S. eriocephala**

1b. Leaf blades cuneate to rounded at the base; stipules smaller or absent; branchlets more or less brittle, except *S. petiolaris*

    2a. Petioles with dark glands at the summit; stomata present on the adaxial [upper] surface; leaves long-sericeous beneath, not darkening in drying; plants trees, spread from cultivation .... **S. alba**

    2b. Petioles eglandular; stomata on abaxial [lower] surface only; leaves silky to felted beneath, commonly darkening in drying; plants shrubs, native

        3a. Branchlets brittle at the base, not glaucous, persistently pubescent, at least at the nodes; leaf blades 4.0–15.0 x 1.0–4.0 cm, toothed to the base; red-brown hairs absent; vigorous shoots with stipules .... **S. sericea**

        3b. Branchlets flexible, sometimes glaucous, glabrous or becoming so; leaf blades 2.5–7.0 x 0.3–2.0 cm, often entire near the base; red-brown hairs sometimes present; stipules wanting .... **S. petiolaris**

Group 6

1a. Leaf blades cordate to rounded at the base; stipules large and persistent; twigs usually flexible at the base

    2a. Leaf blades prominently whitened beneath, tending to be more obovate; plants shrubs, of northern Maine watercourses .... **S. myricoides**

    2b. Leaf blades only thinly glaucous beneath, tending to be more lanceolate; plants widespread .... **S. eriocephala**

1b. Leaf blades cuneate to rounded at the base; stipules inconspicuous or caducous (usually present in *S. sericea*); twigs brittle, except in *S. petiolaris*

    3a. Summit of the petiole bearing dark glands; leaves amphistomic; plants trees, escaped from cultivation, more common around civilization

        4a. Leaf blades lustrous on the adaxial surface, coarsely serrulate, often with only 3–6 serrations per cm, the serrations often more than 0.4 mm high; petiole often glabrous .... **S. fragilis**

        4b. Leaf blades relatively less lustrous, finely serrulate, often with 7 or more serrations per cm, the serrations often 0.4 mm high or less; petiole pubescent .... **S. alba**

    3b. Petiole eglandular; leaves hypostomic; plants large or small shrubs, native

        5a. Branchlets brittle at the base, not glaucous, persistently pubescent, at least at the nodes; leaf blades 4.0–15.0 x 1.0–4.0 cm, toothed to the base; red-brown hairs absent; at least the vigorous shoots with persistent stipules .... **S. sericea**

        5b. Branchlets flexible, not easily broken, sometimes glaucous, essentially glabrous or becoming so; leaf blades 2.5–7.0 x 0.3–2.0 cm, often nearly entire at the base; red-brown hairs sometimes present; stipules minute and caducous ... **S. petiolaris**

Group 7

1a. Bud apex sharp-pointed, the margin free and overlapping; leaf blades narrow-lanceolate or narrower; summit of the petiole glandular; branchlets brittle at the base, breaking off cleanly at the junction of the previous year's growth; plants native, shrubs or trees, to 20.0 m tall .... **S. nigra**

1b. Bud apex blunt, the margin fused; leaf blades and petioles various; branchlets brittle or flexible; plants shrubs, native or introduced, to 6.0 m tall

  2a. Leaves sessile or subsessile with petioles to 3.0 mm long, linear with parallel margins, the margins glandular-dentate; plants shrubs, of northern Maine rivers .... ***S. interior***

  2b. Leaves with petioles 3.0 mm long or more; leaf margins not parallel, entire or serrate; plants of various habitats

    3a. Mature leaves with entire, often revolute, margins; stipules absent; shrub to 1.0 m high, of acid bogs and peaty shores .... ***S. pedicellaris***

    3b. Mature leaves with serrate or serrulate, flat margins; stipules often present; habit and habitat various

      4a. Leaf blades lustrous on the adaxial surface, acuminate to caudate, narrowed to rounded at the base; stipules, when present, glandular at the base; summit of the petiole bearing coarse glands; branchlets brittle at the base

        5a. Emerging leaves often with caducous, red-brown hairs, the blade 5.0–17.0 cm long, caudate at the apex; stipules often persistent; plants native .... ***S. lucida***

        5b. Emerging leaves glabrous, the blade 3.5–10.0 cm long, acuminate at the apex; stipules minute or absent; plants cultivated .... ***S. pentandra***

      4b. Leaf blades not lustrous, acute to subacuminate, cordate to rounded at the base; stipules glandular-serrate; summit of the petiole not bearing coarse glands; branchlets usually flexible at the base

        6a. New branchlets, and often older ones, densely velutinous with gray hairs; leaves tending to be more ovate .... ***S. cordata***

        6b. Branchlets usually glabrous or becoming glabrate; leaves tending to be more lanceolate .... ***S. eriocephala***

Salix alba L.

synonym(s): <*S. a.* 'Caerulea'> = *Salix alba* L. var. *caerulea* (Sm.) Rech. f., *Salix alba* L. var. *calva* G. F. W. Mey.; <*S. a.* 'Vitellina'> = *Salix alba* L. var. *vitellina* (L.) Stokes, *Salix vitellina* L.
common name(s): white willow
habit: tree
range: Eurasia; escaped from cultivation locally
state frequency: occasional
habitat: extensively planted in a wide variety of habitats, usually naturalizing in moist to wet soils of lake and river shores
key to the subspecies:
  1a. Branchlets sericeous .... *S. a.* var. *alba*
  1b. Branchlets promptly glabrate
    2a. Branchlets conspicuously yellow .... *S. a.* 'Vitellina'
    2b. Branchlets brown .... *S. a.* 'Caerulea'
notes: Maine's forms of *Salix alba* with glabrous branchlets are actually horticultural forms and should not be recognized with formal varietal status. Hybridizes with *S. fragilis* (*q.v.*).

Salix arctophila Cockerell *ex* Heller

synonym(s): *Salix groenlandica* Lundstr.
common name(s): arctic willow
habit: depressed, trailing shrub
range: Greenland, s. to ME, w. and n. to Northwest Territories

state frequency: very rare
habitat: alpine areas
notes: This species is ranked S1 by the Maine Natural Areas Program.

Salix argyrocarpa Anderss.
synonym(s): —
common name(s): silvery willow
habit: low shrub
range: Labrador, s. to the higher mountains of ME, NH, and VT
state frequency: very rare; possibly extirpated
habitat: alpine areas
notes: This species is ranked S1 by the Maine Natural Areas Program. Known to hybridize with *Salix planifolia*.

Salix bebbiana Sarg.
synonym(s): *Salix bebbiana* Sarg. var. *capreifolia* (Fern.) Fern.
common name(s): long-beaked willow
habit: shrub or small tree
range: Newfoundland, s. to MD, w. to AZ, n. to AK
state frequency: common
habitat: old fields, roadsides
notes: Known to hybridize with *Salix candida*, *S. petiolaris*, and *S. eriocephala*.

Salix candida Fluegge *ex* Willd.
synonym(s): —
common name(s): hoary willow, sage-leaved willow
habit: low shrub
range: Newfoundland, s. to NJ, w. to ID, n. to AK
state frequency: very rare
habitat: calcareous wetlands
notes: This species is ranked S1 by the Maine Natural Areas Program. Known to hybridize with *Salix bebbiana*, *S. eriocephala* and *S. petiolaris*.

Salix cinerea L. ssp. *oleifolia* (Sm.) Macreight
synonym(s): *Salix atrocinerea* Brot.
common name(s): gray willow, gray florist's willow
habit: shrub to small tree
range: Europe; escaped from cultivation in eastern North America
state frequency: very rare
habitat: areas of cultivation
notes: The subcortical ridges are best observed on peeled branches 1.0–3.0 cm in diameter.

Salix cordata Michx.
synonym(s): *Salix adenophylla* Hook.
common name(s): dune willow, furry willow
habit: shrub
range: Quebec, s. to ME, w. to MI, n. to Hudson Bay
state frequency: very rare
habitat: shores
notes: This species is ranked SH by the Maine Natural Areas Program.

Salix discolor Muhl.

synonym(s): *Salix discolor* Muhl. var. *overi* Ball
common name(s): pussy willow
habit: shrub or small tree
range: Newfoundland, s. to MD, w. to ID, n. to British Columbia
state frequency: common
habitat: damp, open areas
notes: Hybrids with *Salix eriocephala* and *S. humilis*, called *S.* ×*conifera*, Wangenh. occur in Maine. Also known to hybridize with *S. myricoides* (*q.v.*) and *S. pellita*.

Salix eriocephala Michx.

synonym(s): *Salix cordata* Muhl., *Salix cordata* Michx. var. *abrasa* Fern., *Salix rigida* Muhl.
common name(s): red-tipped willow, heart-leaved willow
habit: shrub
range: Newfoundland, s. to NC, w. to KS, n. to Manitoba
state frequency: common
habitat: sandy or gravelly shores
notes: Hybrids with *Salix discolor*, *S. petiolaris*, and *S. sericea*, called *S.* ×*bebbii* Gandog., occur in Maine. Also reported to hybridize with *S. bebbiana*, *S. candida*, *S. humilis*, and *S. nigra*, although some of these combinations are doubtful. This species usually has red or red-tinged carpels, a useful character to separate it from other widespread, weedy species.

Salix fragilis L.

synonym(s): —
common name(s): crack willow
habit: tree
range: Europe; escaped from cultivation to Newfoundland, s. to VA, w. to IL, n. to Ontario
state frequency: uncommon
habitat: lawns, fields, and borders near civilization
notes: Most, and perhaps all, of the material from Maine labeled as *Salix fragilis* is actually the hybrid with *S. alba*, called *S.* ×*rubens* Schrank. Hybrids with *S. babylonica* L., called *S.* ×*pendulina* Wenderoth, also occur in Maine and are known as the weeping willow. These hybrids can be identified by their pendulous branches, glabrous leaves, and stipes *ca.* 0.5 mm long.

Salix herbacea L.

synonym(s): —
common name(s): herb-like willow, dwarf willow
habit: few-leaved, depressed, subshrub
range: circumboreal; s. to mountains of New England and NY
state frequency: very rare; known only from Katahdin
habitat: mossy seeps and lawns of alpine areas
notes: This species is ranked S1 by the Maine Natural Areas Program.

Salix humilis Marsh.

synonym(s): <*S. h.* var. *tristis*> = *Salix humilis* Marsh. var. *microphylla* (Anderss.) Fern.
common name(s): prairie willow, upland willow
habit: shrub
range: Newfoundland, s. to FL, w. to TX, n. to ND and Saskatchewan
state frequency: occasional
habitat: old fields and open, dry woods

key to the subspecies:

    1a. Leaf blades 3.0–10.0 x 1.0–2.5 cm, stipulate; petioles 3.0–7.0 mm long; staminate aments 1.0–2.0 cm long; carpellate aments longer than 2.0 cm .... *S. h.* var. *humilis*
    1b. Leaf blades 2.0–5.0 x 0.7–1.2 cm, exstipulate; petioles 0.5–3.0 mm long; staminate aments 0.5–1.1 cm long; carpellate aments shorter than 2.0 cm .... *S. h.* var. *tristis* (Ait.) Griggs

notes: Hybrids with *Salix discolor* (*q.v.*) occur in Maine. Also known to hybridize with *S. petiolaris*.

Salix interior Rowlee

synonym(s): *Salix exigua* Nutt. *sensu* Gleason and Cronquist (1991), *Salix interior* Rowlee var. *exterior* Fern., *Salix longifolia* Muhl. var. *interior* (Rowlee) M. E. Jones
common name(s): sandbar willow
habit: shrub
range: Nova Scotia, s. to MD, w. to OK, n. to AK
state frequency: very rare
habitat: sandbars and shores
notes: This species is ranked S1 by the Maine Natural Areas Program. The seedling leaves are sometimes pinnately lobed in this species.

Salix lucida Muhl.

synonym(s): *Salix lucida* Muhl. var. *angustifolia* (Anderss.) Anderss., *Salix lucida* Muhl. var. *intonsa* Fern.
common name(s): shining willow
habit: shrub or small tree
range: Newfoundland, s. to MD, w. to SD, n. to Saskatchewan
state frequency: occasional
habitat: low, wet areas
notes: —

Salix myricoides Muhl.

synonym(s): *Salix glaucophylloides* Fern.
common name(s): blueleaf willow
habit: shrub
range: Newfoundland, s. to ME, w. to IL, n. to Ontario
state frequency: very rare; known from northern Maine
habitat: sandy or gravelly shores
notes: This species is ranked S1 by the Maine Natural Areas Program. Known to hybridize with and often confused with *Salix discolor*. *Salix myricoides* has thicker, often subcoriaceous, leaves with 5–9 sharp teeth per cm. *Salix discolor* usually has thinner, membranaceous leaves with 1–3(–5) blunt teeth.

Salix nigra Marsh.

synonym(s): —
common name(s): black willow
habit: shrub or tree
range: New Brunswick, s. to FL, w. to CA, n. to Ontario
state frequency: uncommon; more common in the southern portion of the state
habitat: shores, low woods, and meadows
notes: Hybridizes with *Salix eriocephala*.

Salix pedicellaris Pursh

synonym(s): *Salix pedicellaris* Pursh var. *hypoglauca* Fern., *Salix pedicellaris* Pursh var. *tenuescens* Fern.
common name(s): bog willow

habit: shrub
range: Newfoundland, s. to NJ, w. to OR, n. to Yukon
state frequency: uncommon
habitat: bogs, wet meadows
notes: —

Salix pellita (Anderss.) Anderss. *ex* Schneid.

synonym(s): —
common name(s): silky willow
habit: shrub or small tree
range: Newfoundland, s. to ME, w. to MI, n. to Saskatchewan
state frequency: uncommon; more common in the northern and western parts of the state
habitat: alluvial shores
notes: Hybrids with *Salix candida* occur in Maine. Also known to hybridize with *S. discolor*.

Salix pentandra L.

synonym(s): —
common name(s): bay-leaved willow
habit: shrub
range: Europe; escaped from cultivation to Nova Scotia and ME, s. to MD, w. to IA, n. to Ontario
state frequency: uncommon
habitat: meadows, roadsides, shores
notes: —

Salix petiolaris Sm.

synonym(s): *Salix gracilis* Anderss., *Salix gracilis* Anderss. var. *textoris* Fern., *Salix* ×*subsericea*
    (Anderss.) Schneid.
common name(s): slender willow, meadow willow
habit: shrub to small tree
range: New Brunswick, s. to NJ, w. to CO, n. to Alberta
state frequency: common
habitat: wet meadows and shores
notes: Hybrids with *Salix eriocephala* occur in Maine. Also hybridizes with *S. bebbiana*, *S. candida*,
    and *S. humilis*.

Salix planifolia Pursh

synonym(s): *Salix phylicifolia* L. ssp. *planifolia* (Pursh) Hiitonen
common name(s): tea-leaved willow
habit: shrub
range: Labrador, s. to ME, w. to MN, n. to Yukon
state frequency: very rare
habitat: streamsides and wet seeps in alpine areas
notes: This species is ranked S1 by the Maine Natural Areas Program. Known to hybridize with *Salix
    argyrocarpa*.

Salix purpurea L.

synonym(s): —
common name(s): basket willow, purple osier
habit: shrub
range: Europe; escaped from cultivation to Newfoundland, s. to VA, w. to IA, n. to WI and Ontario
state frequency: uncommon
habitat: field edges, roadsides
notes: —

Salix pyrifolia Anderss.

synonym(s): *Salix balsamifera* (Hook.) Barratt *ex* Anderss.
common name(s): balsam willow
habit: shrub
range: Labrador, s. to ME, NH, and VT, w. to MN, n. to British Columbia
state frequency: uncommon
habitat: wetlands, shores, sometimes in subalpine areas
notes: —

Salix sericea Marsh.

synonym(s): *Salix coactilis* Fern.
common name(s): silky willow
habit: shrub or small tree
range: New Brunswick, s. to GA, w. to AR, n. to WI
state frequency: common
habitat: gravelly shores and river banks
notes: Hybrids with *Salix eriocephala* (*q.v.*) occur in Maine.

Salix uva-ursi Pursh

synonym(s): —
common name(s): bearberry willow
habit: prostrate shrub
range: Greenland, s. to mountains of ME, NH, and NY
state frequency: very rare
habitat: exposed, rocky, alpine areas
notes: This species is ranked S1 by the Maine Natural Areas Program.

Salix viminalis L.

synonym(s): —
common name(s): silky osier
habit: shrub or tree
range: Eurasia; escaped from cultivation in the east to Newfoundland, s. to RI and MA
state frequency: very rare
habitat: near establishments, edges
notes: Hybrids with *Salix caprea* L., called *S.* ×*sericans* Tausch *ex* Kern, are planted in Maine and
    sometimes escape.

VIOLACEAE
(1 genus, 24 species)

Viola
(24 species)

All species that produce cyanic [blue to purple] flowers also produce white-flowered forms [the frequency of the white-flowered forms varies with species]. Collecting the underground portion of the stem [the rhizome] is vital for correct identification. Include the caruncle when measuring seed length, unless noted otherwise in the key. All our species, except *Viola arvensis* and *V. tricolor*, produce cleistogamous flowers, which lack petals, in the summer in addition to the chasmogamous, petaliferous flowers produced in the spring and early summer. Hybrids do occur among closely related species, and some can be identified by abortive pollen and malformed seeds. *Viola pedata* L. is reported to occur in Maine by Gleason and Cronquist (1991), but voucher specimens are unknown.

KEY TO THE SPECIES FOR USE WITH CHASMOGAMOUS FLOWERING MATERIAL

1a. Plants without upright stems, the leaves and peduncles arising directly from stolons or rhizomes
  2a. Corolla yellow .... *V. rotundifolia*
  2b. Corolla purple to blue or white with brown-purple lines
    3a. Plants with long, slender rhizomes 1.0–4.0 mm wide
      4a. Spur at least 2.0 times as long as thick .... *V. selkirkii*
      4b. Spur less than 2.0 times as long as thick
        5a. Style terminating in a downward-oriented long-conic tip .... *V. odorata*
        5b. Style terminating in a forward-pointing, short scoop-shaped tip
          6a. Corolla violet, rarely nearly white; plants of moist areas in alpine
            environments .... *V. palustris*
          6b. Corolla white with brown-purple lines; plants mostly not of alpine areas
            7a. Leaf blades narrow, more than 1.5 times as long as wide, cuneate to subcordate at the base
              8a. Leaf blades lanceolate, 3.0–7.0 times as long as wide, narrowly cuneate at the base .... *V. lanceolata*
              8b. Leaf blades lance-ovate, 1.5–2.5 times as long as wide, broadly cuneate to subcordate at the base .... *V. primulifolia*
            7b. Leaf blades wider, up to 1.2 times as long as wide, distinctly cordate at the base
            9a. Leaf blades orbicular to reniform; plants without stolons ....
              .... *V. renifolia*
            9b. Leaf blades ovate-cordate to orbicular-cordate; plants with stolons (though sometimes not produced until mid-summer)
              10a. Leaf blades glabrous on both surfaces, 1.0–3.0 cm wide at anthesis, crenate to nearly entire; the 2 upper petals not, or slightly, recurved and not twisted .... *V. macloskeyi*
              10b. Leaf blades pubescent on one or both surfaces (though sometimes sparsely and only near the basal lobes), mostly 2.0–4.0 cm wide at anthesis, serrate; the 2 upper petals strongly recurved and twisted .... *V. blanda*
    3b. Plants with short, thick rhizomes 3.0–10.0 mm wide
      11a. Leaf blades crenate-serrate, unlobed and without larger, prominent teeth near the base
        12a. Leaf blades mostly 1.5 times or more as long as wide, truncate to subcordate at the base; plants of dry, sandy habitats .... *V. sagittata*
        12b. Leaf blades mostly less than 1.5 times as long as wide, cordate at the base; plants of moist or wet habitats
          13a. Lateral petals directed forward, therefore, the flowers narrow when living; plants of calcareous wetlands .... *V. nephrophylla*
          13b. Lateral petals spreading, the flowers wider and open; plants of moist soils and shores

14a. Lateral petals pubescent with distally thickened, blunt hairs less than 1.0 mm long; foliage glabrous .... *V. cucullata*

14b. Lateral petals pubescent with distally tapered, pointed hairs 1.0 mm long or longer; foliage glabrous or pubescent

    15a. Leaf blades, petioles, and/or peduncles definitely pubescent

        16a. Sepals ciliate; leaf blades usually as wide or wider than long; plants of moist soils .... *V. sororia*

        16b. Sepals not ciliate; leaf blades longer than wide; plants of rocky shores .... *V. novae-angliae*

    15b. Leaf blades, petioles, and peduncles glabrous (a few hairs may be present on the petioles and peduncles)

        17a. Leaf blades predominantly longer than wide, mostly narrow-ovate, acuminate at the apex; spurred petal usually pubescent .... *V. affinis*

        17b. Leaf blades predominantly as wide or wider than long, mostly ovate to orbicular, obtuse to acute at the apex; spurred petal usually glabrous .... *V. sororia*

11b. Leaf blades lobed, divided, or with larger, prominent teeth near the base

    18a. Leaf blades oblong-lanceolate to triangular-ovate, prominently serrate to slightly divided basally with lobes less than 0.5 times the length of the blade .... *V. sagittata*

    18b. Leaf blades ovate, with prominent lobes more than 0.5 times the length of the blade (except early- and late-season leaves that can be unlobed)

        19a. Leaf blades pubescent, with lobes much shorter than the length of the blade; plants of rich woods .... *V. palmata*

        19b. Leaf blades glabrous, with lobes nearly as long as the blade; plants of sandy or peaty soils near the coast .... *V. brittoniana*

1b. Plants with leafy stems, the peduncles arising from the axils of leaves (stems may be very short at anthesis)

    20a. Stipules entire or slightly toothed, scarious or herbaceous; corolla yellow or white with a yellow center

        21a. Corolla white with a yellow center; stipules lance-acuminate, scarious or subscarious .... *V. canadensis*

        21b. Corolla entirely yellow; stipules broad-ovate, herbaceous .... *V. pubescens*

    20b. Stipules prominently toothed, lobed, or fringed, herbaceous or foliaceous; corolla blue to purple, white-yellow, or multicolored

        22a. Blade of upper leaves truncate to cordate at the base; stipules herbaceous, fringed along margins; plants perennial

            23a. Corolla yellow-white; sepals ciliate, with auricles 2.0–4.0 mm long; leaves crenulate; stipules 1.0–2.5 cm long, with teeth 3.0–5.0 mm long ..... *V. striata*

            23b. Corolla dark blue; sepals glabrous, with auricles up to 1.5 mm long; leaves crenate to subentire; stipules up to 1.5 cm long, entire or toothed with teeth up to 3.0 mm long

                24a. Plants pubescent; blade of lower leaves ovate .... *V. adunca*

                24b. Plants glabrous, sometimes with a few hairs on the adaxial leaf surface; blade of lower leaves suborbicular to reniform .... *V. labradorica*

22b. Blade of upper leaves cuneate at the base; stipules foliaceous, with deep
sinuses; plants annual or sometimes perennial
    25a. Petals shorter than, or up to 2.0 mm longer than, the sepals; flowers up to
1.0 cm wide .... *V. arvensis*
    25b. Petals exceeding the sepals by more than 2.0 mm; flowers 1.5–2.5 cm wide
.... *V. tricolor*

KEY TO THE SPECIES FOR USE WITH CLEISTOGAMOUS FLOWERS OR FRUITING
MATERIAL

1a. Plants without upright stems, the leaves and peduncles arising directly from the stolons
or rhizomes
  2a. Plants lacking stolons
    3a. Leaf blades lobed, divided, or with larger, prominent teeth near the base
      4a. Leaf blades oblong-lanceolate to triangular-ovate, prominently serrate to
slightly divided basally with lobes less than 0.5 times the length of the blade
.... *V. sagittata*
      4b. Leaf blades ovate, with prominent lobes more than 0.5 times the length of the
blade (except early and late season blades, which can be unlobed)
        5a. Leaf blades with lobes much shorter than the length of the blade; seeds,
excluding the caruncle, *ca.* 2.0 mm long; plants of rich woods ....
.... *V. palmata*
        5b. Leaf blades cleft, with lobes nearly as long as the blade; seeds, excluding
the caruncle, 1.3–1.5 mm long; plants of sandy or peaty soils near the
coast .... *V. brittoniana*
    3b. Leaf blades crenate-serrate, unlobed and without larger, prominent teeth near the
base
      6a. Plants with long, slender rhizomes 1.0–4.0 mm wide
        7a. Leaf blades mostly ovate-cordate, membranaceous, with a nearly closed
sinus; auricles glabrous; capsules broad-ellipsoid to globose; seeds
1.5–1.9 x 1.0–1.1 mm, with a short caruncle .... *V. selkirkii*
        7b. Leaf blades orbicular to reniform, firm, with an open sinus; auricles
ciliate; capsules ellipsoid; seeds 1.9–2.4 x 1.0–1.4 mm, the caruncle
elongating at maturity .... *V. renifolia*
      6b. Plants with short, thick rhizomes 3.0–10.0 mm wide
        8a. Cleistogamous flowers borne on a prostrate or subterranean, late
appearing stem, (1–)2–several per peduncle; leaves spreading, often
nearly flat to the ground .... *V. rotundifolia*
        8b. Cleistogamous flowers borne on subterranean to erect peduncles
(however, the peduncles normally becoming erect at the maturing of the
capsule), usually 1 per peduncle, a rare form of *V. sagittata* has 3-
flowered, cleistogamous peduncles); leaves normally ascending to erect,
held off the ground
          9a. Leaf blades mostly 1.5 or more times as long as wide, truncate to
subcordate at the base; plants of dry, sandy habitats .... *V. sagittata*
          9b. Leaf blades mostly less than 1.5 times as long as wide, cordate at the
base; plants of moist to wet habitats

10a. Leaf blades, petioles, and peduncles definitely pubescent
   11a. Sepals ciliate; leaf blades usually as wide or wider than long; plants of moist soils .... *V. sororia*
   11b. Sepals not ciliate; leaf blades longer than wide; plants of rocky shores .... *V. novae-angliae*
10b. Leaf blades, petioles, and peduncles glabrous (a few hairs may be present on the petioles and peduncles)
   12a. Capsule ovoid-cylindric, only slightly exceeding the length of the sepals; auricles of sepals 2.0–6.0 mm long ..... *V. cucullata*
   12b. Capsule ovoid or ellipsoid, 1.5–4.0 times the length of the sepals; auricles of sepals up to 2.0 mm long
      13a. Filiform peduncles of cleistogamous flowers ascending; capsules green; seeds green-black; plants of calcareous wetlands .... *V. nephrophylla*
      13b. Fleshy peduncles of cleistogamous flowers at first prostrate or subterranean, later ascending from the declined base; capsules with purple spots; seeds tan to brown; plants of moist or wet, sometimes disturbed, soils
      14a. Leaf blades predominantly longer than wide, mostly narrow-ovate, acuminate at the apex .... *V. affinis*
      14b. Leaf blades predominantly as wide or wider than long, mostly ovate to orbicular, obtuse to acute at the apex ....
      .... *V. sororia*
2b. Plants producing stolons
  15a. Leaf blades narrow, more than 1.5 times as long as wide, cuneate to subcordate at the base; capsules 5.0–10.0 mm long
   16a. Leaf blades lanceolate, 3.0–7.0 times as long as wide, narrowly cuneate at the base .... *V. lanceolata*
   16b. Leaf blades lance-ovate, 1.5–2.5 times as long as wide, broadly cuneate to subcordate at the base .... *V. primulifolia*
  15b. Leaf blades wider, up to 1.2 times as long as wide, distinctly cordate at the base; capsules 4.0–6.0 mm long
   17a. Stolons leafy; capsule densely pubescent; seeds 3.4–4.0 x 1.75–2.0 mm, with an elongate caruncle .... *V. odorata*
   17b. Stolons leafless or with few leaves near the tip; capsule glabrous; seeds 1.5–2.1 x 0.7–1.3 mm, with a short caruncle
      18a. Leaf blades glabrous on both surfaces; capsules green; seeds green-black, shorter than 1.8 mm
      19a. Seeds 1.0–1.4 x 0.7–0.8 mm; plants of moist soils, only casually alpine .... *V. macloskeyi*
      19b. Seeds 1.5–1.7 x 1.0 mm; plants of alpine habitats .... *V. palustris*
      18b. Leaf blades pubescent on one or both surfaces (though sometimes sparsely and only near the basal lobes); capsules purple to purple-brown; seeds brown, longer than 1.8 mm .... *V. blanda*
1b. Plants with leafy stems, the peduncles arising from the axils of leaves

20a. Blade of upper leaves cuneate at the base; stipules foliaceous, with deep sinuses; plants annual or short-lived perennials, without rhizomes; cleistogamous flowers not produced

    21a. Blade of lower leaves rounded to cordate at the base; capsule ellipsoid to ovoid .... *V. tricolor*

    21b. Blade of lower leaves cuneate to rounded at the base; capsule globose (or sometimes ellipsoid) .... *V. arvensis*

20b. Blade of upper leaves truncate to cordate at the base; stipules scarious to herbaceous, entire, toothed, or fringed along the margins; plants perennial, with rhizomes; cleistogamous flowers often produced

    22a. Stipules entire or slightly toothed, scarious or herbaceous; stems occurring singly or a few together

        23a. Stipules lance-acuminate, scarious or subscarious; capsules (7.0–)10.0–12.0(–14.0) mm long; seeds 1.5–2.0 mm long ...... *V. canadensis*

        23b. Stipules broad-ovate, herbaceous; capsules 5.0–9.0 mm long; seeds 2.0–3.0 mm long .... *V. pubescens*

    22b. Stipules prominently toothed, lobed, or fringed, herbaceous or foliaceous; stems often clustered

        24a. Sepals ciliate, with auricles 2.0–4.0 mm long; leaves crenulate; stipules 1.0–2.5 cm long, with teeth 3.0–5.0 mm long .... *V. striata*

        24b. Sepals glabrous, with auricles up to 1.5 mm long; leaves crenate to subentire; stipules up to 1.5 cm long, entire or toothed with teeth up to 3.0 mm long

            25a. Plants pubescent; blade of lower leaves ovate .... *V. adunca*

            25b. Plants glabrous, sometimes with a few hairs on the adaxial leaf surface; blade of lower leaves suborbicular to reniform .... *V. labradorica*

Viola adunca Sm.

synonym(s): —
common name(s): hook-spurred violet, early blue violet, sand violet, hooked violet
habit: tufted, rhizomatous, perennial herb
range: Greenland and Labrador, s. to NY, w. to CA, n. to AK
state frequency: uncommon
habitat: rocky slopes, gravelly soil, woods
notes: Known to hybridize with *Viola labradorica*.

Viola affinis Le Conte

synonym(s): *Viola papilionacea* Pursh *sensu* American authors
common name(s): —
habit: rhizomatous, perennial herb
range: ME and Quebec, s. to GA, w. to AR, n. to WI
state frequency: rare
habitat: damp woods, thickets, meadows, and shores
notes: Hybrids with *Viola cucullata*, called *V. ×consocia* House, occur in Maine. Known to hybridize with *Viola brittoniana*, *V. nephrophylla*, *V. palmata*, *V. sagittata*, and *V. sororia*.

Viola arvensis Murr.

synonym(s): —
common name(s): European field pansy, wild pansy

habit: annual herb
range: Europe; escaped from cultivation throughout our area
state frequency: very rare
habitat: fields, roadsides
notes: —

Viola blanda Willd.

synonym(s): <*V. b.* var. *palustriformis*> = *Viola incognita* Brainerd, *Viola incognita* Brainerd var.
    forbesii Brainerd
common name(s): sweet white violet
habit: rhizomatous, perennial herb
range: Newfoundland, s. to mountains of GA, w. to TN, n. to ND
state frequency: occasional
habitat: rich woods, cool, moist slopes, shaded ravines
key to the subspecies:
    1a. Leaf blades sharp-acute at the apex, with a narrow basal sinus, usually glabrous on the adaxial
        surface; lateral petals usually glabrous .... *V. b.* var. *blanda*
    1b. Leaf blades obtuse to rounded at the apex, with a wide basal sinus, usually pubescent on the
        adaxial surface; lateral petals usually pubescent .... *V. b.* var. *palustriformis* Gray
notes: Known to hybridize with *Viola renifolia*. All the petals are glabrous or with a small tuft of hairs,
    a useful character to help separate this species from some white-flowered violets that normally
    have purple flowers. These latter species normally have an extensive patch of hairs on at least the
    lateral petals.

Viola brittoniana Pollard

synonym(s): —
common name(s): coast violet
habit: acaulescent, rhizomatous, perennial herb
range: ME, s. to NC
state frequency: very rare
habitat: sandy and/or peaty soils along the coastal plain
notes: This species is ranked SH by the Maine Natural Areas Program. Known to hybridize with *Viola
    affinis*, *V. cucullata*, *V. palmata*, *V. sagittata*, and *V. sororia*.

Viola canadensis L.

synonym(s): —
common name(s): tall white violet, Canada violet
habit: rhizomatous, perennial herb
range: Newfoundland, s. to SC, w. to UT, n. to AK
state frequency: very rare
habitat: rich, moist, deciduous woods
notes: This species is ranked S1 by the Maine Natural Areas Program.

Viola cucullata Ait.

synonym(s): *Viola obliqua* Hill
common name(s): blue marsh violet
habit: tufted, rhizomatous, perennial herb
range: Newfoundland, s. to GA, w. to AR, n. to MN and Ontario
state frequency: occasional
habitat: bogs, swamps, wet meadows
notes: Hybrids with *Viola affinis* (*q.v.*), *V. sagittata* var. *ovata*, and *V. sororia* occur in Maine. Also
    known to hybridize with *V. brittoniana*, *V. nephrophylla*, and *V. palmata*.

Viola labradorica Schrank

synonym(s): *Viola adunca* Sm. var. *minor* (Hook.) Fern., *Viola conspersa* Reichenb.
common name(s): —
habit: tufted, rhizomatous, perennial herb
range: Newfoundland, s. to ME and NH, w. to CA, n. to AK
state frequency: uncommon
habitat: swamps, woods, wet to dry rocks and sand
notes: Known to hybridize with *Viola adunca*. *Viola labradorica* is very similar to *V. adunca*, but with a tendency to grow at higher and/or more northern sites than *V. adunca*.

Viola lanceolata L.

synonym(s): —
common name(s): straw-leaved violet
habit: rhizomatous and stoloniferous, perennial herb
range: New Brunswick, s. to FL, w. to TX, n. to MN and Ontario
state frequency: uncommon
habitat: wet, sandy to peaty soil of open areas
notes: Hybrids with the *Viola primulifolia*, called *V.* ×*modesta* House, and *V. macloskeyi*, called *V.* ×*sublanceolata* House, occur in Maine.

Viola macloskeyi Lloyd ssp. *pallens* (Banks *ex* DC.) M. S. Baker

synonym(s): *Viola pallens* (Banks *ex* DC.) Brainerd
common name(s): wild white violet
habit: rhizomatous and stoloniferous, perennial herb
range: Labrador, s. to mountains of GA, w. to CA, n. to British Columbia
state frequency: occasional
habitat: wet, springy meadows, thickets, and woods, shallow water
notes: Hybrids with *Viola lanceolata* (*q.v.*) occur in Maine. Also known to hybridize with *V. primulifolia*, and *V. renifolia*. All the petals are glabrous or nearly so, a useful character to help separate this species from some white-flowered violets that normally have purple flowers.

Viola nephrophylla Greene

synonym(s): —
common name(s): northern bog violet
habit: tufted, perennial herb
range: Newfoundland, s. to CT, w. to AZ and CA, n. to British Columbia
state frequency: rare
habitat: circumneutral soils of fens, ledges, and peaty shores
notes: Known to hybridize with *Viola affinis*, *V. cucullata*, and *V. sororia*. *Viola nephrophylla* is glabrous and normally has a pubescent spurred petal.

Viola novae-angliae House

synonym(s): —
common name(s): New England violet
habit: short-rhizomatous, perennial herb
range: New Brunswick and ME, w. to MN
state frequency: rare
habitat: moist to wet, gravel and rocks
notes: This species is ranked S2 by the Maine Natural Areas Program.

Viola odorata L.

synonym(s): —
common name(s): sweet violet, English violet

habit: rhizomatous and stoloniferous, perennial herb
range: Europe; escaped from cultivation in our area
state frequency: very rare
habitat: roadsides, open woods
notes: —

Viola palmata L.

synonym(s): *Viola triloba* Schwein.
common name(s): wood violet, three-lobed violet
habit: tufted, perennial herb
range: ME, s. to FL, w. to TX, n. to MN and Ontario
state frequency: very rare
habitat: rich woods and clearings
notes: This species is ranked SH by the Maine Natural Areas Program. Known to hybridize with *Viola affinis*, *V. brittoniana*, *V. cucullata*, *V. sagittata*, and *V. sororia*.

Viola palustris L.

synonym(s): —
common name(s): northern marsh violet, alpine marsh violet
habit: rhizomatous and stoloniferous, perennial herb
range: circumboreal; s. to mountains of ME and NH, w. to CA
state frequency: very rare
habitat: shores of streams and ponds in alpine areas
notes: This species is ranked S1 by the Maine Natural Areas Program. Past collections from Katahdin have been the white-flowered form.

Viola primulifolia L.

synonym(s): *Viola primulifolia* L. var. *acuta* (Bigelow) Torr. & Gray
common name(s): primrose-leaved violet
habit: rhizomatous and stoloniferous, perennial herb
range: ME, s. to FL, w. to TX, n. to MN and Ontario
state frequency: uncommon
habitat: moist to dry stream banks, shores, meadows, and thin woods
notes: Known to hybridize with *Viola lanceolata* (*q.v.*) and *V. macloskeyi*.

Viola pubescens Ait.

synonym(s): <*V. p.* var. *scabriuscula*> = *Viola pensylvanica* Michx. var. *leiocarpon* (Fern. & Wieg.) Fern.
common name(s): yellow forest violet, downy yellow violet
habit: short-rhizomatous, perennial herb
range: New Brunswick, s. to GA, w. to TX, n. to ND and Manitoba
state frequency: uncommon
habitat: rich, deciduous woods, meadows
key to the subspecies:
    1a. The solitary stem with 0 or 1 basal leaves; leaves densely pubescent; stipules broad-ovate to reniform, with an obtuse apex .... *V. p.* var. *pubescens*
    1b. Stems 2 or more, with 1–3 basal leaves; leaves glabrous or nearly so; stipules ovate, with an acute apex .... *V. p.* var. *scabriuscula* Torr. & Gray
notes: —

Viola renifolia Gray

synonym(s): *Viola renifolia* Gray var. *brainerdii* (Greene) Fern.
common name(s): kidney-leaved violet

habit: rhizomatous, non-stoloniferous, perennial herb
range: Newfoundland, s. to CT, w. to CO, n. to British Columbia
state frequency: occasional
habitat: *Thuja* swamps, cold woods
notes: Known to hybridize with *Viola blanda* and *V. macloskeyi*.

Viola rotundifolia Michx.
synonym(s): —
common name(s): round-leaved violet, early yellow violet
habit: rhizomatous, perennial herb
range: ME, s. to mountains of GA, w. to TN, n. to Ontario
state frequency: occasional
habitat: deciduous forests
notes: A vegetatively conspicuous forest violet, with spreading leaves that often lay flat on the ground and have darker green veins.

Viola sagittata Ait.
synonym(s): <*V. s.* var. *ovata*> = *Viola fimbriatula* Sm.
common name(s): arrowhead violet, arrow-leaved violet
habit: short-rhizomatous, perennial herb
range: New Brunswick, s. to FL, w. to TX, n. to MN and Ontario
state frequency: occasional
habitat: moist to dry woods, clearings, meadows, and fields
key to the subspecies:
   1a. Leaf blades ovate-triangular, coarsely serrate at the base, densely pubescent; petioles shorter than the leaf blades; sepals ciliate .... *V. s.* var. *ovata* (Nutt.) Torr. & Gray
   1b. Leaf blades lance-oblong to narrow-lanceolate, with prominent serrations or small, basal lobes, sparsely pubescent; petioles equaling or longer than the leaf blades; sepals not ciliate .... *V. s.* var. *sagittata*
notes: *Viola sagittata* var. *ovata* hybridizes with *V. sororia*, producing *V.* ×*fernaldii* House, and with *V. cucullata*, producing *V.* ×*porteriana* Pollard. Both of these hybrids occur in Maine. Also known to hybridize with *V. affinis*, *V. brittoniana*, *V. cucullata*, *V. palmata*, *V. sagittata* var. *sagittata*, and *V. sororia*.

Viola selkirkii Pursh *ex* Goldie
synonym(s): —
common name(s): great-spurred violet
habit: acaulescent, rhizomatous, perennial herb
range: circumboreal; s. to PA, w. to CO, n. to British Columbia
state frequency: uncommon
habitat: cold, rich woods, shady slopes and ravines
notes: —

Viola sororia Willd.
synonym(s): *Viola septentrionalis* Greene
common name(s): dooryard violet
habit: short-rhizomatous, perennial herb
range: Newfoundland, s. to NC, w. to OK, n. to WA and British Columbia
state frequency: occasional
habitat: woods, clearings, meadows, slopes
notes: Hybrids with *Viola sagittata* (*q.v.*) occur in Maine. Also known to hybridize with *Viola affinis*, *V. brittoniana*, *V. cucullata*, *V. nephrophylla*, and *V. palmata*.

Viola striata Ait.
synonym(s): —
common name(s): cream violet, pale violet
habit: short-rhizomatous, perennial herb
range: ME, s. to GA, w. to AR, n. to WI and Ontario
state frequency: very rare
habitat: low woods, meadows, along ditches and streams
notes: —

Viola tricolor L.
synonym(s): —
common name(s): pansy, Johnny jump-up
habit: annual or short-lived perennial herb
range: Europe; escaped from cultivation to New Brunswick, s. to TX, w. and n. to British Columbia
state frequency: uncommon
habitat: fields, old gardens
notes: —

CLUSIACEAE
(2 genera, 12 species)

KEY TO THE GENERA
1a. Petals pink or flesh-colored; stamens 9, connate into 3 distinct groups, alternating with conspicuous, orange glands .... ***Triadenum***
1b. Petals yellow or copper-colored; stamens 5–many, distinct or connate into 3–5 groups, without hypogynous glands .... ***Hypericum***

Hypericum
(10 species)

KEY TO THE SPECIES
1a. Shrubs; ovary and capsule essentially 3(4)-locular .... ***H. prolificum***
1b. Herbs, sometimes woody at the very base; ovary and capsule 1- to 5-locular
  2a. Styles 5; capsule 5-locular, 1.5–3.0 cm tall; petals 2.5–3.0 cm long .... ***H. ascyron***
  2b. Styles 3(4); capsule 1- to 3(4)-locular, 0.2–0.7 cm tall; petals 0.15–1.2 cm long
    3a. Styles connate or closely connivent, persistent in fruit and appearing as a single, straight beak at the summit of the capsule; stigmas minute; capsule unilocular; flowers with 20–40 stamens .... ***H. ellipticum***
    3b. Styles distinct nearly to the base and often divergent, deciduous in fruit and not appearing as a single beak at the summit of the capsule; stigmas capitate; capsule 1- to 3(4)-locular; flowers with 5–40 stamens
      4a. Ovary and capsule 3(4)-locular; flowers with 20–40 stamens
        5a. Petals conspicuously dotted and streaked with black; sepals obtuse to broad-acute at the apex; seeds less than 1.0 mm long; branches terete, without decurrent ridges .... ***H. punctatum***
        5b. Petals dotted only at margins; sepals acuminate; seeds 1.0–1.3 mm long; branches with sharp, decurrent ridges below the leaves .... ***H. perforatum***

4b. Ovary and capsule unilocular; flowers with 5–22 stamens
   6a. Leaves subulate to linear-subulate, 1-nerved, appressed, 1.0–4.0 mm long; inflorescence not obviously cymose, often with alternate branches ... . *H. gentianoides*
   6b. Leaves linear to elliptic, 1- to 7-nerved, spreading to ascending, 3.0–45.0 mm long; inflorescence clearly cymose, with a terminal flower and 2 opposite branches
      7a. Capsule ellipsoid, abruptly tapered and rounded at the summit; sepals linear to linear-lanceolate, broadest at or near the middle; stems decumbent, leafy-bracted at the base
         8a. Bracts of the cyme elliptic, leaf-like; sepals acute at the apex, subequal in length to the capsule .... *H. mutilum*
         8b. Bracts of the cyme subulate, not resembling the leaves; sepals obtuse at the apex, conspicuously shorter than the capsule .... *H. boreale*
      7b. Capsule ellipsoid-conic to conic, gradually tapered and pointed at the summit; sepals lanceolate, broadest below the middle; stems erect, not leafy-bracted at the base
         9a. Leaves linear to narrow-oblanceolate, mostly 1.0–4.0 mm wide, 1- or 3(5)-nerved, narrowly tapering to the base; petals 2.5–3.0 mm long .... *H. canadense*
         9b. Leaves lanceolate to narrow-elliptic or narrow-oblong, mostly 3.0–9.0 mm wide, 5- or 7-nerved, broadly tapering to rounded at the base; petals 3.5–4.0 mm long .... *H. majus*

Hypericum ascyron L.

synonym(s): *Hypericum pyramidatum* Ait.
common name(s): great St. Johnswort
habit: erect, branched, perennial herb
range: ME, s. to MD, w. to KS, n. to MN and Manitoba
state frequency: very rare
habitat: moist thickets and meadows
notes: This species is ranked SH by the Maine Natural Areas Program.

Hypericum boreale (Britt.) Bickn.

synonym(s): —
common name(s): —
habit: decumbent, rhizomatous, perennial herb
range: Newfoundland, s. to VA, w. to IA, n. to MN and Ontario
state frequency: occasional
habitat: wet, sandy or gravelly soil, shallow water
notes: —

Hypericum canadense L.

synonym(s): —
common name(s): Canada St. Johnswort
habit: slender, annual or perennial herb
range: Newfoundland, s. to GA, w. to AL, n. to MN and Manitoba
state frequency: occasional

habitat: wet meadows, sandy shores, bogs, swamps

notes: Hybrids between *Hypericum canadense* and either *H. boreale* or *H. mutilum* are called *H.* ×*dissimulatum* Bickn. and occur in Maine.

Hypericum ellipticum Hook.

synonym(s): —
common name(s): —
habit: erect, simple, rhizomatous, perennial herb
range: Newfoundland, s. to MD and mountains of NC, w. to IL and IA, n. to ND and Manitoba
state frequency: occasional
habitat: wet, sandy or gravelly shores, marshes
notes: —

Hypericum gentianoides (L.) B. S. P.

synonym(s): —
common name(s): orange-grass, pineweed
habit: erect, branched, annual herb
range: ME, s. to FL, w. to TX, n. to MN and Ontario
state frequency: uncommon
habitat: sterile, sandy or rocky soil
notes: —

Hypericum majus (Gray) Britt.

synonym(s): —
common name(s): —
habit: erect, rhizomatous, perennial herb
range: Newfoundland, s. to DE, w. to CO, n. to WA and British Columbia
state frequency: uncommon
habitat: wet meadows, shores, and open areas
notes: —

Hypericum mutilum L.

synonym(s): *Hypericum mutilum* L. var. *parviflorum* (Willd.) Fern.
common name(s): —
habit: slender, erect, branched, annual or perennial herb
range: Newfoundland, s. to FL, w. to TX, n. to MN and Manitoba
state frequency: uncommon
habitat: low, wet open areas
notes: —

Hypericum perforatum L.

synonym(s): —
common name(s): common St. Johnswort
habit: erect, branched, perennial herb
range: Europe; naturalized throughout much of North America
state frequency: common
habitat: fields, roadsides, waste places
notes: —

Hypericum prolificum L.

synonym(s): *Hypericum spathulatum* (Spach) Steud.
common name(s): shrubby St. Johnswort
habit: shrub

range: NY, s. to GA, w. to LA, n. to MN and Ontario; escaped from cultivation to northern New
   England
state frequency: very rare
habitat: swamps, slopes, woods
notes: This species has conspicuously 2-edged twigs.

Hypericum punctatum Lam.
synonym(s): —
common name(s): spotted St. Johnswort
habit: erect, sparsely branched, perennial herb
range: Quebec, s. to FL, w. to TX, n. to MN and Ontario
state frequency: uncommon
habitat: moist or dry fields and open woods
notes: —

Triadenum
(2 species)

Often recognizable vegetatively due to its oblong to broad-elliptic, cordate-based leaves, usually with
reddish coloration.

KEY TO THE SPECIES
1a. Sepals 5.0–8.0 mm long, acute to acuminate at the apex; petals 8.0–10.0 mm long; styles
   2.0–3.0 mm long; capsule gradually tapering to the apex .... *T. virginicum*
1b. Sepals 3.0–5.0 mm long, obtuse to rounded at the apex; petals 5.0–8.0 mm long; styles
   0.5–1.5 mm long; capsule somewhat abruptly tapering to the apex .... *T. fraseri*

Triadenum fraseri (Spach) Gleason
synonym(s): *Hypericum virginicum* L. var. *fraseri* (Spach) Fern.
common name(s): marsh St. Johnswort
habit: erect, rhizomatous, perennial herb
range: Newfoundland, s. to CT, w. to NE, n. to Saskatchewan
state frequency: occasional
habitat: bogs, marshes, wet shores
notes: —

Triadenum virginicum (L.) Raf.
synonym(s): *Hypericum virginicum* L.
common name(s): marsh St. Johnswort
habit: erect, rhizomatous, perennial herb
range: Nova Scotia, s. to FL, w. to MS, n. to Ontario
state frequency: occasional
habitat: bogs, marshes, wet shores
notes: —

EUPHORBIACEAE
(3 genera, 10 species)

KEY TO THE GENERA

1a. Plants with a milky latex; inflorescence a cyathium; flowers without a calyx, subtended by a cup-like involucre with glands (sometimes with petaloid appendages) at the margin, housing within a single, peduncled, carpellate flower and 1 to several staminate flowers, collectively resembling a single bisexual flower

    2a. Leaves of the stem alternate, the blades with equilateral bases, those subtending the base of the cyathiescence whorled; stipules absent; involucre with 4 glands that lack petaloid appendages or with (4)5 glands that possess petaloid appendages in *Euphorbia corollata*; plants annual or perennial, with ascending to erect stems .... ***Euphorbia***

    2b. Leaves opposite throughout, the blades with inequilateral bases; stipules present; involucre with 4 glands that possess petaloid appendages in most species; plants annual, with prostrate to ascending stems .... ***Chamaesyce***

1b. Plants without latex, the sap watery; inflorescence an axillary cyme; flowers with a calyx, clearly unisexual .... ***Acalypha***

Acalypha
(1 species)

Acalypha virginica L.
synonym(s): <*A. v.* var. *rhomboidea*> = *Acalypha rhomboidea* Raf.
common name(s): three-seeded mercury, Virginia copperleaf
habit: erect, usually branched, annual herb
range: ME, s. to FL, w. to TX, n. to SD
state frequency: uncommon
habitat: dry, open woods, fields, and roadsides
key to the subspecies:
    1a. Carpellate bracts with 5–9 lobes, usually stipitate-glandular, but without long hairs; petioles 0.35–1.0 times as long as the leaf blades, those of the larger leaves commonly more than 0.5 times as long .... *A. v.* var. *rhomboidea* (Raf.) Cooperrider
    1b. Carpellate bracts with 9–15 lobes, without stipitate glands, usually with some long hairs; petioles 0.35–0.5 times as long as the leaf blades .... *A. v.* var. *virginica*
notes: *Acalypha virginica* var. *virginica* is ranked SH by the Maine Natural Areas Program.

Chamaesyce
(4 species)

Members of this genus in Maine have petaloid appendages at the margin of the involucre causing the cyathium to appear to be a single flower. However, inspection of the stamens will reveal a constriction on the filament. This constriction is a joint indicating the location of the receptacle and is where the peduncle stops and the filament begins.

KEY TO THE SPECIES

1a. Leaves with entire margins; seeds with smooth faces; glands with very small or obsolete appendages .... ***C. polygonifolia***

1b. Leaves with serrate or partially serrate margins; seeds with wrinkled, ridged, or pitted faces; glands with small but evident appendages

    2a. Leaves and stem glabrous; leaves serrate only toward the apex and sometimes near the base; faces of the seed with prominent, transverse ridges .... *C. glyptosperma*

    2b. Leaves (at least when young) and stems pubescent; leaves serrate throughout the margin; faces of the seed wrinkled or pitted

        3a. Ovary and capsule glabrous; leaves usually 0.5–0.65 times as wide as long, green or tinged with red; seeds acutely angled .... *C. vermiculata*

        3b. Ovary and capsule strigose; leaves usually 0.35–0.5 times as wide as long, often with red spots or blotches; seeds obtusely angled .... *C. maculata*

Chamaesyce glyptosperma (Engelm.) Small

synonym(s): *Euphorbia glyptosperma* Engelm.
common name(s): ridge-seed spurge
habit: prostrate, branched, annual herb
range: New Brunswick, s. to NY, w. to AZ and CA, n. to British Columbia
state frequency: uncommon
habitat: dry, sandy, open soil
notes: —

Chamaesyce maculata (L.) Small

synonym(s): *Euphorbia maculata* L., *Euphorbia supina* Raf.
common name(s): eyebane, milk-purslane, spotted spurge
habit: prostrate to ascending, simple or branched, annual herb
range: New Brunswick, s. to FL, w. to TX and Mexico, n. to ND and Ontario
state frequency: occasional
habitat: open woods, cultivated ground, roadsides, waste places
notes: —

Chamaesyce polygonifolia (L.) Small

synonym(s): *Euphorbia polygonifolia* L.
common name(s): seaside spurge
habit: prostrate to ascending, mat-forming, simple or branched, annual herb
range: New Brunswick, s. to GA; also the Great Lakes region
state frequency: rare
habitat: sand dunes, sandy or gravelly beaches and shores
notes: —

Chamaesyce vermiculata (Raf.) House

synonym(s): *Chamaesyce rafinesquei* (Greene) Arthur, *Euphorbia vermiculata* Raf.
common name(s): hairy spurge
habit: prostrate to ascending, annual herb
range: Quebec, s. to NJ, w. to IN, n. to WI; also AZ, NM, and British Columbia
state frequency: uncommon
habitat: dry fields, roadsides, waste places
notes: —

Euphorbia
(5 species)

Flowers are similar to *Chamaesyce* (*q.v.*), but the leaves that subtend the cyathiescence differ in arrangement and dimensions. Care must be taken to measure the leaves specified in the key.

KEY TO THE SPECIES

1a. Involucre with (4)5 glands, the glands with evident, white or rarely green, petaloid appendages .... *E. corollata*
1b. Involucre with 4 glands, these without petaloid appendages
   2a. Leaves serrulate; glands of the involucre orbicular to elliptic .... *E. helioscopia*
   2b. Leaves entire; glands of the involucre crescent-shaped, the concave surface facing outwards, often narrowed to horn-like ends
      3a. Umbel-like cyathiescence with 3–5 primary branches; plants annual; involucre 1.0–2.0 mm tall .... *E. peplus*
      3b. Umbel-like cyathiescence usually with 7–15 primary branches; plants perennial from rhizomes; involucre 2.5–3.0 mm tall
         4a. Leaves of the stem 1.0–3.0 x 0.1–0.3 cm; seeds 1.5–2.0 mm long; plants 1.5–4.0 dm tall .... *E. cyparissias*
         4b. Leaves of the stem 3.0–8.0 x 0.4–0.8 cm; seeds 2.0–2.5 mm long; plants 3.0–7.0 dm tall .... *E. esula*

Euphorbia corollata L.

synonym(s): *Tithymalopsis corollata* (L.) Klotzsch
common name(s): flowering spurge, tramp's spurge, wild hippo
habit: erect, perennial herb
range: NY, s. to FL, w. to TX, n. to MN and Ontario; naturalized to ME, s. to CT
state frequency: rare
habitat: dry fields, roadsides, and open woods
notes: —

Euphorbia cyparissias L.

synonym(s): *Tithymalus cyparissias* (L.) Hill
common name(s): cypress spurge
habit: densely tufted, rhizomatous, perennial herb
range: Eurasia; escaped from cultivation to New Brunswick, s. to VA, w. to AR, n. to WA and British Columbia
state frequency: occasional
habitat: roadsides, fields, waste places, old cemeteries
notes: —

Euphorbia esula L.

synonym(s): *Tithymalus esula* (L.) Hill
common name(s): leafy spurge
habit: erect, colonial, perennial herb
range: Eurasia; escaped from cultivation to New Brunswick, s. to MD, w. to CO, n. to British Columbia
state frequency: very rare
habitat: fields, banks, roadsides
notes: —

Euphorbia helioscopia L.
synonym(s): *Tithymalus helioscopius* (L.) Hill
common name(s): wartweed
habit: ascending, annual herb
range: Europe; escaped from cultivation to New Brunswick, s. to MD, w. to KS, n. to British
    Columbia
state frequency: very rare
habitat: dry, waste places
notes: —

Euphorbia peplus L.
synonym(s): *Tithymalus peplus* (L.) Hill
common name(s): petty spurge
habit: branched, annual herb
range: Eurasia; escaped from cultivation to Newfoundland, s. to FL, w. to CA, n. to AK
state frequency: rare
habitat: cultivated ground, waste places
notes: —

LYTHRACEAE
(2 genera, 4 species)

KEY TO THE GENERA
1a. Perianth mostly 4- or 5-merous; flowers with 8 or 10 stamens; leaves more or less
    opposite or whorled; hypanthium about as wide as long; plants shrubs, with upwardly
    arching stems .... ***Decodon***
1b. Perianth mostly 6-merous; flowers with usually 4–6 or 12 stamens; leaves opposite or
    the upper alternate; hypanthium longer than wide; plants herbs, upright .... ***Lythrum***

Decodon
(1 species)

Decodon verticillatus (L.) Ell.
synonym(s): *Decodon verticillatus* (L.) Ell. var. *laevigatus* Torr. & Gray
common name(s): water-willow, water-oleander
habit: shrub
range: New Brunswick, s. to FL, w. to MO, n. to MN and Ontario
state frequency: uncommon
habitat: swamps, shallow water
notes: —

Lythrum
(3 species)

Some of the species with heteromorphic flowers [the stamens and styles of different lengths within
different flowers].

KEY TO THE SPECIES

1a. Inflorescence an elongate, terminal thyrse; flowers with 12 stamens, trimorphic; petals 7.0–12.0 mm long; leaves 3.0–10.0 cm long, 2 or 3 at each node .... *L. salicaria*

1b. Inflorescence composed of solitary or paired flowers in the axils of leaves; flowers with 4–6 stamens, mono- or dimorphic; petals 2.0–6.0 mm long; leaves 1.0–4.0 cm long, usually 1 at each node

    2a. Plants perennial; flowers dimorphic [pin and thrum]; leaves 4.0–10.0 mm wide, rounded to truncate to subcordate at the base .... *L. alatum*

    2b. Plants annual; flowers monomorphic; leaves 1.0–6.0 mm wide, cuneate at the base ...
    .... *L. hyssopifolia*

Lythrum alatum Pursh

synonym(s): —
common name(s): winged loosestrife
habit: erect, virgately branched, perennial herb
range: NY, s. to GA, w. to TX, n. to British Columbia; naturalized to ME, s. to NJ
state frequency: rare
habitat: swamps, wet meadows and ditches
notes: —

Lythrum hyssopifolia L.

synonym(s): —
common name(s): annual loosestrife
habit: simple or branched, annual herb
range: ME, s. in the east to NJ; also OH
state frequency: very rare
habitat: wet, sterile soil, marshes, generally coastal
notes: —

Lythrum salicaria L.

synonym(s): *Lythrum salicaria* L. var. *gracilior* Turcz., *Lythrum salicaria* L. var. *tomentosum* (P. Mill.) DC.
common name(s): purple loosestrife, spiked loosestrife
habit: stout, erect, perennial herb
range: Eurasia; naturalized to Newfoundland, s. to VA, w. to IN, n. to MN
state frequency: occasional
habitat: swamps, bogs, wet meadows and ditches
notes: —

ONAGRACEAE

(4 genera, 20 species)

KEY TO THE GENERA

1a. Perianth 2-merous; fruit indehiscent, covered with uncinate bristles; leaves opposite; petals white .... *Circaea*

1b. Perianth 4- to 6-merous; fruit dehiscent, without uncinate bristles; leaves opposite or alternate; petals absent or white, pink, purple, or yellow

    2a. Petals absent; sepals persistent in fruit; flowers sessile; hypanthium not prolonged beyond the summit of the ovary .... *Ludwigia*

2b. Petals present; sepals deciduous in fruit; flowers pedicellate; hypanthium prolonged beyond the summit of the ovary (except in *Epilobium angustifolium*)

    3a. Petals yellow; seeds not comose; dehiscing fruit usually without recurving valves; leaves alternate; hypanthium extending beyond the summit of the ovary .... ***Oenothera***

    3b. Petals white to pink to purple; seeds comose; dehiscing fruit usually with recurving valves; leaves alternate or, commonly, the lower opposite; hypanthium not, or only shortly, extending beyond the ovary .... ***Epilobium***

Circaea
(2 species)

KEY TO THE SPECIES

1a. Ovary and fruit bilocular; rhizomes slender; inflorescence open; petals 2.5–4.0 mm long; anthers 0.7–1.0 mm long; stems 1.5–7.0 dm tall .... ***C. lutetiana***

1b. Ovary and fruit unilocular; rhizomes tuberous-thickened at the end; inflorescence clustered near the apex; petals 1.0–2.5 mm long; anthers 0.2–0.3 mm long; stems 1.0–3.0 dm tall .... ***C. alpina***

Circaea alpina L.
synonym(s): —
common name(s): alpine enchanter's-nightshade
habit: weak-stemmed, rhizomatous, perennial herb
range: circumboreal; s. to mountains of NC, w. to mountains of NM
state frequency: occasional
habitat: moist to wet woods and openings, bogs
notes: —

Circaea lutetiana L. ssp. *canadensis* (L.) Aschers. & Magnus
synonym(s): *Circaea canadensis* (L.) Hill, *Circaea quadrisulcata* (Maxim.) Franch. & Savigny var. *canadensis* (L.) Hara
common name(s): common enchanter's-nightshade
habit: erect, rhizomatous, perennial herb
range: New Brunswick, s. to GA, w. to OK, n. to ND and Manitoba
state frequency: occasional
habitat: moist woods, thickets, and ravines
notes: —

Epilobium
(9 species)

KEY TO THE SPECIES

1a. Hypanthium not extending beyond the summit of the ovary; petals 1.0–2.0 cm long, entire at the apex; stamens in 1 series, declined .... ***E. angustifolium***

1b. Hypanthium shortly extending beyond the summit of the ovary; petals 0.2–1.5 cm long, notched at the apex; stamens in 2 series, erect to ascending

    2a. Stigma 4-lobed; petals 1.0–1.5 cm long; plants evidently pubescent .... ***E. hirsutum***

    2b. Stigma entire or nearly so; petals 0.2–1.0 cm long; plants short-pubescent

3a. Leaves with entire or undulate, usually revolute, margins; pubescence of the stem not in decurrent lines from the leaf bases
    4a. Plants pubescent with divergent hairs .... *E. strictum*
    4b. Plants pubescent with appressed hairs
        5a. Leaves essentially glabrous on the adaxial surface; inflorescence nodding in bud .... *E. palustre*
        5b. Leaves pubescent on the adaxial surface; inflorescence erect to arching in bud .... *E. leptophyllum*
3b. Leaves with usually toothed, flat margins; pubescence of the stem in decurrent lines from the leaf bases
    6a. Stems matted, decumbent to erect, soboliferous, usually unbranched above the base of the plant, 0.5–4.5 dm tall; plants boreal and alpine
        7a. Seeds smooth on the outer surface; stems mostly 0.5–1.5 dm long; capsule 2.3–4.0 cm long; leaves 1.0–2.0 cm long .... *E. anagallidifolium*
        7b. Seeds papillose on the outer surface; stems mostly 1.5–4.5 dm long; capsule 3.5–7.5 cm long; leaves 1.5–4.5 cm long .... *E. hornemannii*
    6b. Stems usually solitary, erect, without sobols, usually branched above the base of the plant, (0.5–)3.0–10.0 dm tall; plants of various habitats, commonly found at lower elevations
        8a. Coma brown; flower buds pointed at the apex due to the projecting sepal tips; seeds minutely papillose, without a beak; leaves gray-green, rugose-veiny .... *E. coloratum*
        8b. Coma white or nearly so; flower buds obtuse to rounded at the apex; seeds longitudinally striate, with a broad, short beak; leaves without gray color, not or only slightly rugose-veiny .... *E. ciliatum*

Epilobium anagallidifolium Lam.

synonym(s): *Epilobium alpinum* L.
common name(s): alpine willow-herb
habit: dwarf, soboliferous, perennial herb
range: circumboreal; s. in the east to ME, NH, and NY
state frequency: very rare
habitat: moist, seepy spots in alpine areas
notes: This species is ranked S1 by the Maine Natural Areas Program.

Epilobium angustifolium L.

synonym(s): *Chamaenerion angustifolium* (L.) Scop.
common name(s): fireweed, great willow-herb
habit: erect, usually simple, perennial herb
range: circumboreal; s. to mountains of NC, w. to NM
state frequency: occasional
habitat: open woods, thickets, new clearings, burned areas
notes: —

Epilobium ciliatum Raf.

synonym(s): <*E. c.* ssp. *ciliatum*> = *Epilobium glandulosum* Lehm. var. *adenocaulon* (Hausskn.) Fern.; <*E. c.* ssp. *glandulosum*> = *Epilobium glandulosum* Lehm. var. *occidentale* (Trel.) Fern.
common name(s): American willow-herb
habit: erect, mostly solitary and branched, perennial herb

range: Labrador, s. to NC, w. to TX, Mexico, and CA, n. to AK
state frequency: occasional
habitat: wet rocks, springy areas
key to the subspecies:
>   1a. Plants in the fall producing turions at and below ground level; petals 5.0–10.0 mm long, pink to purple (white); leaves narrow-ovate to ovate, rounded to cordate at the base; petioles 0.0–2.0 mm long; coma of seeds persistent ..... *E. c.* ssp. *glandulosum* (Lehm.) Hoch & Raven
>   1b. Plants not producing turions; petals 2.0–6.0 mm long, white (pink); leaves lanceolate to narrow-ovate, cuneate to rounded at the base; petioles (0.0–)2.0–10.0 mm long; coma of seeds caducous .... *E. c.* ssp. *ciliatum*

notes: The turions in *Epilobium ciliatum* ssp. *glandulosum* appear as loose, spherical rosettes of fleshy leaves.

Epilobium coloratum Biehler

synonym(s): —
common name(s): eastern willow-herb
habit: erect, usually solitary and branched, perennial herb
range: New Brunswick, s. to GA, w. to TX, n. to MN and Ontario
state frequency: uncommon
habitat: wet, low ground, springy areas
notes: —

Epilobium hirsutum L.

synonym(s): —
common name(s): hairy willow-herb, great hairy willow-herb
habit: erect, branched, rhizomatous, perennial herb
range: Eurasia; naturalized to New Brunswick, s. to MD, w. to IL, n. to MI and Ontario
state frequency: very rare
habitat: meadows, roadsides, thickets, waste places
notes: —

Epilobium hornemannii Reichenb.

synonym(s): —
common name(s): Hornemann's willow-herb
habit: slender, ascending, simple or forked, soboliferous, perennial herb
range: circumboreal; s. to ME and NH, w. to NV and CA
state frequency: very rare
habitat: damp rocks, stream banks
notes: This species is ranked S1 by the Maine Natural Areas Program.

Epilobium leptophyllum Raf.

synonym(s): —
common name(s): American marsh willow-herb, narrow-leaved willow-herb
habit: simple or branched, perennial herb
range: Quebec, s. to VA, w. to KS, n. to Alberta
state frequency: uncommon
habitat: marshes, bogs, wet meadows and ditches
notes: —

Epilobium palustre L.

synonym(s):*Epilobium oliganthum* Michx., *Epilobium palustre* L. var. *grammadophyllum* Hausskn., *Epilobium palustre* L. var. *oliganthum* (Michx.) Fern.
common name(s): marsh willow-herb, swamp willow-herb

habit: slender, simple or branched, stoloniferous, perennial herb
range: circumboreal; s. to CT, w. to CO and OR
state frequency: uncommon
habitat: bogs, swales, wet, low ground
notes: —

Epilobium strictum Muhl. *ex* Spreng.
synonym(s): —
common name(s): northeastern willow-herb, downy willow-herb
habit: erect, simple or branched, stoloniferous, rhizomatous, perennial herb
range: Quebec, s. to VA, w. to IL, n. to MN and Ontario
state frequency: uncommon
habitat: bogs, swamps, wet thickets and meadows
notes: —

Ludwigia
(1 species)

Ludwigia palustris (L.) Ell.
synonym(s): *Isnardia palustris* L., *Ludwigia palustris* (L.) Ell. var. *americana* (DC.) Fern. & Grisc.
common name(s): common water-purslane
habit: prostrate or floating, perennial herb
range: New Brunswick and Nova Scotia, s. to GA, w. to TX and Mexico, n. to WA
state frequency: uncommon
habitat: muddy shores, shallow water
notes: —

Oenothera
(8 species)

In certain species of *Oenothera*, the sepal appendages are an important morphological character. These are small, narrow, lobe-like tips found near the apex of the sepals. The sepal appendages are sometimes located at the very tip of the sepal, called terminal, and appear to be the true tip of the sepal. In other species, the sepal appendages are located just below the very tip of the sepal on the abaxial surface, called subterminal. In this case, the 4 sepals appear to have small, horn-like tips.

KEY TO THE SPECIES
1a. Stamens alternately of different size; ovary quadrangular or 4-winged, at least in the apical portion; capsule sharply quadrangular or 4-winged
   2a. Petals 3.0–10.0 mm long; style 3.0–10.0 mm long; anthers 1.5–2.5 mm long; inflorescence of scattered flowers, the axis and apex commonly drooping in bud .... *O. perennis*
   2b. Petals 10.0–25.0 mm long; style 10.0–20.0 mm long; anthers 4.0–8.0 mm long; inflorescence relatively more compact, the axis and apex commonly erect in bud
      3a. Sepal appendages mostly up to 2.0 mm long; capsule clavate to obpyramidal, stipitate; plants subglabrous to sparsely pubescent with hairs mostly shorter than 1.0 mm; hypanthium 0.5–1.5 cm long .... *O. fruticosa*
      3b. Sepal appendages mostly 1.0–3.0 mm long; capsule linear-elliptic, elliptic, to narrow-clavate, sessile or short-stipitate; plants conspicuously pubescent with hairs mostly 1.0–3.0 mm long; hypanthium 1.5–2.5 cm long .... *O. pilosella*

1b. Stamens all of equal size; ovary terete or nearly so; capsule terete to obscurely quadrangular

    4a. Leaves entire to dentate; capsule slightly tapering to the apex, widest at the base; ovules and seeds horizontally oriented in the locules, the seeds with angular faces, not pitted; plants biennial or short-lived perennial

        5a. Anthers 7.0–15.0 mm long; seeds mostly 1.2–1.8 mm long; sepal appendages terminal, closely connivent in bud

            6a. Petals 3.0–6.0 cm long; calyx lobes 2.2–5.0 cm long; styles (0.7–)1.8–5.0 cm long .... ***O. grandiflora***

            6b. Petals 1.0–2.5 cm long; calyx lobes 1.0–3.0 cm long; styles 0.3–1.7 cm long

                7a. Abaxial surface of calyx lobes, ovaries, and capsules hidden by a dense pubescence; inflorescence pubescent with few or no gland-tipped hairs; leaves firm, strongly ascending .... ***O. villosa***

                7b. Abaxial surface of calyx lobes, ovaries, and capsules sparsely pubescent to subglabrous; inflorescence pubescent with some gland-tipped hairs; leaves membranaceous, spreading to ascending .... ***O. biennis***

        5b. Anthers 4.0–7.0 mm long; seeds mostly (1.6–)1.8–2.2 mm long; sepal appendages subterminal, separate in bud, the true apex of the sepal represented by a small lobe .... ***O. parviflora***

    4b. Leaves prominently dentate to pinnatifid; capsule not tapering to the apex, essentially of uniform width; ovules and seeds vertically oriented in the locules, the seeds without angular faces, conspicuously pitted; plants annual .... ***O. laciniata***

Oenothera biennis L.

synonym(s): *Oenothera muricata* L.
common name(s): common evening-primrose
habit: stout, erect, simple or branched, biennial or short-lived perennial herb
range: Newfoundland, s. to FL, w. to TX, n. to WA and British Columbia
state frequency: occasional
habitat: fields, roadsides, waste places
notes: —

Oenothera fruticosa L.

synonym(s): <*O. f.* ssp. *fruticosa*> = *Kneiffia fruticosa* (L.) Raimann, *Oenothera tetragona* Roth var. *longistipata* (Pennell) Munz; <*O. f.* ssp. *glauca*> = *Kneiffia glauca* (Michx.) Spach, *Oenothera tetragona* Roth
common name(s): southern sundrops
habit: erect to ascending, simple or branched, perennial herb
range: MA, s. to FL, w. to OK, n. to IL and MI; also Manitoba; naturalized to New Brunswick, Nova Scotia, and ME
state frequency: rare
habitat: open woods, thicket edges, fields, meadows, roadsides
key to the subspecies:

    1a. Pubescence of the inflorescence and capsule, when present, mostly or entirely of gland-tipped hairs; sepal appendages mostly 1.0–2.0 mm long .... *O. f.* ssp. *glauca* (Michx.) Straley

    1b. Pubescence of the inflorescence and capsule mostly or entirely of eglandular hairs; sepal appendages mostly shorter than 1.0 mm .... *O. f.* ssp. *fruticosa*

notes: —

Oenothera grandiflora L'Hér. *ex* Ait.
synonym(s): —
common name(s): large evening-primrose
habit: stout, erect, simple or branched, biennial herb
range: Europe; escaped from cultivation in the northeast
state frequency: rare
habitat: fields, roadsides, waste places
notes: —

Oenothera laciniata Hill
synonym(s): *Raimannia laciniata* (Hill) Rose
common name(s): cut-leaved evening-primrose
habit: decumbent to erect, simple or branched, annual herb
range: NJ, s. to FL, w. to TX, n. to ND; escaped from cultivation in the northeast
state frequency: very rare
habitat: dry, sandy roadsides and waste places
notes: —

Oenothera parviflora L.
synonym(s): *Oenothera cruciata* Nutt. *ex* G. Don var. *sabulonensis* Fern.
common name(s): small-flowered evening-primrose
habit: simple or rarely branched, biennial or short-lived perennial herb
range: Newfoundland, s. to VA, w. to AR, n. to MN and Ontario
state frequency: common
habitat: sandy to gravelly shores, waste places, talus
notes: —

Oenothera perennis L.
synonym(s): *Kneiffia perennis* (L.) Pennell, *Kneiffia pumila* (L.) Spach
common name(s): little sundrops
habit: erect to ascending, mostly simple, fibrous-rooted, perennial herb
range: Newfoundland, s. to VA and mountains of GA, w. to MO, n. to Manitoba
state frequency: occasional
habitat: dry to moist fields, meadows, and open woods
notes: —

Oenothera pilosella Raf.
synonym(s): *Kneiffia pratensis* Small
common name(s): midwestern sundrops
habit: erect, simple or seldom branched, short-rhizomatous, perennial herb
range: Ontario, s. to WV, w. to LA, n. to MI; naturalized to ME, s. to VA
state frequency: rare
habitat: moist fields, meadows, and open woods
notes: —

Oenothera villosa Thunb.
synonym(s): *Oenothera biennis* L. var. *canescens* Torr. & Gray, *Oenothera depressa* Greene
common name(s): —
habit: biennial or short-lived perennial herb
range: Quebec and New Brunswick, s. to NY, w. to OK, n. to Alberta
state frequency: very rare
habitat: fields, roadsides, waste places
notes: —

MELASTOMACEAE
(1 genus, 1 species)

Rhexia
(1 species)

Rhexia virginica L.
synonym(s): —
common name(s): meadow beauty, wing-stem meadow pitcher
habit: simple or branched, fibrous-rooted, perennial herb
range: Nova Scotia, s. to FL, w. to TX, n. to WI and Ontario
state frequency: uncommon
habitat: moist sands and gravels of open places
notes: The petals are usually red-purple in this species.

RESEDACEAE
(1 genus, 2 species)

Reseda
(2 species)

KEY TO THE SPECIES
1a. Flowers green-white, with 15–20 stamens and 4 carpels; leaves regularly lobed ....
.... ***R. alba***
1b. Flowers green-yellow, with 12–15 stamens and 3 carpels; leaves irregularly lobed ....
.... ***R. lutea***

Reseda alba L.
synonym(s): —
common name(s): —
habit: erect, annual or perennial herb
range: Europe; naturalized to ME, s. to DE, w. to IL
state frequency: very rare
habitat: waste places
notes: —

Reseda lutea L.
synonym(s): —
common name(s): —
habit: erect to ascending, biennial or perennial herb
range: Europe; naturalized to ME, s. to MD, w. to MO, n. to MI
state frequency: very rare
habitat: waste places
notes: —

BRASSICACEAE
(33 genera, 66 species)

KEY TO THE GENERA FOR USE WITH FLOWERING MATERIAL

1a. Petals yellow to orange (some petals fading to white with age)............................ **Group 1**
1b. Petals white, ochroleucous, pink, blue, or purple (sometimes with yellow near the base),
    or usually absent in *Cardamine longii* and *Lepidium densiflorum* .... **Group 2**

Group 1

1a. Leaves simple, lacking pronounced lobes (except for basal lobes in cordate- or sagittate-
    shaped leaves)
    2a. At least the lower stem leaves clasping, perfoliate, or with basal lobes that partly
        surround the stem
        3a. Upper stem leaves perfoliate; petals 1.0–1.5 mm long .... *Lepidium*
        3b. Upper stem leaves clasping but not completely surrounding the stem; petals
            2.0–12.0 mm long
            4a. Flower-bearing stems arising from a basal rosette of leaves; style short and
                scarcely differentiated from the ovary
                5a. Ovules in 2 rows in the ovary .... *Turritis*
                5b. Ovules in 1 row in the ovary .... *Arabis*
            4b. Flower-bearing stems lacking conspicuous basal rosettes; style longer and
              clearly differentiated from the summit of the ovary
              6a. Petals 2.0–6.0 mm long; ovary obovoid to globose, abruptly tapering to a
                long, slender style; leaves acute at the apex or nearly so
                7a. Ovary orbicular, becoming reticulate-roughened; ovules 2 per locule
                    .... *Neslia*
                7b. Ovary obovoid or pyriform, smooth; ovules 4–12 per locule ....
                    .... *Camelina*
              6b. Petals 7.0–12.0 mm long; ovary cylindric, gradually tapering to a short,
              thick style; leaves, except the upper in *Brassica*, rounded to obtuse at the
              apex
              8a. Leaves strictly entire; two shorter stamens subtended by a U-shaped
                gland .... *Conringia*
              8b. Leaves with some form of dentition; stamens subtended by 4 round
                glands .... *Brassica*
    2b. Stem leaves narrowed to the base
        9a. Basal leaves usually glabrous; stems lacking leaves in the apical half; pubescence
        absent or entirely of simple hairs .... *Diplotaxis*
        9b. Basal leaves usually pubescent; stems with leaves in the basal and apical
        portions; pubescence partly or entirely of compound hairs
            10a. Plants with a conspicuous rosette of leaves either at the base of the plant or
            on basal offshoots
               11a. Ovary linear-cylindric; anthers linear-oblong .... *Erysimum*
              11b. Ovary ovoid; anthers oval to oblong .... *Draba*
           10b. Plants without a conspicuous rosette of basal leaves
             12a. Petals deeply 2-lobed at the apex .... *Berteroa*

12b. Petals simple, rounded at the apex

    13a. Ovary compressed; petals definitely pubescent .... *Alyssum*

    13b. Ovary cylindric to quadrangular; petals glabrous or sparsely
        pubescent .... *Erysimum*

1b. At least the lower leaves lobed or divided

  14a. Plants glabrous or pubescent with simple [unbranched] hairs

    15a. Inflorescence lacking bracts, except sometimes the lowest pedicel

      16a. Petals 10.0–20.0 mm long

        17a. Ovary with prominent transverse partitions separating the ovules; petals
          conspicuously veined, the veins often of a color different from that of the
          petal .... *Raphanus*

        17b. Ovary without prominent transverse partitions separating the ovules;
          petals uniformly colored, without conspicuous veins

          18a. Sepals spreading to reflexed .... *Sinapis*

          18b. Sepals erect to ascending .... *Brassica*

      16b. Petals 1.5–10.0 mm long

        19a. Stem leaves clasping at the base

          20a. Ovules arranged in 2 rows in the ovary; petals 1.5–5.0 mm long ....
          .... *Rorippa*

          20b. Ovules arranged in 1 row in the ovary; petals (3.0–)6.0–10.0 mm
          long

            21a. Lower pedicels 0.3–0.6 cm long; anthers 0.9–1.3 mm long ....
            .... *Barbarea*

            21b. Lower pedicels 1.0–2.0 cm long; anthers 1.3–2.1 mm long ....
            .... *Brassica*

        19b. Stem leaves narrowed at the base

          22a. Each locule of the ovary with 2 rows of ovules

            23a. Ovary linear; plants mostly of sandy or disturbed areas ....
            .... *Diplotaxis*

            23b. Ovary ellipsoid to cylindric; plants mostly of wet soils ....
            .... *Rorippa*

          22b. Each locule of the ovary with 1 row of ovules

            24a. Ovary with a prominent beak; petals 6.0–12.0 mm long

              25a. Sepals spreading to reflexed .... *Sinapis*

              25b. Sepals erect to ascending .... *Brassica*

            24b. Ovary without a prominent beak; petals 3.0–8.0 mm long

              26a. Plants annual, with fibrous roots, of fields and disturbed areas
              .... *Sisymbrium*

              26b. Plants perennial, with underground, creeping stems, of wet
              soil .... *Rorippa*

    15b. Inflorescence with conspicuous, lobed bracts, only the uppermost pedicel
    sometimes lacking bracts .... *Erucastrum*

  14b. Plants pubescent with compound [branched] hairs

    27a. Stem leaves 1- to 3-times pinnately compound; petals 1.5–2.5 mm long ....
    .... *Descurainia*

    27b. Stem leaves simple; petals 3.0–6.0 mm long .... *Turritis*

Group 2

1a. Leaves simple, lacking pronounced lobes (except for basal lobes in cordate- or sagittate-shaped leaves)

2a. Plants glabrous or pubescent with simple [unbranched] hairs

3a. Leaves subulate, not flat; plants submerged or partly submerged aquatics .... *Subularia*

3b. Leaves not subulate, definitely flat; plants terrestrial

4a. Stem leaves auriculate, sagittate, conspicuously clasping, or perfoliate

5a. Ovary broad-elliptic, ovate, orbicular, or obovate

6a. Ovary notched at the summit; plants annual or biennial

7a. Ovules 1 per locule; plants glabrous or pubescent with short hairs .... *Lepidium*

7b. Ovules 4 or more per locule; plants glabrous .... *Thlaspi*

6b. Ovary without a notch at the summit; plants perennial .... *Cardaria*

5b. Ovary cylindric to linear

8a. Flower-bearing stems arising from a basal rosette of leaves; plants biennial or short-lived perennial; leaves entire or toothed; petals 3.0–9.0 mm long

9a. Ovules in 2 rows in the ovary

10a. Petals 3.0–6.0 mm long, yellow to ochroleucous .... *Turritis*

10b. Petals 5.0–9.0 mm long, usually white .... *Arabis*

9b. Ovules in 1 row in the ovary .... *Arabis*

8b. Flower-bearing stems lacking a conspicuous basal rosette; plants annual; leaves strictly entire; petals 7.0–10.0 mm long .... *Conringia*

4b. Stem leaves not surrounding the stem to any degree

11a. Ovary elongate, linear or cylindric

12a. Plants with a conspicuous rosette of leaves either at the base of the plants or on basal offshoots .... *Arabis*

12b. Plants without a conspicuous rosette of basal leaves

13a. Blade of stem leaves ovate to deltate to reniform, cordate at the base

14a. Petals absent; plants of estuaries .... *Cardamine*

14b. Petals present; plants not of estuaries

15a. Leaves short-petioled or sessile; flowers on pedicels longer than 5.0 mm .... *Lunaria*

15b. Leaves with long petioles; flowers on pedicels up to 5.0 mm long .... *Alliaria*

13b. Blade of stem leaves oblanceolate to obovate or lanceolate to ovate, cuneate to rounded at the base

16a. Petals absent; leaves suborbicular to reniform; plants of estuaries .... *Cardamine*

16b. Petals present; leaves oblanceolate to elliptic; habitats various

17a. Plants fleshy, much branched, halophytic annuals .... *Cakile*

17b. Plants of various habits, not halophytic, annual or perennial

18a. Inflorescence with conspicuous, lobed bracts, only the uppermost pedicel sometimes lacking bracts .... *Erucastrum*

18b. Inflorescence lacking bracts, except sometimes the lowest pedicel

19a. Plants tufted, perennial, 0.3–1.0 dm tall, of alpine areas, elevation exceeding 1000 m; axis of raceme very short; petals 3.0–5.0 mm long ...... *Cardamine*

19b. Plants annual or perennial, 2.0–8.0 dm tall, of disturbed soils, elevation less than 1000 m; axis of raceme elongate; petals much longer ..... *Diplotaxis*

11b. Ovary shorter, lanceolate or oblanceolate to orbicular

20a. Blade of stem leaves cordate at the base

21a. Plants annual; stem leaves short-petioled or sessile .... *Lunaria*

21b. Plants perennial; at least the lower stem leaves long-petioled .... *Armoracia*

20b. Blade of stem leaves narrowed to rounded at the base

22a. Petals dimorphic, the outer pair much larger than the inner pair .... *Iberis*

22b. Petals monomorphic or nearly so

23a. Petals 5.0–8.0 mm long; ovary not flat; plants perennial, mostly 6.0–13.0 dm tall .... *Armoracia*

23b. Petals 1.0–2.0 mm long or absent; ovary flat; plants annual or biennial, mostly 1.0–5.0 dm tall .... *Lepidium*

2b. Plants pubescent partly or entirely with compound [branched] hairs

24a. Petals deeply 2-lobed

25a. Stems leafy; leaves (1.0–)2.0–5.0 cm long; style evident, *ca.* 2.0–3.0 mm long .... *Berteroa*

25b. Stems scapose; leaves 1.0–2.0 cm long; style minute .... *Draba*

24b. Petals not lobed

26a. Ovary ovoid; stems generally unbranched above the base or freely branched in *Lobularia*

27a. Plants perennial, with a conspicuous rosette of leaves either at the base of the plants or on basal offshoots .... *Draba*

27b. Plants annual or perennial, without a conspicuous rosette of leaves

28a. Pubescence stellate; leaves 0.6–1.5 cm long; ovules 2 per locule .... *Alyssum*

28b. Pubescence 2-pronged; leaves 2.0–5.0 cm long; ovules 1 per locule .... *Lobularia*

26b. Ovary cylindric; stems often branched

29a. Plants without a conspicuous rosette of basal leaves; petals 2.0–2.5 cm long .... *Hesperis*

29b. Plants with a conspicuous rosette of leaves either at the base of the plants or on basal offshoots; petals 0.2–0.9 cm long

30a. Petals 2.0–3.0 mm long; sepals often pilose on the abaxial surface; plants annual .... ***Arabidopsis***

30b. Petals 3.0–9.0 mm long; sepals usually glabrous; plants biennial to short-lived perennial

    31a. Ovules in 2 rows in the ovary

        32a. Petals 3.0–6.0 mm long, yellow to ochroleucous .... ***Turritis***

        32b. Petals 5.0–9.0 mm long, usually white .... ***Arabis***

    31b. Ovules in 1 row in the ovary .... ***Arabis***

1b. At least the lower leaves lobed or divided

    33a. Plants glabrous or pubescent with simple [unbranched] hairs

        34a. Stem leaves auricled, sagittate, conspicuously clasping, or perfoliate

            35a. Petals 6.0–8.0 mm long; ovary cylindric; ovules numerous per locule ....
.... ***Arabis***

            35b. Petals 1.0–2.5 mm long; ovary flat; ovules 1 per locule .... ***Lepidium***

        34b. Stem leaves not surrounding the stem to any degree

            36a. Petals dimorphic, the outer pair much larger than the inner pair .... ***Iberis***

            36b. Petals monomorphic or nearly so

                37a. Petals 10.0–20.0 mm long

                    38a. Stems with 2 or 3 palmately lobed or palmately divided leaves ....
.... ***Cardamine***

                    38b. Stems with few to many pinnately lobed or pinnately divided leaves

                        39a. Lower leaves pinnately divided with petiolulate leaflets; plants growing in moist soils of meadows, lawns, and seepy spots ....
.... ***Cardamine***

                        39b. Lower leaves toothed to pinnatifid to pinnate, when pinnate, the leaflets sessile; weeds of disturbed soils

                            40a. Petals uniformly white; ovules in 2 rows in the locules ....
.... ***Diplotaxis***

                            40b. Petals usually with veins of a color different from the rest of the petal; ovules in a single row in the locules .... ***Raphanus***

                37b. Petals absent or up to 8.0 mm long

                    41a. Ovary ovate or elliptic to reniform

                        42a. Petals absent or up to 2.5 mm long; apex of the ovary retuse, the stigma borne in the notch; ovules 1 per locule; leaves 1.0–10.0 cm long

                            43a. Ovary smooth, flat, ovate to orbicular; plants erect to ascending .... ***Lepidium***

                            43b. Ovary rough-textured, plump, reniform; plants depressed to spreading .... ***Coronopus***

                      42b. Petals 6.0–8.0 mm long; apex of the ovary not retuse; ovules numerous per locule; lower leaf blades 10.0–30.0 cm long ....
.... ***Armoracia***

                  41b. Ovary linear or cylindric

                    44a. Petals pink to purple (white); plants fleshy, halophytic, of Atlantic coast shores .... ***Cakile***

                    44b. Petals white; plants non-halophytic, of inland areas

45a. Plants aquatic, submerged or floating (unless stranded), freely rooting from the lower nodes; nectary glands large and horseshoe-shaped; pedicels after anthesis divergent to recurving .... *Rorippa*

45b. Plants terrestrial, not rooting from the lower nodes, emergent when in water; nectary glands semicircular or reniform; pedicels after anthesis erect to ascending

    46a. Petals 1.5–4.0 mm long; leaves produced throughout the stem; plants native, of wet or dry soils .... *Cardamine*

    46b. Petals 6.0–12.0 mm long; leaves confined to the basal half of the stem; plants introduced, weeds of fields and disturbed soils .... *Diplotaxis*

33b. Plants pubescent partly or entirely with compound [branched] hairs .... *Capsella*

KEY TO THE GENERA FOR USE WITH FRUITING MATERIAL

1a. Fruit a silicle [*i.e.*, fruit up to 3.0 times as long as wide] .... **Group 1**

1b. Fruit a silique [*i.e.*, fruit greater than 3.0 times as long as wide] .... **Group 2**

Group 1

1a. Silicle indehiscent, containing 1–4 seeds

  2a. Plants pubescent with compound hairs; silicle firm, evidently reticulate-textured ....
.... *Neslia*

  2b. Plants glabrous or pubescent with simple hairs; silicle often inflated or corky, smooth, papillose, or ribbed

    3a. Silicle 2.5–3.0 mm long, subglobose; plants perennial; leaves auriculate-clasping
.... *Cardaria*

    3b. Silicle longer than 13.0 mm; plants annual

      4a. Silicle 1.3–2.5 cm long, usually with 1 or 2 seeds; plants glabrous, of Atlantic coast shores .... *Cakile*

      4b. Silicle longer than 3.0 cm, with 2 or 3 seeds; plants at least sparsely hispid, of fields and disturbed soils .... *Raphanus*

1b. Silicle dehiscent (with an apical, indehiscent beak in *Brassica* and *Sinapis*), not prominently corky or spongy

  5a. Silicle compressed at right angles to the septum, the septum therefore much narrower than the width of the fruit

    6a. Leaves subulate, not flat; plants submerged or partly submerged aquatics ....
.... *Subularia*

    6b. Leaves not subulate, definitely flat; plants terrestrial

      7a. Silicle with a wrinkled texture; racemes borne laterally .... *Coronopus*

      7b. Silicle with a smooth surface; racemes chiefly terminal

        8a. Silicle with 1 or 2 seeds

          9a. Silicle 2.0–6.0 mm long, with either a short style or a truncate apex or both .... *Lepidium*

          9b. Silicle 5.0–9.0 mm long, with a long style projecting from the prominently retuse apex .... *Iberis*

        8b. Silicle with 4 or more seeds

10a. Fruit oval to orbicular, wing-margined, rounded to obtuse at the
 base, conspicuously retuse at the apex; plants glabrous or pubescent
 with simple hairs .... ***Thlaspi***

10b. Fruit obtriangular to triangular-cordate, not wing-margined, acute at
 the base, convex to slightly concave at the apex; plants pubescent with
 compound hairs .... ***Capsella***

5b. Silicle either not compressed or compressed parallel to the septum, in either case the
septum as wide as the fruit

11a. Silicle thin and flat, wider than 6.0 mm [commonly about 2.5 cm wide], borne
on a gynophore longer than 10.0 mm; leaves cordate to cordate-deltate ....
.... ***Lunaria***

11b. Silicle narrower than 6.0 mm, either sessile at the end of the pedicel or on a very
short gynophore; leaves otherwise

12a. Silicle conspicuously compressed; pubescence, at least in part, compound
[branched]

13a. Flower-bearing stems arising from a basal rosette of leaves .... ***Draba***

13b. Flower-bearing stems lacking a conspicuous basal rosette

14a. Silicle with flat valves; seeds winged; plants 3.0–7.0(–10.0) dm tall
.... ***Berteroa***

14b. Silicle with slightly convex valves; seeds without wings; plants
1.0–3.0 dm tall

15a. Plants pubescent with stellate hairs; leaves 0.6–1.5 cm long;
seeds 2 per locule .... ***Alyssum***

15b. Plants pubescent with 2-pronged hairs; leaves 2.0–5.0 cm long;
seeds 1 per locule .... ***Lobularia***

12b. Silicle with definite, convex valves, scarcely, if at all, compressed;
pubescence, if present, simple [unbranched] or some of the hairs compound
in *Camelina*

16a. Plants pubescent with both simple and compound hairs; leaves clasping
at the base; each valve of the silicle with a prominent midnerve ....
.... ***Camelina***

16b. Plants glabrous or pubescent with simple hairs; leaves narrowed to the
base; each valve of the silicle without a midnerve or with an obscure
midnerve

17a. Leaves simple; seeds rarely maturing; silicle with a persistent style
tip 0.3 mm long .... ***Armoracia***

17b. Leaves pinnately lobed or divided; seeds usually maturing; silicle
with a persistent style tip 0.5–2.0 mm long .... ***Rorippa***

Group 2

1a. Silique indehiscent, separating at maturity into 1-seeded segments, prominently corky or
spongy, at least between the seeds

2a. Silique 1.3–2.5 cm long, usually with 1 or 2 seeds; plants glabrous, of Atlantic coast
shores .... ***Cakile***

2b. Silique 3.0–7.0 cm long, with 2–10 seeds; plants at least sparsely hispid, of fields
and disturbed soils .... ***Raphanus***

1b. Silique longitudinally dehiscent (with an apical, indehiscent beak in *Brassica* and *Sinapis*), the open valves revealing the seeds and septum, not prominently corky or spongy

  3a. Leaves simple, lacking pronounced lobes (except for basal lobes in cordate- or sagittate-shaped leaves)

    4a. Plants pubescent partly or entirely with compound [branched] hairs

      5a. Silique 3.0–10.0 cm long

        6a. Flower-bearing stems usually arising from a basal rosette of leaves; silique with 2 rows of seeds in each locule

          7a. Siliques subterete; seeds with a very narrow wing-margin up to 0.2 mm wide .... ***Turritis***

          7b. Siliques compressed; seeds with a definite wing-margin usually wider than 0.2 mm .... ***Arabis***

        6b. Flower-bearing stems lacking a conspicuous, basal rosette; silique with 1 row of seeds in each locule .... ***Hesperis***

      5b. Silique 0.4–3.0 cm long

        8a. Flower-bearing stems arising from a basal rosette of leaves; silique 0.4–2.0 cm long

          9a. Silique compressed, 0.4–1.2 cm long, up to 5.0 times as long as wide .... ***Draba***

          9b. Silique subterete, 1.0–2.0 cm long, more than 10.0 times as long as wide .... ***Arabidopsis***

        8b. Flower-bearing stems lacking a conspicuous, basal rosette; silique 1.2–3.0 cm long .... ***Erysimum***

    4b. Plants glabrous or pubescent with simple [unbranched] hairs

      10a. Leaves conspicuously clasping at the base

        11a. Plants annual; silique quadrangular, with seeds in 1 row in each locule.... .... ***Conringia***

        11b. Plants biennial or perennial; silique flat or subterete, with seeds in 1 or 2 rows in each locule

          12a. Siliques subterete; seeds with a very narrow wing-margin up to 0.2 mm wide .... ***Turritis***

          12b. Siliques compressed; seeds with a definite wing-margin usually greater than 0.2 mm wide .... ***Arabis***

      10b. Leaves not clasping at the base

        13a. Leaves deltate to suborbicular to reniform, rounded to cordate at the base

          14a. Plants of estuaries; silique 0.5–1.2 cm long .... ***Cardamine***

          14b. Weedy plants of woods, fields, and disturbed soils; silique 4.0–6.0 cm long .... ***Alliaria***

        13b. Leaves lance-linear to elliptic or oblanceolate to obovate, narrowed at the base

          15a. Silique flat

            16a. Replum bordered by a thin wing that is set perpendicular to the plane of the replum, the replum, including the wing, therefore I-beam-shaped in cross-section; siliques 0.5–3.0 cm long, dehiscing

elastically from the base, the valves becoming coiled ....
.... *Cardamine*

16b. Replum bordered, but the border not wing-like, of nearly equal thickness as the replum, the replum, including the wing, therefore appearing flat in cross-section; siliques 2.0–11.0 cm long, not dehiscing elastically, the valves not coiling

17a. Seeds in 1 row in each locule; siliques 7.0–11.0 cm long; lower pedicels 0.7–1.2 cm long .... *Arabis*

17b. Seeds in 2 rows in each locule; siliques 2.0–5.0 cm long; lower pedicels 1.0–3.0 cm long .... *Diplotaxis*

15b. Silique terete or nearly so to angled, but not flat

18a. Seeds in 1 row in each locule; each valve of a silique with 3–5 nerves .... *Sinapis*

18b. Seeds in 2 rows in each locule; each valve of a silique with 1 nerve .... *Diplotaxis*

3b. At least the lower leaves lobed or divided

19a. Principal leaves palmately lobed or palmately divided into 3–5 segments; plants with a stout, fleshy rhizome .... *Cardamine*

19b. Principal leaves pinnately lobed or pinnately divided; plants without a stout, fleshy rhizome

20a. Plants pubescent with compound hairs .... *Descurainia*

20b. Plants glabrous or pubescent with simple hairs

21a. Siliques flat

22a. Replum bordered by a thin wing that is set perpendicular to the plane of the replum, the replum, including the wing, therefore I-beam-shaped in cross-section; silique dehiscing elastically from the base, the valves becoming coiled .... *Cardamine*

22b. Replum bordered, but the border not wing-like, of nearly equal thickness as the replum, the replum, including the wing, therefore appearing flat in cross-section; silique not dehiscing elastically, the valves not coiling

23a. Seeds in 1 row in each locule; siliques 5.0–9.0 cm long ... *Arabis*

23b. Seeds in 2 rows in each locule; siliques 2.0–5.0 cm long ....
.... *Diplotaxis*

21b. Siliques terete or angled

24a. Stem leaves auriculate- or cordate-clasping

25a. Valves of silique nerveless or nearly so; siliques up to 1.5 cm long; seeds rarely maturing .... *Rorippa*

25b. Valves of silique with a prominent midnerve; siliques 1.0–7.0 cm long; seeds usually maturing

26a. Silique tipped with a terete, indehiscent beak 0.8–2.0 cm long; seeds subglobose .... *Brassica*

26b. Silique tipped with a short, thick, persistent style 0.2–0.3 cm long; seeds oblong to quadrate .... *Barbarea*

24b. Stem leaves not clasping

27a. Seeds in 2 rows in each locule

28a. Valves of the silique nerveless or nearly so; pedicels
0.3–1.5 cm long .... ***Rorippa***
28b. Valves of the silique with a prominent midnerve; pedicels
1.0–3.0 cm long .... ***Diplotaxis***
27b. Seeds in 1 row in each locule
29a. Plants native aquatics, submerged or floating (unless
stranded), freely rooting from the lower nodes .... ***Rorippa***
29b. Plants terrestrial weeds, not rooting from the lower nodes
30a. Seeds subglobose; silique tipped by an indehiscent beak
1.0–20.0 mm long
31a. Each valve of the silique with 1 prominent nerve;
indehiscent beak of fruit terete or angled, seedless ....
.... ***Brassica***
31b. Each valve of the silique with 3 prominent nerves;
indehiscent beak of fruit compressed or compressed-
quadrangular, often containing 1(–3) seeds ...... ***Sinapis***
30b. Seeds oblong or ovoid; silique tipped by the persistent
style up to 3.0 mm long
32a. Valves of silique with 3 nerves; pedicels lacking
bracts .... ***Sisymbrium***
32b. Valves of silique with 1 nerve; at least the lower
pedicels subtended by leafy bracts .... ***Erucastrum***

Alliaria
(1 species)

Alliaria petiolata (Bieb.) Cavara & Grande
synonym(s): *Alliaria officinalis* Andrz. *ex* Bieb.
common name(s): garlic mustard
habit: tall, erect, simple to little-branched, biennial herb
range: Europe; naturalized to Quebec, s. to VA, w. to KS, n. to British Columbia
state frequency: uncommon
habitat: fields, edges of thickets, roadsides
notes: —

Alyssum
(1 species)

Alyssum alyssoides (L.) L.
synonym(s): —
common name(s): —
habit: low, simple, annual herb
range: Europe; naturalized to Quebec and ME, s. to NJ, w. to CA, n. to British Columbia
state frequency: very rare
habitat: waste places, roadsides
notes: —

Arabidopsis
(1 species)

Arabidopsis thaliana (L.) Heynh.
synonym(s): *Arabis thaliana* L.
common name(s): mouse-ear cress
habit: slender, erect, simple or branched, annual herb
range: Europe; naturalized throughout much of the United States
state frequency: very rare
habitat: dry fields, roadsides, waste places
notes: —

Arabis
(4 species)

Most of Maine's members of this genus will be transferred to the genus *Boechera* based on chromosome number and DNA sequence data. *Arabis canadensis* L. is reported to occur in Maine by several sources, but this is in error, as the voucher is a collection of *Hesperis matronalis*.

KEY TO THE SPECIES

1a. Ovules in 2 rows in the ovary [therefore, seeds in 2 rows in the silique]; petals 5.0–9.0 mm long .... *A. drummondii*
1b. Ovules in 1 row in the ovary [therefore, seeds in 1 row in the silique]; petals 3.0–8.0 mm long

    2a. Basal leaves evidently pubescent, many of the hairs compound; pedicels erect in fruit; siliques 3.0–5.0 x 0.07–0.11 cm .... *A. hirsuta*
    2b. Basal leaves glabrous to sparsely pubescent, the compound hairs absent or sparse; pedicels ascending to spreading in fruit; siliques 5.0–10.0 x 0.12–0.2 cm
        3a. Petals 3.0–5.0 mm long, equaling to slightly exceeding the length of the sepals; stems glaucous, averaging 13 internodes to the first flower; basal leaves serrate to subentire; midvein of silique extending as much as half way from the base to the apex .... *A. laevigata*
        3b. Petals 6.0–8.0 mm long, about 2.0 times as long as the petals; stems green, averaging 25 internodes to the first flower; basal leaves serrate to pinnately lobed; midvein of silique extending 0.5–0.65 of the way from the base to the apex .... *A. missouriensis*

Arabis drummondii Gray
synonym(s): *Boechera drummondii* (Gray) A. & D. Löve
common name(s): Drummond's rock-cress
habit: erect, biennial herb
range: Labrador, s. to DE, w. to AZ, n. to British Columbia
state frequency: uncommon
habitat: ledges, gravels, rocky thickets
notes: This species has ascending pedicels in fruit.

Arabis hirsuta (L.) Scop. var. **pycnocarpa** (M. Hopkins) Rollins
synonym(s): —
common name(s): hairy rock-cress
habit: erect, stout, biennial herb

range: circumboreal; s. to GA, w. to AZ and CA
state frequency: uncommon
habitat: ledges, gravel, woods
notes: —

Arabis laevigata (Muhl. *ex* Willd.) Poir.
synonym(s): *Turritis laevigata* Muhl. *ex* Willd.
common name(s): smooth rock-cress
habit: simple or branched, biennial herb
range: ME, s. to GA, w. to OK and CO, n. to SD
state frequency: very rare
habitat: rich woods, slopes, ledges, talus
notes: This species is ranked S1 by the Maine Natural Areas Program.

Arabis missouriensis Greene
synonym(s): *Arabis viridis* Harger
common name(s): Missouri rock-cress
habit: biennial herb
range: ME, s. to NJ, w. to OK, n. to WI
state frequency: very rare
habitat: ledges, rocky woods
notes: This species is ranked S1 by the Maine Natural Areas Program.

Armoracia
(1 species)

Armoracia rusticana P. G. Gaertn., B. Mey., & Scherb.
synonym(s): *Armoracia armoracia* (L.) Britt., *Armoracia lapathifolia* Gilib.
common name(s): horseradish
habit: erect, coarse, perennial herb
range: Eurasia; escaped from cultivation to much of North America
state frequency: uncommon
habitat: moist fields and roadsides
notes: —

Barbarea
(3 species)

KEY TO THE SPECIES
1a. Basal leaves simple or with 1–4 pairs of lateral lobes; siliques 1.5–4.0 cm long; fruiting
    pedicels up to 1.0 mm thick
    2a. Silique with a persistent style beak 2.0–3.0 mm long; petals 6.0–8.0 mm long; upper
        leaves usually toothed or with a few, shallow sinuses .... ***B. vulgaris***
    2b. Silique with a persistent style beak 0.5–1.5(–2.0) mm long; petals 3.0–5.0 mm long;
        upper leaves usually coarsely toothed to pinnately lobed .... ***B. orthoceras***
1b. Basal leaves with 4–10 pairs of lateral lobes; siliques 4.5–7.0 cm long; fruiting pedicels
    1.2–1.8 mm thick .... ***B. verna***

Barbarea orthoceras Ledeb.
synonym(s): —
common name(s): American winter-cress, northern winter-cress

habit: simple or branched, biennial or short-lived perennial herb
range: circumboreal; s. to ME, w. to CA
state frequency: very rare
habitat: stream banks, swamps, wet woods, sandy or gravelly strands, rocky shores
notes: This species is ranked SH by the Maine Natural Areas Program.

Barbarea verna (P. Mill.) Aschers.

synonym(s): —
common name(s): early winter-cress
habit: erect, biennial herb
range: Eurasia; naturalized to Newfoundland, s. to FL, w. to CA, n. to WA
state frequency: very rare
habitat: fields, roadsides
notes: —

Barbarea vulgaris Ait. f.

synonym(s): *Barbarea barbarea* (L.) MacM., *Barbarea vulgaris* Ait. f. var. *arcuata* (Opiz *ex* J. & K. Presl) Fries, *Barbarea vulgaris* Ait. f. var. *brachycarpa* Rouy & Foucaud
common name(s): yellow rocket, common winter-cress
habit: erect, branched, biennial or perennial herb
range: Eurasia; naturalized to Newfoundland, s. to VA, w. to KS, n. to British Columbia
state frequency: common
habitat: damp fields, meadows, roadsides, woods, and brooksides
notes: —

Berteroa
(1 species)

Berteroa incana (L.) DC.

synonym(s): —
common name(s): hoary alyssum
habit: erect, branched, annual or perennial herb
range: Europe; naturalized to New Brunswick, s. to NJ, w. to KS, n. to British Columbia
state frequency: occasional
habitat: fields, waste places
notes: —

Brassica
(4 species)

KEY TO THE SPECIES
1a. Upper leaves with petioles, not clasping the stem
    2a. Silique terete or subterete, 1.5–4.0 cm long; fruiting pedicels ascending, 10.0–15.0 mm long; plants glabrous .... *B. juncea*
    2b. Silique quadrangular, 1.0–2.0 cm long; fruiting pedicels erect or appressed, 3.0–4.0 mm long; plants hirsute in the lower portion .... *B. nigra*
1b. Upper leaves sessile and clasping the stem
    3a. Plants glaucous, often sparsely hispid; petals 6.0–7.0 mm long, pale yellow; stem not thickened at the base .... *B. rapa*

3b. Plants green, glabrous; petals less than 6.0 mm long, deep yellow; stem with an
    evident, thickened base .... ***B. napus***

Brassica juncea (L.) Czern.
synonym(s): *Brassica juncea* (L.) Czern. var. *crispifolia* Bailey
common name(s): Chinese mustard, leaf mustard
habit: annual herb
range: Eurasia; naturalized to much of North America
state frequency: uncommon
habitat: cultivated ground
notes: —

Brassica napus L.
synonym(s): —
common name(s): turnip
habit: annual herb
range: Eurasia; escaped from cultivation locally
state frequency: uncommon
habitat: cultivated ground, compost areas
notes: —

Brassica nigra (L.) W. D. J. Koch
synonym(s): —
common name(s): black mustard
habit: annual herb
range: Eurasia; naturalized to much of North America
state frequency: occasional
habitat: cultivated ground, fields, waste places
notes: —

Brassica rapa L.
synonym(s): —
common name(s): bird's rape
habit: succulent, annual herb
range: Eurasia; naturalized to much of North America
state frequency: occasional
habitat: cultivated ground
notes: —

Cakile
(1 species)

Cakile edentula (Bigelow) Hook.
synonym(s): —
common name(s): sea-kale, sea rocket
habit: simple or branched, spreading to ascending, fleshy, annual herb
range: Labrador, s. to NC; also near/at margins of the Great Lakes
state frequency: occasional
habitat: sandy or gravelly beaches
notes: —

Camelina
(2 species)

KEY TO THE SPECIES

1a. Plants rough-pubescent; simple hairs [1.0–2.0 mm long] exceeding the length of the compound hairs; silicles 2.5–5.0 x up to 5.0 mm; fruiting pedicels 4.0–9.0 mm long; seeds up to 1.0 mm long .... *C. microcarpa*

1b. Plants sparsely pubescent to glabrate; simple hairs smaller, not exceeding the length of the compound hairs; silicles 7.0–9.0 x 6.0–7.0 mm; fruiting pedicels 10.0–21.0 mm long; seeds 1.0–1.5 mm long .... *C. sativa*

Camelina microcarpa DC.
synonym(s): —
common name(s): small-seed false flax, small-fruited false flax
habit: erect, branched, annual herb
range: Eurasia; naturalized to Newfoundland, s. to VA, w. to TX, n. to British Columbia
state frequency: uncommon
habitat: fields, roadsides, waste places
notes: —

Camelina sativa (L.) Crantz
synonym(s): —
common name(s): gold of pleasure, large-seed false flax, Dutch-flax
habit: erect, branched, annual herb
range: Eurasia; naturalized to Quebec, s. to SC, w. to CA, n. to British Columbia
state frequency: uncommon
habitat: cultivated ground, roadsides, waste places
notes: —

Capsella
(1 species)

Capsella bursa-pastoris (L.) Medik.
synonym(s): *Bursa bursa-pastoris* (L.) Britt.
common name(s): shepherd's purse
habit: simple or branched, annual herb
range: Europe; naturalized throughout North America
state frequency: common
habitat: cultivated ground, roadsides
notes: —

Cardamine
(8 species)

KEY TO THE SPECIES

1a. Principal leaves palmately lobed or palmately divided into 3–5 segments; plants with a stout, fleshy rhizome; silique with a persistent style beak commonly 6.0–8.0 mm long

2a. Rhizome continuous or with irregular constrictions, not readily separable; stem glabrous in the apical portion [above the leaves]; segments of leaves lance-ovate to ovate or rhombic-ovate, ciliate with appressed hairs *ca.* 0.1 mm long; stems usually with 2 leaves .... *C. diphylla*

2b. Rhizome regularly constricted and readily separable at the constrictions; stem pubescent in the apical portion; segments of leaves linear to oblanceolate, ciliate with spreading hairs 0.1–0.3 mm long; stems usually with 3 leaves .... *C. concatenata*

1b. Principal leaves entire to pinnately lobed or pinnately divided; plants without a stout, fleshy rhizome; silique with a persistent style beak 0.5–2.0 mm long

3a. Flowers without petals; siliques 0.5–1.2 cm long; pedicels mostly 1.0–3.0 mm long; plants of brackish shores .... *C. longii*

3b. Flowers with petals (sometimes absent in *C. impatiens*); siliques 1.5–4.0 cm long; pedicels 2.0–15.0 mm long; plants not of brackish shores

4a. Leaves simple, 0.5–1.0 cm long; axis of raceme very short, scarcely elongating in fruit; alpine plants of elevation exceeding 1000 m .... *C. bellidifolia*

4b. Leaves usually with at least 2 lateral lobes, 2.0–8.0 cm long; axis of raceme longer, elongating in fruit; plants of dry or wet soils of lower elevation

5a. Petals 8.0–15.0 mm long; lower leaves with 7–17 pairs of leaflets; plants perennial .... *C. pratensis*

5b. Petals 1.5–4.0 mm long; lower leaves with 1–9 pairs of leaflets or lobes; plants annual or biennial

6a. Stem leaves sagittate-auriculate; leaflets of the stem leaves sharply toothed to lacerate-margined, commonly acuminate at the apex ....
.... *C. impatiens*

6b. Stem leaves not auriculate at the base; leaflets of the stem leaves entire to toothed, rounded to acute at the apex

7a. Stems glabrous; stem leaves mostly 2.0–4.0 cm long, with linear to oblanceolate, non-decurrent lateral leaflets; style up to 1.0 mm long; seeds 0.7–0.9 mm long .... *C. parviflora*

7b. Stems usually sparsely hispid near the base; stem leaves mostly 4.0–8.0 cm long, with elliptic to suborbicular, decurrent lateral leaflets; style 0.5–2.0 mm long; seeds 1.0–1.5 mm long ....
.... *C. pensylvanica*

Cardamine bellidifolia L.

synonym(s): —
common name(s): alpine bitter-cress, northern bitter-cress
habit: dwarf, tufted, perennial herb
range: circumboreal; s. to mountains of ME and NH, w. to OR
state frequency: very rare
habitat: alpine brooksides and ravines
notes: This species is ranked S1 by the Maine Natural Areas Program.

Cardamine concatenata (Michx.) Sw.

synonym(s): *Dentaria laciniata* Muhl. *ex* Willd.
common name(s): cut-leaved toothwort
habit: rhizomatous, perennial herb
range: New Brunswick, s. to FL, w. to OK, n. to MN and Ontario
state frequency: very rare
habitat: rich woods

notes: This species is ranked S1 by the Maine Natural Areas Program. Hybrids with *Cardamine diphylla*, called *C. ×maxima* (Nutt.) Wood are ranked SH and can be recognized by poorly developed fruits, stems with 2 or 3 leaves, rhizomes with irregular constrictions and enlargements, a middle segment of leaves more or less lance-ovate, and leaf margins ciliate with spreading hairs 0.2–0.3 mm long.

Cardamine diphylla (Michx.) Wood
synonym(s): *Dentaria diphylla* Michx.
common name(s): broad-leaved toothwort
habit: long-rhizomatous, perennial herb
range: Quebec, s. to GA, w. to AL, n. to MN and Ontario
state frequency: uncommon
habitat: rich woods
notes: Hybridizes with *Cardamine concatenata* (*q.v.*).

Cardamine impatiens L.
synonym(s): —
common name(s): —
habit: annual or biennial, weedy herb
range: Europe; naturalized to ME, s. to PA, w. and n. to MI
state frequency: very rare
habitat: shaded, grassy areas
notes: —

Cardamine longii Fern.
synonym(s): —
common name(s): Long's bitter-cress
habit: prostrate to erect, fibrous-rooted, biennial or short-lived perennial herb
range: ME, s. to VA
state frequency: rare
habitat: brackish shores
notes: This species is ranked S2 by the Maine Natural Areas Program.

Cardamine parviflora L. var. *arenicola* (Britt.) O. E. Schulz
synonym(s): —
common name(s): dry land bitter-cress
habit: solitary, erect, simple, annual or biennial herb
range: Newfoundland, s. to FL, w. to TX, n. to MN and Ontario
state frequency: uncommon
habitat: dry ledges and woods
notes: —

Cardamine pensylvanica Muhl. *ex* Willd.
synonym(s): *Cardamine pensylvanica* Muhl. *ex* Willd. var. *brittoniana* Farw.
common name(s): Pennsylvania bitter-cress
habit: erect to trailing, biennial or short-lived perennial herb
range: Newfoundland, s. to FL, w. to CA, n. to AK
state frequency: common
habitat: swamps, wet woods and clearings, stream banks
notes: Submerged stems of this species often lack the hispid pubescence, making it more difficult to distinguish them from those of *Cardamine parviflora*. However, *C. parviflora* would not be found in wet habitats.

Cardamine pratensis L.
synonym(s): —
common name(s): cuckoo flower
habit: erect, simple or branched herb
range: Eurasia; naturalized to Newfoundland, s. to NJ
state frequency: uncommon
habitat: moist meadows, lawns, and roadsides
notes: —

Cardaria
(1 species)

Cardaria draba (L.) Desv.
synonym(s): *Lepidium draba* L.
common name(s): hoary-cress, heart-podded hoary-cress
habit: stout, erect to decumbent, rhizomatous, perennial herb
range: Europe; naturalized in the east to Nova Scotia and ME, s. to DC
state frequency: very rare
habitat: fields, roadsides, waste places
notes: —

Conringia
(1 species)

Conringia orientalis (L.) Andrz.
synonym(s): —
common name(s): hare's-ear mustard
habit: simple or branched, erect, annual herb
range: Eurasia; naturalized throughout much of North America
state frequency: uncommon
habitat: new fields, waste places
notes: —

Coronopus
(1 species)

Coronopus didymus (L.) Sm.
synonym(s): *Carara didyma* (L.) Britt.
common name(s): swine-cress, lesser swine-cress
habit: spreading to ascending, mat-forming, branched, annual or biennial herb
range: South America; naturalized to Newfoundland, s. to FL, w. to TX, n. to British Columbia
state frequency: very rare
habitat: cultivated areas, roadsides, waste places
notes: —

Descurainia
(2 species)

KEY TO THE SPECIES

1a. Fruiting pedicels ascending to spreading, 8.0–14.0 mm long; siliques 1.5–2.5 cm long; seeds 12–25 in each locule; leaves usually 2- or 3-times pinnately divided ...... **D. sophia**

1b. Fruiting pedicels erect to ascending, 3.0–6.0 mm long; siliques 0.5–1.2 cm long; seeds 4–10 in each locule; leaves usually once-pinnately divided, the leaflets entire to lobed .....
.... **D. incana**

Descurainia incana (Bernh. *ex* Fisch. & C. A. Mey.) Dorn
synonym(s): *Descurainia richardsonii* O. E. Schulz
common name(s): Richardson's tansy-mustard, gray tansy-mustard
habit: erect, annual or biennial herb
range: Quebec, s. to CT, w. to NM, n. to AK
state frequency: very rare
habitat: roadsides, river banks, disturbed gravel
notes: This species is ranked SH by the Maine Natural Areas Program.

Descurainia sophia (L.) Webb *ex* Prantl
synonym(s): *Sophia sophia* (L.) Britt.
common name(s): herb Sophia
habit: slender, branched, annual herb
range: Eurasia; naturalized to Quebec, s. to DE, w. to UT and CA, n. to WA
state frequency: rare
habitat: fields, roadsides, waste places
notes: —

Diplotaxis
(2 species)

KEY TO THE SPECIES

1a. Plants perennial; leaves borne throughout the basal half of the stem; silique borne on a stipe 1.0–2.0 mm long; lower pedicels mostly 2.0–3.0 cm long; sepals 5.0–8.0 mm long
.... **D. tenuifolia**

1b. Plants usually annual or biennial; leaves borne chiefly at the base of the plant; silique not stipitate; lower pedicels mostly 1.0–1.5 cm long; sepals 3.0–5.0 mm long ..... **D. muralis**

Diplotaxis muralis (L.) DC.
synonym(s): —
common name(s): sand-rocket, stinking wall-rocket
habit: erect to decumbent, branched near the base, annual or perennial herb
range: Europe; naturalized to New Brunswick, s. to NJ, w. to IL, n. to British Columbia
state frequency: very rare
habitat: sandy roadsides and waste places
notes: —

Diplotaxis tenuifolia (L.) DC.
synonym(s): —
common name(s): slimleaf wall-rocket
habit: erect, suffruticose at the base, perennial herb

range: Europe; naturalized to New Brunswick, s. to VA, w. and n. to MI and Ontario
state frequency: very rare
habitat: roadsides, waste places
notes: —

Draba
(3 species)

Draba allenii Fern. is reported for Maine by Fernald (1950), but voucher specimens are unknown.

KEY TO THE SPECIES
1a. Plants annual or winter-annual; petals 2.0–3.0 mm long, lobed nearly to the middle; stems scapose; fruits essentially planar .... ***D. verna***
1b. Plants perennial; petals 3.0–6.0 mm long, the apex entire or retuse; stems leafy; fruits twisted
    2a. Fruit stellate-pubescent, 1.5–2.5 mm wide; pedicels ascending, at least the lower pedicels commonly with foliaceous bracts .... ***D. cana***
    2b. Fruit glabrous, 2.8–3.8 mm wide; pedicels often spreading, commonly without bracts .... ***D. arabisans***

Draba arabisans Michx.
synonym(s): —
common name(s): rock whitlow-cress
habit: simple or branched, perennial herb
range: Newfoundland, s. to ME, w. to MN, n. to Ontario
state frequency: very rare
habitat: calcareous ledges and rocky areas
notes: This species is ranked S1 by the Maine Natural Areas Program.

Draba cana Rydb.
synonym(s): *Draba lanceolata* Royle *sensu* Fernald (1950), *Draba stylaris* J. Gay *ex* W. D. J. Koch
common name(s): lance-leaved draba, ashy whitlow-grass
habit: loosely mat-forming, perennial herb
range: Quebec, s. to ME, w. to NV, n. to AK
state frequency: very rare
habitat: calcareous ledges and slopes
notes: This species is ranked S1 by the Maine Natural Areas Program.

Draba verna L.
synonym(s): *Draba verna* var. *boerhaavii* van Hall, *Erophila verna* (L.) Bess.
common name(s): spring whitlow-grass
habit: slender, annual or winter-annual herb
range: Eurasia; naturalized to ME, s. to GA, w. to IL, n. to Ontario
state frequency: very rare
habitat: fields, roadsides, open, dry areas
notes: —

Erucastrum
(1 species)

Erucastrum gallicum (Willd.) O. E. Schulz
synonym(s): —
common name(s): dog-mustard
habit: annual or biennial herb
range: Eurasia; naturalized to Newfoundland, s. to PA, w. to MO, n. to British Columbia
state frequency: rare
habitat: fields, roadsides, waste places
notes: —

Erysimum
(2 species)

KEY TO THE SPECIES
1a. Anthers 0.4–0.6 mm long; sepals 2.0–3.5 mm long; petals bright yellow, 3.5–5.5 mm
   long; siliques glabrous to sparsely pubescent .... ***E. cheiranthoides***
1b. Anthers 1.5–2.5 mm long; sepals 5.0–7.0 mm long; petals pale yellow, 6.0–10.0 mm
   long; siliques pubescent .... ***E. inconspicuum***

Erysimum cheiranthoides L.
synonym(s): *Cheirinia cheiranthoides* (L.) Link
common name(s): wormseed-mustard
habit: simple or branched, annual herb
range: circumboreal; s. to NC, w. to OR
state frequency: uncommon
habitat: cultivated fields, meadows, roadsides, waste places
notes: —

Erysimum inconspicuum (S. Wats.) MacM.
synonym(s): *Cheirinia inconspicua* (S. Wats.) Britt.
common name(s): small-flowered treacle-mustard
habit: erect, usually simple, perennial herb
range: Ontario and MI, s. to MO, w. to NV, n. to British Columbia; naturalized to Newfoundland, s. to
   MA
state frequency: very rare
habitat: dry, open soil
notes: —

Hesperis
(1 species)

Hesperis matronalis L.
synonym(s): —
common name(s): dame's rocket
habit: erect, often branched, biennial or perennial herb
range: Eurasia; escaped from cultivation to Newfoundland, s. to GA, w. to KS, n. to MI and Ontario
state frequency: uncommon
habitat: roadsides, thickets, open woods, waste places
notes: —

Iberis
(1 species)

Iberis amara L.
synonym(s): —
common name(s): candytuft
habit: branched, annual herb
range: Europe; escaped from cultivation to ME, MA, and CT
state frequency: very rare
habitat: waste places, cultivated ground
notes: —

Lepidium
(6 species)

KEY TO THE SPECIES

1a. Stem leaves either auriculate- or cordate-clasping
    2a. Stem leaves cordate-clasping, the basal lobes rounded; lower leaves pinnately dissected; petals yellow, 1.0–1.5 mm long; plants sparsely pubescent; silicle about 3.5–4.0 mm long .... *L. perfoliatum*
    2b. Stem leaves auriculate-clasping, the basal lobes acute; lower leaves entire or toothed, not lobed; petals white, 2.0–2.5 mm long; plants densely short-pubescent; silicle 5.0–6.0 mm long .... *L. campestre*
1b. Stem leaves narrowed to the base, not surrounding the stem
    3a. Flowers with 6 stamens; silicles 5.0–7.0 mm long, prominently winged; fruiting pedicels thick, ascending to erect .... *L. sativum*
    3b. Flowers with 2(4) stamens; silicles 2.0–4.0 mm long, without a wing, or winged only in the apical portion; fruiting pedicels slender, spreading to ascending
        4a. Petals typically present, 1.0–2.0 times as long as the sepals; cotyledons accumbent .... *L. virginicum*
        4b. Petals obsolete or rudimentary and shorter than the sepals; cotyledons incumbent
            5a. Racemes densely flowered, with 9–18 fruits per cm; plants inodorous; silicle obovate to obcordate .... *L. densiflorum*
            5b. Racemes relatively sparsely flowered, with 6–10 fruits per cm; plants foul-scented; silicle ovate to elliptic .... *L. ruderale*

Lepidium campestre (L.) Ait. f.
synonym(s): —
common name(s): field-cress, cow-cress
habit: simple or branched, annual or biennial herb
range: Europe; naturalized to Newfoundland, s. to NC, w. and n. to British Columbia
state frequency: uncommon
habitat: fields, roadsides, waste places
notes: —

Lepidium densiflorum Schrad.
synonym(s): —
common name(s): prairie pepperweed
habit: branched, annual or biennial herb

range: Quebec, w. to OR; adventive in the northeast
state frequency: occasional
habitat: roadsides, waste places
notes: —

Lepidium perfoliatum L.

synonym(s): —
common name(s): clasping pepperweed
habit: erect, usually branched, annual herb
range: Europe; naturalized to ME, s. to PA, w. to KS, n. to MI
state frequency: very rare
habitat: fields, roadsides, waste places
notes: —

Lepidium ruderale L.

synonym(s): —
common name(s): stinking pepperweed
habit: malodorous, annual or biennial herb
range: Europe; naturalized to Newfoundland, s. to DE, w. to OH, n. to Saskatchewan
state frequency: rare
habitat: fields, roadsides, waste places
notes: —

Lepidium sativum L.

synonym(s): —
common name(s): garden-cress
habit: annual herb
range: Europe; escaped from cultivation occasionally in the northeast
state frequency: rare
habitat: roadsides, waste places
notes: —

Lepidium virginicum L.

synonym(s): —
common name(s): poor-man's pepper
habit: erect, annual or biennial herb
range: Newfoundland, s. to FL, w. to the Pacific states
state frequency: occasional
habitat: dry fields, gardens, roadsides, waste places
notes: Later flowers sometimes lack petals, confusing the identification of this species with *Lepidium densiflorum* and *L. ruderale*.

Lobularia
(1 species)

Lobularia maritima (L.) Desv.

synonym(s): *Koniga maritima* (L.) R. Br.
common name(s): sweet-alyssum
habit: branched, annual or perennial herb
range: Europe; escaped from cultivation in the eastern United States
state frequency: uncommon
habitat: vacant lots, waste places
notes: A commonly cultivated species with white to blue-purple petals.

Lunaria
(1 species)

Lunaria annua L.
synonym(s): —
common name(s): honesty
habit: annual herb
range: Europe; escaped from cultivation in the northeast
state frequency: very rare
habitat: roadsides, waste places
notes: —

Neslia
(1 species)

Neslia paniculata (L.) Desv.
synonym(s): —
common name(s): ball-mustard
habit: slender, erect, simple or branched, annual or biennial herb
range: Eurasia; naturalized to Newfoundland, s. to NJ, w. to IL, n. to SD and British Columbia
state frequency: very rare
habitat: cultivated fields, railroads, waste places
notes: —

Raphanus
(2 species)

KEY TO THE SPECIES

1a. Petals commonly pale yellow, turning white, with purple veins; silique 4.0–6.0 mm wide, with 4–10 seeds, constricted between the seeds and dividing at maturity into 1-seeded segments; plants with a stout taproot .... *R. raphanistrum*
1b. Petals commonly pale purple; silique 6.0–10.0 mm wide, with 2 or 3 seeds, continuous, not dividing into 1-seeded segments; plants with a conspicuously thickened taproot ....
.... *R. sativus*

Raphanus raphanistrum L.
synonym(s): —
common name(s): wild radish, jointed charlock
habit: coarse, annual herb
range: Eurasia; naturalized to Newfoundland, s. to VA, w. to KS, n. to Manitoba
state frequency: uncommon
habitat: cultivated fields, waste places
notes: Very variable in flower color, the petals can be white, yellow, or purple, and the veins can be purple, yellow, or green. Rarely the veins are of similar color as the petals.

Raphanus sativus L.
synonym(s): —
common name(s): radish
habit: annual or biennial herb
range: Europe; escaped from cultivation occasionally
state frequency: uncommon

habitat: fields, waste places
notes: —

Rorippa
(5 species)

KEY TO THE SPECIES

1a. Plants aquatic, submerged or floating (unless stranded), freely rooting from the lower
nodes; petals white
    2a. Seeds in 1 row in each locule, with about 100–150(–175) polygonal depressions on
each surface; silique 1.0–1.5 mm wide, with a slender, persistent style beak
0.5–2.0 mm long .... *R. microphylla*
    2b. Seeds in 2 rows in each locule, with about 25–50(–60) polygonal depressions on
each surface; silique 1.5–2.5 mm wide, with a short, thick, persistent style beak
0.0–1.0 mm long .... *R. nasturtium-aquaticum*
1b. Plants terrestrial weeds (although sometimes growing in wet environments), not freely
rooting from the lower nodes except sometimes in *R. amphibia*; petals yellow
    3a. Petals 1.5–2.5 mm long, about as long as the sepals; anthers mostly 0.2–0.3 mm
long; plants annual or biennial, without creeping stems or rhizomes .... *R. palustris*
    3b. Petals 4.0–5.0 mm long, about twice as long as the sepals; anthers mostly
0.7–1.3 mm long; plants perennial, with creeping rhizomes and/or stems
        4a. Siliques 3.0–6.0 mm long not including the persistent style beak 1.0–2.0 mm
long, shorter than their pedicels; leaves (except the divided, submerged leaves)
simple and toothed; stems hollow .... *R. amphibia*
        4b. Siliques 10.0–15.0 mm long not including the persistent style beak 0.5–1.0 mm
long, longer than their pedicels; leaves with deep sinuses or divided; stems solid
.... *R. sylvestris*

Rorippa amphibia (L.) Bess.
synonym(s): *Nasturtium amphibium* (L.) Ait. f.
common name(s): amphibious-cress
habit: erect to ascending, emergent or terrestrial, perennial herb
range: Eurasia; naturalized to ME, s. to MA, w. to NY, n. to Quebec
state frequency: very rare
habitat: swamps, shallow water, shores, roadsides
notes: Thought to hybridize with *Rorippa sylvestris*.

Rorippa microphylla (Boenn. *ex* Reichenb.) Hyl. *ex* A. & D. Löve
synonym(s): *Nasturtium microphyllum* Boenn. *ex* Reichenb., *Nasturtium officinale* Ait. f. var.
microphyllum (Boenn. *ex* Reichenb.) Thellung
common name(s): water-cress
habit: submerged to floating, perennial herb
range: Eurasia; naturalized to Newfoundland, s. to FL, w. and n. to MI and Ontario
state frequency: rare
habitat: shallow water, mud
notes: —

Rorippa nasturtium-aquaticum (L.) Hayek
synonym(s): *Nasturtium officinale* Ait. f., *Sisymbrium nasturtium-aquaticum* L.
common name(s): water-cress, true water-cress

habit: submerged to floating, perennial herb
range: Eurasia; naturalized throughout much of North America
state frequency: rare
habitat: shallow water, mud
notes: —

Rorippa palustris (L.) Bess.

synonym(s): <*R. p.* ssp. *fernaldiana*> = *Rorippa islandica* (Oeder) Borbás var. *fernaldiana* Butters &
    Abbe; <*R. p.* ssp. *hispida*> = *Radicula hispida* (Desv.) Britt., *Rorippa islandica* (Oeder) Borbás
    var. *hispida* (Desv.) Butters & Abbe
common name(s): common yellow-cress
habit: erect, simple or branched, annual or biennial herb
range: circumboreal; s. to FL, w. to CA
state frequency: uncommon
habitat: wet shores and waste places
key to the subspecies:
    1a. Abaxial surface of leaves usually hirsute; stem normally hirsute; fruits broad-lanceolate to
       ovoid, 2.0–5.0 x 1.7–4.0 mm .... *R. p.* ssp. *hispida* (Desv.) Rydb.
    1b. Abaxial surface of leaves glabrous; stem glabrous or sparsely hirsute near the base; fruits
       narrow-ellipsoid to subcylindric, 3.0–9.0 x 1.0–2.5 mm ....
       .... *R. p.* ssp. *fernaldiana* (Butters & Abbe) Jonsell
notes: —

Rorippa sylvestris (L.) Bess.

synonym(s): *Radicula sylvestris* (L.) Druce
common name(s): creeping yellow-cress
habit: erect to spreading, simple or branched, perennial herb
range: Eurasia; naturalized to Newfoundland, s. to VA, w. to MO, n. to British Columbia
state frequency: uncommon
habitat: wet shores, meadows, and roadsides
notes: Thought to hybridize with *Rorippa amphibia*.

Sinapis
(2 species)

KEY TO THE SPECIES

1a. Silique glabrous or sparsely bristly, *ca.* 2.0 mm wide, 7- to 13-seeded, with an
    indehiscent, compressed-quadrangular beak 0.35–1.0 times as long as the body; fruiting
    pedicels ascending, 3.0–7.0 mm long; leaves, especially the middle and upper, often
    merely toothed .... ***S. arvensis***
1b. Silique densely bristly, *ca.* 4.0 mm wide, 4- to 8-seeded, with an indehiscent, flat beak
    1.0–2.0 times as long as the body; fruiting pedicels divergent, 6.0–15.0 mm long; most
    of the leaves prominently pinnately lobed .... ***S. alba***

Sinapis alba L.

synonym(s): *Brassica hirta* Moench
common name(s): white-mustard
habit: annual herb
range: Eurasia; naturalized to New Brunswick, s. to DE, w. to IA, n. to British Columbia
state frequency: very rare
habitat: fields, roadsides, waste places
notes: —

Sinapis arvensis L.
synonym(s): *Brassica kaber* (DC.) L. C. Wheeler var. *pinnatifida* (Stokes) L. C. Wheeler
common name(s): charlock
habit: annual herb
range: Eurasia; naturalized throughout much of the United States
state frequency: uncommon
habitat: fields, cultivated ground, shores, waste places
notes: —

Sisymbrium
(2 species)

Members of this genus, as with many Brassicaceae, can lose their pigmentation in the petals with drying and age following collection. Noting on labels fresh petal color can avoid later confusion.

KEY TO THE SPECIES
1a. Pedicels erect to appressed, 2.0–3.0 mm long; silique subulate, 0.8–1.5 cm long; petals 3.0–4.0 mm long; lateral leaf segments oblong to ovate .... *S. officinale*
1b. Pedicels divergent 5.0–10.0 mm long; silique linear-cylindric, 5.0–10.0 cm long; petals 6.0–8.0 mm long; lateral leaf segments linear to lanceolate .... *S. altissimum*

Sisymbrium altissimum L.
synonym(s): *Norta altissima* (L.) Britt.
common name(s): tumble-mustard, tumbling-mustard
habit: tall, erect, branched, annual herb
range: Eurasia; naturalized throughout most of the United States
state frequency: uncommon
habitat: fields, roadsides, waste places
notes: —

Sisymbrium officinale (L.) Scop.
synonym(s): *Erysimum officinale* L., *Sisymbrium officinale* (L.) Scop. var. *leiocarpum* DC.
common name(s): hedge-mustard
habit: branched, annual herb
range: Eurasia; naturalized throughout most of the United States
state frequency: uncommon
habitat: waste places
notes: —

Subularia
(1 species)

Often found in its vegetative state, Maine's sole member of this genus can be identified by its non-flaccid, subulate leaves that are slightly flattened in the distal portion and arranged in a basal tuft, lack of rhizomes, and presence of non-septate roots.

Subularia aquatica L.
synonym(s): —
common name(s): water awlwort
habit: submerged or emergent, annual herb
range: circumboreal; s. to ME, w. to WY and CA

state frequency: rare
habitat: sandy or gravelly margins of lakes, ponds, and slow streams
notes: This species is ranked S2 by the Maine Natural Areas Program.

Thlaspi
(1 species)

Thlaspi arvense L.
synonym(s): —
common name(s): field penny-cress
habit: erect, annual herb
range: Europe; naturalized throughout most of North America
state frequency: uncommon
habitat: fields, roadsides, waste places
notes: —

Turritis
(1 species)

Turritis glabra L.
synonym(s): *Arabis glabra* (L.) Bernh.
common name(s): tower-mustard
habit: erect, stout, biennial or short-lived perennial herb
range: circumboreal; s. to NC, w. to NM and CA
state frequency: uncommon
habitat: ledges, thickets, fields, roadsides
notes: Petals are usually yellow, 3.0–6.0 mm long.

ANACARDIACEAE
(2 genera, 6 species)

KEY TO THE GENERA
1a. Drupe red, conspicuously pubescent; inflorescence terminal or borne laterally on the previous season's twigs, the flowers and fruits crowded; leaves with 7–31 leaflets; petals often pubescent on the adaxial surface .... **Rhus**
1b. Drupe white to yellow, glabrous or sparsely pubescent; inflorescence axillary, relatively more open; leaves with 3–13 leaflets; petals glabrous .... **Toxicodendron**

Rhus
(3 species)

KEY TO THE SPECIES
1a. Leaflets entire or with a few teeth near the apex; rachis of leaf winged in distal portion; bud less than half encircled by the leaf scar .... **R. copallinum**
1b. Leaflets serrate; rachis of leaf not winged; bud nearly encircled by the leaf scar
2a. Twigs and petioles densely hirsute; drupe pubescent with hairs 1.0–2.0 mm long ....
.... **R. hirta**

2b. Twigs and petioles glabrous, the twigs sometimes glaucous; drupe pubescent with hairs *ca.* 0.2 mm long .... ***R. glabra***

Rhus copallinum L. var. *latifolia* Engl.
synonym(s): —
common name(s): shining sumac, dwarf sumac
habit: shrub to small tree
range: ME, s. to FL, w. to TX, n. to WI
state frequency: uncommon; found in southern Maine
habitat: dry woods and open areas
notes: —

Rhus glabra L.
synonym(s): —
common name(s): smooth sumac
habit: shrub
range: ME, s. to FL, w. to TX and Mexico, n. to British Columbia
state frequency: uncommon
habitat: borders, fields, roadsides
notes: Hybrids with *Rhus hirta*, called *R. ×pulvinata* Greene, occur in Maine.

Rhus hirta (L.) Sudworth
synonym(s): *Rhus typhina* L.
common name(s): staghorn sumac
habit: tall shrub to small tree
range: Quebec, s. to GA, w. to AL, n. to MN and Ontario
state frequency: common
habitat: dry borders, open areas, roadsides
notes: Hybrids with *Rhus glabra* (*q.v.*) occur in Maine.

Toxicodendron
(3 species)

This genus is composed of allergenic shrubs and woody vines.

KEY TO THE SPECIES
1a. Leaves with 7–13 leaflets; inflorescence up to 20.0 cm long; buds sessile, covered by scales; plants erect shrubs 2.0–7.0 mm tall .... ***T. vernix***
1b. Leaves with 3 entire to toothed leaflets; inflorescence up to 10.0 cm long; buds usually stalked, naked; plants erect shrubs to mostly 1.0 m tall or woody vines
    2a. Stems erect, without aerial roots; leaves tending to be clustered near the tip of the stem, often slightly folded; inflorescence simple or few-branched, usually with fewer than 25 flowers; drupes glabrous, sessile or subsessile .... ***T. rydbergii***
    2b. Stems climbing or straggling by aerial roots; leaves alternately scattered throughout the stem, usually flat; inflorescence normally branched, usually with more than 25 flowers; drupes sparsely pubescent, scabrous, or papillose, pedicellate ... ***T. radicans***

Toxicodendron radicans (L.) Kuntze
synonym(s): *Rhus radicans* L.
common name(s): poison-ivy, common poison-ivy
habit: trailing to climbing shrub

range: New Brunswick and Nova Scotia, s. to FL, w. to TX, n. to MN and Ontario
state frequency: occasional
habitat: open woods, rocky areas
notes: Maine has 2 species of poison ivy—*Toxicodendron radicans* and *T. rydbergii*. These species are superficially similar and both are often referred to as *T. radicans*. *Toxicodendron radicans* has a more southern distribution and is relatively more common south of the 45th parallel. *Toxicodendron rydbergii*, on the other hand, is the more common poison ivy north of the 45th parallel.

Toxicodendron rydbergii (Small *ex* Rydb.) Greene
synonym(s): *Rhus radicans* (L.) Kuntze var. *rydbergii* (Small *ex* Rydb.) Rehd., *Toxicodendron radicans* (L.) Kuntze var. *rydbergii* (Small *ex* Rydb.) Erskine
common name(s): western poison-ivy
habit: simple or branched, rhizomatous shrub
range: Quebec, s. to VA, w. to AZ, n. to British Columbia
state frequency: occasional
habitat: open woods and rocky slopes
notes: Similar to *Toxicodendron radicans* (*q.v.*).

Toxicodendron vernix (L.) Kuntze
synonym(s): *Rhus vernix* L.
common name(s): poison-sumac
habit: shrub to small tree
range: ME, s. to FL, w. to TX, n. to MN and Ontario
state frequency: rare, known only from southern Maine
habitat: swamps
notes: —

SAPINDACEAE
(2 genera, 11 species)

KEY TO THE GENERA
1a. Perianth zygomorphic; leaves palmately compound; fruit a capsule, commonly prickly; ovaries mostly 3-locular .... *Aesculus*
1b. Perianth actinomorphic; leaves simple or pinnately compound; fruit a samaroid schizocarp, without prickles; ovaries mostly 2-locular .... *Acer*

Acer
(9 species)

KEY TO THE SPECIES FOR USE WITH VEGETATIVE MATERIAL
1a. Leaves pinnately compound with 3–9 leaflets; twigs sometimes glaucous, with a line from each leaf scar meeting in a raised point between the leaf scars .... *A. negundo*
1b. Leaves simple, palmately lobed; twigs not glaucous, with the line from each leaf scar not meeting in a raised point
   2a. Leaves with rounded sinuses, the margins bearing only a few, large teeth
      3a. Broken petioles yielding a milky latex; bark closely furrowed; leaves green on the abaxial surface .... *A. platanoides*

3b. Broken petioles yielding a watery latex; bark roughened in age with large furrows or curling plates; leaves thinly glaucous on the abaxial surface

    4a. Buds red, blunt at the apex; twigs often with collateral buds; leaves with deep, narrow sinuses; central lobe of the leaf more than 0.5 times the length of the blade and narrowed near the base; bark with large, curling plates; plants of moist to wet soil .... *A. saccharinum*

    4b. Buds brown, pointed at the apex; twigs without collateral buds; leaves lobed with open sinuses; central lobe of the leaf up to 0.5 times the length of the blade and not narrowed near the base; bark with deep furrows; plants of rich, mesic soils .... *A. saccharum*

2b. Leaves with angled [V-shaped] sinuses, the margins crenate or serrate with many teeth

    5a. Buds stalked, covered with 2 valvate scales

        6a. Twigs and bud scales pubescent; bark brown or gray-brown; leaves 7.5–12.5 cm long .... *A. spicatum*

        6b. Twigs and bud scales glabrous; bark green or green-brown, with slender, longitudinal, white stripes; larger leaves 15.0–20.0 cm long .... *A. pensylvanicum*

    5b. Buds sessile, covered with 3–10 imbricate scales

        7a. Leaves green on the abaxial surface .... *A. ginnala*

        7b. Leaves thinly glaucous on the abaxial surface

            8a. Leaves crenate; collateral buds absent .... *A. pseudoplatanus*

            8b. Leaves serrate; collateral buds often present .... *A. rubrum*

KEY TO THE SPECIES FOR USE WITH REPRODUCTIVE MATERIAL

1a. Flowers very precocious, borne in dense clusters on very short pedicels, the pedicels elongating at maturity of the fruit

    2a. Petals present; nectary disk present; twigs nearly straight; ovary and body of the mericarp glabrous .... *A. rubrum*

    2b. Petals absent; nectary disk absent or vestigial; twigs commonly arching; ovary pubescent; body of the mericarp sparsely pubescent .... *A. saccharinum*

1b. Flowers subprecocious to serotinous, pedicellate, in some species borne in elongate inflorescences

    3a. Flowers subprecocious to coaetaneous

        4a. Flowers without a nectary disk .... *A. negundo*

        4b. Flowers with a nectary disk

            5a. Pedicels spreading to ascending, glabrous; perianth consisting of sepals and petals; mericarps diverging at an angle of nearly 180°, each mericarp, including the wing, 3.5–5.0 cm long .... *A. platanoides*

            5b. Pedicels drooping, pubescent; perianth consisting of sepals only; mericarps diverging at an angle of less than 120°, each mericarp, including the wing, 2.5–4.0 cm long .... *A. saccharum*

    3b. Flowers serotinous

        6a. Inflorescence pubescent

            7a. Sepals and petals approximately equal in length, 4.0–5.0 mm long; mericarps, including the wing, 3.0–4.0 cm long .... *A. pseudoplatanus*

7b. Sepals 1.0–2.0 mm long; petals *ca.* 3.0 mm long; mericarps, including the wing, 1.8–2.5 cm long .... ***A. spicatum***
6b. Inflorescence glabrous or nearly so
   8a. Petals white; basal portion of leaves with deep sinuses, glabrous on the abaxial surface, sometimes with a few hairs near the junctions of major veins .... ***A. ginnala***
   8b. Petals green to yellow; basal portion of leaves unlobed or with tiny lobes, usually pubescent on the veins on the abaxial surface .... ***A. pensylvanicum***

Acer ginnala Maxim.
synonym(s): —
common name(s): Amur maple
habit: shrub to small tree
range: Asia; escaped from cultivation to New Brunswick, s. to CT, w. and n. to Ontario
state frequency: rare
habitat: hedgerows, open woods
notes: —

Acer negundo L.
synonym(s): <*A. n.* var. *violaceum*> = *Negundo aceroides* (L.) Moench ssp. *violaceum* (Kirchn.) W. A. Weber; <*A. n.* var. *negundo*> = *Negundo aceroides* (L.) Moench
common name(s): box-elder, ash-leaved maple
habit: tree
range: NY, s. to FL, w. to TX, n. to MN and Ontario; naturalized to New Brunswick, s. to CT, w. and n. to Quebec
state frequency: occasional
habitat: moist soil, river banks, waste places
key to the subspecies:
   1a. Twigs glaucous .... *A. n.* var. *violaceum* (Kirchn.) Jaeger
   1b. Twigs green, not glaucous .... *A. n.* var. *negundo*
notes: —

Acer pensylvanicum L.
synonym(s): —
common name(s): striped maple, moosewood, goosefoot maple
habit: small, understory tree
range: Quebec, s. to PA and mountains of GA, w. and n. to MN
state frequency: common
habitat: mesic forests
notes: —

Acer platanoides L.
synonym(s): *Acer platanoides* L. var. *schwedleri* Nichols.
common name(s): Norway maple
habit: tree
range: Europe; escaped from cultivation throughout the northeast
state frequency: occasional
habitat: thickets, roadsides
notes: —

Acer pseudoplatanus L.

synonym(s): —
common name(s): sycamore maple
habit: tree
range: Eurasia; escaped from cultivation to New Brunswick, s. to CT
state frequency: rare
habitat: edges, roadsides
notes: —

Acer rubrum L.

synonym(s): <*A. r.* var. *rubrum*> = *Rufacer rubrum* (L.) Small; <*A. r.* var. *trilobum*> = *Acer carolinianum* Walt., *Rufacer carolinianum* (Walt.) Small
common name(s): red maple, soft maple, swamp maple
habit: tree
range: Newfoundland, s. to FL, w. to TX, n. to Manitoba
state frequency: common
habitat: swamps, fens, moist uplands
key to the subspecies:
    1a. Leaves dark green on the adaxial surface, those of the reproductive branches 3.0–10.0 cm wide, usually with 3 lobes (the lobes sometimes nearly absent), the central lobe 1.0–5.0 cm long, the lateral lobes 0.5–2.0(–3.5) cm long .... *A. r.* var. *trilobum* Torr. & Gray *ex* K. Koch
    1b. Leaves green on the adaxial surface, those of the reproductive branches 6.0–15.0 cm wide, with 3 or 5 lobes, the central lobe 4.0–8.0 cm long, the lateral lobes 2.0–5.0 cm long .... *A. r.* var. *rubrum*
notes: Known to hybridize with *Acer saccharinum* (*q.v.*).

Acer saccharinum L.

synonym(s): *Acer saccharinum* L. var. *laciniatum* Pax, *Argentacer saccharinum* (L.) Small
common name(s): silver maple, river maple, white maple, soft maple
habit: tree
range: New Brunswick, s. to FL, w. to OK, n. to SD, MN, and Ontario
state frequency: occasional
habitat: river banks, low areas
notes: Hybrids with *Acer rubrum* occur in Maine.

Acer saccharum Marsh.

synonym(s): *Saccharodendron saccharum* (Marsh.) Moldenke
common name(s): sugar maple, rock maple, hard maple
habit: tree
range: Quebec, s. to GA, w. to TX, n. to MN and Ontario
state frequency: common
habitat: rich uplands
notes: —

Acer spicatum Lam.

synonym(s): —
common name(s): mountain maple
habit: shrub
range: Newfoundland, s. to mountains of GA, w. to IA, n. to Saskatchewan
state frequency: occasional
habitat: cool, moist woods
notes: —

Aesculus
(2 species)

KEY TO THE SPECIES

1a. Corolla with 4 yellow petals; leaves with 5(–7) leaflets, the leaflets acute to acuminate at the apex; winter buds not glutinous; pubescence of the inflorescence and leaves, when present, white; fruit 3.0–4.0 cm long .... *A. glabra*

1b. Corolla with 5 white petals with yellow or red, basal markings; leaves with 5–9, commonly 7, leaflets, the leaflets abruptly acuminate at the apex; winter buds glutinous; pubescence of the inflorescence and leaves, when present, brown; fruit *ca.* 5.0 cm long .... *A. hippocastanum*

Aesculus glabra Willd.
synonym(s): —
common name(s): Ohio buckeye
habit: small tree
range: Ontario, s. to AL, w. to TX, n. to WI; escaped from cultivation in the northeast
state frequency: rare
habitat: moist, well drained, alluvial woods
notes: —

Aesculus hippocastanum L.
synonym(s): —
common name(s): horse-chestnut
habit: tree
range: Eurasia; escaped from cultivation
state frequency: uncommon
habitat: fields, roadsides, old homesites
notes: —

SIMAROUBACEAE
(1 genus, 1 species)

Ailanthus
(1 species)

Ailanthus altissima (P. Mill.) Swingle
synonym(s): —
common name(s): tree of heaven, copal tree
habit: small tree
range: Asia; escaped from cultivation to MA, s. to FL, w. to IA, n. to Ontario; also ME
state frequency: very rare
habitat: roadsides, areas of cultivation
notes: —

RUTACEAE
(2 genera, 2 species)

KEY TO THE GENERA

1a. Leaves with 5–11 leaflets; plants dioecious; gynoecium with 3–5 carpels, the ovaries distinct; fruit a fleshy follicle; stems prickly .... *Zanthoxylum*

1b. Leaves with 3 leaflets; plants polygamous; gynoecium with 2 connate carpels; fruit an orbicular samara; stems unarmed .... *Ptelea*

Ptelea
(1 species)

Ptelea trifoliata L.
synonym(s): —
common name(s): common hop tree, shrubby trefoil, stinking-ash, wafer-ash
habit: tall shrub to small tree
range: NY, s. to FL, w. to TX and Mexico, n. to NE and Ontario; escaped from cultivation in the northeast
state frequency: very rare
habitat: near establishments, areas of cultivation
notes: —

Zanthoxylum
(1 species)

Zanthoxylum americanum P. Mill.
synonym(s): *Xanthoxylum americanum* P. Mill.
common name(s): common prickly-ash, northern prickly-ash
habit: shrub
range: NY, s. to FL, w. to OK, n. to ND; naturalized to ME, s. to CT
state frequency: very rare
habitat: fields, roadsides, near establishments
notes: —

CISTACEAE
(3 genera, 7 species)

KEY TO THE GENERA

1a. Petaliferous flowers with 5 conspicuous, yellow petals; petals wrinkled in bud; plants lacking a basal rosette of crowded branches

2a. Leaves narrow-oblong to elliptic or elliptic-oblanceolate, 2.0–3.0(–4.0) cm long, widest near or above the middle, not imbricate; plants 1.5–5.0 dm tall, erect to ascending, not forming mats; style very short, not over 1.0 mm long; flowers of 2 types—chasmogamous flowers with large, conspicuous petals and cleistogamous flowers with reduced or absent petals .... *Helianthemum*

2b. Leaves linear-subulate or lance-ovate to triangular, 0.1–0.6 cm long, widest below the middle, densely imbricate; plants 1.0–2.0 dm tall, diffusely branched, forming low mats and mounds; style slender, *ca.* 2.0 mm long; flowers of 1 type—all chasmogamous, with conspicuous petals .... ***Hudsonia***

1b. Petaliferous flowers with 3 inconspicuous, red petals; petals flat in bud; plants overwintering by a basal rosette of crowded, leafy branches .... ***Lechea***

Helianthemum
(2 species)

This genus produces 2 kinds of flowers. The open pollinated [chasmogamous] flowers are borne at the summit of the main axis and have larger petals, larger capsules, and more seeds. The closed pollinated [cleistogamous] flowers are borne in clusters along, or at the apex of, the lateral branches and have smaller or no petals, smaller capsules, and fewer seeds.

KEY TO THE SPECIES

1a. Chasmogamous [petaliferous] flowers solitary or paired at the top of the stem, overtopped later in the season by lateral branches; chasmogamous capsules 6.0–7.0 mm long, with 30–45 seeds; cleistogamous capsules 2.0–3.0 mm long, with 5–10 seeds; seeds papillate .... ***H. canadense***

1b. Chasmogamous flowers in a 2- to 18-flowered inflorescence at the apex of the stem, usually not overtopped later in the season by lateral branches; chasmogamous capsules 3.5–5.0 mm long, with 12–26 seeds; cleistogamous capsules 1.5–2.0 mm long, with 1–3 seeds; seeds indistinctly reticulate .... ***H. bicknellii***

Helianthemum bicknellii Fern.
synonym(s): *Crocanthemum majus* (L.) Britt.
common name(s): frostweed
habit: erect to arched, perennial herb
range: ME, s. to GA, w. to CO, n. to MN and Manitoba
state frequency: uncommon
habitat: dry, sandy or rocky soil
notes: Usually flowers 2 or 3 weeks later than *Helianthemum canadense* when both species are present at the same site.

Helianthemum canadense (L.) Michx.
synonym(s): *Crocanthemum canadense* (L.) Britt.
common name(s): frostweed
habit: erect to ascending, more or less cespitose, perennial herb
range: Nova Scotia, s. to GA, w. to MO, n. to MN and Ontario
state frequency: uncommon
habitat: dry, open woods and sandy soil
notes: Usually flowers 2 or 3 weeks earlier than *Helianthemum bicknellii* when both species are present at the same site.

Hudsonia
(2 species)

Pedicel lengths include the basal, leafy portion.

KEY TO THE SPECIES

1a. Leaves lance-ovate to triangular, closely appressed, densely tomentose, 1.0–3.0 mm long; ovary and capsule glabrous; pedicels 0.0–3.0 mm long .... ***H. tomentosa***

1b. Leaves linear-subulate, erect to spreading, sparsely pubescent, 3.0–6.0 mm long; ovary and capsule pubescent; pedicels 5.0–15.0 mm long .... *H. ericoides*

Hudsonia ericoides L.
synonym(s): —
common name(s): golden heather
habit: low, branched subshrub
range: Newfoundland, s. to ME; also sporadically to SC
state frequency: uncommon
habitat: dry, acidic rocks and pine barrens
notes: Hybrids with *Hudsonia tomentosa*, called *H.* ×*spectabilis* Morse, occur in Maine and can be recognized by their intermediate morphology.

Hudsonia tomentosa Nutt.
synonym(s): —
common name(s): beach heather
habit: low, branched subshrub
range: Labrador, s. to NC, w. to IL, n. to Alberta
state frequency: uncommon
habitat: coastal dunes, inland, sandy areas
notes: Hybrids with *Hudsonia ericoides* (*q.v.*) occur in Maine.

Lechea
(3 species)

KEY TO THE SPECIES
1a. The 2 narrow, outer sepals about half as long as the 3 broad, inner sepals; leaves of the stem 2.0–5.3 mm wide, less than 10.0 times as long as wide
    2a. Leaves green on the abaxial surface, the surface glabrous or pubescent on the midrib and margins only; panicle commonly 0.35–0.5 of the height of the plant, cylindric to narrow-pyramidal, with ascending to erect branches .... *L. intermedia*
    2b. Leaves gray on the abaxial surface due to fine, close pubescence throughout the surface; panicle commonly 0.5–0.65 of the height of the plant, pyramidal to broad-pyramidal, with divergent to ascending branches .... *L. maritima*
1b. The 2 narrow, outer sepals as long or longer than the 3 broad, inner sepals; leaves of the stem 1.0–1.5 mm wide, more than 10.0 times as long as wide .... *L. tenuifolia*

Lechea intermedia Leggett *ex* Britt.
synonym(s): —
common name(s): pinweed
habit: erect, perennial herb
range: New Brunswick, s. to PA and mountains of VA, w. to NE, n. to ND and Saskatchewan
state frequency: occasional
habitat: dry, sterile soil
key to the subspecies:
    1a. Capsule ovoid, *ca.* 1.0 mm wide; inner sepals elliptic-ovate, subacute at the apex, with 3 nerves; panicles densely and closely branched .... *L. i.* var. *juniperina* (Bickn.) B. L. Robins.
    1b. Capsule depressed-globose, *ca.* 2.0 mm wide; inner sepals broad-ovate, obtuse or nearly so at the apex, with 3 or 5 nerves; panicles openly branched .... *L. i.* var. *intermedia*
notes: —

Lechea maritima Leggett *ex* B. S. P.
synonym(s): —
common name(s): seaside pinweed
habit: erect to ascending, perennial herb
range: New Brunswick, s. along the coast to VA
state frequency: uncommon
habitat: dunes, sandy soil
notes: —

Lechea tenuifolia Michx.
synonym(s): —
common name(s): slender pinweed
habit: slender, branched, perennial herb
range: ME, s. to SC, w. to TX, n. to MN
state frequency: very rare
habitat: dry, upland woods
notes: This species is ranked SX by the Maine Natural Areas Program.

THYMELAEACEAE
(2 genera, 2 species)

KEY TO THE GENERA

1a. Sepals well developed, spreading, petaloid; hypanthium pubescent; stamens and style
included in the perianth; flowers white to pink to purple; drupe 0.6–1.0 cm long; leaf
blades commonly oblanceolate to oblong; introduced shrubs escaped from cultivation .....
.... *Daphne*
1b. Sepals minute, represented by tiny lobes at the summit of the glabrous, tubular
hypanthium; stamens and style exserted beyond the perianth; flowers yellow; drupe
1.2–1.5 cm long; leaf blades commonly ovate or obovate; native shrubs of rich woods ....
.... *Dirca*

Daphne
(1 species)

Daphne mezereum L.
synonym(s): —
common name(s): merzereum
habit: shrub
range: Eurasia; escaped from cultivation to Newfoundland and Nova Scotia, s. to CT, w. to OH, n. to
Ontario
state frequency: uncommon
habitat: roadsides, thickets
notes: —

Dirca
(1 species)

Dirca palustris L.
synonym(s): —
common name(s): leatherwood, wicopy
habit: shrub
range: New Brunswick and Nova Scotia, s. to FL, w. to AL and OK, n. to MN and Ontario
state frequency: uncommon
habitat: moist, rich woods
notes: —

MALVACEAE
(5 genera, 11 species)

KEY TO THE GENERA

1a. Plants woody; peduncle of the inflorescence adnate to a conspicuous, elongate bract; stamens pentadelphous; fruit a nut-like drupe .... ***Tilia***
1b. Plants herbaceous; inflorescence without an adnate bract; stamens monadelphous; fruit a capsule or schizocarp
    2a. Gynoecium with 5 wholly connate carpels; fruit a capsule; filaments connate, forming a tube, bearing anthers along the sides of the tube, crowned with 5 teeth at the summit of the filament tube; petals yellow with purple near the base .... ***Hibiscus***
    2b. Gynoecium with 5–20 basally connate carpels; fruit a schizocarp, the carpels separating at maturity into indehiscent or dehiscent mericarps; filaments connate, forming a tube, bearing anthers only at the apex; petals of similar color in basal and apical portions
        3a. Styles with the stigmatic surface along the inner surface; epicalyx present, of 3 bractlets; petals emarginate at the apex, white to pink to purple .... ***Malva***
        3b. Styles with the stigmatic surface at the capitate apex; epicalyx absent; petals entire at the apex, yellow
            4a. Carpels with 3–9 ovules [therefore, the mericarp with 3–9 seeds]; leaves 10.0–15.0 cm long; carpels commonly 10–15 per flower; flowers 1.5–2.5 cm wide .... ***Abutilon***
            4b. Carpels with 1 ovule [therefore, the mericarp with 1 seed]; leaves 2.0–4.0 cm long; carpels 5 per flower; flowers 0.4–0.85 cm wide .... ***Sida***

Abutilon
(1 species)

Abutilon theophrasti Medik.
synonym(s): *Abutilon abutilon* (L.) Rusby
common name(s): velvet-leaf
habit: tall, stout, erect, branched, annual herb
range: Asia; naturalized in the northeast
state frequency: uncommon
habitat: cultivated fields, waste places
notes: —

Hibiscus
(1 species)

Hibiscus trionum L.
synonym(s): —
common name(s): flower of an hour
habit: branched, annual herb
range: Europe; naturalized to New Brunswick, s. to FL, w. to TX, n. to MN and Saskatchewan
state frequency: rare
habitat: waste places, fields, roadsides
notes: This species has ephemeral flowers that are open for only a few hours.

Malva
(5 species)

KEY TO THE SPECIES

1a. Flowers borne on long peduncles, solitary in the axils of the upper leaves, exceeding
   their subtending leaves, with petals 2.0–3.5 cm long; plants perennial; leaves with deep
   sinuses to below the middle
   2a. Plants pubescent with compound [stellate] hairs; bractlets ovate to obovate; carpels
      keeled on the back .... *M. alcea*
   2b. Plants pubescent with simple hairs; bractlets linear to narrow-oblanceolate; carpels
      rounded on the back .... *M. moschata*
1b. Flowers borne on short pedicels, fascicled in the axils of leaves, exceeded by the petioles
   of their subtending leaves, with petals 0.6–2.5 cm long; plants annual or biennial; leaves
   with shallow sinuses, rarely as deep as the middle
   3a. Petals red-purple, 2.0–2.5 cm long; bractlets oblong-ovate .... *M. sylvestris*
   3b. Petals white or tinged with pink or purple, 0.6–1.2 cm long; bractlets linear to
      narrow-lanceolate
      4a. Plants prostrate (sometimes ascending); flowers on pedicels to 3.0 cm long;
         leaves with very shallow sinuses; carpels not rugose-reticulate .... *M. neglecta*
      4b. Plants erect; flowers sessile or subsessile; leaves with shallow sinuses; carpels
         rugose-reticulate on the abaxial surface .... *M. crispa*

Malva alcea L.
synonym(s): —
common name(s): vervain mallow
habit: erect, perennial herb
range: Europe; naturalized to New Brunswick, s. to PA, w. to OH, n. to MI and Ontario
state frequency: very rare
habitat: roadsides, fields
notes: —

Malva crispa (L.) L.
synonym(s): *Malva verticillata* L. var. *crispa* L.
common name(s): curled mallow
habit: erect, crisped-leaved, annual herb
range: Eurasia; naturalized to New Brunswick, s. to CT, w. to IL, n. to Alberta
state frequency: rare
habitat: waste places, roadsides
notes: —

Malva moschata L.

synonym(s): —
common name(s): musk mallow
habit: erect to ascending, perennial herb
range: Europe; escaped from cultivation to Nova Scotia and New Brunswick, s. to VA, w. to MO, n.
   to British Columbia
state frequency: occasional
habitat: fields, roadsides, waste places
notes: —

Malva neglecta Wallr.

synonym(s): —
common name(s): common mallow
habit: prostrate to ascending, basally branched, annual or biennial herb
range: Eurasia; naturalized throughout temperate North America
state frequency: uncommon
habitat: waste places
notes: —

Malva sylvestris L.

synonym(s): *Malva sylvestris* L. var. *mauritiana* (L.) Boiss.
common name(s): high mallow
habit: erect, biennial herb
range: Eurasia; escaped from cultivation to Quebec, w. to British Columbia
state frequency: uncommon
habitat: roadsides, waste places
notes: —

Sida
(1 species)

Sida spinosa L.

synonym(s): —
common name(s): prickly-mallow
habit: branched, annual herb
range: pantropical; n. to MA, w. to NE; also ME
state frequency: very rare
habitat: fields, roadsides, waste places
notes: The specific epithet for this species refers to the spine-like tubercle present at the base of some
   leaves.

Tilia
(3 species)

KEY TO THE SPECIES
1a. Leaf blades 7.0–15.0(–20.0) cm long; flowers with staminodes .... *T. americana*
1b. Leaf blades mostly 5.0–10.0 cm long; flowers without staminodes
    2a. Fruit strongly 5-ribbed; leaves pubescent on the adaxial surface .... *T. platyphyllos*
    2b. Fruit not, or only weakly, ribbed; leaves glabrous on the adaxial surface ... *T. cordata*

Tilia americana L.

synonym(s): <*T. a.* var. *americana*> = *Tilia neglecta* Spach; <*T. a.* var. *heterophylla*> = *Tilia heterophylla* Vent.

common name(s): <*T. a.* var. *americana*> = basswood, American linden; <*T. a.* var. *heterophylla*> = white basswood

habit: tree

range: New Brunswick, s. to FL, w. to TX, n. to Manitoba

state frequency: occasional

habitat: rich woods

key to the subspecies

    1a. Mature leaves green and glabrous or sometimes sparsely stellate-pubescent on the abaxial surface; pedicels glabrous .... *T. a.* var. *americana*

    1b. Leaves permanently stellate-pubescent with white or brown hairs; pedicels pubescent .... *T. a.* var. *heterophylla* (Vent.) Loud.

notes: *Tilia americana* var. *heterophylla*, which is native to the southern and western portions of the stated range, is planted in Maine.

Tilia cordata P. Mill.

synonym(s): —

common name(s): littleleaf linden

habit: tree

range: Europe; escaped from cultivation to New Brunswick and ME

state frequency: rare

habitat: edges, streetsides

notes: Hybrids with *Tilia platyphyllos*, called *Tilia* ×*vulgaris* Hayne, have been planted in Maine.

Tilia platyphyllos Scop.

synonym(s): —

common name(s): bigleaf linden, large-leaved linden

habit: tree

range: Europe; escaped from cultivation in our area

state frequency: very rare

habitat: roadsides

notes: Known to hybridize with *Tilia cordata* (*q.v.*).

CORNACEAE
(2 genera, 8 species)

KEY TO THE GENERA

1a. Flowers of 2 types—staminate with 8–15 stamens and functionally carpellate [the anthers nonfunctional] with 5–10 stamens; pith diaphragmed; petals 5–8; leaves alternate .... ***Nyssa***

1b. Flowers of 1 type—bisexual with 4 stamens; pith solid; petals 4; leaves opposite, whorled, or alternate .... ***Cornus***

Cornus
(7 species)

Easily recognized vegetatively by its entire, arcuate-veined leaves that possess stringy fibers within the veins, visible when the leaf is torn.

KEY TO THE SPECIES

1a. Plants herbs, 0.1–0.2 m tall; leaves appearing whorled .... *C. canadensis*
1b. Plants shrubs or small trees, 1.0–10.0 m tall; leaves opposite or alternate
    2a. Leaves alternate, crowded at the apex of the twig; petioles (0.8–)2.0–5.0 cm long; twigs sometimes yellow with minute red-brown to black dots due to infection of *Cryptodioporthe corni* .... *C. alternifolia*
    2b. Leaves opposite, evenly spaced along the twig; petioles 0.3–1.5(–2.5) cm long; twigs without above-mentioned fungal infection
        3a. Flowers pseudanthial, in a dense cluster, subtended by 4 large, petaloid, white to pink bracts; drupe red; lateral buds concealed; terminal flower bud large, globose, or discoid .... *C. florida*
        3b. Flowers not arranged in pseudanthia, not subtended by petaloid bracts; drupe white to blue; lateral buds visible; terminal flower bud ovoid
            4a. Pith of twigs brown; style conspicuously widened in the apical portion; sepals 1.0–2.0 mm long; drupe blue .... *C. amomum*
            4b. Pith white (sometimes light brown in *C. racemosa*); style of nearly uniform diameter throughout; sepals shorter than 1.0 mm; drupe white to blue
                5a. Winter twigs gray-brown; inflorescence convex or pyramidal, nearly as tall as wide or taller, with red branches and pedicels when in fruit; leaves with 3 or 4(5) pairs of lateral veins .... *C. racemosa*
                5b. Winter twigs not uniformly gray-brown; inflorescence flat-topped, often wider than tall, with branches yellow or gray-brown to brown; leaves with 5–8 pairs of lateral veins
                    6a. Twigs red; drupe white (light blue); leaves lanceolate to ovate; seed brown, with 7–9 yellow stripes; buds without scales, densely pubescent throughout with red-brown and some white hairs .... *C. sericea*
                    6b. Twigs yellow-green, often mottled with purple or red; drupe light blue (white); leaves ovate to suborbicular; seed light brown, without stripes; buds covered about 0.65 of their length by sparsely pubescent scales, only the tips of the densely pubescent leaf primordia visible .... *C. rugosa*

Cornus alternifolia L. f.
synonym(s): *Swida alternifolia* (L. f.) Small
common name(s): alternate-leaved dogwood, pagoda dogwood
habit: shrub to small tree
range: Newfoundland, s. to FL, w. to AR, n. to MN and Manitoba
state frequency: occasional
habitat: woods, slopes, thickets
notes: Known to hybridize with *Cornus sericea*.

Cornus amomum P. Mill.

synonym(s): <*C. a.* ssp. *amomum*> = *Swida amomum* (P. Mill.) Small; <*C. a.* ssp. *obliqua*> =
   Cornus amomum P. Mill. var. *scheutzeana* (C. A. Mey.) Rickett, *Cornus obliqua* Raf.
common name(s): silky dogwood, knob-styled dogwood, red-willow
habit: ascending to spreading shrub
range: ME, s. to GA, w. to OK, n. to ND
state frequency: occasional
habitat: moist to wet thickets, woods, and stream banks
key to the subspecies:
   1a. Leaves mostly 6.0–12.0 cm long, ovate to broad-elliptic, 1.0–2.2 times as long as wide,
      subtruncate to rounded at the base, with 4–6 pairs of veins, green and usually non-papillose on
      the abaxial surface, with white or, more commonly, red-brown pubescence ....
      .... *C. a.* ssp. *amomum*
   1b. Leaves mostly 6.0–9.0 cm long, lanceolate to narrow-acute, 2.2–4.0 times as long as wide,
      cuneate at the base, with 3–5 pairs of veins, usually white and papillose on the abaxial surface,
      with white, or only rarely, red-brown pubescence .... *C. a.* ssp. *obliqua* (Raf.) J. S. Wilson
notes: Known to hybridize with *Cornus racemosa*. The ultimate twigs of *C. amomum* mostly lack
   lenticels, whereas the older branchlets will be streaked by numerous, confluent lenticels.

Cornus canadensis L.

synonym(s): *Chamaepericlymenum canadense* (L.) Aschers. & Graebn.
common name(s): bunchberry, dwarf cornel
habit: low, rhizomatous, perennial herb
range: Greenland, s. to NJ, w. to CA, n. to AK
state frequency: common
habitat: moist woods and thickets, bogs
notes: —

Cornus florida L.

synonym(s): *Cynoxylon floridum* (L.) Raf. *ex* B. D. Jackson
common name(s): flowering dogwood
habit: large shrub to small tree
range: ME, s. to FL, w. to TX, n. to MI and Ontario
state frequency: very rare
habitat: woods
notes: This species is ranked S1 by the Maine Natural Areas Program.

Cornus racemosa Lam.

synonym(s): —
common name(s): northern swamp dogwood
habit: ascending shrub
range: ME, s. to VA, w. to OK, n. to MN and Manitoba
state frequency: uncommon
habitat: moist woods, thickets, roadsides, and stream banks
notes: Known to hybridize with *Cornus amomum*. The ultimate twigs of *C. racemosa* are covered by
   numerous, but minute, lenticels.

Cornus rugosa Lam.

synonym(s): —
common name(s): round-leaved dogwood
habit: ascending shrub
range: New Brunswick, s. to NJ and mountains of VA, w. to IA, n. to Manitoba
state frequency: uncommon

habitat: moist to dry, rocky or sandy woods and slopes
notes: Known to hybridize with *Cornus sericea*.

Cornus sericea L.

synonym(s): *Cornus stolonifera* Michx., *Swida stolonifera* (Michx.) Rydb.
common name(s): red osier dogwood
habit: colonial shrub
range: Labrador, s. to WV, w. to Mexico and CA, n. to AK
state frequency: common
habitat: moist woods, thickets, meadows, stream banks, and shores
notes: Known to hybridize with *Cornus alternifolia* and *C. rugosa*. The twigs of *C. sericea* have
   relatively few lenticels, 8–20 per internode.

Nyssa
(1 species)

Nyssa sylvatica Marsh.

synonym(s): —
common name(s): black tupelo, black gum
habit: small to medium-sized tree
range: ME, s. to FL, w. to TX and Mexico, n. to WI
state frequency: uncommon; found in the southern part of the state
habitat: wet woods and shores, swamps
notes: —

HYDRANGEACEAE
(1 genus, 1 species)

Philadelphus
(1 species)

Philadelphus coronarius L.

synonym(s): —
common name(s): European mock-orange
habit: medium-sized shrub
range: Europe; escaped from cultivation in the northeast
state frequency: uncommon
habitat: roadside thickets
notes: —

DIAPENSIACEAE
(1 genus, 1 species)

Diapensia
(1 species)

Diapensia lapponica L.
synonym(s): —
common name(s): diapensia
habit: low, pulvinate, evergreen subshrub
range: circumboreal; s. in the east to ME, NH, VT, and NY
state frequency: rare
habitat: alpine areas
notes: This species is ranked S2 by the Maine Natural Areas Program.

SARRACENIACEAE
(1 genus, 1 species)

Sarracenia
(1 species)

Sarracenia purpurea L.
synonym(s): —
common name(s): pitcher plant
habit: insectivorous, perennial herb
range: Labrador, s. to FL, w. to IL, n. to MN and Manitoba
state frequency: occasional
habitat: *Sphagnum* bogs
notes: This species is insectivorous by means of retrorsely oriented hairs on the inside of the pitcher-shaped leaves that direct insects to a small volume of water and digestive enzymes at the base where they drown and are digested.

ACTINIDIACEAE
(1 genus, 1 species)

Actinidia
(1 species)

Actinidia arguta Miq.
synonym(s): —
common name(s): —
habit: climbing, glabrous shrub
range: Korea, Japan, Manchuria; naturalized in our area
state frequency: very rare
habitat: disturbed ground, fields
notes: —

CLETHRACEAE
(1 genus, 1 species)

Clethra
(1 species)

Clethra alnifolia L.
synonym(s): —
common name(s): sweet pepperbush, white-alder
habit: small shrub
range: Nova Scotia and ME, s. to FL, w. to TX, n. to NY
state frequency: rare
habitat: swamps and moist, often coastal, woods
notes: This species is ranked S2 by the Maine Natural Areas Program.

POLEMONIACEAE
(3 genera, 7 species)

KEY TO THE GENERA
1a. Leaves simple; filaments straight; calyx not accrescent in fruit
    2a. Leaves mostly opposite; plants perennial; calyx with scarious or hyaline intervals
        between the herbaceous lobes; seeds neither mucilaginous nor releasing spiraled
        threads when wet .... ***Phlox***
    2b. Leaves alternate; plants annual; calyx of uniform texture, herbaceous throughout;
        seeds becoming mucilaginous and releasing spiraled threads when wet ...... ***Collomia***
1b. Leaves pinnately compound; filaments declined; calyx accrescent in fruit .. ***Polemonium***

Collomia
(1 species)

Collomia linearis Nutt.
synonym(s): —
common name(s): collomia
habit: simple or branched, annual herb
range: Ontario and WI, s. to NM, w. to CA, n. to British Columbia; also Quebec, s. to MA
state frequency: very rare
habitat: dry, open areas, gravelly shores
notes: Flowers pink to purple to white.

Phlox
(4 species)

KEY TO THE SPECIES
1a. Petals notched at the apex; stems suffrutescent; leaves 0.5–2.0 cm long, subulate, stiff ....
    .... ***P. subulata***
1b. Petals entire at the apex; stems herbaceous; leaves 1.5–15.0 cm long, variously shaped,
    but not subulate, relatively softer

2a. Long-creeping, sterile, basal offshoots with spatulate leaves present at anthesis;
    inflorescence and calyx stipitate-glandular; corolla 2.5–3.0 cm wide ... *P. stolonifera*
2b. Sterile, basal offshoots absent at anthesis or present and short, upcurving, and
    without spatulate leaves; inflorescence and calyx glabrous or pubescent, but not
    glandular; corolla 1.2–2.2 cm wide
    3a. Leaves ciliolate on the margin, with conspicuous lateral veins that connect near
        the edge to form a submarginal vein around the perimeter of the leaf; anthers pale
        yellow to white; stems without red spots .... *P. paniculata*
    3b. Leaves glabrous on the margin, without a submarginal, connecting vein; anthers
        yellow; stems often red-spotted .... *P. maculata*

Phlox maculata L.

synonym(s): —
common name(s): wild-sweet William, meadow phlox, spotted phlox
habit: erect, perennial herb
range: Quebec and NY, s. to VA and mountains of GA, w. to MO, n. to MN; escaped from cultivation
    in our area
state frequency: very rare
habitat: low woods, wet meadows, stream banks
notes: Flowers purple or red-violet.

Phlox paniculata L.

synonym(s): —
common name(s): fall phlox, summer phlox, perennial phlox
habit: erect, perennial herb
range: NY, s. to GA, w. to AR, n. to IL; escaped from cultivation in our area
state frequency: rare
habitat: rich, moist woods and thickets
notes: —

Phlox stolonifera Sims

synonym(s): —
common name(s): creeping phlox, crawling phlox
habit: ascending, stoloniferous, perennial herb
range: PA, s. to GA, w. and n. to OH; escaped from cultivation in our area
state frequency: very rare
habitat: moist woods
notes: —

Phlox subulata L.

synonym(s): —
common name(s): moss-pink, moss phlox, mountain phlox
habit: prostrate, mat-forming, branched, suffruticose, perennial herb
range: coastal NY, s. to MD and mountains of NC, w. to mountains of TN, n. to MI; escaped from
    cultivation to New Brunswick, s. to CT
state frequency: uncommon
habitat: dry, sandy, gravelly, or rocky soil
notes: —

Polemonium
(2 species)

KEY TO THE SPECIES

1a. Larger leaves with 14–21 leaflets mostly 1.0–2.0 cm apart; rhizome horizontal; stamens long-exserted from the corolla .... *P. vanbruntiae*

1b. Larger leaves with 19–29 leaflets mostly less than 1.0 cm apart; rhizome ascending; stamens not exceeding the corolla .... *P. caeruleum*

Polemonium caeruleum L.
synonym(s): —
common name(s): Greek-valerian, Jacob's ladder
habit: erect, simple, perennial herb
range: Europe; escaped from cultivation to ME, w. to MN
state frequency: very rare
habitat: roadsides, waste places
notes: —

Polemonium vanbruntiae Britt.
synonym(s): —
common name(s): Jacob's ladder, Appalachian Jacob's ladder
habit: erect, rhizomatous, perennial herb
range: New Brunswick, s. to MD, w. to WV, n. to NY and VT
state frequency: very rare
habitat: wooded swamps, *Sphagnum* bogs, stream banks
notes: This species is ranked S1 by the Maine Natural Areas Program.

PRIMULACEAE
(7 genera, 17 species)

KEY TO THE GENERA

1a. Plants aquatic; leaves narrowly dissected; peduncle and axes of the terminal racemes conspicuously inflated, the flowers whorled at the constricted nodes .... *Hottonia*

1b. Plants terrestrial (although sometimes of wet environments); leaves simple; peduncles and axes of inflorescences not inflated; inflorescences various, but not a cluster of racemes with whorled flowers

    2a. Leaves basal; stems scapose; inflorescence an umbel .... *Primula*

    2b. Leaves borne on the stem; inflorescence of axillary flowers, racemes, or panicles

        3a. Calyx petaloid; corolla absent; flowers axillary and subsessile; plants fleshy halophytes .... *Glaux*

        3b. Calyx sepaloid; corolla present; flowers not both axillary and subsessile; plants not fleshy, inland species except *Samolus*

            4a. Ovary partly inferior; leaves alternate; flowers small, 2.0–3.0 mm wide; plants of tidal or brackish shores .... *Samolus*

            4b. Ovary superior; leaves usually opposite or whorled (*Trientalis* with some alternate, scale-like leaves below); flowers larger; plants usually of inland habitats

5a. Corolla red or blue (white); plants annual; pyxis circumscissile near the middle .... *Anagallis*
5b. Corolla white or yellow; plants perennial; capsule dehiscing longitudinally
    6a. Stems with 1 or more small, scale-like leaves below and a single whorl of leaves at the apex; flowers usually 7-merous, with white corollas .... *Trientalis*
    6b. Stems with many nodes of opposite or whorled leaves; flowers 5- or 6-merous, usually with yellow corollas .... *Lysimachia*

Anagallis
(2 species)

KEY TO THE SPECIES
1a. Petals red (white), the margin commonly fringed with minute teeth and stipitate glands; pedicels usually exceeding the length of the leaves .... *A. arvensis*
1b. Petals blue, commonly not fringed on the margin; pedicels usually shorter than the leaves .... *A. foemina*

Anagallis arvensis L.
synonym(s): —
common name(s): scarlet pimpernel, common pimpernel
habit: branched, annual herb
range: Europe; naturalized throughout much of North America and other temperate regions of the world
state frequency: uncommon
habitat: sandy fields and open areas
notes: —

Anagallis foemina P. Mill.
synonym(s): *Anagallis arvensis* L. var. *caerulea* (Schreb.) Gren. & Godr.
common name(s): poor man's weather-glass
habit: branched, annual herb
range: Eurasia; naturalized in our area
state frequency: rare
habitat: sandy fields and open areas
notes: —

Glaux
(1 species)

Glaux maritima L.
synonym(s): —
common name(s): sea milkwort, saltwort
habit: low, erect, usually branched, fleshy, perennial herb
range: Quebec, s. to VA, w. to NM, n. to British Columbia
state frequency: uncommon
habitat: saline to brackish shores and marshes
key to the subspecies:

1a. Stems and branches prostrate to ascending (erect); leaves linear to narrow-oblong, 1.5–6.0 mm wide, obtuse to rounded at the apex; capsules 2.0–2.5 mm wide .... *G. m.* ssp. *maritima*
1b. Stems and branches usually erect; leaves oval or broad-oblong, 4.0–8.0 mm wide, rounded at the apex; capsules 2.5–4.0 mm wide .... *G. m.* ssp. *obtusifolia* (Fern.) Boivin
notes: —

Hottonia
(1 species)

Hottonia inflata Ell.
synonym(s): —
common name(s): featherfoil
habit: submerged to slightly emergent, aquatic, annual herb
range: ME, s. to FL, w. to TX, n. to MO and IL
state frequency: very rare
habitat: quiet, shallow water
notes: This species is ranked S1 by the Maine Natural Areas Program.

Lysimachia
(8 species)

These species can rarely have alternate leaves.

KEY TO THE SPECIES
1a. Leaves minutely punctate; androecium without staminodes, the filaments connate near the base (filaments distinct in *L. thyrsiflora*); lobes of the corolla entire (glandular-ciliate in *L. punctata*), without a cuspidate apex and not wrapped about its associated stamen
  2a. Flowers borne in short, dense racemes originating from the middle leaf axils; filaments distinct, usually of equal length; lobes of the corolla linear, 4.0–5.0 mm long, much shorter than the stamens .... *L. thyrsiflora*
  2b. Flowers solitary in the axils of leaves or in racemes from the apex of the stem or branches; filaments connate at the base, often of unequal length; lobes of the corolla lanceolate to ovate, 5.0–16.0 mm long, as long or longer than the stamens
    3a. Leaves quadrate to suborbicular, 1.0–2.5 cm long; stems extensively creeping over the ground .... *L. nummularia*
    3b. Leaves narrow-lanceolate to lance-ovate, 3.5–12.0 cm long; stems erect or ascending
      4a. Petals not dotted or streaked with black or red; stems pubescent
        5a. Calyx lobes 3.0–5.0 mm long, dark-margined with red; lobes of the corolla with entire margins; flowers borne in terminal or axillary racemes
        .... *L. vulgaris*
        5b. Calyx lobes 5.0–8.0 mm long, entirely green; lobes of the corolla ciliate with stipitate glands; flowers borne in whorls from the axils of the leaves
        .... *L. punctata*
      4b. Petals dotted or streaked with black or red; stems glabrous or sparsely pubescent
        6a. Flowers borne in a terminal raceme; leaves normally opposite, commonly developing bulblets in the axils; pedicels 0.8–1.5 cm long ..... *L. terrestris*

6b. Flowers borne in whorls from the axils of the leaves; leaves normally whorled, without bulblets; pedicels 2.0–5.0 cm long .... *L. quadrifolia*

1b. Leaves not punctate; androecium with staminodes alternating with stamens, the filaments distinct; lobes of the corolla erose on the apical portion of the margin, cuspidate at the apex, wrapped about its associated stamen

7a. Leaf blades lanceolate to ovate, rounded (subcordate) at the base, 1.5–6.0 cm wide; petioles usually ciliate their entire length; plants with long, slender rhizomes .... *L. ciliata*

7b. Leaf blades narrow-lanceolate to lance-ovate, cuneate to rounded at the base, 1.0–2.0 cm wide; petioles without cilia, or ciliate only in the basal half; plants with short, stout rhizomes .... *L. hybrida*

Lysimachia ciliata L.

synonym(s): *Steironema ciliatum* (L.) Baudo
common name(s): fringed loosestrife
habit: erect, rhizomatous, perennial herb
range: Quebec, s. to FL, w. to NM, n. to British Columbia
state frequency: occasional
habitat: moist woods, thickets, and shores
notes: —

Lysimachia hybrida Michx.

synonym(s): *Steironema hybridum* (Michx.) Raf. *ex* B. D. Jackson
common name(s): fringed loosestrife
habit: erect, rhizomatous, perennial herb
range: Quebec and ME, s. to FL, w. to TX, n. to ND and Ontario; also AZ, WA, and Alberta
state frequency: uncommon
habitat: wet woods and shores, swamps
notes: —

Lysimachia nummularia L.

synonym(s): —
common name(s): moneywort
habit: creeping, mat-forming, perennial herb
range: Europe; naturalized to Newfoundland, s. to GA, w. to KS, n. to British Columbia
state frequency: occasional
habitat: moist lawns and roadsides
notes: —

Lysimachia punctata L.

synonym(s): —
common name(s): spotted loosestrife, garden loosestrife
habit: erect, rhizomatous, perennial herb
range: Eurasia; naturalized to Newfoundland, s. to NJ, w. to IL, n. to Ontario; also British Columbia
state frequency: uncommon
habitat: roadsides, waste places
notes: —

Lysimachia quadrifolia L.

synonym(s): —
common name(s): whorled loosestrife
habit: erect, long-rhizomatous, perennial herb

range: New Brunswick and ME, s. to GA, w. to AL, n. to IL, MN, and Ontario
state frequency: occasional
habitat: moist or dry, open woods, thickets, shores
notes: Hybridizes with *Lysimachia terrestris* (*q.v.*).

Lysimachia terrestris (L.) B. S. P.
synonym(s): *Lysimachia terrestris* (L.) B. S. P. var. *ovata* (Rand & Redf.) Fern.
common name(s): swamp candles, yellow loosestrife, swamp loosestrife
habit: erect, rhizomatous, perennial herb
range: Newfoundland, s. to GA, w. to TN, n. to IA and Saskatchewan
state frequency: common
habitat: swamps, wet shores and low areas
notes: Hybrids with *L. quadrifolia*, called *L.* ×*producta* (Gray) Fern., and with *L. thyrsiflora*, called *L.*
  ×*commixta* Fern., occur in Maine and can be recognized by their relatively intermediate
  morphology.

Lysimachia thyrsiflora L.
synonym(s): *Naumburgia thyrsiflora* (L.) Duby
common name(s): tufted loosestrife, swamp loosestrife
habit: erect, rhizomatous, perennial herb
range: circumboreal; s. to NJ, w. to CA
state frequency: uncommon
habitat: swamps and bogs
notes: Hybridizes with *Lysimachia terrestris* (*q.v.*).

Lysimachia vulgaris L.
synonym(s): —
common name(s): garden loosestrife
habit: erect, rhizomatous, perennial herb
range: Eurasia; naturalized to Quebec, s. to MD, w. to IL, n. to Ontario
state frequency: uncommon
habitat: roadsides, waste places
notes: —

Primula
(3 species)

KEY TO THE SPECIES
1a. Corollas yellow; calyx 9.0–15.0 mm long; plants cultivated, sometimes escaping near
  establishments .... *P. veris*
1b. Corollas commonly pink to purple with a yellow base; calyx 3.0–11.0 mm long; plants
  native, of ledges, shores, and meadows, these habitats often calcareous
  2a. Sepals 3.0–6.0 mm long; scape 3.0–21.0 cm tall; largest leaves 1.0–4.0(–6.0) cm
    long; bracts of umbel 3.0–6.0 mm long barely, if at all, saccate at the base; capsules
    2.0–3.0 mm in diameter; seeds rounded, their surface nearly smooth ....
    .... *P. mistassinica*
  2b. Sepals 6.0–9.0 mm long; scape 10.0–40.0 cm tall; largest leaves 4.0–8.0 cm long;
    bracts of umbel 6.0–14.0 mm long, definitely saccate at the base; capsules
    2.5–5.0 mm in diameter; seeds angled, their surface evidently reticulate ....
    .... *P. laurentiana*

Primula laurentiana Fern.

synonym(s): *Primula farinosa* L. var. *macropoda* Fern.
common name(s): bird's-eye primrose, Laurentide primrose
habit: perennial herb
range: Labrador, s. to ME, w. and n. to Ontario
state frequency: rare; found mostly on Atlantic coast shores and islands
habitat: wet and/or calcareous rock of ledges, cliffs, and shores
notes: This species is ranked S2 by the Maine Natural Areas Program.

Primula mistassinica Michx.

synonym(s): —
common name(s): bird's-eye primrose, Mistassini primrose
habit: perennial herb
range: Labrador, s. to ME, w. to IL and IA, n. to AK
state frequency: rare
habitat: calcareous ledges and shores
notes: This species is ranked S3 by the Maine Natural Areas Program.

Primula veris L.

synonym(s): —
common name(s): English cowslip
habit: scapose, perennial herb
range: Europe; escaped from cultivation locally
state frequency: very rare
habitat: abandoned lots, waste places
notes: —

Samolus
(1 species)

Samolus valerandi L. ssp. *parviflorus* (Raf.) Hultén

synonym(s): *Samolus parviflorus* Raf.
common name(s): water-pimpernel, brookweed
habit: slender, branched, perennial herb
range: New Brunswick, s. to FL, w. to Mexico and CA, n. to AK
state frequency: rare
habitat: brackish water of estuaries, shallow water and mud of streams and ditches
notes: This species is ranked S2 by the Maine Natural Areas Program.

Trientalis
(1 species)

Trientalis borealis Raf.

synonym(s): *Trientalis americana* Pursh
common name(s): starflower
habit: simple, stoloniferous, perennial herb
range: Labrador, s. to VA and mountains of GA, w. to CA, n. to British Columbia
state frequency: common
habitat: bogs, forests
notes: —

BALSAMINACEAE
(1 genus, 3 species)

Impatiens
(3 species)

The fruit of *Impatiens* is a capsule, the valves of which dehisce rapidly with any disturbance, projecting the seeds for a short distance.

KEY TO THE SPECIES

1a. Leaves opposite or whorled, sharply serrate; corolla blue or purple to pink or white; capsule obovoid .... *I. glandulifera*
1b. Leaves alternate, crenate-serrate; corolla orange-yellow or pale yellow; capsule slender-clavate
   2a. Corolla orange-yellow (bright yellow or white), spotted with red-brown (unspotted); spur 7.0–10.0 mm long, strongly curved, projecting forward; saccate portion of the spurred sepal longer than wide .... *I. capensis*
   2b. Corolla pale yellow (yellow-white or white), sparingly dotted with red-brown; spur 4.0–6.0 mm long, curved at a right angle, not projecting forward; saccate portion of the spurred sepal wider than long .... *I. pallida*

Impatiens capensis Meerb.
synonym(s): *Impatiens biflora* Walt.
common name(s): orange touch-me-not, jewelweed, spotted touch-me-not
habit: branched, annual herb
range: Newfoundland, s. to FL, w. to OK, n. to AK
state frequency: occasional
habitat: moist woods, brooksides, wet, springy places
notes: —

Impatiens glandulifera Royle
synonym(s): —
common name(s): Himalayan balsam
habit: tall, erect, coarse, annual herb
range: Asia; escaped from cultivation to New Brunswick, s. to MA, w. and n. to Ontario
state frequency: uncommon
habitat: roadside thickets, waste places
notes: —

Impatiens pallida Nutt.
synonym(s): —
common name(s): yellow touch-me-not, jewelweed, pale touch-me-not
habit: branched, annual herb
range: Newfoundland, s. to GA, w. to OK, n. to Saskatchewan
state frequency: rare
habitat: wet woods, meadows, and springy places
notes: This species is ranked S2 by the Maine Natural Areas Program.

ERICACEAE
(21 genera, 46 species)

KEY TO THE GENERA FOR USE WITH FLOWERING MATERIAL

1a. Corolla apopetalous, or absent in *Corema*
  2a. Plants lacking chlorophyll, the stems and capsules white to pink (to red) or yellow to brown; leaves reduced, scale-like .... ***Monotropa***
  2b. Plants with chlorophyll, the stems and capsules variously colored; leaves evident, though sometimes scale-like
    3a. Herbs with basal leaves
      4a. Flower solitary; petals widely spreading, forming a rotate corolla .... ***Moneses***
      4b. Flowers borne in racemes; petals inwardly arching, forming a campanulate or subglobose corolla
        5a. Flowers borne on a secund raceme; petals bituberculate at the base; style straight, exserted beyond the petals, with a 10-lobed hypogynous disk at the base .... ***Orthilia***
        5b. Flowers borne on a spirally arranged raceme; petals without tubercles; style either strongly curved and exserted beyond the petals or straight and about equaling the petals, lacking a 10-lobed hypogynous disk at the base .... ***Pyrola***
    3b. Shrubs or suffruticose herbs with alternate or opposite leaves
      6a. Leaves numerous, crowded, linear to narrow-elliptic, about 1.0 mm wide; perianth inconspicuous; stamens, when present, 2 or 4 per flower
        7a. Flowers on pedicels from the axils of leaves, with 4 sepals, 4 petals, 4 stamens (when present), and 6–9 stigmas (when present) .... ***Empetrum***
        7b. Flowers in small, terminal clusters, with 3 or 4 sepals, 0 petals, 2 stamens (when present), and 2–5 stigmas (when present) .... ***Corema***
      6b. Leaves larger and fewer, much wider than 1.0 mm; perianth conspicuous; stamens 5–10 per flower
        8a. Leaves opposite, sometimes verticillate, serrate; stamens 10 per flower .... ***Chimaphila***
        8b. Leaves alternate, entire; stamens 5–8 per flower
          9a. Corolla 4-merous, pink to red; leaves 0.8–1.8 cm long, dotted with black glands on the abaxial surface; flowers with 8 stamens .... ***Vaccinium***
          9b. Corolla 5-merous, white; leaves 2.0–5.0 cm long, densely villous-tomentose on the abaxial surface; flowers with 5–7 stamens .... ***Rhododendron***
1b. Corolla gamopetalous
  10a. Ovary inferior
    11a. Perianth 4-merous; plants low, commonly prostrate or trailing
      12a. Leaves glabrous; ovary wholly inferior .... ***Vaccinium***
      12b. Leaves pubescent; ovary partly inferior .... ***Gaultheria***
    11b. Perianth 5-merous; plants ascending (depressed in extreme conditions)
      13a. Ovary 4- or 5-locular, with numerous ovules; leaves lacking resin dots and stipitate glands .... ***Vaccinium***

13b. Ovary 10-locular, with 10 ovules; leaves resin-dotted or stipitate-glandular .... *Gaylussacia*

10b. Ovary superior

14a. Corolla campanulate to salverform to rotate, not constricted in the apical portion

15a. Corolla rotate, 10-saccate, the anthers fitting in the corolla sacs and later springing forward .... *Kalmia*

15b. Corolla campanulate to salverform, not saccate

16a. Leaves auriculate, opposite, 2.0–4.0 mm long; inflorescence a secund raceme; plants introduced .... *Calluna*

16b. Leaves without auricles, variously arranged, 2.0–200.0 mm long; inflorescence various, but not a secund raceme; plants native (except *Rhododendron maximum*, though native, extensively planted as well)

17a. Corolla 4-merous; leaves 5.0–10.0 mm long, bristly on the abaxial surface .... *Gaultheria*

17b. Corolla 5-merous; leaves various

18a. Leaf blades 2.0–4.0 mm long, bristle-like; anthers spurred; plants with a moss-like aspect .... *Harrimanella*

18b. Leaf blades 4.0–200.0 mm long, broader, not resembling bristles; anthers lacking spurs; plants not with the aspect of mosses

19a. Leaf blades 0.4–1.0 cm long, opposite; plants of alpine habitats, elevation exceeding 1000 m .... *Loiseleuria*

19b. Leaf blades 1.0–20.0 cm long, alternate; plants not of alpine habitats except *Rhododendron lapponicum*

20a. Calyx ebracteate; shrubs depressed to, more commonly, erect; anthers opening by 2 pores .... *Rhododendron*

20b. Calyx subtended by 2 bracts; shrubs prostrate and trailing; anthers opening by longitudinal slits .... *Epigaea*

14b. Corolla globose, ovoid, urceolate, to tubular, slightly to strongly constricted in the apical portion

21a. Leaves opposite; ovary 2- or 3-locular .... *Loiseleuria*

21b. Leaves alternate; ovary 4- or 5-locular

22a. Inflorescence a solitary flower from the axils of leaves; calyx subtended by 2 bracts

23a. Calyx and abaxial surface of leaves lepidote-scaly; plants 30.0–150.0 cm tall; anthers without spurs; plants without wintergreen flavor .... *Chamaedaphne*

23b. Calyx and leaves without lepidote scales; plants up to 20.0 cm tall; each pollen sac of an anther with 2 spurs; plants with wintergreen flavor .... *Gaultheria*

22b. Inflorescence composed of 2 or more flowers (sometimes solitary in *Phyllodoce*); calyx not subtended by 2 bracts (though sometimes the associated pedicel may possess 2 bracts near its base)

24a. Corolla purple; leaves 0.4–1.0 x 0.2 cm; pedicels stipitate-glandular; anthers without spurs .... *Phyllodoce*

24b. Corolla white to pink; leaves 1.0–7.0 x 0.2–5.0 cm; pedicels without stipitate glands; anthers spurred

25a. Stamens spurred at the junction of the filament and anther; inflorescence borne at the apex of the previous year's twig; leaves deciduous .... *Lyonia*

25b. Stamens with a spur attached to the anther; inflorescence borne at the apex of the current year's twig; leaves persistent or marcescent

26a. Leaves pubescent on the abaxial surface, conspicuously revolute; sepals valvate in bud .... *Andromeda*

26b. Leaves glabrous, with flat margins; sepals imbricate in bud ... . *Arctostaphylos*

KEY TO THE GENERA FOR USE WITH FRUITING MATERIAL

1a. Fruit fleshy and indehiscent

2a. Ovary superior

3a. Fruit a capsule, either fleshy or enclosed by a fleshy calyx

4a. Fruit glabrous, with a wintergreen flavor .... *Gaultheria*

4b. Fruit pubescent, without a wintergreen flavor .... *Epigaea*

3b. Fruit a berry-like drupe

5a. Leaves 1.0–5.0 x 0.4–1.0 cm, oblanceolate to oblong-obovate or oblong-spatulate .... *Arctostaphylos*

5b. Leaves 0.25–0.75 x 0.1 cm, linear to narrow-elliptic

6a. Fruits on pedicels from the axils of leaves; drupe juicy when mature; plants trailing; leaves spreading .... *Empetrum*

6b. Fruits in small, terminal clusters; drupe dry when mature; plants erect; leaves ascending .... *Corema*

2b. Ovary partly or wholly inferior

7a. Ovary partly inferior, maturing as a capsule enclosed by a fleshy calyx with a wintergreen flavor; fruit white .... *Gaultheria*

7b. Ovary wholly inferior, maturing as a berry, sweet or tart, but without a wintergreen flavor; fruit red or blue to black

8a. Ovary 4- or 5-locular; berry with numerous seeds; leaves lacking resin dots and stipitate glands (except *Vaccinium vitis-idaea* with small, black glands) .... *Vaccinium*

8b. Ovary 10-locular; berry with 10 seeds; leaves resin-dotted or stipitate-glandular .... *Gaylussacia*

1b. Fruit dry and dehiscent

9a. Plants lacking chlorophyll, the herbaceous stems and capsules white to pink (to red) or yellow to brown; leaves reduced, scale-like .... *Monotropa*

9b. Plants with chlorophyll, the stems and capsules variously colored, woody or herbaceous; leaves evident (though sometimes scale-like)

10a. Plants herbs; leaves basal

11a. Capsule solitary, dehiscing from apex to base, its valves with glabrous margins .... *Moneses*

11b. Capsules more than 1, borne in a raceme, dehiscing from base to apex, its valves with arachnoid-pubescent margins

12a. Capsules borne on a secund raceme, 3.0–5.0 mm wide .... *Orthilia*
12b. Capsules borne on a spirally-arranged raceme, 4.0–9.0 mm wide ....
.... *Pyrola*
10b. Plants shrubs or suffruticose herbs; leaves alternate, opposite, or whorled
   13a. Leaves alternate (appearing crowded and imbricate in *Harrimanella* and
     Phyllodoce)
     14a. Capsule fleshy or enclosed in a fleshy calyx
       15a. Fruit glabrous, with a wintergreen flavor .... *Gaultheria*
       15b. Fruit pubescent, without a wintergreen flavor .... *Epigaea*
     14b. Capsule dry, not fleshy
       16a. Leaves crowded, imbricate, the blades 0.2–1.0 cm long; alpine plants
        occurring at elevation exceeding 1000 m
        17a. Leaves 2.0–4.0 mm long; capsules loculicidal; capsules and
         pedicels without glands .... *Harrimanella*
        17b. Leaves 4.0–10.0 mm long; capsules septicidal; capsules and
         pedicels minutely glandular .... *Phyllodoce*
       16b. Leaves neither conspicuously crowded nor imbricate, the blades
        1.0–20.0 cm long; plants of elevation less than 1000 m (except the
        alpine *Rhododendron lapponicum*)
        18a. Capsules septicidal
        19a. Capsules as wide as long .... *Kalmia*
        19b. Capsules longer than wide .... *Rhododendron*
        18b. Capsules loculicidal
        20a. Leaves evidently revolute and glaucous .... *Andromeda*
        20b. Leaves neither revolute nor glaucous
         21a. Leaves persistent, lepidote-scaly; suture of the capsule,
          prior to dehiscence, a thin, pale line *ca.* 0.1 mm wide ....
          .... *Chamaedaphne*
         21b. Leaves deciduous, not lepidote-scaly; suture of the
          capsule, prior to dehiscence, a thick, pale strip *ca.* 0.5 mm
          wide .... *Lyonia*
   13b. Leaves opposite or whorled
     22a. Capsules 4-locular; corolla marcescent on the fruit; leaves auriculate;
      plants introduced .... *Calluna*
     22b. Capsules 2-, 3-, or 5-locular, but not 4-locular; corolla deciduous from
      the fruit; leaves without auricles; plants native
      23a. Capsules 5-locular; leaves 10.0–100.0 mm long; plants
       10.0–100.0 cm tall
       24a. Plants shrubs; leaves entire; capsules septicidal .... *Kalmia*
       24b. Plants suffruticose herbs; leaves toothed; capsules loculicidal ......
        .... *Chimaphila*
      23b. Capsules 2- or 3-locular; leaves 5.0–8.0 mm long; plants up to
       10.0 cm tall .... *Loiseleuria*

Andromeda
(1 species)

Andromeda polifolia L. var. **glaucophylla** (Link) DC.
synonym(s): *Andromeda glaucophylla* Link
common name(s): bog rosemary
habit: low, evergreen shrub
range: Labrador, s. to NJ, w. to IN, n. to MN and Manitoba
state frequency: occasional
habitat: bogs
notes: —

Arctostaphylos
(2 species)

KEY TO THE SPECIES

1a. Leaves entire, persistent and living throughout the season, not rugose-reticulate-veined; twigs pubescent; drupe red; pyrenes partly or entirely concrescent; plants of elevation less than 1000 m .... *A. uva-ursi*
1b. Leaves serrate, marcescent, conspicuously rugose-reticulate-veined; twigs glabrous; drupe black; pyrenes distinct; alpine plants of elevation exceeding 1000 m ...... *A. alpina*

Arctostaphylos alpina (L.) Spreng.
synonym(s): *Mairania alpina* (L.) Desv.
common name(s): alpine bearberry
habit: trailing shrub
range: Greenland, s. to higher mountains of ME and NH, w. and n. to AK
state frequency: very rare
habitat: alpine areas
notes: This species is ranked S1 by the Maine Natural Areas Program.

Arctostaphylos uva-ursi (L.) Spreng.
synonym(s): *Arctostaphylos uva-ursi* (L.) Spreng. var. *coactilis* Fern. & J. F. Macbr., *Uva-ursi uva-ursi* (L.) Britt.
common name(s): bearberry
habit: trailing to prostrate, evergreen shrub
range: circumboreal; s. to VA, w. to CA
state frequency: occasional
habitat: exposed, sandy or rocky soil
notes: —

Calluna
(1 species)

Calluna vulgaris (L.) Hull
synonym(s): —
common name(s): heather, Scotch heather
habit: ascending, evergreen shrub
range: Europe; naturalized to Newfoundland, s. along coast to NJ; also MI
state frequency: rare
habitat: sandy areas
notes: —

Chamaedaphne
(1 species)

Chamaedaphne calyculata (L.) Moench
synonym(s): <*C. c.* var. *angustifolia*> = *Cassandra calyculata* (L.) D. Don var. *angustifolia* (Ait.)
Seymour; <*C. c.* var. *latifolia*> = *Cassandra calyculata* (L.) D. Don var. *latifolia* (Ait.) Seymour
common name(s): leatherleaf, Cassandra
habit: branched, evergreen shrub
range: circumboreal; s. to NC, w. to British Columbia
state frequency: occasional
habitat: bogs, pond margins
key to the subspecies:
  1a. Leaves 0.25–0.40 times as wide as long; sepals acute to obtuse at the apex ....
  .... *C. c.* var. *angustifolia* (Ait.) Rehd.
  1b. Leaves *ca.* 0.5 times as wide as long; sepals obtuse to rounded at the apex ....
  .... *C. c.* var. *latifolia* (Ait.) Fern.
notes: —

Chimaphila
(2 species)

KEY TO THE SPECIES
1a. Leaves more or less oblanceolate, acute to obtuse at the apex, tapering to the base, not
    marked with white .... *C. umbellata*
1b. Leaves more or less lanceolate, acute to acuminate at the apex, tapering to rounded at the
    base, marked with white along the midvein and primary veins .... *C. maculata*

Chimaphila maculata (L.) Pursh
synonym(s): —
common name(s): spotted wintergreen
habit: low, evergreen, suffruticose herb
range: ME, s. to GA, w. to AL, n. to MI and Ontario
state frequency: rare
habitat: dry woods, often sandy soil
notes: This species is ranked S2 by the Maine Natural Areas Program.

Chimaphila umbellata (L.) W. Bart. ssp. *cisatlantica* (Blake) Hultén
synonym(s): —
common name(s): pipsissewa, prince's-pine
habit: low, evergreen, suffruticose herb
range: Quebec, s. to GA, w. to IL, n. to MN and Ontario
state frequency: occasional
habitat: dry woods, often sandy soil
notes: —

Corema
(1 species)

Corema conradii (Torr.) Torr. *ex* Loud.
synonym(s): —
common name(s): poverty-grass, broom-crowberry

habit: low, branched, evergreen shrub
range: Newfoundland, s. to NJ
state frequency: uncommon
habitat: sandy or rocky soils, often near the coast
notes: —

Empetrum
(2 species)

Most members of this genus commonly have unisexual flowers. Our species, however, normally have bisexual flowers. Often confused based on similarities in habit, Maine's 2 species of crowberry are easily separated.

KEY TO THE SPECIES
1a. Twigs densely white-villous or -tomentose; drupe purple .... *E. eamesii*
1b. Twigs stipitate-glandular; drupe black .... *E. nigrum*

Empetrum eamesii Fern. & Wieg. ssp. *atropurpureum* (Fern. & Wieg.) D. Löve
synonym(s): *Empetrum atropurpureum* Fern. & Wieg., *Empetrum rubrum* Vahl var. *atropurpureum* (Fern. & Wieg.) R. Good
common name(s): purple crowberry, red crowberry
habit: low, trailing shrub
range: Labrador, s. to ME, w. to NY; also MI and MN
state frequency: uncommon
habitat: exposed granitic rock and gravel
notes: —

Empetrum nigrum L.
synonym(s): —
common name(s): black crowberry
habit: low, trailing shrub
range: circumboreal; s. to coastal NY, w. to CA
state frequency: occasional
habitat: exposed, peaty and rocky areas
notes: —

Epigaea
(1 species)

Epigaea repens L.
synonym(s): *Epigaea repens* L. var. *glabrifolia* Fern.
common name(s): mayflower, trailing-arbutus
habit: prostrate, trailing, evergreen shrub
range: Newfoundland, s. to FL, w. to MS, n. to Saskatchewan
state frequency: uncommon
habitat: sandy woods
notes: —

Gaultheria
(2 species)

Common woodland plants with wintergreen flavor.

KEY TO THE SPECIES

1a. Stems erect; leaves 2.0–5.0 cm long, glabrous; flowers with 5 petals and superior ovary; fruit red .... ***G. procumbens***

1b. Stems prostrate or trailing; leaves 0.5–1.0 cm long, bristly on the abaxial surface; flowers with 4 petals and a partly inferior ovary; fruit white .... ***G. hispidula***

Gaultheria hispidula (L.) Muhl. *ex* Bigelow

synonym(s): *Chiogenes hispidula* (L.) Torr. & Gray
common name(s): creeping snowberry
habit: prostrate or trailing, evergreen shrub
range: Labrador, s. to NJ and mountains of NC, w. to MN, n. to British Columbia
state frequency: occasional
habitat: wet, mossy woods, bogs
notes: —

Gaultheria procumbens L.

synonym(s): —
common name(s): wintergreen, teaberry, checkerberry
habit: low, creeping, rhizomatous subshrub
range: Newfoundland, s. to VA and mountains of GA and AL, w. to IN, n. to Manitoba
state frequency: common
habitat: dry or wet woods
notes: —

Gaylussacia
(2 species)

KEY TO THE SPECIES

1a. Leaves, bracts, sepals, and pedicels with resin dots; plants 3.0–20.0 dm tall; fruit glabrous; leaves not lustrous on the adaxial surface .... ***G. baccata***

1b. Leaves, bracts, sepals, and pedicels with stipitate glands; plants 2.0–5.0 dm tall; fruit stipitate-glandular; leaves lustrous on the adaxial surface .... ***G. dumosa***

Gaylussacia baccata (Wangenh.) K. Koch

synonym(s): —
common name(s): black huckleberry
habit: branched shrub
range: Newfoundland, s. to GA, w. to MO, n. to Saskatchewan
state frequency: occasional
habitat: dry to moist woods
notes: —

Gaylussacia dumosa (Andr.) Torr. & Gray var. ***bigeloviana*** Fern.

synonym(s): —
common name(s): dwarf huckleberry
habit: low, rhizomatous shrub
range: Newfoundland, s. along coastal plain to VA
state frequency: uncommon
habitat: bogs, wet, peaty soils
notes: —

Harrimanella
(1 species)

Harrimanella hypnoides (L.) Coville
synonym(s): *Cassiope hypnoides* (L.) D. Don
common name(s): moss-plant, moss bell-heather
habit: tufted, matted shrub
range: circumboreal; s. to ME, w. to Lake Superior
state frequency: very rare
habitat: wet seeps and crevices in alpine areas
notes: This species is ranked S1 by the Maine Natural Areas Program.

Kalmia
(3 species)

KEY TO THE SPECIES
1a. Leaves whorled [3 at each node]; inflorescence lateral; corolla 0.6–1.2 cm wide;
    capsules 3.0–5.0 mm wide .... *K. angustifolia*
1b. Leaves opposite or alternate; inflorescence terminal; corolla 1.0–2.5 cm wide; capsules
    6.0–8.0 mm wide
    2a. Leaves mostly alternate; leaf blades 5.0–10.0 cm long, lanceolate to elliptic, borne on
        petioles 1.0–2.0 cm long, glabrous; corolla 1.0–1.6 cm wide .... *K. latifolia*
    2b. Leaves opposite; leaf blades 1.0–4.0 cm long, linear to lanceolate, subsessile,
        pubescent on the abaxial surface; corolla 2.0–2.5 cm wide .... *K. polifolia*

Kalmia angustifolia L.
synonym(s): —
common name(s): sheep laurel, lambkill
habit: low, branched shrub
range: Labrador, s. to VA, w. to TN, n. to Manitoba
state frequency: common
habitat: sterile, wet or dry, usually acid, soil
notes: —

Kalmia latifolia L.
synonym(s): —
common name(s): mountain laurel
habit: shrub or small tree
range: ME, s. to FL, w. to LA, n. to IN
state frequency: rare
habitat: rocky or sandy woods
notes: This species is ranked S3 by the Maine Natural Areas Program.

Kalmia polifolia Wangenh.
synonym(s): *Kalmia glauca* Ait.
common name(s): bog laurel, pale laurel
habit: low shrub
range: Labrador, s. to NJ, w. to CA, n. to AK
state frequency: occasional
habitat: bogs, peaty soil, mountain summits
notes: —

Loiseleuria
(1 species)

This genus will be transferred to *Kalmia* based on DNA evidence.

Loiseleuria procumbens (L.) Desv.
synonym(s): *Chamaecistus procumbens* (L.) Kuntze
common name(s): alpine-azalea
habit: low, branched, evergreen shrub
range: circumboreal; s. to ME, w. to Alberta
state frequency: very rare
habitat: exposed alpine areas
notes: This species is ranked S1 by the Maine Natural Areas Program.

Lyonia
(1 species)

Shrubs with 2 red bud scales and shreddy-exfoliating bark.

Lyonia ligustrina (L.) DC.
synonym(s): *Xolisma ligustrina* (L.) Britt.
common name(s): maleberry
habit: shrub
range: ME, s. to FL, w. to TX, n. to OH, KY, and NY
state frequency: occasional
habitat: wet soil, swamps
notes: —

Moneses
(1 species)

Moneses uniflora (L.) Gray
synonym(s): *Pyrola uniflora* L.
common name(s): one-flowered shinleaf, one-flowered pyrola, single delight
habit: perennial herb
range: circumboreal; s. to CT, w. to UT and OR
state frequency: uncommon
habitat: bogs, mossy woods
notes: —

Monotropa
(2 species)

KEY TO THE SPECIES
1a. Inflorescence a solitary flower; stems glabrous, usually white (rarely pink or red); flower with 0–5 sepals; anthers dehiscing by 2 clefts across the top; stigmas glabrous .... *M. uniflora*
1b. Inflorescence a raceme with 2–16 flowers; stems pubescent, usually yellow to light brown (rarely pink or red); terminal flower with 5 sepals, the lower flowers with 3–4 sepals; anthers dehiscing by a single cleft; stigmas villous on the margin .... *M. hypopithys*

Monotropa hypopithys L.
synonym(s): *Hypopitys americana* (DC.) Small
common name(s): pinesap, false beechdrops
habit: achlorophyllous, mycoparasitic, perennial herb
range: interruptedly circumboreal at lower latitudes; s. to FL, w. to Mexico
state frequency: uncommon
habitat: humus of moist or dry woods
notes: —

Monotropa uniflora L.
synonym(s): —
common name(s): Indian pipe
habit: achlorophyllous, mycoparasitic, perennial herb
range: Newfoundland, s. to FL, w. to CA, n. to British Columbia; also Columbia and eastern Asia
state frequency: occasional
habitat: humus of rich woods
notes: The conspicuous nodding flower matures as an erect capsule.

Orthilia
(1 species)

Orthilia secunda (L.) House
synonym(s): *Pyrola secunda* L., *Pyrola secunda* L. var. *obtusata* Turcz.
common name(s): one-sided shinleaf, one-sided pyrola
habit: low, evergreen, perennial herb
range: circumboreal, s. to VA, w. to NM and CA
state frequency: uncommon
habitat: wet, mossy woods and bogs
notes: —

Phyllodoce
(1 species)

This species is sometimes confused vegetatively with *Empetrum*. However, *Phyllodoce* has minutely serrulate leaf margins, whereas those of *Empetrum* are entire.

Phyllodoce caerulea (L.) Bab.
synonym(s): —
common name(s): mountain-heath
habit: low, branched, evergreen subshrub
range: circumboreal, s. in the east to mountains of ME and NH
state frequency: very rare
habitat: alpine areas
notes: This species is ranked S1 by the Maine Natural Areas Program.

Pyrola
(5 species)

Petioles tend to be longer in deeply shaded plants, causing blade to petiole ratios to be less valuable as an identifier under these conditions.

KEY TO THE SPECIES

1a. Styles short and straight; stigmas broadly peltate, 5-lobed, without a subterminal collar; anthers connivent around the style, up to 1.0 mm long; petals 3.0–5.0 mm long; scape 5.0–15.0 cm tall .... *P. minor*

1b. Styles elongate and strongly declined; stigmas capitate, with a subterminal collar; anthers not connivent around the style, 2.0–3.5 mm long; petals 4.0–10.0 mm long; scape 10.0–30.0 cm tall

    2a. Leaf blades 1.0–3.0(–4.0) cm long, usually shorter than the petiole; anther tubes 0.6–0.8 mm long; petals green-white .... *P. chlorantha*

    2b. Leaf blades 2.5–7.0 cm long, equaling to exceeding the length of the petiole; anther tubes 0.2–0.3 mm long; petals white or pink to pale purple

        3a. Leaf blades elliptic or oblong to obovate, usually longer than the petiole; sepals as wide or wider than long, about 0.25 times as long as the petals; bracts below the flowers numbering 0–3, narrow-lanceolate, not clasping the scape .... *P. elliptica*

        3b. Leaf blades broad-elliptic to reniform, usually about equaling the length of the petiole; sepals longer than wide, about 0.35–0.6 times as long as the petals; bracts below the flowers numbering 1–5, broad-lanceolate, clasping the scape

            4a. Petals white; sepals oblong or ovate-oblong, not overlapping at the base, firm, 3- to 5-nerved .... *P. americana*

            4b. Petals pink to pale purple; sepals deltate, overlapping at the base, thin, nerveless or obscurely nerved .... *P. asarifolia*

Pyrola americana Sweet

synonym(s): *Pyrola rotundifolia* L. var. *americana* (Sweet) Fern.
common name(s): round-leaved pyrola
habit: rhizomatous, evergreen, perennial herb
range: circumboreal; s. to NC, w. to MN and Manitoba
state frequency: occasional
habitat: dry or moist woods
notes: —

Pyrola asarifolia Michx.

synonym(s): *Pyrola asarifolia* Michx. var. *purpurea* (Bunge) Fern.
common name(s): pink pyrola, liver-leaf pyrola
habit: rhizomatous, evergreen, perennial herb
range: Newfoundland, s. to MA, w. to NM, n. to AK
state frequency: rare
habitat: moist woods, bogs
notes: This species is ranked S3 by the Maine Natural Areas Program.

Pyrola chlorantha Sw.

synonym(s): *Pyrola virens* Schreb.
common name(s): greenish-flowered pyrola
habit: rhizomatous, evergreen, perennial herb
range: circumboreal; s. to MD, w. to AZ
state frequency: uncommon
habitat: dry, often coniferous, woods
notes: —

Pyrola elliptica Nutt.
synonym(s): —
common name(s): shinleaf
habit: rhizomatous, evergreen, perennial herb
range: Newfoundland, s. to DE, w. to NM, n. to British Columbia
state frequency: occasional
habitat: dry or moist woods
notes: —

Pyrola minor L.
synonym(s): —
common name(s): little shinleaf
habit: rhizomatous, evergreen, perennial herb
range: circumboreal; s. in the east to ME, NH, and VT
state frequency: rare
habitat: cool, moist woods
notes: This species is ranked S3 by the Maine Natural Areas Program.

Rhododendron
(5 species)

KEY TO THE SPECIES

1a. Corolla apopetalous, white; capsules opening from base to apex; leaves densely white-tomentose on the abaxial surface, turning rusty-tomentose .... *R. groenlandicum*
1b. Corolla gamopetalous, white to purple, often spotted or striped; capsules opening from apex to base; leaves variously scaly, scurfy, or pubescent on the abaxial surface, but not white- or rusty-tomentose
  2a. Leaves lepidote-scaly; capsules 4.0–8.0 mm long; plants 1.0–3.0 dm tall, depressed, alpine .... *R. lapponicum*
  2b. Leaves not lepidote-scaly; capsules 7.0–20.0 mm long; plants 1.5–30.0 dm tall, erect, non-alpine
    3a. Leaves coriaceous, persistent and living throughout the year, the blade 10.0–20.0 cm long; sepals 2.0–4.0 mm long .... *R. maximum*
    3b. Leaves membranaceous, deciduous, the blade 2.0–6.0 cm long; sepals less than 1.0 mm long
      4a. Corolla strongly zygomorphic, divided into 3 portions—the upper portion 3-lobed, the 2 lower portions narrow and unlobed; flowers each with 10 stamens, appearing with or before the leaves; styles 1.5–2.0 cm long; capsules puberulent, sparsely, if at all, setose; twigs glabrous or soft-puberulent .... *R. canadense*
      4b. Corolla weakly zygomorphic; flowers each with 5 stamens, appearing after the leaves; styles 4.0–5.0 cm long; capsules evidently strigose-setose; twigs coarsely setose .... *R. viscosum*

Rhododendron canadense (L.) Torr.
synonym(s): *Rhodora canadensis* L.
common name(s): rhodora
habit: branched shrub to 1.0 m
range: Newfoundland, s. to NJ, w. to PA, n. to Ontario

state frequency: occasional
habitat: bogs, moist thickets
notes: —

Rhododendron groenlandicum (Oeder) Kron & Judd
synonym(s): *Ledum groenlandicum* Oeder
common name(s): Labrador-tea
habit: branched, evergreen shrub
range: Greenland, s. to NJ, w. to OR, n. to AK
state frequency: occasional
habitat: bogs, rocky slopes
notes: —

Rhododendron lapponicum (L.) Wahlenb.
synonym(s): —
common name(s): Lappland rosebay
habit: low, mat-forming, evergreen shrub
range: circumboreal; s. to ME and NH, w. to WI
state frequency: very rare
habitat: alpine areas
notes: This species is ranked S1 by the Maine Natural Areas Program.

Rhododendron maximum L.
synonym(s): —
common name(s): great-laurel, rosebay, great rhododendron
habit: evergreen shrub or small tree
range: Nova Scotia and ME, s. to GA, w. to AL, n. to Ontario
state frequency: very rare
habitat: pond margins, moist woods, swamps
notes: This species is ranked S1 by the Maine Natural Areas Program. *Rhododendron maximum* is commonly planted and therefore is seen more commonly than the actual frequency of the wild populations.

Rhododendron viscosum (L.) Torr.
synonym(s): *Azalea viscosa* L.
common name(s): clammy azalea
habit: branched shrub
range: ME, s. to FL, w. and n. to OH
state frequency: very rare
habitat: wet woods, swamps, shores
notes: This species is ranked S1 by the Maine Natural Areas Program.

Vaccinium
(10 species)

Gleason and Cronquist (1991) report *Vaccinium stramineum* present in Maine, but voucher specimens are unknown.

KEY TO THE SPECIES
1a. Leaves coriaceous, lustrous, evergreen; shrubs prostrate or trailing, to 2.0 dm tall; corolla with 4 lobes; fruit usually red or dark red, tart at maturity

2a. Corolla divided to near the middle; leaves with erect, black glands on the abaxial surface; pedicels less than 0.5 cm long .... *V. vitis-idaea*

2b. Corolla divided to below the middle; leaves without glands on the abaxial surface; pedicels 1.0–5.0 cm long

  3a. Leaf blades elliptic-oblong, rounded at the apex, with flat or slightly revolute margins; pedicels with green, leaf-like bracteoles 2.0–4.0 x 1.0–2.0 mm; styles 5.0–7.0 mm long; berry 1.0–1.5 cm in diameter .... *V. macrocarpon*

  3b. Leaf blades ovate to triangular, acute at the apex, with strongly revolute margins; pedicels with red, scale-like bracteoles 0.5–2.0 x less than 1.0 mm; styles 3.0–4.0 mm long; berry 0.6–1.2 cm in diameter .... *V. oxycoccos*

1b. Leaves herbaceous, not lustrous, deciduous; shrubs prostrate or upright, never trailing, 0.5–30.0 dm tall; corolla with 4 or, more commonly, 5 lobes; fruit blue to blue-black, sweet at maturity

  4a. Twigs papillose; anthers lacking spurs

    5a. Shrubs 1.0–3.0 m high; leaves commonly longer than 4.5 cm ..... *V. corymbosum*

    5b. Shrubs shorter than 1.0 m; leaves seldom as long as 4.5 cm

      6a. Leaves entire or nearly so

        7a. Twigs and leaves evidently pubescent; bud scale apex acute; leaf blades not glaucous, elliptic to narrow-elliptic .... *V. myrtilloides*

        7b. Twigs and leaves sparsely pubescent at most; bud scale apex rounded; leaf blades glaucous, ovate to broad-elliptic .... *V. pallidum*

      6b. Leaves finely serrulate

        8a. Leaves 2.0–6.0 mm wide; plants rarely reaching 9.0 cm tall; corolla 3.0–4.0 mm long .... *V. boreale*

        8b. Leaves 6.0–22.0 mm wide; plants 9.0–51.0 cm tall; corolla 4.0–8.0 mm long

          9a. Leaf blades regularly serrulate, not glaucous, mostly 1.0–3.0 cm long, *ca.* 0.35 times as wide .... *V. angustifolium*

          9b. Leaf blades often irregularly serrulate, glaucous, the larger ones 2.5–5.0 cm long, *ca.* 0.5 times as wide .... *V. pallidum*

  4b. Twigs not papillose; anthers with a pair of upward-projecting spurs in addition to the terminal tubules

    10a. Flowers mostly 5-merous; bud scales 2; leaves finely serrulate .... *V. cespitosum*

    10b. Flowers mostly 4-merous; bud scales more than 2; leaves entire .... *V. uliginosum*

Vaccinium angustifolium Ait.

synonym(s): *Vaccinium angustifolium* Ait. var. *hypolasium* Fern., *Vaccinium angustifolium* Ait. var. *laevifolium* House, *Vaccinium angustifolium* Ait. var. *nigrum* (Wood) Dole
common name(s): low sweet blueberry, common lowbush blueberry
habit: low shrub
range: Labrador, s. to VA and mountains of NC, w. to IL and IA, n. to Manitoba
state frequency: common
habitat: dry sandy or rocky soil of barrens, ledges, and mountain tops
notes: —

Vaccinium boreale Hall & Aalders

synonym(s): —
common name(s): sweet hurts, alpine blueberry
habit: low, diminutive shrub
range: Newfoundland, s. to mountains of ME, w. to NY, n. to Quebec
state frequency: very rare
habitat: exposed, often high elevation, rocky and gravelly areas
notes: This species is ranked S1 by the Maine Natural Areas Program. *Vaccinium boreale* usually
 flowers 10–20 days earlier than *V. angustifolium* at the same site.

Vaccinium cespitosum Michx.

synonym(s): —
common name(s): dwarf bilberry
habit: depressed to ascending shrub
range: circumboreal; s. to ME, NH, and VT, w. to CA
state frequency: uncommon
habitat: gravelly or rocky shores and clearings, casually alpine
notes: —

Vaccinium corymbosum L.

synonym(s): *Vaccinium atrococcum* (Gray) Heller, *Vaccinium caesariense* Mackenzie, *Vaccinium
 corymbosum* L. var. *albiflorum* (Hook.) Fern., *Vaccinium corymbosum* L. var. *glabrum* Gray
common name(s): highbush blueberry
habit: shrub
range: New Brunswick, s. to FL, w. to TX, n. to WI
state frequency: uncommon
habitat: bogs, swamps, low woods
notes: —

Vaccinium macrocarpon Ait.

synonym(s): —
common name(s): large cranberry, American cranberry
habit: low, trailing, evergreen shrub
range: Newfoundland, s. to mountains of NC, w. to IL, n. to MN and Manitoba
state frequency: uncommon
habitat: bogs and wet, low areas
notes: —

Vaccinium myrtilloides Michx.

synonym(s): *Vaccinium canadense* Kalm *ex* A. Rich.
common name(s): velvet-leaf blueberry, sour-top blueberry
habit: shrub
range: Newfoundland, s. to mountains of VA, w. to IA, n. to British Columbia
state frequency: occasional
habitat: moist woods and clearings, bogs, swamps
notes: —

Vaccinium oxycoccos L.

synonym(s): *Vaccinium oxycoccos* L. var. *ovalifolium* Michx., *Oxycoccos oxycoccos* (L.) MacM.
common name(s): small cranberry, bog cranberry
habit: slender, diminutive, creeping, evergreen shrub

range: circumboreal; s. to mountains of NC, w. to OR
state frequency: occasional
habitat: bogs
notes: —

Vaccinium pallidum Ait.

synonym(s): *Vaccinium vacillans* Kalm *ex* Torr.
common name(s): hillside blueberry
habit: low shrub
range: ME, s. to GA, w. to OK, n. to MN
state frequency: uncommon
habitat: dry, upland woods and fields
notes: —

Vaccinium uliginosum L.

synonym(s): *Vaccinium uliginosum* L. var. *alpinum* Bigelow
common name(s): alpine bilberry, bog bilberry
habit: low, depressed to ascending, mat-forming, stoloniferous shrub
range: circumboreal; s. to ME, NH, and VT, w. to CA
state frequency: uncommon
habitat: rocky barrens, mountain slopes
notes: —

Vaccinium vitis-idaea L. ssp. *minus* (Lodd.) Hultén

synonym(s): *Vitis-idaea vitis-idaea* (L.) Britt.
common name(s): mountain cranberry, lingonberry, ling berry
habit: prostrate, creeping, evergreen shrub
range: Newfoundland, s. to ME, w. to MN, n. to British Columbia
state frequency: occasional
habitat: dry, rocky areas, bogs, peaty soil
notes: —

CONVOLVULACEAE
(4 genera, 11 species)

KEY TO THE GENERA

1a. Plants parasitic, with haustoria, lacking chlorophyll; stems pink-yellow to orange, with scale-like leaves .... *Cuscuta*
1b. Plants autotrophic, without haustoria, with chlorophyll; stems more or less green, with foliaceous leaves
    2a. Stigmas 1, terminated by 3 small lobes; ovary and capsule 3-locular .... *Ipomoea*
    2b. Stigmas 2, linear to ovoid, unlobed; ovary and capsule 1-, 2-, or 4-locular
        3a. Pedicels subtended by 2 foliaceous bracts that are positioned near and partly conceal the calyx; corolla 2.0–7.0 cm long; stigmas ovoid to ellipsoid .... *Calystegia*
        3b. Pedicels subtended by small bracts that are positioned well below the calyx and do not conceal it; corolla 1.5–2.5 cm long; stigmas linear .... *Convolvulus*

Calystegia
(3 species)

KEY TO THE SPECIES

1a. Stems erect, only the tip twining; leaf blades broad-cuneate to subcordate at the base; leaves subtending the flowers with petioles less than 0.5 times the length of the leaf midvein, those of the basal portion of the plant conspicuously reduced; bracts subtending the calyx cuneate at the base .... ***C. spithamaea***
1b. Stems twining or trailing; leaf blades sagittate to hastate at the base leaves subtending the flowers with petioles more than 0.5 times the length of the leaf midvein, those of the basal portion of the plant only slightly reduced; bracts subtending the calyx rounded to cordate at the base
   2a. Bracts subtending the calyx mostly 1.5–3.5 cm long; stamens 2.0–3.3 cm long, unmodified; corolla 4.0–7.0 cm long .... ***C. sepium***
   2b. Bracts subtending the calyx mostly 0.8–1.5(–2.5) cm long; stamens, if present, 1.2–1.6 cm long, often modified, forming a second whorl of connate petals; corolla 2.0–4.0 cm long .... ***C. pellita***

Calystegia pellita (Ledeb.) G. Don
synonym(s): *Convolvulus pellitus* Ledeb.
common name(s): —
habit: climbing to trailing, rhizomatous, perennial herb
range: Asia; escaped from cultivation to ME, s. to VA, w. to MO, n. to MI
state frequency: rare
habitat: fields, roadsides, waste places
notes: —

Calystegia sepium (L.) R. Br.
synonym(s): *Convolvulus sepium* L.
common name(s): hedge-bindweed, wild-morning glory
habit: twining to trailing, rhizomatous, perennial herb
range: Newfoundland, s. to FL, w. to NM and CA, n. to British Columbia
state frequency: occasional
habitat: thickets, shores, especially in coastal areas
notes: —

Calystegia spithamaea (L.) Pursh
synonym(s): *Convolvulus spithamaeus* L.
common name(s): upright-bindweed, low-bindweed
habit: erect to ascending, simple or branched, rhizomatous, perennial herb
range: ME, s. to VA and mountains of GA, w. to MO, n. to MN and Ontario
state frequency: very rare
habitat: rocky or sandy, open areas, fields, open woods
notes: This species is ranked S1 by the Maine Natural Areas Program.

Convolvulus
(1 species)

Convolvulus arvensis L.
synonym(s): —
common name(s): field-bindweed

habit: slender, trailing or climbing, perennial herb
range: Europe; naturalized throughout much of North America
state frequency: uncommon
habitat: fields, roadsides, waste places
notes: —

Cuscuta
(5 species)

Twining, parasitic herbs lacking chlorophyll, with reduced, attenuate, alternate, scale-like leaves.

KEY TO THE SPECIES

1a. Stigmas filiform, elongate-tapering, much taller than wide; fruit a pyxis, circumscissile near the base

    2a. Perianth mostly 5-merous; lobes of the calyx acute at the apex; scale-like appendages near the stamens simple and fringed; seeds up to 1.0 mm long; plants usually parasitic on *Trifolium* .... *C. epithymum*

    2b. Perianth mostly 4-merous; lobes of the calyx obtuse at the apex; scale-like appendages near the stamens bifid and lacking a fringe; seeds *ca.* 1.5 mm long; plants parasitic on various herbs and shrubs .... *C. europaea*

1b. Stigmas capitate, dilated near the apex, not or only slightly taller than wide; fruit a utricle or an irregularly rupturing capsule

    3a. Perianth mostly 5-merous; ovary and fruit subglobose; styles 1.0–1.5 mm long; fruit tipped by a thickened stylopodium .... *C. gronovii*

    3b. Perianth mostly 4-merous; ovary and fruit depressed-globose; styles 0.5–1.5 mm long; fruit without a thickened summit

        4a. Lobes of the corolla acute; calyx equaling or exceeding the length of the connate portion of the corolla; marcescent corolla cupping the base of the fruit; fruit 2- to 4-seeded; plants commonly parasitic on members of the Polygonaceae .... *C. polygonorum*

        4b. Lobes of the corolla obtuse; calyx shorter than the length of the connate portion of the corolla; marcescent corolla capping the summit of the fruit; fruit 1- or 2-seeded; plants parasitic on various herbs and shrubs .... *C. cephalanthi*

Cuscuta cephalanthi Engelm.
synonym(s): —
common name(s): buttonbush dodder
habit: coarse, climbing, annual herb
range: New Brunswick, s. to FL, w. to NM, n. to WA and British Columbia
state frequency: very rare
habitat: saltmarshes, river banks
notes: —

Cuscuta epithymum L.
synonym(s): —
common name(s): clover dodder, legume dodder
habit: slender, branched, annual herb
range: Europe; naturalized throughout much of North America
state frequency: rare
habitat: clover fields
notes: —

Cuscuta europaea L.

synonym(s): —
common name(s): —
habit: twining, perennial herb
range: Europe; adventive locally
state frequency: very rare
habitat: hedgerows
notes: —

Cuscuta gronovii Willd. *ex* J. A. Schultes

synonym(s): —
common name(s): common dodder
habit: slender to coarse, annual herb
range: New Brunswick, s. to FL, w. to AZ, n. to Manitoba
state frequency: uncommon; more abundant in coastal areas
habitat: meadows, shores, low areas
notes: —

Cuscuta polygonorum Engelm.

synonym(s): —
common name(s): smartweed dodder
habit: slender, annual herb
range: ME, s. to MD, w. to KS, n. to ND and Ontario
state frequency: very rare
habitat: low areas
notes: —

Ipomoea
(2 species)

KEY TO THE SPECIES

1a. Leaves usually entire; sepals mostly 1.0–1.5 cm long, acute to acuminate at the apex; corolla 4.0–6.0 cm long .... *I. purpurea*
1b. Leaves commonly 3-lobed, sometimes entire; sepals mostly 1.5–2.5 cm long, with an elongate, linear tip; corolla 3.0–5.0 cm long .... *I. hederacea*

Ipomoea hederacea Jacq.

synonym(s): —
common name(s): ivy-leaved morning glory
habit: slender, twining, annual herb
range: tropical America; naturalized to ME, s. to FL, w. to AZ, n. to ND
state frequency: very rare
habitat: fields, roadsides, waste places
notes: —

Ipomoea purpurea (L.) Roth

synonym(s): —
common name(s): common morning glory
habit: twining, high-climbing, annual herb
range: tropical America; naturalized in our area
state frequency: very rare
habitat: fields, roadsides, waste places
notes: —

BORAGINACEAE
(13 genera, 23 species)

KEY TO THE GENERA

1a. Ovary unlobed or inconspicuously lobed; style terminating the summit of the ovary; stigma peltate-annular, commonly surmounted by an entire or bifid cone-like process .... ***Heliotropium***

1b. Ovary with deep sinuses, appearing as 4 distinct carpels; style arising from between the lobes; stigma not peltate-annular, terminal

  2a. Corolla rotate; anthers 5.0–9.0 mm long, with conspicuous linear appendages *ca.* 3.0 mm long .... ***Borago***

  2b. Corolla tubular, funnelform, salverform, or campanulate; anthers smaller, without appendages

    3a. Corolla weakly zygomorphic, the tubular, connate portion sometimes bent

      4a. Apex of corolla tube with fornices, these appearing to close off the tube; stamens exserted beyond the corolla; mericarps without a stipe-like process, attaching directly to the receptacle; receptacle flat, without a pit .... ***Echium***

      4b. Apex of corolla tube lacking fornices, the tube appearing open; stamens included within the corolla tube; mericarps attached to the receptacle by a stipe-like process, this fitting into a pit in the receptacle .... ***Anchusa***

    3b. Corolla actinomorphic

      5a. Ovary and mericarps with barbed prickles, at least on the marginal or abaxial surfaces

        6a. Plants annual; each flower closely subtended by a bract .... ***Lappula***

        6b. Plants biennial or perennial; all or most of the flowers without closely associated bracts

          7a. Corolla 5.0–8.0 mm wide; mericarps attached to one another by the apical third of their inner faces; mericarps 3.5–5.0 mm long .... ***Cynoglossum***

          7b. Corolla 2.0–3.0 mm wide; mericarps attached to one another along the middle third of their inner faces; mericarps 2.0–3.0 mm long .... ***Hackelia***

      5b. Ovary and mericarps smooth or textured, sometimes muricate, but never with barbed prickles

        8a. Corolla yellow to orange, the apex of the tube without fornices .... ***Amsinckia***

        8b. Corolla white, yellow-white, pink, or blue (with a yellow center in some *Myosotis*), the apex of the tube with fornices (lacking fornices in *Buglossoides arvensis* and *Mertensia virginica*)

          9a. Corolla tubular or tubular-campanulate, the lobes usually erect to ascending

            10a. Leaves and stems glabrous; mericarps without a stipe-like process, attaching directly to the receptacle; receptacle flat, without a pit .... ***Mertensia***

10b. Leaves and stems hispid-hirsute; mericarps attached to the receptacle by a stipe-like process, this fitting into a pit in the receptacle .... **Symphytum**

9b. Corolla salverform or funnelform, the lobes spreading

11a. Flowers, except sometimes the lower, not subtended by bracts; mericarps smooth .... **Myosotis**

11b. Flowers all or mostly all subtended by bracts; mericarps smooth to wrinkled or pitted

12a. Limb of the corolla 6.0–11.0 mm wide; mericarps attached to the receptacle by a stipe-like process, this fitting into a pit in the receptacle .... **Anchusa**

12b. Limb of the corolla up to 4.0 mm wide; mericarps without a stipe-like process, attaching directly to the receptacle; receptacle flat, without a pit

13a. Plants perennial; corolla 4.0–5.0 mm long; leaves with 2 or 3 conspicuous, lateral veins; mericarps lustrous, white to white-brown, smooth .... **Lithospermum**

13b. Plants annual (biennial); corolla 5.0–8.0 mm long; leaves without conspicuous, lateral veins; mericarps dull brown, wrinkled and pitted .... **Buglossoides**

Amsinckia
(2 species)

KEY TO THE SPECIES

1a. Leaves entire; corolla funnelform; mericarps irregularly rugulose .... *A. douglasiana*
1b. Leaves sparingly denticulate; corolla salverform; mericarps transversely rugulose ....
.... *A. menziesii*

Amsinckia douglasiana A. DC.
synonym(s): —
common name(s): —
habit: annual herb
range: western United States; adventive in the east
state frequency: very rare
habitat: fields, roadsides, waste places
notes: —

Amsinckia menziesii (Lehm.) A. Nels. & J. F. Macbr.
synonym(s): —
common name(s): —
habit: annual herb
range: western United States; adventive in the east
state frequency: very rare
habitat: fields, roadsides, waste places
notes: —

Anchusa
(2 species)

KEY TO THE SPECIES

1a. Corolla weakly zygomorphic; plants annual; leaves 4.0–8.0 cm long; stamens inserted at the middle or in the basal half of the corolla tube .... *A. arvensis*
1b. Corolla actinomorphic; plants perennial; leaves 6.0–20.0 cm long; stamens inserted on the apical half of the corolla tube .... *A. officinalis*

Anchusa arvensis (L.) Bieb.
synonym(s): *Lycopsis arvensis* L.
common name(s): small bugloss
habit: erect to ascending, annual herb
range: Europe; naturalized to Newfoundland, s. to VA, w. to NE, n. to Ontario
state frequency: very rare
habitat: waste places, sandy beaches
notes: —

Anchusa officinalis L.
synonym(s): —
common name(s): common bugloss, common alkanet
habit: erect, biennial herb
range: Europe; naturalized to ME, s. to NJ, w. to OH, n. to MI
state frequency: very rare
habitat: roadsides, waste places
notes: —

Borago
(1 species)

Borago officinalis L.
synonym(s): —
common name(s): borage
habit: erect, annual herb
range: Europe; Newfoundland, s. to VA, w. to ND, n. to Ontario
state frequency: rare
habitat: roadsides, waste places
notes: —

Buglossoides
(1 species)

Buglossoides arvensis (L.) I. M. Johnston
synonym(s): *Lithospermum arvense* L.
common name(s): corn gromwell, common gromwell
habit: pubescent, annual or biennial herb
range: Eurasia; introduced to Quebec, s. to FL, w. to CA, n. to British Columbia
state frequency: rare
habitat: dry, waste places
notes: —

Cynoglossum
(1 species)

Cynoglossum virginianum L. var. **boreale** (Fern.) Cooperrider
synonym(s): *Cynoglossum boreale* Fern.
common name(s): wild comfrey, northern wild comfrey
habit: erect, perennial herb
range: Newfoundland, s. to CT, w. to IA, n. to British Columbia
state frequency: rare
habitat: rich, upland woods
notes: This species is ranked SH by the Maine Natural Areas Program.

Echium
(2 species)

KEY TO THE SPECIES
1a. Lower branches of the inflorescence 2.5–6.0 cm long, bearing flowers from base to apex
.... *E. vulgare*
1b. Lower branches of the inflorescence 12.0–22.0 cm long, not bearing flowers in the basal
portion .... *E. creticum*

Echium creticum L.
synonym(s): *Echium australe* Lam.
common name(s): —
habit: erect, bristly, biennial herb
range: western Mediterranean; escaped from cultivation in the northeast
state frequency: very rare
habitat: fields, waste places, gardens
notes: —

Echium vulgare L.
synonym(s): —
common name(s): blueweed, blue devils
habit: erect, bristly, taprooted, biennial herb
range: Europe; naturalized to Newfoundland, s. to MA, w. and n. to Ontario; also British Columbia
state frequency: uncommon
habitat: meadows, fields, roadsides, waste places
notes: —

Hackelia
(1 species)

Hackelia deflexa (Wahlenb.) Opiz var. **americana** (Gray) Fern. & I. M. Johnston
synonym(s): *Hackelia americana* (Gray) Fern.
common name(s): northern stickseed
habit: branched, biennial herb
range: circumboreal; s. to ME, w. to WA
state frequency: very rare
habitat: moist, rocky woods, thickets, and shores
notes: This species is ranked S1 by the Maine Natural Areas Program.

Heliotropium
(1 species)

Heliotropium curassavicum L.
synonym(s): —
common name(s): seaside heliotrope
habit: prostrate to ascending, mat-forming, perennial herb
range: tropical America; naturalized as far north as ME
state frequency: very rare
habitat: sandy seashores, borders of fresh to saline marshes
notes: —

Lappula
(2 species)

KEY TO THE SPECIES
1a. Mericarps 2.0–3.0 mm long, with a single row of marginal prickles; prickles with more
    or less confluent bases .... *L. occidentalis*
1b. Mericarps 3.0–4.0 mm long, with 2(3) rows of marginal prickles; prickles distinct ....
    .... *L. squarrosa*

Lappula occidentalis (S. Wats.) Greene
synonym(s): *Lappula redowskii auct. non* (Hornem.) Greene var. *occidentalis* (S. Wats.) Rydb.
common name(s): western stickseed
habit: erect, annual herb
range: Eurasia and western North America; adventive in the east
state frequency: very rare
habitat: railways, roadsides, waste places
notes: —

Lappula squarrosa (Retz.) Dumort.
synonym(s): *Lappula echinata* Gilib., *Lappula lappula* (L.) Karst.
common name(s): two-row stickseed
habit: erect, annual herb
range: Eurasia; naturalized to Newfoundland, s. to NJ, w. to TX, n. to AK
state frequency: uncommon
habitat: fields, roadsides, waste places
notes: —

Lithospermum
(1 species)

Sometimes only 1 mericarp matures in the fruits of *Lithospermum*.

Lithospermum officinale L.
synonym(s): —
common name(s): common gromwell, European gromwell
habit: erect, branched, perennial herb
range: Eurasia; naturalized to New Brunswick, s. to NJ, w. to IL, n. to MN and Ontario
state frequency: very rare
habitat: thickets, pastures, roadsides, waste places
notes: —

Mertensia
(2 species)

KEY TO THE SPECIES

1a. Corolla 6.0–9.0 mm long; stems spreading to decumbent; plants native, of Atlantic coast beaches .... *M. maritima*
1b. Corolla 18.0–25.0 mm long; stems erect to ascending; plants naturalized, not of coastal beaches .... *M. virginica*

Mertensia maritima (L.) S. F. Gray
synonym(s): *Pneumaria maritima* (L.) Hill
common name(s): seaside lungwort, seaside bluebell, oysterleaf
habit: spreading to decumbent, perennial herb
range: Greenland, s. to MA, w. to British Columbia, n. to AK
state frequency: uncommon
habitat: rocks, sandy or gravelly sea beaches
notes: —

Mertensia virginica (L.) Pers. *ex* Link
synonym(s): —
common name(s): Virginia cowslip, bluebell, eastern bluebell
habit: erect to ascending, perennial herb
range: NY, s. to SC and AL, w. to KS, n. to WI; adventive in our area
state frequency: very rare
habitat: moist to wet woods and clearings
notes: —

Myosotis
(5 species)

KEY TO THE SPECIES

1a. Calyx pubescent with erect to appressed hairs that are not hooked at the tip; plants of moist to wet soils, even growing in shallow water
  2a. Limb of the corolla 2.0–5.0 mm wide; mericarps taller than the style; stems terete, not stoloniferous .... *M. laxa*
  2b. Limb of the corolla 5.0–10.0 mm wide; mericarps shorter than the style; stems angled, often stoloniferous .... *M. scorpioides*
1b. Calyx pubescent, at least in part with spreading, uncinate hairs; plants of dry to moist soils
  3a. Corolla white, with a limb 1.0–2.0 mm wide; calyx weakly zygomorphic, 3 of the lobes shorter than the other 2 .... *M. verna*
  3b. Corolla blue or, less often, white, with a limb 2.0–8.0 mm wide; calyx actinomorphic
    4a. Limb of the corolla 5.0–8.0 mm wide, horizontally spreading .... *M. sylvatica*
    4b. Limb of the corolla 2.0–4.0 mm wide, spreading-ascending .... *M. arvensis*

Myosotis arvensis (L.) Hill
synonym(s): —
common name(s): field scorpion-grass
habit: erect to ascending, often branched, biennial or annual herb
range: Eurasia; naturalized to Newfoundland, s. to NJ, w. to WV, n. to MN

state frequency: occasional
habitat: fields, roadsides, waste places
notes: —

Myosotis laxa Lehm.

synonym(s): —
common name(s): smaller forget-me-not
habit: slender, decumbent, short-lived perennial or annual herb
range: Newfoundland, s. to GA, w. to TN, n. to Ontario; also British Columbia, s. to CA
state frequency: uncommon
habitat: wet shores, shallow water
notes: —

Myosotis scorpioides L.

synonym(s): —
common name(s): water scorpion-grass, true forget-me-not
habit: creeping, often stoloniferous, fibrous-rooted, perennial herb
range: Europe; naturalized to Newfoundland, s. to GA, w. to LA, n. to Ontario; also Manitoba, British
  Columbia, and AK
state frequency: uncommon
habitat: wet soil, shallow, quiet water
notes: —

Myosotis sylvatica Ehrh. *ex* Hoffmann

synonym(s): —
common name(s): garden forget-me-not
habit: tufted, annual, biennial, or short-lived perennial herb
range: Eurasia; escaped from cultivation to Quebec, s. to RI, w. to MI, n. to Ontario
state frequency: uncommon
habitat: wet meadows, waste places
notes: —

Myosotis verna Nutt.

synonym(s): —
common name(s): early scorpion-grass
habit: simple or branched, annual herb
range: ME, s. to FL, w. to TX and CA, n. to British Columbia
state frequency: uncommon
habitat: upland fields, banks, and rocky woods
notes: —

Symphytum
(2 species)

KEY TO THE SPECIES
1a. Leaves decurrent on the stem; pubescence of the stem composed of subterete hairs;
    mericarps smooth; filaments nearly as wide as the anthers .... *S. officinale*
1b. Leaves not decurrent; pubescence of the stem composed of flattened hairs; mericarps
    conspicuously roughened; filaments narrower than the anthers .... *S. asperum*

Symphytum asperum Lepechin

synonym(s): —
common name(s): prickly comfrey, rough comfrey
habit: coarse, branched, perennial herb
range: Asia; naturalized to Newfoundland, s. to MD, w. to MI, n. to Manitoba; also British Columbia
state frequency: uncommon
habitat: roadsides, waste places
notes: —

Symphytum officinale L.

synonym(s): —
common name(s): common comfrey
habit: coarse, branched, perennial herb
range: Eurasia; naturalized to Newfoundland, s. to GA, w. to LA, n. to Ontario; also Alberta and British Columbia
state frequency: uncommon
habitat: roadsides, waste places
notes: —

SOLANACEAE

(7 genera, 17 species)

KEY TO THE GENERA FOR USE WITH FLOWERING MATERIAL

1a. Corolla rotate, valvate in bud; anthers dehiscing by terminal pores or small clefts, connivent around the style .... *Solanum*
1b. Corolla campanulate, funnelform, tubular, or urceolate, not valvate in bud; anthers dehiscing by longitudinal slits, separate
  2a. One anther conspicuously smaller than the other four; calyx connate only at the very base .... *Petunia*
  2b. All five anthers similar in size; calyx connate for at least half its length (except *Nicandra*)
    3a. Flowers sessile or subsessile; leaves of the apical half of the stem sessile or nearly so; corolla weakly zygomorphic .... *Hyoscyamus*
    3b. Flowers borne on pedicels; leaves with petioles; corolla actinomorphic
      4a. Flowers 7.0–20.0 cm long .... *Datura*
      4b. Flowers 0.9–2.5 cm long
        5a. Flowers with 3–5 carpels, solitary on the stem; corolla 2.0–2.5 cm long ...
        .... *Nicandra*
        5b. Flowers with 2 carpels, usually more than 1 per stem; corolla 0.6–2.0 cm long
          6a. Plants woody; corolla blue to purple .... *Lycium*
          6b. Plants herbaceous; corolla yellow to white .... *Physalis*

KEY TO THE GENERA FOR USE WITH FRUITING MATERIAL

1a. Fruit a capsule or pyxis
  2a. Fruit solitary on the stem, 4-locular (at least in the basal portion), prickly, 3.0–5.0 cm long .... *Datura*

2b. Fruit usually more than 1, 2-locular, without prickles, up to 1.5 cm long
    3a. Fruit a pyxis, circumscissile above the middle, borne in the axils of alternately arranged bracts; leaves 5.0–20.0 cm long .... *Hyoscyamus*
    3b. Fruit a capsule, longitudinally dehiscent, borne in the axils of oppositely arranged bracts; leaves 3.0–8.0 cm long .... *Petunia*
1b. Fruit a berry
    4a. Plants woody; leaves borne on short petioles 3.0–10.0 mm long, entire .... *Lycium*
    4b. Plants herbaceous; leaves either borne on petioles longer than 10.0 mm or toothed to lobed or both
        5a. Calyx not accrescent in fruit, subtending, but not enveloping, the fruit .. *Solanum*
        5b. Calyx accrescent in fruit, partly or completely enveloping the fruit
            6a. Calyx glabrous or pubescent; plants unarmed, glabrous or pubescent with simple hairs
                7a. Berry with 3–5 locules; calyx connate at the very base only, with basal auricles .... *Nicandra*
                7b. Berries with 2 locules; calyx prominently connate in the basal portion, without basal auricles .... *Physalis*
            6b. Calyx spiny; plants armed with slender prickles, pubescent with compound hairs .... *Solanum*

Datura
(2 species)

KEY TO THE SPECIES
1a. Corolla 5-toothed at the apex, 7.0–10.0 cm long; capsule erect; plants glabrous or inconspicuously pubescent .... *D. stramonium*
1b. Corolla 10-toothed at the apex, 15.0–20.0 cm long; capsule nodding; plants pubescent ...
.... *D. inoxia*

Datura inoxia P. Mill.
synonym(s): —
common name(s): —
habit: coarse, malodorous, branched, perennial herb
range: Mexico and tropical America; adventive locally
state frequency: very rare
habitat: roadsides, waste places
notes: —

Datura stramonium L.
synonym(s): *Datura stramonium* L. var. *tatula* (L.) Torr.
common name(s): jimsonweed, stramonium, thorn-apple
habit: coarse, malodorous, branched, annual herb
range: Asia; naturalized to New Brunswick, s. to PA, w. to OH, n. to Ontario; also Saskatchewan and British Columbia
state frequency: uncommon
habitat: dry waste places
notes: —

Hyoscyamus
(1 species)

Hyoscyamus niger L.
synonym(s): —
common name(s): black henbane
habit: coarse, malodorous, biennial or annual herb
range: Europe; naturalized to New Brunswick, s. to NY, w. to SD, n. to Ontario
state frequency: very rare
habitat: roadsides, waste places
notes: —

Lycium
(1 species)

Lycium chinense P. Mill.
synonym(s): —
common name(s): Chinese matrimony-vine
habit: shrub
range: Asia; escaped from cultivation to ME, s. to VA, w. to OK, n. to MI
state frequency: very rare
habitat: roadsides, waste places
notes: —

Nicandra
(1 species)

Nicandra physalodes (L.) Gaertn.
synonym(s): *Physalodes physalodes* (L.) Britt.
common name(s): apple-of-Peru
habit: coarse, annual herb
range: Peru; escaped from cultivation in the northeast
state frequency: rare
habitat: old gardens, roadsides, waste places
notes: The petals are usually blue in this species.

Petunia
(1 species)

Annual herbs with flowers that have a weakly zygomorphic calyx and are borne in the axils of opposite leaves.

Petunia axillaris (Lam.) B. S. P.
synonym(s): *Petunia hybrida* Vilm., *Petunia violacea* Lindl.
common name(s): common petunia, garden petunia
habit: viscid-pubescent, annual herb
range: South America; escaped from cultivation in our area
state frequency: very rare
habitat: roadsides, fields, gardens
notes: Our escaped, cultivated forms usually have purple flowers.

Physalis
(5 species)

KEY TO THE SPECIES

1a. Plants annual from taproots; corolla 6.0–10.0 mm long .... *P. pruinosa*
1b. Plants perennial from rhizomes; corolla 10.0–20.0 mm long
    2a. Corolla white, with 5 evident lobes; fruiting calyx bright red .... *P. alkekengi*
    2b. Corolla yellow, often with 5 purple spots within, scarcely lobed; fruiting calyx green, yellow, or brown
        3a. Apical half of the stem pubescent with long, soft, spreading hairs; anthers yellow to light blue; leaf blades ovate to rhombic .... *P. heterophylla*
        3b. Apical half of the stem pubescent with short, stiff, appressed or decurved hairs; anthers blue; leaf blades linear to lanceolate to ovate
            4a. Pubescence of the stem appressed; connate portion of the calyx pubescent in 10 vertical strips along the nerves with minute, appressed hairs up to 0.5 mm long; anthers 3.0–4.0 mm long; berry yellow .... *P. longifolia*
            4b. Pubescence of the stem decurved; connate portion of the calyx pubescent throughout with long, spreading hairs usually 0.5–1.5 mm long; anthers 2.0–3.0 mm long; berry orange-red .... *P. virginiana*

Physalis alkekengi L.
synonym(s): —
common name(s): Chinese lantern plant, winter-cherry
habit: erect, simple, rhizomatous, perennial herb
range: Eurasia; escaped from cultivation locally
state frequency: very rare
habitat: waste places
notes: —

Physalis heterophylla Nees
synonym(s): *Physalis heterophylla* Nees var. *ambigua* (Gray) Rydb.
common name(s): clammy ground-cherry
habit: erect to spreading, rhizomatous, perennial herb
range: ME, s. to SC, w. to TX and CO, n. to Saskatchewan
state frequency: uncommon
habitat: dry, upland woods and clearings
notes: —

Physalis longifolia Nutt. var. *subglabrata* (Mackenzie & Bush) Cronq.
synonym(s): *Physalis virginiana* P. Mill. var. *subglabrata* (Mackenzie & Bush) Waterfall
common name(s): longleaf ground-cherry, Virginia ground-cherry
habit: slender, erect to ascending, usually branched, rhizomatous, perennial herb
range: VT and Ontario, s. to VA, w. to AZ, n. to MT; adventive in our area
state frequency: very rare
habitat: open woods, fields, cultivated ground
notes: —

Physalis pruinosa L.
synonym(s): *Physalis pubescens* L. var. *grisea* Waterfall
common name(s): downy ground-cherry
habit: branched, taprooted, annual herb

range: ME, s. to FL, w. to KS, n. to WI
state frequency: very rare
habitat: fields, waste places
notes: This species is ranked SH by the Maine Natural Areas Program.

Physalis virginiana P. Mill.

synonym(s): —
common name(s): Virginia ground-cherry
habit: branched, rhizomatous, perennial herb
range: CT, s. to FL, w. to AZ, n. to CO and Manitoba; naturalized in our area
state frequency: very rare
habitat: woods, fields, clearings
notes: —

Solanum
(6 species)

KEY TO THE SPECIES

1a. Leaves glabrous or pubescent with simple hairs; stems unarmed; anthers not tapering to the apex

    2a. Plants woody near the base, climbing or twining; corolla blue to purple (white); berry red, ellipsoid to ovoid .... *S. dulcamara*

    2b. Plants herbaceous, not climbing; corolla white, sometimes tinged with blue; berry black, yellow, or red, globose

        3a. Stems glabrous or sparsely pubescent; berry black

            4a. Leaves thin, membranaceous, translucent; anthers 1.3–2.0 mm long; berries 5.0–9.0 mm long, lustrous; seeds 1.2–1.8 mm long; flowers mostly 1–4 in an umbel-like inflorescence .... *S. americanum*

            4b. Leaves thick, opaque or nearly so; anthers 1.8–2.6 mm long; berries 8.0–13.0 mm long, dull; seeds 1.7–2.3 mm long; flowers mostly 5–10 in a raceme-like inflorescence .... *S. nigrum*

        3b. Stems conspicuously pubescent; berry yellow or red .... *S. villosum*

1b. Leaves pubescent with compound hairs; stems armed with slender prickles; anthers tapering to the apex

    5a. Calyx beset with spines; leaves deeply 1- or 2-times pinnately lobed; plants annual ... . *S. rostratum*

    5b. Calyx pubescent, or sometimes with 1 or 2 spines; leaves with 2–5 pairs of large teeth or shallow lobes; plants perennial .... *S. carolinense*

Solanum americanum P. Mill.

synonym(s): —
common name(s): —
habit: slender, branched, annual herb
range: ME, s. to FL, w. to TX, n. to ND
state frequency: uncommon
habitat: woods, thickets, shores, cultivated ground, waste places
notes: —

Solanum carolinense L.

synonym(s): —
common name(s): horse-nettle
habit: coarse, erect, simple or branched, rhizomatous, perennial herb
range: VA, s. to FL, w. to TX; naturalized to ME, s. to MD, w. to MN, n. to Ontario
state frequency: very rare
habitat: sandy areas, fields, waste places
notes: —

Solanum dulcamara L.

synonym(s): —
common name(s): bittersweet nightshade
habit: climbing, rhizomatous, perennial shrub
range: Eurasia; naturalized throughout our area
state frequency: occasional
habitat: thickets, clearings
key to the subspecies:
    1a. Branches and leaves glabrous or nearly so .... *S. d.* var. *dulcamara*
    1b. Branches and leaves pubescent .... *S. d.* var. *villosissimum* Desv.
notes: This species is woody near the base.

Solanum nigrum L.

synonym(s): —
common name(s): black nightshade
habit: branched, annual herb
range: Europe; naturalized to New Brunswick and Nova Scotia, s. to FL
state frequency: uncommon
habitat: roadsides, waste places
notes: —

Solanum rostratum Dunal

synonym(s): —
common name(s): buffalo bur
habit: coarse, branched, annual herb
range: Great Plains; naturalized in the eastern United States
state frequency: uncommon
habitat: dry fields, waste places
notes: —

Solanum villosum L.

synonym(s): —
common name(s): —
habit: branched, annual herb
range: Eurasia; adventive locally
state frequency: very rare
habitat: waste places
notes: —

OLEACEAE
(3 genera, 6 species)

KEY TO THE GENERA

1a. Leaves pinnately compound; fruit a samara; corolla absent; bud scales scurfy-pubescent; terminal bud present .... *Fraxinus*
1b. Leaves simple; fruit a capsule or drupe; corolla present, showy; bud scales or pubescent, but not scurfy; terminal bud absent
    2a. Anthers exserted beyond the corolla tube; flowers white, 2.5–8.0 mm long; fruit a black drupe .... *Ligustrum*
    2b. Anthers included within the corolla tube; flowers white to purple, usually exceeding 10.0 mm in length; fruit a capsule .... *Syringa*

Fraxinus
(3 species)

Maine's members of this genus of trees have inconspicuous, unisexual flowers.

KEY TO THE SPECIES

1a. Samara with wings extending nearly to the base of the compressed body; leaflets numbering 7–11, sessile; terminal bud separated from the most distal pair of lateral buds; calyx deciduous .... *F. nigra*
1b. Samara with wings extending to 0.35–0.6 of the distance to the base of the terete or subterete body; leaflets numbering 5–9, with petiolules; terminal bud adjacent to the most distal pair of lateral buds; calyx persistent, recognizable as a lobed cupule at the base of the samara
    2a. Twigs and leaflets pubescent, sometimes only sparsely so; twigs gray-brown; terminal bud acute; leaves not papillose on the abaxial surface; wing of samara extending *ca.* 0.5 of the distance to the base of the body; leaf scars with a straight or slightly concave, upper margin; petiolules commonly winged .... *F. pennsylvanica*
    2b. Twigs and leaflets glabrous; twigs brown to blue-brown; terminal bud rounded; leaves papillose on the abaxial surface; wing of samara extending *ca.* 0.35 of the distance to the base of the body; leaf scars with a concave upper margin; petiolules not winged .... *F. americana*

Fraxinus americana L.
synonym(s): —
common name(s): white ash
habit: tree
range: New Brunswick, s. to FL, w. to TX, n. to MN and Ontario
state frequency: common
habitat: moist, rich woods
notes: —

Fraxinus nigra Marsh.
synonym(s): —
common name(s): black ash
habit: tree
range: Newfoundland, s. to VA, w. to IA, n. to ND and Manitoba

state frequency: occasional
habitat: swamps, wet woods
notes: —

Fraxinus pennsylvanica Marsh.
synonym(s): *Fraxinus pennsylvanica* Marsh. var. *austinii* Fern., *Fraxinus pennsylvanica* Marsh. var. *subintegerrima* (Vahl) Fern.
common name(s): green ash
habit: tree
range: Quebec, s. to FL, w. to TX, n. to MN and Manitoba
state frequency: common
habitat: river banks, low, wet woods
notes: —

Ligustrum
(2 species)

KEY TO THE SPECIES
1a. Basal, tubular portion of the corolla 5.0–8.0 mm long, about 2.0 times as long as the corolla lobes; anthers *ca.* 3.0 mm long; calyx usually pubescent; twigs pubescent, some of the hairs *ca.* 0.5 mm long .... *L. amurense*
1b. Basal, tubular portion of the corolla 2.5–3.0 mm long, about as long as the corolla lobes; anthers *ca.* 2.0 mm long; calyx glabrous; twigs minutely pubescent with hairs *ca.* 0.1 mm long .... *L. vulgare*

Ligustrum amurense Carr.
synonym(s): —
common name(s): —
habit: branched, deciduous to semi-evergreen shrub
range: Japan, China; escaped from cultivation in our area
state frequency: rare
habitat: fields, edges, roadsides
notes: —

Ligustrum vulgare L.
synonym(s): —
common name(s): common privet
habit: branched, deciduous shrub
range: Europe; escaped from cultivation in our area
state frequency: uncommon
habitat: fields, edges, roadsides
notes: —

Syringa
(1 species)

This is a commonly planted shrub with fragrant, light purple to white flowers.

Syringa vulgaris L.
synonym(s): —
common name(s): lilac, common lilac
habit: shrub to small tree

range: Europe; introduced to our area
state frequency: common
habitat: abandoned homesites, fields, roadsides
notes: The hybrid of this species with *Syringa* ×*persica* L., called *S.* ×*chinensis* Willd., sometimes escapes from cultivation in Maine. It can be identified by its cuneate- to round-based leaves *vs.* truncate- to cordate-based in *S. vulgaris*.

PHRYMACEAE
(1 genus, 1 species)

Phryma
(1 species)

Phryma leptostachya L.
synonym(s): —
common name(s): lopseed
habit: erect, simple or branched, perennial herb
range: New Brunswick, s. to FL, w. to TX, n. to Manitoba
state frequency: very rare
habitat: moist, rich woods and thickets
notes: This species is ranked SH by the Maine Natural Areas Program.

VERBENACEAE
(1 genus, 3 species)

Verbena
(3 species)

Spike widths used in the following key are for the flowering portions of the inflorescence.

KEY TO THE SPECIES
1a. Bracts commonly shorter than the calyx; calyx 2.0–3.0 mm long; leaves toothed, often coarsely so, sometimes hastately lobed near the base
  2a. Corolla white, the limb usually *ca.* 2.0 mm wide; spikes slender, up to 5.0 mm wide, not crowded, the flowers not contiguous .... *V. urticifolia*
  2b. Corolla usually blue, the limb 2.5–4.5 mm wide; spikes thicker, mostly 6.0–8.0 mm wide, crowded, the flowers contiguous .... *V. hastata*
1b. Bracts 8.0–15.0 mm long, much longer than the calyx; calyx 3.0–4.0 mm long; leaves pinnately incised or lobed .... *V. bracteata*

Verbena bracteata Lag. & Rodr.
synonym(s): *Verbena bracteosa* Michx.
common name(s): prostrate vervain, bracted vervain
habit: decumbent to ascending, branched, annual or perennial herb
range: VA, s. to FL, w. to TX and Mexico, n. to British Columbia; adventive in the northeast
state frequency: very rare
habitat: sandy fields, roadsides, and waste places
notes: —

Verbena hastata L.

synonym(s): —
common name(s): blue vervain, common vervain
habit: erect, branched, perennial herb
range: New Brunswick, s. to FL, w. to AZ and CA, n. to British Columbia
state frequency: occasional
habitat: damp fields, meadows, thickets, and shores, swales, swamps
notes: Known to hybridize with *Verbena urticifolia*.

Verbena urticifolia L.

synonym(s): —
common name(s): white vervain
habit: erect, branched, annual or perennial herb
range: New Brunswick, s. to FL, w. to TX, n. to ND and Saskatchewan
state frequency: very rare
habitat: borders, thickets, moist fields, meadows, waste places
notes: This species is ranked SH by the Maine Natural Areas Program. Known to hybridize with
   Verbena hastata.

LAMIACEAE

(27 genera, 47 species)

KEY TO THE GENERA

1a. Calyx with a small, but evident, protuberance on the upper side .... *Scutellaria*
1b. Calyx without a protuberance on the upper side
   2a. Ovary with deep sinuses, appearing as 4 distinct carpels; style gynobasic; mericarps
     attached near their base
      3a. Flowers with only 2 stamens, the other 2 stamens absent or represented by small
        staminodes
        4a. Calyx strongly zygomorphic; corolla blue; plants annual (perennial in *Salvia*
          officinalis)
          5a. Inflorescence of whorled flowers in the axils of foliage leaves; anthers
            composed of 2 divergent pollen sacs, without connectives .... *Hedeoma*
          5b. Inflorescence of whorled flowers in the axils of reduced, bract-like leaves;
            anthers attached to a thread-like connective that is articulated to the
            filament, the upper end of the connective bearing a single pollen sac, the
            lower end of the connective either with a deformed pollen sac or without a
            pollen sac .... *Salvia*
        4b. Calyx actinomorphic or nearly so; corolla white, red, red-purple, or purple;
          plants perennial
          6a. Corolla white, weakly zygomorphic, up to 5.0 mm long; flowers in dense
            whorls in the axils of leaves .... *Lycopus*
          6b. Corolla pink to purple, strongly zygomorphic, 20.0–45.0 mm long;
            flowers in dense, capitate clusters at the tips of the stems and branches,
            sometimes also in dense, axillary clusters .... *Monarda*
      3b. Flowers with 4 stamens

7a. Connate portion of the corolla [the tube] longer than and somewhat
concealing the stamens

    8a. Stems conspicuously white-tomentose; corolla white; lobes of the calyx
numbering 10, with reflexed and spinulose apices; plants from a taproot ...
.... ***Marrubium***

    8b. Stems retrorse-scabrous, hirsute, or glabrous; corolla blue-purple; lobes
of the calyx numbering 5, awned at the apex; plants from rhizomes and
stolons .... ***Glechoma***

7b. Connate portion of the corolla shorter than the stamens, the stamens readily
visible

    9a. Inflorescence axillary, composed of few to many flowers in the axils of
foliage leaves

        10a. Calyx actinomorphic or nearly so

            11a. Corolla 4-lobed, nearly actinomorphic, the upper lobe wider than
the others and often emarginate at the apex (rarely with 5 subequal
lobes) .... ***Mentha***

            11b. Corolla zygomorphic [with an upper and lower lip]

                12a. Flowers with short pedicels; upper pair of stamens longer
than the lower pair .... ***Glechoma***

                12b. Flowers sessile; upper pair of stamens equal to or shorter
than the lower pair

                    13a. Leaves palmately lobed; corolla 8.0–12.0 mm long, the
upper lip conspicuously villous .... ***Leonurus***

                    13b. Leaves simple; corolla 10.0–30.0 mm long, the upper lip
glabrous or sparsely pilose

                        14a. Apex of sepals acuminate, but not spiny; anthers
dehiscing by longitudinal valves; leaf blades orbicular
to ovate, truncate to cordate at the base; mericarps
truncate at the apex; lower lip of the corolla without a
pair of projections near the base .... ***Lamium***

                        14b. Apex of sepals terminated by short, stiff spines;
anthers dehiscing by latitudinal valves; leaf blades
rhombic-elliptic to ovate, cuneate to rounded at the
base; mericarps rounded at the apex; lower lip of the
corolla with a pair of projections near the base on the
adaxial [upper] surface .... ***Galeopsis***

        10b. Calyx zygomorphic [with an upper and lower lip]

            15a. Corolla 4-lobed, nearly actinomorphic, the upper lobe wider than
the others and often emarginate at the apex (rarely with 5 subequal
lobes) .... ***Mentha***

            15b. Corolla zygomorphic, with an upper and lower lip

                16a. Upper calyx lip with broad-triangular lobes, the lobes wider
than long; corolla white or yellow (rarely becoming pink), the
connate portion curved upward; leaf blades ovate to deltate-
ovate; plants citrus-scented .... ***Melissa***

16b. Upper calyx lip with narrow-triangular lobes, the lobes longer than wide; corolla pale purple, red-purple, or pink (rarely white), the connate portion straight; leaf blades ovate, elliptic, oblong, or narrower; plants without a citrus scent

 17a. Flowers in cymules of 2–6 in the axils of many of the foliage leaves, each cymule subtended by minute bracts shorter than the calyx; calyx pubescent inside the connate tube; plants annual, without stolons, 1.0–2.0 dm tall .... *Acinos*

 17b. Flowers numerous in dense clusters at the stem tip and 1–3 of the upper axils, each flower subtended by ciliate, linear-subulate bracts about as long as the calyx; calyx glabrous inside the connate tube; plants perennial from short stolons, 2.0–5.0 dm tall .... *Clinopodium*

9b. Inflorescence a spike, raceme, or panicle-like cyme, terminal or axillary, but not in a conspicuous whorl from the axils of normal foliage leaves

 18a. Inflorescence an open panicle-like cyme .... *Trichostema*

 18b. Inflorescence a spike (sometimes capitate) or raceme

  19a. Each node of the inflorescence with 2 flowers, 1 each in the axil of a bract .... *Physostegia*

  19b. Each node of the inflorescence with 4 or more flowers

   20a. Upper lip of the corolla concavely arched, appearing as a small hood

    21a. Calyx zygomorphic [with an upper and lower lip]; filaments bifid near the apex, one of the tooth-like branches bearing the anther; bracts broadly rounded at the apex .... *Prunella*

    21b. Calyx actinomorphic or nearly so; filaments not bifid at the summit; bracts obtuse to acuminate at the apex

     22a. Upper pair of stamens slightly shorter than the lower pair; calyx with 5–10 nerves; lower whorls of flowers sessile .... *Stachys*

     22b. Upper pair of stamens slightly longer than the lower pair; calyx with 15 nerves; lower whorls of flowers often borne on peduncles .... *Nepeta*

   20b. Upper lip of the corolla flat or upwardly curved, not hood-like

    23a. Stamens, at least the longer pair, upwardly curved, closely ascending under the upper corolla lip and not surpassing it

     24a. Uppermost lobe of the calyx wider than the other 4; bracteal leaves aristately toothed; calyx with 15 nerves .... *Dracocephalum*

     24b. Upper 3 lobes of the calyx differing from the other 2; bracteal leaves entire to crenate-serrate; calyx with 10–13 nerves

25a. Corolla 5.0–7.0 mm long; leaves linear to narrow-oblanceolate; each flower subtended by minute bracts shorter than the calyx; plants annual, without stolons, 1.0–3.0 dm tall .... ***Satureja***

25b. Corolla 12.0–15.0 mm long; leaves lance-ovate to ovate to ovate-oblong; each flower subtended by ciliate, linear-subulate bracts about as long as the calyx; plants perennial from short stolons, 2.0–5.0 dm tall .... ***Clinopodium***

23b. Stamens straight or nearly so, at least the longer pair protruding from the corolla

26a. Leaves with entire margins

27a. Plants diffusely branched and mat-forming; leaf blades 0.5–1.0 cm long; calyx pubescent inside the connate tube .... ***Thymus***

27b. Plants upright; leaf blades 1.0–7.0 cm long; calyx glabrous inside the connate tube

28a. Calyx 10- to 13-nerved; anthers with parallel pollen sacs; flowers in crowded, capitate cymes at the apex of the stems and branches ....
.... ***Pycnanthemum***

28b. Calyx 15-nerved; anthers with divergent pollen sacs; flowers borne in clusters in the axils of bracts, forming a terminal, spike-like inflorescence .... ***Hyssopus***

26b. Leaves with toothed margins

29a. Inflorescence secund; corolla 5-lobed; anthers with divergent pollen sacs; plants annual, without rhizomes .... ***Elsholtzia***

29b. Inflorescence not secund; corolla 4-lobed; anthers with parallel pollen sacs; plants rhizomatous perennials

30a. Corolla nearly actinomorphic, the upper lobe wider than the others and often emarginate at the apex (rarely with 5 subequal lobes); leaves with toothed margins .... ***Mentha***

30b. Corolla zygomorphic [with an upper and lower lip]; leaves with entire or toothed margins
.... ***Pycnanthemum***

2b. Ovary lobed, the carpels obviously connate near the base for some distance; style terminal, emerging from between the lobes of the ovary; mericarps attached by their sides

31a. Corolla clearly composed of an upper and lower lip; stamens long-exserted from the corolla, with arcuate filaments .... ***Trichostema***

31b. Corolla apparently composed of a lower lip only, the upper lip either very small or part of the lower lip; stamens not conspicuously surpassing the corolla, the filaments straight or nearly so

    32a. Lower lip 5-lobed, the 2 small, lateral lobes representing the laterally displaced lobes of the upper lip; inflorescence an elongate raceme 5.0–20.0 cm tall, leafy-bracted only at the base .... ***Teucrium***

    32b. Lower lip 3- or 4-lobed, the upper lip present, but very small; inflorescence either whorled flowers in the axils of foliage leaves or in leafy-bracted spikes 4.0–8.0 cm tall .... ***Ajuga***

Acinos
(1 species)

Acinos arvensis (Lam.) Dandy
synonym(s): *Satureja acinos* (L.) Scheele
common name(s): mother of thyme, basil thyme
habit: erect, basally branched, annual to short-lived perennial herb
range: Europe; naturalized to ME, s. to NJ, w. to WI, n. to Ontario
state frequency: very rare
habitat: old fields, pastures, roadsides, waste places
notes: —

Ajuga
(2 species)

KEY TO THE SPECIES
1a. Plants upright, not forming mats; upper lip of the corolla minutely 2-lobed; stems commonly pubescent throughout; flowers in whorls of 6 or more at each node ....
.... ***A. genevensis***
1b. Plants with stolons, mat-forming, the flowering stems upright; upper lip of the corolla entire; stems commonly pubescent in lines of decurrence from the leaf bases; flowers in whorls of usually 6 at each node .... ***A. reptans***

Ajuga genevensis L.
synonym(s): —
common name(s): standing bugle, blue bugle
habit: erect, tufted, rhizomatous, perennial herb
range: Eurasia; escaped from cultivation to ME, s. to PA, w. to IL, n. to Ontario
state frequency: very rare
habitat: fields, gardens, lawns, roadsides, waste places
notes: —

Ajuga reptans L.
synonym(s): —
common name(s): carpet bugle, common bugle
habit: stoloniferous, mat-forming, perennial herb
range: Eurasia; escaped from cultivation to Newfoundland, s. to PA, w. to OH, n. to WI
state frequency: rare
habitat: fields, roadsides, lawns
notes: —

Clinopodium
(1 species)

Clinopodium vulgare L.
synonym(s): *Satureja vulgaris* (L.) Fritsch, *Satureja vulgaris* (L.) Fritsch var. *neogaea* Fern.
common name(s): wild basil, dogmint
habit: creeping, simple or branched, perennial herb
range: Europe; naturalized to Newfoundland, s. to DE and mountains of NC, w. to NM, n. to Manitoba
state frequency: uncommon
habitat: woods, thickets, slopes, shores, waste places
notes: —

Dracocephalum
(1 species)

Dracocephalum parviflorum Nutt.
synonym(s): *Moldavica parviflora* (Nutt.) Britt.
common name(s): American dragonhead
habit: erect, simple or branched, taprooted, annual or biennial herb
range: western Quebec, s. to NY, w. to AZ, n. to AK; naturalized to New Brunswick, Nova Scotia, and ME
state frequency: uncommon
habitat: dry, rocky or gravelly soil, often calcareous
notes: —

Elsholtzia
(1 species)

Elsholtzia ciliata (Thunb.) Hyl.
synonym(s): —
common name(s): crested late summer mint
habit: branched, annual herb
range: Asia; naturalized to New Brunswick, s. to NJ, w. and n. to WI
state frequency: very rare
habitat: old fields, roadsides
notes: —

Galeopsis
(2 species)

KEY TO THE SPECIES
1a. Corolla 15.0–24.0 mm long; middle lobe of the lower lip truncate at the apex; leaves commonly rounded at the base; lobes of the calyx 7.5–11.0 mm long in fruit .. **G. tetrahit**
1b. Corolla 13.0–16.0 mm long; middle lobe of the lower lip retuse or cleft at the apex; leaves commonly cuneate at the base; lobes of the calyx 5.0–8.0 mm long in fruit ....
.... **G. bifida**

Galeopsis bifida Boenn.

synonym(s): *Galeopsis tetrahit* L. var. *bifida* (Boenn.) Lej. & Court.
common name(s): splitlip hempnettle
habit: slender, branched or simple, taprooted, annual herb
range: Europe; naturalized to Newfoundland, s. to NC, w. to OH, n. to SD
state frequency: uncommon
habitat: waste places
notes: —

Galeopsis tetrahit L.

synonym(s): —
common name(s): brittlestem hempnettle
habit: branched or simple, taprooted, annual herb
range: Eurasia; naturalized to Newfoundland, s. to CT, w. to IA, n. to WI and Manitoba
state frequency: uncommon
habitat: cultivated fields, roadsides, waste places
notes: —

Glechoma
(1 species)

Glechoma hederacea L.

synonym(s): *Glechoma hederacea* L. var. *micrantha* Moric., *Nepeta hederacea* (L.) Trevisan
common name(s): gill over the ground, run away robin
habit: creeping, stoloniferous, fibrous-rooted, perennial herb
range: Eurasia; naturalized to Newfoundland, s. to VA, w. to MO, n. to Ontario
state frequency: occasional
habitat: moist woods, lawns, roadsides, waste places
notes: This species has dimorphic flowers—smaller flowers, 8.0–15.0 mm long, that are functionally
carpellate with minute, included stamens, and larger flowers, 13.0–23.0 mm long, that are bisexual
with shortly exserted stamens.

Hedeoma
(2 species)

KEY TO THE SPECIES

1a. Leaves linear, with entire margins; sterile filaments minute and inconspicuous; lobes of
the calyx subulate, ciliate on the margin .... *H. hispida*
1b. Leaves lanceolate to ovate or obovate, entire to serrulate; sterile filaments evident; lower
lobes of the calyx subulate, the upper triangular, all the lobes glabrous on the margin ......
.... *H. pulegioides*

Hedeoma hispida Pursh

synonym(s): —
common name(s): rough false pennyroyal
habit: erect, simple or branched, annual herb
range: Ontario and NY, s. to MS, w. to TX, n. to Saskatchewan; naturalized to ME, s. to CT
state frequency: very rare
habitat: dry, sandy or rocky areas
notes: —

Hedeoma pulegioides (L.) Pers.
synonym(s): —
common name(s): American-pennyroyal, American false pennyroyal
habit: erect, branched, annual herb
range: New Brunswick and Nova Scotia, s. to FL, w. to AR, n. to SD and MN
state frequency: uncommon
habitat: dry, gravelly soil and ledges, woods
notes: —

Hyssopus
(1 species)

Hyssopus officinalis L.
synonym(s): —
common name(s): hyssop
habit: erect, rhizomatous, perennial herb
range: Eurasia; naturalized as far north as ME, w. to MN and Ontario
state frequency: very rare
habitat: dry pastures, roadsides, waste places
notes: —

Lamium
(4 species)

KEY TO THE SPECIES

1a. Corolla 2.0–3.0 cm long, with an upper lip 7.0–12.0 mm long; plants perennial
    2a. Corolla white; leaves commonly with a white blotch bordering each side of the midrib; terminal tooth of the upper leaves usually obtuse at the apex .. *L. maculatum*
    2b. Corolla pink-purple (white); leaves without a white blotch; terminal tooth of the upper leaves acute at the apex .... *L. album*
1b. Corolla 1.0–1.8 cm long, with an upper lip 2.0–5.0 mm long; plants annual
    3a. Upper leaves and bracts sessile; calyx 5.0–8.0 mm long, the lobes erect [oriented straight forward]; flowers without a subtending bractlet; connate portion of the corolla [the tube] glabrous on the inner surface .... *L. amplexicaule*
    3b. All the leaves petiolate; calyx 5.0–6.0 mm long, the lobes spreading; flowers commonly with a subtending bractlet; connate portion of the corolla often pubescent on the inner surface .... *L. purpureum*

Lamium album L.
synonym(s): —
common name(s): snowflake, white dead-nettle
habit: decumbent, rhizomatous, perennial herb
range: Eurasia; escaped from cultivation to New Brunswick, s. to VA, w. and n. to MN and Ontario
state frequency: very rare
habitat: lawns, roadsides, waste places
notes: —

Lamium amplexicaule L.
synonym(s): —
common name(s): henbit

habit: decumbent to ascending, branched, annual herb
range: Eurasia; naturalized to Labrador, s. to CT, w. to MI, n. to Ontario
state frequency: rare
habitat: fields, roadsides, waste places
notes: Plants of this species sometimes produce small, cleistogamous flowers.

Lamium maculatum L.

synonym(s): —
common name(s): spotted dead-nettle, spotted henbit
habit: creeping, rhizomatous, perennial herb
range: Eurasia; escaped from cultivation to ME, s. to NC, w. to TN, n. to Ontario
state frequency: very rare
habitat: roadsides, waste places
notes: —

Lamium purpureum L.

synonym(s): —
common name(s): purple dead-nettle, red dead-nettle
habit: prostrate to decumbent, basally branched, hollow-stemmed, annual herb
range: Eurasia; naturalized to Newfoundland, s. to SC, w. to MO, n. to MI
state frequency: very rare
habitat: fields, roadsides, waste places
notes: —

Leonurus
(1 species)

Leonurus cardiaca L.

synonym(s): —
common name(s): common motherwort
habit: tall, erect, fibrous-rooted, perennial herb
range: Eurasia; naturalized throughout much of North America
state frequency: uncommon
habitat: waste places
notes: —

Lycopus
(3 species)

The abaxial surface of the leaves in Maine species is glandular-punctate, a useful feature for distinguishing this genus.

KEY TO THE SPECIES

1a. Lobes of the calyx narrow-triangular, 1.0–2.0 mm long, less than 0.5 times as wide as long, acuminate at the apex, surpassing the mericarps at maturity; lower leaves coarsely serrate to pinnately lobed .... *L. americanus*

1b. Lobes of the calyx broad-triangular, shorter than 1.0 mm, more than 0.5 times as wide as long, obtuse to subacute at the apex, surpassed by the mericarps at maturity; lower leaves serrate to coarsely serrate

2a. Perianth 5-merous; plants with tubers; interior angles of the mericarps shorter than the 2 lateral angles, therefore, the apex of the combined mericarps concave; leaves and stem glabrous to sparsely pubescent; corolla with flaring lobes .... ***L. uniflorus***

2b. Perianth 4-merous; plants without tubers; interior angles of the mericarps as long as the 2 lateral angles, therefore, the apex of the combined mericarps flat across the apex; leaves and stem pubescent; corolla with forward-oriented lobes, not flaring ......
.... ***L. virginicus***

Lycopus americanus Muhl. *ex* W. Bart.

synonym(s): —
common name(s): American water-horehound
habit: slender, erect, stoloniferous, rhizomatous, perennial herb
range: Newfoundland, s. to FL, w. to CA, n. to British Columbia
state frequency: occasional
habitat: low ground, seepy areas
notes: —

Lycopus uniflorus Michx.

synonym(s): —
common name(s): northern water-horehound, northern bugleweed
habit: slender, stoloniferous, perennial herb
range: Newfoundland, s. to MD and mountains of NC, w. to AR and CA, n. to AK
state frequency: occasional
habitat: wet woods, shores, and low areas
notes: —

Lycopus virginicus L.

synonym(s): —
common name(s): Virginia water-horehound
habit: stoloniferous, perennial herb
range: ME, s. to FL, w. to TX, n. to MN
state frequency: uncommon
habitat: rich, moist soil
notes: —

Marrubium
(1 species)

Marrubium vulgare L.

synonym(s): —
common name(s): horehound, common horehound
habit: prostrate to ascending, taprooted, perennial herb
range: Eurasia; naturalized throughout much of North America
state frequency: very rare
habitat: waste places
notes: —

Melissa
(1 species)

Melissa officinalis L.
synonym(s): —
common name(s): common balm, lemon balm
habit: erect, branched, perennial herb
range: Asia; naturalized to ME, s. to FL, w. to AR and KS, n. to OH
state frequency: very rare
habitat: open woods, roadsides, waste places
notes: —

Mentha
(4 species)

Hybridization is common in *Mentha*, the progeny identified by their intermediate morphology. *Mentha suaveolens* Ehrh. is reported from Maine by Gleason and Cronquist (1991), but voucher specimens are unknown. It is described here due to its role in hybridization in the state: leaves sessile, broad-elliptic to round-ovate, cordate at the base, villous-tomentose on the abaxial surface; calyx tomentose; inflorescence tall and slender, 3.0–15.0 x 0.5–1.0 cm

KEY TO THE SPECIES

1a. Flowers in whorls in the axils of ordinary foliage leaves
  2a. Lower and middle leaves ovate to elliptic, 1.25–2.5 times as long as wide, often rounded at the base; whorls of flowers about as long as the petioles .... **M. arvensis**
  2b. Lower and middle leaves lanceolate to narrow-oblong, 2.0–4.0 times as long as wide, cuneate at the base; whorls of flowers usually shorter than the petioles ....
  .... **M. canadensis**
1b. Flowers in terminal spikes, subtended by leaves of reduced size
  3a. Inflorescence a globose or ovoid spike, seldom exceeding 2.5 cm tall in fruit; connate portion of the calyx pubescent; stems pubescent; leaf blades usually broad-ovate, up to 2.0 times as long as wide, borne on petioles .... **M. aquatica**
  3b. Inflorescence a narrow, cylindric spike, 3.0–12.0 cm tall in fruit; connate portion of the calyx glabrous; stems glabrous or nearly so; leaf blades lanceolate to elliptic, 2.0–3.5 times as long as wide, sessile or subsessile .... **M. spicata**

Mentha aquatica L.
synonym(s): —
common name(s): water mint
habit: perennial herb
range: Eurasia; escaped from cultivation to Nova Scotia, s. to DE
state frequency: very rare
habitat: wet places
notes: Sterile hybrids with *Mentha spicata*, called *M.* ×*piperita* L., occur in Maine. Also known to hybridize with *M. arvensis*.

Mentha arvensis L.
synonym(s): —
common name(s): field mint, common mint, wild mint
habit: ascending to erect, branched or simple, perennial herb

range: circumboreal; s. to NC, w. to CA
state frequency: occasional
habitat: moist streamsides, shores, and open areas
notes: Hybrids with *Mentha spicata*, called *M.* ×*gracilis* Sole and previously known as *Mentha cardiaca* (S. F. Gray) Gerarde *ex* Baker, occur in Maine. *Mentha arvensis* is also involved in a triparental hybrid, *Mentha arvensis* × *M. spicata* × *M. suaveolens*, that occurs Maine. Also known to hybridize with *Mentha aquatica*.

Mentha canadensis L.

synonym(s): *Mentha arvensis* L. var. *villosa* (Benth.) S. R. Stewart
common name(s): Canada mint
habit: ascending to erect, branched or simple, perennial herb
range: circumboreal; s. to NC, w. to CA
state frequency: uncommon
habitat: streamsides, shores
notes: —

Mentha spicata L.

synonym(s): —
common name(s): spearmint
habit: usually branched, rhizomatous, perennial herb
range: Europe; naturalized throughout much of North America
state frequency: uncommon
habitat: stream banks, wet meadows and ditches, waste places
notes: Hybridizes with *Mentha arvensis* (*q.v.*) in a biparental and triparental hybrid, with *M. aquatica* (*q.v.*), and with *M. suaveolens*, producing *M.* ×*villosa* Huds. All these hybrids occur in Maine. The sterile hybrid, *M.* ×*villosa* can be identified by its broad-ovate leaves that are subcordate (to rounded) at the base and its tomentose calyx lobes.

Monarda
(3 species)

Maine's species of *Monarda* have conspicuous, showy corollas that are dotted with resin glands and are pubescent to some extent with septate hairs.

KEY TO THE SPECIES

1a. Corolla 3.0–4.5 cm long, glabrous to sparsely pubescent, red (red-purple); bractlets commonly tinged with red; leaf blades 7.0–15.0 cm long, borne on petioles 1.0–4.0 cm long; calyx 10.0–14.0 mm long .... **M. didyma**
1b. Corolla 2.0–3.5 cm long, pubescent, pink to purple or red-purple; bractlets green, gray, purple, or purple-tinged; leaf blades 6.0–12.0 cm long, borne on petioles 1.0–2.0 cm long; calyx 7.0–10.0 mm long
    2a. Corolla pink to purple; upper lip of the corolla densely villous near the apex, forming a tuft of hairs; pubescence of the corolla with pale, inconspicuous septa; bractlets green or gray, sometimes tinged with purple; leaves green to gray-green ....
    .... **M. fistulosa**
    2b. Corolla red-purple; upper lip of the corolla sparsely pubescent near the apex; pubescence of the corolla with conspicuous, red-purple septa; bractlets partly or entirely purple; leaves dark green .... **M. media**

Monarda didyma L.

synonym(s): —
common name(s): bee-balm, Oswego-tea, scarlet bee-balm
habit: tall, erect, simple or branched, perennial herb
range: NY, s. to mountains of GA, w. to mountains of TN, n. to MI and Ontario; escaped from cultivation to New Brunswick and ME
state frequency: rare
habitat: moist, rich woods, thickets, and bottomlands
notes: —

Monarda fistulosa L.

synonym(s): <*M. f.* var. *mollis*> = *Monarda mollis* L.
common name(s): wild bergamot
habit: tall, erect, often branched, rhizomatous, perennial herb
range: Quebec, s. to MD and mountains of GA, w. to TX, n. to MN and Saskatchewan; escaped from cultivation in the northeast
state frequency: uncommon
habitat: upland woods, borders of woods, thickets, and clearings
key to the subspecies:
    1a. Leaves pubescent on the abaxial surface with long hairs commonly exceeding 1.0 mm in length; hairs of the stem decurved to spreading .... *M. f.* var. *fistulosa*
    1b. Leaves pubescent on the abaxial surface with short hairs up to 0.5 mm long or nearly glabrous; hairs of the stem decurved .... *M. f.* var. *mollis* (L.) Benth.
notes: Our most common form is *Monarda fistulosa* var. *mollis*.

Monarda media Willd.

synonym(s): —
common name(s): purple bergamot
habit: tall, erect, usually branched, perennial herb
range: NY, w. to IN, n. to Ontario; escaped from cultivation to ME, s. to VA and NC
state frequency: uncommon
habitat: woods, roadside thickets
notes: —

Nepeta
(1 species)

Nepeta cataria L.

synonym(s): —
common name(s): catnip
habit: erect, taprooted, perennial herb
range: Eurasia; naturalized throughout our area
state frequency: uncommon
habitat: roadsides, dooryards, waste places
notes: —

Physostegia
(1 species)

Physostegia virginiana (L.) Benth.
synonym(s): *Dracocephalum virginianum* L., *Physostegia virginiana* (L.) Benth. var. *granulosa* (Fassett) Fern., *Physostegia virginiana* (L.) Benth. var. *speciosa* (Sweet) Gray
common name(s): false dragonhead, obedient plant
habit: erect, simple or branched, stoloniferous, perennial herb
range: New Brunswick and Nova Scotia, s. to FL, w. to Mexico, n. to ND and Manitoba
state frequency: uncommon
habitat: river banks, damp thickets, meadows, shores, and waste places
notes: —

Prunella
(1 species)

Prunella vulgaris L.
synonym(s): —
common name(s): selfheal, heal-all
habit: tufted, prostrate to ascending, simple or branched, perennial herb
range: Eurasia; naturalized to Newfoundland, s. to NC, w. to NM and CA, n. to AK
state frequency: common
habitat: fields, lawns, shores, roadsides, waste places
key to the subspecies:
    1a. Middle stem leaves ovate to ovate-oblong, usually rounded at the base, 2.0–5.0 times as long as wide .... *P. v.* ssp. *lanceolata* (W. Bart.) Hultén
    1b. Middle stem leaves lanceolate to lance-oblong, usually cuneate at the base, 1.5–2.5 times as long as wide .... *P. v.* ssp. *vulgaris*
notes: —

Pycnanthemum
(3 species)

KEY TO THE SPECIES
1a. Leaves lance-oblong to oblong or lance-ovate to ovate, obscurely serrate, the larger 1.5–4.0 cm wide .... *P. muticum*
1b. Leaves linear to lance-oblong, 0.2–1.0 cm wide
    2a. Leaves with 3 or 4 pairs of lateral veins, 3.0–10.0 mm wide, usually pubescent on the midvein of the abaxial surface; stem pubescent; inner bractlets acute to short-acuminate at the apex; lobes of the calyx 0.5–1.0 mm long .... *P. virginianum*
    2b. Leaves with 1 or 2(3) pairs of lateral veins, 2.0–4.0 mm wide, glabrous; stem glabrous; inner bractlets long-acuminate at the apex, the midvein prolonged into a stiff point; lobes of the calyx 1.0–1.5 mm long .... *P. tenuifolium*

Pycnanthemum muticum (Michx.) Pers.
synonym(s): *Koellia mutica* (Michx.) Kuntze
common name(s): short-tooth mountain-mint, blunt mountain-mint
habit: branched, rhizomatous, perennial herb
range: ME, s. to FL, w. to TX, n. to IL and MI
state frequency: very rare

habitat: woods, thickets, meadows, clearings
notes: This species is ranked SH by the Maine Natural Areas Program.

Pycnanthemum tenuifolium Schrad.
synonym(s): *Koellia flexuosa auct. non* (Walt.) MacM., *Pycnanthemum flexuosum auct. non* (Walt.)
   B. S. P.
common name(s): narrow-leaved mountain-mint
habit: branched, rhizomatous, perennial herb
range: ME, s. to FL, w. to TX, n. to MN and Ontario
state frequency: uncommon
habitat: thickets, open areas, bogs
notes: Hybrids with *Pycnanthemum virginianum* occur in Maine.

Pycnanthemum virginianum (L.) T. Dur. & B. D. Jackson *ex* B. L. Robins. & Fern.
synonym(s): *Koellia virginiana* (L.) MacM.
common name(s): Virginia mountain-mint
habit: branched, rhizomatous, perennial herb
range: New Brunswick, s. to VA and mountains of GA, w. to OK, n. to ND
state frequency: uncommon
habitat: woods, thickets, meadows, shores
notes: Hybrids with *Pycnanthemum tenuifolium* (*q.v.*) occur in Maine.

Salvia
(1 species)

Salvia officinalis L.
synonym(s): —
common name(s): garden sage
habit: suffrutescent, perennial herb
range: old world; escaped from cultivation in our area
state frequency: very rare
habitat: waste places, areas of cultivation
notes: —

Satureja
(1 species)

Satureja hortensis L.
synonym(s): —
common name(s): summer savory
habit: branched, annual herb
range: Europe; escaped from cultivation locally
state frequency: very rare
habitat: roadsides, waste places
notes: —

Scutellaria
(3 species)

Scutellaria galericulata L.
synonym(s): *Scutellaria epilobiifolia* A. Hamilton
common name(s): common skullcap, marsh skullcap
habit: ascending to erect, simple or forked, rhizomatous, perennial herb
range: circumboreal; s. to DE, w. to AZ and CA
state frequency: occasional
habitat: moist to wet thickets, meadows, and sandy, gravelly, or rocky shores
notes: Hybrids with *Scutellaria lateriflora*, called *S.* ×*churchilliana* Fern., occur in Maine.

Scutellaria lateriflora L.
synonym(s): —
common name(s): mad-dog skullcap, blue skullcap
habit: solitary, usually simple, slender-rhizomatous, perennial herb
range: Newfoundland, s. to GA, w. to CA, n. to British Columbia
state frequency: occasional
habitat: alluvial thickets, swampy woods, moist bottomlands, meadows
notes: Known to hybridize with *Scutellaria galericulata* (*q.v.*).

Scutellaria parvula Michx.
synonym(s): —
common name(s): little skullcap, Leonard's skullcap
habit: slender, erect, simple or forked from the base, stoloniferous, perennial herb
range: ME, s. to GA, w. to TX, n. to ND and Ontario
state frequency: very rare
habitat: sandy or gravelly soil, woods, ledges
key to the subspecies:
    1a. Leaves lance-ovate to deltate, 2.0–3.0 times as long as wide, the principal ones with 1 or 2 pairs
        of lateral veins; stems and calyx without glandular hairs .... *S. p.* var. *leonardii* (Epling) Fern.
    1b. Leaves ovate to orbicular, 1.3–2.0 times as long as wide, the principal ones with 3–5 pairs of
        lateral veins; stems and calyx with glandular hairs .... *S. p.* var. *parvula*
notes: This species is ranked SH by the Maine Natural Areas Program.

Stachys
(4 species)

KEY TO THE SPECIES

1a. Plants annual from a taproot; leaf blades 2.0–4.0 cm long, obtuse to rounded at the apex; corolla 6.0–8.0 mm long .... *S. arvensis*
1b. Plants perennial from rhizomes; leaf blades 3.0–15.0 cm long, acute to acuminate at the apex; corolla 10.0–16.0 mm long

    2a. Stems glabrous, sometimes pubescent only on the angles with short, stout hairs; calyx without glandular hairs

        3a. Leaves sessile or on petioles up to 8.0 mm long, the blades mostly 3.0–7.0 cm long .... *S. hyssopifolia*
        3b. Leaves on petioles 8.0–25.0 mm long, the blades mostly 6.0–15.0 cm long .... *S. tenuifolia*

    2b. Stems pubescent on the angles and faces; calyx pubescent in part with glandular hairs .... *S. palustris*

Stachys arvensis (L.) L.
synonym(s): —
common name(s): field hedge-nettle, staggerweed
habit: decumbent to ascending, basally branched, annual herb
range: Europe; naturalized to ME, s. and w. to VA, n. to NY
state frequency: very rare
habitat: fields, waste places
notes: —

Stachys hyssopifolia Michx. var. *ambigua* Gray
synonym(s): *Stachys ambigua* (Gray) Britt., *Stachys aspera* Michx.
common name(s): hyssopleaf hedge-nettle
habit: erect, rhizomatous, perennial herb
range: PA, s. to GA, w. to MO, n. to IA; introduced to ME
state frequency: very rare
habitat: roadsides, fields
notes: —

Stachys palustris L.
synonym(s): <*S. p.* ssp. *arenicola*> = *Stachys palustris* L. var. *homotricha* Fern.; <*S. p.* ssp. *palustris*> = *Stachys palustris* L. var. *nipigonensis* Jennings; <*S. p.* ssp. *pilosa*> = *Stachys palustris* L. var. *segetum* (Mutel) Grogn.
common name(s): woundwort, marsh hedge-nettle
habit: simple or branched, rhizomatous, perennial herb
range: circumboreal; s. to CT, w. to AZ
state frequency: uncommon
habitat: wet fields and meadows, shores
key to the subspecies:

    1a. Calyx short-pubescent with glandular and eglandular hairs of nearly similar length; lobes of the calyx abruptly tapering to the apex; corolla purple; angles of the stem retrorse-pubescent with stout, pustulose-based hairs, relatively longer and wider than the hairs of the stem faces .... *S. p.* ssp. *palustris*
    1b. Calyx pubescent with glandular hairs and long, stout, eglandular hairs; lobes of the calyx gradually tapering to the apex; corolla pink to pale purple; stem pubescence variable

2a. Leaves linear-oblong to lance-oblong, 0.4–2.0(–3.0) cm wide; angles and faces of the stem evidently pubescent with hairs of similar dimensions .... *S. p.* ssp. *arenicola* (Britt.) Gill

2b. Leaves oblong or lance-ovate, 1.5–5.0 cm wide; angles of the stem retrorse-pubescent with pustulose-based hairs, relatively longer than the hairs of the stem faces ..... *S. p.* ssp. *pilosa* (Nutt.) Epling

notes: —

Stachys tenuifolia Willd.

synonym(s): *Stachys tenuifolia* Willd. var. *hispida* (Pursh) Fern., *Stachys tenuifolia* Willd. var. *platyphylla* Fern.
common name(s): smooth hedge-nettle
habit: erect, usually branched, rhizomatous, perennial herb
range: New Brunswick, s. to SC, w. to TX, n. to MN
state frequency: very rare
habitat: moist shores, meadows, and low woods
notes: —

Teucrium
(1 species)

Teucrium canadense L.

synonym(s): <*T. c.* var. *occidentale*> = *Teucrium boreale* Bickn., *Teucrium occidentale* Gray, *Teucrium occidentale* Gray var. *boreale* (Bickn.) Fern.
common name(s): American germander, wood-sage, Canada germander
habit: stiff, erect, simple or branched, rhizomatous, perennial herb
range: New Brunswick and Nova Scotia, s. to FL, w. to NM and CA, n. to British Columbia
state frequency: uncommon
habitat: moist to wet woods, thickets, and shores
key to the subspecies:

1a. Much of the pubescence of the plant, especially in the inflorescence and surface of the calyx, glandular; abaxial surface of the leaves pubescent with straight hairs; upper 3 calyx lobes obtuse at the apex (the middle lobe sometimes acute) .... *T. c.* var. *occidentale* (Gray) McClintock & Epling

1b. Pubescence of the plant entirely eglandular; abaxial surface of the leaves pubescent with curled or bent hairs; upper 3 calyx lobes acute at the apex (the middle lobe sometimes acuminate)

   2a. Leaves lanceolate to ovate, 2.0–6.0 cm wide, sparsely pubescent on the abaxial surface, membranaceous .... *T. c.* var. *virginicum* (L.) Eat.

   2b. Leaves lanceolate to lance-ovate, 1.5–3.0 cm wide, densely pubescent on the abaxial surface, firm .... *T. c.* var. *canadense*

notes: —

Thymus
(1 species)

Thymus pulegioides L.

synonym(s): —
common name(s): lemon thyme
habit: branched, mat-forming shrub
range: Europe; escaped from cultivation
state frequency: uncommon
habitat: lawns, edges, areas of cultivation
notes: —

Trichostema
(1 species)

Trichostema dichotomum L.
synonym(s): —
common name(s): blue curls, forked blue curls
habit: branched, glandular, annual herb
range: ME, s. to FL, w. to MO, n. to MI
state frequency: uncommon
habitat: dry, sandy woods and open fields
notes: —

LENTIBULARIACEAE
(2 genera, 10 species)

KEY TO THE GENERA
1a. Leaves simple, with a viscid, adaxial surface, without traps; flowers solitary, without bracts; calyx with 5 lobes; plants terrestrial, of wet cliffs .... *Pinguicula*
1b. Leaves simple to dissected, not viscid, bearing bladder-like traps on the leaves or on separate branches; flowers solitary or borne in racemes, each subtended by a bract; calyx with 2 lobes; plants aquatic, of wetlands .... *Utricularia*

Pinguicula
(1 species)

Pinguicula vulgaris L.
synonym(s): —
common name(s): violet butterwort
habit: acaulescent, viscid-leaved, perennial herb
range: Greenland and Labrador, s. to New Brunswick and ME, interruptedly w. to OR, n. to AK
state frequency: very rare
habitat: wet, alpine cliffs
notes: This species is ranked S1 by the Maine Natural Areas Program.

Utricularia
(9 species)

KEY TO THE SPECIES FOR USE WITH VEGETATIVE MATERIAL
1a. Plants aquatic or amphibious, either free-floating or sometimes creeping over the substrate; leaves mostly divided into 3 or more segments (except *U. gibba* with mostly 2 segments)
    2a. Leaf-like branches whorled; traps located at the tips of branches .... *U. purpurea*
    2b. Leaf-like branches alternate; traps scattered along the branches
        3a. Branch segments terete or capillary (sometimes flattened in pressing)
            4a. Branches divided 6 or more times; winter buds variously sized; plants found in shallow water

5a. Branches repeatedly dichotomous [*i.e.*, they are many times forked with no main axis]; winter buds diminutive, up to 1.0 mm thick ...... ***U. radiata***

5b. Branches forked at the base, each fork with a straightish or flexuous rachis with segments emerging from its sides; winter buds larger, 2.0 mm wide or more

    6a. Plants large; branches 1.0–6.0 cm long, with spicules along the margins; winter buds 10.0–20.0 mm long, ciliate with stout, gray hairs .... ***U. macrorhiza***

    6b. Plants small; branches 1.0–2.0 cm long, with spicules only near the tips of the divisions; branches that comprise the winter buds glabrous or with a few cilia; winter buds globose, 2.0–5.0 mm in diameter .... ***U. geminiscapa***

4b. Branches divided at most 3 times, usually twice; winter buds diminutive, up to 1.0 mm thick; plants creeping over the ground or in shallow water .... ***U. gibba***

3b. Branch segments flat

    7a. Plants with 2 types of branches—green, photosynthetic branches without traps, and trap-bearing, non-photosynthetic branches lacking chlorophyll; margins of the terminal branchlets minutely serrate; winter buds ciliate with stout, gray hairs .... ***U. intermedia***

    7b. Plants with only 1 type of branch, which is both photosynthetic and trap-bearing; margins of the terminal branchlets entire; winter buds red-green, glabrous or with a few cilia .... ***U. minor***

1b. Plants terrestrial, the stem definitely anchored in the substrate [when in a vegetative state usually found only by careful sod sampling]; protruding leaves simple or slightly parted

8a. Branch-bearing nodes of the rhizome each consistently with 2 lateral, root-like runners; leaves with a few lobes or sometimes simple; plants of wet, sandy shores .... ***U. resupinata***

8b. Branch-bearing nodes of the rhizome with 0–4 lateral, root-like runners, not consistently 2; leaves simple; plants of wet, organic soils, rarely of wet sands .... ***U. cornuta***

KEY TO THE SPECIES FOR USE WITH REPRODUCTIVE MATERIAL

1a. Peduncles subtended by a whorl of conspicuously inflated branches that hold the peduncle above the surface of the water .... ***U. radiata***

1b. Peduncles not subtended by a whorl of inflated branches

  2a. Corolla yellow

    3a. Bracts subtending the pedicels accompanied by a pair of bractlets; calyx enclosing the capsule .... ***U. cornuta***

    3b. Bracts subtending the pedicels without additional bractlets; calyx not enclosing the capsule

      4a. Lower lip of the corolla *ca.* 2.0 times as long as the upper lip

        5a. Lower lip of the corolla 8.0–12.0 mm long, with a well developed palate; spur nearly as long as the lower lip .... ***U. intermedia***

5b. Lower lip of the corolla 4.0–8.0 mm long, without a palate or with a poorly developed one; spur only about 0.5 times as long as the lower lip .... *U. minor*

4b. Lower lip of the corolla only slightly, if at all, longer than the upper lip

6a. Spur thick, conic, *ca.* 0.5 times as long as the lower lip; peduncles with 1–3 flowers .... *U. gibba*

6b. Spur longer, 0.65–1.0 times as long as the lower lip; peduncles with 2–12 flowers

7a. Peduncles with 2–5 flowers; bracts without basal lobes; cleistogamous flowers [without petals] commonly produced on the submersed branches .... *U. geminiscapa*

7b. Peduncles with 6–12 flowers; bracts with basal lobes; cleistogamous flowers not produced .... *U. macrorhiza*

2b. Corolla purple or red-purple (white), with a yellow spot in *U. purpurea*

8a. Lateral lobes of the lower corolla lip saccate; bracts occurring singly; lower lip with a well developed palate; peduncles 1- to 4-flowered .... *U. purpurea*

8b. Lateral lobes of the lower corolla lip not saccate; bracts occurring in pairs, fused basally; lower lip with a poorly developed palate; peduncles with a single flower .... *U. resupinata*

Utricularia cornuta Michx.

synonym(s): *Stomoisia cornuta* (Michx.) Raf.
common name(s): naked bladderwort
habit: terrestrial, scapose, perennial herb
range: Newfoundland, s. FL, w. to TX, n. MN and Manitoba
state frequency: uncommon
habitat: bogs, wet, peaty shores
notes: —

Utricularia geminiscapa Benj.

synonym(s): *Utricularia clandestina* Nutt.
common name(s): mixed bladderwort
habit: free-floating, aquatic, perennial herb
range: Newfoundland, s. to VA, w. and n. to WI
state frequency: uncommon
habitat: quiet water, slow streams
notes: —

Utricularia gibba L.

synonym(s): —
common name(s): creeping bladderwort
habit: creeping or free-floating, perennial herb
range: New Brunswick and Nova Scotia, s. to FL, w. to Mexico and CA, n. to MN and Ontario
state frequency: uncommon
habitat: shallow water, bogs
notes: —

Utricularia intermedia Hayne

synonym(s): —
common name(s): northern bladderwort

habit: free-floating or creeping, perennial herb
range: circumboreal; s. to DE, w. to CA
state frequency: uncommon
habitat: shallow water, slow streams, shores
notes: —

Utricularia macrorhiza Le Conte
synonym(s): *Utricularia vulgaris* L.
common name(s): common bladderwort
habit: free-floating, perennial herb
range: circumboreal; s. to FL, w. to CA
state frequency: occasional
habitat: quiet water
notes: —

Utricularia minor L.
synonym(s): —
common name(s): lesser bladderwort
habit: mostly creeping, perennial herb
range: circumboreal; s. to NJ, w. to CA
state frequency: uncommon
habitat: shallow water, bogs, wet meadows and shores
notes: —

Utricularia purpurea Walt.
synonym(s): *Vesiculina purpurea* (Walt.) Raf.
common name(s): spotted bladderwort, greater purple bladderwort
habit: free-floating, perennial herb
range: New Brunswick and Nova Scotia, s. to FL, w. to LA, n. to MN
state frequency: uncommon
habitat: quiet water, slow streams
notes: —

Utricularia radiata Small
synonym(s): *Utricularia inflata* Walt. var. *minor* Chapman
common name(s): floating bladderwort
habit: free-floating, perennial herb
range: Nova Scotia and ME, s. to FL, w. to TX, n. to IN
state frequency: uncommon
habitat: ponds near coast
notes: —

Utricularia resupinata B. D. Greene *ex* Bigelow
synonym(s): *Lecticula resupinata* (B. D. Greene *ex* Bigelow) Barnh.
common name(s): resupinate bladderwort, lesser purple bladderwort
habit: terrestrial, perennial herb
range: New Brunswick and Nova Scotia, s. to FL, w. to GA, n. to MN and Ontario
state frequency: very rare
habitat: muddy shores, shallow water
notes: This species is ranked S1 by the Maine Natural Areas Program.

OROBANCHACEAE
(11 genera, 21 species)

KEY TO THE GENERA

1a. Plants without chlorophyll, the stems yellow to brown or purple; leaves reduced to scales; ovary unilocular; placentation parietal

2a. Stems branched, usually many times; flowers dimorphic—the upper with a well developed corolla and appearing bisexual [although actually staminate], the lower carpellate, with a small and unopening corolla that is forced off by the developing ovary; plants parasitic on *Fagus grandifolia* .... ***Epifagus***

2b. Stems simple or rarely few-branched; flowers monomorphic; plants parasitic on various species of plants

3a. Flowers 1–3 per plant, borne on long, slender pedicels; calyx nearly actinomorphic; plants parasitic on various species of plants .... ***Orobanche***

3b. Flowers numerous, borne in a thick, fleshy spike; calyx zygomorphic, with a deep split on the lower side; plants parasitic on several species of the genus *Quercus* ...
.... ***Conopholis***

1b. Plants with chlorophyll, with green stems and/or leaves in life; leaves foliaceous; ovary bilocular; placentation axile

4a. Foliage leaves of the stem alternate; anthers glabrous

5a. Anther sacs similar; corolla 2.0 or more times as long as the calyx; bracts herbaceous .... ***Pedicularis***

5b. Anther sacs dissimilar—one attached to the filament near its middle, the other suspended from near its apex; corolla less than 2.0 times as long as the calyx; bracts petaloid .... ***Castilleja***

4b. Foliage leaves of the stem opposite; anthers pubescent (glabrous in *Odontites*)

6a. Corolla weakly zygomorphic, not galeate, 1.0–5.0 cm long

7a. Corolla yellow (sometimes marked with brown or red-brown), 2.5–5.0 cm long; leaves evidently lobed; capsules ovoid to ellipsoid, acute at the apex; anther locules awned at the base .... ***Aureolaria***

7b. Corolla pink to purple (white), 1.0–2.0 cm long; leaves entire; capsules globose to subglobose, rounded at the apex (except for the mucro, if present); anther locules obtuse to cuspidate at the base .... ***Agalinis***

6b. Corolla zygomorphic and galeate [with an upper, hooded lip that invests the stamens], 0.25–1.4 cm long

8a. Leaves and bracts palmately veined; corolla white, blue, or purple, usually with darker purple lines, the sides of the upper lip reflexed .... ***Euphrasia***

8b. Leaves and bracts pinnately veined; corolla white, yellow, or red, the sides not reflexed

9a. Calyx partly inflated in flower, prominently enlarged and inflated in fruit; corolla yellow; foliage leaves prominently toothed .... ***Rhinanthus***

9b. Calyx not, or only scarcely, inflated; corolla red or both white and yellow; foliage leaves entire or weakly toothed

10a. Corolla red, pubescent; anthers glabrous; capsule shorter than the persistent calyx, with more than 4 seeds .... ***Odontites***

10b. Corolla white with a yellow palate, glabrous; anthers pubescent; capsule longer than the persistent calyx, with 1–4 seeds ....*Melampyrum*

Agalinis
(4 species)

KEY TO THE SPECIES

1a. Lobes of the calyx oblong or semiorbicular, obtuse to rounded at the apex; plants fleshy, of coastal saltmarshes .... *A. maritima*
1b. Lobes of the calyx triangular to subulate, acute to acuminate at the apex; plants not fleshy, of inland wetlands, forests, and open areas
  2a. Lobes of the calyx 1.8–5.0 mm long, 0.75–2.0 times as long as the basal, connate portion; pollen sacs *ca.* 1.0 mm long; style 4.0–5.0 mm long .... *A. neoscotica*
  2b. Lobes of the calyx 0.3–2.8 mm long, 0.2–1.0 times as long as the basal, connate portion; pollen sacs 1.4–2.2 mm long; style 8.0–14.0 mm long
    3a. Pedicels 0.1–0.4 cm long at anthesis; corolla pubescent on the adaxial surface [inside], the lobes of the upper lip reflexed-spreading; style 0.8–1.0 mm long.......
    .... *A. paupercula*
    3b. Pedicels 1.0–2.0 cm long at anthesis; corolla glabrous on the adaxial surface, the upper lip arched forward over the stamens; style 1.0–1.4 mm long .. *A. tenuifolia*

Agalinis maritima (Raf.) Raf.
synonym(s): *Gerardia maritima* Raf.
common name(s): saltmarsh false foxglove, saltmarsh agalinis
habit: fleshy, simple or branched, fibrous-rooted, annual herb
range: New Brunswick and Nova Scotia, s. to FL, w. to TX and Mexico
state frequency: rare
habitat: saltmarshes
notes: This species is ranked S2 by the Maine Natural Areas Program.

Agalinis neoscotica (Greene) Fern.
synonym(s): *Agalinis purpurea* (L.) Pennell var. *neoscotica* (Greene) Boivin, *Gerardia neoscotica* Greene, *Gerardia purpurea* L. var. *neoscotica* (Greene) Gleason
common name(s): Nova Scotia false foxglove, smooth agalinis
habit: simple or branched, fibrous-rooted, annual herb
range: Nova Scotia and ME
state frequency: very rare
habitat: bogs, sandy shores
notes: This species is ranked S1 by the Maine Natural Areas Program.

Agalinis paupercula (Gray) Britt.
synonym(s): *Agalinis purpurea* (L.) Pennell var. *parviflora* (Benth.) Boivin *sensu* Gleason and Cronquist (1991), *Gerardia paupercula* (Gray) Britt.
common name(s): small-flowered agalinis
habit: simple or branched, fibrous-rooted, annual herb
range: New Brunswick, s. to NJ, w. to IA, n. to MN
state frequency: uncommon
habitat: damp shores and open areas, bogs
notes: —

Agalinis tenuifolia (Vahl) Raf.
synonym(s): *Gerardia tenuifolia* Vahl
common name(s): common agalinis
habit: branched, fibrous-rooted, annual herb
range: ME, s. to GA, w. to LA, n. to MI
state frequency: uncommon
habitat: dry woods, thickets, and fields
notes: —

Aureolaria
(2 species)

KEY TO THE SPECIES
1a. Calyx, pedicels, capsules, and axis of the inflorescence glabrous; corolla yellow, glabrous on the outside; anthers larger, the bodies 5.0–6.0 mm long, the basal awns 1.0–1.5 mm long; seeds 2.0–2.7 mm long; plants perennial ....*A. flava*
1b. Calyx, pedicels, capsules, and axis of the inflorescence pubescent; corolla yellow with purple-brown markings, glandular-pubescent on the outside; anthers smaller, the bodies 3.0–3.4 mm long, the basal awns *ca.* 0.8 mm long; seeds 0.8–1.0 mm long; plants annual ....*A. pedicularia*

Aureolaria flava (L.) Farw.
synonym(s): *Agalinis flava* (L.) Boivin, *Dasistoma flava* (L.) Wood, *Gerardia flava* L.
common name(s): smooth false foxglove
habit: simple or branched, perennial herb
range: ME, s. to FL, w. to LA, n. to WI
state frequency: uncommon
habitat: dry, deciduous woods
notes: —

Aureolaria pedicularia (L.) Raf.
synonym(s): *Agalinis pedicularia* (L.) Blake, *Dasistoma pedicularia* (L.) Benth., *Gerardia pedicularia* L.
common name(s): annual false foxglove, fern-leaved false foxglove
habit: branched, perennial herb
range: ME, s. to FL, w. to LA, n. to MN
state frequency: rare
habitat: dry woods
notes: This species is ranked S2 by the Maine Natural Areas Program.

Castilleja
(2 species)

KEY TO THE SPECIES
1a. Foliage leaves entire to 3- to 5-cleft; bracts 3- to 5-lobed, red or rarely yellow; calyx 17.0–25.0 mm long; plants annual or biennial, of moist sands, gravels, and meadows ....*C. coccinea*
1b. Foliage leaves entire; bracts entire or 1- to 3-toothed, white to yellow (purple-tinged); calyx 15.0–20.0 mm long; plants perennial, of alpine areas and northern river shores ....*C. septentrionalis*

Castilleja coccinea (L.) Spreng.
synonym(s): —
common name(s): scarlet painted-cup
habit: usually simple, annual or biennial herb
range: ME, s. to SC and MS, w. to OK, n. to Manitoba
state frequency: very rare
habitat: damp, sandy meadows
notes: This species is ranked SX by the Maine Natural Areas Program.

Castilleja septentrionalis Lindl.
synonym(s): —
common name(s): northern painted-cup, northeastern painted-cup
habit: simple or branched, perennial herb
range: Labrador, s. to ME, w. to MI and UT, n. to Alberta
state frequency: rare
habitat: damp, often calcareous, rocky or gravelly soil
notes: This species is ranked S3 by the Maine Natural Areas Program.

Conopholis
(1 species)

Conopholis americana (L.) Wallr. f.
synonym(s): —
common name(s): squawroot
habit: stout, erect, achlorophyllous, parasitic herb
range: Nova Scotia and ME, s. to FL, w. to AL, n. to WI
state frequency: uncommon
habitat: oak roots in rich woods
notes: —

Epifagus
(1 species)

Parasitic herbs with dimorphic flowers, the larger flowers with a calyptrate corolla.

Epifagus virginiana (L.) W. Bart.
synonym(s): —
common name(s): beechdrops
habit: slender, often branched, achlorophyllous, parasitic herb
range: New Brunswick and Nova Scotia, s. to FL, w. to LA, n. to WI and Ontario
state frequency: occasional
habitat: under beech trees
notes: —

Euphrasia
(5 species)

KEY TO THE SPECIES
1a. Bracts with subulate or bristle-tipped teeth; corollas 5.0–10.0 mm long
2a. Bracts erect, cuneate at the base, with ascending teeth .... ***E. stricta***

2b. Bracts spreading, rounded to truncate at the base, with spreading teeth ....
.... *E. nemorosa*
1b. Bracts with obtuse to acute teeth; corollas 2.5–6.0 mm long
   3a. Internodes elongate, at least some more than 3.0 times as long as the leaves; calyx
      lobes 2.5–3.0 mm long at maturity; corollas 4.5–6.0 mm long .... *E. disjuncta*
   3b. Internodes shorter, none of them exceeding 3.0 times as long as the leaves; calyx
      lobes 1.5–2.0 mm long at maturity; corollas 2.5–4.5 mm long
      4a. Stems with 2–5 pairs of leaves, the leaves increasing in size apically;
         inflorescence densely crowded in flower, 0.5–2.5 cm long in fruit; plants of
         alpine areas, elevation exceeding 1000 m .... *E. oakesii*
      4b. Stems usually with more than 5 pairs of leaves, the leaves decreasing in size
         apically; inflorescence relatively more elongate in flower, 2.0–14.0 cm long in
         fruit; plants of moist, often exposed, peats and gravels, but not alpine ... *E. randii*

Euphrasia disjuncta Fern. & Wieg.

synonym(s): —
common name(s): arctic eyebright
habit: slender, simple or branched, annual herb
range: Labrador, s. to ME, w. to MT, n. to AK
state frequency: very rare
habitat: stream banks, damp, open areas, bogs
notes: This species is ranked SX by the Maine Natural Areas Program.

Euphrasia nemorosa (Pers.) Wallr.

synonym(s): *Euphrasia americana* Wettst., *Euphrasia canadensis auct. non* Townsend
common name(s): —
habit: simple or branched, annual herb
range: Europe; naturalized to Newfoundland, s. to ME
state frequency: uncommon
habitat: fields, roadsides
notes: —

Euphrasia oakesii Wettst.

synonym(s): —
common name(s): white mountain eyebright
habit: low, simple, annual herb
range: Labrador, s. to mountains of ME and NH
state frequency: very rare
habitat: alpine areas
notes: This species is ranked S1 by the Maine Natural Areas Program.

Euphrasia randii B. L. Robins.

synonym(s): *Euphrasia randii* B. L. Robins. var. *farlowii* B. L. Robins., *Euphrasia randii* B. L.
   Robins. var. *reeksii* Fern.
common name(s): Rand's eyebright, Nova Scotian eyebright
habit: simple or branched, annual herb
range: Newfoundland, s. to Nova Scotia and ME
state frequency: uncommon
habitat: moist, grassy or peaty slopes, headlands, brackish shores
notes: —

Euphrasia stricta D. Wolff *ex* J. F. Lehm.
synonym(s): *Euphrasia rigidula* Jord.
common name(s): —
habit: simple or branched, annual herb
range: Europe; naturalized to Newfoundland, s. to RI and PA
state frequency: occasional
habitat: thickets, pastures, fields
notes: —

Melampyrum
(1 species)

Melampyrum lineare Desr.
synonym(s): <*M. l.* var. *lineare*> = *Melampyrum lineare* Desr. var. *americanum* (Michx.) Beauverd
common name(s): cowwheat
habit: erect, commonly branched, annual herb
range: Labrador, s. to VA and mountains of GA, w. to WA, n. to British Columbia
state frequency: occasional
habitat: woods, bogs, rocky barrens
key to the subspecies:
  1a. Leaves linear to narrow-lanceolate, 1.0–6.0 mm wide; lowest 2 nodes separated by an
      internode usually 0.5–3.0 cm long .... *M. l.* var. *lineare*
  1b. Leaves lanceolate to ovate, 0.5–30.0 mm wide; lowest 2 nodes separated by an internode
      usually 3.0–4.5 cm long .... *M. l.* var. *latifolium* Bart.
notes: —

Odontites
(1 species)

Odontites vernus (Bellardi) Dumort. ssp. **serotinus** (Dumort.) Corb.
synonym(s): *Odontites serotinus* Dumort., *Odontites rubra* Gilib.
common name(s): red bartsia
habit: branched, annual herb
range: Newfoundland, s. to MA, w. to WI, n. to Quebec
state frequency: uncommon
habitat: fields, roadsides, waste places
notes: —

Orobanche
(1 species)

Orobanche uniflora L.
synonym(s): *Thalesia uniflora* (L.) Britt.
common name(s): one-flowered cancerroot
habit: low, achlorophyllous, parasitic herb
range: Newfoundland, s. to FL, w. to CA, n. to Yukon
state frequency: uncommon
habitat: near to below ground in moist woods and thickets, stream banks
notes: —

Pedicularis
(2 species)

KEY TO THE SPECIES

1a. Calyx split on the upper and lower side, the lower split larger; upper lip of the corolla with 2 slender teeth near the apex; capsule 2.0–3.0 times as long as the persistent calyx; stems 1.5–4.0 dm tall .... *P. canadensis*
1b. Calyx subequally 5-lobed; upper lip of the corolla with 4 very short teeth near the apex; capsule barely exceeding the persistent calyx; stems 4.0–9.0 dm tall .... *P. furbishiae*

Pedicularis canadensis L.
synonym(s): —
common name(s): common lousewort, wood-betony
habit: erect, simple, short-rhizomatous, perennial herb
range: New Brunswick and ME, s. to FL, w. to NM, n. to Manitoba
state frequency: uncommon
habitat: woods, clearings, ledges
notes: —

Pedicularis furbishiae S. Wats.
synonym(s): —
common name(s): Furbish's lousewort
habit: erect, perennial herb
range: New Brunswick and ME
state frequency: very rare
habitat: river banks
notes: This species is ranked S2 by the Maine Natural Areas Program.

Rhinanthus
(1 species)

Rhinanthus minor L.
synonym(s): *Rhinanthus crista-galli* L., *Rhinanthus crista-galli* L. var. *fallax* (Wimmer & Grab.) Druce
common name(s): yellow rattle
habit: erect, simple or branched, annual herb
range: circumboreal; s. to ME, w. to CO and OR
state frequency: occasional
habitat: thickets, old fields, clearings, roadsides
notes: —

SCROPHULARIACEAE
(2 genera, 5 species)

KEY TO THE GENERA

1a. Foliage leaves all basal or alternate; corolla weakly zygomorphic, rotate; androecium composed of 5 stamens, all of which bear pollen; plants biennial .... *Verbascum*
1b. Foliage leaves opposite; corolla zygomorphic, bilabiate; androecium composed of 5 stamens—4 pollen-bearing, 1 sterile; plants perennial .... *Scrophularia*

Scrophularia
(2 species)

KEY TO THE SPECIES

1a. Corollas 5.0–8.0 mm long; sterile filament dark brown or purple-brown, usually wider than long; petioles without a winged margin; capsule 4.0–7.0 mm long ... **S. marilandica**
1b. Corollas 7.0–11.0 mm long; sterile filament yellow-green, usually longer than wide; petioles with a winged margin; capsule 6.0–10.0 mm long .... **S. lanceolata**

Scrophularia lanceolata Pursh
synonym(s): —
common name(s): American figwort
habit: erect, perennial herb
range: Quebec, New Brunswick, and Nova Scotia, s. to SC, w. to NM and CA, n. to British Columbia
state frequency: uncommon
habitat: open woods, borders of woods, thickets, fields, roadsides
notes: —

Scrophularia marilandica L.
synonym(s): —
common name(s): eastern figwort, carpenter's square
habit: erect, perennial herb
range: ME, s. to GA, w. to LA and OK, n. to MN
state frequency: very rare
habitat: rich, open woods, thickets
notes: This species is ranked SX by the Maine Natural Areas Program.

Verbascum
(3 species)

KEY TO THE SPECIES

1a. Plants pubescent with compound, eglandular hairs; 3 upper [shorter] filaments pubescent with white to yellow hairs; 2 lower [longer] filaments glabrous or nearly so; capsule ellipsoid to ovoid, densely pubescent
    2a. Leaves decurrent on the stem nearly or completely to the next leaf; inflorescence at maturity crowded, the flowers concealing the axis of the spike; stigma capitate .... **V. thapsus**
    2b. Leaves not, or only very shortly, decurrent on the stem; inflorescence at maturity less dense, the flowers spaced enough to expose the axis of the spike; stigma spatulate, decurrent on the sides of the style .... **V. phlomoides**
1b. Plants pubescent with simple, glandular hairs; filaments pubescent with purple, knob-tipped hairs; capsule subglobose, pubescent with minute hairs or glabrous ... **V. blattaria**

Verbascum blattaria L.
synonym(s): —
common name(s): moth mullein
habit: slender, simple or branched, glandular, biennial herb
range: Eurasia; naturalized throughout much of North America
state frequency: uncommon
habitat: old fields, roadsides, waste places
notes: —

Verbascum phlomoides L.

synonym(s): —
common name(s): clasping mullein
habit: biennial herb
range: Europe; naturalized to ME, s. to NC, w. to IA, n. to MN
state frequency: very rare
habitat: clearings, roadsides
notes: —

Verbascum thapsus L.

synonym(s): —
common name(s): common mullein, flannel plant
habit: tall, stout, erect, simple, biennial herb
range: Europe; naturalized throughout much of North America
state frequency: common
habitat: fields, gravelly or rocky banks, roadsides, waste places
notes: —

VERONICACEAE

(17 genera, 51 species)

KEY TO THE GENERA

1a. Leaves 6–12 at each node; flowers with a single stamen borne at the summit of the ovary; calyx a minute rim around the apex of the ovary .... ***Hippuris***

1b. Leaves 1 or 2 at each node (but crowded in the apical portion in *Callitriche*) or all basal; flowers with 1–5 stamens, these not borne at the summit of the ovary; calyx absent or present and borne near the base of the ovary

    2a. Perianth absent; plants aquatic, with opposite leaves that are crowded in the apical portion forming a floating rosette .... ***Callitriche***

    2b. Perianth present; plants terrestrial (aquatic in *Limosella* and *Littorella*); leaves not forming a floating rosette

        3a. Perianth inconspicuous, 3- or 4-merous, with an actinomorphic, scarious corolla; leaves all basal (opposite and borne on a stem in *Plantago psyllium*)

            4a. Scape 1- to 3-flowered; fruit an achene, 1-locular, 1-seeded; plants aquatic ....
            .... ***Littorella***

            4b. Scape many-flowered; fruit usually a pyxis, 2-locular, 2- to many-seeded; plants terrestrial .... ***Plantago***

        3b. Perianth conspicuous and showy, with a zygomorphic, 5-merous corolla; leaves borne on a stem (clustered at the nodes of a horizontal stem in *Limosella*)

            5a. Corolla with a slender, basal spur protruding between the 2 lower calyx lobes, and with a palate

                6a. Inflorescence a terminal raceme; capsule symmetrical, dehiscing by terminal teeth, glabrous; plants without glandular hairs

                    7a. Palate composed of 2 short, white ridges; seeds wingless, smooth or nearly so .... ***Nuttallanthus***

7b. Palate composed of a single, usually yellow or orange, ridge; seeds with wings or wing-like angles, or reticulate-patterned .... ***Linaria***

6b. Inflorescence a solitary flower in the axils of leaves; capsule asymmetrical, one of the locules larger, dehiscing by 2 terminal pores, pubescent; plants with glandular hairs .... ***Chaenorrhinum***

5b. Corolla lacking a slender, basal spur, only *Chelone*, *Misopates*, *Mimulus*, and 1 species of *Penstemon* with a palate

8a. Corolla nearly actinomorphic; stems prostrate, bearing a cluster of leaves, roots, and 1-flowered peduncles at the nodes; plants aquatic, of tidal shores .... ***Limosella***

8b. Corolla zygomorphic (weakly zygomorphic in *Penstemon tubiflorus*); stems usually upright, not bearing leaves, roots, and peduncles at a single node; plants of various habits and habitats

9a. Corolla broadly saccate at the base; capsule dehiscing by subterminal pores .... ***Misopates***

9b. Corolla without a saccate swelling near the base; capsule dehiscing by valves or slits

10a. Foliage leaves alternate; inflorescence a secund raceme .... ***Digitalis***

10b. Foliage leaves opposite; inflorescence various, but not a secund raceme

11a. Androecium composed of 2 pollen-bearing stamens and 2 sterile stamens represented by filaments; inflorescence a terminal spike, raceme, or panicle

12a. Corolla apparently composed of 4 lobes due to fusion of the upper 2 petals; calyx composed of 4 sepals (5 sepals in *Veronica officinalis*)

13a. Leaves opposite, sometimes the upper alternate; lobes of the corolla longer than the basal tube; capsule, before dehiscence, emarginate or lobed at the apex .... ***Veronica***

13b. Leaves mostly in whorls of 3–7; lobes of the corolla shorter than the basal tube; capsule, before dehiscence, tapering to the apex, not lobed .... ***Veronicastrum***

12b. Corolla composed of 2 lips—the upper lip 2-lobed, the lower lip 3-lobed; calyx composed of 5 sepals

14a. Calyx subtended by bractlets; second pair of stamens represented by minute filaments .... ***Gratiola***

14b. Calyx without bractlets; second pair of stamens represented by a pair of slender filaments .... ***Lindernia***

11b. Androecium composed of 4 pollen-bearing stamens, with or without 1 sterile stamen; inflorescence of solitary flowers in the axils of normal foliage leaves

15a. Androecium composed of 4 pollen-bearing stamens and 1 sterile stamen; sepals distinct nearly to the base

16a. Sterile stamen shorter than the pollen-bearing stamens; sepals subtended by bractlets; seeds winged ... . ***Chelone***

16b. Sterile stamen of length nearly equal to the pollen-bearing stamens; sepals not subtended by bractlets; seeds without wings .... ***Penstemon***

15b. Androecium composed of 4 pollen-bearing stamens; sepals evidently connate, forming a basal tube .... ***Mimulus***

Callitriche
(2 species)

KEY TO THE SPECIES

1a. Schizocarp *ca.* 0.2 mm longer than wide; mericarps winged, at least in the distal margin, with an evident groove between the wings; surface markings of mericarps in vertical rows .... ***C. palustris***

1b. Schizocarp nearly or fully as wide as long, no more than 0.1 mm longer than wide; mericarps lacking a wing-margin, without or with an inconspicuous groove between the wings; surface markings of mericarps not or obscurely in vertical rows .. ***C. heterophylla***

Callitriche heterophylla Pursh
synonym(s): *Callitriche anceps* Fern.
common name(s): water starwort
habit: submerged to floating, aquatic, annual herb
range: Greenland, s. to FL, w. to TX and Mexico, n. to AK
state frequency: uncommon
habitat: quiet, shallow water, muddy shores
notes: This species is ranked S1 by the Maine Natural Areas Program.

Callitriche palustris L.
synonym(s): *Callitriche verna* L.
common name(s): —
habit: submerged to emergent, aquatic, annual herb
range: circumboreal; s. to VA, w. to CA
state frequency: occasional
habitat: quiet water, wet shores
notes: —

Chaenorrhinum
(1 species)

Chaenorrhinum minus (L.) Lange
synonym(s): —
common name(s): lesser-toadflax, dwarf-snapdragon
habit: erect, branched, annual herb
range: Europe; naturalized to New Brunswick and Nova Scotia, s. to VA, w. to MO, n. to Ontario
state frequency: uncommon
habitat: roadsides, waste places
notes: —

Chelone
(2 species)

KEY TO THE SPECIES

1a. Leaves sessile or with a winged and obscure petiole, cuneate to rounded at the base; corolla white or pale, sometimes pink to purple or green-yellow near the apex; palate pubescent with white or pale yellow hairs; staminode green .... *C. glabra*
1b. Leaves with a distinct petiole 1.0–3.0 cm long, rounded at the base; corolla purple throughout; palate pubescent with bright yellow hairs; staminode white .... *C. lyonii*

Chelone glabra L.
synonym(s): *Chelone glabra* L. var. *dilatata* Fern. & Wieg.
common name(s): white turtlehead
habit: erect, simple or branched, perennial herb
range: Newfoundland, s. to GA, w. to AL and MO, n. to MN and Manitoba
state frequency: occasional
habitat: stream margins, wet thickets, shores, and ditches
notes: —

Chelone lyonii Pursh
synonym(s): —
common name(s): —
habit: erect, simple or branched, perennial herb
range: southeastern United States; escaped from cultivation in New England
state frequency: very rare
habitat: shores, thickets
notes: —

Digitalis
(3 species)

KEY TO THE SPECIES

1a. Lobes of the calyx broad-elliptic to obovate, mostly 6.0–9.0 mm wide .... *D. purpurea*
1b. Lobes of the calyx narrowed, 1.0–2.5 mm wide
  2a. Leaves narrowed at the base, not clasping the stem; stem glabrous; inflorescence tomentose .... *D. lanata*
  2b. Upper leaves broad at the base, clasping the stem; stem pubescent; inflorescence glabrous .... *D. grandiflora*

Digitalis grandiflora P. Mill.
synonym(s): *Digitalis ambigua* Murr.
common name(s): yellow foxglove
habit: biennial or perennial herb
range: Eurasia; escaped from cultivation
state frequency: very rare
habitat: fields, areas of cultivation
notes: —

Digitalis lanata Ehrh.
synonym(s): —
common name(s): Grecian foxglove

habit: perennial or biennial herb
range: Europe; escaped from cultivation to ME, w. to IN
state frequency: very rare
habitat: roadsides, waste places
notes: —

Digitalis purpurea L.
synonym(s): —
common name(s): common foxglove, purple foxglove
habit: stout, biennial herb
range: Europe; escaped from cultivation to Newfoundland, s. to CT
state frequency: rare
habitat: fields, clearings
notes: —

Gratiola
(2 species)

KEY TO THE SPECIES
1a. Leaves lanceolate to ovate, with a broad base, glandular-punctate, mostly 1.0–2.5 cm
  long; corolla 12.0–16.0 mm long, commonly yellow throughout; capsule 2.0–3.0 mm
  long; plants perennial .... ***G. aurea***
1b. Leaves lanceolate to oblanceolate, with a narrow base, not glandular-punctate, mostly
  2.0–5.0 cm long; corolla 8.0–10.0(–12.0) mm long, commonly white at the base with
  yellow lobes; capsule 3.0–5.0 mm long; plants annual .... ***G. neglecta***

Gratiola aurea Pursh
synonym(s): *Gratiola lutea* Raf.
common name(s): golden pert, yellow hedge-hyssop
habit: basally creeping to ascending, stoloniferous, rhizomatous, perennial herb
range: Newfoundland, s. to FL, w. to IL, n. to ND
state frequency: uncommon
habitat: sandy, gravelly, or peaty shores, swamps
notes: The corolla in this species is sometimes white throughout.

Gratiola neglecta Torr.
synonym(s): —
common name(s): —
habit: simple or branched, annual herb
range: New Brunswick and Nova Scotia, s. to GA, w. to AZ and CA, n. to British Columbia
state frequency: uncommon
habitat: muddy soil, shallow water
notes: —

Hippuris
(1 species)

This genus is easily distinguished from other aquatic genera by its simple leaves, produced 6–12 at
each node.

Hippuris vulgaris L.
synonym(s): —
common name(s): mare's tail
habit: erect, simple, rhizomatous, emergent, aquatic, perennial herb
range: circumboreal; s. to ME, w. to NE, NM, and CA
state frequency: rare
habitat: shallow, quiet, fresh to brackish water
notes: This species is ranked S? by the Maine Natural Areas Program.

Limosella
(1 species)

Limosella australis R. Br.
synonym(s): *Limosella subulata* Ives
common name(s): Atlantic mudwort
habit: prostrate, creeping, mat-forming, annual herb
range: Newfoundland, s. to VA
state frequency: rare
habitat: fresh to brackish, muddy or sandy shores
notes: This species is ranked S3 by the Maine Natural Areas Program.

Linaria
(4 species)

KEY TO THE SPECIES
1a. Corolla blue to violet, or white and striped with violet, with a yellow or orange palate
    2a. Spur 2.0–5.0 mm long; plants perennial by creeping rhizomes; seeds reticulate-patterned .... *L. repens*
    2b. Spur 8.0–15.0 mm long; plants annual; seeds with 4–6 ring-like ribs ....
    .... *L. maroccana*
1b. Corolla yellow, with an orange palate
    3a. Leaves linear, 0.2–0.4 cm wide .... *L. vulgaris*
    3b. Leaves lance-ovate to ovate, 1.0–2.0 cm wide .... *L. dalmatica*

Linaria dalmatica (L.) P. Mill.
synonym(s): —
common name(s): dalmatian toadflax
habit: erect, coarse, simple or branched, perennial herb
range: Europe; naturalized to Nova Scotia and ME, s. to PA, w. to OH, n. to Ontario
state frequency: very rare
habitat: fields, roadsides, waste places
notes: —

Linaria maroccana Hook. f.
synonym(s): —
common name(s): —
habit: erect, annual herb
range: northern Africa; escaped from cultivation locally in New England
state frequency: very rare
habitat: fields, roadsides, waste places
notes: —

Linaria repens (L.) P. Mill.

synonym(s): —
common name(s): striped toadflax
habit: ascending, branched, colonial, perennial herb
range: Europe; naturalized to Newfoundland, s. to MA
state frequency: very rare
habitat: thickets, fields, roadsides, waste places
notes: Known to hybridize with *Linaria vulgaris*.

Linaria vulgaris P. Mill.

synonym(s): *Linaria linaria* (L.) Karst.
common name(s): butter and eggs, common toadflax
habit: ascending, colonial, perennial herb
range: Europe; naturalized throughout most of North America
state frequency: occasional
habitat: fields, roadsides, waste places
notes: Known to hybridize with *Linaria repens*.

Lindernia

(1 species)

Lindernia dubia (L.) Pennell

synonym(s): <*L. d.* var. *anagallidea*> = *Ilysanthes inequalis* (Walt.) Pennell, *Lindernia anagallidea*
(Michx.) Pennell; <*L. d.* var. *dubia*> = *Ilysanthes dubia* (L.) Barnh., *Lindernia dubia* (L.) Pennell
var. *riparia* (Raf.) Fern.
common name(s): false pimpernel
habit: low, depressed to ascending, branched or simple, annual herb
range: New Brunswick and Nova Scotia, s. to FL, w. to TX and Mexico, n. to MN and Ontario; also
    WA, s. to CA
state frequency: uncommon
habitat: sandy or muddy shores, shallow water
key to the subspecies:
    1a. Pedicels mostly 1.0–5.0 mm long, bearing only cleistogamous flowers [corollas unexpanded];
        bracts rounded at the apex; plants usually of tidal shores ....
        .... *L. d.* var. *inundata* (Pennell) Pennell
    1b. Pedicels mostly 5.0–25.0 mm long, bearing chasmogamous [expanded corollas] or
        cleistogamous flowers; bracts obtuse to acute at the apex; plants of various wet places and
        shores, but usually not of tidal habitats
        2a. Foliage leaves up to 3.0 cm long, not much larger than the bracts, at least the lower
            narrowed to the base; pedicels 5.0–15.0 mm long, not conspicuously surpassing the
            subtending bracts; seeds 2.0–3.0 times as long as wide .... *L. d.* var. *dubia*
        2b. Foliage leaves mostly 6.0–15.0 cm long, much larger than the bracts, all broadly rounded at
            the base; pedicels 10.0–25.0 mm long, conspicuously surpassing the bracts (except
            sometimes the lowest); seeds 1.5–2.0 times as long as wide ....
            .... *L. d.* var. *anagallidea* (Michx.) Cooperrider
notes: *Lindernia dubia* var. *anagallidea* is ranked SX by the Maine Natural Areas Program.

Littorella
(1 species)

Littorella uniflora (L.) Aschers.
synonym(s): *Littorella americana* Fern.
common name(s): shoreweed, American littorella, American shore-grass
habit: fibrous-rooted, perennial herb
range: Newfoundland, s. to ME, w. to MN, n. to Ontario
state frequency: rare
habitat: sandy, gravelly, or muddy shores, shallow water
notes: This species is ranked S2 by the Maine Natural Areas Program.

Mimulus
(2 species)

KEY TO THE SPECIES
1a. Corolla blue (white), 2.0–3.0 cm long; stems glabrous, erect .... **M. ringens**
1b. Corolla yellow, 1.7–2.2 cm long; stems viscid-villous, creeping at the base, ascending ....
.... **M. moschatus**

Mimulus moschatus Dougl. *ex* Lindl.
synonym(s): —
common name(s): muskflower, musky monkeyflower
habit: spreading to ascending, perennial herb
range: Newfoundland, w. to MI and Ontario; escaped from cultivation in ME
state frequency: very rare
habitat: springy places, streambeds, ditches
notes: —

Mimulus ringens L.
synonym(s): <*M. r.* var. *ringens*> = *Mimulus ringens* L. var. *minthodes* (Greene) A. L. Grant
common name(s): Allegheny monkeyflower
habit: erect, perennial herb
range: New Brunswick and Nova Scotia, s. to GA, w. to TX and CO, n. to Saskatchewan
state frequency: occasional
habitat: wet woods, meadows, and shores, swamps, tidal mud
key to the subspecies:
    1a. Lobes of the calyx 3.0–8.0 mm long; internodes mostly 3.0–7.0 cm long; lower pedicels
    2.0–6.0 cm long; principal foliage leaves 5.0–13.0 cm long; plants mostly of inland wetlands
    and shores .... *M. r.* var. *ringens*
    1b. Lobes of the calyx 1.5–2.5 mm long; internodes 1.5–2.5 cm long; pedicels 1.0–1.7 cm long;
    principal foliage leaves 2.5–5.0 cm long; plants of tidal shores ...... *M. r.* var. *colpophilus* Fern.
notes: *Mimulus ringens* var. *colpophilus* is ranked S2 by the Maine Natural Areas Program.

Misopates
(1 species)

Misopates orontium (L.) Raf.
synonym(s): *Antirrhinum orontium* L.
common name(s): lesser snapdragon
habit: annual herb

range: Europe; rarely escaped from cultivation
state frequency: very rare
habitat: fields, waste places
notes: Flowers pink-purple, purple, or white.

Nuttallanthus
(1 species)

Nuttallanthus canadensis (L.) D. A. Sutton
synonym(s): *Linaria canadensis* (L.) Chaz.
common name(s): old-field toadflax
habit: slender, annual or biennial herb
range: New Brunswick and Nova Scotia, s. to FL, w. to TX and Mexico, n. to SD, MN, and Ontario
state frequency: occasional
habitat: dry, sandy or sterile soil
notes: —

Penstemon
(5 species)

KEY TO THE SPECIES
1a. Corolla weakly zygomorphic, the upper and lower lips not well developed, pubescent on
   both surfaces with minute, glandular hairs .... *P. tubiflorus*
1b. Corolla zygomorphic, with conspicuous upper and lower lips, pubescent on the inner
   surface with eglandular hairs
   2a. Lower lip with an upward swelling, forming a palate that closes the orifice of the
      tubular corolla; corolla without purple lines on the inside of the tube; sterile filament
      densely pubescent .... *P. hirsutus*
   2b. Lower lip without a palate, the orifice of the corolla open; corolla marked with fine
      purple lines on the inside of the tube; sterile filament sparsely to moderately
      pubescent
      3a. Corolla with a slender, basal tube and an evident, abrupt, dilated, distal portion,
         only moderately ridged within the tube; leaves glabrous (sometimes with some
         hairs along the midvein); stems glabrous or pubescent only in lines of decurrence
         from the leaf bases
         4a. Anthers pubescent; leaves subcoriaceous, entire to denticulate; lobes of the
            calyx up to 1.0 cm long in fruit; capsule 9.0–12.0 mm long .... *P. digitalis*
         4b. Anthers glabrous; leaves relatively more herbaceous, with sharper teeth;
            lobes of the calyx up to 1.2 cm long in fruit; capsule 7.0–8.0 mm long ....
            .... *P. calycosus*
      3b. Corolla gradually widened distally, strongly ridged within; leaves pubescent on
         both surfaces; stems pubescent .... *P. pallidus*

Penstemon calycosus Small
synonym(s): *Penstemon laevigatus* Ait. ssp. *calycosus* (Small) Bennett
common name(s): eastern beard-tongue
habit: erect, perennial herb

range: ME, s. to FL, w. to AR, n. to MI
state frequency: rare
habitat: woods, meadows, rocky slopes
notes: —

Penstemon digitalis Nutt. *ex* Sims
synonym(s): —
common name(s): tall white beard-tongue, foxglove beard-tongue
habit: erect, branched, perennial herb
range: New Brunswick and Nova Scotia, s. to VA and AL, w. to TX, n. to SD, MN, and Ontario
state frequency: uncommon
habitat: open woods, meadows, fields, clearings, roadsides
notes: —

Penstemon hirsutus (L.) Willd.
synonym(s): —
common name(s): northeastern beard-tongue
habit: erect, perennial herb
range: Quebec and ME, s. to VA, w. to TN, n. to WI and Ontario
state frequency: rare
habitat: dry woods and fields
notes: —

Penstemon pallidus Small
synonym(s): —
common name(s): eastern white beard-tongue
habit: erect, perennial herb
range: ME, s. to GA, w. to AR and KS, n. to MN; thought to be naturalized in the northeastern portion
    of its range
state frequency: uncommon
habitat: woods, fields, openings, roadsides
notes: —

Penstemon tubiflorus Nutt.
synonym(s): —
common name(s): tube beard-tongue
habit: solitary or clustered, erect, perennial herb
range: WI and IN, s. to MS, w. to TX, n. to NE; adventive to ME, s. to PA
state frequency: uncommon
habitat: open woods, fields, roadsides
key to the subspecies:
    1a. Lowest cymules of peduncles 0.2–1.5(–3.0) cm long; larger leaves 2.0–4.0 cm wide, subacute
        at the apex, membranaceous; nodes distant .... *P. t.* var. *tubiflorus*
    1b. Lowest cymules on peduncles 1.5–8.0 cm long; larger leaves 1.0–2.5 cm wide, subacuminate
        at the apex, subcoriaceous; nodes approximate .... *P. t.* var. *achoreus* Fern.
notes: —

Plantago
(9 species)

KEY TO THE SPECIES

1a. Leaves borne on a stem, opposite, linear, 1.0–3.0 mm wide; peduncles arising from the axils of leaves .... *P. psyllium*
1b. Leaves all basal, linear to ovate, 1.0–110.0 mm wide; peduncles arising from basal rosettes of leaves
  2a. Bracts and sepals glabrous or ciliate on the apical portion of the keel; plants perennial
    3a. Leaves linear, 1.0–12.0 mm wide, fleshy; corolla tube pubescent on the abaxial [outside] surface; plants of Atlantic coast shores .... *P. maritima*
    3b. Leaves narrow-elliptic to ovate or obovate, 6.0–110.0 mm wide, definitely not fleshy; corolla tube glabrous on the abaxial surface; plants not of Atlantic coast shores
      4a. Bracts and sepals conspicuously keeled; lobes of the corolla up to 1.0 mm long; pyxis 4- to 30-seeded; spikes 5.0–30.0 cm tall
        5a. Pyxis circumscissile near the middle, containing 6–30 seeds; bracts broad-ovate; seeds 1.0–1.7 mm long, reticulate-patterned; petiole usually green near the base .... *P. major*
        5b. Pyxis circumscissile near the base, containing 4–10 seeds; bracts narrow-triangular; seeds 1.5–2.0(–2.5) mm long, not reticulate-patterned; petiole usually anthocyanic near the base .... *P. rugelii*
      4b. Bracts and sepals flat or slightly keeled; lobes of the corolla exceeding 1.0 mm; pyxis 2- to 4-seeded; spikes 1.5–10.0 cm tall
        6a. Calyx composed of 3 sepals, 1 of the sepals with 2 midveins and formed by fusion of 2 sepals; leaves lanceolate to narrow-elliptic; flowers only slightly fragrant; seeds deeply concave on the inner face ..... *P. lanceolata*
        6b. Calyx composed of 4 distinct sepals; leaves elliptic to ovate or obovate; flowers fragrant; seeds slightly concave on the inner surface ...... *P. media*
  2b. Bracts and/or sepals conspicuously pubescent; plants annual or biennial
    7a. Plants subdioecious; lobes of the corolla of carpel-bearing, chasmogamous flowers erect after anthesis, connivent, closing over the fruit; leaves oblanceolate to obovate .... *P. virginica*
    7b. Plants synoecious; lobes of the corolla spreading to reflexed after anthesis; leaves linear
      8a. Bracts conspicuous, exceeding the flowers and fruits, the lower ones extending 5.0–25.0 mm beyond the flowers; inflorescence sparsely tomentose .... *P. aristata*
      8b. Bracts inconspicuous, often not exceeding 5.0 mm long, scarcely, if at all, extending beyond the flowers and fruits; inflorescence evidently tomentose .... *P. patagonica*

Plantago aristata Michx.
synonym(s): —
common name(s): bracted plantain, buckhorn
habit: taprooted, annual or short-lived perennial herb

range: IL, s. to LA, w. to TX; naturalized to eastern United States and Canada
state frequency: uncommon
habitat: dry, open soil, waste places
notes: —

Plantago lanceolata L.

synonym(s): *Plantago altissima auct. non* L., *Plantago lanceolata* L. var. *sphaerostachya* Mert. &
  Koch
common name(s): English plantain, ribgrass, ripplegrass
habit: fibrous-rooted or rhizomatous, perennial herb
range: Eurasia; naturalized throughout North America
state frequency: occasional
habitat: lawns, roadsides
notes: Occasionally this species produces inflorescences composed partly or entirely of carpellate
  flowers.

Plantago major L.

synonym(s): <*P. m.* var. *intermedia*> = *Plantago major* L. var. *scopulorum* Fries & Broberg
common name(s): common plantain
habit: fibrous-rooted, perennial or annual herb
range: Eurasia; naturalized to United States and southern Canada
state frequency: common
habitat: lawns, dooryards, roadsides, waste places
key to the subspecies:
  1a. Pyxis rounded at the apex, circumscissile in a line below the tips of the sepals; leaves
    subcoriaceous-fleshy, usually pubescent, with broad, short petioles; plants of saline shores ....
    .... *P. m.* var. *intermedia* (DC.) Pilger
  1b. Pyxis pointed at the apex, circumscissile in a line above or below the sepals; leaves
    membranaceous to fleshy-herbaceous, glabrous or pubescent, with longer, more slender
    petioles; plants of inland habitats
    2a. Pyxis circumscissile in a line below the tips of the sepals; leaves fleshy-herbaceous, usually
      pubescent on one or both surfaces .... *P. m.* var. *major*
    2b. Pyxis circumscissile in a line at the tips of the sepals; leaves membranaceous, usually
      glabrous .... *P. m.* var. *pilgeri* Domin
notes: —

Plantago maritima L. var. *juncoides* (Lam.) Gray

synonym(s): *Plantago juncoides* Lam. var. *decipiens* (Barneoud) Fern., *Plantago juncoides* Lam. var.
  glauca (Hornem.) Fern., *Plantago oliganthos* Roemer & J. A. Schultes, *Plantago oliganthos*
  Roemer & J. A. Schultes var. *fallax* Fern.
common name(s): seaside plantain, goosetongue
habit: fleshy, deep-rooted, perennial herb
range: circumboreal; s. along the Atlantic coast to NJ
state frequency: uncommon
habitat: saltmarshes, coastal beaches and rocks
notes: —

Plantago media L.

synonym(s): —
common name(s): hoary plantain
habit: fibrous-rooted, perennial herb
range: Eurasia; naturalized to New Brunswick and Nova Scotia, s. to CT, w. to MI, n. to Ontario and
  Manitoba

state frequency: rare
habitat: fields, lawns, waste places
notes: —

Plantago patagonica Jacq.
synonym(s): *Plantago purshii* Roemer & J. A. Schultes
common name(s): woolly plantain
habit: annual herb
range: South America; naturalized to IN, s. and w. to TX, n. to British Columbia; also ME, s. to NJ
state frequency: very rare
habitat: dry, open areas
notes: —

Plantago psyllium L.
synonym(s): *Plantago arenaria* Waldst. & Kit., *Plantago indica* L.
common name(s): leafy-stemmed plantain
habit: annual herb
range: Europe; naturalized to the eastern United States
state frequency: rare
habitat: waste places
notes: —

Plantago rugelii Dcne.
synonym(s): —
common name(s): American plantain
habit: fibrous-rooted, perennial herb
range: New Brunswick and Nova Scotia, s. to FL, w. to TX, n. to MT
state frequency: uncommon
habitat: shores, lawns, gardens, roadsides, waste places
notes: —

Plantago virginica L.
synonym(s): —
common name(s): hoary plantain, pale-seed plantain
habit: taprooted, annual or biennial herb
range: ME, s. to FL, w. to CA, n. to OR
state frequency: very rare
habitat: dry, open areas
notes: Occasionally this species produces cleistogamous flowers in addition to the normal, chasmogamous flowers. The cleistogamous flowers have spreading corolla lobes, which may cause confusion in the identification key.

Veronica
(14 species)

KEY TO THE SPECIES
1a. Racemes borne in the axils of leaves, subtended by very small bracts; leaves opposite throughout; plants perennial
    2a. Stems and usually the leaves conspicuously pubescent; styles 2.5–8.0 mm long; plants of mesic habitats

3a. Flowering stems creeping near the base, the tip ascending; leaves thick, subcoriaceous, permanently pubescent; corolla 4.0–8.0 mm wide; capsule distinctly exceeding the calyx .... *V. officinalis*

3b. Flowering stems ascending to erect; leaves thin, membranaceous, becoming glabrate; corolla 8.0–13.0 mm wide; capsule not, or only slightly, exceeding the calyx

    4a. Leaf blades ovate, 1.0–2.0 times as long as wide; style 3.0–5.0 mm long; sepals subequal in length; styles 3.0–5.0 mm long; stems 1.0–4.0 dm tall .... *V. chamaedrys*

    4b. Leaf blades lanceolate to ovate, 2.0–4.0 times as long as wide; style 6.0–8.0 mm long; upper sepals much shorter than the lower sepals; styles 6.0–8.0 mm long; stems 3.0–8.0 dm tall .... *V. austriaca*

2b. Stems and leaves essentially glabrous; styles 1.5–4.0 mm long; plants of hydrophytic habitats

    5a. Leaf blades mostly 1.5–4.0 times as long as wide; rachis of raceme straight; seeds up to 0.5 mm long

        6a. Leaves borne on short petioles; raceme 4- to 30-flowered; style 2.5–3.5 mm long .... *V. americana*

        6b. Leaves of the flowering stems sessile; raceme 20- to 54-flowered; style 1.5–2.5 mm long .... *V. anagallis-aquatica*

    5b. Leaf blades mostly 4.0–20.0 times as long as wide; rachis of raceme flexuous; seeds 1.2–1.8 mm long .... *V. scutellata*

1b. Racemes terminating the main stem, subtended by foliaceous bracts (subtended by very small bracts in *V. longifolia*); upper leaves [those subtending flowers] commonly alternate; plants annual or perennial

7a. Plants annual; bracteal leaves smaller than or similar in size to foliage leaves

    8a. Pedicels very short, up to 2.0 mm long; seeds 0.4–1.2 mm long; bracteal leaves smaller than or foliage leaves

        9a. Corolla largely or entirely white; leaves linear-oblong to oblanceolate, 3.0–10.0 times as long as wide; style 0.1–0.3 mm long; seeds 0.4–0.8 mm long .... *V. peregrina*

        9b. Corolla blue; leaves ovate to broad-elliptic, 1.0–2.0 times as long as wide; style 0.4–1.0 mm long; seeds 0.8–1.2 mm long .... *V. arvensis*

    8b. Pedicels longer, 6.0–40.0 mm long; seeds 1.2–2.5 mm long; bracteal leaves similar in size to foliage leaves

        10a. Pedicels 15.0–40.0 mm long in fruit; corolla mostly 8.0–11.0 mm wide; style 1.8–3.0 mm long; capsule 5.0–9.0 mm wide .... *V. persica*

        10b. Pedicels 6.0–15.0 mm long in fruit; corolla 4.0–8.0 mm wide; style mostly 1.0–1.5 mm long; capsule 3.0–6.0 mm wide

            11a. Corolla blue; style up to 1.5 mm long, extending beyond the apical notch of the capsule; capsule 2.5–4.0 x 3.5–6.0 mm, with 9–12 seeds per locule .... *V. polita*

            11b. Corolla blue or white; style up to 1.0 mm long, not extending beyond the apical notch of the capsule; capsule 4.0–6.0 x 3.0–4.0 mm, with 3–8 seeds per locule .... *V. agrestis*

7b. Plants perennial; bracteal leaves smaller than foliage leaves

12a. Leaves 4.0–10.0 cm long, sharply serrate; style about 2.0 times as long as the capsule; capsule pubescent with eglandular hairs; stems 3.0–15.0 dm tall .... *V. longifolia*
12b. Leaves 1.0–4.0 cm long, entire or weakly toothed; style shorter than to about as long as the capsule; capsule pubescent with glandular hairs; stems 0.5–3.0 dm tall
    13a. Capsule 4.0–7.0 mm tall, taller than wide; stems erect; plants of alpine habitats, elevation exceeding 1000 m .... *V. wormskjoldii*
    13b. Capsule 3.0–4.0 mm tall, wider than tall; stems creeping at the base or producing prostrate lower branches; plants of fields and roadsides, elevation less than 1000 m .... *V. serpyllifolia*

Veronica agrestis L.

synonym(s): —
common name(s): field speedwell
habit: prostrate to ascending, basally forked, annual herb
range: Eurasia; naturalized to Newfoundland, s. to CT, w. to PA, n. to MI
state frequency: very rare
habitat: fields, lawns, roadsides, waste places
notes: —

Veronica americana Schwein. *ex* Benth.

synonym(s): —
common name(s): American speedwell, American brooklime
habit: fleshy, erect to ascending, simple, rhizomatous, perennial herb
range: Newfoundland, s. to NC, w. to Mexico and CA, n. to AK
state frequency: uncommon
habitat: seepy areas, stream banks, swamps, shallow water
notes: —

Veronica anagallis-aquatica L.

synonym(s): —
common name(s): water speedwell, brook-pimpernel
habit: erect, fibrous-rooted, biennial or short-lived perennial herb
range: Eurasia; naturalized to ME, s. to NC, w. to NM and AZ, n. to WA
state frequency: very rare
habitat: springs, slow water, wet ditches and shores, shallow water
notes: —

Veronica arvensis L.

synonym(s): —
common name(s): corn speedwell
habit: erect to ascending, simple or branched, annual herb
range: Eurasia; naturalized to Newfoundland, s. to CT, w. to MN, n. to Ontario; also British Columbia
state frequency: uncommon
habitat: open woodlands, ledges, fields, gardens, lawns, waste places
notes: —

Veronica austriaca L. ssp. *teucrium* (L.) D. A. Webb

synonym(s): *Veronica latifolia* L., *Veronica teucrium* L.
common name(s): —
habit: stiff, erect, perennial herb

range: Eurasia; escaped from cultivation to ME, s. to MD, w. to IN, n. to SD
state frequency: very rare
habitat: roadsides
notes: —

Veronica chamaedrys L.

synonym(s): —
common name(s): bird's eye, germander speedwell
habit: prostrate to ascending, rhizomatous, perennial herb
range: Europe; naturalized to Newfoundland, s. to MD, w. to IL and MI, n. to Ontario
state frequency: uncommon
habitat: borders of woods, fields, lawns, roadsides
notes: —

Veronica longifolia L.

synonym(s): —
common name(s): long-leaved speedwell
habit: erect, perennial herb
range: Europe; naturalized to Newfoundland, s. to MD, w. to IL, n. to ND
state frequency: uncommon
habitat: thickets, fields, roadsides, waste places
notes: —

Veronica officinalis L.

synonym(s): —
common name(s): common speedwell
habit: creeping, branched, perennial herb
range: Europe; naturalized to Newfoundland, s. to NC, w. to TN, n. to WI and Ontario
state frequency: occasional
habitat: dry hills, fields and open woods
notes: —

Veronica peregrina L.

synonym(s): —
common name(s): purslane speedwell, neckweed
habit: fleshy, erect, simple or branched, annual herb
range: Newfoundland, s. to FL, w. to TX and Mexico, n. to AK
state frequency: occasional
habitat: moist to wet open areas, marshes
key to the subspecies:
    1a. Stems and commonly the sepals and capsules glandular-pubescent ....
    .... *V. p.* ssp. *xalapensis* (Kunth) Pennell
    1b. Stem, sepals, and capsules glabrous .... *V. p.* ssp. *peregrina*
notes: —

Veronica persica Poir.

synonym(s): —
common name(s): bird's-eye speedwell
habit: ascending, simple or branched, annual herb
range: Asia; naturalized throughout much of North America
state frequency: uncommon
habitat: lawns, gardens, roadsides, waste places
notes: —

Veronica polita Fries

synonym(s): —
common name(s): wayside speedwell
habit: prostrate to ascending, annual herb
range: Eurasia; naturalized as far north as New Brunswick, w. to Manitoba
state frequency: very rare
habitat: lawns, roadsides, waste places
notes: —

Veronica scutellata L.

synonym(s): —
common name(s): marsh speedwell, narrow-leaved speedwell
habit: erect to ascending, rhizomatous, perennial herb
range: Labrador, s. to VA, w. to CA, n. to AK
state frequency: occasional
habitat: wet shores, swamps, bogs
notes: —

Veronica serpyllifolia L.

synonym(s): <*V. s.* ssp. *humifusa*> = *Veronica tenella* All.
common name(s): thyme-leaved speedwell
habit: creeping to ascending, simple or branched, rhizomatous, perennial herb
range: Newfoundland, s. to CT, w. to MN, n. to Ontario
state frequency: occasional
habitat: open woods, fields, meadows, lawns, roadsides
key to the subspecies:
    1a. Corolla pale blue with blue lines; pedicels puberulent; corolla mostly 2.0–4.0 mm wide;
        filaments 1.0–2.5 mm long; capsules mostly 3.0–4.0 mm wide .... *V. s.* ssp. *serpyllifolia*
    1b. Corolla blue; pubescence of pedicels in part of glandular hairs; corolla mostly 5.0–8.0 mm
        wide; filaments 2.0–4.0 mm long; capsules mostly 4.0–6.0 mm wide ....
        .... *V. s.* ssp. *humifusa* (Dickson) Syme
notes: —

Veronica wormskjoldii Roemer & J. A. Schultes

synonym(s): *Veronica alpina* L. var. *unalaschcensis* Cham. & Schlecht.
common name(s): northern speedwell
habit: erect, simple, rhizomatous, perennial herb
range: Greenland, s. to ME and NH, w. to NM, n. to AK
state frequency: very rare
habitat: seeps and gullies in alpine areas
notes: This species is ranked S1 by the Maine Natural Areas Program.

Veronicastrum
(1 species)

Veronicastrum virginicum (L.) Farw.

synonym(s): *Leptandra virginica* (L.) Nutt., *Veronica virginica* L.
common name(s): Culver's root, Culver's psychic
habit: tall, erect, usually branched, perennial herb
range: Asia; naturalized to ME, s. to FL, w. to TX, n. to Manitoba
state frequency: very rare
habitat: upland woods, thickets, meadows
notes: —

GENTIANACEAE
(7 genera, 11 species)

KEY TO THE GENERA

1a. Leaves of the stem reduced to scales; flowers 0.25–0.4(–0.7) cm long; lobes of the corolla imbricate in bud .... ***Bartonia***
1b. Leaves of the stem foliaceous; flowers 0.5–5.5 cm long; lobes of the corolla convolute in bud
    2a. Style evident, slender; corolla pink, 1.0–1.3 cm long, its lobes 3.0–4.0 mm long; anthers becoming spirally coiled .... ***Centaurium***
    2b. Style absent, inconspicuous, or short and stout; corolla blue to white or green to purple-green, the corolla and/or its lobes longer; anthers not spirally coiled
        3a. Lobes of the corolla with conspicuous, basal spurs in the larger flowers; corolla green to purple-green .... ***Halenia***
        3b. Lobes of the corolla lacking spurs; corolla blue to white
            4a. Corolla rotate; calyx with deep sinuses; stigmas decurrent along the sides of the ovary; plants 1.0–2.0 dm tall .... ***Lomatogonium***
            4b. Corolla funnelform, campanulate, or salverform; calyx with an elongate, basal, connate portion; stigmas sessile or on a short style, but not decurrent on the sides of the ovary; plants 1.5–8.0 dm tall
                5a. Sinuses of the corolla folded or plaited, often of a color or texture different from that of the corolla lobes; plants perennial; calyx with a continuous membrane or rim on the inside at the summit .... ***Gentiana***
                5b. Sinuses of the corolla neither folded nor plaited, of a color and texture similar to that of the lobes; plants annual or biennial; calyx without a membrane or with a membrane confined to the base of each sinus
                    6a. Margins of the corolla lobes conspicuously fringed in the apical portion; corolla 3.5–5.5 cm long, usually 4-merous; ovules covering much of the inner surface of the ovary; calyx with a membrane on the inside at the base of each sinus .... ***Gentianopsis***
                    6b. Margins of the corolla lobes not fringed; corolla 1.0–2.3 cm long, commonly 5-merous; ovules confined to bands along each of the 2 sutures of the ovary; calyx without an internal membrane ....
.... ***Gentianella***

Bartonia
(2 species)

KEY TO THE SPECIES

1a. Leaves below the inflorescence opposite; petals oblong, rounded to obtuse at the often mucronulate apex; stigmas, including the basal lobes, 1.5–2.3 mm long; flowers borne on erect to ascending pedicels .... ***B. virginica***
1b. Leaves alternate, sometimes the lower ones opposite; petals lanceolate, obtuse to acuminate at the apex; stigmas 0.8–1.5 mm long; flowers borne on ascending to spreading pedicels .... ***B. paniculata***

Bartonia paniculata (Michx.) Muhl.
synonym(s): <*B. p.* ssp. *iodandra*> = *Bartonia paniculata* (Michx.) Muhl. var. *intermedia* Fern.
common name(s): screwstem
habit: erect to trailing, sometimes twining, annual or biennial herb
range: Newfoundland, s. to FL, w. to TX, n. to MI and Ontario
state frequency: very rare
habitat: wet peat or sand, swamps, bogs, wet meadows
key to the subspecies:
   1a. Anthers commonly purple; calyx connate in the basal portion; petals obtuse to acute at the
       apex; stems often purple .... *B. p.* ssp. *iodandra* (B. L. Robins.) J. Gillett
   1b. Anthers yellow; calyx distinct; petals acuminate at the apex; stems green or yellow-green ....
       .... *B. p.* ssp. *paniculata*
notes: This species is ranked S1 by the Maine Natural Areas Program.

Bartonia virginica (L.) B. S. P.
synonym(s): —
common name(s): —
habit: usually simple, erect, annual or biennial, herb
range: Quebec, s. to FL, w. to LA, n. to MN and Ontario
state frequency: uncommon
habitat: dry to wet, often acid soil of meadows, bogs, and gravels
notes: —

Centaurium
(1 species)

Centaurium pulchellum (Sw.) Druce
synonym(s): —
common name(s): branching centaury
habit: erect, branched, annual herb
range: Europe; naturalized to ME and VT, s. to VA, w. and n. to IL
state frequency: very rare
habitat: fields, waste places
notes: —

Gentiana
(3 species)

KEY TO THE SPECIES
1a. Leaves and lobes of the calyx minutely ciliolate; lobes of the corolla about as long as the
    appendages [plaits] in the sinuses of the corolla .... *G. clausa*
1b. Leaves and lobes of the calyx without cilia; lobes of the corolla 1.5–5.0 times as long as
    the appendages in the sinuses of the corolla
    2a. Involucral leaves ascending, linear to linear-lanceolate, 0.3–1.5 cm wide; lobes of the
        corolla 1.5–3.0 times as long as the appendages in the sinuses of the corolla; leaves
        dark green .... *G. linearis*
    2b. Involucral leaves spreading, lance-ovate to ovate, 1.5–3.0 cm wide; lobes of the
        corolla 3.0–5.0 times as long as the appendages in the sinuses of the corolla; leaves
        light green .... *G. rubricaulis*

Gentiana clausa Raf.

synonym(s): —
common name(s): bottle gentian
habit: tufted, ascending, usually simple, perennial herb
range: Quebec and ME, s. to MD and mountains of NC, w. to MO
state frequency: uncommon
habitat: moist borders, thickets, meadows, and stream banks
notes: —

Gentiana linearis Froel.

synonym(s): *Dasystephana linearis* (Froel.) Britt.
common name(s): closed gentian, narrow-leaved gentian
habit: slender, perennial herb
range: Labrador, s. to MD, w. to NE, n. to MN
state frequency: uncommon
habitat: wet woods, thickets, and meadows, bogs
notes: —

Gentiana rubricaulis Schwein.

synonym(s): *Dasystephana grayi* (Kusnez.) Britt., *Gentiana linearis* Froel. var. *latifolia* Gray
common name(s): red-stemmed gentian, closed gentian, Great Lakes gentian
habit: perennial herb
range: Ontario, s. to MI, w. to NE, n. to Saskatchewan; also ME and New Brunswick
state frequency: very rare
habitat: moist woods, meadows, and shores
notes: This species is ranked SH by the Maine Natural Areas Program.

Gentianella
(2 species)

KEY TO THE SPECIES

1a. Corolla 1.0–1.5 cm long, with a fringe of hairs on the inside around the base of the lobes; lobes of the corolla 3.0–5.0 mm long, obtuse to acute at the apex .... *G. amarella*
1b. Corolla 1.5–2.3 cm long, without a fringe at the base of the lobes; lobes of the corolla 4.0–7.0 mm long, acute to acuminate at the apex .... *G. quinquefolia*

Gentianella amarella (L.) Boerner ssp. *acuta* (Michx.) J. Gillett

synonym(s): *Gentiana amarella* L.
common name(s): northern gentian, felwort
habit: simple or branched, annual or biennial herb
range: interruptedly circumboreal; s. to ME, w. to Mexico and CA
state frequency: very rare
habitat: moist rocks, gravel, and sand
notes: This species is ranked S1 by the Maine Natural Areas Program.

Gentianella quinquefolia (L.) Small

synonym(s): *Gentiana quinquefolia* L.
common name(s): stiff gentian, ague weed
habit: simple or branched, annual or biennial herb
range: ME, s. and w. to FL, n. to Ontario
state frequency: very rare
habitat: woods, wet banks and fields
notes: This species is ranked SH by the Maine Natural Areas Program.

Gentianopsis
(1 species)

Gentianopsis crinita (Froel.) Ma
synonym(s): *Gentiana crinita* Froel.
common name(s): fringed gentian
habit: simple or branched, annual or biennial herb
range: ME, s. to MD and mountains of GA, w. to IA, n. to Manitoba
state frequency: uncommon
habitat: low woods, wet thickets, brooksides, and meadows
notes: —

Halenia
(1 species)

Halenia deflexa (Sm.) Griseb.
synonym(s): —
common name(s): spurred gentian
habit: erect, simple or branched, annual or biennial herb
range: Labrador, s. to ME, w. to IL, n. to SD and Ontario; also Mexico and MT
state frequency: uncommon
habitat: moist to wet, cool woods, bogs
notes: —

Lomatogonium
(1 species)

Lomatogonium rotatum (L.) Fries *ex* Fern.
synonym(s): —
common name(s): marsh-felwort
habit: simple or branched, annual or biennial herb
range: circumboreal; s. to ME, w. to CO and AK
state frequency: very rare
habitat: edges of pools and peats of rocky, coastal shores
notes: This species is ranked S1/S2 by the Maine Natural Areas Program.

APOCYNACEAE
(4 genera, 8 species)

KEY TO THE GENERA
1a. Stems climbing or twining; corolla purple-black, the lobes pubescent on the adaxial surface; corona present, forming a fleshy, lobed cup .... **Cynanchum**
1b. Stems depressed to erect, or trailing, but not climbing; corolla variously colored, but not purple-black, the lobes glabrous; corona absent or present and forming a whorl of hoods and horns
    2a. Perianth composed of 2 whorls—the calyx and the corolla; pollen distinct, not aggregated in pollinia; carpels connate in the basal portion

3a. Corolla blue-purple (white); stems trailing over the ground and forming extensive mats; flowers solitary in the axils of leaves; seeds without a coma .... ***Vinca***

3b. Corolla white, pink, yellow, or green-white; stems depressed to erect, not trailing; flowers in terminal cymes; seeds comose .... ***Apocynum***

2b. Perianth composed of 3 whorls—calyx, corolla, and corona—the corona forming a conspicuous whorl of hoods, each hood with a protruding horn; pollen aggregated in pollinia; carpels distinct in the basal portion, connate near the apex .... ***Asclepias***

Apocynum
(2 species)

KEY TO THE SPECIES

1a. Corolla pink or white and pink-striped, 6.0–10.0 mm long, its lobes spreading or recurving; seeds 2.5–3.0 mm long; leaves wide-spreading to drooping ....
.... ***A. androsaemifolium***

1b. Corolla white, green-white, or yellow, 3.0–6.0 mm long, its lobes erect or slightly spreading; seeds 4.0–6.0 mm long; leaves erect to spreading .... ***A. cannabinum***

Apocynum androsaemifolium L.
synonym(s): —
common name(s): spreading dogbane
habit: simple or branched, spreading to ascending, perennial herb
range: Newfoundland, s. to GA, w. to AZ, n. to British Columbia
state frequency: common
habitat: dry thickets, borders, fields, and roadsides
notes: Hybrids with *Apocynum cannabinum*, called *A.* ×*floribundum* Greene, occur in Maine and can be recognized by their intermediate morphology—corolla white or tinged with pink, 4.0–7.0 mm long, the lobes not recurving; seeds 3.0–4.0 mm long; leaves wide-spreading.

Apocynum cannabinum L.
synonym(s): *Apocynum sibiricum* Jacq.
common name(s): hemp dogbane, clasping dogbane, Indian-hemp
habit: erect to prostrate, branched, perennial herb
range: Newfoundland, s. to FL, w. to CA, n. to AK
state frequency: occasional
habitat: thickets, borders of woods, stream margins
notes: Known to hybridize with *Apocynum androsaemifolium* (q.v.).

Asclepias
(4 species)

Members of this genus have unusual and highly specialized flowers. Milkweeds have a third whorl of the perianth, called a corona, that appears as a whorl of 5 hoods, each of which has a central horn. The stamens have been highly modified to form 2 pollen sacs connected by a translator. The pollen grains from each pollen sac are united and form a single unit, called a pollinium. A sticky gland, the corpusculum, attaches the translator and pollinia to an insect pollinator. The center of the flower is composed of the gynostegium, which is formed from fused anther and stigma material. This drum-shaped structure bears the stigmatic slits to which pollen is transferred during pollination.

KEY TO THE SPECIES

1a. Leaves of the stem alternate; corolla yellow or orange .... *A. **tuberosa***
1b. Leaves of the stem opposite; corolla white, pink, red, or purple
    2a. Follicle remotely covered with conic processes 1.0–3.0 mm tall; hoods of the corona
        surpassing the gynostegium; horns not surpassing the hoods .... *A. **syriaca***
    2b. Follicle not covered with conic processes; hoods of the corona approximately
        equaling the gynostegium; horns evidently exceeding the hoods
        3a. Lobes of the corolla 4.5–6.0 mm long; lateral margins of the hoods divergent
            from near the base, the hood, therefore, upwardly flaring; follicles 5.0–9.0 cm
            long; plants of wetlands .... *A. **incarnata***
        3b. Lobes of the corolla 7.0–10.0 mm long; lateral margins of the hoods parallel, the
            hood, therefore, tubular-shaped; follicles 10.0–15.0 cm long; plants of upland
            habitats .... *A. **exaltata***

Asclepias exaltata L.

synonym(s): —
common name(s): tall milkweed
habit: erect, perennial herb
range: ME, s. to VA and mountains of GA, w. to IA, n. to MN
state frequency: uncommon
habitat: rich, moist, upland woods, clearings
notes: —

Asclepias incarnata L.

synonym(s): —
common name(s): swamp milkweed
habit: solitary or clustered, erect, stout, branched, perennial herb
range: New Brunswick, s. to FL, w. to NM, n. to Manitoba
state frequency: uncommon
habitat: swamps, wet thickets, meadows, ditches, and shores
key to the subspecies:
    1a. Stems often branched, glabrous or nearly so; leaves lanceolate to lance-oblong, usually tapering
        to the base, glabrous or nearly so on the abaxial surface .... *A. i.* ssp. *incarnata*
    1b. Stems usually few-branched, short-pilose; leaves broad-lanceolate to elliptic, usually rounded at
        the base, short-pilose on the abaxial surface .... *A. i.* ssp. *pulchra* (Ehrh. *ex* Willd.) Woods.
notes: —

Asclepias syriaca L.

synonym(s): —
common name(s): common milkweed
habit: erect, stout, rhizomatous, perennial herb
range: New Brunswick, s. to GA, w. to OK, n. to Saskatchewan
state frequency: common
habitat: thickets, fields, meadows, roadsides
notes: —

Asclepias tuberosa L.

synonym(s): —
common name(s): butterflyweed
habit: ascending to erect, simple or branched, perennial herb
range: ME, s. to FL, w. to Mexico and AZ, n. to SD

state frequency: very rare
habitat: dry, sandy soil
notes: This species is ranked SX by the Maine Natural Areas Program.

Cynanchum
(1 species)

Cynanchum louiseae Kartesz & Gandhi
synonym(s): *Cynanchum nigrum* (L.) Pers., *Vincetoxicum nigrum* (L.) Moench
common name(s): black swallowwort
habit: twining, perennial herb
range: Europe; naturalized to ME, s. to PA, w. to KS
state frequency: uncommon
habitat: fields, thickets, roadsides, waste places
notes: —

Vinca
(1 species)

Vinca minor L.
synonym(s): —
common name(s): common periwinkle
habit: creeping, mat-forming, suffrutescent, perennial herb
range: Europe; escaped from cultivation in our area
state frequency: uncommon
habitat: open woods, roadsides, waste places
notes: —

RUBIACEAE
(5 genera, 23 species)

KEY TO THE GENERA

1a. Plants woody; inflorescence a densely flowered, spherical cluster; corolla slender-funnelform .... ***Cephalanthus***
1b. Plants herbaceous; inflorescence not a spherical cluster; corolla salverform, funnelform, or rotate
    2a. Leaves 3–12 at a node [*i.e.*, whorled] (rarely only 2 leaves at some of the nodes in *Galium palustre*); fruit dry, consisting of 2 indehiscent, 1-seeded carpels; corolla glabrous (pubescent on the abaxial surface in *Galium circaezans*)
        3a. Sepals obsolete; corolla white, green-white, yellow, yellow-green, green-purple, or purple, usually rotate; inflorescence not subtended by an involucre .... ***Galium***
        3b. Sepals triangular; corolla blue or pink, funnelform; inflorescence subtended by an involucre .... ***Sherardia***
    2b. Leaves 2 at each node [*i.e.*, opposite]; fruit fleshy or dehiscent; corolla pubescent on the adaxial [inside] surface
        4a. Fruit a berry; ovules 1 in each locule; plants creeping, with evergreen leaves .... ***Mitchella***

4b. Fruit a capsule; ovules numerous in each locule; plants erect, often tufted, with deciduous leaves .... ***Houstonia***

Cephalanthus
(1 species)

Cephalanthus occidentalis L.
synonym(s): —
common name(s): buttonbush
habit: shrub
range: Quebec, s. to FL, w. to Mexico, n. to MN; also CA
state frequency: occasional
habitat: swamps, margins of streams and ponds
notes: —

Galium
(17 species)

The fruit in *Galium* is a schizocarp, composed of 2 fleshy or dryish mericarps. Fruit measurements in the key are for each individual mericarp.

KEY TO THE SPECIES
1a. Primary axis and main branches of plants with 2–4 leaves at each node
  2a. Principal leaves with 3 or 5 nerves; fruits bristly (sometimes glabrous in *G. boreale*)
    3a. Corolla white or cream-white; leaves lance-linear, 0.1–0.25 cm wide; bristles of the fruits, when present, not hooked at the apex; angles of the stem commonly retrorse-scabrous .... ***G. boreale***
    3b. Corolla green-purple, purple, or yellow-green; leaves ovate to obovate or elliptic to oval, 0.5–2.5 cm wide; bristles of the fruits hooked at the apex; angles of the stem glabrous or pubescent, but not retrorse-scabrous
      4a. Flowers all pedicellate; leaves 1.0–3.0 cm long .... ***G. kamtschaticum***
      4b. Some of the flowers sessile; leaves 2.0–8.0 cm long
        5a. Leaves oval to elliptic, usually widest near the middle, obtuse at the apex; petals commonly pubescent on the abaxial surface, acute at the apex ....
        .... ***G. circaezans***
        5b. Leaves elliptic to lanceolate, usually widest below the middle, acute to acuminate at the apex; petals glabrous, acuminate at the apex ....
        .... ***G. lanceolatum***
  2b. Principal leaves with 1 nerve; fruits without bristles
    6a. Cymes branched 3 or more times, bearing 5 or more flowers; some of the pedicels spreading, especially in fruit .... ***G. palustre***
    6b. Cymes branched 1 or 2 times, bearing 2–4 flowers; pedicels mostly erect to ascending
      7a. Corollas mostly 3-merous, up to 1.8 mm wide, the obtuse lobes as wide or wider than long; stems often retrorse-scabrous on the angles, usually not densely short-bearded just below the nodes; plants developing prostrate, basal offshoots late in the season

8a. Leaves 4–6 at each node, usually some of the nodes with at least 5; each
peduncle with 2 or 3 flowers .... *G. tinctorium*
8b. Leaves mostly 4 at each node; peduncles with 1 or 2(3) flowers
    9a. Leaves 2.0–10.0 mm long; mature mericarps 0.8–1.0 mm in diameter;
    peduncles 0.4–4.0 mm long .... *G. brevipes*
    9b. Leaves 7.0–20.0 mm long; mature mericarps 1.2–2.0 mm in diameter;
    peduncles 5.0–30.0 mm long .... *G. trifidum*
7b. Corollas 4-merous, 2.0–3.5 mm wide, the acute lobes longer than wide;
stems not scabrous on the angles, densely short-bearded just below the nodes;
plants not developing basal offshoots in late season
    10a. Leaves 10.0–30.0 x 1.0–6.0 mm, ascending to spreading; mericarps
    4.0–5.0 mm in diameter, tuberculate, usually only 1 developing; stems
    2.0–8.0 dm tall .... *G. obtusum*
    10b. Leaves 8.0–15.0 x 1.0–3.0 mm, recurved to deflexed; mericarps
    2.0–3.0 mm in diameter, smooth, usually both developing; stems
    1.0–4.0 dm tall .... *G. labradoricum*
1b. Primary axis and main branches with 5–12 leaves at each node
  11a. Corollas yellow; leaves linear, (6–)8–12 per node
    12a. Panicle densely flowered, its lower branches longer than its internodes; flowers
    inodorous .... *G. verum*
    12b. Panicle loosely flowered, its lower branches shorter than its internodes; flowers
    fragrant .... *G. wirtgenii*
  11b. Corollas white or green-white; leaves linear to elliptic, 4–8 per node
    13a. Leaves sharply pointed to cuspidate at the apex
      14a. Inflorescence a many-flowered, relatively large, open, cymose panicle; stems
      usually erect, not scabrous on the angles
        15a. Leaves 1.0–2.5 cm long, broadest above the middle, the surfaces
        concolored .... *G. mollugo*
        15b. Leaves 2.5–5.0 cm long, broadest near or below the middle, the abaxial
        surface paler than the adaxial surface .... *G. sylvaticum*
      14b. Inflorescence a 2- to 5-flowered, once- or twice-branched cyme; stems
      reclining to erect, usually scabrous on the angles, at least near the apex
        16a. Leaves (6–)8 per node; plants annual from a short taproot; mericarps
        2.0–4.0 mm long .... *G. aparine*
        16b. Leaves 4–6 per node; plants perennial, with creeping rhizomes;
        mericarps 1.5–2.0 mm long
          17a. Fruits smooth; leaves 0.8–2.0 x 0.2–0.6 cm .... *G. asprellum*
          17b. Fruits uncinate-bristly; leaves 1.5–8.5 x 0.4–1.3 cm ..... *G. triflorum*
    13b. Leaves blunt at the apex
      18a. Cymes branched 3 or more times, bearing 5 or more flowers; some of the
      pedicels spreading, especially in fruit; corollas 4-merous, 2.5–3.3 mm wide,
      the lobes longer than wide .... *G. palustre*
      18b. Cymes branched 1 or 2 times, bearing 2–4 flowers; pedicels mostly erect or
      ascending; corollas mostly 3-merous, 1.0–1.8 mm wide, the lobes as wide or
      wider than long .... *G. tinctorium*

Galium aparine L.

synonym(s): —
common name(s): spring cleavers
habit: weak, reclining, annual herb
range: circumpolar; s. in eastern North America to FL
state frequency: uncommon
habitat: woods, thickets, fields, waste places, seashores
notes: The fruit of *Galium aparine* is hispid with uncinate hairs.

Galium asprellum Michx.

synonym(s): —
common name(s): rough bedstraw
habit: branched, perennial herb
range: Newfoundland, s. to NC, w. to MO and NE, n. to MN and Ontario
state frequency: occasional
habitat: wet woods, thickets, and meadows, marshes
notes: —

Galium boreale L.

synonym(s): *Galium boreale* L. var. *hyssopifolium* (Hoffmann) DC., *Galium boreale* L. var. *intermedium* DC.
common name(s): northern bedstraw
habit: erect, perennial herb
range: circumboreal; s. to DE, w. to CA
state frequency: uncommon
habitat: open woods, gravelly or rocky banks, thickets, meadows, shores
notes: —

Galium brevipes Fern. & Wieg.

synonym(s): —
common name(s): —
habit: branched, mat-forming, perennial herb
range: Quebec and New Brunswick, s. to ME, w. to MN, n. to Ontario
state frequency: very rare
habitat: swamps, wet shores, often calcareous
notes: This species is ranked S? by the Maine Natural Areas Program.

Galium circaezans Michx.

synonym(s): —
common name(s): forest bedstraw, wild-licorice
habit: simple or basally branched, erect to ascending, perennial herb
range: ME, s. to FL, w. to TX, n. to MN
state frequency: uncommon
habitat: woods, thickets
key to the subspecies:
    1a. Leaves pubescent on the abaxial surface, 2.0–5.0 x 1.0–2.5 cm; stems pubescent on the angles
    .... *G. c.* var. *hypomalacum* Fern.
    1b. Leaves glabrous or sparsely hispid on the abaxial surface, 1.5–4.0 x 0.7–1.8 cm; stems
        glabrous or sparsely pubescent on the angles .... *G. c.* var. *circaezans*
notes: —

Galium kamtschaticum Steller *ex* J. A. & J. H. Schultes
synonym(s): —
common name(s): northern wild-licorice, boreal bedstraw
habit: erect, glabrous, perennial herb
range: Newfoundland, s. to ME, w. to NY and Lake Superior; also AK, s. to WA
state frequency: rare
habitat: cool thickets, brooksides, and mossy woods
notes: This species is ranked S? by the Maine Natural Areas Program.

Galium labradoricum (Wieg.) Wieg.
synonym(s): —
common name(s): bog bedstraw, Labrador bedstraw
habit: slender, erect, ascending, simple or branched, perennial herb
range: Labrador, s. to NJ, w. to IA, n. to MN and Alberta
state frequency: uncommon
habitat: cool bogs, thickets, and mossy woods
notes: This species is ranked S? by the Maine Natural Areas Program.

Galium lanceolatum Torr.
synonym(s): —
common name(s): wild-licorice
habit: slender, erect to ascending, basally branched, perennial herb
range: ME, s. to NC, w. to TN, n. to MN
state frequency: uncommon
habitat: dry woods and thickets
notes: —

Galium mollugo L.
synonym(s): *Galium erectum* Huds.
common name(s): white bedstraw, smooth bedstraw
habit: decumbent to erect, perennial herb
range: Eurasia; naturalized to Newfoundland, s. to VA, w. to IN, n. to Ontario
state frequency: occasional
habitat: fields, roadsides, waste places
notes: Known to hybridize with *Galium verum*.

Galium obtusum Bigelow
synonym(s): —
common name(s): bluntleaf bedstraw
habit: matted, branched, perennial herb
range: New Brunswick and Nova Scotia, s. to FL, w. to AZ, n. to MN and Ontario
state frequency: very rare
habitat: wet thickets, meadows, shores, and low woods
notes: This species is ranked SX by the Maine Natural Areas Program.

Galium palustre L.
synonym(s): —
common name(s): marsh bedstraw
habit: slender, branched, perennial herb
range: Newfoundland, s. to NJ, w. to WI, n. to Ontario
state frequency: uncommon
habitat: wet banks, meadows, and shores
notes: —

Galium sylvaticum L.

synonym(s): —
common name(s): Scotch mist
habit: erect to ascending, perennial herb
range: Europe; naturalized to ME, s. to CT, w. to NY
state frequency: uncommon
habitat: thickets, fields, roadsides
notes: —

Galium tinctorium (L.) Scop.

synonym(s): *Galium claytonii* Michx., *Galium trifidum* L. ssp. *tinctorium* (L.) Hara
common name(s): southern three-lobed bedstraw, Clayton's bedstraw
habit: slender, ascending to reclining, branched, perennial herb
range: Newfoundland, s. to FL, w. to TX and Mexico, n. to MN and Ontario
state frequency: occasional
habitat: swamps, low, wet places
notes: Similar to *Galium trifidum* (*q.v.*), but also with shorter pedicels—up to 0.8 cm long.

Galium trifidum L.

synonym(s): *Galium subbiflorum* Rydb., *Galium tinctorium* (L.) Scop. var. *subbiflorum* (Wieg.)
    Fern., *Galium trifidum* L. ssp. *hallophilum* (Fern. & Wieg.) Puff, *Galium trifidum* L. ssp.
    subbiflorum (Wieg.) Piper
common name(s): northern three-lobed bedstraw
habit: slender, reclining, branched, perennial herb
range: circumpolar; s. in eastern North America to NY and PA
state frequency: occasional
habitat: swamps, fresh to saline shores
notes: Similar to *Galium tinctorium* (*q.v.*), but also with longer pedicels—0.5–3.0 cm long.

Galium triflorum Michx.

synonym(s): *Galium triflorum* Michx. var. *asprelliforme* Fern.
common name(s): sweet-scented bedstraw
habit: simple or forked, prostrate to scrambling, perennial herb
range: circumboreal; s. to FL, w. to Mexico and CA
state frequency: occasional
habitat: woods, thickets
notes: —

Galium verum L.

synonym(s): —
common name(s): yellow bedstraw
habit: stiff, creeping to erect, perennial herb
range: Eurasia; naturalized to Newfoundland, s. to VA, w. to KS, n. to ND and Ontario
state frequency: uncommon
habitat: fields, roadsides, waste places
notes: Known to hybridize with *Galium mollugo*.

Galium wirtgenii F. W. Schultz

synonym(s): —
common name(s): yellow bedstraw
habit: creeping to erect, perennial herb

range: Europe; naturalized to ME and Quebec, s. to PA, w. to IN, n. to MI
state frequency: very rare
habitat: fields, meadows
notes: —

Houstonia
(3 species)

KEY TO THE SPECIES

1a. Lower leaves with a petiole; corollas salverform; inflorescence a peduncled, single
flower; corolla light purple with a yellow center, or entirely white; capsule compressed;
seeds globose, with a deep, round cavity on the inner face .... *H. caerulea*
1b. Leaves sessile; corollas funnelform; inflorescence a cyme; corolla purple to white;
capsule not compressed; seeds meniscoid, with a ridge across the hollowed inner face
    2a. Leaves of the stem rounded to subcordate at the base; sepals 1.7–6.5 mm long at
    anthesis; capsules wider than tall .... *H. purpurea*
    2b. Leaves of the stem narrowed to the base; sepals 1.0–2.0 mm long at anthesis;
    capsules as tall or taller than wide .... *H. longifolia*

Houstonia caerulea L.
synonym(s): *Hedyotis caerulea* (L.) Hook.
common name(s): bluets, Quaker ladies
habit: tufted, erect, slender-rhizomatous, perennial herb
range: New Brunswick and Nova Scotia, s. to GA, w. to AR, n. to WI and Ontario
state frequency: common
habitat: woods, thickets, meadows, fields, lawns
notes: —

Houstonia longifolia Gaertn.
synonym(s): *Hedyotis longifolia* (Gaertn.) Hook.
common name(s): long-leaved bluets
habit: tufted, erect, simple or branched, fibrous-rooted, perennial herb
range: ME, s. to GA, w. to OK, n. to Saskatchewan
state frequency: rare
habitat: rocky or gravelly soil, commonly of river shores
notes: This species is ranked S2 by the Maine Natural Areas Program.

Houstonia purpurea L. var. *calycosa* Gray
synonym(s): *Hedyotis lanceolata* Poir., *Houstonia lanceolata* (Poir.) Britt.
common name(s): large houstonia
habit: erect, simple or branched, fibrous-rooted, perennial herb
range: ME, s. to GA, w. to TX, n. to MN and Ontario
state frequency: very rare
habitat: dry, open woods, slopes, pastures
notes: —

Mitchella
(1 species)

Mitchella repens L.
synonym(s): —
common name(s): patridgeberry
habit: trailing, mat-forming, evergreen, perennial herb
range: Newfoundland, s. to FL, w. to TX, n. to MN and Ontario
state frequency: occasional
habitat: woods
notes: The round-ovate to ovate leaves have a conspicuous white-green midstripe.

Sherardia
(1 species)

Sherardia arvensis L.
synonym(s): —
common name(s): field madder
habit: slender, basally procumbent, branched, annual herb
range: Eurasia; naturalized to Nova Scotia and ME, s. to NC, w. to MO, n. to Quebec
state frequency: very rare
habitat: fields, cultivated areas, waste places
notes: —

ADOXACEAE
(2 genera, 10 species)

KEY TO THE GENERA
1a. Leaves pinnately divided; drupe with 3–5 pyrenes; terminal bud absent .... *Sambucus*
1b. Leaves simple; drupe with a single pyrene; terminal bud present .... *Viburnum*

Sambucus
(2 species)

KEY TO THE SPECIES
1a. Inflorescence with a main axis; pith orange-brown; winter buds large, ovoid; drupes red;
plants in anthesis in May and June .... *S. racemosa*
1b. Inflorescence branched from near the base and lacking a prominent, main axis; pith
white; winter buds small, conical; drupes purple-black; plants in anthesis in July and
August .... *S. canadensis*

Sambucus canadensis L.
synonym(s): —
common name(s): common elder
habit: shrub
range: New Brunswick, s. to FL, w. to OK, n. to Manitoba
state frequency: occasional
habitat: moist woods, fields, roadsides
notes: —

Sambucus racemosa L. ssp. ***pubens*** (Michx.) House
synonym(s): *Sambucus pubens* Michx.
common name(s): red-berried elder, stinking elder
habit: shrub
range: circumboreal; s. to mountains of NC and GA, w. to OR
state frequency: occasional
habitat: rich woods, openings
notes: —

Viburnum
(8 species)

KEY TO THE SPECIES FOR USE WITH VEGETATIVE MATERIAL
1a. Leaves (except sometimes the most apical pair of the new shoot) 3-lobed
   2a. Bud scales 3 or more, generally at least sparsely pubescent; twigs pubescent with
     stellate hairs .... *V. acerifolium*
   2b. Bud scales 2 (sometimes appearing as 1), glabrous; twigs glabrous or stipitate-
     glandular, not stellate-pubescent
     3a. Leaves without stipules; buds dark purple .... *V. edule*
     3b. Leaves with slender stipules; buds red .... *V. opulus*
1b. Leaves not lobed
   4a. Buds naked; twigs, buds, leaves, and inflorescence stellate-pubescent; leaves
     5.0–18.0 cm long at maturity
     5a. Leaf blades 5.0–12.0 cm long; pubescence gray; plants introduced, erect shrubs
     .... *V. lantana*
     5b. Leaf blades 10.0–18.0 cm long; pubescence red-brown; plants native, straggling
     or loosely ascending .... *V. lantanoides*
   4b. Buds covered with scales; twigs, buds, leaves, and inflorescence glabrous to
     pubescent or scurfy; stellate hairs, if present, few and on the leaves; leaves
     4.0–12.0 cm long
     6a. Buds linear, with 2 scales, scurfy-pubescent; leaves with many-forked, less
      conspicuous, pinnate veins that anastomose near the margin and do not extend to
      the tip of each tooth; abaxial surface of leaves dotted with small, red-brown
      scales
      7a. Leaf margin entire to undulate to crenate; petioles lacking a prominent wing
       margin; buds brown .... *V. nudum*
      7b. Leaf margin sharply serrulate; petioles often with a conspicuous, undulate,
       winged margin; buds purple .... *V. lentago*
     6b. Buds ovoid, with 4 scales, glabrous or pubescent; leaves with prominent,
      subpalmate, simple to twice-forked veins that extend to the tip of each tooth;
      abaxial surface of leaves dotted with scales .... *V. dentatum*

KEY TO THE SPECIES FOR USE WITH REPRODUCTIVE MATERIAL
1a. Marginal flowers of the cyme neutral, with greatly enlarged, somewhat zygomorphic
   corollas; drupes red or dark red
   2a. Cyme stellate-pubescent, sessile; seeds with several grooves .... *V. lantanoides*

2b. Cyme not stellate-pubescent, on a peduncle 2.0–5.0 cm long; seeds not grooved .... *V. opulus*

1b. All the flowers of the cyme similar, fertile and with relatively small corollas; drupes blue to blue-black (red in *V. edule*)

   3a. Leaves (except sometimes the most apical pair of the new shoot) 3-lobed

      4a. Filaments 3.0–4.0 mm long; drupes blue-black .... *V. acerifolium*

      4b. Filaments up to 1.0 mm long; drupes red .... *V. edule*

   3b. Leaves not lobed

      5a. Cyme stellate-pubescent with light gray hairs; drupes red turning dark; pyrenes with 3 deep grooves .... *V. lantana*

      5b. Cyme glabrous or scurfy-pubescent, the stellate hairs usually few and red to brown; drupes blue to black; pyrenes with 1 or no grooves

         6a. Stylopodium pubescent; cyme glabrous or nearly so; pyrenes deeply grooved on one side .... *V. dentatum*

         6b. Stylopodium glabrous; cyme scurfy-pubescent; pyrenes without deep grooves

            7a. Cyme with a peduncle 2.0–5.0 cm long .... *V. nudum*

            7b. Cyme sessile .... *V. lentago*

Viburnum acerifolium L.
synonym(s): —
common name(s): dockmackie, maple-leaved viburnum
habit: shrub
range: Quebec, s. to FL, w. to LA, n. to MN
state frequency: occasional
habitat: moist or dry, rocky woods
notes: —

Viburnum dentatum L. var. *lucidum* Ait.
synonym(s): *Viburnum recognitum* Fern.
common name(s): arrowwood
habit: shrub
range: New Brunswick, s. to FL, w. to OH, n. to MI and Ontario
state frequency: occasional
habitat: moist to dry woods, thickets, and borders, swamps
notes: —

Viburnum edule (Michx.) Raf.
synonym(s): *Viburnum pauciflorum* La Pylaie *ex* Torr. & Gray
common name(s): squashberry, mooseberry
habit: straggling to erect shrub
range: Labrador, s. to PA, w. to OR, n. to AK
state frequency: rare
habitat: moist woods and shores in subalpine areas
notes: This species is ranked S2 by the Maine Natural Areas Program.

Viburnum lantana L.
synonym(s): —
common name(s): wayfaring-tree
habit: shrub to small tree

range: Eurasia; escaped from cultivation in the northeastern United States and adjacent Canada
state frequency: very rare
habitat: roadsides
notes: —

Viburnum lantanoides Michx.

synonym(s): *Viburnum alnifolium* Marsh.
common name(s): hobblebush, witch hobble
habit: shrub
range: New Brunswick, s. to mountains of NC, w. to TN, n. to MI and Ontario
state frequency: occasional
habitat: moist forests
notes: —

Viburnum lentago L.

synonym(s): —
common name(s): nannyberry, sheepberry
habit: shrub or small tree
range: Quebec, s. to VA, w. to WY, n. to MT and Saskatchewan
state frequency: occasional
habitat: woods, roadsides, stream banks
notes: —

Viburnum nudum L. var. *cassinoides* (L.) Torr. & Gray

synonym(s): *Viburnum cassinoides* L.
common name(s): witherod, wild-raisin
habit: shrub
range: Newfoundland, s. to GA, w. to AL, n. to WI
state frequency: occasional
habitat: wet woods, swamps
notes: —

Viburnum opulus L.

synonym(s): <*V. o.* var. *americanum*> = *Viburnum trilobum* Marsh.
common name(s): <*V. o.* var. *americanum*> = highbush-cranberry; <*V. o.* var. *trilobum*> = Guelder-
rose
habit: shrub
range: Newfoundland, s. to PA, w. to WA, n. to British Columbia; also Eurasia
state frequency: occasional
habitat: moist forests, borders, thickets
key to the subspecies:
  1a. Marginal flowers of the cyme neutral, with greatly enlarged, somewhat zygomorphic corollas;
    glands of the petiole mostly sessile, wider than tall, concave at the apex .... *V. o.* var. *opulus*
  1b. All the flowers of the cyme similar, fertile, and with relatively small corollas; glands of the
    petiole stalked, taller than wide, convex at the apex .... *V. o.* var. *americanum* Ait.
notes: *Viburnum opulus* var. *opulus* is a Eurasian species that has escaped from cultivation in our
  area. The cultivated snowball shrub is a horticultural form of *V. opulus* var. *opulus* that has all of
  the flowers of the cyme enlarged and neutral.

AQUIFOLIACEAE
(2 genera, 4 species)

KEY TO THE GENERA

1a. Perianth 4- or 5-merous; sepals minute, deciduous in fruit; petals linear to linear-oblong, yellow, distinct; stamens distinct; leaves without stipules; twigs and buds dark purple .... ***Nemopanthus***

1b. Perianth 4- to 8-merous; sepals evident, persistent in fruit; petals obovate, white, connate at the base; stamens adnate to the corolla tube for a short distance; leaves with a stipular vestige on each side; twigs and buds gray to brown .... ***Ilex***

Ilex
(3 species)

Bundle scars that appear as 1 and minute stipular projections serve as identification aids for winter material of the deciduous-leaved species.

KEY TO THE SPECIES

1a. Leaves coriaceous, evergreen, punctate on the abaxial surface, the margins toothed only above the middle; drupe black .... ***I. glabra***

1b. Leaves thin, deciduous, not punctate, the margins toothed throughout; drupe red, orange-red, or yellow

    2a. Sepals with ciliate margins, obtuse at the apex; leaves serrate, acuminate at the apex, dull on the adaxial surface, commonly with some pubescence on the abaxial surface; all the flowers borne on a short pedicel 1.0–4.0 mm long .... ***I. verticillata***

    2b. Sepals with entire margins (rarely with a few cilia), acute or nearly so at the apex; leaves serrulate, acute at the apex, lustrous on the adaxial surface, usually glabrous on the abaxial surface; staminate flowers borne on pedicels 8.0–16.0 mm long .... ***I. laevigata***

Ilex glabra (L.) Gray
synonym(s): —
common name(s): inkberry
habit: evergreen shrub
range: Nova Scotia, s. along the coastal plain to FL and LA
state frequency: very rare
habitat: bogs, swamps, wet woods, near the coast
notes: This species is ranked S1 by the Maine Natural Areas Program.

Ilex laevigata (Pursh) Gray
synonym(s): —
common name(s): smooth winterberry
habit: shrub
range: ME, s. to GA
state frequency: rare
habitat: swamps, wet woods
notes: This species is ranked S2/S3 by the Maine Natural Areas Program.

Ilex verticillata (L.) Gray
synonym(s): *Ilex verticillata* (L.) Gray var. *fastigiata* (Bickn.) Fern., *Ilex verticillata* (L.) Gray var. *padifolia* (Willd.) Torr. & Gray *ex* S. Wats., *Ilex verticillata* (L.) Gray var. *tenuifolia* (Torr.) S. Wats.
common name(s): winterberry, black-alder
habit: dioecious shrub
range: Newfoundland, s. to GA, w. to MO, n. to MN and Ontario
state frequency: occasional
habitat: swamps, wet woods, thickets
notes: —

Nemopanthus
(1 species)

Shrubs with dark purple twigs and buds.

Nemopanthus mucronatus (L.) Loes.
synonym(s): *Nemopanthus fascicularis* Raf.
common name(s): mountain holly, common mountain holly
habit: erect, much branched shrub
range: Newfoundland, s. to VA, w. to IL, n. to MN and Ontario
state frequency: occasional
habitat: damp woods, thickets, and shores, swamps, bogs
notes: —

CAPRIFOLIACEAE
(8 genera, 18 species)

KEY TO THE GENERA

1a. Plants herbaceous or suffrutescent; stamens fewer than the lobes of the corolla or as many as the lobes of the corolla in *Triosteum*
  2a. Flowers and fruits sessile in the axils of leaves; stamens 5 per flower, as many as the lobes of the corolla .... ***Triosteum***
  2b. Flowers and fruits borne on a peduncle; stamens 3 or 4 per flower, as many or fewer than the lobes of the corolla
    3a. Flowers borne in a capitulum-like cluster; corolla 4-lobed .... ***Knautia***
    3b. Flowers borne separately in open inflorescences; corolla 5-lobed
      4a. Stems trailing; leaves 1.0–2.0 cm long; corolla 10.0–15.0 mm long; flowers nodding, paired .... ***Linnaea***
      4b. Stems upright; leaves 1.0–20.0 cm long; corolla 1.5–7.0 mm long; flowers neither nodding nor paired
        5a. Stem leaves pinnately divided; calyx composed of 5–20 narrow segments, these inconspicuous at anthesis, but unrolling and forming a plumose, pappus-like crown in fruit; plants perennial .... ***Valeriana***
        5b. Stem leaves simple; calyx obsolete or minute; plants annual ....
          .... ***Valerianella***
1b. Plants definitely woody; stamens as many as the lobes of the corolla

6a. Leaves regularly serrate; fruit a capsule .... *Diervilla*
6b. Leaves entire or with a few, irregular, tooth-like lobes; fruit fleshy
    7a. Corolla 10.0–50.0 mm long; ovary with 2 or 3 locules, each locule bearing
    several ovules; fruit red, blue, or black; leaves strictly entire; leaf scars with 3
    vascular bundles .... *Lonicera*
    7b. Corolla 3.0–8.0 mm long; ovary 4-locular, 2 of the locules with several abortive
    ovules, the other 2 locules with a solitary, fertile ovule; fruit white; leaves entire
    or with irregular, tooth-like lobes; leaf scars with 1 vascular bundle ....
    .... *Symphoricarpos*

Diervilla
(1 species)

***Diervilla lonicera* P. Mill.**
synonym(s): —
common name(s): bush-honeysuckle
habit: low, arching shrub
range: Newfoundland, s. to NC, w. to IA, n. to Saskatchewan
state frequency: occasional
habitat: dry woods and rocky places
notes: —

Knautia
(1 species)

The calyx in *Knautia* is composed of 8–12 setaceous appendages.

***Knautia arvensis* (L.) Coult.**
synonym(s): *Scabiosa arvensis* L.
common name(s): blue buttons
habit: pubescent, taprooted, perennial herb
range: Europe; naturalized to Newfoundland, s. to WV, w. and n. to Quebec
state frequency: very rare
habitat: fields, roadsides, waste places
notes: —

Linnaea
(1 species)

***Linnaea borealis* L. ssp. *longiflora* (Torr.) Hultén**
synonym(s): *Linnaea borealis* L. var. *americana* (Forbes) Rehd.
common name(s): twinflower
habit: creeping, trailing, evergreen, suffrutescent perennial
range: circumpolar; s. in North America to NJ, w. to UT and CA
state frequency: occasional
habitat: moist or dry, cool woods, cold bogs
notes: —

Lonicera
(10 species)

KEY TO THE SPECIES

1a. Flowers paired, borne on axillary peduncles; plants upright (twining or climbing in *L. japonica*); leaves distinct
   2a. Plants twining or climbing; corolla 3.0–5.0 cm long; fruit black .... ***L. japonica***
   2b. Plants upright; corolla 1.0–2.2 cm long; fruit red or blue
      3a. Style glabrous; corolla essentially actinomorphic
         4a. Ovaries separate; fruit blue; bractlets subtending the flowers mostly 4.0–6.0 mm long; winter buds with 2 scales; plants of wet, low areas .... ***L. villosa***
         4b. Ovaries definitely united; fruit red; bractlets subtending the flowers mostly 1.2–3.0 mm long; winter buds with more than 2 scales; plants of upland, dry to moist soils .... ***L. canadensis***
      3b. Style pubescent; corolla weakly to strongly zygomorphic
         5a. Twigs with solid, white pith; bractlets subtending the flower obsolete; plants native, of circumneutral fens .... ***L. oblongifolia***
         5b. Twigs with hollow, darker pith; bractlets subtending the flowers 2.0–8.0 mm long; plants introduced, of fields, roadsides, and disturbed areas
            6a. Upper lip of the corolla with deep sinuses, the sinuses extending nearly to the lip's base; filaments glabrous; bractlets eglandular; winter buds conical or ovoid, glabrous or pubescent
               7a. Leaves pubescent on the abaxial surface; peduncles 5.0–15.0 mm long; corolla pubescent on the abaxial [outside] surface, saccate at the base; twigs pubescent .... ***L. morrowii***
               7b. Leaves glabrous on the abaxial surface; peduncles 15.0–25.0 mm long; corolla glabrous on the abaxial surface, not saccate at the base; twigs glabrous .... ***L. tatarica***
            6b. Upper lip of the corolla with shallower sinuses, the sinuses extending *ca.* half the distance to the base; filaments pubescent; bractlets glandular; winter buds fusiform, the scales conspicuously ciliate .... ***L. xylosteum***
1b. Flowers in 3-flowered cymules, sessile in the axils of leaves; plants twining or climbing; upper 1 or 2 pairs of leaves gamophyllous (all distinct in *L. periclymenum*)
   8a. All the leaves distinct, with petioles; basal connate tube of the corolla glandular-pubescent; distal end of the twigs usually glandular-pubescent .... ***L. periclymenum***
   8b. Upper 1 or 2 pairs of leaves gamophyllous; basal connate tube of the corolla eglandular; twigs usually glabrous
      9a. Corolla 3.0–5.0 cm long, weakly zygomorphic, the 5 lobes nearly equal; seeds 4.0–4.5 mm long .... ***L. sempervirens***
      9b. Corolla 1.5–2.5 cm long, strongly zygomorphic, with an upper and lower lip; seeds 3.0–3.5 mm long .... ***L. dioica***

Lonicera canadensis Bartr. *ex* Marsh.
synonym(s): —
common name(s): fly honeysuckle
habit: straggling shrub

range: Quebec, s. to NJ and mountains of NC, w. to IN, n. to Saskatchewan
state frequency: occasional
habitat: forests
notes: —

Lonicera dioica L.

synonym(s): —
common name(s): wild honeysuckle, mountain honeysuckle
habit: climbing to reclining shrub
range: ME, s. to GA, w. to MO, n. to Manitoba
state frequency: very rare
habitat: thickets, rocky banks, woods
notes: This species is ranked S1 by the Maine Natural Areas Program.

Lonicera japonica Thunb.

synonym(s): —
common name(s): Japanese honeysuckle
habit: trailing to climbing shrub
range: Asia; naturalized throughout the eastern United States
state frequency: very rare
habitat: borders of woods, thickets, fields, roadsides
notes: —

Lonicera morrowii Gray

synonym(s): —
common name(s): Morrow's honeysuckle
habit: shrub
range: Eurasia; escaped from cultivation to New Brunswick, s. to NJ, w. to PA, n. to MI
state frequency: occasional
habitat: roadsides, thickets
notes: Hybrids with *Lonicera tatarica*, called *L.* ×*bella* Zabel, occur in Maine. This hybrid can be
   identified by its sparsely pubescent leaves, glabrous and scarcely saccate corolla, and peduncles
   5.0–15.0 mm long.

Lonicera oblongifolia (Goldie) Hook.

synonym(s): —
common name(s): swamp fly honeysuckle
habit: erect shrub
range: New Brunswick, s. to PA, w. to OH, n. to Manitoba
state frequency: rare
habitat: circumneutral fens
notes: This species is ranked S2 by the Maine Natural Areas Program.

Lonicera periclymenum L.

synonym(s): —
common name(s): woodbine, European woodbine
habit: climbing to trailing shrub
range: Europe; escaped from cultivation to Newfoundland, s. to ME
state frequency: rare
habitat: thickets, roadsides
notes: —

Lonicera sempervirens L.

synonym(s): —
common name(s): trumpet honeysuckle, coral honeysuckle
habit: twining shrub
range: ME, s. to FL, w. to TX, n. to NE
state frequency: very rare
habitat: swamps, forested banks
notes: This species is ranked S1 by the Maine Natural Areas Program. Though native, this species is
   also planted and occasionally escapes.

Lonicera tatarica L.

synonym(s): —
common name(s): Tatarian honeysuckle
habit: shrub
range: Eurasia; escaped from cultivation to New Brunswick, s. to NJ, w. to KY and IA, n. to Ontario
state frequency: uncommon
habitat: borders of woods, thickets, shores
notes: Known to hybridize with *Lonicera morrowii* (*q.v.*) and *L. xylosteum*.

Lonicera villosa (Michx.) J. A. Schultes

synonym(s): <*L. v.* var. *villosa*> = *Lonicera caerulea* L. var. *villosa* (Michx.) Torr. & Gray
common name(s): mountain fly honeysuckle
habit: depressed to erect shrub
range: Labrador, s. to CT, w. to PA, n. to Manitoba
state frequency: occasional
habitat: bogs, swamps, rocky or peaty areas
key to the subspecies:
   1a. Leaves densely villous-tomentose on both surfaces; branchlets densely pubescent; calyx
      pubescent, at least at anthesis; corolla pubescent; plants depressed .... *L. v.* var. *villosa*
   1b. Leaves moderately pubescent to glabrous on one or both surfaces; branchlets pubescent to
      glabrous; calyx glabrous; corolla glabrous or rarely pubescent; plants usually upright
      2a. Twigs glabrous; leaves sparsely pubescent to glabrous .... *L. v.* var. *tonsa* Fern.
      2b. Twigs pubescent; leaves moderately to sparsely pubescent, rarely glabrous
         3a. Twigs pubescent with short, fine and longer, coarser hairs; leaves pilose on the abaxial
            surface .... *L. v.* var. *solonis* (Eat.) Fern.
         3b. Twigs pubescent with only short, fine hairs; leaves pilose to glabrate on the abaxial
            surface .... *L. v.* var. *calvescens* (Fern. & Wieg.) Fern.
notes: —

Lonicera xylosteum L.

synonym(s): —
common name(s): European fly honeysuckle
habit: shrub
range: Europe; escaped from cultivation to New Brunswick, s. to NJ, w. to OH, n. to MI
state frequency: rare
habitat: open woods, thickets, roadsides
notes: Known to hybridize with *Lonicera tatarica*.

Symphoricarpos
(1 species)

Symphoricarpos albus (L.) Blake
synonym(s): *Symphoricarpos racemosus* Michx. var. *laevigatus* (Fern.) Blake
common name(s): snowberry
habit: ascending shrub
range: western North America; introduced to Quebec, s. to VA, w. and n. to MN
state frequency: uncommon
habitat: roadsides, banks, fields
notes: With hollow pith, as many of the introduced *Lonicera*, but with 1 bundle scar and lacking superposed buds.

Triosteum
(1 species)

Triosteum aurantiacum Bickn.
synonym(s): —
common name(s): wild-coffee, horse-gentian, feverwort
habit: coarse, perennial herb
range: New Brunswick, s. to GA, w. to OK, n. to MN and Ontario
state frequency: very rare
habitat: rich woods and thickets
notes: This species is ranked S1 by the Maine Natural Areas Program. The fruit is a bright orange drupe.

Valeriana
(2 species)

KEY TO THE SPECIES
1a. Basal leaves simple, or sometimes with a single pair of basal lobes; stem leaves with 3–13 leaflets, the rachis and abaxial surfaces glabrous .... *V. uliginosa*
1b. Basal leaves pinnately divided; stem leaves with 11–21 leaflets, the rachis and abaxial surface sparsely pubescent .... *V. officinalis*

Valeriana officinalis L.
synonym(s): —
common name(s): garden-heliotrope
habit: coarse, erect, fibrous-rooted, perennial herb
range: Eurasia; naturalized to New Brunswick, s. to NJ, w. to OH, n. to MN; also Manitoba and British Columbia
state frequency: uncommon
habitat: roadsides, thickets
notes: —

Valeriana uliginosa (Torr. & Gray) Rydb.
synonym(s): *Valeriana sitchensis* Bong. var. *uliginosa* (Torr. & Gray) Boivin
common name(s): marsh valerian
habit: fibrous-rooted, rhizomatous, perennial herb

range: Quebec and New Brunswick, s. to MA, w. to OH, n. to MI and Ontario
state frequency: rare; known only from northern Maine
habitat: wet woods and meadows, swamps, bogs
notes: This species is ranked S2 by the Maine Natural Areas Program.

Valerianella
(1 species)

Valerianella locusta (L.) Lat.
synonym(s): *Valerianella olitoria* (L.) Pollich
common name(s): European corn salad
habit: branched, annual herb
range: Europe; naturalized to ME, s. to NC, w. to TN, n. to IN
state frequency: very rare
habitat: moist fields, roadsides, waste places
notes: —

APIACEAE
(25 genera, 37 species)

KEY TO THE GENERA FOR USE WITH FLOWERING MATERIAL
1a. Gynoecium composed of 2–5 carpels; plants woody or herbaceous; leaves whorled, alternate, or appearing all basal
    2a. Leaves alternate or appearing all basal; inflorescence composed of 2 or more umbels; gynoecium composed of 5 carpels .... *Aralia*
    2b. Leaves in a single whorl; inflorescence a single umbel; gynoecium composed of 2 or 3 carpels .... *Panax*
1b. Gynoecium composed of 2 carpels; plants herbaceous; leaves alternate
    3a. Plants with slender, creeping stems; leaves simple; umbels simple
        4a. Leaf blades broad-ovate to orbicular, neither septate nor hollow; plants inland, of wet soil .... *Hydrocotyle*
        4b. Leaf blades linear-spatulate, septate, hollow; plants coastal, of brackish soil ....
        .... *Lilaeopsis*
    3b. Plants with upright stems; leaves evidently lobed or compound to decompound (simple in a rare brackish water form of *Sium suave*); umbels mostly compound
        5a. Leaves palmately divided into 3–7 broad segments; plants polygamous, with bisexual flowers and unisexual [staminate] flowers in separate umbellets or intermixed with the bisexual flowers; stylopodium absent, the styles arising from a ring-like disk .... *Sanicula*
        5b. Leaves compound to decompound; plants mostly synoecious; stylopodium present, the styles found at the apex of the stylopodium (stylopodium wanting in *Zizia*)
            6a. Ultimate segments of larger leaves narrow-linear to filiform, up to 1.0 mm wide
                7a. Petals yellow; styles short and stout; umbels with 10–40 primary branches

8a. Plants annual; petiolar sheaths of larger leaves 1.0–3.0 cm long ....
.... *Anethum*

8b. Plants perennial; petiolar sheaths of larger leaves 3.0–10.0 cm long ....
.... *Foeniculum*

7b. Petals white; styles slender; umbels with 7–14 primary branches ....
.... *Carum*

6b. Ultimate segments of larger leaves linear to orbicular, wider than 1.0 mm

9a. Leaves with 3 leaflets, these leaflets [the ultimate segments] simple or lobed, but not again divided

10a. Upper leaf sheaths not dilated, less than 1.0 cm wide; umbel subtended by 0 or 1 bract; corollas actinomorphic; plants 0.3–1.0 m tall .... *Cryptotaenia*

10b. Upper leaf sheaths conspicuously dilated, 1.0 cm or more wide; umbel subtended by more than 1 bract; corollas of the marginal flowers of the umbellets weakly zygomorphic; plants robust, 1.0–5.0 m tall .... *Heracleum*

9b. Leaves with 5 or more ultimate segments

11a. Leaves with clearly defined leaflets of consistently similar shape, the ultimate segments commonly wider than 2.0 cm

12a. Principal leaves once-compound (twice-compound in the submerged leaves of *Sium suave*)

13a. Upper leaf sheaths conspicuously dilated, 1.0 cm or more wide; stems pubescent .... *Heracleum*

13b. Upper leaf sheaths not dilated, less than 1.0 cm wide; stems glabrous or minutely pubescent

14a. Umbels and umbellets usually lacking bracts and bractlets, respectively; plants taprooted, weeds of fields and waste places

15a. Flowers yellow; stems grooved .... *Pastinaca*

15b. Flowers white; stems without grooves ...... *Pimpinella*

14b. Umbels and umbellets with bracts and bractlets, respectively; plants fibrous-rooted, aquatic .... *Sium*

12b. Principal leaves twice- or thrice-compound

16a. Petals yellow or green-yellow

17a. Central flower of each umbellet sessile, the others pedicellate; leaf divisions and leaflets ternately arranged ....
.... *Zizia*

17b. All the flowers of the umbellet pedicelled; leaf divisions and leaflets not ternately arranged

18a. Umbels and umbellets with 0–2 inconspicuous, linear bracts and bractlets, respectively; plants taprooted biennials .... *Pastinaca*

18b. Umbels and umbellets with a few, conspicuous, lanceolate bracts and bractlets, respectively; plants perennial .... *Levisticum*

16b. Petals white

19a. Upper leaf sheaths conspicuously dilated, 1.0 cm or more
wide .... ***Angelica***
19b. Upper leaf sheaths not dilated, less than 1.0 cm wide
20a. Veins of the leaves directed to the sinuses; base of the
stem thickened; some of the roots tuberous-thickened;
plants of wet soil .... ***Cicuta***
20b. Veins of the leaves directed to the teeth; base of the
stem not thickened; roots without tubers (although
thick in *Osmorhiza*); plants not of wet soil
21a. Ovary bristly; plants of rich woods ..... ***Osmorhiza***
21b. Ovary glabrous; plants of coastal shores or of
fields and disturbed areas
22a. Sepals absent; styles much longer than the
stylopodia, deflexed; umbellets lacking
bractlets; plants of fields and disturbed areas .....
.... ***Aegopodium***
22b. Sepals present; styles shorter than the
stylopodia, ascending; umbellets with bractlets;
plants of coastal shores .... ***Ligusticum***
11b. Leaves dissected to decompound, without clearly defined leaflets
and/or with irregular branching, the ultimate segments commonly less
than 1.0 cm wide
23a. Stem streaked or spotted with purple .... ***Conium***
23b. Stem not marked with purple
24a. Ovary bristly; bracts of the umbel pinnatifid; central flower of
the inflorescence purple or pink .... ***Daucus***
24b. Ovary glabrous; bracts of the umbel entire or absent; central
flower of the inflorescence white
25a. Bractlets narrow-ovate, with conspicuously ciliate or
fimbriate margins .... ***Anthriscus***
25b. Bractlets absent or linear to lanceolate, their margins
entire or minutely ciliolate
26a. Axils of upper leaves with bulbils .... ***Cicuta***
26b. Axils of leaves without bulbils
27a. Central umbellet not, or only slightly, overtopped
by the lateral umbellets; plants perennial, of
wetlands and forests .... ***Conioselinum***
27b. Central umbellet overtopped by the lateral
umbellets; plants annual, of fields and disturbed
areas .... ***Aethusa***

KEY TO THE GENERA FOR USE WITH FRUITING MATERIAL

1a. Fruit a fleshy, berry-like drupe; gynoecium composed of 2–5 carpels; plants woody or
herbaceous; leaves whorled, alternate, or appearing all basal
2a. Leaves alternate or appearing all basal; infructescence composed of 2 or more
umbels; gynoecium composed of 5 carpels; drupes dark purple to black .... ***Aralia***

2b. Leaves in a single whorl; infructescence a single umbel; gynoecium composed of 2 or
    3 carpels; drupes red or yellow .... *Panax*

1b. Fruit dry, a schizocarp; gynoecium composed of 2 carpels; plants herbaceous; leaves
  alternate

  3a. Plants with slender, creeping stems; leaves simple; umbels simple

    4a. Leaf blades broad-ovate to orbicular, neither septate nor hollow; plants inland, of
      wet soil .... *Hydrocotyle*

    4b. Leaf blades linear-spatulate, septate, hollow; plants coastal, of brackish soil ....
      .... *Lilaeopsis*

  3b. Plants with upright stems; leaves evidently lobed or compound to decompound
    (simple in a rare, brackish water form of *Sium suave*); umbels mostly compound

    5a. Fruits pubescent, bristly, or prickly

      6a. Fruits without ribs, sessile or subsessile; stylopodium absent, the styles
        arising from a ring-like disk .... *Sanicula*

      6b. Fruits with longitudinal ribs, at least on the beak, pedicellate; stylopodium
        present, the styles (if persistent) found at the apex of the stylopodium

        7a. Principal leaves once-divided, the ultimate segments 5.0–40.0 cm wide;
          oil tubes of the fruit large, conspicuous, extending 0.5–0.65 times the
          distance from the stylopodium to the base of the fruit .... *Heracleum*

        7b. Principal leaves 2 or more times divided, the ultimate segments
          commonly less than 5.0 cm wide; oil tubes of the fruit slender, obscure,
          absent, or extending nearly the entire distance from the stylopodium to the
          base

          8a. Fruit with both primary and secondary ribs, the primary ribs smaller
            and bristly, the secondary ribs larger, wing-like, covered with
            prickles; umbels with pinnatifid bracts .... *Daucus*

          8b. Fruit with evident, primary ribs, the secondary ribs weak and
            inconspicuous; umbels either lacking bracts or with entire bracts

            9a. Mericarps linear to linear-oblong, 12.0–22.0 mm long; umbel with
              2–8 primary branches .... *Osmorhiza*

            9b. Mericarps lanceolate to ovoid, oblong, or elliptic, 2.0–9.0 mm
              long; umbel with 8–20 primary branches .... *Pimpinella*

    5b. Fruits glabrous

      10a. Leaves with clearly defined leaflets of consistently similar shape, the
        ultimate segments commonly wider than 2.0 cm (submersed leaves irregularly
        dissected in *Sium*)

        11a. Principal leaves once-compound (twice-compound in the submerged
          leaves of *Sium suave*)

          12a. Schizocarp 7.0–12.0 mm long; upper leaf sheaths conspicuously
            dilated, 1.0 cm or more wide .... *Heracleum*

          12b. Schizocarp 2.0–7.0 mm long; upper leaf sheaths not dilated, less
            than 1.0 cm wide

            13a. Leaves usually with 3 leaflets; pedicels irregularly of unequal
              length; umbels with 2–7 primary branches; plants of rich forests ...
              .... *Cryptotaenia*

13b. Longer leaves with 5–19 leaflets; pedicels of similar length; umbels with 6–25 primary branches; plants not of rich forests

   14a. Plants with fibrous roots; umbels with bracts; plants of wetlands .... ***Sium***

   14b. Plants with taproots; umbels usually without bracts; plants of fields and disturbed areas

      15a. Mericarps conspicuously compressed, wing-margined, 5.0–7.0 mm long; stems coarse and corrugated; oil tubes of the fruit large, conspicuous, extending up to 0.75 times the distance from the stylopodium to the base of the fruit; leaves not confined to a basal rosette .... ***Pastinaca***

      15b. Mericarps nearly terete in cross-section, not wing-margined, 2.0–2.5 mm long; stems slender and terete; oil tubes of the fruit slender, obscure, absent and/or extending nearly the entire distance from the stylopodium to the base; leaves mostly confined to a basal rosette, the cauline leaves progressively reduced .... ***Pimpinella***

11b. Principal leaves twice- or thrice-compound

   16a. Upper leaf sheaths conspicuously dilated, 1.0 cm or more wide; umbel with 20–45 primary branches .... ***Angelica***

   16b. Upper leaf sheaths not dilated, less than 1.0 cm wide; umbel with 8–25 primary branches

      17a. Veins of the leaves directed to the sinuses; base of the stem thickened; some of the roots tuberous-thickened; plants of wet, inland soils .... ***Cicuta***

      17b. Veins of the leaves directed to the teeth; base of the stem not thickened; roots without tubers; plants not of wet, inland soil

         18a. Mericarps with wing-like ribs, somewhat to strongly compressed

            19a. Leaves divided into 3 portions; plants of Atlantic coast shores .... ***Ligusticum***

            19b. Leaves not clearly divided into 3 portions; plants not of Atlantic coast shores

               20a. Both the marginal and dorsal ribs wing-like; umbel with 0–2 inconspicuous, linear bracts .... ***Pastinaca***

               20b. Only the marginal ribs wing-like, the ribs of the dorsal surface elevated, but not wing-like; umbel with a few, conspicuous, lanceolate bracts .... ***Levisticum***

         18b. Mericarps without wing-like ribs, only somewhat compressed

            21a. Stylopodium present; all the fruits of the umbellets pedicelled; stems arising from stolons and/or rhizomes; umbellets usually without bractlets .... ***Aegopodium***

            21b. Stylopodium wanting; central fruit of each umbellet sessile, the others pedicellate; stems arising from a cluster of fibrous roots; umbellets with bractlets .... ***Zizia***

   10b. Leaves dissected to decompound, without clearly defined leaflets and/or
       with irregular branching, the ultimate segments commonly less than 1.0 cm
       wide

       22a. Stems streaked or spotted with purple .... ***Conium***

       22b. Stems not marked with purple

          23a. Schizocarp with a ribbed beak 1.0–3.0 mm long, the body of the
             schizocarp without ribs; bractlets with conspicuously ciliate to
             fimbriate margins .... ***Anthriscus***

          23b. Schizocarp without a beak, with ribs; bractlets with entire or
             minutely ciliolate margins or absent

             24a. Axils of some of the upper leaves with bulbils; schizocarps rarely
                maturing .... ***Cicuta***

             24b. Axils of the leaves without bulbils; schizocarps normally
                maturing

                25a. Umbellets usually without bracts; ultimate leaf segments of
                   larger leaves narrow-linear to filiform, up to 1.0 mm wide

                   26a. Plants perennial; petiolar sheath of larger leaves
                      3.0–10.0 cm long .... ***Foeniculum***

                   26b. Plants annual; petiolar sheath 1.0–3.0(–4.0) cm long

                     27a. Schizocarp compressed perpendicular to the
                      commissure; umbel with 7–14 primary branches; styles
                      persistent in fruit .... ***Carum***

                     27b. Schizocarp compressed parallel with the commissure;
                      umbel with 10–40 primary branches; styles deciduous
                      in fruit .... ***Anethum***

                25b. Umbellets with bracts; ultimate segments of larger leaves
                  linear to linear-oblong, wider than 1.0 mm

                   28a. Plants perennial; schizocarp compressed; lateral ribs of
                    the fruit more prominent and with larger wings than the
                    ribs of the abaxial surface .... ***Conioselinum***

                   28b. Plants annual; schizocarp subterete; lateral and abaxial
                    ribs equally prominent .... ***Aethusa***

Aegopodium
(1 species)

Aegopodium podagraria L.
synonym(s): *Aegopodium podagraria* L. var. *variegatum* Bailey
common name(s): goutweed
habit: erect, branched, rhizomatous, perennial herb
range: Eurasia; escaped from cultivation to Newfoundland, s. to NC, w. and n. to MI and Ontario; also
    Manitoba and British Columbia
state frequency: uncommon
habitat: roadsides, waste places, moist, shaded areas
notes: The commonly cultivated form of this species has doubly serrate leaflets with white markings.

Aethusa

(1 species)

A monotypic genus with poisonous herbage.

Aethusa cynapium L.
synonym(s): —
common name(s): fool's parsley
habit: branched, annual herb
range: Eurasia; naturalized to New Brunswick, s. to DE, w. to OH, n. to MN and Ontario; also British Columbia
state frequency: very rare
habitat: fields, waste places, cultivated ground
notes: —

Anethum

(1 species)

Strongly scented annuals with filiform-dissected leaves.

Anethum graveolens L.
synonym(s): —
common name(s): dill
habit: erect, usually branched, annual herb
range: Eurasia; escaped from cultivation to ME, s. to NJ, w. and n. to MN
state frequency: very rare
habitat: roadsides, waste places
notes: —

Angelica

(2 species)

KEY TO THE SPECIES
1a. Lateral ribs of the mericarps with thin, flat wings; upper sheaths much longer than the leaf blade or lacking blades altogether; plants of moist to wet, inland soils, 1.0–3.0 m tall .... *A. atropurpurea*
1b. Lateral and dorsal ribs of the mericarps thick and corky; upper sheaths shorter than the leaf blades; plants of coastal beach habitats, 0.5–1.5 m tall .... *A. lucida*

Angelica atropurpurea L.
synonym(s): —
common name(s): Alexanders, purple-stem angelica
habit: tall, very stout, perennial herb
range: Labrador, s. to DE, w. to IL, n. to MN and Ontario
state frequency: uncommon
habitat: swamps, bottomlands, wet woods, rich thickets
notes: —

Angelica lucida L.
synonym(s): *Coelopleurum actaeifolium* (Michx.) Coult. & Rose, *Coelopleurum lucidum* (L.) Fern.
common name(s): sea beach angelica, sea coast angelica
habit: stout, perennial herb

range: Greenland, s. to coastal NY; also AK, s. to CA
state frequency: uncommon
habitat: coastal beaches, rocks, and borders of woods
notes: —

Anthriscus
(1 species)

Anthriscus sylvestris (L.) Hoffmann
synonym(s): —
common name(s): wild chervil, cow-parsley
habit: coarse, branched, biennial or short-lived perennial herb
range: Europe; naturalized to Newfoundland, s. to NJ, w. and n. to Ontario
state frequency: very rare
habitat: fields, roadsides, waste places
notes: —

Aralia
(4 species)

KEY TO THE SPECIES
1a. Plants scapose, the single leaf arising from a very short, woody stem at the surface of the
     ground .... *A. nudicaulis*
1b. Plants with evident, above-ground stems and few to many leaves
   2a. Plants armed with stout prickles, up to 12.0 m tall; styles distinct .... *A. spinosa*
   2b. Plants unarmed or armed with bristles near the base of the plant, up to 2.0 m tall;
      styles connate about half their length
      3a. Inflorescence composed of a few to several umbels in a cluster resembling a
         corymb; plants armed near the base with bristles; plants of dry, sandy, and/or
         sterile soils; leaflets rounded to cuneate at the base .... *A. hispida*
      3b. Inflorescence composed of numerous umbels in a cluster resembling a panicle;
         plants unarmed; plants of rich soils; leaflets cordate at the base .... *A. racemosa*

Aralia hispida Vent.
synonym(s): —
common name(s): bristly sarsaparilla
habit: rhizomatous shrub or suffrutescent herb
range: Labrador, s. to NC, w. to IN, n. to MN and Manitoba
state frequency: occasional
habitat: dry, sandy, open woods
notes: —

Aralia nudicaulis L.
synonym(s): —
common name(s): wild sarsaparilla
habit: erect, rhizomatous, perennial herb
range: Labrador, s. to mountains of GA, w. to CO, n. to ID
state frequency: common
habitat: dry to wet forests
notes: —

Aralia racemosa L.

synonym(s): —
common name(s): spikenard
habit: erect, perennial herb
range: Quebec, s. to GA, w. to KS, n. to Manitoba
state frequency: uncommon
habitat: rich woods
notes: —

Aralia spinosa L.

synonym(s): —
common name(s): Hercules' club
habit: shrub to small tree
range: NJ, s. to FL, w. to TX, n. to MO; introduced to ME, w. to MI
state frequency: rare
habitat: moist, rich woods
notes: —

Carum

(1 species)

Carum carvi L.

synonym(s): —
common name(s): caraway
habit: branched, biennial herb
range: Eurasia; escaped from cultivation to Newfoundland, s. to CT, w. to IL, n. to British Columbia
state frequency: uncommon
habitat: fields, roadsides, waste places
notes: —

Cicuta

(2 species)

Deadly poisonous herbs of wet soils.

KEY TO THE SPECIES
1a. Axils of the upper leaves bearing bulbils; leaflets linear, up to 5.0 mm wide; schizocarp 1.5–2.0 mm long .... *C. bulbifera*
1b. Axils of the leaves without bulbils; leaflets linear to lanceolate, commonly wider than 5.0 mm; schizocarp 2.0–4.0 mm long .... *C. maculata*

Cicuta bulbifera L.

synonym(s): —
common name(s): bulbiferous water-hemlock, bulblet-bearing water-hemlock
habit: slender, perennial herb
range: Newfoundland, s. to VA, w. to OR, n. to British Columbia
state frequency: uncommon
habitat: swamps, marshes, wet thickets
notes: —

Cicuta maculata L.

synonym(s): —
common name(s): common water-hemlock, spotted cowbane
habit: stout, branched, perennial herb
range: Quebec, s. to FL, w. to Mexico and CA, n. to AK
state frequency: uncommon
habitat: swamps, marshes, wet meadows and thickets
notes: —

Conioselinum
(1 species)

Conioselinum chinense (L.) B. S. P.

synonym(s): —
common name(s): hemlock-parsley
habit: stout to slender, perennial herb
range: Labrador, s. to mountains of NC, w. to MO, n. to MN and Ontario
state frequency: uncommon
habitat: swamps, bogs, wet meadows, thickets, and forests
notes: —

Conium
(1 species)

Conium maculatum L.

synonym(s): —
common name(s): poison-hemlock
habit: large, branched, biennial herb
range: Eurasia; naturalized to Quebec, s. to FL, w. to IA, n. to British Columbia
state frequency: uncommon
habitat: waste places
notes: —

Cryptotaenia
(1 species)

Cryptotaenia canadensis (L.) DC.

synonym(s): *Deringa canadensis* (L.) Kuntze
common name(s): honewort
habit: single-stemmed, perennial herb
range: New Brunswick, s. to GA, w. to TX, n. to Manitoba
state frequency: very rare
habitat: rich woods and thickets
notes: This species is ranked SH by the Maine Natural Areas Program.

Daucus
(1 species)

Daucus carota L.
synonym(s): —
common name(s): Queen Anne's lace, wild-carrot
habit: erect, biennial herb
range: Eurasia; naturalized to most of North America
state frequency: common
habitat: dry fields and waste places
notes: —

Foeniculum
(1 species)

Foeniculum vulgare P. Mill.
synonym(s): —
common name(s): fennel
habit: stout, short-lived, perennial herb
range: Europe; naturalized to Australia and North America from ME, s. to FL, w. to NE, n. to MI; also CA
state frequency: very rare
habitat: dry fields and roadsides
notes: —

Heracleum
(3 species)

Robust, stout-stemmed herbs with relatively large fruits.

KEY TO THE SPECIES
1a. Principal umbel mostly with 50–150 primary branches, up to 5.0 dm wide; leaflets up to 13.0 dm long; schizocarp on a pedicel 1.5–4.0 cm long .... **H. mantegazzianum**
1b. Principal umbel mostly with 15–45 primary branches, rarely exceeding 2.0 dm wide; leaflets 0.5–3.0(–6.0) dm long; schizocarp on a pedicel 0.6–2.0 cm long
    2a. Leaves with 3 ternately arranged, sharply serrate leaflets; umbels with 5–10 bracts; larger leaflets 10.0–30.0(–60.0) cm long; stylopodium not, or only slightly, extending above a deep notch in the mericarp .... **H. maximum**
    2b. Leaves with 3–7 pinnately arranged, commonly bluntly toothed leaflets; umbels with fewer than 5 bracts, or lacking bracts altogether; leaflets mostly 5.0–10.0 cm long; stylopodium extending above a notch in the mericarp .... **H. sphondylium**

Heracleum mantegazzianum Sommier & Levier
synonym(s): —
common name(s): giant hogweed
habit: robust, biennial or monocarpic perennial herb
range: Asia; naturalized to northeastern United States
state frequency: very rare
habitat: fields, roadsides, waste places
notes: —

Heracleum maximum Bartr.

synonym(s): *Heracleum lanatum* Michx.
common name(s): cow-parsnip, masterwort
habit: tall, stout, perennial herb
range: Labrador, s. to mountains of GA, w. to NM and CA, n. to AK
state frequency: uncommon
habitat: low, damp areas
notes: —

Heracleum sphondylium L.

synonym(s): —
common name(s): eltrot, hogweed
habit: tall, stout, perennial herb
range: Eurasia; naturalized to Newfoundland, s. to NY
state frequency: rare
habitat: fields, roadsides, waste places
notes: —

Hydrocotyle
(1 species)

Hydrocotyle americana L.

synonym(s): —
common name(s): marsh pennywort
habit: slender, creeping, stoloniferous, perennial herb
range: Newfoundland, s. to NC, w. to IN, n. to MN
state frequency: uncommon
habitat: moist woods, meadows, thickets, and low ground
notes: —

Levisticum
(1 species)

Levisticum officinale W. D. J. Koch

synonym(s): *Hipposelinum levisticum* (L.) Britt. & Rose
common name(s): lovage
habit: tall, stout, branched, perennial herb
range: Europe; escaped from cultivation occasionally in the northeast
state frequency: rare
habitat: waste places, fields
notes: —

Ligusticum
(1 species)

Ligusticum scothicum L.

synonym(s): —
common name(s): Scotch-lovage
habit: stout, simple or branched, perennial herb
range: Greenland, s. to coastal NY
state frequency: uncommon

habitat: saline marshes, sandy or rocky seashores
notes: —

Lilaeopsis
(1 species)

Lilaeopsis chinensis (L.) Kuntze
synonym(s): *Lilaeopsis lineata* (Michx.) Greene
common name(s): lilaeopsis
habit: creeping, rhizomatous, aquatic, perennial herb
range: Nova Scotia, s. to FL, w. to MS
state frequency: very rare
habitat: mud of brackish marshes and tidal shores
notes: This species is ranked S1 by the Maine Natural Areas Program.

Osmorhiza
(3 species)

Determining presence/absence of the bractlets subtending the umbellets is important for identifying species in this genus. The bractlets are sometimes deciduous in fruit, which could lead to incorrect character assessment. Like leaves, the falling bractlets leave a scar that indicates their earlier presence. Therefore, specimens of *Osmorhiza* collected late in the season lacking bractlets should be checked carefully for bractlet scars to be sure they truly lack bractlets. Include the stylopodium when measuring style length.

KEY TO THE SPECIES

1a. Bractlets absent; styles at maturity strongly outcurved, 0.3–1.0 mm long .... *O. berteroi*
1b. Bractlets present; styles at maturity nearly straight, 1.0–4.0 mm long
    2a. Styles 1.0–1.5 mm long; roots without a conspicuous, anise taste; bractlets linear-attenuate .... *O. claytonii*
    2b. Styles 2.0–4.0 mm long; roots tasting of anise; bractlets lance-attenuate ....
    .... *O. longistylis*

Osmorhiza berteroi DC.
synonym(s): *Osmorhiza chilensis* Hook. & Arn.
common name(s): mountain sweet cicely, tapering sweet cicely
habit: erect, perennial herb
range: Newfoundland, s. to ME, w. to AZ and CA, n. to AK
state frequency: rare
habitat: moist to dry woods and clearings
notes: This species is ranked S2 by the Maine Natural Areas Program.

Osmorhiza claytonii (Michx.) C. B. Clarke
synonym(s): —
common name(s): bland sweet cicely
habit: slender, erect, perennial herb
range: Newfoundland, s. to NC, w. to AL, AR, and KS, n. to Saskatchewan
state frequency: occasional
habitat: moist woods, wooded slopes, and clearings
notes: —

Osmorhiza longistylis (Torr.) DC.

synonym(s): —
common name(s): long-styled sweet cicely, anise-root
habit: erect, perennial herb
range: Quebec, s. to VA, w. to OK, n. to Alberta
state frequency: uncommon
habitat: rich, moist, clearings, thickets, and alluvial woods
notes: —

Panax
(2 species)

KEY TO THE SPECIES

1a. Leaflets with long petiolules, acuminate at the apex; plants 2.0–6.0 dm tall, in anthesis from mid-June through early August; flowers mostly bisexual; drupe red ....
.... *P. quinquefolius*
1b. Leaflets sessile or nearly so, obtuse to acute at the apex; plants 1.0–2.0 dm tall, in anthesis from mid-May through mid-June; flowers often unisexual [staminate]; drupe yellow .... *P. trifolius*

Panax quinquefolius L.

synonym(s): —
common name(s): American ginseng
habit: solitary, erect, perennial herb
range: ME, s. to FL, w. to OK, n. to Manitoba
state frequency: rare
habitat: rich, cool woods
notes: This species is ranked S2 by the Maine Natural Areas Program.

Panax trifolius L.

synonym(s): —
common name(s): dwarf ginseng, groundnut
habit: solitary, erect, perennial herb
range: New Brunswick, s. to mountains of GA, w. to NE, n. to MN and Ontario
state frequency: uncommon
habitat: rich or damp forests
notes: —

Pastinaca
(1 species)

Pastinaca sativa L.

synonym(s): —
common name(s): parsnip
habit: tall, stout, taprooted, biennial herb
range: Eurasia; naturalized throughout most of North America
state frequency: uncommon
habitat: fields, roadsides, waste places
notes: —

Pimpinella
(1 species)

Pimpinella saxifraga L.
synonym(s): —
common name(s): burnet-saxifrage
habit: branched, perennial herb
range: Eurasia; naturalized to Newfoundland, s. to DE, w. to IN, n. to WI
state frequency: uncommon
habitat: fields, roadsides, waste places
notes: —

Sanicula
(3 species)

KEY TO THE SPECIES
1a. Styles much exceeding the bristles of the fruit; staminate flowers numbering 12–25 per
   umbellet; sepals 0.5–2.0 mm long, not exceeding the bristles of the fruit; branches of the
   inflorescence erect to ascending; plants perennial
   2a. Sepals 1.0–2.0 mm long, rigid, lance-subulate; schizocarps subsessile, slightly
      exceeded by the persistent, staminate flower remnants .... *S. marilandica*
   2b. Sepals 0.5–1.0 mm long, soft, deltate; schizocarps stipitate, slightly exceeding the
      persistent, staminate flower remnants .... *S. odorata*
1b. Styles exceeded by the bristles of the fruit; staminate flowers 2–7 per umbellet; sepals
   2.0–2.5 mm long, forming a beak that exceeds the bristles of the fruit; branches of the
   inflorescence spreading to ascending; plants biennial .... *S. trifoliata*

Sanicula marilandica L.
synonym(s): —
common name(s): black snakeroot
habit: rhizomatous, perennial herb
range: Newfoundland, s. to FL, w. to NM, n. to British Columbia
state frequency: uncommon
habitat: open woods, thickets, meadows, shores
notes: —

Sanicula odorata (Raf.) K. M. Pryer & L. R. Phillippe
synonym(s): *Sanicula gregaria* Bickn.
common name(s): cluster sanicle
habit: slender, perennial herb
range: Quebec and Nova Scotia, s. to FL, w. to TX, n. to MN
state frequency: very rare
habitat: rich forests
notes: —

Sanicula trifoliata Bickn.
synonym(s): —
common name(s): beaked sanicle
habit: biennial herb

range: New Brunswick, s. to NC, w. to IA, n. to MN and Ontario
state frequency: very rare
habitat: rich forests
notes: —

Sium
(1 species)

Sium suave Walt.
synonym(s): —
common name(s): water-parsnip
habit: solitary, erect, branched, emergent to terrestrial, perennial herb
range: Newfoundland, s. to FL, w. to CA, n. to AK
state frequency: occasional
habitat: wet meadows and thickets, swamps, muddy shores
notes: Sometimes producing simple, emersed leaves when growing in tidal habitats.

Zizia
(1 species)

Zizia aurea (L.) W. D. J. Koch
synonym(s): —
common name(s): golden Alexanders
habit: branched, perennial herb
range: New Brunswick, s. to FL, w. to TX, n. to Saskatchewan
state frequency: uncommon
habitat: moist to wet woods, thickets, fields, meadows, and shores
notes: —

MENYANTHACEAE
(2 genera, 2 species)

KEY TO THE GENERA
1a. Leaves compound, with 3 leaflets, emersed; petals pubescent on the adaxial [inner] surface, lacking glands; inflorescence a raceme or panicle .... ***Menyanthes***
1b. Leaves simple, floating; petals glabrous on the adaxial surface, bearing a yellow gland near the base of each lobe; inflorescence an umbel .... ***Nymphoides***

Menyanthes
(1 species)

Menyanthes trifoliata L.
synonym(s): *Menyanthes trifoliata* L. var. *minor* Raf.
common name(s): buckbean, bogbean
habit: rhizomatous, emergent, aquatic, perennial herb
range: circumboreal; s. to VA, w. to CA
state frequency: uncommon
habitat: cold bogs, pond margins, and quiet, shallow water
notes: —

Nymphoides
(1 species)

Nymphoides cordata (Ell.) Fern.
synonym(s): *Nymphoides lacunosa* (Vent.) Kuntze
common name(s): little floating heart
habit: stout-rhizomatous, aquatic, perennial herb
range: Newfoundland, s. to FL, w. to LA, n. to Ontario
state frequency: occasional
habitat: quiet water of ponds and slow streams
notes: Some of the flowers are often replaced by tuberous-thickened, spur-like roots, these forming a cluster beneath the leaf near the base of the umbel.

CAMPANULACEAE
(3 genera, 14 species)

KEY TO THE GENERA
1a. Corolla zygomorphic; stamens connate; gynoecium composed of 2 carpels; capsules opening near the top .... *Lobelia*
1b. Corolla actinomorphic; stamens separate or merely connivent; gynoecium composed of 3 or 5 carpels; capsules opening by lateral pores
  2a. Flowers and fruits solitary, sessile or subsessile in the axils of leaves; capsules slender-cylindric; plants annual .... *Triodanis*
  2b. Flowers and fruits solitary or borne in a spike, raceme, or glomerule, sessile or pedicellate; capsules globose to ovoid or obconic; plants perennial .... *Campanula*

Campanula
(7 species)

KEY TO THE SPECIES
1a. Corolla 0.4–1.3 cm long; stems 3-angled, slender, weak, leaning on other vegetation; native wetland plants .... *C. aparinoides*
1b. Corolla 1.0–5.0 cm long; stems terete or obscurely angled, erect to ascending; native or introduced plants of terrestrial habitats
  2a. Flowers sessile in axillary and terminal glomerules .... *C. glomerata*
  2b. Flowers borne on pedicels
    3a. Flowers borne in conspicuous, secund racemes
      4a. Corolla 2.0–3.0 cm long; stems and leaves glabrous or sparsely pubescent; plants with well developed rhizomes; upper leaves sessile, but not clasping ...
      .... *C. rapunculoides*
      4b. Corolla 1.0–2.0 cm long; stems and leaves tomentulose; plants with poorly developed rhizomes; upper leaves clasping the stem .... *C. bononiensis*
    3b. Flowers borne in open clusters or loose racemes
      5a. Blade of stem leaves linear or linear-lanceolate; corolla 1.5–3.0 cm long; plants native .... *C. rotundifolia*

5b. Blade of stem leaves lanceolate to ovate or triangular; corolla 2.5–5.0 cm long; plants introduced
    6a. Corolla 2.5–4.0 cm long; peduncles with (1–)2 or 3 flowers; calyx and hypanthium bristly-pubescent .... *C. trachelium*
    6b. Corolla 4.0–5.0 cm long; peduncles with 1 flower; calyx and hypanthium glabrous .... *C. latifolia*

Campanula aparinoides Pursh
synonym(s): *Campanula aparinoides* Pursh var. *grandiflora* Holz., *Campanula uliginosa* Rydb.
common name(s): marsh bellflower
habit: simple or branched, slender-rhizomatous, perennial herb
range: New Brunswick and Nova Scotia, s. to GA, w. to MO and CO, n. to Saskatchewan
state frequency: uncommon
habitat: wet meadows and shores, swales
notes: —

Campanula bononiensis L.
synonym(s): —
common name(s): —
habit: slender, perennial herb
range: Europe; escaped from cultivation locally
state frequency: very rare
habitat: old homesteads, compost areas
notes: —

Campanula glomerata L.
synonym(s): —
common name(s): clustered bellflower
habit: stout, erect, perennial herb
range: Eurasia; escaped from cultivation to ME and Quebec, s. to MA, w. and n. to MN
state frequency: very rare
habitat: old fields, roadsides, waste places
notes: —

Campanula latifolia L.
synonym(s): —
common name(s): —
habit: erect, perennial herb
range: Eurasia; escaped from cultivation locally
state frequency: very rare
habitat: waste places, old homesteads, compost areas
notes: —

Campanula rapunculoides L.
synonym(s): *Campanula rapunculoides* L. var. *ucranica* (Bess.) K. Koch
common name(s): rover bellflower, creeping bellflower
habit: slender, usually simple, rhizomatous, perennial herb
range: Eurasia; naturalized to Newfoundland, s. to DE and MD, w. to MO, n. to ND
state frequency: uncommon
habitat: thickets, lawns, roadsides, waste places
notes: —

Campanula rotundifolia L.

synonym(s): —
common name(s): harebell, bluebell
habit: slender, erect, simple or branched, perennial herb
range: circumboreal; s. to NJ, w. to Mexico and CA
state frequency: occasional
habitat: rocky areas, woods, meadows, shores, beaches
notes: These plants have dimorphic leaves—the basal ones with orbicular blades and the cauline with linear to linear-lanceolate blades.

Campanula trachelium L.

synonym(s): —
common name(s): nettle-leaved bellflower, throatwort
habit: erect, perennial herb
range: Eurasia and northern Africa; naturalized to ME and Quebec, s. to MA, w. to OH
state frequency: very rare
habitat: thickets, roadsides, waste places
notes: —

Lobelia

(6 species)

The androecium in *Lobelia* has a similar pollen presentation mechanism as the Asteraceae. It is composed of connate anthers that form a ring or tube. As the style grows up through the middle of the anthers, it pushes out the pollen.

KEY TO THE SPECIES

1a. Flowers 2.0–4.5 cm long; corolla fenestrate
  2a. Corolla red (pink or white); flowers 3.0–4.5 cm long; filament tube 2.4–3.3 cm long
    .... *L. cardinalis*
  2b. Corolla blue (white); flowers 2.0–3.0 cm long; filament tube 1.2–1.5 cm long ....
    .... *L. siphilitica*
1b. Flowers 0.7–1.8 cm long; corolla lacking open slits
  3a. Leaves all basal, with 2 hollow tubes extending their entire length; all the anthers
    pubescent at the apex; plants aquatic .... *L. dortmanna*
  3b. Leaves borne on a stem, flat, not hollow; only the 2 smaller anthers pubescent at the
    apex; plants terrestrial
    4a. Stem leaves linear to linear-lanceolate, mostly 0.5–4.0 mm wide; lower lip of the
      corolla glabrous on the adaxial [inside] surface .... *L. kalmii*
    4b. Stem leaves lanceolate or oblanceolate to ovate or obovate, mostly wider than
      5.0 mm; lower lip of the corolla pubescent on the adaxial surface near the base
      5a. Corolla 6.0–8.0 mm long; hypanthium nearly as long as the corolla, evidently
        inflated in fruit; middle stem leaves typically ovate-oblong; stems long-
        pubescent, at least below; plants annual .... *L. inflata*
      5b. Corolla 7.0–11.0 mm long; hypanthium shorter than the corolla, scarcely
        inflated in fruit; middle stem leaves typically oblanceolate; stems short-
        pubescent; plants biennial or perennial .... *L. spicata*

Lobelia cardinalis L.

synonym(s): —
common name(s): cardinal flower
habit: erect, usually simple, perennial herb
range: New Brunswick and ME, s. to FL, w. to TX, n. to MN and Ontario
state frequency: occasional
habitat: damp shores and meadows, swamps
notes: —

Lobelia dortmanna L.

synonym(s): —
common name(s): water lobelia
habit: glabrous, aquatic, perennial herb
range: Newfoundland, s. to NJ, w. to PA, n. to MN; also British Columbia, s. to OR
state frequency: occasional
habitat: pond margins
notes: —

Lobelia inflata L.

synonym(s): —
common name(s): Indian-tobacco
habit: erect, branched, annual herb
range: Newfoundland, s. to GA, w. to KS, n. to MN and Saskatchewan
state frequency: occasional
habitat: open woods, fields, roadsides, waste places
notes: —

Lobelia kalmii L.

synonym(s): —
common name(s): Kalm's lobelia
habit: slender, simple or branched, perennial herb
range: Newfoundland, s. to NJ, w. to CO and WA, n. to British Columbia
state frequency: uncommon
habitat: wet ledges and shores, swamps, often calcareous
notes: —

Lobelia siphilitica L.

synonym(s): —
common name(s): great lobelia, blue cardinal flower
habit: stout, erect, simple, perennial herb
range: ME, s. to VA and mountains of NC, w. to TX and CO, n. to MN and Manitoba
state frequency: very rare
habitat: swamps, wet ground
notes: This species is ranked SX by the Maine Natural Areas Program.

Lobelia spicata Lam.

synonym(s): —
common name(s): spiked lobelia, pale-spike lobelia
habit: erect, usually simple, perennial herb
range: New Brunswick and Nova Scotia, s. to GA, w. to AR and KS, n. to Alberta
state frequency: uncommon
habitat: woods, thickets, clearings, meadows

key to the subspecies:
1a. Stem, bracts, and calyx lobes hirtellous .... *L. s.* var. *hirtella* Gray
1b. Stems glabrous except sometimes near the base; bracts glabrous; calyx lobes glabrous or merely ciliate
  2a. Flowers 7.0–9.0 mm long, blue; anthers white; inflorescence usually 10- to 30-flowered; capsules more or less globose .... *L. s.* var. *campanulata* McVaugh
  2b. Flowers 9.0–12.0 mm long, light blue to white; anthers light blue-gray; inflorescence often with more flowers; capsules short-hemispherical .... *L. s.* var. *spicata*
notes: *Lobelia spicata* var. *hirtella* is ranked S3 by the Maine Natural Areas Program.

Triodanis
(1 species)

Species of *Triodanis* have dimorphic flowers. The lower flowers are cleistogamous, with 3 or 4 calyx lobes and reduced corollas. The upper flowers are chasmogamous, with 5 calyx lobes and normal, open corollas.

Triodanis perfoliata (L.) Nieuwl.
synonym(s): *Specularia perfoliata* (L.) A. DC.
common name(s): round-leaved triodanis
habit: simple or branched, annual herb
range: New Brunswick and ME, s. to FL, w. to Mexico and CA, n. to British Columbia
state frequency: rare
habitat: shores, roadsides, waste places
notes: —

ASTERACEAE
(71 genera, 220 species)

KEY TO THE GENERA
1a. Capitula composed entirely of bisexual ray flowers; plants with a milky latex ... **Group 1**
1b. Capitula composed entirely of disk flowers or of both ray and disk flowers, the ray flowers, when present, restricted to the margin of the capitulum and unisexual or sterile; plants commonly without a milky latex
  2a. Capitula without ray flowers, composed entirely of disk flowers
    3a. Pappus composed of capillary bristles, either simple or pinnately branched .... **Group 2**
    3b. Pappus composed of scales [that are divided at the apex into many bristles in *Dyssodia*], awns, or entirely absent .... **Group 3**
  2b. Capitula with ray flowers [these sometimes minute and inconspicuous, but can be identified by their zygomorphic corollas]
    4a. Corollas of ray flowers yellow or orange (brown or purple-brown at the base in some *Coreopsis*, *Helenium*, and *Rudbeckia*)
      5a. Pappus composed partly or entirely of capillary bristles (short scales may also be present) .... **Group 4**
      5b. Pappus composed entirely of scales [that are divided at the apex into many bristles in *Dyssodia*], awns, a crown, or completely absent .... **Group 5**
    4b. Corollas of ray flowers of various colors, but not yellow or orange .... **Group 6**

Group 1

1a. Pappus absent; plants annual; flowers yellow
  2a. Phyllaries membranaceous to herbaceous, not prominently keeled; peduncles not swollen; stems with leaves .... ***Lapsana***
  2b. Phyllaries becoming enlarged with an indurate, keeled midrib after anthesis; peduncles conspicuously swollen; stem with minute, bracteal leaves .... ***Arnoseris***
1b. Pappus present in some form; plants annual or perennial; flowers of various colors
  3a. Pappus composed partly or entirely of scales (bristles may also be present)
    4a. Pappus composed entirely of scales, the scales numerous; flowers blue (rarely pink or white); involucre 9.0–15.0 mm tall .... ***Cichorium***
    4b. Pappus of 5 scales and 5–10 slender bristles; flowers yellow; involucre 4.0–7.0 mm tall .... ***Krigia***
  3b. Pappus composed entirely of slender bristles
    5a. Pappus bristles simple
      6a. Cypsela terete or several-angled [this can often be successfully assessed with the flowering ovary]
        7a. Cypsela muricate, at least in the apical portion, also tipped by a long, slender beak .... ***Taraxacum***
        7b. Cypsela without sharp projections, with or without an apical beak
          8a. Flowers pink, purple, white, yellow-white, or green-white; capitula with 5–16 flowers .... ***Prenanthes***
          8b. Flowers yellow, orange, or red-orange; capitula with 8–100 flowers
            9a. Plants taprooted annuals or biennials .... ***Crepis***
            9b. Plants fibrous-rooted perennials from short or long rhizomes or a caudex .... ***Hieracium***
      6b. Cypsela evidently compressed
        10a. Cypsela without an enlarged disk at the apex where the pappus attaches, also lacking a beak; capitula with 80–250 flowers .... ***Sonchus***
        10b. Cypsela with an enlarged disk at the apex where the pappus attaches, with or without a beak; capitula with 5–56 flowers
          11a. Capitula with 5 flowers .... ***Mycelis***
          11b. Capitula with 13–55 flowers .... ***Lactuca***
    5b. Pappus bristles pinnately branched
      12a. Involucre composed of 1 series of phyllaries of equal length; leaves long and slender, grass-like; body of the cypsela, excluding the beak, 10.0–25.0 mm long .... ***Tragopogon***
      12b. Involucre calyculate or with phyllaries of differing lengths; leaves relatively wider, not grass-like; body of the cypsela 2.0–7.5 mm long
        13a. Stems not resembling a scape, definitely leafy; body of the cypsela, excluding the beak, 2.0–4.0 mm long .... ***Picris***
        13b. Stems resembling a scape, without leaves or with few, very small leaves; body of the cypsela 4.0–7.5 mm long
          14a. Receptacle without chaff; cypselas rugulose, without beaks ....
          .... ***Leontodon***
          14b. Receptacle with chaff; cypselas muricate, the inner with slender beaks .... ***Hypochaeris***

Group 2

1a. Leaves spiny-margined; phyllaries with a spine-tip (except *Cirsium muticum*); receptacle densely bristly between the disk flowers

    2a. Pappus bristles simple; stems spiny-winged .... ***Carduus***

    2b. Pappus bristles pectinately branched; stems not spiny-winged except in *Cirsium vulgare* .... ***Cirsium***

1b. Leaves without spines; phyllaries not spine-tipped (except *Aster ericoides*); receptacle lacking bristles (except *Centaurea*)

    3a. Most or all of the phyllaries fimbriately or pectinately fringed at the apex; receptacle densely hairy; cypselas attached obliquely to the receptacle .... ***Centaurea***

    3b. Phyllaries with entire or ciliate margins; receptacle lacking bristles; cypselas attached basally to the receptacle

        4a. Plants pubescent with white or gray tomentum, at least on the abaxial leaf surface

            5a. Plants dioecious or monoecious, none of the flowers bisexual [central flowers of carpellate *Anaphalis* appearing bisexual but have an undivided style and are functionally staminate]

                6a. Stems resembling a scape, appearing before the leaves; leaves all basal, 5.0–40.0 cm wide, palmately lobed .... ***Petasites***

                6b. Stems resembling a scape or not, in either case, appearing after the leaves; leaves basal or produced on a stem, 0.2–5.5 cm wide, entire

                    7a. Plants stoloniferous, with conspicuous rosettes of basal leaves; staminate pappus apically clavate; carpellate pappus connate at the base and falling together; stems subscapose, with reduced, bract-like leaves .... ***Antennaria***

                    7b. Plants not stoloniferous, without conspicuous rosettes of basal leaves; pappus slender at the apex and distinct, falling separately; stems leafy, the leaves 7.5–12.0 cm long .... ***Anaphalis***

            5b. Plants polygamous—the numerous outer flowers unisexual [carpellate], the few inner bisexual

                8a. Pappus bristles connate at the base, falling together

                    9a. Plants annual or biennial, weedy, of fields and disturbed ground; cypselas smooth or papillate .... ***Gamochaeta***

                    9b. Plants perennial from a caudex, of boreal areas; cypselas sparsely strigose .... ***Omalotheca***

                8b. Pappus bristles distinct, falling separately

                  10a. Capitulescence resembling a spike; plants perennial, of alpine areas .... ***Omalotheca***

                  10b. Capitulescence resembling a corymb or panicle or clusters of capitula; plants annual or winter-annual, of disturbed areas

                    11a. Involucre 2.0–3.0 mm tall, brown; capitulescence of small, axillary and terminal clusters, overtopped by its subtending leaves; plants 0.5–2.5 dm tall .... ***Gnaphalium***

                    11b. Involucre 5.0–7.0 mm tall, white or yellow-white, sometimes tinged with brown; capitulescence terminal, overtopping its subtending leaves; plants 1.0–10.0 dm tall .... ***Pseudognaphalium***

        4b. Plants glabrous or pubescent, but not conspicuously tomentose

12a. Leaves opposite or whorled (those of the upper stem rarely alternate)

   13a. Lower and middle stem leaves opposite

      14a. Leaves with a distinct petiole .... *Ageratina*

      14b. Leaves lacking a distinct petiole .... *Eupatorium*

   13b. Lower and middle stem leaves whorled .... *Eupatorium*

12b. Leaves alternate throughout the stem

   15a. Plants vines; capitula with 4 principal phyllaries and 5 flowers ....
      .... *Mikania*

   15b. Plants not vines; capitula with more than 4 phyllaries and more than 4 flowers

      16a. All flowers of the capitula bisexual

         17a. Flowers yellow or orange; cypsela 1.5–2.5 mm long;
            capitulescence resembling a panicle or corymb .... *Senecio*

         17b. Flowers pink-purple (white); cypsela 6.0–7.0 mm long;
            capitulescence tall and slender, resembling a raceme .... *Liatris*

      16b. At least the marginal flowers of the capitula unisexual [carpellate]

         18a. Involucre 1.0–1.5 cm tall, in 1 series of long, nearly equal
            phyllaries, sometimes calyculate, turbinate-cylindric,
            conspicuously swollen at the base before anthesis .... *Erechtites*

         18b. Involucre 0.25–1.2 cm tall, usually in 2 or more series of
            different length phyllaries, saucer-shaped to cylindric, but without
            a basal swelling

            19a. Capitula on leafy-bracted branches; plants flowering
               mid-August to late September; style appendages acute to
               acuminate at the apex; leaves somewhat fleshy; plants of saline
               sand and mud .... *Symphyotrichum*

            19b. Capitula on bracteate branches; plants mostly flowering June
               to July; style appendages acute to, more commonly, obtuse;
               leaves not fleshy; plants not of saline habitats

               20a. Rays 0.5–1.0 mm long; plants annual; involucre
                  3.0–4.0 mm long, its phyllaries glabrous or sparsely
                  pubescent .... *Conyza*

               20b. Rays up to 5.0 mm long; plants biennial or perennial;
                 involucre 5.0–12.0 mm long, its phyllaries glandular and
                 often hirsute .... *Erigeron*

Group 3

1a. Capitula 1-flowered, aggregated in clusters forming a false, secondary capitulum that is subtended by a common involucre; true involucre commonly spine-tipped and subtended by a tuft of capillary bristles; leaves spiny-margined .... *Echinops*

1b. Capitula with more than 1 flower, not conspicuously aggregated in secondary capitula; involucre with or without awn-tips; leaves not spiny-margined

   2a. Receptacle bristly or chaffy, at least near the margin of the capitulum

      3a. Pappus absent

         4a. Receptacle densely bristly or long-hairy or naked

            5a. Cypselas attached obliquely or laterally to the receptacle; style branches
              with a thickened, often pubescent, ring, the texture of the branches

changing to papillate distal to the ring; involucre 10.0–25.0 mm tall; most or all the phyllaries fimbriate or pectinately fringed at the apex .... *Centaurea*

5b. Cypselas attached basally to the receptacle; style branches truncate and penicillate; involucre 1.0–7.5 mm tall; phyllaries with entire margins .... *Artemisia*

4b. Receptacle with flattened or filiform chaff

6a. Staminate and carpellate flowers in separate capitula (staminate usually the uppermost capitula and possessing an undivided style); involucre armed with tubercles, spines, or prickles; carpellate flowers lacking a corolla

7a. Staminate involucre of distinct phyllaries; carpellate involucre a conspicuous, prickly bur, 0.8–3.5 cm tall .... *Xanthium*

7b. Staminate involucre of united phyllaries; carpellate involucre with 1 or more series of tubercles or spines, 0.3–1.0 cm tall .... *Ambrosia*

6b. Staminate and carpellate flowers in the same capitula; involucre unarmed; all the flowers with a corolla (except in *Iva xanthifolia*, in which the carpellate flowers sometimes lack a corolla)

8a. Leaves alternate; plants polygamous—the outer flowers unisexual [carpellate], the inner flowers bisexual [with a divided style] ... *Madia*

8b. Leaves opposite (except sometimes the uppermost alternate); plants either monoecious with all the flowers of the capitula unisexual [the outer carpellate, the inner staminate with an undivided style and abortive ovary] or synoecious with all the flowers bisexual

9a. Anthers scarcely united; involucre with 1–3 series of monomorphic, herbaceous phyllaries; leaves simple .... *Iva*

9b. Anthers united most of their length; involucre with 2 series of dimorphic phyllaries—the outer series larger, herbaceous to foliaceous, the inner series smaller and membranaceous; leaves (except the uppermost) pinnately trifoliate .... *Bidens*

3b. Pappus of some form present

10a. Most or all of the leaves opposite; pappus of 2–6 retrorsely barbed (rarely antrorsely barbed) awns (sometimes the awns nearly obsolete in *Bidens discoidea*); receptacle with flattened scales .... *Bidens*

10b. Leaves alternate; pappus of bristles or scales; receptacle densely bristly

11a. Phyllaries simple, with attenuate, hooked tips; leaves usually simple, with blades rounded to cordate at the base .... *Arctium*

11b. Phyllaries fimbriate to pectinate-fringed, without hooked tips; leaves pinnately lobed or with blades cuneate at the base or both .... *Centaurea*

2b. Receptacle without bristles or chaff

12a. Leaves opposite; pappus composed of 10–20 scales that are cleft into 5–10 bristles at the apex .... *Dyssodia*

12b. Leaves alternate; pappus of awns, a short crown, or none

13a. Pappus of 2–8 firm, but deciduous, awns; phyllaries squarrose and heavily resinous .... *Grindelia*

13b. Pappus a short crown or none; phyllaries neither squarrose nor resinous

14a. Capitulescence resembling a spike, raceme, or panicle .... *Artemisia*
14b. Capitulescence a solitary capitulum or resembling a corymb
    15a. Leaves toothed, sometimes with a few basal lobes as well; cypselas with 5–10 rib-like nerves .... *Chrysanthemum*
    15b. Leaves 1- to 3-times pinnatifid; cypselas with 5 or fewer rib-like nerves
        16a. Receptacle flat or low-convex; disk corollas 5-lobed; plants 1.0–15.0 dm tall .... *Tanacetum*
        16b. Receptacle high-convex, pointed; disk corollas 4-lobed; plants 0.5–4.0 dm tall .... *Matricaria*

Group 4

1a. Leaves opposite, except sometimes the uppermost .... *Arnica*
1b. Leaves alternate throughout the stem
    2a. Involucre with 1 series of long phyllaries of equal length, sometimes calyculate as well
        3a. Stems with small, bract-like leaves only, appearing and flowering before the cordate and suborbicular blades of the basal leaves are produced; disk flowers sterile .... *Tussilago*
        3b. Stems with leaves, present during flowering; disk flowers fertile .... *Senecio*
    2b. Involucre with 2 or more series of long phyllaries of distinctly unequal lengths
        4a. Anthers sagittate-tailed at the base; phyllaries 2.0–2.5 cm long; leaves densely velvety-tomentose on the abaxial surface .... *Inula*
        4b. Anthers cuneate to sagittate at the base, but not tailed; phyllaries 0.3–1.1 cm long; leaves glabrous to pubescent, but not velvety-tomentose
            5a. Leaves resinous-punctate; capitulescence usually resembling a corymb ....
            .... *Euthamia*
            5b. Leaves not resinous-punctate; capitulescence not resembling a corymb (except in the alpine *Solidago multiradiata*) .... *Solidago*

Group 5

1a. Receptacle naked
    2a. Leaves opposite; pappus composed of 10–20 scales that are cleft into 5–10 bristles at the apex .... *Dyssodia*
    2b. Leaves alternate; pappus of awns, a short crown, or none
        3a. Pappus a tiny crown or completely absent
            4a. Rays 8.0–15.0 mm long .... *Chrysanthemum*
            4b. Rays up to 4.0 mm long .... *Tanacetum*
        3b. Pappus composed of distinct awns or scales
            5a. Phyllaries resinous; pappus composed of 2–8 firm, but deciduous, awns; capitula with 25–40 rays .... *Grindelia*
            5b. Phyllaries not resinous; pappus composed of 5–10 lanceolate to ovate, awn-tipped scales; capitula with 8–21 rays .... *Helenium*
1b. Receptacle bristly or chaffy, at least toward the margin of the capitulum
    6a. Leaves opposite, at least the upper connate-perfoliate; stem square; disk flowers unisexual [staminate], with an undivided style and abortive ovary .... *Silphium*

6b. Leaves alternate or opposite, not connate-perfoliate; stem terete or angled; disk flowers bisexual, the style divided and ovary fertile

    7a. Leaves regularly alternate throughout the stem

        8a. Receptacle bristly or setose; pappus composed of 6–10 conspicuously awned scales .... ***Gaillardia***

        8b. Receptacle chaffy with flattened scales; pappus various but not composed of awned scales

            9a. Phyllaries in 1 series, laterally compressed and enfolding a ray cypsela; receptacle chaffy only near the margin; plants glandular-aromatic .. ***Madia***

            9b. Phyllaries in 1 or more series, flat, not enfolding a ray cypsela; receptacle chaffy throughout; plants variously aromatic or not

                10a. Capitula with 20–40 ray flowers; phyllaries dry and scarious-margined .... ***Anthemis***

                10b. Capitula with 6–21 ray flowers; phyllaries herbaceous or foliaceous, commonly without scarious margins

                    11a. Cypselas compressed; chaff of receptacle flat or nearly so, not enfolding the disk flowers; receptacle convex to low-conic .... ***Coreopsis***

                    11b. Cypselas quadrangular; chaff of receptacle partially enfolding the disk flowers; receptacle conic to columnar .... ***Rudbeckia***

    7b. Leaves opposite or whorled, except sometimes the upper, which are alternate

        12a. Plants aquatic, with flaccid stems; leaves whorled, finely dissected into narrow, limp segments; capitulescence commonly a solitary capitulum; cypselas terete .... ***Bidens***

        12b. Plants terrestrial or of wetlands, with erect stems; leaves opposite, entire or divided into broad, firmer segments; capitulescence commonly composed of 2 or more capitula; cypselas compressed, quadrangular, or triangular

            13a. Ray flowers carpellate and fertile, becoming chartaceous in fruit and persistent on the quadrangular or triangular cypsela .... ***Heliopsis***

            13b. Ray flowers neutral, or carpellate in *Madia*, deciduous at or before maturity of the usually compressed cypsela

                14a. Phyllaries in 1 series, laterally compressed and enfolding a ray cypsela; receptacle chaffy only near the margin; plants glandular-aromatic .... ***Madia***

                14b. Phyllaries in 1 or more series, flat, not enfolding a ray cypsela; receptacle chaffy throughout; plants variously aromatic or not

                    15a. Phyllaries more or less monomorphic; chaff of the receptacle partially enfolding the disk flowers; cypselas compressed at right angles to the phyllaries .... ***Helianthus***

                    15b. Phyllaries dimorphic, in 2 series—the outer series larger and more foliaceous than the inner; chaff flat or nearly so, not enfolding the disk flowers; cypselas compressed parallel to the phyllaries

                        16a. Pappus of 2–6 retrorsely barbed awns (rarely antrorsely barbed or sometimes obsolete in *Bidens discoidea*); cypsela not or scarcely wing-margined .... ***Bidens***

16b. Pappus of 2 short teeth or none; cypsela wing-margined ....
.... *Coreopsis*

Group 6

1a. Pappus composed of capillary bristles (also with an additional series of minute, slender scales in some *Erigeron*)

  2a. Plants subdioecious, each capitulum composed almost entirely of unisexual flowers; stems scaly-bracteate; leaves basal, palmately lobed .... *Petasites*

  2b. Plants polygamous, each capitulum with the outer [ray] flowers unisexual [carpellate] and the inner [disk] flowers bisexual; stems with leaves; leaves various, but neither basal nor palmately lobed

    3a. Plants annual, with a taproot; rays up to 3.0 mm long

      4a. Involucre 3.0–4.0 mm long; capitula on bracteate branches; plants flowering spring–summer; style appendages acute to, more commonly, obtuse; leaves not fleshy; plants not of saline habitats .... *Conyza*

      4b. Involucre 5.0–8.0 mm long; capitula on leafy-bracted branches; plants flowering late summer and fall; style appendages acute to acuminate at the apex; leaves somewhat fleshy; plants of saline sand and mud ....
.... *Symphyotrichum*

    3b. Plants annual to perennial; rays 2.0–20.0 mm long

      5a. Phyllaries scarcely to fully herbaceous, but not foliaceous or chartaceous-based with a green tip; style appendages lanceolate or wider, up to 0.3 mm long; plants mostly flowering in June to July (sometimes later) .... *Erigeron*

      5b. Phyllaries either foliaceous or, more commonly, with a chartaceous base and an evident green tip; style appendages lanceolate or narrower, commonly more than 0.3 mm long; plants mostly flowering August to September

        6a. Rays 2.0–3.0 mm long, numbering 7–14 per capitulum; capitulescence elongate, slender, resembling a thyrse .... *Solidago*

        6b. Rays 3.0–20.0 mm long, numbering 9–100 per capitulum; capitulescence generally not resembling a narrow, elongate thyrse

          7a. Pappus in 2 series—an outer ring of short [less than 1.0 mm] bristles and an inner series of long bristles

            8a. Inner [longer] series of pappus bristles tapering to the apex; leaves rigid, linear, 1-nerved, 1.2–4.0 mm wide; rays commonly light purple .... *Ionactis*

            8b. Inner series of pappus bristles clavate-thickened at the apex; leaves herbaceous, more or less narrow-elliptic, several-nerved, 7.0–35.0 mm wide; rays white .... *Doellingeria*

          7b. Pappus in 1 series of long bristles

            9a. Blade of the basal and lower stem leaves evidently cordate and borne on a petiole

              10a. Capitulescence resembling a corymb, usually flat- or round-topped; outer phyllaries usually less than 2.5 times as long as wide; rays white or purple; cypselas only slightly compressed
.... *Aster*

              10b. Capitulescence resembling a panicle, usually elongate and taller than wide; outer phyllaries usually more than 3.0 times

as long as wide; rays blue or purple; cypselas compressed ....
.... *Symphyotrichum*
9b. None of the leaf blades both cordate and borne on a petiole
    11a. Ray flowers mostly 3–8; cypselas densely silky-pubescent;
    phyllaries cartilaginous, with green tips; pappus light brown to
    red-brown .... *Sericocarpus*
    11b. Ray flowers mostly 8–100; cypselas glabrous to pubescent,
    but not densely silky-pubescent; phyllaries not cartilaginous;
    pappus commonly white
        12a. Ovaries and cypselas glandular; phyllaries with scarious
        tips .... *Oclemena*
        12b. Ovaries and cypselas eglandular; phyllaries with
        herbaceous tips
            13a. Phyllaries oblong, rounded to obtuse at the apex;
            leaves usually scabrous on the adaxial surface, rugose-
            veiny on the abaxial surface .... *Aster*
            13b. Phyllaries linear to oblong, obtuse to attenuate at the
            apex; leaves glabrous to pubescent on the adaxial
            surface, not rugose-veiny on the abaxial surface ....
            .... *Symphyotrichum*
1b. Pappus composed entirely of scales, awns, or a short crown, or absent
  14a. Leaves opposite; capitula with 3–6 [commonly 5] ray flowers .... *Galinsoga*
  14b. Leaves alternate or all basal; capitula with 4–many ray flowers
    15a. Receptacle densely bristly or setose; pappus composed of 6–10 conspicuously
    awned scales .... *Gaillardia*
    15b. Receptacle not densely bristly, either with chaff or naked; pappus various, but
    not composed of awned scales
      16a. Receptacle chaffy, at least toward the middle
        17a. Phyllaries dry and scarious-margined, appressed; leaves pinnately
        divided (except *Achillea ptarmica* with simple leaves); chaff without a
        spine-like tip
          18a. Rays 5.0–11.0 mm long, numbering 10–16 per capitulum; disk
          5.0–10.0 mm wide; capitulescence not resembling a corymb, the
          capitula located at the tips of branches .... *Anthemis*
          18b. Rays 2.0–5.0 mm long, mostly numbering 4–10 per capitulum
          (sometimes more in cultivated forms of *Achillea ptarmica*); disk
          2.0–8.0 mm wide; capitulescence resembling a corymb .... *Achillea*
        17b. Phyllaries herbaceous, without a scarious margin, spreading to reflexed;
        leaves entire; chaff with a stout, spinescent tip .... *Echinacea*
      16b. Receptacle without chaff
        19a. Plants with basal leaves, scapose stems, and solitary capitula .... *Bellis*
        19b. Plants with leafy stems and commonly more than 1 capitulum
          20a. Leaves simple, with entire margins; style branches with a short,
          lanceolate, externally pubescent appendage; pappus of the disk
          flowers composed of several small scales and 2 or 4 longer bristles .....
          .... *Boltonia*

20b. Leaves simple and with toothed margins or pinnately divided; style branches truncate and penicillate; pappus either a short crown or absent

21a. Leaves twice-pinnatifid, with linear to filiform ultimate segments; receptacle convex, rounded, or pointed .... *Matricaria*

21b. Leaves crenate to once- or sometimes twice-pinnatifid, the teeth or ultimate segments broader; receptacle flat or low-convex

22a. Capitula solitary at the tips of branches; disk 1.0–2.0 cm wide; rays 1.0–2.0 cm long .... *Leucanthemum*

22b. Capitula several to many in a capitulescence, resembling a corymb; disk 0.4–0.9 cm wide; rays up to 0.8 cm long

23a. Rays commonly shorter than 4.0 mm .... *Chrysanthemum*

23b. Rays 4.0–8.0 mm long .... *Tanacetum*

Achillea
(2 species)

Our species have white, or sometimes pink, flowers.

KEY TO THE SPECIES

1a. Leaves shallowly serrate to subentire; rays 4.0–5.0 mm long, 5–15 [commonly 8–10] per capitulum .... *A. ptarmica*

1b. Leaves pinnately dissected; rays 1.0–4.0 mm long, 4–6 [commonly 5] per capitulum ......
.... *A. millefolium*

Achillea millefolium L.

synonym(s): <*A. m.* var. *borealis*> = *Achillea borealis* Bong., *Achillea millefolium* L. var. *nigrescens* E. Mey. *sensu* Gleason and Cronquist (1991); <*A. m.* var. *occidentalis*> = *Achillea lanulosa* Nutt.
common name(s): common yarrow, milfoil
habit: erect, aromatic, rhizomatous, perennial herb
range: Eurasia; naturalized to Newfoundland, s. to FL, w. to Mexico and CA, n. to AK
state frequency: common
habitat: fields, roadsides
key to the subspecies:

1a. Phyllaries with conspicuous, dark brown to black margins; rays 2.5–4.0 mm long; stems with 4–9 leaves .... *A. m.* var. *borealis* (Bong.) Farw.

1b. Phyllaries with pale brown margins (sometimes the uppermost with dark margins); rays 1.0–2.5 mm long; stems with mostly 5–20 leaves

2a. Stems glabrate to arachnoid-pubescent; capitulescence flat-topped or nearly so, 6.0–30.0 cm wide; leaves flat, the relatively broader segments all aligned nearly parallel to each other .... *A. m.* var. *millefolium*

2b. Stems tomentose; capitulescence convex-topped, 2.0–10.0 cm wide; leaves more 3-dimensional, the relatively narrower segments aligned in numerous planes ....
.... *A. m.* var. *occidentalis* DC.

notes: *Achillea millefolium* var. *borealis* is ranked SH by the Maine Natural Areas Program. *Achillea m.* var. *millefolium* is noted to have dark-margined, upper phyllaries when in exposed habitats, thus making its identification and that of *A. m.* var. *borealis* sometimes difficult.

Achillea ptarmica L.
synonym(s): —
common name(s): sneezeweed
habit: erect, rhizomatous, perennial herb
range: Eurasia; naturalized to Labrador, s. to NY, w. to MO, n. to MN and Ontario
state frequency: uncommon
habitat: roadsides, fields, waste places, shores
notes: Cultivated forms often have numerous ray flowers in each capitula.

Ageratina
(1 species)

Ageratina altissima (L.) King & H. E. Robins.
synonym(s): *Eupatorium rugosum* Houtt.
common name(s): white snakeroot
habit: erect, simple, perennial herb
range: Nova Scotia, s. to GA, w. to TX, n. to Saskatchewan
state frequency: uncommon
habitat: rich, moist woods, borders, thickets, rocky places
notes: —

Ambrosia
(3 species)

Ragweeds have unisexual capitula. The staminate ones are borne in a capitulescence resembling a spike or raceme. The carpellate capitula are found in the axils of leaves or bracts below the staminate capitula.

KEY TO THE SPECIES

1a. Plants annual, 0.5–5.0 m high; leaves entire to palmately lobed, opposite throughout; receptacle without chaff; staminate involucre with 3 evident nerves on one side; carpellate involucre 5.0–10.0 mm long .... *A. trifida*
1b. Plants annual or perennial, 0.2–1.0(–2.5) m high; leaves once- or twice-pinnatifid, usually opposite below and alternate above; receptacle with chaff; staminate involucre inconspicuously nerved; carpellate involucre mostly 3.0–5.0 mm long
    2a. Plants usually perennial from a creeping rootstock; leaves relatively thicker; carpellate involucre with *ca.* 4 tubercles near the apex, these sometimes inconspicuous .... *A. coronopifolia*
    2b. Plants annual; leaves relatively thinner; carpellate involucre with 4–7 sharp spines near or above the middle .... *A. artemisiifolia*

Ambrosia artemisiifolia L.
synonym(s): —
common name(s): common ragweed
habit: simple or branched, annual herb
range: Newfoundland, s. to FL, w. to TX and Mexico, n. to British Columbia
state frequency: common
habitat: waste places, roadsides, fields
key to the subspecies:
    1a. Staminate involucre 3.0–7.0 mm wide; leaves simple to once-pinnatifid (rarely twice-pinnatifid) .... *A. a.* var. *artemisiifolia*

1b. Staminate involucre 1.5–5.0 mm wide; leaves 2- or 3-times pinnatifid
    2a. Staminate involucre 1.5–2.5 mm wide .... *A. a.* var. *paniculata* (Michx.) Blank.
    2b. Staminate involucre 2.5–5.0 mm wide .... *A. a.* var. *elatior* (L.) Descourtils
notes: —

Ambrosia coronopifolia Torr. & Gray
synonym(s): *Ambrosia psilostachya* DC. var. *coronopifolia* (Torr. & Gray) Farw.
common name(s): western ragweed
habit: erect, perennial herb
range: Nova Scotia, s. to LA, w. to TX and Mexico, n. to MT and Saskatchewan
state frequency: very rare
habitat: dry, sandy waste places
notes: —

Ambrosia trifida L.
synonym(s): —
common name(s): giant ragweed, great ragweed
habit: erect, simple or branched, robust, annual herb
range: Quebec, s. to FL, w. to AZ and Mexico, n. to British Columbia
state frequency: uncommon
habitat: moist waste places
notes: —

Anaphalis
(1 species)

Herbs with the aspect of *Pseudognaphalium*.

Anaphalis margaritacea (L.) Benth. & Hook. f.
synonym(s): *Anaphalis margaritacea* (L.) Benth. & Hook. f. var. *angustior* (Miq.) Nakai, *Anaphalis margaritacea* (L.) Benth. & Hook. f. var. *intercedens* Hara, *Anaphalis margaritacea* (L.) Benth. & Hook. f. var. *subalpina* Gray
common name(s): pearly everlasting
habit: erect, rhizomatous, perennial herb
range: Newfoundland, s. to VA, w. to CA, n. to AK
state frequency: common
habitat: dry woods, open areas, roadsides
notes: —

Antennaria
(4 species)

Pussytoes are dioecious plants, with separate carpellate and staminate individuals. Many of our species are apomictic, and the often smaller, staminate plants are missing from a given population or are unknown altogether. Because the staminate plants of a species often have had comparatively less study, fewer characters exist for them, and they are more difficult to assign to species. Therefore, collections should include carpellate plants.

KEY TO THE SPECIES
1a. Rosette leaves small, 0.2–2.1 cm wide, 1-veined or sometimes with 2 additional, evanescent, lateral veins

2a. Middle and upper stem leaves tipped by a flat or involute-margined, scarious appendage
    3a. New rosette leaves bright green and promptly glabrous on the adaxial surface .... ***A. howellii***
    3b. New rosette leaves white or gray-green and tomentose on the adaxial surface .... ***A. neglecta***
2b. Middle and upper stem leaves blunt- to aristate-tipped, only the leaves of the capitulescence with a scarious appendage .... ***A. howellii***
1b. Rosette leaves larger, 0.7–5.5 cm wide, with 3 or 5 prominent veins
    4a. Carpellate involucre 5.0–7.0 mm tall; central corollas 2.5–4.3 mm long; cypselas 1.0–1.5 mm long; mature pappus 4.0–5.5 mm long .... ***A. plantaginifolia***
    4b. Carpellate involucre (6.0–)7.0–10.0 mm tall; central corollas 4.5–7.0 mm long; cypselas 1.3–2.2 mm long; mature pappus 6.0–9.0 mm long .... ***A. parlinii***

Antennaria howellii Greene
synonym(s): <*A. h.* ssp. *canadensis*> = *Antennaria canadensis* Greene; <*A. h.* ssp. *neodioica*> = *Antennaria neodioica* Greene, *Antennaria neodioica* Greene var. *attenuata* Fern., *Antennaria neodioica* Greene var. *chlorophylla* Fern., *Antennaria neodioica* Greene var. *grandis* Fern., *Antennaria rupicola* Fern.; <*A. h.* ssp. *petaloidea*> = *Antennaria petaloidea* (Fern.) Fern., *Antennaria petaloidea* (Fern.) Fern. var. *scariosa* Fern., *A. petaloidea* (Fern.) Fern. var. *subcorymbosa* (Fern.) Fern.
common name(s): —
habit: stoloniferous, perennial herb
range: Newfoundland, s. to VA, w. to IN, n. to MN and Manitoba
state frequency: common
habitat: open woods and fields
key to the subspecies:
    1a. Middle and upper stem leaves tipped by a flat or involute-margined, scarious appendage; new rosette leaves bright green and promptly glabrous on the adaxial surface .... *A. h.* ssp. *canadensis* (Greene) Bayer
    1b. Middle and upper stem leaves blunt- to aristate-tipped, only the leaves of the capitulescence with a scarious appendage; new rosette leaves white or gray-green and tomentose on the adaxial surface or sometimes bright green and promptly glabrous in forms of *A. howellii* var. *neodioica*
        2a. Stolons and basal offshoots short, leafy, terminated by rosettes; rosette leaves tending to have defined petioles .... *A. h.* ssp. *neodioica* (Greene) Bayer
        2b. Stolons elongate, cord-like, with few, small leaves, only tardily developing terminal rosettes; rosette leaves with ill defined petioles .... *A. h.* ssp. *petaloidea* (Fern.) Bayer
notes: Staminate plants are very rare except in certain forms of *Antennaria howellii* ssp. *neodioica*.

Antennaria neglecta Greene
synonym(s): —
common name(s): field pussytoes
habit: stoloniferous, perennial herb
range: Nova Scotia, s. to VA, w. to OK, n. to Northwest Territories
state frequency: common
habitat: open woods and fields
notes: Staminate plants are relatively common.

Antennaria parlinii Fern.
synonym(s): <*A. p.* ssp. *fallax*> = *Antennaria brainerdii* Greene, *Antennaria fallax* Greene, *Antennaria munda* Fern.; <*A. p.* ssp. *parlinii*> = *Antennaria parlinii* Fern. var. *arnoglossa* (Greene) Fern.
common name(s): Parlin's pussytoes
habit: larger-leaved, stoloniferous, perennial herb
range: Nova Scotia, s. to NC, w. to TX, n. to MN and Ontario
state frequency: occasional
habitat: open woods and fields
key to the subspecies:
    1a. New rosette leaves bright green and promptly glabrous or glabrate on the adaxial surface; stem and phyllaries often purple .... *A. p.* ssp. *parlinii*
    1b. New rosette leaves gray-green and tomentose on the adaxial surface; stem and phyllaries rarely purple .... *A. p.* ssp. *fallax* (Greene) Bayer & Stebbins
notes: Staminate plants are relatively uncommon.

Antennaria plantaginifolia (L.) Richards.
synonym(s): *Antennaria plantaginifolia* (L.) Richards. var. *petiolata* (Fern.) Heller
common name(s): plantain-leaved pussytoes
habit: larger-leaved, stoloniferous, perennial herb
range: New Brunswick, s. to GA, w. to AL, n. to MN
state frequency: uncommon
habitat: dry, open woods and fields
notes: Staminate plants are relatively common.

Anthemis
(3 species)

A group of species superficially similar to, and often hard to separate from, *Matricaria* in the field.

KEY TO THE SPECIES
1a. Ray flowers yellow, 20–30 per capitulum; leaves deeply once-pinnately lobed; disk 12.0–18.0 mm wide .... *A. tinctoria*
1b. Ray flowers white, 6–16 per capitulum; leaves finely twice- to thrice-pinnately divided; disk 5.0–12.0 mm wide
    2a. Ray flowers staminate or neutral; receptacle chaffy only near the middle; cypselas glandular-tuberculate; plants ill scented .... *A. cotula*
    2b. Ray flowers carpellate; receptacle chaffy throughout; cypselas not tuberculate; plants not ill scented .... *A. arvensis*

Anthemis arvensis L.
synonym(s): —
common name(s): corn chamomile
habit: erect, annual herb
range: Europe; naturalized to Newfoundland, s. to GA, w. to MO, n. to British Columbia
state frequency: uncommon
habitat: roadsides, waste places, fields
key to the subspecies:
    1a. Chaff shorter than the disk flowers .... *A. a.* var. *agrestis* (Wallr.) DC.
    1b. Chaff exceeding the disk flowers .... *A. a.* var. *arvensis*
notes: —

Anthemis cotula L.
synonym(s): —
common name(s): stinking chamomile, dogfennel
habit: erect, annual herb
range: Europe; naturalized throughout much of North America
state frequency: occasional
habitat: roadsides, waste places, fields
notes: —

Anthemis tinctoria L.
synonym(s): —
common name(s): yellow chamomile
habit: branched or simple, short-lived, perennial herb
range: Europe; naturalized to Newfoundland, s. to NJ, w. to MN, n. to British Columbia
state frequency: uncommon
habitat: roadsides, fields, waste places
notes: —

Arctium
(4 species)

An easily recognized genus of European weeds with bur-like involucres.

KEY TO THE SPECIES

1a. Capitulescence resembling a corymb; capitula with a peduncle mostly 3.0–10.0 cm long; petioles strongly angled; leaf blades broad-ovate to orbicular-ovate, rounded at the apex
  2a. Involucre 1.5–2.5 cm wide, its arachnoid-tomentose phyllaries not surpassing the flowers; petioles hollow .... *A. tomentosum*
  2b. Involucre 2.5–4.0 cm wide, its glabrous to sparsely arachnoid phyllaries often surpassing the flowers; petioles solid .... *A. lappa*
1b. Capitulescence resembling a raceme; capitula sessile or with short peduncles mostly shorter than 3.0 cm; petioles weakly angled; leaf blades narrow- to broad-ovate, acute at the apex
  3a. Involucre 2.5–3.5 cm wide, its phyllaries equaling or exceeding the flowers; cypselas (5.5–)6.0–7.5 mm long .... *A. vulgare*
  3b. Involucre 1.5–2.5 cm wide, its phyllaries shorter than the flowers; cypselas 5.0–6.0 mm long .... *A. minus*

Arctium lappa L.
synonym(s): —
common name(s): great burdock
habit: large/tall, branched, biennial herb
range: Eurasia; naturalized to New Brunswick, s. to PA, w. to IL, n. to British Columbia
state frequency: uncommon
habitat: roadsides, waste places
notes: Known to hybridize with *Arctium minus*.

Arctium minus Bernh.
synonym(s): *Lappa minor* Hill
common name(s): common burdock
habit: medium to large, branched, biennial herb

range: Eurasia; naturalized to Newfoundland, s. to VA, w. to CA, n. to British Columbia
state frequency: common
habitat: disturbed soil of waste places and roadsides
notes: Known to hybridize with *Arctium lappa* and *A. tomentosum.*

Arctium tomentosum P. Mill.

synonym(s): —
common name(s): cotton burdock
habit: medium-sized, branched, biennial herb
range: Eurasia; naturalized to Quebec, s. to PA, w. to MO, n. to Alberta
state frequency: very rare
habitat: waste places
notes: Known to hybridize with *Arctium minus.*

Arctium vulgare (Hill) Evans

synonym(s): *Arctium minus* Bernh. ssp. *nemorosum* (Lej. & Court.) Syme, *Arctium nemorosum* Lej.
& Court.
common name(s): —
habit: medium to large, biennial herb
range: Europe; naturalized to Newfoundland, s. to VA, w. to CA, n. to British Columbia
state frequency: uncommon
habitat: waste places
notes: Intermediate between *Arctium lappa* and *A. minus,* and perhaps produced by hybridization
between them.

Arnica
(1 species)

Superficially similar to *Helianthus,* the sunflowers. *Arnica* has carpellate ray flowers and lacks chaff,
whereas *Helianthus* has neutral ray flowers and possesses chaff on the receptacle.

Arnica lanceolata Nutt.

synonym(s): *Arnica mollis* Hook. var. *petiolaris* Fern.
common name(s): New England arnica, hairy arnica
habit: erect, perennial herb
range: Quebec, s. to mountains of ME and NH, w. to CO and CA, n. to British Columbia
state frequency: rare
habitat: rocky shores, wet ledges, seepy areas
notes: This species is ranked S2 by the Maine Natural Areas Program.

Arnoseris
(1 species)

A monotypic genus, containing only the listed species, adventive in our part of the world.

Arnoseris minima (L.) Schweig. & Koerte

synonym(s): —
common name(s): dwarf nipplewort, lamb succory
habit: erect, branched, annual herb
range: Europe; naturalized to Nova Scotia, s. to PA, w. to OH, n. to MI
state frequency: very rare
habitat: fields, roadsides, waste places
notes: —

Artemisia
(9 species)

When assessing leaf division, use the lower or middle leaves.

KEY TO THE SPECIES

1a. Central flowers of disk functionally unisexual [staminate], the gynoecium of these flowers with an abortive ovary and undivided style; leaves 2- or 3-times pinnatifid into linear or linear-filiform segments up to 2.0 mm wide .... *A. campestris*
1b. Central flowers of disk bisexual, the gynoecium of these flowers with a fertile ovary and divided style; leaves various
  2a. Receptacle densely hairy; leaves 2- or 3-times pinnatifid into approximately oblong segments 1.5–4.0 mm wide .... *A. absinthium*
  2b. Receptacle without hairs; leaves various
    3a. Plants perennial, arising from a rhizome or woody caudex; leaves tomentose on one or both surfaces, sometimes sparsely so; phyllaries tomentose, at least sparsely so
      4a. Plants shrubby or suffrutescent; leaves 2- or 3-times pinnatifid into linear or filiform segments 0.5–1.5 mm wide; involucre 2.0–3.5 mm tall
        5a. Leaves 3.0–6.0 cm long, thinly tomentose on the abaxial surface, glabrous on the adaxial surface; plants shrubby, 0.5–2.0 m tall .... *A. abrotanum*
        5b. Leaves 1.0–3.0 cm long, tomentose on both surfaces; plants suffrutescent, 0.4–1.0 m tall .... *A. pontica*
      4b. Plants herbaceous [dieing back to the ground each year]; leaves entire or 1- or 2-times pinnatifid into broader lobes wider than 1.5 mm; involucre 2.5–7.5 mm tall
        6a. Leaves entire or toothed, the lower sometimes shallowly lobed, but the primary lobes without additional teeth or lobes; involucre 2.5–3.5 mm tall
        .... *A. ludoviciana*
        6b. Leaves 1- or 2-times pinnatifid, the primary lobes either entire or with additional teeth or lobes; involucre 3.5–7.5 mm tall
          7a. Involucre 3.5–4.5 mm tall; disk corollas 2.0–2.8 mm long; stem below the capitulescence glabrous or nearly so .... *A. vulgaris*
          7b. Involucre 6.0–7.5 mm tall; disk corollas 3.2–4.0 mm long; stem tomentose .... *A. stelleriana*
    3b. Plants annual or biennial, arising from a taproot; leaves glabrous on both surfaces; phyllaries glabrous
      8a. Capitulescence dense, resembling a spike, with inconspicuous peduncles and erect or ascending capitula; involucre 2.0–3.0 mm tall; plants inodorous or nearly so; leaves 1- or 2-times pinnatifid .... *A. biennis*
      8b. Capitulescence open, resembling a panicle, with conspicuous peduncles and nodding capitula; involucre 1.0–2.0 mm tall; plants sweet scented; leaves 2- or 3-times pinnatifid .... *A. annua*

Artemisia abrotanum L.
synonym(s): —
common name(s): southernwood
habit: shrub to 2.0 m

range: Europe; naturalized over much of the United States and northeastern Canada
state frequency: very rare
habitat: roadsides, waste places
notes: —

Artemisia absinthium L.

synonym(s): —
common name(s): common wormwood, absinthium
habit: coarse, perennial herb
range: Europe; naturalized to Newfoundland, s. to NC, w. to OH, n. to British Columbia
state frequency: rare
habitat: roadsides, fields, waste places
notes: —

Artemisia annua L.

synonym(s): —
common name(s): annual wormwood, sweet wormwood
habit: sweet scented, annual herb
range: Eurasia; naturalized to New Brunswick, s. to VA, w. to AL and AR, n. to Ontario
state frequency: rare
habitat: waste places, roadsides, fields
notes: —

Artemisia biennis Willd.

synonym(s): —
common name(s): biennial wormwood
habit: coarse, simple, annual or biennial herb
range: mid- to western North America; naturalized in the east
state frequency: uncommon
habitat: waste places, stream banks, open areas
notes: —

Artemisia campestris L.

synonym(s): <*A. c.* ssp. *borealis*> = *Artemisia canadensis* Michx.; <*A. c.* ssp. *caudata*> = *Artemisia caudata* Michx.
common name(s): sagewort wormwood, Canadian wormwood
habit: slightly scented, biennial or perennial herb
range: circumboreal; s. to FL, w. to AZ and CA
state frequency: rare
habitat: open, sandy soils, rocks, river shores
key to the subspecies:
   1a. Involucre 2.0–3.0 x 2.0–3.0 mm; capitula with 14–25 flowers; disk corollas 1.4–2.0 mm long; plants biennial (rarely perennial), often with a single stem, of open, often coastal, sandy soils in southern Maine .... *A. c.* ssp. *caudata* (Michx.) Hall & Clements
   1b. Involucre 3.0–4.0 x 3.5–6.0 mm; capitula with 23–45 flowers; disk corollas 2.2–3.0 mm long; plants perennial (rarely biennial), often with many stems, of circumneutral cliffs and river shores in northern Maine .... *A. c.* ssp. *borealis* (Pallas) Hall & Clements
notes: *Artemisia campestris* ssp. *borealis* is ranked SH by the Maine Natural Areas Program.

Artemisia ludoviciana Nutt.

synonym(s): *Artemisia ludoviciana* Nutt. var. *brittonii* (Rydb.) Fern., *Artemisia ludoviciana* Nutt. var. *latifolia* (Bess.) Torr. & Gray
common name(s): white sage, western mugwort

habit: rhizomatous, perennial herb
range: Manitoba and MN, s. to IL and KS, w. to NM, n. to British Columbia; naturalized to Quebec, s. to CT
state frequency: rare
habitat: dry waste places
notes: —

Artemisia pontica L.

synonym(s): —
common name(s): Roman wormwood
habit: rhizomatous, suffrutescent, perennial herb
range: Europe; naturalized to Quebec, s. to DE, w. to OH, n. to Manitoba
state frequency: rare
habitat: dry, open waste places
notes: —

Artemisia stelleriana Bess.

synonym(s): —
common name(s): dusty miller, beach wormwood
habit: rhizomatous, perennial herb
range: Asia; naturalized to New Brunswick, s. to VA; also NY, w. to MN
state frequency: uncommon
habitat: sandy, coastal beaches and dunes
notes: —

Artemisia vulgaris L.

synonym(s): —
common name(s): common mugwort
habit: rhizomatous, simple or branched, perennial herb
range: Eurasia; naturalized to Newfoundland, s. to GA, w. to MN, n. to Ontario; also Manitoba to British Columbia
state frequency: occasional
habitat: fields, roadsides, waste places, thickets
key to the subspecies:
    1a. Ultimate leaf segments acuminate at the apex; leaves relatively thicker and more deeply divided .... *A. v.* var. *vulgaris*
    1b. Ultimate leaf segments rounded to acute at the apex; leaves relatively thinner and less deeply divided .... *A. v.* var. *latiloba* Ledeb.
notes: —

Aster
(4 species)

KEY TO THE SPECIES
1a. Blade of basal and lower stem leaves evidently cordate and borne on a petiole
    2a. Branches of the capitulescence glandular; rays purple or at least purple-tinged; plants widespread .... ***A. macrophyllus***
    2b. Branches of the capitulescence without glands; rays white; plants of southern Maine
        3a. Plants with vegetative tufts of leaves produced separately from the flowering stem; involucre slender-cylindric, mostly 8.0–9.0 mm tall; leaves at the base of the stem firmer, often with rectangle-shaped sinuses .... ***A. schreberi***

3b. Plants without separate vegetative tufts of leaves; involucre campanulate, mostly 6.0–8.0 mm tall; leaves at the base of the stem thin, with V-shaped sinuses .... *A. divaricatus*

1b. None of the leaf blades both cordate and borne on a petiole .... *A. radula*

Aster divaricatus (Nutt.) Torr. & Gray
synonym(s): *Eurybia divaricata* (L.) Nesom
common name(s): common white heart-leaved aster, white wood aster
habit: rhizomatous, perennial herb
range: ME, s. to mountains of GA, w. to AL, n. to Ontario
state frequency: very rare
habitat: dry woods and clearings
notes: This species is ranked S1 by the Maine Natural Areas Program.

Aster macrophyllus L.
synonym(s): *Aster macrophyllus* L. var. *apricensis* Burgess, *Aster macrophyllus* L. var. *excelsior* Burgess, *Aster macrophyllus* L. var. *ianthinus* (Burgess) Fern., *Aster macrophyllus* L. var. *pinguifolius* Burgess, *Aster macrophyllus* L. var. *sejunctus* Burgess, *Aster macrophyllus* L. var. *velutinus* Burgess, *Eurybia macrophylla* (L.) Cass.
common name(s): big-leaved aster, large-leaved aster
habit: rhizomatous, perennial herb
range: New Brunswick and Nova Scotia, s. to mountains of GA, w. to IL, n. to MN and Manitoba
state frequency: common
habitat: woods, thickets, clearings
notes: —

Aster radula Sol. *ex* Ait.
synonym(s): *Aster radula* Ait. var. *strictus* (Pursh) Gray, *Eurybia radula* (Sol. *ex* Ait.) Nesom
common name(s): low rough aster, file-blade aster
habit: rhizomatous, perennial herb
range: Labrador, s. to VA, w. to KY, n. to Quebec
state frequency: uncommon
habitat: moist to wet soils of wetlands, shores, and forests
notes: —

Aster schreberi Nees
synonym(s): *Eurybia schreberi* (Nees) Nees
common name(s): Schreber's aster
habit: rhizomatous, perennial herb
range: ME, s. to VA, w. to IL, n. to WI; also AL and TN
state frequency: very rare
habitat: damp woods, thickets
notes: This species is ranked SX by the Maine Natural Areas Program.

Bellis
(1 species)

The stems of this genus resemble a scape, bearing very reduced leaves, if any.

Bellis perennis L.
synonym(s): —
common name(s): English daisy

habit: perennial herb
range: Europe; escaped from cultivation to much of the northern United States
state frequency: rare
habitat: lawns, waste places
notes: —

Bidens
(10 species)

In some species, the outer cypselas of a capitulum will have smaller dimensions and shorter awns than cypselas from the inner portion of the disk. Cypsela measurements in the key do not include the awns.

KEY TO THE SPECIES

1a. Plants aquatic, with flaccid stems; leaves whorled, finely dissected into narrow, limp segments; capitulescence commonly a solitary capitulum; cypselas terete .... **B. beckii**
1b. Plants terrestrial or of wetlands, with erect stems; leaves opposite, entire or divided into broad, firmer segments; capitulescence commonly composed of 2 or more capitula; cypselas compressed, quadrangular, or triangular
  2a. Leaves (except the uppermost) 1- to 3-times divided into distinct leaflets, at least the terminal leaflet with a petiolule
    3a. Capitula with ray flowers; rays 1.0–2.5 cm long; leaflets coarsely serrate to pinnately divided .... **B. aristosa**
    3b. Capitula with or without rays; rays, if present, up to 0.5 cm long; leaflets serrate
      4a. Capitula with 3–5 outer [larger] phyllaries; phyllaries glabrous-margined, at least near the base; cypselas 3.0–6.5 mm long, with awns 0.0–2.4 mm long
        .... **B. discoidea**
      4b. Capitula with 5–21 outer phyllaries; phyllaries ciliate-margined, at least near the base; cypselas 5.3–11.3 mm long, with awns 2.0–9.5 mm long
        5a. Capitula with 5–10 [averaging 8] outer phyllaries; outer and inner cypselas 5.3–7.0 and 7.0–10.0 mm long, respectively; anthers exserted from the corolla .... **B. frondosa**
        5b. Capitula with 10–21 [averaging 13] outer phyllaries; outer and inner cypselas 6.5–11.3 and 9.0–17.0 mm long, respectively; anthers included within the corolla .... **B. vulgata**
  2b. Leaves simple or lobed into 3–7 segments, the segments broad and without petiolules
    6a. Leaves sessile, or sometimes the lowermost with a short, narrowed base; cypselas either with a convex, cartilaginous apex or with a truncate apex in *B. tripartita*
      7a. Outer phyllaries ascending to erect; disk corollas 3- to 5-lobed, usually 4-lobed; anthers included within the corolla; capitula erect
        8a. Cypselas with a convex, cartilaginous apex; capitula cylindric to campanulate; plants of estuaries .... **B. hyperborea**
        8b. Cypselas with a truncate apex; capitula campanulate to hemispherical; plants of moist to dry soils, only rarely of estuaries .... **B. tripartita**
      7b. Outer phyllaries spreading; disk corollas 5-lobed; anthers slightly exserted from the corolla; capitula nodding in age .... **B. cernua**
    6b. Leaves with distinct petioles 1.0–4.0 cm long; cypselas with a truncate apex

9a. Capitula cylindric to narrow-campanulate, each with 7–30 flowers; plants of estuaries .... *B. eatonii*

9b. Capitula campanulate to hemispherical, each with 30–150 flowers; plants of wet to dry soils, only rarely of estuaries

    10a. Disk corollas usually 4-lobed, yellow; cypselas very flat, smooth, with inconspicuous midribs, and usually with 3 awns; leaves serrate .... *B. tripartita*

    10b. Disk corollas 5-lobed, orange to orange-yellow; cypselas compressed-quadrangular, often tuberculate and sparsely strigose, with evident midribs (at least near the apex), and with 2–6 [usually 4] awns; leaves serrate or sometimes lobed as well .... *B. connata*

Bidens aristosa (Michx.) Britt.

synonym(s): *Bidens aristosa* (Michx.) Britt. var. *fritcheyi* Fern.
common name(s): midwestern tickseed-sunflower
habit: annual or biennial herb
range: ME, s. to VA, w. to LA and TX, n. to MN
state frequency: very rare
habitat: damp, shady, roadsides and waste places
notes: —

Bidens beckii Torr. *ex* Spreng.

synonym(s): *Megalodonta beckii* (Torr. *ex* Spreng.) Greene
common name(s): water-marigold, water beggar ticks
habit: aquatic, perennial herb
range: New Brunswick and Nova Scotia, s. to NJ, w. to MO, n. to Manitoba; also British Columbia, s. to OR
state frequency: uncommon
habitat: ponds, slow streams, shallow, quiet water
notes: *Bidens beckii* can be separated vegetatively from other similar aquatics by its whorled, dichotomously forked leaves that lack marginal spicules.

Bidens cernua L.

synonym(s): *Bidens cernua* L. var. *elliptica* Wieg., *Bidens cernua* L. var. *minima* (Huds.) Pursh
common name(s): nodding beggar ticks, bur-marigold
habit: simple or branched, annual herb
range: Eurasia; naturalized to New Brunswick and Nova Scotia, s. to NC, w. to CA, n. to British Columbia
state frequency: occasional
habitat: low, wet places
notes: —

Bidens connata Muhl. *ex* Willd.

synonym(s): *Bidens connata* Muhl. *ex* Willd. var. *gracilipes* Fern., *Bidens connata* Muhl. *ex* Willd. var. *petiolata* (Nutt.) Farw.
common name(s): purple stem beggar ticks
habit: branched, annual herb
range: New Brunswick, s. to NC, w. to KS, n. to ND
state frequency: uncommon
habitat: wet shores and waste places, swamps
notes: —

Bidens discoidea (Torr. & Gray) Britt.

synonym(s): —
common name(s): few-bracted beggar ticks
habit: slender, simple or branched, annual herb
range: New Brunswick and Nova Scotia, s. to AL, w. to TX, n. to MN and Ontario
state frequency: rare
habitat: swamps, shores, wet places
notes: —

Bidens eatonii Fern.

synonym(s): *Bidens eatonii* Fern. var. *kennebecensis* Fern., *Bidens eatonii* Fern. var. *mutabilis* Fassett
common name(s): New England estuarine beggar ticks, Eaton's bur-marigold
habit: slender, simple or branched, annual herb
range: ME, s. to CT
state frequency: very rare
habitat: estuaries, tidal shores
notes: *Bidens eatonii* is ranked S1 by the Maine Natural Areas Program. It is similar to, and often confused with, *B. hyperborea*. *Bidens eatonii* has petiolate leaves, truncate-apexed cypselas [that average larger than those of *B. hyperborea*], and mostly 3–5 outer phyllaries. *Bidens hyperborea* has sessile leaves, cartilaginous, convex-apexed cypselas, and mostly 4–8 outer phyllaries.

Bidens frondosa L.

synonym(s): —
common name(s): devil's beggar ticks
habit: annual herb
range: Newfoundland, s. to GA, w. to CA, n. to WA and British Columbia
state frequency: common
habitat: damp waste places
notes: —

Bidens hyperborea Greene

synonym(s): <*B. h.* var. *hyperborea*> = *Bidens hyperborea* Greene var. *colpophila* (Fern. & St. John) Fern.; <*B. h.* var. *svensonii*> = *Bidens hyperborea* Greene var. *cathancensis* Fern.
common name(s): estuary beggar ticks, northern estuarine beggar ticks, estuary bur-marigold
habit: simple or branched, erect to depressed, annual herb
range: Quebec, s. to NJ
state frequency: rare
habitat: estuaries
key to the subspecies:
    1a. Outer cypselas 4.0–5.0 mm long; inner cypselas 5.0–7.0 mm long, with marginal awns [the longer pair] 1.8–3.0 mm long .... *B. h.* var. *hyperborea*
    1b. Outer cypselas 5.0–8.5 mm long; inner cypselas 7.5–10.0 mm long, with marginal awns 3.0–5.0 mm long .... *B. h.* var. *svensonii* Fassett
notes: *Bidens hyperborea* is ranked S1/S2 by the Maine Natural Areas Program. This species is similar to *B. eatonii* (*q.v.*).

Bidens tripartita L.

synonym(s): *Bidens comosa* (Gray) Wieg.
common name(s): strawstem beggar ticks
habit: stout, simple or branched, annual herb

range: Eurasia; naturalized to ME, s. to NC, w. to NM, n. to ND
state frequency: rare
habitat: moist thickets and waste places
notes: —

Bidens vulgata Greene
synonym(s): —
common name(s): tall beggar ticks, stick tight
habit: coarse, branched, annual herb
range: New Brunswick and Nova Scotia, s. to NC, w. to CA, n. to British Columbia
state frequency: occasional
habitat: wet to dry waste places
notes: —

Boltonia
(1 species)

A small genus of American herbs with the aspect of *Aster*. The longer pappus bristles, present on the disk flowers, are commonly lacking from the ray flowers.

Boltonia asteroides (L.) L'Hér.
synonym(s): <*B. a.* var. *asteroides*> = *Boltonia asteroides* (L.) L'Hér. var. *glastifolia* (Hill) Fern.;
    <*B. a.* var. *latisquama*> = *Bidens latisquama* Gray
common name(s): boltonia
habit: slender, simple or branched, perennial herb
range: ME, s. to FL, w. to TX, n. to ND
state frequency: very rare
habitat: moist or wet, gravelly shores and sandy thickets
key to the subspecies:
    1a. Capitulescence normally composed of fewer than 25 capitula; disk 6.0–8.0 mm wide; cypselas
        1.5–2.0 mm long; plants averaging smaller, 2.0–7.0 dm tall .... *B. a.* var. *asteroides*
    1b. Capitulescence normally composed of more than 25 capitula; disk 7.0–14.0 mm wide; cypselas
        2.5–3.0 mm long; plants averaging larger, up to 15.0 dm tall ....
        .... *B. a.* var. *latisquama* (Gray) Cronq.
notes: —

Carduus
(2 species)

A genus very similar to *Cirsium*, consistently differing only in its pappus.

KEY TO THE SPECIES
1a. Capitula 1.5–2.5 cm wide; outer phyllaries herbaceous, erect to spreading; plants very
    spiny, with tough stems; corollas *ca.* 18.0 mm long .... *C. acanthoides*
1b. Capitula 1.0–1.3 cm wide; outer phyllaries rigid, mostly erect; plants sparsely spiny,
    with brittle stems; corollas *ca.* 14.0 mm long .... *C. crispus*

Carduus acanthoides L.
synonym(s): —
common name(s): plumeless thistle
habit: spiny, biennial or annual herb
range: Europe; naturalized to Nova Scotia and ME, s. to VA, w. to NE

state frequency: very rare
habitat: roadsides, pastures, waste places
notes: —

Carduus crispus L.
synonym(s): —
common name(s): welted thistle
habit: brittle-stemmed, spiny, biennial herb
range: Eurasia; naturalized to New Brunswick and Nova Scotia, s. to CT, w. to MO, n. to MN and Ontario
state frequency: very rare
habitat: roadsides, waste places, fields
notes: —

Centaurea
(6 species)

Most species of knapweeds are known to have falsely radiate capitula where the marginal flowers of the disk are enlarged [and sterile]. However, the corollas of these marginal flowers are actinomorphic and do not resemble the typical ray flowers of the Asteraceae.

KEY TO THE SPECIES
1a. Leaves (or some of them) pinnatifid, with narrow lobes
    2a. Pappus 3.0–6.0 mm long; involucre 16.0–25.0 mm tall; dark, pectinate, scarious portion of phyllaries 4.0–6.0 mm long .... *C. scabiosa*
    2b. Pappus 3.0 mm long or less; involucre 10.0–13.0 mm tall; dark, pectinate, scarious portion of phyllaries 1.0–2.0 mm long .... *C. maculosa*
1b. Leaves entire or toothed, sometimes the larger with a few lobes
    3a. Plants annual or winter-annual, white-tomentose when young; leaves linear to lanceolate, often entire, up to 1.0 cm wide, often persistently tomentose on the abaxial surface .... *C. cyanus*
    3b. Plants biennial to perennial, glabrous or pubescent, but not white-tomentose; leaves oblanceolate to elliptic, at least the lower toothed, generally wider than 1.0 cm, not tomentose on the abaxial surface
        4a. Phyllaries light brown to dark brown at the apex, the middle and outer entire to irregularly lacerate at the apex, the inner dilated and bifid at the apex ..... *C. jacea*
        4b. Phyllaries black (at least in part) at the apex, at least the middle and outer regularly pectinate at the apex, rarely any of them conspicuously bifid at the apex
            5a. Dark, pectinate, scarious portion of phyllaries (3.0–)4.0–6.0 mm long; involucre wider than tall; marginal disk flowers typically not enlarged .... *C. nigra*
            5b. Dark, pectinate, scarious portion of phyllaries 1.0–3.0 mm long; involucre taller than wide; marginal disk flowers usually enlarged and falsely appearing as ray flowers .... *C. nigrescens*

Centaurea cyanus L.
synonym(s): —
common name(s): bachelor's button, bluebottle, cornflower
habit: slender, branched, annual herb
range: Europe; escaped from cultivation to Newfoundland, s. to VA, w. to NE, n. to British Columbia

state frequency: uncommon
habitat: fields, roadsides, waste places
notes: —

Centaurea jacea L.

synonym(s): —
common name(s): brown knapweed
habit: simple or branched, perennial herb
range: Europe; naturalized to Quebec, s. to VA, w. to IA, n. to British Columbia
state frequency: uncommon
habitat: fields, roadsides, waste places
notes: Known to hybridize with *Centaurea nigra*. These hybrids have falsely radiate capitula, as with
   C. jacea, but have phyllaries that more closely resemble *C. nigra* [described in the key].

Centaurea maculosa Lam.

synonym(s): —
common name(s): spotted knapweed
habit: branched, biennial or perennial herb
range: Europe; naturalized to New Brunswick and Nova Scotia, s. to VA, w. to KS, n. to British
   Columbia
state frequency: uncommon
habitat: fields, roadsides, waste places
notes: —

Centaurea nigra L.

synonym(s): *Centaurea nigra* L. var. *radiata* DC.
common name(s): black knapweed, Spanish buttons
habit: coarse, branched, perennial herb
range: Europe; naturalized to Newfoundland, s. to DE, w. to OH, n. to MI and Ontario; also British
   Columbia
state frequency: occasional
habitat: fields, roadsides, waste places
notes: Known to hybridize with *Centaurea jacea* (*q.v.*).

Centaurea nigrescens Willd.

synonym(s): *Centaurea vochinensis* Bernh. *ex* Reichenb.
common name(s): short-fringed knapweed
habit: coarse, branched, perennial herb
range: Europe; escaped from cultivation to Nova Scotia and ME, s. to VA, w. to MO, n. to Ontario
state frequency: rare
habitat: fields, roadsides
notes: —

Centaurea scabiosa L.

synonym(s): —
common name(s): hard heads
habit: taprooted, perennial herb
range: Europe; naturalized to Quebec, s. to ME, w. to IA, n. to Ontario
state frequency: very rare
habitat: fields, roadsides, waste places
notes: —

Chrysanthemum
(3 species)

A genus often with dimorphic fruits—the cypselas of the ray flowers often have 3 wing-angles whereas the cypselas of the disk flowers are subterete with 5–10 low ribs.

KEY TO THE SPECIES

1a. Rays present, conspicuous, 8.0–15.0 mm long, yellow; disk 10.0–20.0 mm wide; capitula few, terminating the branches
   2a. Leaves toothed to pinnatifid, clasping the stem .... *C. segetum*
   2b. Leaves bipinnatifid, not clasping the stem .... *C. coronarium*
1b. Rays absent or present and shorter than 1.0 cm, white; disk 4.0–7.0 mm wide; capitula numerous, aggregated in a capitulescence resembling a corymb .... *C. balsamita*

Chrysanthemum balsamita L.
synonym(s): —
common name(s): costmary, mint-geranium
habit: coarse, aromatic, perennial herb
range: Eurasia; escaped from cultivation to ME, s. to DE, w. to KS, n. to Ontario
state frequency: uncommon
habitat: roadsides, waste places
notes: —

Chrysanthemum coronarium L.
synonym(s): —
common name(s): garland chrysanthemum
habit: erect, simple or branched, annual herb
range: Europe; escaped from cultivation to New Brunswick and ME, s. to MA, w. and n. to Ontario
state frequency: very rare
habitat: waste places
notes: —

Chrysanthemum segetum L.
synonym(s): —
common name(s): corn chrysanthemum, corn-marigold
habit: branched, annual herb
range: Europe; escaped from cultivation to Newfoundland, s. to FL, w. and n. to Ontario
state frequency: very rare
habitat: fields, roadsides, waste places
notes: —

Cichorium
(1 species)

Weedy plants with biseriate involucres—the inner series of 8–10 phyllaries, the outer series of 5 short and spreading phyllaries.

Cichorium intybus L.
synonym(s): —
common name(s): common chicory, blue sailors
habit: perennial herb

range: Europe; escaped from cultivation to Newfoundland, s. to Bermuda, w. to CA, n. to British
   Columbia
state frequency: occasional
habitat: roadsides, fields, waste places
notes: —

Cirsium
(5 species)

Spiny-leaved plants with white or pink to, more commonly, purple flowers.

KEY TO THE SPECIES
1a. Involucre 1.0–2.0 cm tall; capitula with, or mostly with, unisexual flowers [by abortion];
   plants colonial from deep-seated, creeping roots .... *C. arvense*
1b. Involucre 2.0–5.0 cm tall; capitula with bisexual flowers; plants not colonial
   2a. Upper stem and branches with conspicuous, undulate- and prickly-margined wings
      formed by decurrent leaf bases .... *C. vulgare*
   2b. Neither the stem nor the branches with decurrent wings
      3a. Leaves persistently white-tomentose on the abaxial surface .... *C. discolor*
      3b. Leaves thinly tomentose or arachnoid-pubescent when young, becoming glabrate
         in age
         4a. Outer phyllaries blunt or tipped with a vestigial spinule up to 0.5 mm long .....
            .... *C. muticum*
         4b. Outer phyllaries tipped with a coarse spine 4.0–6.0 mm long ..... *C. pumilum*

Cirsium arvense (L.) Scop.
synonym(s): *Cirsium arvense* (L.) Scop. var. *vestitum* Wimmer & Grab.
common name(s): Canada thistle
habit: perennial herb
range: Eurasia; naturalized to much of the United States and adjacent Canada
state frequency: common
habitat: fields, roadsides, waste places
notes: —

Cirsium discolor (Muhl. *ex* Willd.) Spreng.
synonym(s): —
common name(s): field thistle
habit: robust, branched, biennial or perennial herb
range: ME, s. to GA, w. to KS, n. to MN and Manitoba
state frequency: very rare
habitat: fields, thickets, river banks, waste places
notes: —

Cirsium muticum Michx.
synonym(s): —
common name(s): swamp thistle
habit: coarse, branched, biennial herb
range: Newfoundland, s. to DE and mountains of NC and TN, w. to MO, n. to Saskatchewan
state frequency: occasional
habitat: low woods, swamps, thickets, wet meadows
notes: —

Cirsium pumilum (Nutt.) Spreng.
synonym(s): —
common name(s): pasture thistle, bull thistle
habit: simple or branched, biennial herb
range: ME, s. to NC
state frequency: uncommon
habitat: dry fields, pastures, and open woods
notes: —

Cirsium vulgare (Savi) Ten.
synonym(s): *Cirsium lanceolatum* (L.) Hill
common name(s): common thistle, bull thistle
habit: biennial herb
range: Eurasia; naturalized to Newfoundland, s. to GA, w. to CA, n. to British Columbia
state frequency: common
habitat: pastures, fields, roadsides, waste places
notes: —

Conyza
(1 species)

Most members of this genus lack rays. However, our American species possess small, inconspicuous rays.

Conyza canadensis (L.) Cronq.
synonym(s): *Erigeron canadensis* L.
common name(s): horseweed
habit: erect, mostly simple, annual herb
range: Newfoundland, s. to FL, w. to Mexico and CA, n. to British Columbia
state frequency: common
habitat: waste places, fields
notes: —

Coreopsis
(3 species)

A group of showy composites, all of ours escaped from cultivation.

KEY TO THE SPECIES
1a. Disk corollas 4-lobed; style appendages short and blunt; outer phyllaries *ca.* 2.0 mm long; inner phyllaries 5.0–8.0 mm long; cypselas linear to linear-oblong .... *C. tinctoria*
1b. Disk corollas 5-lobed; style appendages evidently acute; outer phyllaries 5.0–10.0 mm long; inner phyllaries 13.0–30.0 mm long; cypselas oblong to orbicular
    2a. Leaves usually pinnately lobed into linear-filiform to narrow-lanceolate, lateral segments rarely over 5.0 mm wide; stems with leaves nearly to the summit; chaff 6.0–7.0 mm long .... *C. grandiflora*
    2b. Leaves linear to lance-linear or spatulate, entire or with 1 or 2 lateral lobes near the base; stems leafy mainly in the basal half; chaff 4.0–6.0 mm long .... *C. lanceolata*

Coreopsis grandiflora Hogg *ex* Sweet
synonym(s): —
common name(s): big-flowered tickseed

habit: tall, perennial or annual herb
range: GA, s. to FL, w. to TX and NM, n. to KS; escaped from cultivation as far north as ME, w. to MI
state frequency: very rare
habitat: sandy or rocky thickets, open areas
notes: —

Coreopsis lanceolata L.

synonym(s): —
common name(s): longstalk tickseed
habit: branched, perennial herb
range: Ontario, s. to FL, w. to NM; escaped from cultivation n. and e. to ME
state frequency: very rare
habitat: dry, sandy, rocky, or gravelly places
notes: —

Coreopsis tinctoria Nutt.

synonym(s): —
common name(s): plains tickseed
habit: branched, annual herb
range: MN, s. to TX, w. to CA, n. to WA; escaped from cultivation in the northeast
state frequency: rare
habitat: moist, low, disturbed areas
notes: The ray flowers in this species can rarely be entirely red-brown.

Crepis
(2 species)

Taprooted, lactiferous herbs with basally disposed leaves.

KEY TO THE SPECIES

1a. Inner phyllaries pubescent on the adaxial surface; mature cypselas purple-brown; stem leaves auriculate or not, with revolute margins .... *C. tectorum*
1b. Inner phyllaries glabrous on the adaxial surface; mature cypselas pale brown; stem leaves clasping and auriculate, with flat margins .... *C. capillaris*

Crepis capillaris (L.) Wallr.

synonym(s): —
common name(s): —
habit: branched, ascending, annual or biennial herb
range: Europe; naturalized to New Brunswick and ME, s. to PA
state frequency: very rare
habitat: fields, lawns, waste places
notes: —

Crepis tectorum L.

synonym(s): —
common name(s): —
habit: annual herb
range: Eurasia; naturalized to New Brunswick and Nova Scotia, s. to NJ, w. to NE, n. to MI and Ontario

state frequency: very rare
habitat: cultivated ground, waste places
notes: —

Doellingeria
(1 species)

Doellingeria umbellata (P. Mill.) Nees
synonym(s): *Aster umbellatus* P. Mill.
common name(s): flat-topped white aster
habit: rhizomatous, perennial herb
range: Newfoundland, s. to NC and mountains of GA, w. to mountains of AL, n. to MN
state frequency: occasional
habitat: moist to dry meadows, thickets, borders, and low areas
notes: —

Dyssodia
(1 species)

The leaves and phyllaries of this genus are embedded with oil-glands that are responsible for the ill scent of these plants.

Dyssodia papposa (Vent.) A. S. Hitchc.
synonym(s): *Boebera papposa* (Vent.) Rydb.
common name(s): stinking-marigold, fetid-marigold
habit: branched, annual or biennial herb
range: Ontario and MN, s. to LA, w. to Mexico and AZ, n. to MT; naturalized to ME and other eastern
    states
state frequency: very rare
habitat: dry roadsides and waste places
notes: —

Echinacea
(1 species)

Showy composites with *ca.* 13 purple to white, drooping rays and a purple disk.

Echinacea pallida (Nutt.) Nutt.
synonym(s): —
common name(s): prairie coneflower
habit: simple or occasionally branched, perennial herb
range: WI, s. to LA, w. to TX, n. to MT; naturalized in the northeast
state frequency: very rare
habitat: dry, open places
notes: —

Echinops
(1 species)

Unique among our species, this genus has 1-flowered capitula. With the aspect of thistles, these plants commonly have clasping, spiny-margined leaves with white-tomentose abaxial surfaces.

Echinops sphaerocephalus L.
synonym(s): —
common name(s): —
habit: coarse, branched, perennial herb
range: Eurasia; escaped from cultivation to Quebec, s. to VA, w. to IA, n. to British Columbia
state frequency: rare
habitat: cultivated areas, waste places
notes: —

Erechtites
(1 species)

Erechtites is variable in both morphology and habitat.

Erechtites hieraciifolia (L.) Raf. *ex* DC.
synonym(s): *Erechtites hieraciifolia* (L.) Raf. *ex* DC. var. *intermedia* Fern., *Erechtites hieraciifolia* (L.) Raf. *ex* DC. var. *praealta* (Raf.) Fern.
common name(s): fireweed, pilewort
habit: annual herb
range: New Brunswick and Nova Scotia, s. to FL, w. to TX, n. to MN and Ontario
state frequency: occasional
habitat: dry to damp woods, clearings, shores, marshes
notes: —

Erigeron
(6 species)

A genus with the aspect of *Symphyotrichum* but typically flowering earlier and with characteristically different phyllaries [as specified in the key].

KEY TO THE SPECIES
1a. Rays up to 5.0 x up to 0.5 mm, becoming filiform and inconspicuous in drying; involucre 5.0–12.0 mm tall; capitula with a series of rayless, carpellate flowers within the series of ray flowers .... *E. acris*
1b. Rays evident even in drying; involucre 2.0–7.0 mm tall; capitula with carpellate ray flowers and bisexual disk flowers
    2a. Leaves linear to linear-oblanceolate, 0.1–0.5 cm wide; stems tufted and bearing axillary, sterile, leafy branches; capitula with 20–50 ray flowers; capitulescence a single capitulum at the tip of a branch .... *E. hyssopifolius*
    2b. Leaves linear to suborbicular, up to 7.0 cm wide; stems not tufted, usually without sterile, axillary branches; capitula with 50–400 ray flowers; capitulescence resembling a corymb, only rarely a single capitulum

3a. Pappus of 2 lengths—the inner of long, slender bristles, the outer of short scales; pappus of ray flowers less than 1.0 mm long; stem leaves gradually tapering to the base

4a. Stem leaves linear to lanceolate, mostly entire; pubescence of mid-stem appressed to spreading; rays up to 6.0 mm long; stems with few leaves .... *E. strigosus*

4b. Stem leaves broad-lanceolate or ovate, toothed; pubescence of mid-stem spreading; rays 4.0–10.0 mm long; stems with relatively many leaves .... *E. annuus*

3b. Pappus of 1 length—long, slender bristles; pappus of all the flowers much longer than 1.0 mm; stem leaves rounded to clasping at the base

5a. Capitula with 50–100 ray flowers; corollas of disk flowers 4.0–6.0 mm long; rays 1.0 mm wide or wider; plants with slender, elongate, stoloniform rhizomes .... *E. pulchellus*

5b. Capitula with 150–400 ray flowers; corollas of disk flowers 2.5–3.2 mm long; rays up to 0.5 mm wide; plants without stoloniform rhizomes .... *E. philadelphicus*

Erigeron acris L. var. *kamtschaticus* (DC.) Herder

synonym(s): *Erigeron angulosus* Gaudin var. *kamtschaticus* (DC.) Hara, *Trimorpha acris* (L.) Nesom var. *kamtschatica* (DC.) Nesom

common name(s): trimorphic-daisy, bitter fleabane

habit: stout, erect, biennial or short-lived perennial herb

range: circumboreal; s. to ME, w. to UT

state frequency: very rare

habitat: damp, sandy banks, thickets, and clearings

notes: This species is ranked SH by the Maine Natural Areas Program. Between the bisexual disk flowers and the unisexual ray flowers is a ring of unisexual [carpellate] disk flowers.

Erigeron annuus (L.) Pers.

synonym(s): —

common name(s): daisy fleabane, annual fleabane

habit: coarse, annual or rarely biennial herb

range: semicosmopolitan

state frequency: common

habitat: fields, waste places

notes: —

Erigeron hyssopifolius Michx.

synonym(s): —

common name(s): hyssop-leaved fleabane, hyssop-daisy

habit: cespitose, perennial herb

range: Newfoundland, s. to ME, w. to MI, n. to Yukon

state frequency: rare

habitat: calcareous, rocky shores and banks

notes: This species is ranked S2 by the Maine Natural Areas Program.

Erigeron philadelphicus L.

synonym(s): —

common name(s): Philadelphia-daisy

habit: biennial or short-lived perennial herb

range: Newfoundland, s. to FL, w. to TX, n. to Yukon
state frequency: occasional
habitat: rich thickets, shores, moist meadows
notes: —

Erigeron pulchellus Michx.
synonym(s): —
common name(s): robin's-plantain
habit: rhizomatous, stoloniferous, perennial herb
range: ME, s. to GA, w. to TX, n. to MN and Ontario
state frequency: uncommon
habitat: woods, meadows, stream banks
notes: —

Erigeron strigosus Muhl. *ex* Willd.
synonym(s): —
common name(s): rough fleabane, daisy fleabane
habit: annual or rarely biennial herb
range: Newfoundland, s. to SC, w. to OK, n. to British Columbia
state frequency: common
habitat: fields, roadsides, waste places
key to the subspecies:
  1a. Phyllaries with conspicuously flattened hairs over 1.0 mm long; pubescence of mid-stem longer
      and usually spreading .... *E. s.* var. *septentrionalis* (Fern. & Wieg.) Fern.
  1b. Phyllaries with terete or obscurely flattened hairs less than 1.0 mm long; pubescence of mid-
      stem shorter and usually appressed .... *E. s.* var. *strigosus*
notes: —

Eupatorium
(5 species)

Certain species complexes are noted to hybridize with each other [the whorled-leaved species forming one complex and the opposite-leaved species another]. Distinct morphologies have been observed in the eastern United States and are yet to be named.

KEY TO THE SPECIES
1a. Leaves whorled; flowers pink to purple; involucre 6.5–9.0 mm tall
    2a. Stems viscid-puberulent, at least near the apex; leaves with 3 prominent, somewhat
        parallel veins, abruptly contracted to the petiole, 5.0–15.0 cm long .... *E. dubium*
    2b. Stems not viscid-puberulent; leaves commonly pinnately veined, gradually tapering
        to the petiole, 6.0–30.0 cm long
        3a. Capitula with 8–22 flowers; capitulescence, or its divisions, commonly flat-
            topped; stems commonly spotted or streaked with purple and not glaucous ....
            .... *E. maculatum*
        3b. Capitula with 4–7 flowers; capitulescence convex; stems purple throughout, but
            not spotted, glaucous .... *E. fistulosum*
1b. Leaves usually opposite; flowers white; involucre 4.0–6.6 mm tall
    4a. Leaves connate at the base, the stem appearing to pierce through; capitula with 9–23
        flowers .... *E. perfoliatum*
    4b. Leaves not connate at the base; capitula with 5–7 flowers .... *E. rotundifolium*

Eupatorium dubium Willd. *ex* Poir.

synonym(s): —
common name(s): three-nerved joe-pye weed
habit: tall, coarse, perennial herb
range: Nova Scotia and ME, s. to SC
state frequency: very rare
habitat: damp thickets and shores
notes: This species is ranked S1 by the Maine Natural Areas Program. Similar to *Eupatorium maculatum*, this species often has purple-streaked stems. *Eupatorium dubium*, however, has fewer flowers [5–10] per capitulum.

Eupatorium fistulosum Barratt

synonym(s): —
common name(s): hollow-stemmed joe-pye weed, hollow joe-pye weed, trumpet weed
habit: tall, coarse, perennial herb
range: ME, s. to FL, w. to TX, n. to IA
state frequency: very rare
habitat: moist woods, thickets, meadows, and bottomlands
notes: This species is ranked S1 by the Maine Natural Areas Program.

Eupatorium maculatum L.

synonym(s): —
common name(s): spotted joe-pye weed
habit: tall, coarse, perennial herb
range: Newfoundland, s. to NC, w. to UT, n. to British Columbia
state frequency: common
habitat: moist thickets, meadows, and shores
key to the subspecies:
    1a. Leaves of the upper node large and elongate, 8.0–30.0 x 3.0–8.0 cm, commonly surpassing the inflorescence .... *E. m.* var. *foliosum* (Fern.) Wieg.
    1b. Leaves of the upper node smaller, 3.0–18.0 x 0.5–5.5 cm, usually not surpassing the inflorescence .... *E. m.* var. *maculatum*
notes: Habitat is also useful in separating the varieties of *Eupatorium maculatum*. *Eupatorium m.* var. *foliosum* often occurs on higher elevation sites than *E. m.* var. *maculatum*.

Eupatorium perfoliatum L.

synonym(s): —
common name(s): boneset, estuary boneset
habit: tall, coarse, perennial herb
range: Nova Scotia and ME, s. to FL, w. to TX, n. to Manitoba
state frequency: occasional
habitat: moist or wet, low woods, thickets, and shores
key to the subspecies:
    1a. Leaves and phyllaries pubescent; stem coarsely pubescent, relatively thicker; leaves relatively wider; plants of inland habitats .... *E. p.* var. *perfoliatum*
    1b. Leaves and phyllaries glabrous or glabrate; stem minutely pubescent to glabrous, relatively thinner; leaves relatively narrower; plants of tidal shores .... *E. p.* var. *colpophilum* Fern. & Grisc.
notes: *Eupatorium perfoliatum* var. *colpophilum* is ranked SH by the Maine Natural Areas Program. This species will very rarely have whorled leaves [in 3s], purple flowers, or distinct leaves.

Eupatorium rotundifolium L. var. **ovatum** (Bigelow) Torr.
synonym(s): *Eupatorium pubescens* Muhl. *ex* Willd.
common name(s): round-leaved eupatorium, hairy boneset
habit: tall, perennial herb
range: ME, s. to VA and mountains of GA, w. to mountains of AL and KY
state frequency: very rare
habitat: dry to seldom moist, open woods, thickets, and clearings
notes: This species is ranked SH by the Maine Natural Areas Program. *Eupatorium rotundifolium* var. *ovatum* will sometimes have alternate upper leaves.

Euthamia
(2 species)

KEY TO THE SPECIES
1a. Leaves with 3 evident nerves and occasionally with 2 or 4 fainter nerves, 3.0–12.0 mm wide; capitula with 20–35(–45) flowers .... ***E. graminifolia***
1b. Leaves with 1 evident nerve and occasionally with 2 fainter nerves, 1.0–4.0 mm wide; capitula with 10–21 flowers (or with as many as 50 in a very rare variety) ....
.... ***E. tenuifolia***

Euthamia graminifolia (L.) Nutt.
synonym(s): <*E. g.* var. *graminifolia*> = *Solidago graminifolia* (L.) Salisb.; <*E. g.* var. *nuttallii*> = *Solidago graminifolia* (L.) Salisb. var. *nuttallii* (Greene) Fern.
common name(s): common flat-topped goldenrod, grass-leaved goldenrod
habit: erect. rhizomatous, perennial herb
range: Newfoundland, s. to NC, w. to NM, n. to British Columbia
state frequency: common
habitat: moist to dry thickets, shores, and open areas
key to the subspecies:
　　1a. Stems, branches, and leaves glabrous or occasionally with some lines of scabrous pubescence
　　... *E. g.* var. *graminifolia*
　　1b. Branches and also often the stems and leaves pubescent ....
　　... *E. g.* var. *nuttallii* (Greene) W. Stone
notes: Reported to hybridize with *Euthamia tenuifolia*, but further study is required to document interspecific mating.

Euthamia tenuifolia (Pursh) Nutt.
synonym(s): <*E. t.* var. *pycnocephala*> = *Euthamia galetorum* (Greene) Friesner, *Solidago tenuifolia* Pursh var. *pycnocephala* Fern.; <*E. t.* var. *tenuifolia*> = *Solidago tenuifolia* Pursh
common name(s): coastal-plain flat-topped goldenrod, narrow-leaved goldenrod
habit: slender, erect, perennial herb
range: Nova Scotia and ME, s. to FL, w. to IN, n. to MI
state frequency: very rare
habitat: damp to dry, open, sandy soil
key to the subspecies:
　　1a. Stems usually unbranched and without axillary fascicles of leaves; leaves thick or subcoriaceous; capitulescence 1.0–10.0 cm wide ....
　　... *E. t.* var. *pycnocephala* (Fern.) C. & J. Taylor
　　1b. Stems commonly branched and/or with axillary fascicles of leaves; leaves thin; capitulescence 3.0–30.0 cm wide .... *E. t.* var. *tenuifolia*
notes: This species is ranked S1/S2 by the Maine Natural Areas Program. Reported to hybridize with *Euthamia graminifolia*.

Gaillardia
(1 species)

Plants with showy, often bicolored, 3-cleft, ray flowers.

Gaillardia pulchella Foug.
synonym(s): —
common name(s): rosering, blanket flower, gaillardia
habit: branched, annual herb
range: VA, s. to FL, w. to Mexico and NM, n. to CO and MN; escaped from cultivation to ME
state frequency: very rare
habitat: dry, sandy, open areas
notes: —

Galinsoga
(2 species)

Maine's species of *Galinsoga* originated from tropical America and arrived in New England in the mid-1800s.

KEY TO THE SPECIES
1a. Pappus of the disk flowers with a short awn-tip; pappus of ray flowers well developed
    and equaling the length of the ray flower tube; cypselas pubescent .... *G. quadriradiata*
1b. Pappus of the disk flowers without an awn-tip; pappus of ray flowers absent or poorly
    developed; cypselas sparsely pubescent or glabrous .... *G. parviflora*

Galinsoga parviflora Cav.
synonym(s): —
common name(s): lesser quickweed
habit: branched, annual herb
range: southwest United States and Mexico, s. to South America; naturalized throughout the world
state frequency: rare
habitat: gardens, waste places
notes: —

Galinsoga quadriradiata Ruiz & Pavón
synonym(s): *Galinsoga ciliata* (Raf.) Blake
common name(s): common quickweed
habit: branched, annual herb
range: Central and South America; naturalized throughout the world
state frequency: uncommon
habitat: gardens, waste places
notes: —

Gamochaeta
(1 species)

Gamochaeta purpurea (L.) Cabrera
synonym(s): *Gnaphalium purpureum* L.
common name(s): purple cudweed
habit: simple or seldom branched, annual or biennial herb
range: ME, s. to FL, w. to CA, n. to British Columbia

state frequency: very rare
habitat: sandy soil, waste places
notes: This species is ranked SX by the Maine Natural Areas Program.

Gnaphalium
(1 species)

Gnaphalium uliginosum L.
synonym(s): —
common name(s): low cudweed
habit: branched, annual herb
range: Newfoundland, s. to VA, w. to UT, n. to British Columbia
state frequency: common
habitat: stream banks, damp clearings
notes: —

Grindelia
(1 species)

Showy composites with resinous-punctate leaves.

Grindelia squarrosa (Pursh) Dunal
synonym(s): —
common name(s): curly-top gumweed
habit: branched, biennial or perennial herb
range: MN, s. to TX, w. to NV, n. to British Columbia; naturalized in the northeast
state frequency: very rare
habitat: open areas, waste places
notes: —

Helenium
(2 species)

Plants with glandular-punctate leaves that are decurrent on the stem.

KEY TO THE SPECIES
1a. Ray flowers carpellate; disk flowers yellow; leaves coarsely toothed .... *H. autumnale*
1b. Ray flowers neutral, rarely absent; disk flowers red-brown or purple-brown; leaves
     entire or subentire .... *H. flexuosum*

Helenium autumnale L.
synonym(s): —
common name(s): common sneezeweed
habit: simple or branched, fibrous-rooted, perennial herb
range: Quebec, s. to FL, w. to AZ, n. to British Columbia
state frequency: very rare
habitat: moist, low ground, shores
notes: Known to hybridize with *Helenium flexuosum*.

Helenium flexuosum Raf.
synonym(s): *Helenium nudiflorum* Nutt.
common name(s): southern sneezeweed

habit: perennial herb
range: ME, s. to FL, w. to TX, n. to WI
state frequency: rare
habitat: moist ground, waste places
notes: Known to hybridize with *Helenium autumnale*.

Helianthus
(12 species)

Hybridization is thought to be common in mixed populations.

KEY TO THE SPECIES

1a. Plants annual, with fibrous roots; leaves alternate (except the lowermost); receptacle flat or nearly so; disk flowers usually red-purple
  2a. Phyllaries lance-oblong to ovate, abruptly narrowed to the apex, conspicuously long-ciliate; cypselas usually glabrous except near the pubescent apex, 4.0–8.0 mm wide; leaves dentate; plants 0.5–3.0 m tall .... *H. annuus*
  2b. Phyllaries lanceolate, gradually narrowed to the apex, either not ciliate or ciliate with short hairs of similar length to the hairs of the abaxial surface of the phyllaries; cypselas pubescent, 1.2–2.5 mm wide; leaves entire to undulate-dentate; plants usually not exceeding 1.0 m in height
    3a. Chaff near the center of the disk inconspicuously short-pubescent; cypselas mottled, pubescent with spreading-ascending hairs; stems frequently mottled; leaves dark green and scabrous .... *H. debilis*
    3b. Chaff near the center of the disk conspicuously pubescent near the apex with long, white hairs; cypselas not mottled, pubescent with appressed-ascending hairs; stems not mottled; leaves pale green and strigose .... *H. petiolaris*
1b. Plants perennial, with rhizomes, stolons, tuberous roots, and/or tough, overwintering bases; leaves opposite (except often the upper); receptacle usually convex to some degree; disk flowers yellow (red-purple in *H. pauciflorus*)
  4a. Disk flowers red-purple; phyllaries acute to obtuse at the apex, broad, appressed, of several, conspicuously different lengths .... *H. pauciflorus*
  4b. Disk flowers yellow; phyllaries acuminate to attenuate at the apex, narrow, at least some merely ascending, of several, inconspicuously different lengths
    5a. Stem and capitulescence conspicuously pubescent
      6a. Stem and abaxial surface of leaves and phyllaries densely pubescent with short, soft hairs; leaves sessile, subcordate to cordate at the base ..... *H. mollis*
      6b. Stem and abaxial surface of leaves and phyllaries with sparser and/or coarser pubescence; leaves at least short-petioled (sometimes sessile in *H. giganteus*), tapering to rounded at the base
        7a. Leaves 1.0–3.5 cm wide, with a petiole up to 1.5 cm long; roots fleshy and merely thickened
          8a. Stem spreading-hirsute; leaves flat, usually with 3 prominent, parallel nerves near the base .... *H. giganteus*
          8b. Stem strigose; leaves sometimes folded and falcate-arched [especially in drying], pinnately veined near the base .... *H. maximiliani*

7b. Leaves 4.0–12.0 cm wide, with a petiole (1.5–)2.0–8.0 cm long; roots forming tubers .... *H. tuberosus*

5b. Stem glabrous (or very nearly so), only the capitulescence pubescent

9a. Leaves sessile or with short petioles to 0.5(1.0) cm long

10a. Leaves narrowly tapering to the base, usually alternately arranged on the upper portion of the stem; disk 1.5–2.5 cm wide; plants of moist to wet soils .... *H. giganteus*

10b. Leaves truncate to broadly rounded at the base, usually oppositely arranged throughout the stem; disk 1.0–1.5 cm wide; plants of dry soils .... *H. divaricatus*

9b. Well developed leaves with petioles (0.5–)1.0–4.0 cm long

11a. Leaves generally opposite throughout the stem (except the very uppermost), at least somewhat triple-veined, those near the middle of the stem 2.5–10.0 cm wide, broad-lanceolate to ovate, and usually less than 3.0 times as long as wide; capitula with 8–15 ray flowers

12a. Phyllaries spreading, evidently surpassing the disk; lobes of the disk corollas pubescent; petioles 1.5–6.0 cm long; leaves thin and membranaceous, prominently serrate, subglabrous to scabrous on the abaxial surface .... *H. decapetalus*

12b. Phyllaries ascending, approximately equaling the height of the disk; lobes of the disk corollas glabrous; petioles 0.5–3.0 cm long; leaves thick and firm, subentire to shallowly serrate, scabrous-hispid on the abaxial surface .... *H. strumosus*

11b. Leaves usually alternate on the upper half of the stem, pinnately veined, those near the middle of the stem 1.0–4.0 cm wide, mostly lanceolate, and usually 3.0–8.0 times as long as wide; capitula with 10–20 ray flowers

13a. Lobes of the disk corollas glabrous or glabrate; leaves slightly, if at all, scabrous on the adaxial surface and pubescent on the abaxial surface with hairs less than 1.0 mm long .... *H. grosseserratus*

13b. Lobes of the disk corollas pubescent; leaves strongly scabrous on the adaxial surface and pubescent on the abaxial surface with hairs that commonly equal or exceed 1.0 mm .... *H. giganteus*

Helianthus annuus L.

synonym(s): —
common name(s): common sunflower
habit: simple or branched, coarse, annual herb
range: MN, s. to TX, w. to CA, n. to Manitoba; escaped from cultivation in the eastern United States
state frequency: uncommon
habitat: moist, low areas, waste places, roadsides
notes: Wild forms have several, relatively smaller capitula. Known to hybridize with *Helianthus decapetalus.*

Helianthus debilis Nutt.

synonym(s): —
common name(s): —
habit: decumbent to suberect, annual herb
range: TX; escaped from cultivation in our area

state frequency: very rare
habitat: waste places, roadsides
notes: —

Helianthus decapetalus L.

synonym(s): —
common name(s): forest sunflower, thin-leaved sunflower
habit: simple or branched, rhizomatous, perennial herb
range: New Brunswick and ME, s. to GA, w. to MO, n. to MN
state frequency: uncommon
habitat: woods, thickets, meadows, stream banks, river banks
notes: Known to hybridize with *Helianthus annuus*.

Helianthus divaricatus L.

synonym(s): *Helianthus divaricatus* L. var. *angustifolius* Kuntze
common name(s): divaricate sunflower, rough sunflower
habit: simple or branched, erect, rhizomatous, perennial herb
range: ME, s. to FL, w. to OK, n. to Saskatchewan
state frequency: rare
habitat: dry, open thickets and woods, openings
notes: Known to hybridize with *Helianthus grosseserratus* and *H. giganteus*.

Helianthus giganteus L.

synonym(s): *Helianthus borealis* E. E. Wats.
common name(s): swamp sunflower, tall sunflower, giant sunflower
habit: simple or branched, rhizomatous, perennial herb
range: New Brunswick and Nova Scotia, s. to GA, w. to NE, n. to Alberta
state frequency: rare
habitat: wet thickets and woods, swamps, borders of saltmarshes
notes: Known to hybridize with *Helianthus divaricatus*, *H. grosseserratus*, and *H. mollis*.

Helianthus grosseserratus Martens

synonym(s): —
common name(s): sawtooth sunflower
habit: rhizomatous, perennial herb
range: OH, s. to AR, w. to TX, n. to ND; escaped from cultivation in the east
state frequency: rare
habitat: rich thickets, roadsides
notes: Hybrids with *Helianthus salicifolius* A. Dietr., called *H.* ×*kellermanii* Britt., occur in Maine. These hybrids are somewhat similar to *H. grosseserratus*, except their narrow-lanceolate leaves are only 1.0–2.0 cm wide and their corolla lobes are pubescent with white hairs. Hybrids with *H. maximiliani*, called *H.* ×*intermedia*, R. W. Long, also occur in Maine. Also known to hybridize with *Helianthus divaricatus*, *H. giganteus*, and *H. mollis*.

Helianthus maximiliani Schrad.

synonym(s): —
common name(s): Maximilian's sunflower
habit: stout, rhizomatous, perennial herb
range: MN, s. to MO, w. to TX, n. to British Columbia; escaped from cultivation in the east
state frequency: very rare
habitat: waste places
notes: Hybrids with *Helianthus grosseserratus* (*q.v.*) occur in Maine.

Helianthus mollis Lam.
synonym(s): —
common name(s): ashy sunflower, hairy sunflower
habit: simple or branched, rhizomatous, perennial herb
range: MI and OH, s. to GA, w. to TX, n. to IA; escaped from cultivation in the east
state frequency: very rare
habitat: dry, open areas
notes: Known to hybridize with *Helianthus giganteus* and *H. grosseserratus.*

Helianthus pauciflorus Nutt.
synonym(s): <*H. p.* ssp. *pauciflorus*> = *Helianthus laetiflorus* Pers. var. *rigidus* (Cass.) Fern.; <*H. p.*
    ssp. *subrhomboideus*> = *Helianthus laetiflorus* Pers. var. *subrhomboideus* (Rydb.) Fern.
common name(s): stiff sunflower, rhombic-leaved sunflower
habit: simple or branched, rhizomatous, perennial herb
range: MI, s. to GA, w. to NM, n. to Alberta; escaped from cultivation to ME, s. to MA
state frequency: rare
habitat: roadsides, waste places
key to the subspecies:
    1a. Stems 1.0–2.0 m tall, with 9–15 nodes below the capitulescence; leaves lanceolate to lance-
        ovate, acuminate at the apex, 8.0–27.0 cm long .... *H. p.* ssp. *pauciflorus*
    1b. Stems 0.3–1.2 m tall, with 5–10 nodes below the capitulescence; leaves lance-linear to
        rhombic-ovate, acute to obtuse at the apex, 5.0–12.0 cm long ....
        .... *H. p.* ssp. *subrhomboideus* (Rydb.) O. Spring & E. Schilling
notes: Hybrids with *Helianthus tuberosus,* called *H.* ×*laetiflorus* Pers., occur in Maine. These hybrids
resemble most closely *H. pauciflorus,* but usually have yellow disk flowers.

Helianthus petiolaris Nutt.
synonym(s): —
common name(s): plains sunflower, prairie sunflower
habit: simple or branched, annual herb
range: Manitoba and MN, s. to LA, w. to AZ, n. to WA; escaped from cultivation to ME, s. to VA
state frequency: rare
habitat: dry roadsides and waste places
notes: —

Helianthus strumosus L.
synonym(s): —
common name(s): rough-leaved sunflower, pale-leaved wood sunflower
habit: often branched, rhizomatous, perennial herb
range: New Brunswick and ME, s. to GA, w. to AR, n. to Saskatchewan
state frequency: uncommon
habitat: open woods, thickets, clearings
notes: —

Helianthus tuberosus L.
synonym(s): —
common name(s): Jerusalem artichoke
habit: branched, rhizomatous, perennial herb
range: New Brunswick and ME, s. to GA, w. to AR, n. to Saskatchewan
state frequency: uncommon
habitat: moist or wet, rich thickets
notes: Known to hybridize with *Helianthus pauciflorus* (*q.v.*).

Heliopsis
(1 species)

Plants with dimorphic cypselas—those of the disk flowers quadrangular, those of the ray flowers triangular.

Heliopsis helianthoides (L.) Sweet var. ***scabra*** (Dunal) Fern.
synonym(s): *Heliopsis scabra* Dunal
common name(s): ox eye, sunflower-everlasting
habit: robust, perennial herb
range: NY, s. to GA, w. to NM, n. to British Columbia; naturalized n. and e. to ME and New Brunswick
state frequency: rare
habitat: dry woods and thickets, waste places
notes: —

Hieracium
(14 species)

Those species that lack a conspicuous cluster of basal leaves at anthesis [identified in the key] do so because they have caducous, lower leaves. The lower leaves can sometimes be found, but will be brown and senescing.

KEY TO THE SPECIES
1a. Rhizome long and slender; plants stoloniferous, with a conspicuous cluster of basal leaves at anthesis; stems resembling a scape, without leaves or with 1 or 2 very small leaves
    2a. Capitulescence composed of 1(–3) capitula; stem 0.3–25.0(–40.0) dm tall ....
    .... ***H. pilosella***
    2b. Capitulescence composed of 2–many (commonly 5 or more) capitula; stem 1.0–9.0 dm tall
        3a. Corolla orange-red to red .... ***H. aurantiacum***
        3b. Corolla yellow .... ***H. caespitosum***
1b. Rhizome short and thick; plants without stolons, with or without a conspicuous cluster of basal leaves at anthesis; stems with or without leaves
    4a. Plants with a conspicuous cluster of basal leaves at anthesis
        5a. Cypselas 2.2–4.0(–5.0) mm long; basal leaves mostly 1.7–5.0 times as long as wide; leaves sometimes red-purple or with red-purple mottling or red-purple veins
            6a. Each capitulum with 40–80 flowers; involucre composed of phyllaries of 2 or 3 subequal lengths; pappus bristles numerous, of unequal lengths; cypselas truncate at the apex
                7a. Blades of basal leaves rounded to cordate at the base; stems resembling a scape, with 0–2 small leaves borne near the base .... ***H. murorum***
                7b. Blades of basal leaves gradually tapering to the base; stems with 2–12 scattered leaves
                    8a. Peduncles and phyllaries evidently stipitate-glandular as well as hispid- and/or stellate-pubescent; basal leaves 1.5–5.5 cm wide;

plants 1.5–10.0 dm tall, weeds, of disturbed fields and roadsides ....
.... *H. lachenalii*

    8b. Peduncles and phyllaries hispid- or stellate-pubescent, with few or no
        stipitate glands; basal leaves 0.7–2.0 cm wide; plants 1.0–3.5 dm tall,
        native, of ledges and rocky shores .... *H. robinsonii*

  6b. Each capitulum with 15–40 flowers; involucre composed of phyllaries of 2
      distinctly unequal lengths—a long, inner set and a minute, outer set; pappus
      bristles fewer, of equal length; cypselas gradually tapered to the apex (or
      sometimes truncate in *H. venosum*)

      9a. Capitulescence resembling a corymb, the lower branches elongate; midrib
        and main veins of the leaf commonly red-purple; stems glabrous, except
        sometimes at the very base .... *H. venosum*

      9b. Capitulescence resembling a panicle, the lower branches not elongate;
        midrib and main veins of the leaf not red-purple; stems pubescent, except
        sometimes in the apical half .... *H. gronovii*

5b. Cypselas 1.5–2.0 mm long; basal leaves mostly 5.0–12.0 times as long as wide;
  leaves green, without red-purple mottling or red-purple veins

  10a. Stems and leaves thinly glaucous; leaves glabrous to sparsely setose

    11a. Leaves sparsely setose; plants with slender, ascending to spreading,
      often numerous, sterile branches arising from the base .... *H. praealtum*

    11b. Leaves glabrous or nearly so; plants without sterile branches arising
      from the base .... *H. piloselloides*

  10b. Stems and leaves not glaucous; leaves evidently setose .... *H. caespitosum*

4b. Plants lacking a conspicuous cluster of basal leaves at anthesis

  12a. Ovary and cypsela gradually tapering to the apex .... *H. gronovii*

  12b. Ovary and cypsela truncate at the apex

    13a. Each capitulum with 8–30 flowers; peduncles glabrous ...... *H. paniculatum*

    13b. Each capitulum with 40–100 flowers; peduncles pubescent

      14a. Peduncles and phyllaries evidently stipitate-glandular; involucre
        composed of phyllaries of 2 distinctly unequal lengths—a long, inner set
        and a minute, outer set; pappus bristles fewer, of equal length ....
        .... *H. scabrum*

      14b. Peduncles and phyllaries with no or few stipitate glands; involucre
        composed of phyllaries of 2 or 3 subequal lengths; pappus bristles
        numerous, of unequal lengths

        15a. Stem leaves 2–10; phyllaries acuminate to caudate at the apex, with
          some stellate hairs; plants 1.0–3.5 dm tall .... *H. robinsonii*

        15b. Stem leaves 5–50; phyllaries rounded to obtuse at the apex, without
          stellate hairs; plants 1.5–15.0 dm tall

          16a. Styles pale yellow; capitulescence open, with slender peduncles,
            to 10.0 cm long; stems with 5–30 thin to membranaceous leaves;
            involucre gray-brown to black; middle phyllaries oblong-
            lanceolate, entirely gray-brown or with a wide, dark central band
            .... *H. canadense*

          16b. Styles darkened by minute, brown scabrules; capitulescence
            more dense, with stout peduncles 2.0–4.0 cm long; stems with

25–50 firm to coriaceous leaves; involucre gray-green; middle phyllaries narrow-lanceolate, entirely pale green or with a narrow, green, central band and apex .... **H. kalmii**

Hieracium aurantiacum L.

synonym(s): —
common name(s): orange hawkweed, devil's paintbrush, Indian paintbrush
habit: rhizomatous, perennial herb
range: Europe; naturalized to Newfoundland, s. to NC, w. to IA, n. to MN; also British Columbia
state frequency: common
habitat: fields, roadsides, lawns, meadows
notes: Known to hybridize with *Hieracium caespitosum*, *H. pilosella*, and *H. piloselloides*.

Hieracium caespitosum Dumort.

synonym(s): *Hieracium pratense* Tausch
common name(s): yellow king devil
habit: rhizomatous, perennial herb
range: Europe; naturalized to Quebec, s. to GA, w. and n. to TN
state frequency: occasional
habitat: fields, pastures, roadsides
notes: Hybrids with *Hieracium lactucella* Wallr. and *H. pilosella*, called *H.* ×*floribundum* Wimmer & Grab. and *H.* ×*flagellare* Willd., respectively, occur in Maine. Also known to hybridize with *H. aurantiacum*.

Hieracium canadense Michx.

synonym(s): —
common name(s): Canada hawkweed
habit: erect to ascending, perennial herb
range: Labrador, s. to ME, w. to OR, n. to British Columbia
state frequency: occasional
habitat: thickets, borders of woods, fields
notes: Hybrids with *Hieracium scabrum*, called *H.* ×*fernaldii* Lepage, occur in Maine. Also known to hybridize with *H. lachenalii*.

Hieracium gronovii L.

synonym(s): —
common name(s): beaked hawkweed, hairy hawkweed
habit: perennial herb
range: MA, s. to FL, w. to TX, n. to MI and Ontario; also ME
state frequency: very rare
habitat: dry, open woods, fields, pastures
notes: This species is ranked SX by the Maine Natural Areas Program. *Hieracium gronovii* doubtfully remains part of our flora. Hybrids with *H. paniculatum* and *H. venosum*, called *H.* ×*marianum* Willd., occur in Maine.

Hieracium kalmii L.

synonym(s): *Hieracium canadense* Michx. var. *fasciculatum* (Pursh) Fern.
common name(s): —
habit: erect, leafy, perennial herb
range: Quebec, s. to NJ, w. to IL, n. to MN
state frequency: very rare
habitat: clearings, fields, shores
notes: Known to hybridize with *Hieracium scabrum*.

Hieracium lachenalii K. C. Gmel.

synonym(s): *Hieracium vulgatum* Fries
common name(s): European hawkweed
habit: rhizomatous, perennial herb
range: Europe; naturalized to Newfoundland, s. to NJ, w. and n. to MN
state frequency: uncommon
habitat: fields, roadsides, waste places
notes: The leaves are sometimes spotted or veined with purple. Known to hybridize with *Hieracium canadense*.

Hieracium murorum L.

synonym(s): —
common name(s): wall hawkweed
habit: rhizomatous, perennial herb
range: Europe; naturalized to Newfoundland, s. to NJ, w. and n. to MN; also British Columbia
state frequency: rare
habitat: fields, roadsides, thickets, waste places
notes: —

Hieracium paniculatum L.

synonym(s): —
common name(s): panicled hawkweed
habit: solitary, perennial herb
range: New Brunswick, s. to mountains of GA, w. to AL, n. to MN and Ontario
state frequency: occasional
habitat: open woods, thickets
notes: Known to hybridize with *Hieracium gronovii* and *H. venosum*.

Hieracium pilosella L.

synonym(s): —
common name(s): mouse-ear hawkweed
habit: rhizomatous, carpet-forming, perennial herb
range: Europe; naturalized to Newfoundland, s. to NC, w. to OH, n. to MN
state frequency: common
habitat: pastures, fields, lawns
key to the subspecies:
    1a. Leaves conspicuously white-tomentose on the abaxial surface only when young, at maturity the pubescence thinning .... *H. p.* var. *pilosella*
    1b. Leaves permanently and conspicuously white-tomentose on the abaxial surface .... *H. p.* var. *niveum* Muell.-Arg.
notes: Hybrids with *Hieracium caespitosum* (*q.v.*) occur in Maine. Also known to hybridize with *H. aurantiacum*.

Hieracium piloselloides Vill.

synonym(s): *Hieracium florentinum* All.
common name(s): glaucous king devil
habit: rhizomatous, perennial herb
range: Europe; naturalized to Newfoundland, s. to VA, w. to IA, n. to MI and Ontario
state frequency: uncommon
habitat: fields, clearings, roadsides
notes: Known to hybridize with *Hieracium aurantiacum*.

Hieracium praealtum Vill. *ex* Gochnat var. *decipiens* W. D. J. Koch
synonym(s): —
common name(s): king devil
habit: branched, perennial herb
range: Europe; naturalized to Newfoundland, s. to NY
state frequency: rare
habitat: pastures, fields, roadsides
notes: —

Hieracium robinsonii (Zahn) Fern.
synonym(s): —
common name(s): northeastern hawkweed, Robinson's hawkweed
habit: rhizomatous, perennial herb
range: Newfoundland, s. to ME and NH
state frequency: very rare
habitat: ledges, rocky margins of streams
notes: This species is ranked SH by the Maine Natural Areas Program.

Hieracium scabrum Michx.
synonym(s): —
common name(s): rough hawkweed, sticky hawkweed
habit: stout, coarse, solitary, perennial herb
range: New Brunswick and Nova Scotia, s. to mountains of GA, w. to MO, n. to MN and Ontario
state frequency: occasional
habitat: dry, open woods and clearings
key to the subspecies:
    1a. Lower portion of stem, petioles, and abaxial midribs pubescent with hairs 2.0–3.0 mm long; adaxial surface of leaf sparsely pubescent .... *H. s.* var. *scabrum*
    1b. Lower portion of stem, petioles, and abaxial midribs pubescent with hairs 0.2–0.6 mm long; adaxial surface of leaf glabrous or sparsely pubescent .... *H. s.* var. *tonsum* Fern. & St. John
notes: Hybrids with *Hieracium canadense* (*q.v.*) occur in Maine. Also known to hybridize with *H. kalmii*.

Hieracium venosum L. var. *nudicaule* (Michx.) Farw.
synonym(s): —
common name(s): veiny hawkweed, rattlesnake hawkweed
habit: rhizomatous, perennial herb
range: ME, s. to GA, w. to MO, n. to MI and Ontario
state frequency: very rare
habitat: dry, open woods, clearings
notes: *Hieracium venosum* is ranked S1 by the Maine Natural Areas Program. Known to hybridize with *H. gronovii* (*q.v.*) and *H. paniculatum*.

Hypochaeris
(1 species)

Hypochaeris glabra L.
synonym(s): —
common name(s): smooth cat's ear
habit: simple or branched, annual herb

range: Europe; naturalized in the eastern United States
state frequency: very rare
habitat: disturbed areas, waste places
notes: —

Inula
(1 species)

Plants with dimorphic phyllaries—the inner narrow and scarious, the outer broader and herbaceous.

Inula helenium L.
synonym(s): —
common name(s): elecampane
habit: stout, perennial herb
range: Europe; escaped from cultivation in the United States
state frequency: uncommon
habitat: moist to wet roadsides, meadows, and clearings
notes: —

Ionactis
(1 species)

A distinctive segregate from the genus *Aster*, *Ionactis* is readily separable vegetatively and florally.

Ionactis linariifolius (L.) Greene
synonym(s): *Aster linariifolius* L.
common name(s): stiff aster, spruce aster
habit: wiry, stiff-stemmed, perennial herb
range: New Brunswick and ME, s. to FL, w. to TX, n. to MN
state frequency: uncommon
habitat: dry, open, sandy soil, ledges, rocky banks
notes: —

Iva
(3 species)

Members of this genus are monoecious, with separate carpellate and staminate flowers in the same capitulum. The flowers near the margin of the disk are carpellate and possess a branched style. The more numerous central flowers have an abortive ovary and are functionally staminate. They can be identified by their undivided style.

KEY TO THE SPECIES
1a. Capitula not subtended by leafy bracts; carpellate corollas absent or up to 0.5 mm long; leaf blades broad-ovate to suborbicular, 2.5–15.0 cm wide; cypselas glabrous or hispidulous near the apex .... *I. xanthifolia*
1b. Capitula subtended by leafy bracts; carpellate corollas 1.0–1.5 mm long; leaf blades lanceolate to broad-ovate, 1.0–7.0 cm wide; cypselas resin-dotted
    2a. Plants suffrutescent perennials of maritime marshes and shores; leaves short-petiolate
    .... *I. frutescens*
    2b. Plants herbaceous annuals of moist, often disturbed, inland soils; leaves evidently petiolate .... *I. annua*

Iva annua L.
synonym(s): *Iva ciliata* Willd.
common name(s): rough marsh-elder, sumpweed
habit: annual herb
range: IN, s. to LA, w. to NM, n. to ND; naturalized to ME, s. to MA
state frequency: very rare
habitat: alluvial or moist soil, waste places
notes: —

Iva frutescens L. ssp. *oraria* (Bartlett) R. C. Jackson
synonym(s): —
common name(s): maritime marsh-elder
habit: suffrutescent perennial
range: Nova Scotia, s. to VA
state frequency: very rare
habitat: saline marshes
notes: This species is ranked S1 by the Maine Natural Areas Program.

Iva xanthifolia Nutt.
synonym(s): —
common name(s): big marsh-elder
habit: tall, coarse, annual herb
range: WI, s. to TX, w. to NM, n. to British Columbia; also Quebec, s. to NJ
state frequency: uncommon
habitat: rich, alluvial soil, moist waste places
notes: —

Krigia
(1 species)

Krigia virginica (L.) Willd.
synonym(s): —
common name(s): Virginia dwarf-dandelion
habit: slender, annual herb
range: ME, s. to FL, w. to TX, n. to WI and Ontario
state frequency: very rare
habitat: dry, sterile, sandy soil
notes: This species is ranked SH by the Maine Natural Areas Program.

Lactuca
(5 species)

Nerve number on the body of the cypsela is an important character for identification. In dried specimens, the shrunken cypsela may have crease lines that are not nerves. The nerves are relatively straight and raised above the surface of the ovary.

KEY TO THE SPECIES

1a. Leaves very wide, about as broad as long, toothed .... *L. sativa*
1b. Leaves narrower, much longer than broad, with deep sinuses (sometimes not lobed in *L. canadensis*)

2a. Cypselas with a single, prominent nerve on each face, sometimes with an additional pair of faint nerves, with a slender, firm beak 0.5–1.0 times as long as the body of the cypsela; capitula with 13–22 flowers; corollas yellow (sometimes purple in age)

    3a. Involucre 10.0–15.0 mm tall in fruit; cypselas, including the beak, 4.5–6.0 mm long; pappus 5.0–7.0 mm long .... *L. canadensis*

    3b. Involucre 15.0–22.0 mm tall in fruit; cypselas, including the beak, 7.0–10.0 mm long; pappus 8.0–12.0 mm long .... *L. hirsuta*

2b. Cypselas with 3 prominent nerves on each face, beaked or beakless; capitula with 15–55 flowers; corollas blue, white, or sometimes yellow in *L. biennis*

    4a. Cypselas beakless or with a short, stout beak; plants biennial from a tap root; involucre 10.0–14.0 mm tall in fruit .... *L. biennis*

    4b. Cypselas with a long, slender beak up to as long as the body of the cypsela; plants perennial from deep-seated, creeping roots; involucre 15.0–20.0 mm tall in fruit
.... *L. tatarica*

Lactuca biennis (Moench) Fern.

synonym(s): *Lactuca spicata* (Lam.) A. S. Hitchc.
common name(s): tall blue lettuce
habit: robust, biennial or annual herb
range: Newfoundland, s. to mountains of NC, w. to CA, n. to AK
state frequency: occasional
habitat: moist or rich thickets and openings
notes: Known to hybridize with *Lactuca canadensis* (*q.v.*).

Lactuca canadensis L.

synonym(s): —
common name(s): tall lettuce
habit: biennial or annual herb
range: New Brunswick and Nova Scotia, s. to FL, w. to TX, n. to Saskatchewan
state frequency: occasional
habitat: thickets, borders, fields, waste places
key to the subspecies:

    1a. Leaves unlobed, sometimes the very lower leaves lobed

        2a. Main stem leaves lanceolate to lance-ovate, those near the middle of the stem usually entire
.... *L. c.* var. *canadensis*

        2b. Main stem leaves oblanceolate to narrow-obovate, usually denticulate ....
.... *L. c.* var. *obovata* Wieg.

    1b. Leaves lobed, sometimes the very upper leaves unlobed

        3a. Lobes of leaves linear-falcate, usually entire; unlobed upper leaves, if present, linear to linear-lanceolate .... *L. c.* var. *longifolia* (Michx.) Farw.

        3b. Lobes of leaves broad-falcate or obovate, entire to toothed; unlobed upper leaves, if present, lanceolate or oblanceolate to narrow-obovate .... *L. c.* var. *latifolia* Kuntze

notes: Hybrids with *Lactuca biennis*, called *L.* ×*morssii* B. L. Robins., occur in Maine. In their common form, these hybrids have the habit, vegetative characters, and flower color [blue] of *L. biennis*, but the pappus color [cream] and cypselas are intermediate between the two parents.

Lactuca hirsuta Muhl. *ex* Nutt.

synonym(s): —
common name(s): hairy tall lettuce
habit: annual or biennial, pubescent herb
range: Nova Scotia and ME, s. to FL, w. to TX, n. to MI and Ontario

state frequency: uncommon
habitat: dry, open woods and clearings
key to the subspecies:
    1a. Lower half of the stem pubescent; leaves pubescent on both surfaces .... *L. h.* var. *hirsuta*
    1b. Lower half of the stem glabrous or nearly so; leaves essentially glabrous except for the often
        villous midrib on the abaxial surface .... *L. h.* var. *sanguinea* (Bigelow) Fern.
notes: —

Lactuca sativa L.

synonym(s): *Lactuca scariola* L. var. *integrifolia* (Bogenh.) G. Beck
common name(s): garden lettuce
habit: annual or biennial herb
range: Eurasia; escaped from cultivation in the United States
state frequency: uncommon
habitat: garden waste, compost areas
notes: —

Lactuca tatarica (L.) C. A. Mey. var. *pulchella* (Pursh) Breitung

synonym(s): *Lactuca pulchella* (Pursh) DC.
common name(s): blue lettuce
habit: perennial herb
range: MI, s. to MO, w. to AZ and CA, n. to AK; adventive in eastern North America
state frequency: very rare
habitat: meadows, thickets, river banks
notes: —

Lapsana
(1 species)

Annuals with yellow ray flowers and minutely calyculate involucres.

Lapsana communis L.

synonym(s): —
common name(s): nipplewort
habit: branched, annual herb
range: Eurasia; naturalized to New Brunswick and Nova Scotia, s. to VA, w. to MO, n. to British
    Columbia
state frequency: uncommon
habitat: woods, fields, roadsides, waste places
notes: —

Leontodon
(1 species)

Plants with the aspect of *Taraxacum*—well developed basal rosettes and scapose stems.

Leontodon autumnalis L.

synonym(s): —
common name(s): fall-dandelion
habit: scapose, perennial herb
range: Eurasia; naturalized to Newfoundland, s. to DE, w. to MI, n. to Ontario
state frequency: common

habitat: fields, meadows, roadsides, waste places
key to the subspecies:
  1a. Phyllaries tomentose-puberulent to glabrous, but without any hirsute pubescence ....
  .... *L. a.* ssp. *autumnalis*
  1b. Phyllaries hirsute-pubescent, otherwise also tomentose-puberulent to glabrous ....
  .... *L. a.* ssp. *pratensis* (Link) Arcang.
notes: —

Leucanthemum
(1 species)

Leucanthemum vulgare Lam.
synonym(s): *Chrysanthemum leucanthemum* L., *Chrysanthemum leucanthemum* L. var.
    pinnatifidum (Lecoq & Lamotte) Moldenke
common name(s): ox-eye daisy, white daisy
habit: simple or occasionally branched, rhizomatous, perennial herb
range: Eurasia; naturalized to much of North America
state frequency: common
habitat: fields, roadsides, waste places
notes: —

Liatris
(1 species)

A genus with showy disk flowers, usually arranged in a spike- or raceme-like capitulescence.

Liatris scariosa (L.) Willd. var. *novae-angliae* Lunell
synonym(s): *Liatris borealis* Nutt., *Liatris novae-angliae* (Lunell) Shinners
common name(s): northern blazing star
habit: perennial herb
range: ME, s. to NJ, w. and n. to NY
state frequency: very rare
habitat: dry, open places
notes: This species is ranked S1 by the Maine Natural Areas Program.

Madia
(2 species)

Strongly aromatic herbs with stipitate-glandular stems [at least near the apex].

KEY TO THE SPECIES
  1a. Leaves 2.0–7.0 x 0.1–0.4 cm; stems stipitate-glandular only near the summit; capitula
      2.0–5.0 mm wide [after pressing], with 0–5 ray flowers; rays *ca.* 2.0 mm long ....
      .... *M. glomerata*
  1b. Leaves 4.0–18.0 x 0.4–1.2 cm; stems conspicuously stipitate-glandular throughout;
      capitula 6.0–12.0 mm wide [after pressing], with 5–13 ray flowers; rays 2.0–7.0 mm
      long .... *M. sativa*

Madia glomerata Hook.
synonym(s): —
common name(s): mountain tarweed

habit: annual herb
range: western North America; naturalized to New England
state frequency: very rare
habitat: open places
notes: —

Madia sativa Molina

synonym(s): *Madia capitata* Nutt., *Madia sativa* Molina var. *congesta* Torr. & Gray
common name(s): Chile tarweed
habit: coarse, simple or ascending-branched, annual herb
range: western North America; naturalized to the eastern United States
state frequency: very rare
habitat: roadsides, waste places
notes: —

Matricaria
(4 species)

KEY TO THE SPECIES

1a. Capitula with conspicuous, white ray flowers; disk corollas 5-lobed; plants aromatic or without scent

    2a. Receptacle conic, acute at the apex; cypselas with 5 slightly raised ribs, smooth over its surface; pappus none; plants aromatic .... *M. recutita*

    2b. Receptacle hemispherical, rounded at the apex; cypselas with 3 thickened, wing-like ribs, minutely roughened on its surface and between the ribs; pappus a short crown; plants nearly without scent

        3a. Plants ascending to erect; capitula 3.0–4.0 cm wide; ultimate segments of the leaves filiform, 4.0–20.0 mm long; rays 10.0–20.0 mm long .... *M. perforata*

        3b. Plants depressed to spreading; capitula 1.5–3.0 cm wide; ultimate segments of the leaves linear, 1.5–5.0 mm long; rays 7.0–12.0 mm long .... *M. maritima*

1b. Capitula without ray flowers; disk corollas 4-lobed; plants aromatic [suggesting pineapple] .... *M. discoidea*

Matricaria discoidea DC.

synonym(s): *Matricaria matricarioides* (Less.) Porter
common name(s): pineapple-weed
habit: low, erect, branched, annual herb
range: western North America; naturalized to Newfoundland, s. to DE, w. to MO, n. to Manitoba
state frequency: common
habitat: roadsides, waste places
notes: —

Matricaria maritima L.

synonym(s): —
common name(s): scentless chamomile
habit: low, annual herb
range: Europe; naturalized to Newfoundland, s. to PA; also British Columbia
state frequency: rare
habitat: roadsides, fields, waste places
notes: —

Matricaria perforata Mérat
synonym(s): *Matricaria maritima* L. var. *agrestis* (Knaf) Wilmott
common name(s): scentless chamomile
habit: ascending to erect, annual herb
range: Europe; naturalized to Newfoundland, s. to PA, w. to KY and KS, n. to British Columbia
state frequency: uncommon
habitat: fields, shores, waste places
notes: —

Matricaria recutita L.
synonym(s): *Matricaria chamomilla* L., *Matricaria chamomilla* L. var. *coronata* (J. Gay) Coss. & Germ.
common name(s): chamomile
habit: branched, annual herb
range: Europe; naturalized to Newfoundland, s. to PA, w. and n. to MN
state frequency: very rare
habitat: roadsides, waste places
notes: —

Mikania
(1 species)

Twining herbs, ours with fleshy, fascicled roots.

Mikania scandens (L.) Willd.
synonym(s): —
common name(s): climbing hempweed
habit: twining, vining herb
range: ME, s. to FL, w. to TX, n. to MI and Ontario
state frequency: very rare
habitat: moist thickets, swamps, stream banks
notes: This species is ranked SH by the Maine Natural Areas Program.

Mycelis
(1 species)

Mycelis muralis (L.) Dumort.
synonym(s): *Lactuca muralis* (L.) Fresen.
common name(s): wall-lettuce
habit: slender, annual or biennial herb
range: Europe; naturalized to ME, s. to NY, w. to MI, n. to Quebec
state frequency: very rare
habitat: moist roadsides, and waste places
notes: —

Oclemena
(2 species)

1a. Rays white or white tinged with pink; leaves membranaceous, prominently toothed, flat-margined, those of the stem below the capitulescence mostly numbering 10–22 ....
.... ***O. acuminata***

1b. Rays pink to purple; leaves firm, entire or nearly so, often revolute-margined, those of
the stem below the capitulescence 40–75 .... *O. nemoralis*

Oclemena acuminata (Michx.) Nesom
synonym(s): *Aster acuminatus* Michx.
common name(s): whorled aster
habit: rhizomatous, perennial herb
range: Newfoundland, s. to PA and along mountains to GA, w. to TN, n. to Ontario
state frequency: common
habitat: woods, clearings
notes: Hybrids with *Oclemena nemoralis*, called *O.* ×*blakei* (Porter) Nesom, occur in Maine and can
be recognized by intermediate morphology—leaves 5.0–24.0 mm wide, slightly toothed, somewhat
firm, numbering 25–40 below the capitulescence.

Oclemena nemoralis (Ait.) Greene
synonym(s): *Aster nemoralis* Ait.
common name(s): bog aster, leafy bog aster
habit: rhizomatous, perennial herb
range: Newfoundland, s. to DE, w. to MI, n. to Ontario
state frequency: uncommon
habitat: *Sphagnum* bogs, peaty shores
notes: Hybrids with *Oclemena acuminata* (*q.v.*) occur in Maine.

Omalotheca
(2 species)

Our species are woolly, perennial herbs of boreal and alpine areas.

KEY TO THE SPECIES
1a. Pappus bristles distinct and falling separately; leaves 1.2–2.5 cm long; alpine plants
0.2–1.0 dm tall, growing at elevation exceeding 1000 m .... *O. supina*
1b. Pappus bristles united at the base and falling together; leaves 2.5–5.0 cm long; boreal
plants 1.0–6.0 dm tall, growing at elevation less than 1000 m .... *O. sylvatica*

Omalotheca supina (L.) DC.
synonym(s): *Gnaphalium supinum* L.
common name(s): alpine cudweed
habit: dwarf, tufted, perennial herb
range: circumboreal; s. in the east to ME and NH
state frequency: very rare
habitat: moist meadows and gullies in alpine areas
notes: This species is ranked S1 by the Maine Natural Areas Program.

Omalotheca sylvatica (L.) Schultz-Bip. & F. W. Schultz
synonym(s): *Gnaphalium sylvaticum* L.
common name(s): woodland cudweed
habit: erect, simple, perennial herb
range: circumboreal; s. to ME, w. to WI
state frequency: rare
habitat: open woods, borders, fields, waste places
notes: This species is ranked S3 by the Maine Natural Areas Program.

Petasites
(1 species)

Species of *Petasites* are subdioecious. The staminate plants have capitula composed entirely of staminate flowers, or sometimes some of the marginal flowers will be carpellate as well. Similarly, the carpellate plants have capitula composed entirely of carpellate flowers or sometimes with a few staminate flowers near the center of the capitula.

Petasites frigidus (L.) Fries var. *palmatus* (Ait.) Cronq.
synonym(s): *Petasites palmatus* (Ait.) Gray
common name(s): northern sweet-coltsfoot
habit: creeping, rhizomatous, perennial herb
range: circumboreal; s. to MA, w. to CA
state frequency: uncommon
habitat: moist meadows and low woods, swampy places
notes: —

Picris
(1 species)

Inner flowers of capitula with a pappus of plumose bristles and narrow, lustrous, brown cypselas. Outer flowers of capitula with pappus reduced or absent and crescent-shaped, pale, woolly cypselas.

Picris echioides L.
synonym(s): —
common name(s): ox tongue, bristly ox tongue
habit: coarse, tall, branched, annual herb
range: Eurasia; naturalized to New Brunswick and Nova Scotia, s. to NJ, w. to OH, n. to Ontario; also Saskatchewan and Alberta
state frequency: very rare
habitat: fields, roadsides, waste places
notes: —

Prenanthes
(6 species)

Leaf shape in non-alpine species is highly variable and should not be used for identification.

KEY TO THE SPECIES
1a. Stem leaves sessile; phyllaries long-hirsute .... *P. racemosa*
1b. At least the lower stem leaves petioled; phyllaries glabrous or minutely puberulent near the apex
    2a. Each capitulum with 4–6 phyllaries and 5 or 6 flowers .... *P. altissima*
    2b. Each capitulum with 7–13 phyllaries and 9–18 flowers
        3a. Phyllaries green-black to black; plants mostly 1.0–4.0 dm tall, of alpine areas, elevation exceeding 1000 m
            4a. Upper stem and branches of capitulescence villous-puberulent; leaves entire or toothed; outer [shorter] phyllaries spreading .... *P. boottii*
            4b. Stem and branches of capitulescence glabrous; at least the lower leaves usually lobed; outer phyllaries appressed .... *P. nana*

3b. Phyllaries green to purple, sometimes black-spotted in *P. trifoliolata*; plants mostly 4.0–17.0 dm tall, of woods and open areas at lower elevation (except *P. trifoliolata*, which is casually subalpine)

5a. Pappus deep red-brown .... *P. alba*

5b. Pappus white to light yellow-brown .... *P. trifoliolata*

Prenanthes alba L.

synonym(s): —
common name(s): rattlesnake root
habit: stout, perennial herb
range: ME, s. to GA, w. to MO, n. to Saskatchewan
state frequency: very rare
habitat: rich woods and thickets
notes: —

Prenanthes altissima L.

synonym(s): —
common name(s): tall white-lettuce
habit: slender, perennial herb
range: Newfoundland, s. to GA, w. to AR, n. to Manitoba
state frequency: common
habitat: moist woods
notes: —

Prenanthes boottii (DC.) Gray

synonym(s): —
common name(s): dwarf simple white-lettuce, simple dwarf white-lettuce, Boott's rattlesnake root
habit: perennial herb
range: ME, w. to NY
state frequency: very rare
habitat: alpine areas
notes: This species is ranked S1 by the Maine Natural Areas Program.

Prenanthes nana (Bigelow) Torr.

synonym(s): *Prenanthes trifoliolata* (Cass.) Fern. var. *nana* (Bigelow) Fern.
common name(s): dwarf cut-leaf white-lettuce, gall of the earth, dwarf rattlesnake root
habit: perennial herb
range: Labrador, s. to ME, w. to NY
state frequency: very rare
habitat: rocky or mossy, alpine areas
notes: This species is ranked S1 by the Maine Natural Areas Program.

Prenanthes racemosa Michx.

synonym(s): —
common name(s): glaucous white-lettuce, glaucous rattlesnake root
habit: perennial herb
range: Quebec, s. to NJ, w. to CO, n. to British Columbia
state frequency: very rare
habitat: stream banks, river banks, shores
notes: This species is ranked S2 by the Maine Natural Areas Program.

Prenanthes trifoliolata (Cass.) Fern.
synonym(s): —
common name(s): gall of the earth
habit: perennial herb
range: Newfoundland, s. to NC and mountains of GA, w. to OH, n. to Quebec
state frequency: common
habitat: dry, sandy clearings, thickets, and woods
notes: —

Pseudognaphalium
(3 species)

Our species typically have entire, white-tomentose leaves and white to yellow flowers.

KEY TO THE SPECIES
1a. Leaves decurrent on the stem as thin wings, acuminate at the apex; stem glandular-pubescent .... *P. viscosum*
1b. Leaves not decurrent on the stem, obtuse to acute at the apex; stem glandular-pubescent or not
   2a. Stem tomentose, usually lacking glandular hairs (except sometimes near the base) .... *P. obtusifolium*
   2b. Stem not tomentose (except near the apex), glandular-pubescent throughout .... *P. helleri*

Pseudognaphalium helleri (Britt.) A. Anderb. ssp. *micradenium* (Weatherby) Kartesz
synonym(s): *Gnaphalium helleri* Britt. var. *micradenium* (Weatherby) Mahler, *Gnaphalium obtusifolium* L. var. *micradenium* Weatherby
common name(s): small rabbit-tobacco
habit: erect, fragrant, annual herb
range: New Brunswick, s. to SC, w. to MO, n. to MI
state frequency: very rare
habitat: sandy woods, thickets, and clearings
notes: This species is ranked S? by the Maine Natural Areas Program.

Pseudognaphalium obtusifolium (L.) Hilliard & Burtt
synonym(s): *Gnaphalium obtusifolium* L.
common name(s): fragrant cudweed, catfoot
habit: erect, fragrant, annual herb
range: New Brunswick and Nova Scotia, s. to FL, w. to TX, n. to WI and Ontario
state frequency: uncommon
habitat: dry fields, clearings, and borders
notes: —

Pseudognaphalium viscosum (Kunth) W. A. Weber
synonym(s): *Gnaphalium macounii* Greene, *Gnaphalium viscosum* Kunth
common name(s): clammy cudweed, povertyweed, sticky-everlasting
habit: erect, annual herb
range: New Brunswick, s. to NC, w. to Mexico and AZ, n. to British Columbia
state frequency: occasional
habitat: clearings, pastures, borders
notes: —

Rudbeckia
(2 species)

A well known group of plants with conspicuous, yellow ray flowers that lack both stamens and carpels [*i.e.*, they are neutral].

KEY TO THE SPECIES
1a. Pappus absent; stems pubescent; style appendages elongate and pointed; chaff acute at the apex .... ***R. hirta***
1b. Pappus present; stems glabrous; style appendages short and blunt; chaff blunt at the apex .... ***R. laciniata***

Rudbeckia hirta L.
synonym(s): <*R. h.* var. *pulcherrima*> = *Rudbeckia serotina* Nutt., *Rudbeckia serotina* Nutt. var. *sericea* (T. V. Moore) Fern. & Schub.
common name(s): black-eyed Susan
habit: short-lived perennial or biennial herb
range: Newfoundland, s. to FL, w. to Mexico, n. to British Columbia
state frequency: common
habitat: woods, thickets, fields, roadsides, waste places
key to the subspecies:
    1a. Leaves entire to finely toothed; basal leaves lanceolate to oblanceolate, 1.0–2.5(–5.0) cm wide, (3.0–)4.0–5.0 times as long as wide; stem leaves linear to oblanceolate or spatulate .... *R. h.* var. *pulcherrima* Farw.
    1b. Leaves coarsely toothed; basal leaves ovate to rhombic-oval, 2.5–7.0 cm wide, about 2.0 times as long as wide; stem leaves lance-ovate to ovate or pandurate .... *R. h.* var. *hirta*
notes: Habitat is also useful in separating the 2 varieties found in Maine. *Rudbeckia hirta* var. *hirta* is a native plant of relatively undisturbed habitats, found in woods and thickets. *Rudbeckia hirta* var. *pulcherrima* is an aggressive weed of disturbed habitats such as fields and roadsides.

Rudbeckia laciniata L.
synonym(s): *Rudbeckia laciniata* L. var. *hortensis* Bailey
common name(s): cut-leaf coneflower
habit: coarse, biennial or short-lived perennial herb
range: Quebec, s. to FL, w. to AZ, n. to MT
state frequency: uncommon
habitat: rich, moist areas
notes: —

Senecio
(6 species)

A large and cosmopolitan genus with yellow to orange flowers.

KEY TO THE SPECIES
1a. Plants annual or biennial (rarely a short-lived perennial in *S. jacobaea*), often with evident taproots; stems leafy throughout
    2a. Rays 4.0–8.0 mm long; leaves relatively more divided, usually 2- or 3-times pinnatifid .... ***S. jacobaea***
    2b. Rays absent or up to 2.0 mm long; leaves relatively less divided, usually toothed to pinnatifid

   3a. Rays present [though inconspicuous]; phyllaries without a black apex; involucre
       with about 13 principal phyllaries .... *S. sylvaticus*
   3b. Rays absent; phyllaries with a black apex; involucre with about 21 principal
       phyllaries .... *S. vulgaris*
1b. Plants perennial, with rhizomes, stolons, and/or a caudex; stems with evident tufts of
    basal leaves, the middle and upper leaves conspicuously reduced
   4a. Blade of basal leaves tapering to the petiole, usually cuneate at the base; plants of
       gravels, ledges, or cliffs .... *S. pauperculus*
   4b. Blade of basal leaves abruptly contracted to the petiole, usually truncate to cordate at
       the base; plants of moist to wet soil of forests, meadows, and wetlands
      5a. Blade of basal leaves usually 1.75–3.5 times as long as wide, finely toothed,
          rounded to subcordate at the base, acute to obtuse at the apex ....
          .... *S. schweinitzianus*
      5b. Blade of basal leaves usually 0.75–1.5(–1.75) times as long as wide, crenate,
          strongly cordate at the base, rounded at the apex .... *S. aureus*

Senecio aureus L.
synonym(s): *Senecio aureus* L. var. *aquilonius* Fern., *Senecio aureus* L. var. *intercursus* Fern.
common name(s): heart-leaved groundsel, golden ragwort, squawweed
habit: rhizomatous, perennial herb
range: Labrador, s. to FL, w. to AR, n. to MN and Manitoba
state frequency: uncommon
habitat: moist meadows, woods, and shores, swampy places
notes: —

Senecio jacobaea L.
synonym(s): —
common name(s): tansy ragwort, stinking Willie
habit: coarse, biennial or short-leaved perennial herb
range: Europe; naturalized to Newfoundland, s. to NJ; also Ontario and British Columbia
state frequency: very rare
habitat: dry fields, roadsides, and waste places
notes: —

Senecio pauperculus Michx.
synonym(s): *Senecio pauperculus* Michx. var. *balsamitae* (Muhl. *ex* Willd.) Fern.
common name(s): northern meadow groundsel
habit: occasionally rhizomatous, perennial herb
range: Labrador, s. to GA, w. to NE and WY, n. to AK
state frequency: uncommon
habitat: meadows, ledges, rocky stream banks and shores
notes: —

Senecio schweinitzianus Nutt.
synonym(s): *Senecio robbinsii* Oakes *ex* Rusby
common name(s): New England groundsel, swamp ragwort
habit: rhizomatous, perennial herb
range: Quebec, s. to NC
state frequency: occasional
habitat: wet meadows and thickets, swampy woods
notes: —

Senecio sylvaticus L.
synonym(s): —
common name(s): woodland groundsel
habit: generally simple, annual herb
range: Europe; naturalized to Newfoundland, s. to NJ, w. to OH, n. to WI; also British Columbia
state frequency: uncommon
habitat: clearings, open woods, rocky slopes, waste places
notes: —

Senecio vulgaris L.
synonym(s): —
common name(s): common groundsel
habit: branched, annual herb
range: Europe; naturalized to Labrador, s. to NC, w. to CA, n. to AK
state frequency: uncommon
habitat: cultivated land, waste places
notes: —

Sericocarpus
(1 species)

Sericocarpus asteroides (L.) B. S. P.
synonym(s): *Aster paternus* Cronq.
common name(s): white-topped aster, toothed white-topped aster
habit: rhizomatous, perennial herb
range: ME, s. to FL, w. to AL, n. to MI
state frequency: very rare
habitat: dry woods and clearings
notes: This species is ranked S1 by the Maine Natural Areas Program.

Silphium
(1 species)

The disk flowers of *Silphium* are functionally staminate. The carpels of these flowers are sterile and have a simple style.

Silphium perfoliatum L.
synonym(s): —
common name(s): cup plant
habit: tall, stout, quadrangular-stemmed, perennial herb
range: Ontario, s. to GA, w. to OK, n. to ND; naturalized to ME, s. to PA
state frequency: very rare
habitat: rich woods, thickets, low ground, river banks
notes: —

Solidago
(19 species)

Pubescence of the cypsela can often be assessed by examining the ovary while plants are in flower. Some herbarium specimens of *Solidago* can be difficult to distinguish from *Symphyotrichum* (*q.v.*). Color of fresh ray flowers is important to note in the field as these flowers can fade and discolor with age. Hybridization is reported to be common between closely related species, and many combinations are possible.

KEY TO THE SPECIES

1a. Capitulescence terminal and resembling a panicle that nods at the summit and/or has branches with secund capitula

  2a. Plants often with conspicuous tufts of basal leaves; leaves of the stem progressively reduced upward, those of the apical portion of the stem smaller, often of different shape, and less petiolate than those of the basal portion

    3a. Stems and often leaves closely and minutely pubescent; rays pale yellow .... *S. nemoralis*

    3b. Stems and commonly the leaves glabrous, or the stems sometimes with long, scattered hairs; rays bright yellow

      4a. Leaves fleshy, entire; capitula with 12–17 rays; plants of coastal habitats .... *S. sempervirens*

      4b. Leaves not fleshy, serrate to subentire; capitula with 1–12 rays; plants not of coastal habitats

        5a. Lower leaves abruptly contracted to a petiole

          6a. Abaxial surface of leaves hirsute; capitulum with 4–7 disk flowers; cypselas minutely pubescent .... *S. ulmifolia*

          6b. Abaxial surface of leaves glabrous; capitulum with 8–20 disk flowers; cypselas glabrous .... *S. arguta*

        5b. Lower leaves gradually tapered to the base

          7a. Lower leaves conspicuously sheathing, covering 0.5–0.75 of the circumference of the stem; capitulescence taller than wide; plants of fens and bogs .... *S. uliginosa*

          7b. Lower leaves not conspicuously sheathing, covering less than 0.5 of the circumference of the stem; capitulescence as wide as or wider than tall; plants of fields, woods, and open, sandy areas .... *S. juncea*

  2b. Plants lacking conspicuous tufts of basal leaves; leaves of the apical portion of the stem of nearly similar shape and not dramatically reduced in size relative to the leaves of the basal portion of the stem

    8a. Leaves with 3 conspicuous, parallel nerves [veins]—a midrib and 2 evident and prolonged, lateral nerves

      9a. Stem glaucous and glabrous below the capitulescence; phyllaries obtuse to acute at the apex, green .... *S. gigantea*

      9b. Stem not glaucous, pubescent in at least the apical half; phyllaries acuminate at the apex, yellow-green

        10a. Plants less pubescent; basal half of stem usually glabrous; abaxial surface of leaves normally pubescent only on the midrib and major veins; leaves thin, usually sharply serrate; involucre 2.0–3.0 mm tall .... *S. canadensis*

        10b. Plants more pubescent; basal half [as well as the apical half] of the stem usually pubescent; abaxial surface of leaves normally pubescent on and between veins; leaves firm, subentire to remotely serrate; involucre 3.0–4.0(–5.0) mm tall .... *S. altissima*

    8b. Leaves without 3 conspicuous, parallel nerves, the lateral veins pinnately branched and not aligned parallel to the midrib

      11a. Plants with long, creeping rhizomes; leaves sessile or nearly so .... *S. rugosa*

11b. Plants from a caudex, without creeping rhizomes; lower leaves abruptly tapered to a petiole .... *S. ulmifolia*
1b. Capitulescence consisting of clusters of axillary capitula or resembling a corymb or thyrse [simple or compound], neither nodding at the summit nor with secund capitula
  12a. Plants often with conspicuous tufts of basal leaves; leaves of the stem progressively reduced upward, those of the apical portion of the stem smaller, often of different shape, and less petiolate than those of the basal portion
    13a. Involucre 8.0–11.0 mm long, composed of acuminate- to attenuate-tipped phyllaries; cypsela 4.0–5.0 mm long; lower leaves with an acuminate apex .... *S. macrophylla*
    13b. Involucre 3.0–9.0 mm long, composed of round- to acuminate-tipped phyllaries; cypsela less than 4.0 mm long; lower leaves usually with an obtuse to acute apex
      14a. Stem (at least the apical portion) and often the leaves pubescent with minute, viscidulous hairs; phyllaries narrow, acuminate at the apex, up to 0.5(0.75) mm wide at the middle .... *S. puberula*
      14b. Stem and leaves glabrous or pilose, not pubescent with minute hairs; phyllaries wider, rounded to acute at the apex, (0.75–)1.0–1.5 mm wide at the middle
        15a. Capitula with 1–8 ray flowers; leaves conspicuously sheathing, covering 0.5–0.75 the circumference of the stem; cypselas glabrous; plants of fens and bogs .... *S. uliginosa*
        15b. Capitula with 7–17 ray flowers; leaves not conspicuously sheathing, covering less than 0.5 the circumference of the stem; cypselas glabrous or pubescent; plants of dry, open or rocky areas, and alpine summits
          16a. Phyllaries, especially the outer, with squarrose tips; involucre 5.0–9.0 mm tall; cypselas glabrous .... *S. squarrosa*
          16b. Phyllaries appressed to spreading; involucre 3.0–6.0 mm tall (up to 9.0 mm tall in *S. multiradiata*); cypselas glabrous or pubescent
            17a. Cypselas pubescent; leaves glabrous or nearly so
              18a. Capitula composed of 30–50 flowers [both ray and disk, ray flowers 10–13]; cypselas 3.0–3.5 mm long; lobes of the disk corolla 1.2–2.0 mm long .... *S. multiradiata*
              18b. Capitula composed of 10–20 flowers [both ray and disk, ray flowers 7–10]; cypselas 2.0–2.6 mm long; lobes of the disk corolla 0.8–1.4 mm long .... *S. simplex*
            17b. Cypselas glabrous; leaves pilose on one or both surfaces
              19a. Corolla of ray flowers white or yellow-white; involucre 3.0–5.0 mm long; phyllaries white-brown or light yellow-brown at the base, green at the apex, the colors sharply contrasting .... *S. bicolor*
              19b. Corolla of ray flowers yellow; involucre 4.0–6.0 mm long; phyllaries light green or yellow-green at the base, green at the apex, the colors not strongly contrasting .... *S. hispida*
  12b. Plants lacking conspicuous tufts of basal leaves; leaves of the apical portion of the stem of nearly similar shape and not dramatically reduced in size relative to the leaves of the basal portion of the stem

20a. Capitula with 30–50 flowers [both ray and disk, ray flowers 10–13]; leaves of the stem below the capitulescence numbering 2–7; plants 0.5–3.5 dm, of alpine summits .... **S. multiradiata**

20b. Capitula with 10–25 flowers [both ray and disk, ray flowers 3–9]; leaves of the stem below the capitulescence numbering 7–60; plants 3.0–15.0 dm, of woods and rocky or sandy, open areas

    21a. Leaves elliptic to ovate, 3.0–10.0 cm wide, 1.2–2.5 times as long as wide; stem angled, not glaucous .... **S. flexicaulis**

    21b. Leaves lanceolate to narrow-elliptic, 1.0–3.0 cm wide, 3.0–10.0 times as long as wide; stem terete, glaucous .... **S. caesia**

Solidago altissima L.

synonym(s): *Solidago canadensis* L. var. *scabra* (Muhl. *ex* Willd.) Torr. & Gray, *Solidago scabra* Muhl. *ex* Willd.

common name(s): tall goldenrod

habit: erect, rhizomatous, perennial herb

range: Quebec and ME, s. to FL, w. to TX, n. to Ontario

state frequency: very rare

habitat: fields, open areas

notes: This species is ranked SH by the Maine Natural Areas Program. Reported to hybridize with *Solidago canadensis*.

Solidago arguta Ait.

synonym(s): —

common name(s): forest goldenrod

habit: mostly solitary, perennial herb

range: ME, s. to NC and AL, w. to MO, n. to Ontario

state frequency: uncommon

habitat: open woods, clearings, dry meadows

notes: Reported to hybridize with *Solidago juncea* and *S. uliginosa*.

Solidago bicolor L.

synonym(s): —

common name(s): silverrod, white goldenrod

habit: simple or sparsely branched, perennial herb

range: New Brunswick and Nova Scotia, s. to GA, w. to AR, n. to MI and Ontario

state frequency: occasional

habitat: fields

notes: Hybrids with *Solidago puberula* occur in Maine. Also reported to hybridize with *S. caesia, S. hispida, S. nemoralis*, and *S. rugosa*.

Solidago caesia L.

synonym(s): —

common name(s): blue-stem goldenrod, axillary goldenrod

habit: slender, short- to long-rhizomatous, perennial herb

range: New Brunswick and Nova Scotia, s. to FL, w. to TX, n. to WI and Ontario

state frequency: uncommon

habitat: rich woods, thickets, clearings

notes: Reported to hybridize with *Solidago bicolor, S. canadensis, S. flexicaulis, S. rugosa*, and *S. ulmifolia*.

Solidago canadensis L.

synonym(s): —
common name(s): Canada goldenrod, common goldenrod
habit: perennial herb
range: Newfoundland, s. to VA, w. to IL, n. to Ontario
state frequency: common
habitat: dry thickets, clearings, fields, roadsides, and waste places
notes: Hybrids with *Solidago macrophylla* (*q.v.*) occur in Maine. *Solidago canadensis* is reported to hybridize with *S. altissima*, *S. caesia*, *S. gigantea*, *S. macrophylla*, *S. rugosa*, *S. simplex*, and *S. uliginosa*.

Solidago flexicaulis L.

synonym(s): —
common name(s): zigzag goldenrod
habit: slender, erect to ascending, rhizomatous, perennial herb
range: Quebec, s. to NC and mountains of GA, w. to AR and KS, n. to ND and Ontario
state frequency: common
habitat: rich woods, thickets, and slopes
notes: Reported to hybridize with *Solidago caesia*.

Solidago gigantea Ait.

synonym(s): *Solidago gigantea* Ait. var. *leiophylla* Fern.
common name(s): smooth goldenrod
habit: clustered or occasionally solitary, rhizomatous, perennial herb
range: New Brunswick, s. to GA, w. to NM, n. to British Columbia
state frequency: common
habitat: moist, rich, thickets and open places
notes: Reported to hybridize with *Solidago canadensis* var. *canadensis*.

Solidago hispida Muhl. *ex* Willd.

synonym(s): <*S. h.* var. *hispida*> = *Solidago bicolor* L. var. *concolor* Torr. & Gray; <*S. h.* var. *lanata*> = *Solidago bicolor* L. var. *lanata* (Hook.) Seymour
common name(s): hairy goldenrod
habit: simple or occasionally branched, perennial herb
range: Newfoundland, s. to GA, w. to AR, n. to Saskatchewan
state frequency: uncommon
habitat: dry woods and fields, rocky, open areas
key to the subspecies:
    1a. Stem pubescent with hairs both curved and distinct [non-intermeshed] .... *S. h.* var. *hispida*
    1b. Stem pubescent with either straight or intermeshed hairs .... *S. h.* var. *lanata* (Hook.) Fern.
notes: Reported to hybridize with *Solidago bicolor*.

Solidago juncea Ait.

synonym(s): —
common name(s): early goldenrod
habit: solitary or few-stemmed, perennial herb
range: New Brunswick, s. to mountains of GA, w. to MO, n. to MN and Saskatchewan
state frequency: common
habitat: dry, sandy soil of woods and fields, shores
notes: Reported to hybridize with *Solidago arguta*, *S. nemoralis*, *S. sempervirens*, and *S. uliginosa*. Hybrids with *S. puberula*, *S. rugosa*, and *S. simplex* var. *randii* are occur in Maine.

Solidago macrophylla Pursh

synonym(s): *Solidago macrophylla* Pursh var. *thyrsoidea* (E. Mey.) Fern.

common name(s): large-leaved goldenrod, big-leaved goldenrod

habit: perennial herb

range: Labrador, s. to MA, w. to NY, n. to Ontario

state frequency: uncommon

habitat: moist, cool woods, thickets, and peaty areas

notes: Hybrids with *Solidago canadensis* var. *canadensis*, called *S.* ×*calcicola* Fern., occur in Maine. This hybrid, ranked SX by the Maine Natural Areas Program, can be recognized by its narrow-elliptic to narrow-ovate leaves 1.5–3.5 cm wide, involucre 4.0–6.0 mm tall, weakly triple-nerved leaves, and barely, if at all, secund capitula. Also reported to hybridize with *S. squarrosa* and *S. rugosa*.

Solidago multiradiata Ait. var. *arctica* (DC.) Fern.

synonym(s): *Solidago cutleri* Fern.

common name(s): New England goldenrod, Cutler's goldenrod

habit: tufted or solitary, perennial herb

range: Newfoundland, s. to ME and NY; also AK, s. to British Columbia

state frequency: very rare

habitat: alpine areas

notes: This species is ranked S1/S2 by the Maine Natural Areas Program. Reported to hybridize with *Solidago simplex*.

Solidago nemoralis Ait.

synonym(s): —

common name(s): gray goldenrod

habit: solitary or tufted, simple or branched, erect to depressed, perennial herb

range: New Brunswick and Nova Scotia, s. to FL, w. to AZ, n. to British Columbia

state frequency: occasional

habitat: dry, sterile soil of woodlands, fields, and roadsides

notes: Reported to hybridize with *Solidago bicolor*, *S. juncea*, and *S. puberula*.

Solidago puberula Nutt.

synonym(s): —

common name(s): downy goldenrod, dusty goldenrod

habit: simple or ascending-branched, perennial herb

range: New Brunswick and Nova Scotia, s. to FL, w. to LA, n. to Quebec

state frequency: occasional

habitat: dry, sandy, acid or rocky soil of open places

notes: Hybrids with *Solidago bicolor* and *S. juncea* occur in Maine. Also known to hybridize with *S. nemoralis* and *S. simplex*.

Solidago rugosa P. Mill.

synonym(s): —

common name(s): rough-stemmed goldenrod, wrinkle-leaved goldenrod

habit: clumped or solitary, rhizomatous, perennial herb

range: Newfoundland, s. to FL, w. to TX, n. to MI and Ontario

state frequency: common

habitat: moist or dry woods, thickets, borders, and open areas

key to the subspecies:

  1a. Capitula with 6–8 ray flowers; phyllaries rounded to obtuse at the apex; leaves thick and firm, conspicuously rugose-veiny, crenate to subentire, usually acute at the apex; pubescence of the plant short and stiff .... *S. r.* ssp. *aspera* (Ait.) Cronq.

1b. Capitula with 8–11 ray flowers; phyllaries acute to obtuse at the apex; leaves thin, not conspicuously rugose-veiny, sharply serrate, usually acuminate at the apex; pubescence of the plant long and soft or absent

    2a. Stem pubescent, without decurrent lines from leaf bases; leaves pubescent, at least on major veins on the abaxial surface

        3a. Capitulescence with long, divergent branches that strongly exceed the small [1.0–7.0 x 0.5–1.5 cm], subtending leaves .... *S. r.* ssp. *r.* var. *rugosa*

        3b. Capitulescence with short, ascending branches that equal or slightly exceed the large [5.0–10.0 x 1.0–3.5 cm], subtending leaves .... *S. r.* ssp. *r.* var. *villosa* (Pursh) Fern.

    2b. Stem essentially glabrous, commonly with decurrent lines from the leaf bases; leaves glabrous or nearly so .... *S. r.* ssp. *r.* var. *sphagnophila* Graves

notes: Hybrids with *Solidago juncea, S. macrophylla,* and *S. sempervirens* (*q.v.*) occur in Maine. Also reported to hybridize with *S. bicolor, S. caesia, S. canadensis* var. *canadensis,* and *S. ulmifolia.*

Solidago sempervirens L.

synonym(s): —
common name(s): seaside goldenrod
habit: stout, fleshy, perennial herb
range: Newfoundland, s. to NJ and locally to VA
state frequency: occasional
habitat: sandy or rocky, coastal shores and marshes
notes: Hybrids with *Solidago rugosa,* called *S.* ×*asperula* Desf., occur in Maine. Also reported to hybridize with *S. juncea.*

Solidago simplex Kunth. ssp. *randii* (Porter) Ringius

synonym(s): <*S. s.* ssp. *r.* var. *monticola*> = *Solidago randii* (Porter) Britt., *Solidago randii* (Porter) Britt. var. *monticola* (Porter) Fern., *Solidago simplex* Kunth. ssp. *randii* (Porter) Ringius var. *randii* (Porter) Kartesz & Gandhi; <*S. s.* ssp. *r.* var. *racemosa*> = *Solidago racemosa* Greene, *Solidago spathulata* DC. var. *racemosa* (Greene) Gleason
common name(s): Rand's goldenrod
habit: tufted or occasionally solitary, perennial herb
range: Nova Scotia, s. to VA, w. to IN, n. to MN and Ontario
state frequency: occasional
habitat: dry ledges, granitic areas, and rocky banks
key to the subspecies:

    1a. Basal leaves mostly 3.0–8.0 times as long as wide, often sharply toothed, with a broad midrib 0.7–1.0 mm wide; capitulescence compact, with short peduncles up to 4.0 mm long; plants mostly of cliffs and ledges .... *S. s.* ssp. *r.* var. *monticola* (Porter) Ringius

    1b. Basal leaves mostly 7.0–10.0 times as long as wide, often subentire, with a slender midrib 0.1–0.2 mm wide; capitulescence more open, with longer peduncles 5.0–15.0 mm long; plants mostly of river shores .... *S. s.* ssp. *r.* var. *racemosa* (Greene) Ringius

notes: Reported to hybridize with *Solidago canadensis* var. *canadensis, S. juncea, S. multiradiata* var. *arctica,* and *S. puberula.*

Solidago squarrosa Muhl.

synonym(s): —
common name(s): squarrose goldenrod, stout ragged goldenrod
habit: sometimes rhizomatous, perennial herb
range: New Brunswick, s. to NC, w. to TX, n. to NE and Ontario
state frequency: uncommon
habitat: dry, rocky woods, clearings, fields
notes: Reported to hybridize with *Solidago macrophylla.*

Solidago uliginosa Nutt.

synonym(s): <*S. u.* var. *linoides*> = *Solidago chrysolepis* Fern., *Solidago purshii* Porter; <*S. u.* var. *uliginosa*> = *Solidago uniligulata* (DC.) Porter

common name(s): northern bog goldenrod

habit: solitary or few-stemmed, perennial herb

range: Newfoundland, s. to NC, w. to TN, n. to MI and Ontario

state frequency: occasional

habitat: bogs, swamps, wet meadows

key to the subspecies:

  1a. Plants robust; stems 6.0–15.0 dm tall; leaves of the stem numbering 20–40, the lower leaves 3.0–8.0 cm wide; capitulescence 3.0–25.0 cm wide .... *S. u.* var. *uliginosa*
  1b. Plants smaller; stems 2.0–9.0 dm tall; leaves of the stem numbering 5–20(–30), the lower 0.7–3.0 cm wide; capitulescence 1.0–14.0 cm wide
    2a. Capitulescence slender, narrow-cylindric, usually with ascending branches, 1.0–10.0 cm wide .... *S. u.* var. *linoides* (Torr. & Gray) Fern.
    2b. Capitulescence wider, more spherical, usually with recurved-spreading branches, 2.5–14.0 cm wide .... *S. u.* var. *terra-novae* (Torr. & Gray) Fern.

notes: Reported to hybridize with *Solidago arguta*, *S. canadensis* var. *canadensis*, and *S. juncea*.

Solidago ulmifolia Muhl. *ex* Willd.

synonym(s): —

common name(s): elm-leaved goldenrod

habit: slender, usually solitary, perennial herb

range: Nova Scotia, s. to FL, w. to TX, n. to MN

state frequency: very rare

habitat: dry, rocky woods and thickets

notes: This species is ranked SX by the Maine Natural Areas Program. Reported to hybridize with *Solidago caesia* and *S. rugosa*.

Sonchus
(3 species)

Lactiferous herbs with yellow ray flowers.

KEY TO THE SPECIES

1a. Plants perennial, with deep-seated, creeping roots; capitula 3.0–5.0 cm wide in flower; involucre 14.0–22.0 mm tall in fruit .... *S. arvensis*
1b. Plants annual, with taproots; capitula 1.5–2.5 cm wide in flower; involucre 9.0–13.0 mm tall in fruit
  2a. Cypselas with 3(–5) ribs on each face; basal auricles of leaves rounded .... *S. asper*
  2b. Cypselas with 3–5 ribs on each face and transversely rugulose; basal auricles of leaves acutely pointed .... *S. oleraceus*

Sonchus arvensis L.

synonym(s): <*S. a.* ssp. *uliginosus*> = *Sonchus arvensis* L. var. *glabrescens* Guenth., Grab., & Wimmer

common name(s): field sow-thistle, perennial sow-thistle

habit: perennial herb

range: Europe; naturalized to Newfoundland, s. to DE, w. to UT, n. to AK

state frequency: occasional

habitat: fields, roadsides, shores, waste places

key to the subspecies:
    1a. Peduncles and phyllaries copiously pubescent with gland-tipped hairs .... *S. a.* ssp. *arvensis*
    1b. Peduncles and phyllaries without gland-tipped hairs .... *S. a.* ssp. *uliginosus* (Bieb.) Nyman
notes: —

Sonchus asper (L.) Hill
synonym(s): —
common name(s): prickly-leaved sow-thistle, spiny-leaved sow-thistle
habit: annual herb
range: Europe; naturalized throughout much of the world
state frequency: occasional
habitat: fields, roadsides, waste places
notes: —

Sonchus oleraceus L.
synonym(s): —
common name(s): common sow-thistle
habit: annual herb
range: Europe; naturalized to much of the world
state frequency: occasional
habitat: cultivated fields, roadsides, waste places
notes: —

Symphyotrichum
(18 species)

A difficult genus containing several species complexes that often do not satisfactorily separate into their respective groups. Hybridization is known to occur in the genus. The disk flowers in *Symphyotrichum* have a slender, basal tube and an apical, expanded portion called the limb. The limb is composed of two parts—the connate, basal region and the distinct teeth or lobes that crown the apex of the disk corolla. The ratio of the length of the lobes to that of the limb is an important character in some groups of species. Some species of *Symphyotrichum* with small, non-anthocyanic rays can be confused with certain species of *Solidago* when the specimens have aged and discerning yellow and white colors is difficult. However, these non-anthocyanic species of *Symphyotrichum* often have more ray flowers [15–40 per capitulum] and/or disk corollas that often become red to purple, especially later in the season. *Solidago* have 1–17 ray flowers per capitulum and yellow disk flowers. *Symphyotrichum dumosum* (L.) Nesom is reported to occur in Maine by Gleason and Cronquist (1991), but valid voucher specimens are unknown.

KEY TO THE SPECIES
1a. Annuals from a short taproot; rays very short and inconspicuous, scarcely or not
    exceeding the mature pappus ....*S. subulatum*
1b. Perennials from a rhizome, caudex, or crown; rays elongate and conspicuous
    2a. Blade of basal and lower stem leaves evidently cordate and borne on a petiole
        3a. Middle and upper stem leaves sessile and conspicuously clasping the stem; leaves
            usually with low, crenate teeth .... *S. undulatum*
        3b. Middle and upper stem leaves petiolate or sessile but not clasping the stem;
            leaves serrate
            4a. Leaves serrate with low teeth; petioles winged; stems usually few-flowered,
                often with fewer than 50 capitula; phyllaries usually acuminate, green at the
                tip; peduncles scarcely bracteate .... *S. ciliolatum*

4b. Leaves serrate with prominent, often arching teeth; petioles commonly not winged; stems usually with more than 50 capitula; phyllaries obtuse to acute, usually with a purple tip; peduncles copiously bracteate .... *S. cordifolium*

2b. None of the leaf blades both cordate and borne on a petiole

  5a. Leaves strongly cordate- or auriculate-clasping

    6a. Phyllaries glandular; plants produced from a short, thick rhizome or branching caudex, sometimes with short, creeping rhizomes

      7a. Rays 45–100, red-purple .... *S. novae-angliae*

      7b. Rays 15–30, commonly blue .... *S. patens*

    6b. Phyllaries eglandular; plants produced from a long, creeping rhizome (except *S. laeve* with a short rhizome or caudex)

      8a. Phyllaries long-acuminate to attenuate, of nearly equal length; stems commonly hispid with spreading hairs, at least in the lower portion .... *S. puniceus*

      8b. Phyllaries mostly obtuse to acute, often of different lengths; stems glabrous or pubescent

        9a. Plants produced from a short, thick rhizome or branching caudex, sometimes with short, creeping rhizomes; stem and leaves below the capitulescence glabrous, commonly glaucous .... *S. laeve*

        9b. Plants produced from a long, creeping rhizome; stem and leaves below the capitulescence glabrous or, more commonly, the stem with some pubescence in lines of decurrence from the leaves, not glaucous

          10a. Rhizomes mostly less than 2.0 mm thick; leaves linear, 2.0–9.0 mm wide; stems mostly less than 2.5 mm thick at the base; plants of cool bogs and fens .... *S. borealis*

          10b. Rhizomes mostly more than 2.0 mm thick; leaves linear-lanceolate to oblanceolate, 4.0–25.0 mm wide; stems mostly more than 2.5 mm thick at the base; plants of river shores, meadows, damp openings, and coastal areas

            11a. Leaves tending to be oblanceolate, narrowing to the base, herbaceous or fleshy; many of the peduncles shorter than 3.0 cm; phyllaries herbaceous nearly to base ...... *S. novi-belgii*

            11b. Leaves tending to be linear-lanceolate, not noticeably narrowing to the base, coriaceous; many of the peduncles longer than 3.0 cm; phyllaries herbaceous only in the apical half .... *S. anticostense*

  5b. Leaves not auriculate-clasping, narrowed to the petiole or stem

    12a. Phyllaries, especially the outer, with a prominent, often squarrose, spinulose tip, the margins coarsely ciliolate; stems pubescent; leaves rigid, coarsely ciliate-margined .... *S. ericoides*

    12b. Phyllaries without a spinulose tip, usually with glabrous, fimbriate, or minutely ciliolate margins; stems glabrous or pubescent; leaves herbaceous to coriaceous, the margins various

      13a. Some of the phyllaries involute in the apical portion; plants from a stout caudex or very short, thick rhizome .... *S. pilosum*

      13b. Phyllaries flat; plants from long, creeping rhizomes

14a. Involucre 5.0–11.0 mm long; phyllaries usually over 1.0 mm wide, often with a pronounced, green portion and appearing herbaceous; rays 6.0–20.0 mm long, usually blue or violet

    15a. Rhizomes mostly less than 2.0 mm thick; leaves linear, 2.0–9.0 mm wide; stems mostly less than 2.5 mm thick at the base; plants of cool bogs and fens .... *S. boreale*

    15b. Rhizomes mostly more than 2.0 mm thick; leaves linear-lanceolate to oblanceolate, 4.0–25.0 mm wide; stems mostly more than 2.5 mm thick at the base; plants of river shores, meadows, damp openings, and coastal areas

        16a. Leaves tending to be oblanceolate, narrowing to the base, herbaceous or fleshy; many of the peduncles shorter than 3.0 cm; phyllaries nearly herbaceous to the base .... *S. novi-belgii*

        16b. Leaves tending to be linear-lanceolate, not noticeably narrowing to the base, coriaceous; many of the peduncles longer than 3.0 cm; phyllaries herbaceous only in the apical half .... *S. anticostense*

14b. Involucre 3.0–7.0 mm long; phyllaries usually less than 1.0 mm wide, the green portion short or slender; rays 3.0–15.0 mm long, usually white (blue to purple in *S. praealtum*)

    17a. Lobes of the disk corolla comprising 0.5–0.75 of the limb; leaves often pubescent on the abaxial surface along the midvein .... · .... *S. lateriflorum*

    17b. Lobes of the disk corolla comprising 0.15–0.40 of the limb; leaves commonly glabrous

        18a. Involucre 2.5–3.5(–4.0) mm long; rays 3.0–6.0 mm long; capitula often secund .... *S. racemosum*

        18b. Involucre 3.5–7.0 mm long; rays 3.0–15.0 mm long; capitula often not secund

            19a. The subcoriaceous leaves with evident, reticulate veinlets; veinlets forming nearly isodiametric areolae; lobes of the disk corolla comprising 0.17–0.25 of the limb .... *S. praealtum*

            19b. The herbaceous leaves usually with obscure veinlets; veinlets forming areolae longer than broad; lobes of the disk corolla 0.15–0.45 of the limb

                20a. Capitulescence strictly ascending, commonly composed of fewer than 15 capitula; rhizomes mostly thinner than 2.0 mm; rays 7.0–15.0 mm long; lobes of the disk corolla comprising 0.15–0.30 of the limb; plants of cool bogs and fens .... *S. boreale*

                20b. Capitulescence ascending to spreading, commonly composed of more than 15 capitula; rhizomes mostly thicker than 2.0 mm; rays 3.0–12.0 mm long; lobes of the disk corolla comprising 0.30–0.45 of the limb; plants not of cool bogs and fens

21a. Plants smaller, 1.0–6.0 dm tall; stems 1.0–3.0 mm thick at the base; leaves 3.0–10.0 x 0.3–1.0 cm; rays 3.0–8.0 mm long, 15–30 per capitulum ....
.... ***S. tradescantii***
21b. Plants larger, 5.0–15.0 dm tall; stems 3.0–6.0 mm thick at the base; leaves 8.0–15.0 x 0.3–3.5 cm; rays 4.5–12.0 mm long, 20–40 per capitulum ....
.... ***S. lanceolatum***

Symphyotrichum anticostense (Fern.) Nesom
synonym(s): *Aster anticostensis* Fern., *Aster gaspensis* Victorin
common name(s): Anticosti aster
habit: rhizomatous, perennial herb
range: Quebec, s. to ME
state frequency: very rare
habitat: calcareous, gravelly river banks
notes: This species is ranked SX by the Maine Natural Areas Program.

Symphyotrichum boreale (Torr. & Gray) Nesom
synonym(s): *Aster borealis* (Torr. & Gray) Prov., *Aster junciformis* Rydb.
common name(s): northern bog aster
habit: rhizomatous, perennial herb
range: Quebec, s. to NJ, w. to CO, n. to AK
state frequency: rare
habitat: open and wooded fens, shores
notes: —

Symphyotrichum ciliolatum (Lindl.) A. & D. Löve
synonym(s): *Aster ciliolatus* Lindl.
common name(s): northern heart-leaved aster
habit: rhizomatous, perennial herb
range: New Brunswick, s. to NY, w. to WY, n. to British Columbia
state frequency: uncommon
habitat: woods, clearings, thickets, roadsides, shores
notes: —

Symphyotrichum cordifolium (L.) Nesom
synonym(s): *Aster cordifolius* L., *Aster cordifolius* L. var. *furbishiae* Fern., *Aster cordifolius* L. var. *polycephalus* Porter
common name(s): common blue heart-leaved aster
habit: perennial herb
range: New Brunswick, s. to GA, w. to MO, n. to MN and Ontario
state frequency: common
habitat: woods, thickets, clearings, roadsides
notes: —

Symphyotrichum ericoides (L.) Nesom
synonym(s): *Aster ericoides* L.
common name(s): squarrose white aster
habit: rhizomatous, perennial herb
range: ME, s. to GA, w. to AZ, n. to British Columbia
state frequency: rare
habitat: dry, open places, thickets
notes: Sometimes confused with *Symphyotrichum pilosum*, which has a short rhizome or caudex, entire or serrate leaves, rays 5.0–10.0 mm long, and phyllaries with entire to fimbriate margins. *Symphyotrichum ericoides* has a long, creeping rhizome, rays 3.0–6.0 mm long, and coarsely ciliate-margined leaves and phyllaries.

Symphyotrichum laeve A. & D. Löve
synonym(s): *Aster laevis* L.
common name(s): smooth aster
habit: perennial herb
range: ME, s. to GA, w. to NM, n. to British Columbia; introduced to Quebec and New Brunswick
state frequency: rare
habitat: dry, open places
notes: —

Symphyotrichum lanceolatum (Willd.) Nesom
synonym(s): *Aster lanceolatus* Willd., *Aster simplex* Willd., *Aster simplex* Willd. var. *ramosissimus* (Torr. & Gray) Cronq.
common name(s): eastern lined aster
habit: rhizomatous, perennial herb
range: Newfoundland, s. to NC, w. to KS, n. to ND and Saskatchewan
state frequency: common
habitat: damp thickets, meadows, shores, roadsides
notes: Hybrids with *Symphyotrichum lateriflorum* and *S. novi-belgii* occur in Maine. Also known to hybridize with *S. puniceum*.

Symphyotrichum lateriflorum (L.) A. & D. Löve
synonym(s): <*S. l.* var. *angustifolium*> = *Aster lateriflorus* (L.) Britt. var. *angustifolium* (Wieg.) Nesom; <*S. l.* var. *hirsuticaulis*> = *Aster hirsuticaulis* Lindl. *ex* DC., *Aster lateriflorus* (L.) Britt. var. *tenuipes* Wieg.; <*S. l.* var. *horizontale*> = *Aster lateriflorus* (L.) Britt. var. *pendulus* (Ait.) Burgess; <*S. l.* var. *lateriflorum*> = *Aster lateriflorus* (L.) Britt.
common name(s): calico aster, starved aster, goblet aster
habit: perennial herb
range: New Brunswick and Nova Scotia, s. to FL, w. to TX, n. to Manitoba
state frequency: common
habitat: various habitats, open areas to closed forests, dry to wet soils
key to the subspecies:
   1a. Peduncle of the capitulum usually shorter than the involucre or absent; midrib of leaf usually pubescent on the abaxial [lower] surface
      2a. Larger leaves 7.0–9.0 x 0.5–3.0 cm, 2.0–6.0 times as long as wide ... *S. l.* var. *lateriflorum*
      2b. Larger leaves 6.5–16.0 x 0.5–1.5 cm, 7.0–15.0 times as long as wide ....
      .... *S. l.* var. *angustifolium* (Wieg.) Nesom
   1b. Peduncle of the capitulum 1.0–4.0 times the length of the involucre; midrib of leaf glabrous or pubescent on the abaxial surface
      3a. Primary branches of the capitulescence mostly not again divided; midrib of leaf usually pubescent on the abaxial surface .... *S. l.* var. *horizontale* (Desf.) Nesom

3b. Primary branches of the capitulescence mostly again divided; midrib of leaf usually glabrous on the abaxial surface .... *S. l.* var. *hirsuticaulis* (Lindl. *ex* DC.) Porter

notes: The disk flowers of *Symphyotrichum lateriflorum* material are quite distinctive, described as goblet-shaped due to the wide-spreading lobes. Hybrids with *S. lanceolatum*, *S. novi-belgii*, *S. puniceum*, *S. racemosum*, and *S. tradescantii* occur in Maine.

Symphyotrichum novae-angliae (L.) Nesom
synonym(s): *Aster novae-angliae* L.
common name(s): New England aster
habit: rhizomatous, perennial herb
range: New Brunswick, s. to NC and AL, w. to NM, n. to ND and Alberta
state frequency: occasional
habitat: moist, open areas, woods
notes: —

Symphyotrichum novi-belgii (L.) Nesom
synonym(s): <*S. n.* var. *elodes*> = *Aster novi-belgii* L. var. *elodes* (Torr. & Gray) Gray; <*S. n.* var. *novi-belgii*> =*Aster foliaceus* L., *Aster novi-belgii* L., *Aster novi-belgii* L. var. *litoreus* Gray, *Aster novi-belgii* L. var. *tardiflorus* (L.) A. G. Jones, *Aster tardiflorus* L.; <*S. n.* var. *villicaule*> = *Aster johannensis* Fern., *Aster johannensis* Fern. var. *villicaulis* (Gray) Fern., *Aster tardiflorus* L. var. *vestitus* Fern.
common name(s): New York aster
habit: sometimes fleshy and/or halophytic, perennial herb
range: Newfoundland, s. to GA
state frequency: common
habitat: moist thickets, meadows, shores, and roadsides
key to the subspecies:
  1a. Leaves linear-lanceolate, barely clasping at the base, the larger 4.0–12.0 mm wide ....
  .... *S. n.* var. *elodes* (Torr. & Gray) Nesom
  1b. Leaves oblanceolate to lance-ovate or oblong, usually clasping at the base, the larger 6.5–25.0 mm wide
    2a. Phyllaries appressed-ascending, the tips not recurving; corolla and pappus of the disk florets mostly 6.0–8.0 mm long; plants mostly of inland areas ....
    .... *S. n.* var. *villicaule* (Gray) Lebrecque & Brouillet
    2b. Phyllaries ascending, at least the outer with squarrose tips; corolla and pappus of the disk florets mostly 4.0–6.0 mm long; plants mostly near the coastal region ....
    .... *S. n.* var. *novi-belgii*
notes: Hybrids with *Symphyotrichum lanceolatum* and *S. lateriflorum* occur in Maine.

Symphyotrichum patens (Ait.) Nesom
synonym(s): *Aster patens* Ait.
common name(s): clasping aster, late purple aster
habit: slender, perennial herb
range: ME, s. to FL, w. to TX, n. to MN
state frequency: very rare
habitat: dry, open woods and fields
notes: This species is ranked SX by the Maine Natural Areas Program.

Symphyotrichum pilosum (Willd.) Nesom
synonym(s): <*S. p.* var. *pilosum*> = *Aster pilosus* Willd.; <*S. p.* var. *pringlei*> = *Aster pilosus* Willd. var. *demotus* Blake, *Aster pilosus* Willd. var. *pringlei* (Gray) Blake
common name(s): awl aster
habit: branched, fibrous-rooted, perennial herb
range: New Brunswick, s. to FL, w. to AR, n. to MN and Ontario

state frequency: uncommon
habitat: dry, open places, sometimes coastal
key to the subspecies:
    1a. Plants pubescent on the stem, branches, and often the leaves .... *S. p.* var. *pilosum*
    1b. Plants glabrous .... *S. p.* var. *pringlei* (Gray) Nesom
notes: Sometimes confused with *Symphyotrichum ericoides* (*q.v.*).

Symphyotrichum praealtum (Poir.) Nesom var. *angustior* (Wieg.) Nesom
synonym(s): *Aster praealtus* Poir. var. *angustior* Wieg.
common name(s): veiny-lined aster
habit: rhizomatous, perennial herb
range: ME, s. to GA, w. to AZ, n. to Manitoba
state frequency: rare
habitat: moist, low thickets and meadows
notes: —

Symphyotrichum puniceum (L.) A. & D. Löve
synonym(s): *Aster puniceus* L., *Aster puniceus* L. var. *firmus* (Nees) Torr. & Gray, *Aster puniceus* L.
    var. *oligocephalus* Fern., *Aster puniceus* L. var. *perlongus* Fern.
common name(s): bristly aster, purple-stemmed aster
habit: stout-stemmed, rhizomatous, perennial herb
range: Newfoundland, s. to mountains of GA, w. to mountains of AL, n. to Saskatchewan
state frequency: occasional
habitat: wet fields, shores
notes: Hybrids with *Symphyotrichum lateriflorum* occur in Maine. Also known to hybridize with *S.*
    lanceolatum.

Symphyotrichum racemosum (Ell.) Nesom
synonym(s): *Aster fragilis* Willd. var. *subdumosus* (Wieg.) A. G. Jones, *Aster racemosus* Ell., *Aster*
    vimineus Lam., *Symphyotrichum racemosum* (Ell.) Nesom var. *subdumosum* (Wieg.) Nesom
common name(s): small-headed aster
habit: erect, rhizomatous, perennial herb
range: ME, s. to FL, w. and n. to MO
state frequency: uncommon
habitat: dry to moist, sandy soils of fields, woodlands, and shores
notes: Hybrids with *Symphyotrichum lateriflorum* occur in Maine. Occasionally this species has
    relatively elongate and bracteate peduncles, superficially similar to *S. dumosum* (L.) Nesom,
    instead of the typical crowded, short-pedunculate capitula. This form is likely, in part, responsible
    for the records of *S. dumosum* in Maine.

Symphyotrichum subulatum (L.) Nesom
synonym(s): *Aster subulatus* Michx.
common name(s): small saltmarsh aster
habit: somewhat fleshy, annual herb
range: ME, s. to DE, w. to MI, n. to Ontario
state frequency: very rare
habitat: saltmarshes
notes: This species is ranked S1 by the Maine Natural Areas Program.

Symphyotrichum tradescantii (L.) Nesom
synonym(s): *Aster tradescantii* L.
common name(s): shore aster
habit: slender, erect, rhizomatous, perennial herb
range: Quebec and Nova Scotia, s. to NJ, w. to NY
state frequency: occasional
habitat: stream and lake shores, usually rocky
notes: Hybrids with *Symphyotrichum lateriflorum* occur in Maine.

Symphyotrichum undulatum (L.) Nesom
synonym(s): *Aster undulatus* L.
common name(s): clasping heart-leaved aster
habit: rhizomatous, perennial herb
range: Nova Scotia and ME, s. to FL, w. to LA, n. to MN and Ontario
state frequency: occasional
habitat: dry, open woods, thickets, and clearings
notes: —

Tanacetum
(3 species)

A group very close to *Chrysanthemum*; the generic limits uncertain and some of the species moved back and forth between genera.

KEY TO THE SPECIES

1a. Rays 4.0–8.0 mm long, white; cypselas with 7–10 ribs .... *T. parthenium*
1b. Rays absent or up to 4.0 mm long and yellow; cypselas with 5 ribs
    2a. Leaves glabrous or nearly so; disk 5.0–10.0 mm wide; capitulescence composed of 20–200 capitula .... *T. vulgare*
    2b. Leaves villous; disk 10.0–18.0 mm wide; capitulescence composed of 1–15(–30) capitula .... *T. bipinnatum*

Tanacetum bipinnatum (L.) Schultz-Bip. ssp. *huronense* (Nutt.) Breitung
synonym(s): *Tanacetum huronense* Nutt. var. *johannense* Fern.
common name(s): eastern tansy, Huron tansy, Lake Huron tansy
habit: rhizomatous, perennial herb
range: Newfoundland, s. to ME, w. to MI, n. to Ontario
state frequency: very rare
habitat: sandy or gravelly strands and river banks
notes: This species is ranked S2 by the Maine Natural Areas Program.

Tanacetum parthenium (L.) Schultz-Bip.
synonym(s): *Chrysanthemum parthenium* (L.) Bernh.
common name(s): feverfew
habit: branched, perennial herb
range: Eurasia; escaped from cultivation to New Brunswick and Nova Scotia, s. to MD, w. to MO, n. to Ontario
state frequency: very rare
habitat: borders, roadsides, waste places
notes: —

Tanacetum vulgare L.
synonym(s): —
common name(s): common tansy, golden buttons
habit: erect, coarse, rhizomatous, perennial herb
range: Eurasia; naturalized throughout most of the United States
state frequency: occasional
habitat: roadsides, fields, waste places
notes: —

Taraxacum
(2 species)

A genus complicated by apomixis, polyploidy, and taxonomic confusion.

KEY TO THE SPECIES
1a. Inner phyllaries mostly corniculate-appendaged; body of the cypselas red to purple- or
  brown-red at maturity; beak of the cypselas 0.5–3.0 times as long as the body ....
  .... *T. laevigatum*
1b. Inner phyllaries mostly not corniculate-appendaged; body of the cypselas light brown or
  gray-brown at maturity; beak of the cypselas 2.5–4.0 times as long as the body ....
  .... *T. officinale*

Taraxacum laevigatum (Willd.) DC.
synonym(s): *Leontodon erythrospermum* (Andrz. *ex* Bess.) Britt., *Taraxacum erythrospermum*
  Andrz. *ex* Bess.
common name(s): red-seeded dandelion
habit: slender, scapose, perennial herb
range: Eurasia; naturalized to Quebec, s. to VA, w. to NM, n. to British Columbia
state frequency: uncommon
habitat: fields, lawns, roadsides, waste places
notes: —

Taraxacum officinale Wiggers
synonym(s): <*T. o.* ssp. *officinale*> = *Taraxacum officinale* G. H. Weber *ex* Wiggers var. *palustre*
  (Lyons) Blytt; <*T. o.* ssp. *vulgare*> = *Leontodon latiloba* (DC.) Britt., *Leontodon taraxacum* L.,
  Taraxacum latilobum DC.
common name(s): common dandelion
habit: scapose, perennial herb
range: Eurasia; naturalized to most of the temperate world
state frequency: common
habitat: lawns, fields, roadsides, waste places
key to the subspecies:
  1a. Cypselas muricate only in the apical half, sometimes muricate in bands in the basal half as well;
    outer phyllaries reflexed; leaves generally with more and sharper lobes .... *T. o.* ssp. *officinale*
  1b. Cypselas muricate in both the apical and basal halves, never muricate in bands; outer phyllaries
    spreading to ascending; leaves generally with fewer and larger lobes ....
    .... *T. o.* ssp. *vulgare* (Lam.) Schinz & R. Keller
notes: —

Tragopogon
(2 species)

A distinctive genus with clasping, grass-like leaves.

KEY TO THE SPECIES

1a. Flowers purple; cypselas 2.5–4.0 cm long; involucre 2.5–4.0 cm tall in flower,
   4.0–7.0 cm tall in fruit; phyllaries exceeding ray flowers in length .... *T. porrifolius*
1b. Flowers yellow; cypselas 1.5–2.5 cm long; involucre 1.2–2.4 cm tall in flower,
   1.9–3.8 cm tall in fruit; phyllaries shorter than to equaling the ray flowers in length ....
   .... *T. pratensis*

Tragopogon porrifolius L.
synonym(s): —
common name(s): salsify, oyster plant, vegetable oyster
habit: biennial herb
range: Europe; escaped from cultivation to New Brunswick and Nova Scotia, s. to GA, w. to KS, n. to
   British Columbia
state frequency: very rare
habitat: moist soil of roadsides, fields, and waste places
notes: —

Tragopogon pratensis L.
synonym(s): —
common name(s): goat's beard, showy goat's beard
habit: biennial herb
range: Europe; naturalized to Quebec, s. to GA, w. to KS, n. to Ontario
state frequency: occasional
habitat: roadsides, fields, rocky banks, waste places
notes: —

Tussilago
(1 species)

In *Tussilago*, the inner flowers of the capitulum are staminate. The carpellate portion of these flowers is sterile and has a simple or merely lobed style.

Tussilago farfara L.
synonym(s): —
common name(s): coltsfoot
habit: rhizomatous, perennial herb
range: Eurasia; naturalized to Newfoundland, s. to VA, w. and n. to MN and British Columbia
state frequency: occasional
habitat: damp soil of brooksides, ditches, and waste places
notes: —

Xanthium
(2 species)

Species of *Xanthium* have unisexual capitula. The staminate capitula are borne on the upper portion of the plant, are many-flowered, and possess a poorly developed involucre. The carpellate capitula are borne on the lower portion of the plant, are 2-flowered, and have phyllaries that are modified into prickles and have coalesced into a conspicuous bur.

KEY TO THE SPECIES

1a. Leaf blades lanceolate, tapering to the base; plants with 3-forked, axillary spines ....
.... *X. spinosum*
1b. Leaf blades ovate to reniform, truncate to cordate at the base; plants without axillary
spines .... *X. strumarium*

Xanthium spinosum L.

synonym(s): —
common name(s): spiny cocklebur
habit: spiny, taprooted, annual herb
range: Europe; naturalized to New Brunswick and ME, s. to MA
state frequency: very rare
habitat: waste places
notes: —

Xanthium strumarium L.

synonym(s): <*X. s.* var. *canadense*> = *Xanthium echinatum* Murr., *Xanthium italicum* Moretti,
Xanthium pensylvanicum Wallr., *Xanthium speciosum* Kearney
common name(s): common cocklebur
habit: erect to ascending, taprooted, annual herb
range: Europe; naturalized to much of the world
state frequency: uncommon
habitat: fields, shores, waste places
key to the subspecies:
    1a. Carpellate involucre brown or yellow-brown, 2.0–3.5 cm long; phyllaries with relatively longer
       pubescence .... *X. s.* var. *canadense* (P. Mill.) Torr. & Gray
    1b. Carpellate involucre yellow-green, 0.8–2.0 cm long; phyllaries with relatively shorter
       pubescence .... *X. s.* var. *strumarium*
notes: —

LITERATURE CITED

Angiosperm Phylogeny Group. In prep. An Ordinal Classification for the Families of Flowering Plants. *Annals of the Missouri Botanical Garden.*

Bremer, K., B. Bremer, and M. Thulin. 1998. Classification of Flowering Plants. Internet www.systbot.uu.se/classification/overview.html.

Britton, Nathaniel Lord., and Hon. Addison Brown. 1913. *An Illustrated Flora of the Northeastern United States, Canada, and the British Possessions.* Vols. I and II. [of 3]. Charles Scribner's Sons, New York, New York.

Campbell, Christopher S., Heman P. Adams, Patricia Adams, Alison C. Dibble, Leslie M. Eastman, Susan C. Gawler, Linda L. Gregory, Barbara A. Grunden, Arthur D. Haines, Ken Jonson, Sally C. Rooney, Thomas F. Vining, Jill E. Weber, and Wesley A. Wright. 1995. *Checklist of the Vascular Plants of Maine*, third revision. Bulletin 844, Maine Agricultural and Forest Experiment Station, Orono, Maine.

Clark, Josephine F. 1930. Ferns of the Red River Country, Maine. *Rhodora* 32(379):133–136.

Fernald, Merritt Lyndon. 1950. *Gray's Manual of Botany*, eighth edition. American Book Company, Boston, Massachusetts.

Gleason, Henry A., and Arthur Cronquist. 1991. *Manual of the Vascular Plants of Northeastern United States and Adjacent Canada.* The New York Botanical Garden, New York, New York.

Hauke, Richard L. 1993. Equisetaceae, pp. 76–84 *in* Flora of North America Editorial Committee (eds.), *Flora of North America: North of Mexico*, volume 2, Oxford University Press, New York, New York.

Judd, Walter, Christopher Campbell, Elizabeth Kellogg, Peter Stevens, and Michael Donoghue. In prep. *Plant Systematics: a Phylogenetic Approach.*

Källersjö, M., J. S. Farris, M. W. Chase, B. Bremer, M. F. Fay, C. J. Humphries, G. Petersen, O. Seberg, and K. Bremer. In press. Simultaneous parsimony jackknife analysis of 2538 *rbc*L DNA sequences reveals support for major clades of green plants, land plants, seed plants, and flowering plants. *Plant Systematics and Evolution.*

Kartesz, John T. 1994. *A Synonymized Checklist of the Vascular Flora of the United States, Canada, and Greenland.* Vols. I and II. Timber Press, Portland, Oregon.

Ogden, E. C., F. H. Steinmetz, and F. Hyland. 1948. *Check-list of the Vascular Plants of Maine.* Spaulding-Moss Company, Boston, Massachusetts.

Pryer, Kathleen M. 1993. *Gymnocarpium*, p. 258 *in* Flora of North America Editorial Committee (eds.), *Flora of North America: North of Mexico*, volume 2, Oxford University Press, New York, New York.

KEY TO THE GENERA FOR USE WITH VEGETATIVE, NON-EMERGENT, AQUATIC PLANT MATERIAL

1a. Plants thalloid, not differentiated into stems and leaves
  2a. Thallus broad-ellipsoid to globose, without roots .... ***Wolffia***
  2b. Thallus flat, with roots
    3a. Each thallus with 2 or more roots .... ***Spirodela***
    3b. Each thallus with 1 root .... ***Lemna***
1b. Plants not thalloid, with stems and leaves
  4a. Plants attaching to the substrate by fleshy disks, usually on rock or ledge of rapidly flowing water .... ***Podostemum***
  4b. Plants without fleshy disk attachments, usually of sandy- or mucky-bottomed, still to slow-moving water
    5a. Leaves clustered at the base of a stem or borne directly from a stolon or rhizome
      6a. Leaves, or some of them, floating on the surface of the water
        7a. Leaves divided into 4 leaflets .... ***Marsilea***
        7b. Leaves simple to lobed
          8a. Leaves simple
            9a. Leaf blades linear, up to 120.0 cm long, with parallel venation ...... ***Sparganium***
            9b. Leaf blades elliptic, 4.0–12.0 cm long, with pinnate venation .... ***Brasenia***
          8b. Leaves with 2 basal lobes
            10a. Leaves with crenate margins .... ***Nymphoides***
            10b. Leaves with entire margins
              11a. Roots conspicuously septate .... ***Sagittaria***
              11b. Roots not septate
                12a. Leaf blades with angular basal lobes; petioles with 4 large air cavities .... ***Nymphaea***
                12b. Leaf blades with rounded basal lobes; petioles with many small air cavities .... ***Nuphar***
      6b. Leaves all submersed
        13a. Leaves pinnately dissected .... ***Sium***
        13b. Leaves simple
          14a. Leaves truncate to broad-rounded at the apex, with 2 large air cavities in cross-section .... ***Lobelia***
          14b. Leaves tapering to the apex, without 2 conspicuous air cavities
            15a. Leaves thin and flat
              16a. Roots conspicuously septate; leaf blades 4.0–30.0 cm long .... ***Sagittaria***
              16b. Roots not septate; leaf blades commonly longer, up to 200.0 cm
                17a. Leaves with an evident midvein and a central band of lacunae that is more densely reticulate than the marginal bands .... ***Vallisneria***
                17b. Leaves without an evident midvein, uniformly reticulate-checkered throughout .... ***Sparganium***
            15b. Leaves capillary, terete, keeled, or compressed

18a. Plants with filiform, horizontal or arching stolons
    19a. Stolons green, arching above the substrate .. ***Ranunculus***
    19b. Stolons white to green, not arching
        20a. Leaves usually blunt at the apex; plants of tidal
        habitats .... ***Limosella***
        20b. Leaves usually pointed at the apex; plants of non-tidal
        habitats .... ***Littorella***
18b. Plants without stolons, the horizontal stem, if present,
subterranean
    21a. Roots conspicuously septate .... ***Eriocaulon***
    21b. Roots not septate
        22a. Leaves linear, with 4 air spaces in cross-section and
        abruptly expanded at the base .... ***Isoetes***
        22b. Leaves subulate or capillary, not regularly with 4 air
        spaces in cross-section, not abruptly expanded at the
        base
            23a. Leaves mostly 30.0–60.0 cm long, septate
            (sometimes obscurely so in *Eleocharis robbinsii*)
            24a. Septa of leaves extending the entire width of
            the blade .... ***Juncus***
            24b. Septa of leaves not extending the entire width
            of the blade, usually connecting the veins only
                25a. Rhizome white-brown to yellow-brown
                (red-brown); erect stems with leaf-bearing,
                open, white-brown sheaths ....
                .... ***Schoenoplectus***
                25b. Rhizome usually red-brown; erect stems
                with leafless, closed, often red-brown,
                sheaths .... ***Eleocharis***
            23b. Leaves 1.0–12.0 cm (up to 30.0 cm in deep
            water), without septa
            26a. Stems [appearing to be leaves] capillary
            throughout, enclosed at the base by a delicate,
            tubular cataphyll; plants with slender rhizomes
            .... ***Eleocharis***
            26b. Leaves definitely flattened at the base, not
            enclosed by a cataphyll; plants without
            rhizomes .... ***Subularia***
5b. Leaves borne on upright or floating stems
    27a. Leaf blades peltate .... ***Brasenia***
    27b. Leaf blades attached at the base
        28a. Plants with bladder-like traps, these traps borne on the photosynthetic,
        leaf-like branches or on separate achlorophyllous branches .... ***Utricularia***
        28b. Plants without traps
            29a. Leaves pectinately divided

30a. Leaves clearly alternate .... *Proserpinaca*

30b. Leaves crowded, irregularly scattered, or whorled

    31a. Leaves densely crowded along a few cm of swollen stem ....
.... *Hottonia*

    31b. Leaves remote to crowded on longer, slender stems ....
.... *Myriophyllum*

29b. Leaves simple to divided, but not pectinately divided

    32a. Nodes with 1 leaf

        33a. Leaves simple

            34a. Leaves terete and with septa that extend the entire width
of the blade .... *Juncus*

            34b. Leaves capillary to flat, without septa that extend the
entire width of the blade

                35a. Leaves appearing as scale-like bumps ....
.... *Myriophyllum*

                35b. Leaves not scale-like

                    36a. Leaves pinnately veined .... *Polygonum*

                    36b. Leaves parallel-veined

                        37a. Midrib of leaves not more prominent than the
lateral ribs

                        38a. Leaves very thin, translucent, without a
keel .... *Zosterella*

                        38b. Leaves thicker, opaque or nearly so, often
somewhat keeled on the abaxial surface ....
.... *Sparganium*

                        37b. Midrib of leaves more prominent than the
lateral ribs

                        39a. Stems hollow; floating leaves not
differentiated into a petiole and blade

                          40a. Margins of the leaf sheath fused for
more than half the length of the sheath;
plants rhizomatous .... *Glyceria*

                          40b. Margins of the leaf sheath distinct and
overlapping for most of its length; plants
without rhizomes .... *Zizania*

                        39b. Stems solid; floating leaves (if present)
with a petiole and expanded blade

                          41a. Stipular sheath completely connate to
the blade, with no free portion .. *Ruppia*

                          41b. Stipular sheath either entirely distinct
from the blade or partly connate and
then with a free, ligule-like tip

42a. Leaves narrow-linear, 0.2–1.5 mm wide, cross-septate throughout; stipules connate to the blade for 0.4–3.0 cm .... ***Stuckenia***

42b. Leaves filiform to orbicular, 0.1–75.0 mm wide, septate only in the lacunar bands (if present); stipules distinct or connate to the blade for less than 1.0 cm (except *Potamogeton robbinsii*) .... ***Potamogeton***

33b. Leaves evidently lobed to compound

43a. Leaves ternately lobed to dissected .... ***Ranunculus***

43b. Leaves not divided in a ternate fashion

44a. Leaves without a central axis, divided into capillary segments .... ***Ranunculus***

44b. Leaves with a central axis [the midvein], with broader segments .... ***Sium***

32b. Nodes with 2 or more leaves

45a. Leaves divided into numerous, narrow segments (emersed leaves of *Megalodonta* lobed to entire)

46a. Leaves many-times irregularly branched, the segments mostly entire .... ***Megalodonta***

46b. Leaves 2- to 4-times dichotomously forked, the segments minutely serrate with tiny spicules near the apex .... ***Ceratophyllum***

45b. Leaves simple

47a. Nodes with 6–12 leaves .... ***Hippuris***

47b. Nodes with 3 or fewer leaves (the lower nodes sometimes with only 2 leaves in *Elodea*)

48a. Nodes with 3 leaves

49a. Leaves 3.0–10.0 cm long; plants of brackish water .... ***Zannichellia***

49b. Leaves 0.6–1.7 cm long; plants of fresh water .... ***Elodea***

48b. Nodes with 2 leaves

50a. Leaves punctate with glands or pellucid dots

51a. Leaves punctate with pellucid dots, strictly entire .... ***Hypericum***

51b. Leaves glandular-punctate, entire to slightly toothed .... ***Gratiola***

50b. Leaves not punctate

52a. Leaves minutely serrate with tiny spicules .... ***Najas***

52b. Leaves entire

53a. Leaves 3.0–10.0 cm long, filiform to linear
.... ***Zannichellia***

53b. Leaves 0.07–4.0 cm long, linear to ovate

  54a. Leaves dimorphic—the lower linear
with a retuse apex, the upper spatulate
to obovate with a rounded apex, closely
crowded to form a floating rosette ....
.... ***Callitriche***

  54b. Leaves monomorphic, all usually wider
than linear, never crowded to form a
floating rosette

    55a. Leaves 1.0–4.0 cm long, narrowed
to a distinct petiole .... ***Ludwigia***

    55b. Leaves 0.07–0.8 cm long, with an
obscure petiole

      56a. Some of the leaves
gamophyllous .... ***Tillaea***

      56b. Leaves all distinct .... ***Elatine***

INDEX

Main entries in the Flora are in **bold face**; synonyms are in light face. Subspecies (ssp.) Are indexed before varieites (var.). Hybrids are listed last under a genus. Subspecific taxa in the Flora are of three kinds—the only representative of a species, a synonym, and a recognized subspecies or variety of a species. The latter are not indexed but may be found in the key to the subspecies under the species that contain them. Common names are indexed under the name that seems to represent the genus. For example, the marginal wood fern (*Dryopteris marginalis*) is indexed as "wood fern, marginal" because *Dryopteris* are the wood ferns. Common names that are hyphenated (*e.g.*, cotton-grass [*Eriophorum*]) indicate that the implied relation is false; *i.e.*, cotton-grass is not a grass (It is a sedge.). Therefore, hyphenated names are treated as a single name and indexed accordingly. Equivalent to a hyphenated name is a name beginning with the word *false*. For example, "narrow false oat" (*Trisetum spicatum*) is not an oat (*Avena*), so is indexed as "false oat, narrow." When there is uncertainty, try each word in the common name.

FLORA OF MAINE ORDER FORM

(Please photocopy)

Name: _____

Address: _____

Town/city: _____ State: _____

Zip code: _____

Please send me _____ copies @ $45.00 _____

Maine 6% sales tax ($2.70 per book)
(for addresses in Maine) _____

Shipping
first book - $6.00
each additional book - $2.00 _____

TOTAL _____

Thank you for your order. Please make checks payable to:

V. F. Thomas Co.
P. O. Box 281
Bar Harbor, Maine 04069-0281